Springer Series in Surface Sciences

Volume 75

This series covers the whole spectrum of surface sciences, including structure and dynamics of clean and adsorbate-covered surfaces, thin films, basic surface effects, analytical methods and also the physics and chemistry of interfaces. Written by leading researchers in the field, the books are intended primarily for researchers in academia and industry and for graduate students.

Zhong Lin Wang

Contact-Electrification of Matter

Springer

Zhong Lin Wang
Beijing Institute of Nanoenergy
and Nanosystems
Chinese Academy of Sciences
Beijing, China

Georgia Institute of Technology
Atlanta, USA

ISSN 0931-5195 ISSN 2198-4743 (electronic)
Springer Series in Surface Sciences
ISBN 978-3-031-88503-7 ISBN 978-3-031-88504-4 (eBook)
https://doi.org/10.1007/978-3-031-88504-4

This Springer imprint is published by the registered company Springer Nature Switzerland AG
The registered company address is: Gewerbestrasse 11, 6330 Cham, Switzerland

Preface

Contact-electrification (CE) refers to a process by which materials become electrically charged when they come into direct contact with each other and then are separated. This phenomenon is commonly known as triboelectrification (TE) in our daily life and is a fundamental aspect of static electricity. The key feature of CE is the transfer of charges between two objects when they come into contact, resulting in one object becoming positively charged and the other negatively charged. In fact, CE is a common phenomenon of matter between any two partners of liquids, solid and gases.

Although TE has been known for over 2600 years, the study about its mechanisms has been cumbersome and rather sluggish, possibly because the phenomenon was viewed from different angles by different communities of scientists and lacking of proper techniques. Because TE is always treated as a negative effect so there is a little motivation for studying it. In the last two decades, fundamental studies about CE has excited a lot of interest because the invention of the triboelectric nanogenerators (TENGs), which is a paradigm shift technology for harvesting mechanical energy with the use of contact-electrification effect. This book is to summarize the progresses made in the field for systematically covering the CE in systems made of liquid, solid and/or gas. We intend to give an updated review in understanding the CE and its various applications. Part I of the book from Chaps. 1–8 is about the fundamental science related to CE, as outlined in follows:

Chapter 1 by Zhong Lin Wang and Aurelia C. Wang is to introduce the brief history about the study of TE. The chapter also summarizes the major contributions made by TE for human civilization. By introducing the new developments in understanding CE and associated applications, there is a desperate need for authoring a text book, so that the science and technologies associated with CE can be advanced faster. This chapter serves as the basis for future chapters.

Chapter 2 by Haiyang Zou, Shiquan Lin, Xiangtian Ding and Zhong Lin Wang is on the methods and techniques for quantifying CE. The basic techniques used for quantifying the density of surface charges from nano-scale to macroscopic scale are presented. The standards set up for quantifying CE are also covered. The establishments of these techniques are critical for studying interface charge transfers at various interfaces.

Chapter 3 by Lok C. Lew Yan Voon and Morten Willatzen is on the theory for understanding the CE. Both semi-classical and first principle quantum theories are introduced. Theoretical understanding about CE is at the early phase in comparison to experimental progress. Much more work is required in the future.

Chapter 4 by Xiangyu Chen and Zhong Lin Wang is on the CE at metal-insulator interface, by mainly introducing Schottky barrier related charge transfer process. This study is possible by the using of atomic force microscopy related probe techniques. By either varying the sample temperature or applying a bias voltage, the dominant charge carriers are proved to be electrons.

Chapter 5 by Cheng Xu and Zhong Lin Wang is about the CE between insulator and insulator. The energy band model is first introduced, which is useful for understanding the electron transfer between materials that can be described using energy band diagram with the introduce of surface states. Then, the electron cloud model first proposed by Wang in 2018 are presented in details, which is believed to be a more general understanding about CE in matters. By assuming that the electron transfer is the dominant process for CE, charge transfer is only possible in the areas there are close contact between atoms with a strong overlap of electron wave functions. Various charge releasing mechanisms are also presented.

Chapter 6 by Shiquan Lin, Xiangtian Ding and Zhong Lin Wang is to introduce the fundamentals about CE at semiconductor interfaces, with a focus on the newly discovered effect—the tribovoltaic effect, which was first proposed by Wang in 2019. By sliding an n-type semiconductor on a p-type semiconductor as an example, the binding between two atoms at the interface would release a binding energy, called bindington, which may exist in a form of internal photons and can excite local electron-hole pairs, which are then separated by the build-in electric field at the pn junction, resulting in a DC output. The effect is now a new area of studies for dynamic semiconductor interfaces for sending and energy harvesting.

Chapter 7 by Di Wei systematically describes the CE at a liquid-solid interface, emphasizing charge transfer processes across interfaces involving conductors, dielectrics, and semiconductors. It emphasizes the formation and dynamics of the electrical double layer (EDL), a fundamental structure regulating charge behavior at a liquid-solid interface. Wang's two-step model is introduced to elucidate the role played by CE in EDL formation on dielectric surfaces, which describe the key factor contributed by electron transfer at the interface and its initial contribution to the formation of EDL.

Chapter 8 by Ziming Wang, Mulin He, Xueyan Yang and Zhong Lin Wang introduces the newly discovered contact-electro-catalysis (CEC), first discovered by Wang's group in 2022, which is about the contact-electrification-driven electron transfer at a liquid-solid interface to facilitate local redox reactions. This chapter provides a comprehensive overview of the theoretical foundation of CEC, strategies for improving CEC, and the unique advantages of CEC. Additionally, a roadmap for future research and development of CEC has been proposed.

Part II of this book from Chaps. 9–12 is about the various applications of CE:

Chapter 9 by Ying-Chih Lai and Jie Wang systematically describes the fundamental approach and technological applications of triboelectric nanogenerators (TENGs). Starting with the introduction of the two primary types of TENGs: AC-type and DC-type, the chapter presents the five working modes of AC-type TENGs and the two material types used in DC TENGs. Key aspects of performance optimization, material selection, environmental control, and charge excitation in solid-solid TENGs are also discussed. TENG applications have been reviewed in the fields of as micro/nano power sources (e.g., from body motion energy, mechanical energy under in vivo conditions, vibration energy, and wind energy), self-powered sensors (including tactile, biomechanical, acoustic, human physiological, and chemical sensing), blue energy harvesting, and high-voltage sources.

Chapter 10 by Gang Cheng, Jinyang Zhang, and Huaifang Qin is about the applications of TENG based on CE between liquid and solid, range from water energy harvesting to sensing systems. TENG has been demonstrated as a probe for measuring the charge transfer and a liquid-solid interface, which is believed to be important for fundamental chemistry. In this chapter, recent advancements in the applications of TENGs utilizing liquid-solid CE are introduced in detail, the applications mainly focus on their roles as energy harvesters, sensors, and TENG probes, which can realize liquid kinetic energy converted into electricity, self-powered microfluidic or chemical sensing, and the liquid-solid interface study.

Chapter 11 by Chi Zhang and Zhi Zhang is about the tribovoltaic nanogenerator (TVNG) as an innovative energy conversion device using the tribovoltaic effect, with great application potential in areas such as mechanical components, multi-source energy harvesting, intelligent flexible devices, and self-powered sensing systems. The chapter also demonstrate the broad adaptability and high efficiency of the tribovoltaic effect across different scenarios but also guide the development of future intelligent systems and new energy technologies.

Chapter 12 by Wei Tang and Peng Jiang is about the applications of contact-electro-catalysis (CEC), including pollutant degradation, CO_2 capture, the synthesis of important chemicals (such as hydrogen peroxide and ammonia), resource recovery (e.g., recycling of lithium-ion batteries and precious metals), and even cancer therapy. CEC offers high reaction efficiency under mild conditions, which sets it apart from more demanding catalytic processes. A particularly significant advantage of CEC lies in its broad catalyst selections including many conventional organic and inorganic materials. Moreover, the scope of CEC extends beyond solid catalysts to liquid-liquid interfaces, such as oil-water boundaries, where liquid-liquid contact electrification can induce electron transfer to drive redox reactions, similar to solid-liquid systems. This opens up new possibilities for catalytic processes at liquid interfaces, where traditional catalytic methods may be less effective.

I have a long wish to get this book written up because I believe that there is an urgent need for such a book, which can be useful for a broad range of readers to understand CE, a phenomenon that is most well-known, available anywhere and happening all the time in nature, but lacking of in-depth of scientific understanding. I believe that the study of CE has been a "forgotten" corner in science in the past centuries possibly due to its negative effects in many cases, such as causing fire and discharge, but now TENG possibly is the 5th major applications of CE in broad, therefore, a systematic understanding is very important. The book is intended as a text and reference book for students and researchers in the fields of physics, chemistry, materials science and electric engineering. Finally, I like to thank all of the authors who have contributed to this book. The book is impossible without the support of many colleagues and collaborators, to whom I am very grateful.

Zhong Lin Wang, Ph.D.
Chief Scientist
Chair Professor and Director
Beijing Institute of Nanoenergy
and Nanosystems
Chinese Academy of Sciences
Beijing, China
wangzhonglin@binn.cas.cn

Georgia Institute of Technology
Atlanta, USA

Competing Interests The author has no competing interests to declare that are relevant to the content of this manuscript.

Contents

Contributors

Xiangyu Chen Beijing Institute of Nanoenergy and Nanosystems, Chinese Academy of Sciences, Beijing, China

Gang Cheng Key Lab for Special Functional Materials of Ministry of Education, School of Nanoscience and Materials Engineering, Henan University, Kaifeng, China

Xiangtian Ding School of Mechanical Engineering, Beijing Institute of Technology, Beijing, China;
Hebei Key Laboratory of Intelligent Assembly and Detection Technology, Tangshan Research Institute, Beijing Institute of Technology, Tangshan, China

Mulin He Beijing Institute of Nanoenergy and Nanosystems, Chinese Academy of Sciences, Beijing, China

Peng Jiang School of Pharmaceutical Sciences, Wuhan University, Wuhan, China

Ying-Chih Lai Department of Materials Science and Engineering, National Chung Hsing University, Taichung, Taiwan;
Innovation and Development Center of Sustainable Agriculture, i-Center for Advanced Science and Technology, National Chung Hsing University, Taichung, Taiwan;
Department of Physics, National Chung Hsing University, Taichung, Taiwan

Shiquan Lin School of Mechanical Engineering, Beijing Institute of Technology, Beijing, China;
Hebei Key Laboratory of Intelligent Assembly and Detection Technology, Tangshan Research Institute, Beijing Institute of Technology, Tangshan, China

Huaifang Qin Key Lab for Special Functional Materials of Ministry of Education, School of Nanoscience and Materials Engineering, Henan University, Kaifeng, China

Lok C. Lew Yan Voon College of Sciences, University of Findlay, Findlay, OH, USA

Aurelia C. Wang School of Materials Science and Engineering, Georgia Institute of Technology, Atlanta, GA, USA

Jie Wang Beijing Institute of Nanoenergy and Nanosystems, Chinese Academy of Sciences, Beijing, China

Zhong Lin Wang Beijing Institute of Nanoenergy and Nanosystems, Chinese Academy of Sciences, Beijing, China

Ziming Wang Beijing Institute of Nanoenergy and Nanosystems, Chinese Academy of Sciences, Beijing, China

Di Wei Beijing Institute of Nanoenergy and Nanosystems, Chinese Academy of Sciences, Beijing, China

Tang Wei Beijing Institute of Nanoenergy and Nanosystems, Chinese Academy of Sciences, Beijing, China

Morten Willatzen Department of Mathematical Sciences, Aalborg University, Aalborg, Denmark

Cheng Xu School of Materials Science and Physics, China University of Mining and Technology, Xuzhou, Jiangsu, China

Xueyan Yang Beijing Institute of Nanoenergy and Nanosystems, Chinese Academy of Sciences, Beijing, China

Chi Zhang Beijing Key Laboratory of Micro-nano Energy and Sensor, CAS Center for Excellence in Nanoscience, Beijing Institute of Nanoenergy and Nanosystems, Chinese Academy of Sciences, Beijing, China

Jinyang Zhang Beijing Institute of Nanoenergy and Nanosystems, Chinese Academy of Sciences, Beijing, China

Zhi Zhang Beijing Key Laboratory of Micro-nano Energy and Sensor, CAS Center for Excellence in Nanoscience, Beijing Institute of Nanoenergy and Nanosystems, Chinese Academy of Sciences, Beijing, China

Haiyang Zou College of Materials Science and Engineering, Sichuan University, Chengdu, China

Part I
Fundamental Science of Contact-Electrification

Chapter 1
Introduction About Contact-Electrification (Triboelectrification)

Abstract This chapter introduces the history for studying triboelectricity (contact-electrification) and its development in the last few centuries. Some recent advances are briefly reviewed, such as the tribovoltaic effect, the charge transfer at liquid–solid interfaces, the Wang's two-step model about the formation of the electric double layer, and the contact-electro-catalysis.

1.1 Triboelectric Effect and Contact Electrification

Triboelectricity (TE) is about the electric charges generated when two different materials or matters come into contact and are then separated [1]. This phenomenon is a form of static charges. It can occur with different materials, such as the sole of a shoe on a carpet, or between two pieces of the same material. TE occurs for all phase of materials between combinations of solids, liquids and gases, for instance liquid flowing in a solid tube or an aircraft flying through air. When TE occurs also when the two materials are in contact without sliding with each other, in this case it is called *contact-electrification* (CE). In general, triboelectricity is more often used in tribology and engineering, and contact-electrification is more a scientific term.

TE has been known for more than ~ 2600 years since ancient Greek civilization, and it exists everywhere, anywhere and at any time in our nature. When children learn physics in middle school, TE is the first effect introduced about electricity (Fig. 1.1a): by contacting rubbing a glass tube on an animal fur, the glass tube is negatively changed, and the fur would be positively charged. Although this universal phenomenon occurs for all known matter in all phases, solid, liquid and even gas (Fig. 1.1b), the understanding of the mechanism of TE is still not complete, and there are few text books on TE [2]. *Nature* published an article on triboelectric charging nearly a century ago, noting "This class of research is simple-seeming. But those who have spent time on the subject will allow that it is very baffling; those who have not done so will at least remember that despite great efforts by physicists the subject has not yet passed the pioneer stage" [3]. Studying of TE has been rather slow because it is considered as a negative effect, because it creates electric fires and

Z. L. Wang, *Contact-Electrification of Matter*, Springer Series in Surface Sciences 75,
https://doi.org/10.1007/978-3-031-88504-4_1

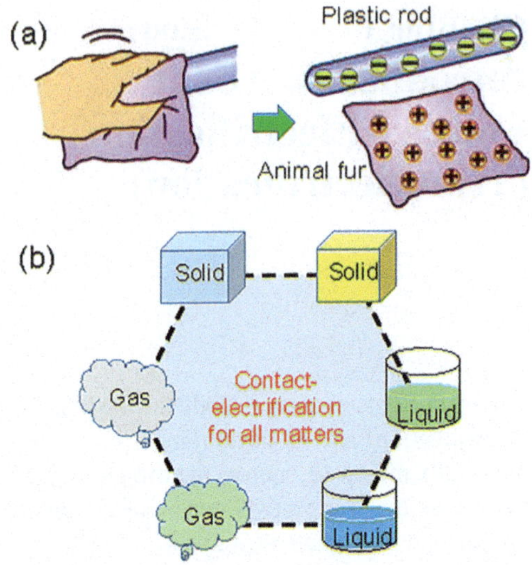

Fig. 1.1 Schematic illustration of triboelectrification **a** by rubbing a fur against a plastic rod, which is normally the first class for teaching kids about electricity. **b** Triboelectrification is a universal phenomenon that occurs at interfaces between solids, liquids and gases

discharge, disrupts flow [4, 5] and blending [6] in pharmaceutical processing, and leads to increased friction [7] and energy losses [8]. So the question is what is new about our recent understanding, why do we still care about TE, and what will be the outcome and impact if we fully understand TE?

CE and TE are conventionally treated the same by many readers, but the two have major differences. TE is a convolution of CE and friction. CE occurs in TE, but friction is not necessary for charge transfer. Friction includes both production of debris and a friction coefficient. Therefore, CE is a scientific effect in which no friction is necessary, while TE is a practical operation. Given that CE occurs for solid, liquid and gas phases of almost any materials, and is present in the process of friction, there has been a long debate regarding whether CE is due to electron transfer [9], ion transfer [10, 11] or even material species transfer [12]. The idea for a role for ion transfer in CE is due to the presence of liquid phases in some cases and the presence of moisture in others, while that for materials species transfer is probably due to the practice of mechanical rubbing.

Lighting in nature is due to the discharge of rain drops, which is a result of CE between air and water droplets. Once water is vaporized, it slow forming small water droplets when rising in air. The collision between air molecules and the water surface make the water droplet charged. The repulsive between the water droplets leads to the formation of cloud. Once the air blows, the small water droplets recombine, forming large droplets and increased surface charge density. Once the surface charge density reaches a threshold, air breakdown would occur, resulting in lighting.

1.2 Brief History for Studying Triboelectric Effect

TE is probably the most widely known, widely used and ever since studied phenomenon in scientific history, because is present in all environment, for all phases of materials and at all time. People's major understanding about TE can be briefly summarized as follows since 2600 years ago (Fig. 1.2).

1. **Ancient Greece time (circa 600 BCE)**: The earliest recorded mention of triboelectricity comes from the Greek philosopher *Thales of Miletus*, who discovered that rubbing amber with wool caused the amber to attract small objects like feathers. This phenomenon was noticed but not understood in terms of electrical principles. The term "amber" is where the word *electron* (for electricity) originates.
2. **17th Century—Early Scientific Inquiry**: In 1600, *William Gilbert*, an English scientist, coined the term "electricus" from the Greek word *ēlektron* (amber). He is often regarded as the father of electrical studies and made significant contributions to understanding static electricity. By the mid-1600 s, other scientists like *Robert Boyle* and *Otto von Guericke* began experimenting with static electricity, using materials like glass and sulfur. The term "triboelectric" was later coined to describe the effects of friction between different materials.
3. **18th Century—the Rise of Electrical Understanding**:
   Benjamin Franklin (1750s): Franklin's famous kite experiment and his work on electricity led to the development of concepts like positive and negative charge. Although Franklin was focused on lightning, his work laid a foundation for understanding static electricity, including the role of triboelectric effects.
   Charles François du Fay (1733): French physicist du Fay discovered that there are two kinds of electric charge—what we now call positive and negative—and that materials could attract or repel based on their charges, including through triboelectric interactions.
   Pierre-Simon Laplace and others: By the late eighteenth century, European scientists began to develop more sophisticated experiments involving electrostatic machines like the *electrophorus*, which were designed to generate static electricity. These machines took advantage of triboelectric effects to accumulate charge.
4. **19th Century—Triboelectricity in Electrostatics**:
   Alessandro Volta (1800): The invention of the first chemical battery by Volta shifted the study of electricity away from static charge to steady current. However, the study of static electricity and triboelectric effects continued to evolve in parallel with these developments.
   Michael Faraday (1830s): Faraday made significant contributions to the understanding of electrical phenomena, including concepts of induction and electric fields, which were essential for broader electrical theories, but the study of triboelectricity was mostly relegated to its own field of electrostatics.

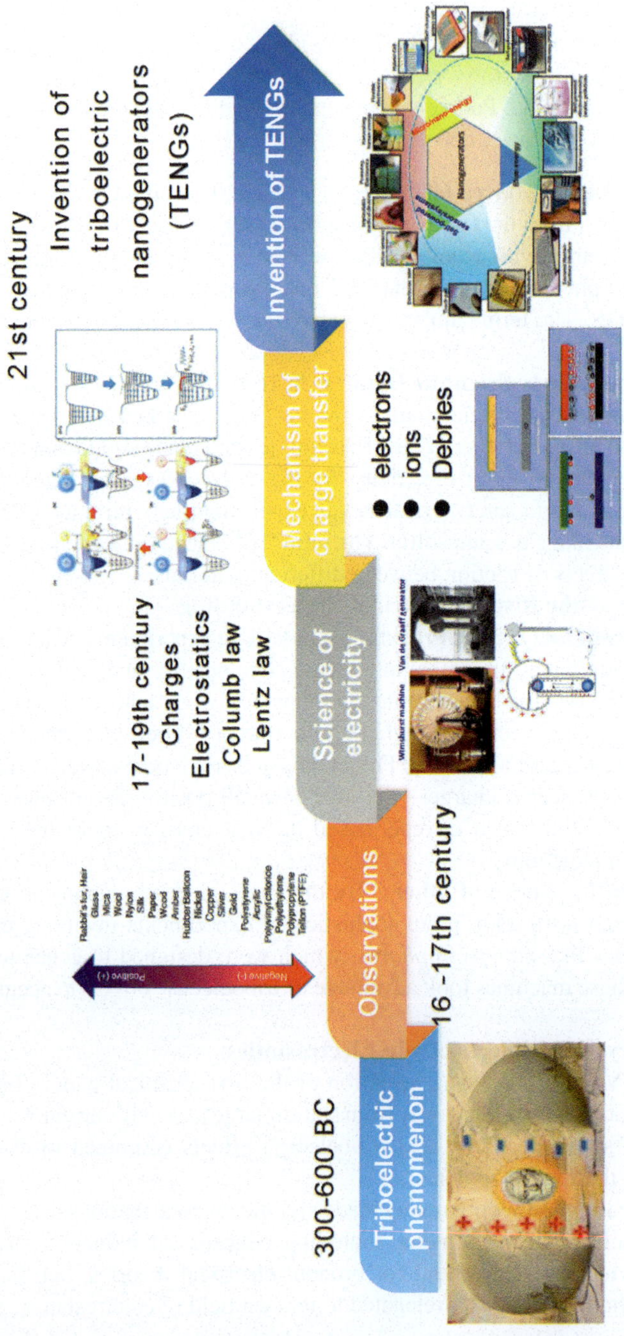

Fig. 1.2 Brief history about triboelectricity

The Electrostatic Generator (1860s): In the mid-1800s, devices like the *Wimshurst machine* and *Van de Graaff generator* were developed, which used triboelectric effects to generate high voltages. These machines were essential for early experiments in high-voltage electricity and paved the way for later advancements in electrostatics.

5. **20th Century—Theoretical Advances and Applications**:
Triboelectric Series: In the early twentieth century, researchers began creating tables of materials and their ability to gain or lose electrons when rubbed against each other, forming what is now known as the *triboelectric series* (Fig. 1.3). This series orders materials based on their tendency to become positively or negatively charged when in contact with other materials. The most famous of these tables was developed by *J.J. Thomson* in 1927. This table gives a guidance about TE, but no quantitative data are provided.
Technological Applications: By the late 19th and early twentieth centuries, triboelectricity found applications in devices like the *electrophotographic process* (precursor to photocopying) and static-sensitive devices. However, it wasn't until the late twentieth century that triboelectricity became widely explored in areas such as materials science and nanotechnology.

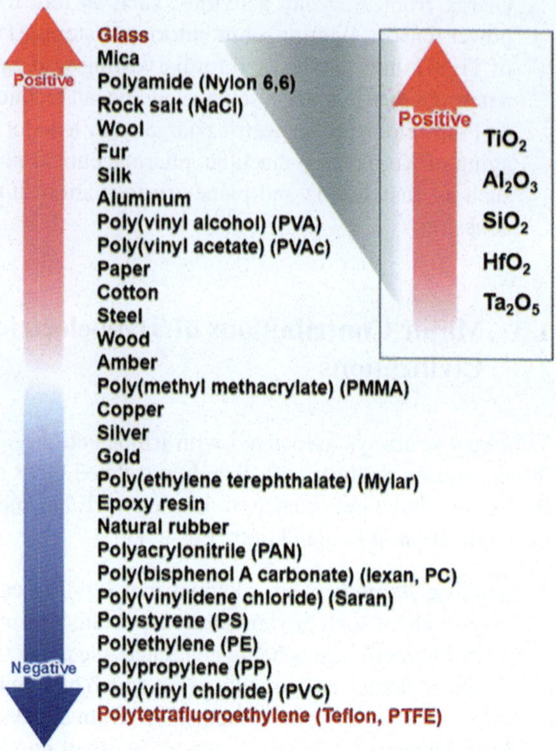

Fig. 1.3 A conventional triboelectric series and an experimentally determined triboelectric series of oxide dielectric materials

6. **21st Century—Modern Research and Energy Harvesting**:
   Triboelectric Nanogenerators (TENGs): In 2012, *Zhong Lin Wang* invented triboelectric nanogenerators that could efficiently convert small mechanical movements into usable electrical power [13]. TENGs have a broad range of applications, especially in energy harvesting and sensing, where they convert mechanical motion into usable electrical energy [14]. Some specific applications include: (a) **Self-Powered Devices**: TENGs can be used to power small, low-energy electronics without external batteries. For example, sensors in remote locations or wearable electronics that are powered by body movement. (b) **Wearable Electronics**: TENGs embedded in clothing or accessories can harvest energy from daily motion. These can be used to power health monitoring devices or small gadgets. (c) **Environmental Energy Harvesting**: TENGs can capture mechanical energy from environmental sources, such as wind, rain, ocean waves, or even vibrations from infrastructure. This has the potential to generate renewable energy from ambient sources. (d) **Biomedical Applications**: TENGs can be used in medical devices, especially for wearable or implantable sensors. These devices could operate autonomously without requiring external battery replacements, which is especially beneficial for long-term health monitoring. (e) **Smart Cities**: TENG technology can be integrated into urban environments to capture energy from everyday activities, such as foot traffic or vehicle movement, to power sensors, lighting, or monitoring systems. (f) **Blue energy**: Networks made of TENG units can be used for harvesting water wave energy in ocean, opening a new chapter in marine science and ocean technologies.

   Furthermore, triboelectric charges play a major role in industries such as packaging of Xerox copy machine, pharmaceutical powders, and in many processes such as dust storms and planetary formation. It can also increase friction and adhesion.

1.3 Major Contributions of Triboelectric Effect to Human Civilizations

Tribology is always associated with triboelectricity. Tribology is vitally important in engineering applications. Tribology related GDP accounts to 2–3% of the total. Besides the huge importance of tribology, TE has made five major contributions to the civilization of human kinds (Fig. 1.4):

1. **Making flame/fire**: It is known that lighting can cause fire, but the site and magnitude of such fire are hardly controlled. Using a wood stick to drill another piece of wood, heat is being generated due to friction. As a result, controlled and localized flame can be made, using which human learnt how to convert raw meat into cooked meat, protect themselves from attacks of wild animals and receiving heat. This is a major step in the advances of human civilization.

Fig. 1.4 Major contributions of triboelectricity to human civilization: **a** Making fire; **b** rolling based transportation; **c** realization of electricity; and **d** purification of water in nature

2. **Transportation**: It is known that rolling friction is much less than sliding friction. Such a principle has been used by human for transportations for thousands of years.
3. **First known about electricity**: Human first realized electricity is through triboelectricity, using which basic knowledge about static electricity has been acquired. This is the beginning of electric power era, which driove the second industrial revolution.
4. **Purification of water in nature**: In nature, it is known that flowing water does not get polluted, while stationary water is easily to be polluted. The background scientific principle is because of the contact-electrification induced catalysis process between water and solids such as stones and soils with the presence of oxygen. Electron exchange between water molecules and solids results in redox reaction, and the produced radicals will degrade the pollutions in water. This process is called the contact-electro-catalysis (CEC) [15]. The rotation of earth is a key for the formation of our planetary environment, and CEC is likely the key process about the chemical processes occurred inside and on the earth.
5. **Triboelectric nanogenerators (TENGs)** [16]: TENG could be the fifth important contribution of TE to the sustainable energy for humankind, but it needs to be verified by future developments. TENG is a technology that harvests energy from our living environment with a broad range of applications from

wearable electronics, robotics, medical science, environment protection, security and large-scale blue energy. Chapters 9 and 10 give a comprehensive review on TENGs.

1.4 Mechanisms of Contact-Electrification of Solid Materials

Although TE is a basic effect of electricity, studying the mechanisms of TE is rather cumbersome, for several reasons. First, TE is a very complex process that not only involves the basic processes occurring in tribology, but also interfacial charge exchange and tunneling. Neglecting the distinction between TE and CE has made the study rather difficult. Second, the TE effect occurs for all matters, so it is rather challenging to propose a unified physical model that covers such a broad range of materials and phases. Third, the lack of tools for probing nano-scale TE was a major limitation for the field until the invention of Kevin probe force microscopy (KPFM) [17]. Last, TE has always been regarded as a negative effect due to the fact that it can create electric fires and discharge. The complexity for studying TE is also due to the charge carriers are suggested as electrons transfer, ion adsorption and/or even materials species transfer during contact by various communities. Figure 1.5 presents a summary of major advances made in understanding triboelectricity until the end of last century [18].

Development of new methods are essential for understanding CE. Quantitative materials database serves as a valuable resource for systematically analyzing the relationship between triboelectrification and the properties of different materials. Numerous approaches have been developed to assess and measure the triboelectric properties of materials. Chapter 2 focuses on the methods used to characterize the triboelectric charge density of various materials. These quantitative triboelectric charge densities could act as a benchmark reference for advancing the use of triboelectrification in energy harvesting and self-powered sensing applications.

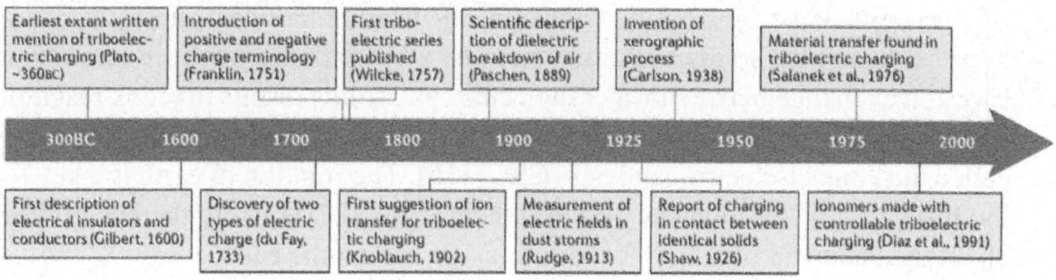

Fig. 1.5 Major developments in the history of studying triboelectric charging through the year 2000. From Lacks and Shinbrot [11]

1.4.1 At Metal–Dielectric Interface

In our daily life, rubbing a metal rod on a rock surface results in sparks, which is a result of TE and related air discharges in a metal-dielectric (M-D) case. If the electronic structure of the dielectric can be described using the energy band structure, the tendency of electron transfer between the metal tip and the dielectric depends on the relative positions of the Fermi level of the metal and the unoccupied surface states of the dielectric. The relative height of the Fermi level can be controlled by applying a bias voltage [19] or by raising the sample temperature [20]. In this case, electron transfer is the dominant charge carriers [21, 22].

Chapter 4 provides a comprehensive overview of the mechanisms underlying CE at metal–insulator interface, emphasizing the interaction between metals and insulators leads to the formation of potential barriers that govern charge flow, with significant implications for electronic device performance. The theoretical model discussed in this chapter includes the Schottky contact effect and work function difference, to elucidate the principles of charge transfer and the dynamics of electron movement at these interfaces. Advanced techniques such as KPFM are highlighted for their ability to provide nanoscale insights into charge distribution and transfer processes, while the influence from CE to the measurements of AFM is also discussed.

1.4.2 At Dielectric-Dielectric Interface—The Surface States Model

For materials whose electronic structures can be represented by energy bands, CE between two materials can be presented as in Fig. 1.6 by using the surface state model [23, 24]. This is reasonable because the presence of a surface means an interruption of the crystal periodicity, e.g., an abrupt change of the periodical crystal lattice, leading to the formation of energy states within the bandgap. The presence of defects (surface defect structures) could result in the occupation of surface states by electrons. Owing to the different valance band and conduction band structures of each material, the occupied surface states in material A could have higher energy than that of the unoccupied surface states of material B (Fig. 1.6a). Once the two materials physically contact, some of the electrons could transfer from material A's surface to the surface of material B due to an energy drive, leading to CE. Such electrons will not transfer back to material A even after the two materials are separated if no thermal or phonon excitation is applied, resulting in net positive electrostatic charges on material A and negative electrostatic charges on material B (Fig. 1.6b). Such charges are surface state bound charges and cannot freely flow in general cases if the conductivity of the materials is rather low, thus they are expected to remain on surfaces permanently. However, in the presence of thermal energy fluctuations and raised temperatures, these bound electrons can be released from surface states according to the thermionic emission model. The surface states model is mostly easy

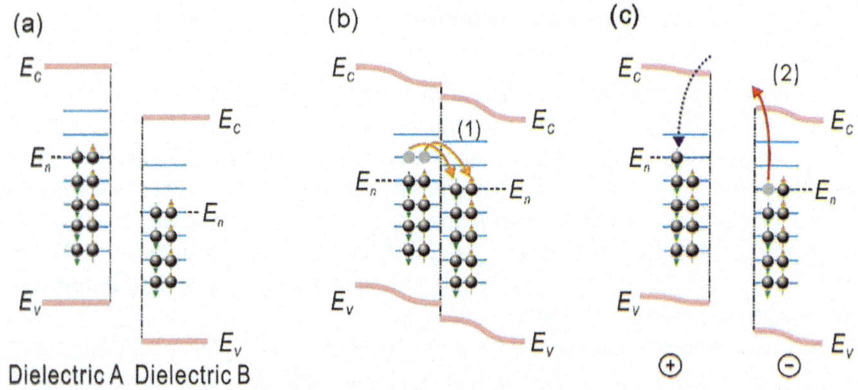

Fig. 1.6 Charge transfer **a** before contact, **b** in contact, and **c** after contact between two different insulator dielectrics for a case that E_n of the former is higher than that of the latter and the band structure model holds

and acceptable for understanding the charge transfer between two solids, provided each of the solids can be described by the energy band diagram.

1.4.3 At Dielectric-Dielectric Interface—The Overlapped Electron Cloud Model

The surface state model is valid if the electronic structure of the dielectric can be represented by a band diagram with the presence of surface and defect states. As for general materials such as polymers and rubber materials, their electronic structures are represented by molecular orbitals and chain alignment rather than the energy band diagram. Furthermore, for materials that may not have a well characterized molecular structure or in the presence of composite phases, as in natural materials such as wool, wood, animal fur, and human hair, there is no simple electronic structure model that can represent these materials. We do know that all of these existing materials exhibit CE, and therefore a general model is needed for explaining CE on atomic and molecular levels.

We first start from experimental observations based on KPFM study [22]. The first question is: how close must the two atoms be before CE occurs? This question was answered by KPFM study [21, 25]. By correlating the phased and amplitude measured for the AFM tip at resonance with the measured surface potential difference (ΔV), the interaction force between the tip and the surface can be probed. When the tip experiences a net attractive force from the surface, it is in the attractive regime, and there is an increase of the phase shift ($\Delta\varphi > 0$). In contrast, if the tip experiences a net repulsive interaction force from the surface, it will be in the repulsive force regime, and there will be a decrease from the original phase shift ($\Delta\varphi < 0$). Therefore, the sign of $\Delta\varphi$ could be regarded as the symbol of the net tip—sample interaction force for each vibration cycle. The result shows that the distance between the tip and the sample for electron transfer to occur should be smaller than the interatomic distance at equilibrium, where the long range attractive force equals to the short-range repulsive force.

The repulsive and attraction interactions between atoms can be easily understood from the interatomic interaction potential [29]. For two atoms that form a bond, which means some sort of overlap in electron clouds or wave functions, an equilibrium distance a is established, which is called the bond length or interatomic distance. If the interatomic distance x is larger than a, the two atoms tend to attract with each other (Fig. 1.7a1), in which case, the overlap between the electron clouds of the two atoms is negligible, thus no electron transfer is possible. If the interatomic distance x is shorter than a, the two atoms tend to repel each other owing to the increased overlap of electron clouds (Fig. 1.7b1). It is in such a range that electron transfer occurs.

One of the key factors in CE is an external force that has to be applied in order for the two surfaces to be contacted. The role played by such an externally applied force is to create local high pressure at certain contacting points even at atomic and nano-scales, where the interatomic distance x is forced to be shorter than the bonding length between the local contacting points, resulting in local repulsive force. This is why that CE is indispensable from friction in practice.

The experimental results indicate that the distance between the AFM tip and the surface has to be shorter than the atomic bonding length in order for the CE to occur. In the repulsive region, the increased overlap in local electron clouds results in electron transition from one material to the other. Therefore, Wang's group first proposed an atomic-scale charge transfer mechanism for CE [26]. Figure 1.7a2 shows a case in which, prior to the atomic-scale contact of the two materials, their respective electron clouds remain separated without overlap. This is the attractive force region. The potential well bounds the electrons tightly in specific orbitals and stops them from freely escaping, which is the case for non-conducting materials. When the two atoms belonging to two materials, respectively, get close and contact with each other, the electron clouds overlap between the two atoms to form a bond. The bonding lengths are shortened even more if an external compression force is applied. In this case, the initial single potential wells become an asymmetric double-well potential, and the energy barrier between the two is lowered as a result of strong electron cloud overlap. Then electrons can then transfer from one atom to the other, resulting in CE. The role played by mechanical contact of the two materials is to shorten the distance between the atoms and cause a strong overlap of their electron clouds in the repulsive region, at least in the area at which the atomic-scale contact occurs, despite the samples could be larger. The process presented in Fig. 1.7b2 is referred as the Wang transition that has been observed experimentally [27].

From the schematic shown in Fig. 1.7, it appears that CE may occur for all of atoms/matters. In reality, it is important to note that only a very small fraction of surface areas of the two materials will reach atomic scale contact with considering the atomic roughness of the surfaces This also explains why more charges are transferred if one

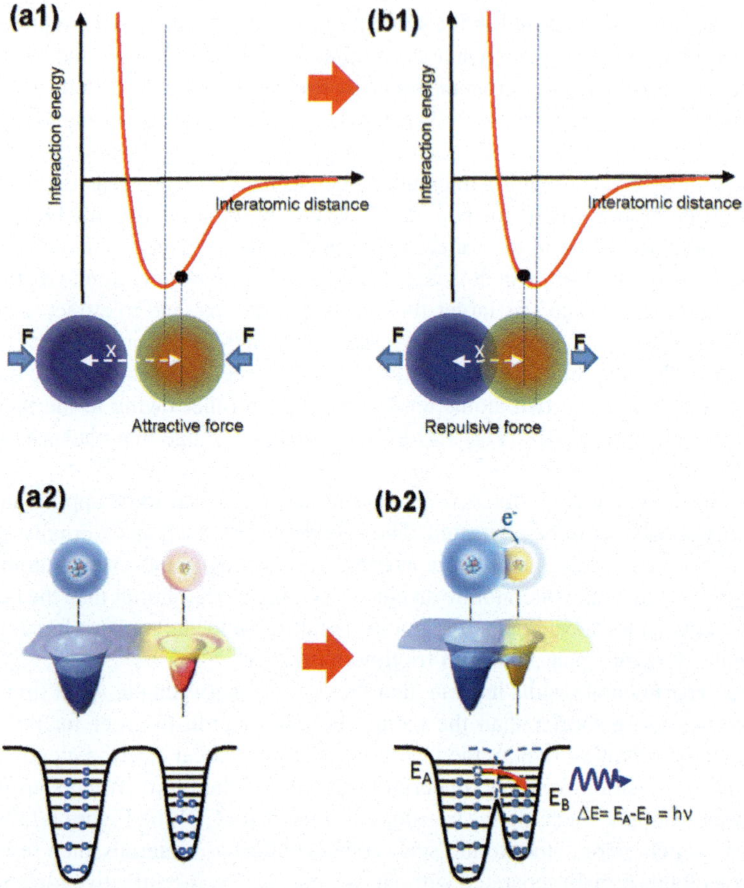

Fig. 1.7 The electron cloud/wave-function overlap model for CE. Interatomic interaction potential between two atoms and the force between the two when they are **a1** with little electron cloud/wave-function overlap, **b1** with strong electron cloud/wave-function overlap in the repulsive region. **a2** and **b2** are the corresponding (3D and 2D) atomic potential energy model and the electronic states of two atoms belonging to two materials A and B, respectively, when they are before contact and in contact, respectively, showing electron transfer from one atom to the other after being forced to have electron cloud overlap. Detailed illustration to show that the increased electron cloud overlap results in a lower potential barrier between the two atoms, leading to interatomic electron transition and possible photon emission, which is indicated in **b2**. The transition shown in **b2** is called Wang transition [27]

material is rubbed harder against the other owing to the strongly overlapped electron clouds under the applied pressure. On the other hand, applied force can even cause local fractures and plastic deformation, which is why CE is always convoluted with friction process possibly with the generation of debris. After being separated, the transferred electrons remain as static charges on the surfaces.

The last process for energy release is the energy from forming a bond. It is generally known that forming a bond places the two atoms into a lower energy state, resulting in energy release. This released "energy quantum" can generate electron–hole pairs at semiconductor surface/interface, e.g., tribovoltaic effect, which is to be presented in the following section.

1.4.4 At Metal–Semiconductor Interface—The Tribovoltaic Effect

It is known that the energy required to separate the charged surfaces is comparable to the fracture energies of materials, which reveals that the electrification is associated with the interactions in electron clouds. Figure 1.8a shows the band diagram for a metal—semiconductor (p type) interface. The energy released by forming a bond is equivalent to a "energy quantum" (so called "bindington") that could excite electron–hole pairs at the semiconductor side, and the electrons will be separated from the holes due to the presence of the Schottky barrier, possibly resulting in a DC current [28]. This is similar to the result of photovoltaic effect except the energy is provided by the tribo-process rather than external photon. This may be simply referred to as the *tribovoltaic effect* first proposed by Wang [29–31]. Chapter 6 gives a comprehensive coverage on the tribovoltatic effect and recent studies of CE involving semiconductors. The mechanism of the tribovoltaic effect is presented.

1.4.5 At Semiconductor–Semiconductor Interface—The Tribovoltaic Effect

Sliding an n-type semiconductor on a p-type semiconductorsurface, a DC current has been received to flow from p-side to the n-side in the external circuit. The observed current strongly depends on the nature of the semiconductor and its conductivity. This effect can be explained as a result of tribovoltaic effect (Fig. 1.8) [29, 30, 32]. Contacting a n-type on a p-type semiconductor, an internal electric field would be generated and an equilibrium is reached. Once the n-type slides on the p-type, new atomic bonds would be formed at the newly formed interface area, which would release energy owing to the formation of new atom bonds. Such energy could be released in a form of internal photon that would excite electron–hole pairs at the pn interface region, the separation of which by the pn junction would generate a DC current. In comparison to photovoltaic effect, which occurs because of the excitation from external photons, the tribovoltaic effect is due to the excitation of internal photons. The output voltage is in the order of semiconductor bandgap.

On other hand, CE can also occur at a sliding pn junction. In this case, the received output voltage is much more than that generated by the tribovoltaic effect in the order

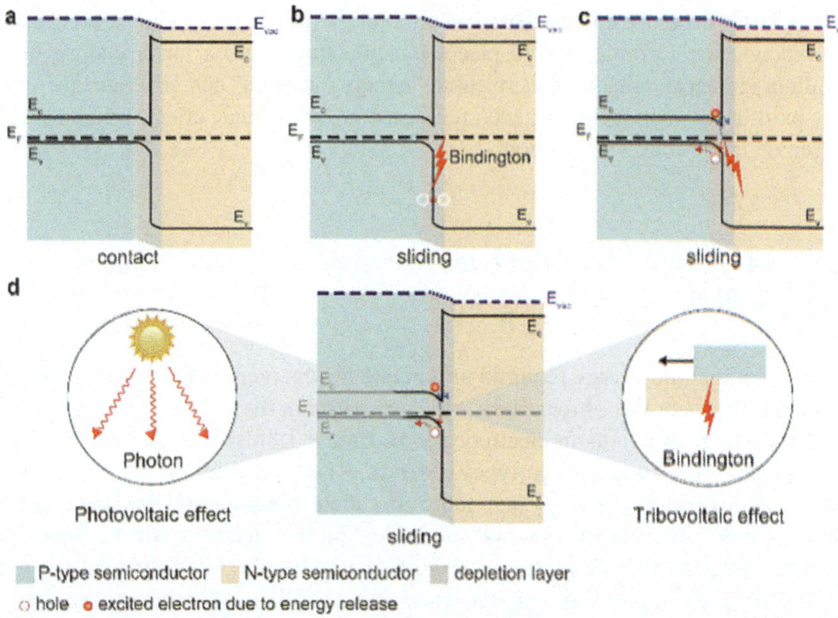

Fig. 1.8 The tribovoltaic effect at a sliding dynamic pn junction by sliding an n-type semiconductor on a p-type semiconductor. (E_c is the bottom of the conduction band, E_f is the Fermi level of the semiconductors, and E_v is the top of the valence band)

of bandgap. The sliding speed should be slow to avoid a rising of local temperature due to friction, which may make the data interpretation complex. A suggestion is that the experiments should done first using a scanning tip as in AFM, so that the above two factors can be minimized.

Chapter 11 presents the applications of using the tribovoltaic effect for various sceneries. The tribovoltaic nanogenerator (TVNG), as an innovative energy conversion device, has demonstrated great application potential in areas such as mechanical components, multi-source energy harvesting, intelligent flexible devices, and self-powered sensing systems in recent years. TVNG converts mechanical vibration and frictional energy into electrical energy, enhancing the energy efficiency and self-driving capabilities of traditional mechanical systems. In self-powered sensing systems, the direct current signal from the TVNG can reflect various operational and environmental parameters, making it suitable for real-time sensing. This provides efficient and reliable solutions for fields such as intelligent environmental monitoring, health monitoring, and industrial automation. These applications not only demonstrate the broad adaptability and high efficiency of the tribovoltaic effect across different scenarios, but also guide the development of future intelligent systems and new energy technologies.

1.5 Contact-Electrification at Liquid–Solid Interfaces

Interface between a liquid and a solid is fundamentally important for catalysis, surface chemical reaction, electrochemistry, chemical engineering and cellular level biology, because charge transfer at the interface dictates many chemical and biological processes. As for the case of liquid–solid CE, it is usually assumed to be ion transfer without concreate studies, simply because ions are often present in solution, such as H^+ and OH^- in water. CE between liquid–solid is the root about the formation of the electric double layer (EDL). The classical model for EDL is an adsorption of a layer of ions on the solid surface, which tends to attract the ions of opposite sign of charges, while repel the ions of the same signs of charges in the solution, forming a distribution of electric potential near the liquid–solid interface. This model is entirely based on there are only ion adsorptions at the liquid–solid interface without including the involvement of electron transfer. Based on the study presented in the last few sections, the formation of EDL has to be revised.

Based on the classical model, the EDL refers to two parallel layers of charges surrounding a solid surface. The first layer consists of ions physically adsorbed onto the surface. The second layer is composed of ions attracted adjacent to the surface via Coulomb interaction to electrically screening the first layer. With considering the liquid property of the solution, this second layer is made of free ions that move in the liquid under the influence of electric attraction and thermal motion rather than being firmly anchored. Although the existence of EDL is a fact, its structure is clear and it has been used for understanding many liquid–solid interface related chemical processes, a fundamental question is what is the cause for the first layer becoming charged at the very beginning when it first contacts with a liquid? A possibility is the initial electron exchange between liquid molecules and solid surface atoms at the very initial contact.

Recently, it was proposed by Wang and Wang [29] that the formation of EDL hay have two-steps (Fig. 1.9). The first step is an electron exchange process between liquid and solid surfaces via CE, which makes the atoms on the solid surface to be ions. This is the first process because electron transfer occurs in picosecond or femtosecond time scale, rather fast. The second step is the interaction of the ions with the ions in the liquid, resulting in a gradient distribution of cations and anions near the interface. Ion migration is rather slow and it is several orders of magnitudes slower than that of electron transfer. The transferred electrons would electrostatically interact with ions in the liquid to form a layer of ion adsorption. The traditional model ignores the first step, just considering the second step. Such a revision about the model for EDL could subsequent affect some related understanding about the interface chemistry, electrochemistry and even cellular-level interactions. Therefore, for simple referring, the two step model is called Wang model for EDL [34].

In the first step, when the molecules in the solution first approach a virgin surface that has no pre-existing surface charges (Fig. 1.9a), it may be possible that the atoms/ molecules in the solution directly interact with the atoms on the solid surface to form strongly overlap of electron clouds, thus, electron transfer is possible as a result of CE

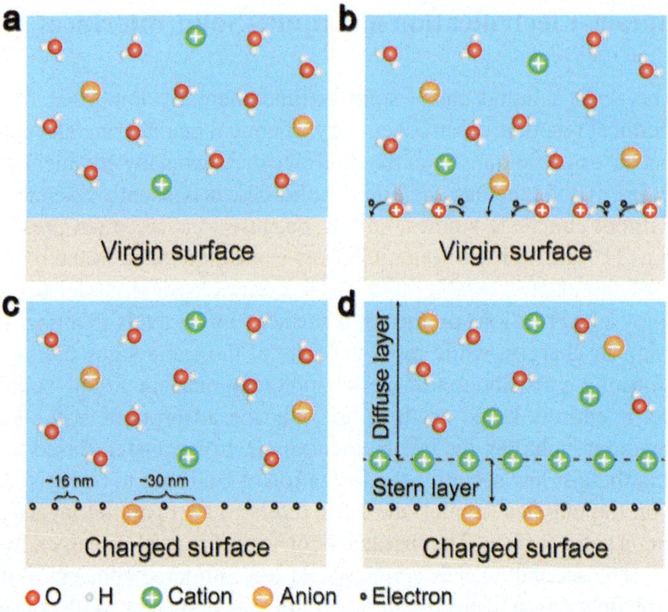

Fig. 1.9 The formation of EDL is a result of contact-electrification at a liquid–solid interface. This diagram shows the proposed two-step model (Wang model) on the procedures for forming an electric double layer at a liquid–solid interface. The first step: formation of the very first layer of electrostatic charges on the solid surface. **a** Schematic representation of molecules in solution; **b** The adsorption of molecules on the surface and electron transfers result in the surface being charged; **c** the adsorbed molecules being pushed off the adsorbed sites due to pressure of the liquid. The second step: segregation of ions in the liquid. **d** the presence of the charged surface as a result of the first step; **e** adsorption of ions in liquid on charged solid surface, forming the EDL. From Lin and Xu et al. [33]

at liquid–solid interface. Electron transfer occurs first to make the "neutral" atoms on solid surface become charged, i.e., the formation of ions (Fig. 1.9b). Water molecule could loss an electron and becomes H_2O^+ ion. Recently, using time resolved X-ray absorption spectroscopy, it has been found that the ionization of H_2O Produces an H_2O^+ ion [35]. But the lifetime of H_2O^+ is about 200 ps before it décomposes into $OH + H_3O^+$. This result supports the proposed model and confirmes that electron transfer is possible at liquid–solid interface. This study suggests that water is not just a solvent as we understand from traditional chemistry, it in fact participates in chemical reaction!

In the second step, if there are ions existing in the liquid, such as H^+ and OH^-, the loosely distributed negative ions in the solution would be attracted to migration toward the surface bonded ions due to electrostatic interactions from the transferred electrons, forming an EDL. *Our suggestion here is that the origin of forming the EDL is likely the result of CE due to electron transfer at the very first step.* Once the EDL is formed, many discussions and models in the literatures regarding to related potential distribution and surface chemistry are applicable. With considering the large differences in mass, diffusion rate and sizes, the transport or leakage behavior of the electrons and ions on the surface can be hugely different, which eventually affect the chemical processes at surfaces for the purpose of energy storage using supercapacitor, corrosion, redox reaction and ion exchange. Chapter 7 gives a systematic description about the mechanisms of CE at liquid–solid interfaces, emphasizing charge transfer processes across interfaces involving conductors, dielectrics, and semiconductors. It emphasizes the formation and dynamics of the electrical double layer, a fundamental structure regulating charge behavior at interfaces. Wang's two-step model is introduced to elucidate the role of CE in EDL formation on dielectric surfaces.

1.6 Contact-Electro-Catalysis

Contact-electro-catalysis (CEC), first proposed in 2022 by Wang's group [15, 36, 37], is an emerging field that utilizes CE-driven electron exchange at the liquid–solid and even liquid–liquid interfaces to stimulate redox reactions. The energy source of CEC is external mechanical agitations, and the solid to be used is general organic (such as FEP, PTFE) and in-organic materials[47] (such as SiO_2, Al_2O_3 etc.) despite they are chemically inert. CEC is proposed based on the coupling of the physical and chemical processes in electron transfer during solid–liquid CE, representing a typical interdisciplinary research field. Figure 1.10a shows the difference of CEC from traditional catalysis, and the basic mechanism for CEC is shown in Fig. 1.10b. The traditional catalysis is an exchange of electron between two reacting molecules right at the surface of the catalysis, which means that the two type of molecules have to meet simultaneously at the same site for the electron exchange to take place. While for CEC, the electron transfer occurs step by step without the direct meeting between the two molecules at the same time. Taking the interaction between FEP particles and an aqueous solution as an example (Fig. 1.10b), when water molecules come into contact with FEP particles, electrons are transferred from the water molecules to the FEP surface. Subsequently, dissolved oxygen in the water gains electrons from the FEP surface, which is a physical process. From the chemical perspective, the water molecules lose electrons, converting water into radical cations, which then rapidly transfer protons to form hydroxide ions and hydroxyl radicals. Then, under the excitation of mechanical agitation such as ultrasonic wave or collisions, the transferred electrons on the surface of the FEP particles would be released and then captured by O_2 molecules in water, superoxide radicals are thus formed, restoring the FEP particles to their original uncharged state, and the water molecule is recovered. This cycle continues with the persistence of mechanical stimulation, generating a large number of reactive radicals in the solution, thereby promoting the occurrence of related chemical reactions, which is the basic principle of contact electro-catalysis, which is believed how does catalysis occur in nature in the absence of light.

CEC might be the potential mechanism for illustrating why the running water can self-clean. Since the CE phenomenon during the flow of water could catalyze the production of ROS for degrading organic pollutants. Besides, CEC has enabled

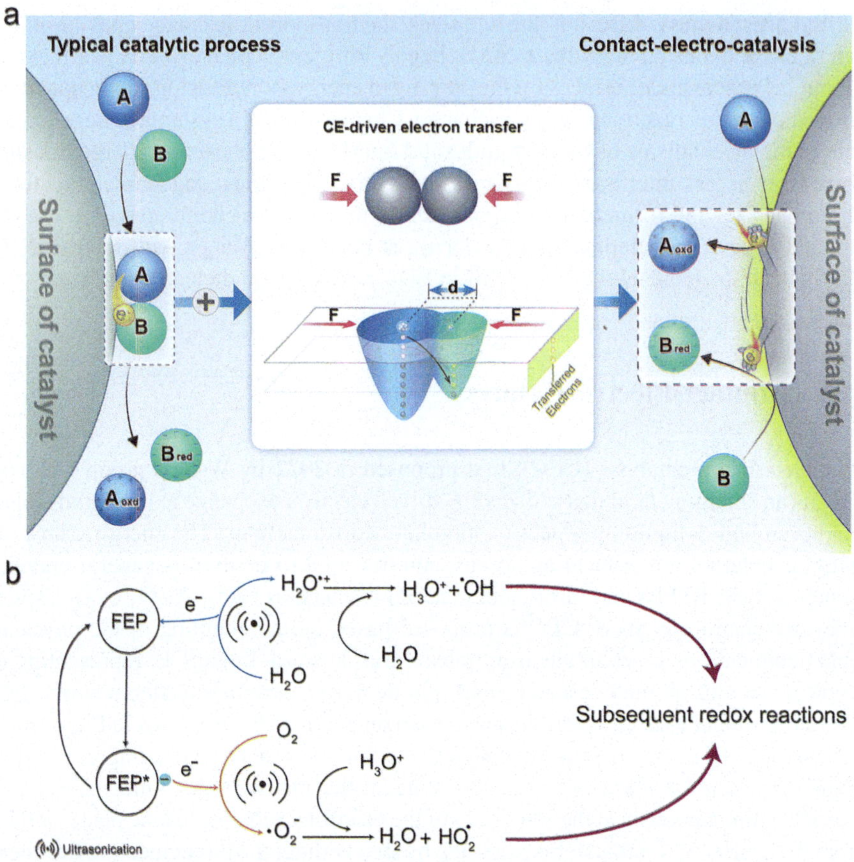

Fig. 1.10 **a** A comparison between traditional catalysis (left-hand side) and contact-electro-catalysis (right-hand side). In traditional catalysis, the reacting molecues A and B have to be met simultaneously at the same site on the catalysis surface before reaction occurs. While in CEC, the two types of molecules are not required to meet at the same time, but arrive sequentially. **b** The mechanism of contact-electro-catalysis

direct synthesis of H_2O_2 under ambient conditions and even anaerobic condition with rather low cost, no harmful intermediates or byproducts, and the potential for scalable production. The third representative application of CEC focuses on recycling cathode materials in spent lithium-ion batteries (LIBs). High leaching efficiencies of both Co and Li were achieved at mild conditions without using toxic and expensive inorganic strong acids, which could not only simplify the process for waste management, but also enable considerable economic benefits. Moreover, we envision the CEC could also be applied in the research frontier of biological sciences such as cancer therapy and longevity studies, which are closely related to the production of ROS. Chapter 8 systematically introduces the fundamental science of CEC, and Chap. 12 covers the various applications of CEC.

1.7 Summary

In summary, triboelectricity is evolving from a mere curiosity of electrostatics to a transformative force in modern technology. Its versatility in energy harvesting, environmental applications, and material science has unlocked new opportunities for sustainable, low-cost, and distributed power generation systems. With continued research and innovation, triboelectricity could play a critical role in the future of energy harvesting and self-powered technologies. This chapter introduces the brief history of studying triboelectricity. The book is composed of two parts. The first part focuses on the fundamental science related to CE for all of the interfaces in matters. The second part is about the applications of CE in various fields, such as contact-electro catalysis and triboelectric nanogenerators. The book emphasizes more on the science side of CE by illustrating various mechanisms and associated applications, rather than from the tribology point of view that is more about friction and debris generation. We anticipate that the book will give a comprehensive review about recent progresses in the field in the last two decades. The objective is to accelerate the advances in science and technologies related to CE for the benefit of humankind.

References

1. https://en.wikipedia.org/wiki/Triboelectric_effect
2. W.R. Harper, *Contact and frictional electrification* (Clarendon P, Oxford, 1967)
3. P. Shaw, The electrical charges from like solids. Nature **118**, 659–660 (1926)
4. K.C. Pingali, S.V. Hammond, F.J. Muzzio, T. Shinbrot, Use of a static eliminator to improve powder flow. Int. J. Pharm. **369**(1–2), 2–4 (2009)
5. F. Salama, A. Sowinski, K. Atieh, P. Mehrani, Investigation of electrostatic charge distribution within the reactor wall fouling and bulk regions of a gas–solid fluidized bed. J. Electrost. **71**(1), 21–27 (2013)
6. Y. Pu, M. Mazumder, C. Cooney, Effects of electrostatic charging on pharmaceutical powder blending homogeneity. J. Pharm. Sci. **98**(7), 2412–2421 (2009)
7. T. Burgo, C. Silva, L. Balestrin, F. Galembeck, Friction coefficient dependence on electrostatic tribocharging. Sci. Rep. **3**(1), 2384 (2013)
8. K. Sayfidinov, S.D. Cezan, B. Baytekin, H.T. Baytekin, Minimizing friction, wear, and energy losses by eliminating contact charging. Sci. Adv. **4**(11), eaau3808 (2018)
9. C. Liu, A.J. Bard, Electrons on dielectrics and contact electrification. Chem. Phys. Lett. **480**(4–6), 145–156 (2009)
10. L.S. McCarty, G.M. Whitesides, Electrostatic charging due to separation of ions at interfaces: contact electrification of ionic electrets. Angew. Chem., Int. Ed. **47**, 2188–2207 (2008)
11. A. Diaz, D. Wollmann, D. Dreblow, Contact electrification: ion transfer to metals and polymers. Chem. Mater. **3**(6), 997–999 (1991)
12. J. Lowell, The role of material transfer in contact electrification. J. Phys. D Appl. Phys. **10**(17), L233 (1977)
13. F. Fan, Z. Tian, Z.L. Wang, Flexible triboelectric generator. Nano Energy **1**, 328–334 (2012)
14. Z.L. Wang, From contact electrification to triboelectric nanogenerators (Review). Rep. Prog. Phys. **84**, 096502 (2021)
15. Z. Wang, X. Dong, W. Tang, Z.L. Wang, Contact-electro-catalysis (CEC). Chem. Soc. Rev. **53**, 4349–4373 (2024)

16. Z. L. Wang, Y. Yang, J. Zhai, J. Wang (eds.), *Handbook of Triboelectric Nanogenerators*. Springer Nature (2023)
17. B. Terris, J. Stern, D. Rugar, H. Mamin, Contact electrification using force microscopy. Phys. Rev. Lett. **63**(24), 2669 (1989)
18. D.J. Lacks, T. Shinbrot, Long-standing and unresolved issues in triboelectric charging. Nat. Rev. Chem. **3**(8), 465–476 (2019)
19. S. Lin, L. Xu, L. Zhu, X. Chen, Z.L. Wang, Electron transfer in nanoscale contact electrification: photon excitation effect. Adv. Mater. **31**, 1901418 (2019)
20. S. Lin, L. Xu, C. Xu, X. Chen, A.C. Wang, B. Zhang, P. Lin, Y. Yang, H. Zhao, Z.L. Wang, Electron transfer in nanoscale contact electrification: effect of temperature in the metal–dielectric case. Adv. Mater. **31**, 1808197 (2019)
21. Y.S. Zhou, S. Wang, Y. Yang, G. Zhu, S. Niu, Z.-H. Lin, Y. Liu, Z.L. Wang, Manipulating nanoscale contact electrification by an applied electric field. Nano Lett. **14**(3), 1567–1572 (2014)
22. S. Li, Y. Zhou, Y. Zi, G. Zhang, Z.L. Wang, Excluding contact electrification in surface potential measurement using kelvin probe force microscopy. ACS Nano **10**(2), 2528–2535 (2016)
23. G. Castle, L. Schein, General model of sphere-sphere insulator contact electrification. J. Electrost. **36**(2), 165–173 (1995)
24. L. Schein, M. LaHa, D. Novotny, Theory of insulator charging. Phys. Lett. A **167**(1), 79–83 (1992)
25. Y. Zhou, S. Li, S. Niu, Z.L. Wang, Effect of contact- and sliding-mode electrification on nanoscale charge transfer for energy harvesting. Nano Res. **9**(12), 3705 (2016)
26. C. Xu, Y. Zi, A.C. Wang, H. Zou, Y. Dai, X. He, P. Wang, Y.C. Wang, P. Feng, D. Li, Z.L. Wang, On the electron-transfer mechanism in the contact-electrification effect. Adv. Mater. **30**, 1706790 (2018)
27. D. Li, C. Xu, Y. Liao, W. Cai, Y. Zhu, Z. L. Wang, Interface Inter-atomic Electron-transition Induced Photon Emission in contact-electrification. Sci. Adv. **7**(39), eabj0349 (2021)
28. S. Lin, Z.L. Wang, The tribovoltaic effect (Review). Mater. Today **62**, 111–128 (2023)
29. Z.L. Wang, A.C. Wang, On the origin of contact-electrification. Mater. Today **30**, 34–51 (2019)
30. M. Zheng, S. Lin, Z. Tang, Y. Feng, Z.L. Wang, Photovoltaic effect and tribovoltaic effect at liquid-semiconductor interface. Nano Energy **83**, 105810 (2021)
31. Z. Zhang, D. Jiang, J. Zhao, G. Liu, T. Bu, C. Zhang, Z.L. Wang, Tribovoltaic Effect on Metal-Semiconductor Interface for Direct-Current Low-Impedance Triboelectric Nanogenerators. Adv. Energy Mater. **10**(9), 1903713 (2020)
32. R. Xu, Q. Zhang, J.Y. Wang, D. Liu, J. Wang, Z.L. Wang, Direct current triboelectric cell by sliding an n-type semiconductor on a p-type semiconductor. Nano Energy **66**, 104185 (2019)
33. S. Lin, L. Xu, A.C. Wang, Z.L. Wang, Quantifying electron-transfer in liquid-solid contact electrification and the formation of electric double-layer. Nat. Commun. **11**, 1–8 (2020)
34. S. Lin, X. Chen, Z.L. Wang, Contact-electrification at liquid-solid interface. (Review), Chem. Rev. **122**, 5209–5232 (2022)
35. Z. H. Loh, G. Doumy, C. Arnold, L. Kjellsson, S. Southworth, A. Al Haddad, Y. Kumagai, M.F. Tu, P. Ho, A. March, Observation of the fastest chemical processes in the radiolysis of water. Science **367**, 179–182 (2020)
36. Z. Wang, A. Berbille, Y. Feng, S. Li, L. Zhu, W. Tang, Z.L. Wang, Contact-electro-catalysis for the degradation of organic pollutants using pristine dielectric powder. Nat. Commu. **13**, 130 (2022)
37. Z. Wang, X. Dong, X. Li, Y. Feng, S. Li, W. Tang, Z.L. Wang, A universal contact-electro-catalysis process for producing reactive oxygen species by ball milling of triboelectric materials. Nat. Commu. **15**, 757 (2024)

Chapter 2
Methods for Quantifying Contact-Electrification

Abstract Contact electrification (CE) has been one of the most important physical phenomena in surface science for more than 2600 years. Although a large number of experiments have been conducted to detect CE and theories about electron, ion, and material transfer have been proposed, the mechanism of CE remains a subject of ongoing debate and continues to stimulate the interest of scientists. A key issue in charge transfer research is how to accurately measure charge transfer in various situations. Developments in charge transfer measurements have consistently led to a deeper understanding of CE. In this review, we explore the field of CE with a focus on methods for measuring charge transfer. And a quantitative materials database serves as a valuable resource for systematically analyzing the relationship between triboelectrification and the properties of different materials. Numerous approaches have been developed to assess and measure the triboelectric properties of materials. This chapter focuses on the methods used to characterize the triboelectric charge density of various materials. These quantitative triboelectric charge densities could act as a benchmark reference for advancing the use of triboelectrification in energy harvesting and self-powered sensing applications.

2.1 Introduction

Contact energization (CE) is a common physical phenomenon [1] that occurs at almost any interface, including solid–solid [2–5], liquid–solid [6–9], and liquid–liquid interfaces [10, 11]. As a fundamental scientific topic, the identity of charge carriers (electron transfer [12–15], ion transfer [16, 17] or matter transfer [18, 19]) in CE has been attracting generations of researchers. However, even after decades of debate, the academic community has not reached a final consensus on the mechanism of CE. In addition to its scientific significance, revealing the mechanism of CE is also important for technological development. Although the amount of charge transferred in CE is limited, it can easily generate an electric field of tens of thousands of volts, and the resulting electrostatic attraction/repulsion is quite significant on the microscopic scale.

Z. L. Wang, *Contact-Electrification of Matter*, Springer Series in Surface Sciences 75,
https://doi.org/10.1007/978-3-031-88504-4_2

In the study of CE, the development of charge measurement methods is fundamental. In order to reveal the mechanism of CE, various charge measurement methods have been developed and applied, including macroscopic and microscopic measurement methods, liquid and solid matter charge measurement methods, and surface and body charge measurement methods. Based on these methods, the mechanism of CE is continuously updated. Theories such as electron transfer [12–15], ion transfer [16, 17], and material transfer [18, 19] have been proposed successively, leading to the increasing of our understanding of CE. However, there is still a long way to go to completely reveal the mechanism of CE. Given the importance of charge measurement in CE research, the design and development of charge measurement methods is one of the key issues in CE research. With the emergence of new measurement tools, new effects in CE are usually discovered. Therefore, it is necessary for researchers to have a clear understanding of the existing charge measurement methods, including their testing principles, advantages and disadvantages, and applications.

In this paper, we systematically review the charge measurement methods, especially their applications in CE research. The charge measurement methods are categorized, including their working principles, advantages and disadvantages, applications and their contributions to the development of charge measurement theory. In addition, the paper introduces the relevant theoretical research methods and their implications for CE research.

2.2 Macroscale KPFM

The presence of surface charge on a solid affects the surface potential. Therefore, a way of thinking about obtaining the surface charge is to measure the surface potential, which can be accomplished by using a Kelvin probe. Kelvin probes were originally used to measure the contact potential difference (CPD) between two metal surfaces. When two metals with different surface functionalities are connected by a wire, a CPD exists between the two metal surfaces. As the two metals are in close proximity to each other, a surface charge is induced due to the presence of the CPD and capacitance. The relationship between induced charge, capacitance between two metals and CPD can be given by the equation

$$Q = CV_{CPD} \tag{2.1}$$

where Q is the induced charge on the surface of the metal, C is the capacitance between the two metals, and V_{CPD} is the CPD between these two metals.

If we vibrate one of the plates, the capacitance C in Eq. (2.1) is a varying value, which causes the charge on the surface of the metal plate to oscillate, generating an alternating current (AC) in the wire connecting the two plates. To counteract the effect of the CPD between the two metal plates, we apply a voltage source in the loop. When the applied compensating voltage and CPD are equal and opposite, we have

$$Q = C(V_{CPD} - V_c) = 0 \tag{2.2}$$

where V_c is the compensation voltage.

At this point, no matter how to change the capacitance, the charge on the surface of the pole plate is 0, and of course the current in the wire is equal to 0. As long as we know the value of the applied compensation voltage, we can get the CPD between the two metals.

When a Kelvin probe is used to measure the surface charge of an insulator, it is necessary to install a metal electrode on the backside of the insulator layer. In the absence of charge on the surface of the insulator, the Kelvin probe gives the CPD between the metal electrode and the probe. however, when there is a charge on the surface of the insulator, the solid surface charge also contributes to the induced charge on the surface of the probe. At this point, an additional voltage is required to compensate for the current generated by this charge, which results in a change in the Kelvin potential, known as the compensation potential (V_c). In general, the Kelvin potential becomes positive when there is a positive charge on the surface and negative when there is a negative charge on the surface. There is a precise mathematical relationship between the insulator surface charge density and the Kelvin potential (V_c). This relationship needs to be calculated by modeling a parallel plate capacitor that is similar to the KPFM's computational model for measuring surface charge density. More details on this will be described later in this paper. A nice series of studies investigating various aspects of CE using Kelvin probes were carried out by Thiago et al. These studies examined the effect of atmospheric water on the generation and dissipation of polymer surface charge, the formation of charge patterns due to self-aligned ions on the polymer surface, and the effect of solvents on tribocharge patterns [20–22]. Figure 2.1a shows the surface potentials recorded using macroscopic KPFM scanning of PTFE and PMMA surfaces that were previously charged by friction rubbing with a rotating PE foam disk at controlled pressures and speeds. The electrostatic potentials on many pixels exceeded ± 3 kv, indicating that the PTFE surfaces contain separated positive and negative fixed charges that are arranged in macroscopic (positive or negative) domains in the millimeter-centimeter range, which is quite counterintuitive and unanticipated within the conceptual framework of the friction electric series. The charge surface concentration that produces a 3 kv potential at a 2 mm electrode-sample distance is 254 charges/μm^2. this means that the average distance between ions accounting for the excess local charge is in the 60 nm range. Assuming that the area occupied by the ions is in the range of 0.1 ~ 1 nm2, the proportion of surface area occupied by ions in the macroscopic domains producing excess charge is less than 0.03%, and the volume concentration of the net excess charge in the surface layer is about 10 ~ 7 mol L^{-1}. At this low concentration, the electrostatic repulsive interactions are very low, following the classical Debye—Huckel theory for electrolyte solutions. Potential maps also show PE films rubbing against glass spheres and PTFE particles on the tabletop of a planetary mixer. An uneven charge is observed: positive domains dominate on the PTFE-scratched PE surface, but are scarce on the PE in contact with the glass balls [22]. Figure 2.1b show some Kelvin probe measurements, the surface potential distributions of the two polymers after CE, and the surface potential distribution of PTFE obtained by the Kelvin probe, which can be seen to have a resolution of roughly a few millimetres [20, 21].

Figure 2.2 give an example for the studying of contact electrification between liquid and solid by using macroscale KPFM. Qin et al. placed the TENG on a support frame with an inclination angle of 11.5° and the back electrode of the TENG was connected to the ground via an electrostatic meter [23]. A glass dropper was placed on the TENG with a ground electrode at its lower end. Oil droplets first fall from the dropper to contact (or collide with) the surface of the TENG, then gradually roll or slide to the lower end of the TENG, and finally leave the TENG and fall into the sink. As shown in Fig. 2.2a, b, the use of the macroscopic KPFM method is directly related to the surface charge density linearly by measuring the surface potential.

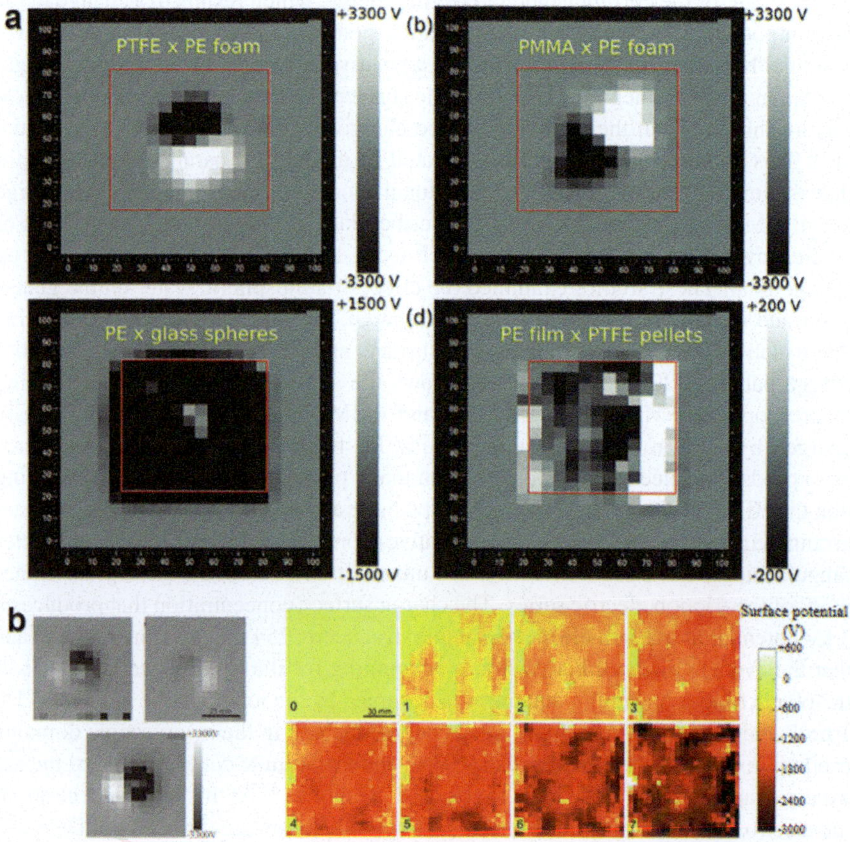

Fig. 2.1 The surface potential of the insulator samples captured by using macroscale KPFM. **a** The contact electrification mosaic effect on the insulator surface. **b** The change of the surface potential of insulator sample. Figure with permission [20], copyright © 2012 The Chemical Society of Japan. Figure with permission [21], copyright © 2013, The Author(s)

As the number of drops increases, the surface potential gradually increases, and up to 400 drops, the surface potential at the sample point is linearly related to the number of drops. Figure 2.2d shows the surface potential at different drop heights, where the potential was imaged after 20 drops of glycerol were dropped. As shown in Fig. 2.2, the potential of the impact zone gradually enhanced with the increase of drop height, showing a trap-like shape of increasing area and rising depth on the 3D image. This indicates that the increase in drop height mainly enhances the impact zone interaction by strengthening the collision. As shown in Fig. 2.2, the minimum value of the surface potential is almost linearly related to the drop height until 7.5 cm, at which the slope becomes flat. Figure 2.2 shows the surface potential of the impact zone after dropping different types of liquids on the TENG surface. When 400 drops of each liquid were applied, the potentials of squalane, castor oil, glycerol, and deionized water were −11.61, −29.44, −60.63, and −70.97 V, respectively. In the experimental setup, the DI water showed a stronger energizing ability in the impact zone. The low viscosity of the water enhances the contact with the surface during the fall. The state of the back electrode during the fall also affects the surface energization. As shown in Fig. 2.2f, whether the back electrode is open or grounded has no significant effect on the energization, and the applied voltage can adjust the section energization. Applying a voltage of −1000 V to the back electrode during oil drop results in a significant increase in the surface potential generated in the impact zone, and a voltage of 1000 V leads to a decrease in the surface potential, which means that more negative charges are generated on the surface during energization. The results show that the charged charge can be regulated by changing the potential of the back electrode during energization.

Macroscale KPFM can also be used to investigate the effect of pre-charge on the contact electrification between liquid and solid. Tang et al. investigated the effect of pre-charging on contact electrification at liquid–solid interface, as shown in Fig. 2.3, and they used the macroscopic KPFM method to analyse the surface potential distribution of PTFE under different solid–liquid contact conditions. Figure 2.3a (i–iii) shows the experimental schematic. The PTFE film was bonded to the backed copper electrode and the surface potential was detected with a potentiometric probe, and the scanning path is shown in Fig. 2.3a (iii). The surface potential distributions of PTFE in the initial state, after contact with droplets, after treatment with electric field and CE under electric field were investigated. As can be seen in Fig. 2.3b, the initial PTFE surface has a slightly negative potential, and when rubbing with water droplets, the potential in the CE region decreases significantly (Fig. 2.3c), indicating that the PTFE acquires more negative charge. In order to exclude the possibility that the polymer is polarized in an electric field and has an intrinsic surface potential, the PTFE film was subjected to both a forward and a reverse electric field, and Fig. 2.3d, e show that the surface potential changed negligibly compared to the initial state. Subsequently, when the polytetrafluoroethylene was rubbed against a water droplet under the electric field, it was unexpectedly found that the surface potential in the CE region became positive (Fig. 2.3f), indicating that the polytetrafluoroethylene lost

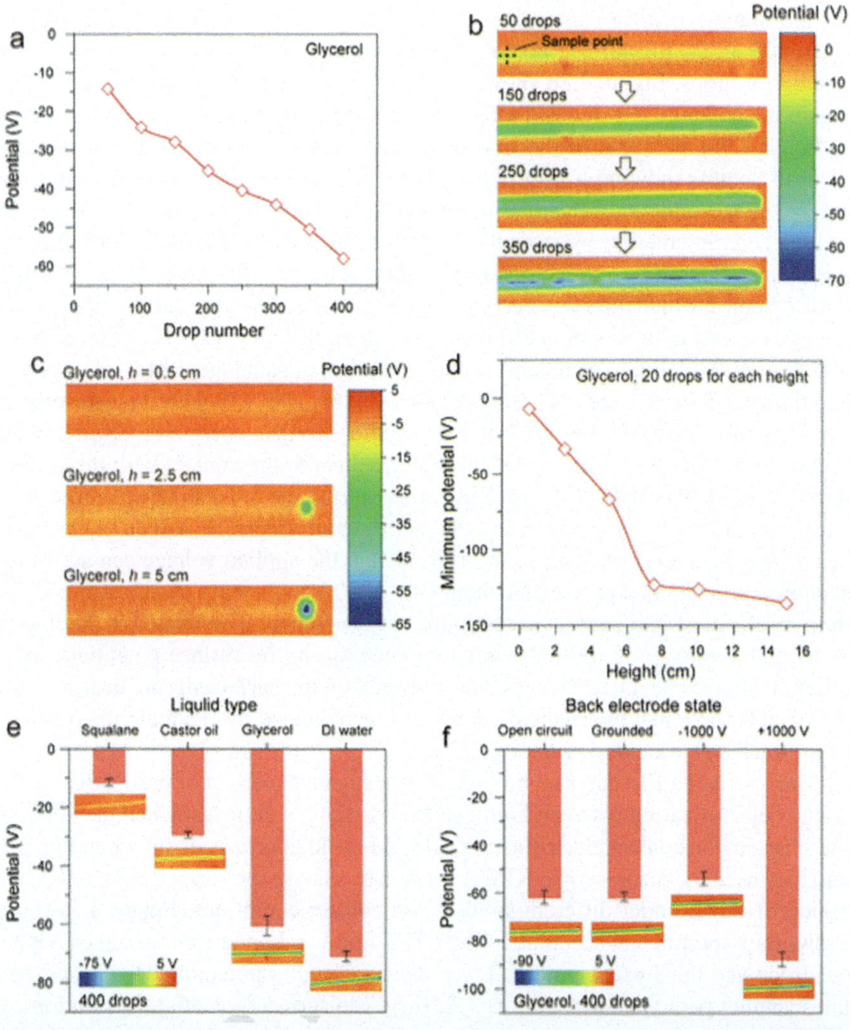

Fig. 2.2 The study of contact electrification between liquid and solid by using macroscale KPFM. **a, b** Surface potential at the sample point **a** and potential images of the motion trace **b** for different numbers of oil droplets. **c, d** Potential images of the motion trace **c** and the minimum potential **d** for different dropping heights. **e** Surface potential with different types of liquids. Figure with permission [23], copyright © 2023, Elsevier Ltd.

electrons, and when the direction of the electric field was changed, the surface potential of the polytetrafluoroethylene decreased, proving that the polytetrafluoroethylene gained electrons (Fig. 2.3g) [24].

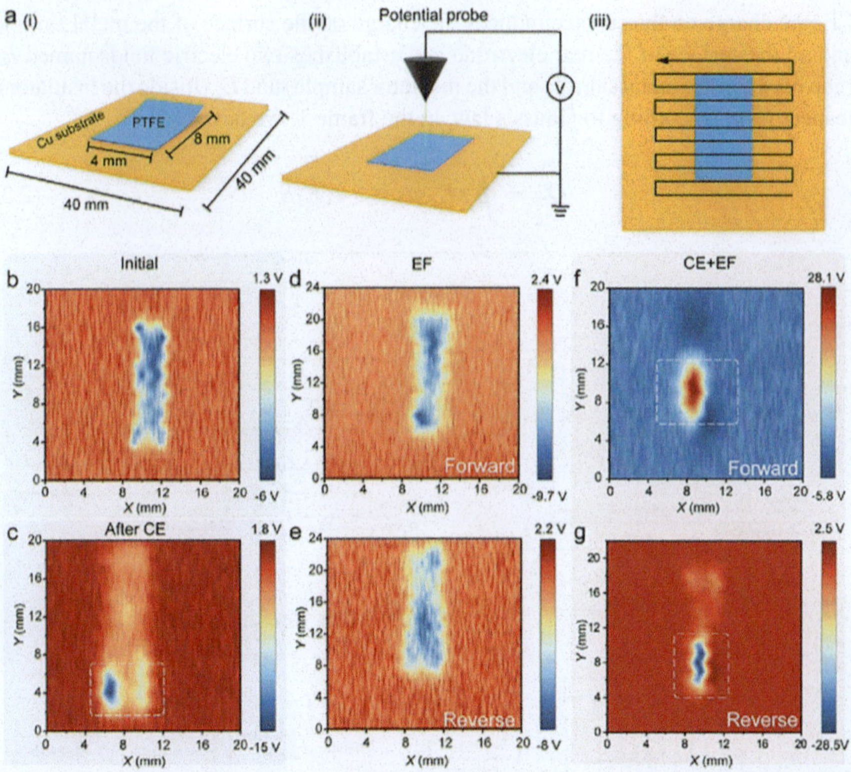

Fig. 2.3 The effect of pre-charging on the contact electrification studied by using macroscale KPFM. **a** Schematic diagram of the surface potential measurement process. **b** The initial state. **c** After contact with the water droplet. Under positive **d** and reverse **e** electric field treatment. After contact with water, when positive **f** or reverse **g** electric field is applied, respectively. The white wire frame represents the area in contact with the water droplet. Figure with permission [24], copyright © 2022, American Chemical Society

2.3 Capacitor Probe

The parallel plate system is a classical, simple and efficient device for CE measurements [25–27], which can be considered as one kind of capacitor probe. In the parallel plate system, two samples are fixed on two opposite back electrodes and a linear electrode is used to bring one sample into contact with the other. When the two samples are separated, the charge generated by the CE remains on the surfaces of the two samples and induces a charge on the surfaces of the two back electrodes, which generates a potential difference between the two back electrodes, which can be measured by an electrostatic meter. In addition, there is a precise relationship between the charge transferred on the sample surface and the potential difference between the two rear electrodes [26]. As shown in Fig. 2.4a, taking the CE between the metal and insulator as an example, when the insulator surface is charged by the

CE, the charge on the surface induces the charge on the surface of the metal sample and on the surface of the rear electrode and establishes two electric fields named E_1 (gap between the metal sample and the insulator sample) and E_2 (inside the insulator), respectively. According to Gauss's law, in the frame 1, we have

$$Q_m = \oint \varepsilon E \cdot ds = -\varepsilon_0 E_1 S \tag{2.3}$$

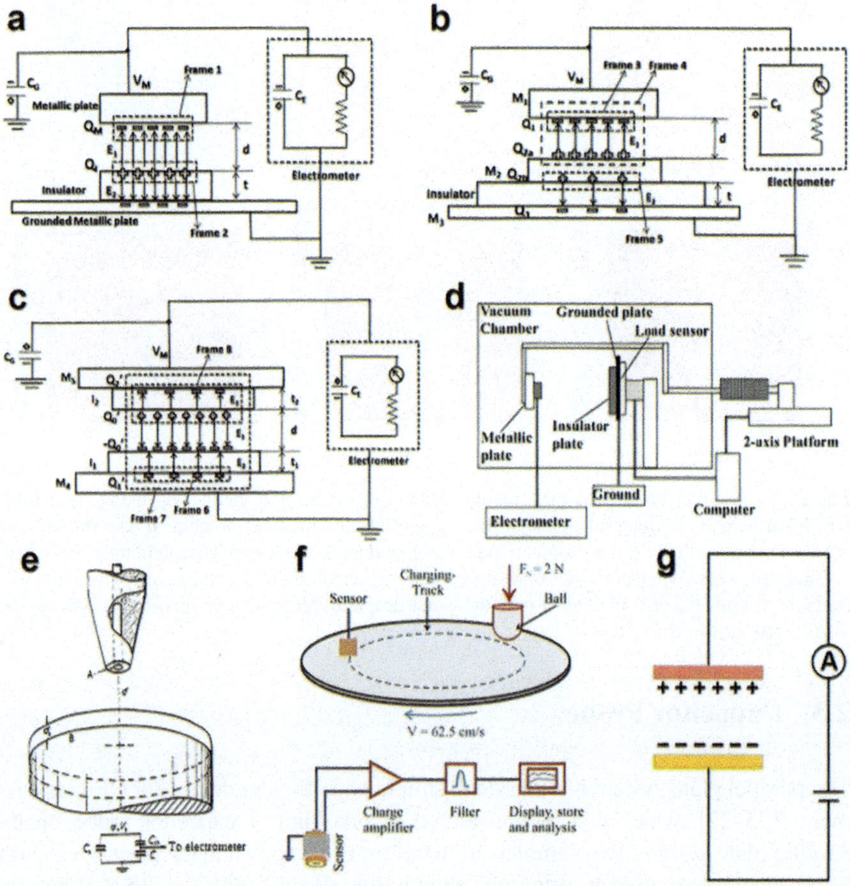

Fig. 2.4 Schematic diagram of the capacitor probe method. Macroscale capacitor probe for **a** metal–insulator CE, **b** metal–metal CE and **c** insulator-insulator CE. **d** The setup of the capacitor probe with environmental control chamber. **e** Illustration of a capacitor probe and its working principle. **f** An experimental setup for measuring the triboelectrification using a capacitor probe. **g** Parallel capacitor to describe the working principle of Kelvin probe. Figure with permission [26], copyright © 2013 Elsevier B.V. Figure with permission [30], copyright © AIP Publishing

where E is the electric field strength, Q_m is the induced charge on the surface of the metal specimen, S is the contact area, and ε_0 is the vacuum dielectric constant.

In the frame 2, we have

$$Q_i = \oint \varepsilon E \cdot ds = \varepsilon_0 E_1 S + \varepsilon_0 \varepsilon_r E_2 S \tag{2.4}$$

where Q_i is the CE charge on the insulator surface and ε_r is the relative dielectric constant of the insulator sample.

The charge on the electrostatic meter (Q_e) and the charge on the grounded capacitor (Q_g) are given by

$$Q_e + Q_g = V_m(C_e + C_g) \tag{2.5}$$

where C_e is the capacitance of the electrostatic meter, C_g is the capacitance to ground, and V_m is the value of the potential difference displayed by the electrostatic meter.

According to the conservation of charge, the relationship between charges should be

$$Q_e + Q_g + Q_m + Q_i = 0 \tag{2.6}$$

In addition, the potential difference between the metal sample and the back electrode can be expressed as

$$V_m = E_2 t - E_1 d \tag{2.7}$$

where t is the thickness of the insulator specimen and d is the gap between the metal sample and the insulator specimen at the time of measurement.

Combining Eqs. (2.3) and (2.7), we get

$$Q_i = -V_m \frac{(C_e + C_g)\varepsilon_r d + (C_e + C_g)t + \varepsilon_0 \varepsilon_r S}{\varepsilon_r d} \tag{2.8}$$

Equation (2.8) gives the relationship between the CE charge on the surface of the insulator and the potential between the metal sample and the back electrode given by the electrostatic meter. The parallel plate system can also be used to study metal–metal CE and insulator-insulator CE with the experimental setups shown in Fig. 2.4b, c, respectively. Details of the transferred charge calculations for both cases in Fig. 2.4b, c can be found in the literature [26].

The advantage of the parallel plate method is that it gives the amount of charge transferred at the moment of separation of the two samples, thus eliminating the effect of charge escape on the charge measurement. On this basis, the effects of load, temperature, humidity and atmosphere on contact energization can be studied. For example, Zhang et al. investigated the effect of load on CE under vacuum conditions by parallel plate system experiments in a vacuum chamber, and proposed a mechanism by which deformation affects the density of surface states of a material and hence the CE [27].

As shown in Fig. 2.4e, the wire probe is encased in a metal shell to form a concentric cylindrical structure. The outer metal shielding shell is grounded and the centre probe is connected to an electrostatic meter to form a capacitive probe [28, 29]. When the probe is close to the solid surface, a capacitor C_{ps} is formed between the circular electrode and the charged surface, a capacitor C is formed between the electrode and the shielding shell, and a third capacitor C_1 is formed between the shielding shell and the charged surface. The specific circuit diagram is shown in Fig. 2.4e [28]. And all these three capacitors can be calculated based on the geometrical structure. Based on the relationship between the three capacitors, the relationship between the induced charge on the electrode surface and the charge density on the solid surface can be obtained, see reference [28]. This method usually has a millimetre resolution and can be used to observe the charge at a point in the friction track. As shown in Fig. 2.4f, a ball is loaded on a rotating insulator disk so that it rubs against the insulator disk against each other. The resulting friction track is a circle. A capacitive probe is placed on the other side of the circle to enable real-time observation of the friction initiation. This experimental setup was very popular for a long time, and the relation between the capacitor and the charge induced on the surface is shown in Fig. 2.4g [30–34].

A grounded metal is one of the simplest capacitor probes, which we call an electrode capacitor probe. When a grounded metal is near a charged surface, an induced charge is generated on the metal surface. An electrostatic meter can be used to measure the current flowing from the ground to the metal surface to get the amount of charge induced on the metal surface. In this method, the induced charge on the metal surface is generally not equal to the electrostatic charge on the solid surface. This is because the metal surface is open, and not only can the solid surface charge induce a charge on the surface of the metal electrode, but all charges around the metal electrode affect the amount of charge induced on the metal surface. Due to their simple structure, they have been widely used in CE research, especially before 1980 [35, 36]. Of course, based on some ingenious designs, the electrode capacitor probe method can still bring surprises, such as a series of studies of CE between ferromagnetic balls and insulators (Fig. 2.5a) [37–40].

As shown in Fig. 2.5b, the metal electrode probe is mounted under the insulator disk. At the bottom of the disk, there is a rotating magnet capable of driving a ferromagnetic ball to roll along a circular path on the insulator surface. During the rolling process, CE occurs between the ball and the insulator disk. As the ball moves away from the electrode, the electrode measures the charge on the disk close to the electrode. As the sphere moves over the electrode, the electrode measures the total charge on the sphere and the disk near the electrode. Thus, when the sphere moves for one cycle, a peak signal appears on the electrostatic meter which represents the sum of the charge on the sphere and the charge on the disk near the electrode. The background signal of charge on the disk, as shown in Fig. 2.5c. By subtracting the background, which corresponds to the charge on the surface of the disk, the amount

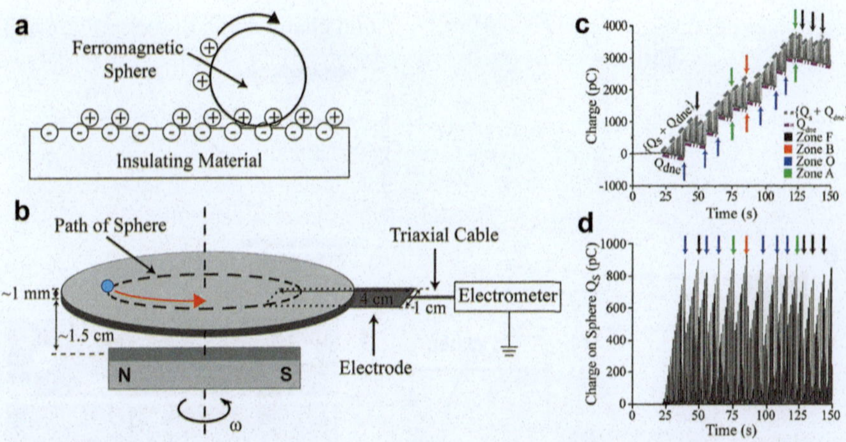

Fig. 2.5 Electrode probe for CE measurement. **a** Schematic illustration of the charge transfer between a rolling steel sphere and an insulating surface. **b** An experimental setup for measuring the dynamics of CE and electrical discharge using an electrode probe. **c** and **d** Data showing the charge sensed by the electrode probe during CE between steel sphere and insulating surface. Figure with permission [37], copyright © 2010, American Chemical Society

of charge on the surface of the sphere, as shown in Fig. 2.5d. Based on this method, the kinetic process between the CE ferromagnetic sphere and the insulator can be investigated, and the charging and discharging of the sphere, and of the insulator by the CE can be conveniently observed.

2.4 Faraday Cup

The structure of a Faraday cup is given in Fig. 2.6a. When a charged object enters a Faraday cup, it generates an electric charge on the inner and outer surfaces of the metal cup. According to Gauss's law, the charge on the inner wall of the inner cup (Q_{ii}) is equal to the charge carried by the object (Q_o) but in the opposite direction, while the charge on the outer wall of the inner cup (Q_{io}) is equal to the charge carried by the object. For the same reason, the induced charge on the inner wall of the outer cup, Q_{oi}, should be equal to the charge carried by the object, and instead, we have

$$Q_o = Q_{oi} = -Q_{io} = -Q_{ii} \tag{2.9}$$

Here, the outer wall of the inner cup and the inner wall of the outer cup form a capacitor and we have

$$Q_o = Q_{oi} = C\Delta V \tag{2.10}$$

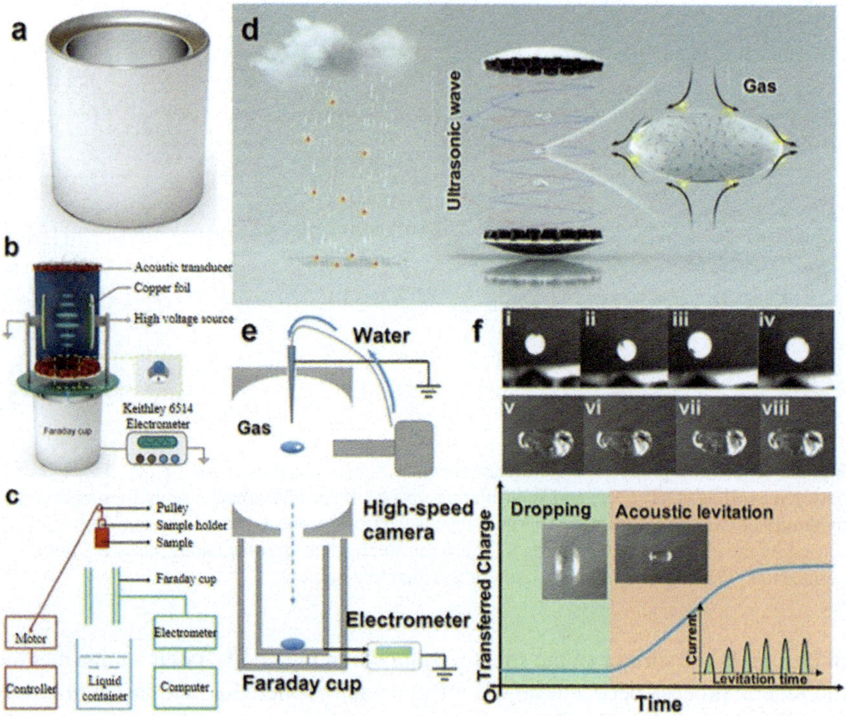

Fig. 2.6 Faraday cup for CE measurement. **a** schematic diagram of a Faraday cup. **b** A droplet charge measuring device based on Faraday cup. **c** A device for liquid–solid contact electrification measurement based on Faraday ring. **d–f** The contact electrification between air and liquid tested by using Faraday cup. Figure with permission [24], copyright © 2022, American Chemical Society. Figure with permission [50], copyright © 2023, The author(s)

where C is the capacitance capacity formed by the outer wall of the inner cup and the inner wall of the outer cup, the magnitude of which depends on the structure of the Faraday cup. ΔV is the potential difference between the inner and outer cups, which can be measured by an electrostatic meter or voltmeter.

From Eq. (2.10), we can increase the detection sensitivity by decreasing the capacity of the Faraday cup if the object carries a certain charge. For example, reduce the size of the Faraday cup and choose a proper space between the inner and outer cups.

There are two general designs for the application of Faraday cups in CE experiments. One is to perform CE experiments outside the Faraday cup and then put the object into the Faraday cup to realize the charge measurement [41]. As shown in Fig. 2.6b, a water droplet is lifted by the acoustic field and contacted with other solid balls, and then the acoustic field is turned off and the water droplet is dropped into the Faraday cup to obtain the charge on the water droplet [24]. However, this method has the disadvantage that the process of placing the object into the Faraday cup may come into contact with air or dust, leading to additional charge generation and inaccurate measurements. Another idea is to perform a CE experiment in a Faraday cup and then remove the object involved in the CE for measurement [42]. This method avoids the extra electricity generated by contact with objects such as air or dust, but makes the design of the device more difficult and increases the size of the Faraday cup. Faraday cups played a crucial role in early CE research and have long been a common device for CE studies. It is usually used to measure the CE behaviour of certain particles by rolling them on a plate and then letting them fall into the Faraday cup to detect charge transfer [43–46]. Some researchers have also used Faraday cups for CE charge dissipation experiments by placing charged samples into the Faraday cup and observing the charge dissipation [47, 48].

Measuring charge with a Faraday cup requires the object to be placed completely in the cup, and there is generally no way to avoid the object coming into contact with the inside wall of the Faraday cup, which would result in CE between the object and the Faraday cup. However, in some cases we do not want our charged sample to come into contact with the wall of the cup, or we may wish to avoid disrupting the motion of the object itself. In such cases, we need to hollow out the bottom of the Faraday cup to allow our sample to pass through the Faraday cup, also known as the Faraday ring [49]. The Faraday ring works similarly to the Faraday cup. However, the Faraday ring generates more leakage fields and is not as accurate as the Faraday cup. Nevertheless, this makes the design of the CE experimental setup more flexible. Of course, the leakage field can be reduced by increasing the depth-to-diameter ratio, thus improving the measurement accuracy. Since the CE process is not disrupted, the Faraday ring allows for the kinetic study of CE. For example, a Faraday ring is placed on top of a solid surface, and particles are dropped from a high place and bounced repeatedly on the surface. The Faraday ring can record the charge acquired by the particle in electrical contact at each bounce [49]. In addition, the Faraday ring can avoid mutual interference between the experimental components. As shown in Fig. 2.3c, Chen et al. suspended a solid sample above a liquid and realized CE between liquid and solid by stretching a string, at which time the bottom of the Faraday cup interfered with the string [50]. To avoid this interference, Chen et al. chose a Faraday ring for their measurements. They passed a string through the Faraday ring and stretched the solid sample in contact with the liquid and passed it through the Faraday ring to achieve charge measurement.

Figures 2.6d–f give a classic work about the CE between liquid and air by using Faraday cup. As shown in Fig. 2.6d, raindrops usually carry some electrostatic charge due to their frictional motion with the air. In the study by Wang et al. an acoustic field was used to suspend droplets, which were generated by two ultrasonic phased arrays (see Fig. 2.6d). The focal point of the sound was formed by precisely adjusting the acoustic frequency to 40 kHz while a Faraday cup was placed below the suspended droplets (see Fig. 2.6e). A precision stepper motor was used to push the syringe to place droplets of precisely controlled volume at the focal point of the sound field (20 μL per drop). The droplets were rotated in the acoustic field at very high speeds while a high-speed camera was used to measure the radius, height, and state of the droplets from the side. An ultra-sensitive electrostatic meter with a Faraday cup made of an aluminium plate shielded from external interference was used to quantify the charge of the drop. The charge on the suspended droplets was measured by integrating the induced short-circuit current generated by the Faraday cup when the droplets were dropped into the Faraday cup. The kinetic behaviour of the suspended droplets during the stable suspension phase was investigated using a high-speed camera, as shown in Fig. 2.6f. It was found that the suspended droplets could rotate steadily at 7200 rpm, thus generating sufficient frictional motion with the air (see Fig. 2.6fi–iv). The detailed rotation of the droplet is shown in Fig. 2.6a, f. The internal part of the droplet moves outward to achieve complete contact between the liquid and air (see Fig. 2.6fv–viii). The high-speed rotation of suspended droplets in an acoustic field can provide sufficient frictional motion between water and gas molecules to lead to a high probability of contact energization. As described in the experimental section, we have done our best to clean the air and remove the floating charges. The amount of charge on the droplets increased with suspension time while the droplets were fixed in the same position and there was no significant airflow in the chamber. Therefore, Wang et al. suggest that charge separation between liquid and gas is the main reason for this electrification process [51].

2.5 Vision Assisted Methods

By using a camara to observe the movement of the charged objects in the electric field is a way to measure the charge carried by the objects. The measurement of droplet charge has always been an important topic. In addition to the use of Faraday cups, an effective way to measure droplet charge is to place a droplet in a uniform electric field and then obtain the droplet charge by observing the electric field force applied to the droplet, as in the famous Milligan experiment and the recent studies on CE between water and hydrophobic surfaces. Similar to these experiments, Lin et al. devised a method to measure the amount of charge carried by fog microdroplets. In Lin et al.'s method, an ultrasonic nebulizer was placed below the gap between two parallel electrodes (50 cm apart), as shown in Fig. 2.7a. Deionized water (DI water) is atomized and the fog microdroplets will move upward and into the gap between the two electrodes. When a constant voltage is applied to the electrodes, a stable electric field will be established between the two electrodes and the charged micro droplets will be subjected to electrostatic force. As shown in Fig. 2.7b, the electrostatic forces applied to the negatively charged microdroplets and the positively charged microdroplets are directed toward the positive and negative electrodes, respectively. Therefore, the polarity of the charge carried by the microdroplets can be obtained by observing the direction of motion of the microdroplets in the electric field. When a fog microdroplet starts to move under the electrostatic force, the microdroplet is subjected to a resistance force opposite to the electrostatic force. The results show that the charged particles and microdroplets will accelerate in the electric field and quickly reach a stable velocity (terminal electrostatic velocity) due to the balance between the electric field force and the resistance. Based on the terminal electrostatic velocity and the size of the particles, the charge carried by the microdroplets can be calculated. Here, the terminal electrostatic velocity (VTE) is measured and calculated using a high-speed camera. As shown in Fig. 2.7c, when 8 and − 8 kV potentials were applied, the fog microdroplets were separated into two parts and attracted by the electrodes. This indicates that there are two types of microdroplets, positive and negative, in the ultrasonic atomization process. By observing the video, the distance travelled by the fog microdroplets in a certain time was measured, and the terminal electrostatic velocity of the microdroplets was obtained by dividing the distance travelled by the time. As shown in Fig. 2.7d, the diameter of the circular charge pattern, which is the diameter of the contact area between the water microdroplet and the PTFE membrane, can be directly measured using KPFM. Based on the geometric conservation and volume conservation of the microdroplets before and after adsorption, the diameter of the microdroplets before adsorption can be calculated [11].

Another example for the camera assisted method for contact electrification measurement is shown in Fig. 2.8 Tang et al. measured the charged energy of the droplets through ultrasonic suspension and visual aid methods. As shown in Fig. 2.8a, the acoustic suspension device composed of 72 ultrasonic transducers can generate standing waves, and the droplets can be trapped in the wave node of the sound field, so that they can be freely suspended without contact with environmental objects. According to the position of the ultrasonic transducer, the distribution of sound pressure in the sound field can be calculated, as shown in Fig. 2.8b. Further, the Gorkov potential in the sound field can be obtained, and the sound field force is the negative gradient of the Gorkov potential (Fig. 2.8c). A pair of parallel flat electrodes are placed on the left and right sides of the droplet to generate an approximately uniform electric field. If the droplet is charged, an electric field force will be generated on the droplet. According to the balance of forces, the electric and sound fields in magnitude and the liquid drop should equal the opposite direction. Figure 2.8d–f shows a schematic diagram of the measurement of the droplet charge. As shown in Fig. 2.8e, the droplet will be trapped in the center of the sound field, and when a voltage is applied to the right electrode and the other electrode is grounded, an electric field will be formed, as shown in Fig. 2.8d. In this case, if the droplet is positively charged, it will move to the left under the pull of the electric field forces until it reaches a position where the electric field forces and the sound forces are in balance. The greater the number of positive charges, the greater the electric field force on the droplet, and the farther the droplet moves to the left. If the droplet is negatively charged, the droplet will move to the right and eventually reach the equilibrium position, as shown in Fig. 2.8f. The sound force and the electric force can be obtained by measuring the movement distance of the droplet. Figure 2.8g–i shows the movement of droplets with different electric charges under an electric field. In Fig. 2.8 g, a droplet of deionized water (DI water) is positively charged by a metal ring with a voltage of –5 kV and trapped in a sound field. When a voltage of + 15kv is applied to the electrode on the right, positively charged deionized water droplets are dragged to the left. The distance between the two electrodes is set to 38 mm, and the movement distance of the droplets measured by the camera is 1.2 mm. As can be seen from Fig. 2.8c, the sound force is 7.05μN. It can be calculated that the positive

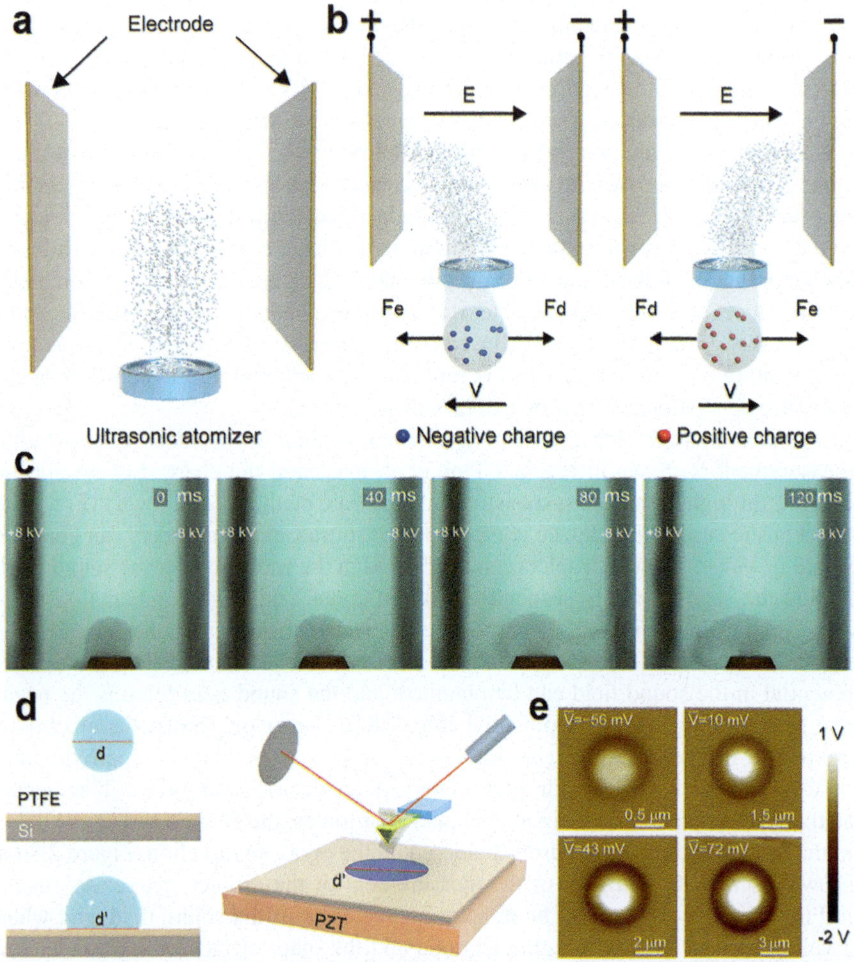

Fig. 2.7 The setup of the experiments. **a** The schematic diagram of the experiment set. An ultrasonic atomizer is placed below the gap of two parallel electrodes, which are 50 cm apart. **b** The schematic for the movement and force analysis of the negatively charged and positively charged water microdroplet in the uniformed electric field. The negatively charged microdroplet will move toward the positive electrode while the positively charged will move toward the negative electrode owing to the opposite's attraction. *Fe* denotes the electrostatic force on the microdroplet, and *Fd* denotes the drag force on the microdroplet. **c** The movement of the fog microdroplets taking by the high-speed camera, when the two electrode potentials are 8 kV and − 8 kV, respectively. **d** The measurement of the diameter of circular charge patterns on the PTFE surface by using KPFM, and the schematic of the relation between the diameter of the microdroplet before adsorption and the diameter of circular charge patterns. **e** The charge pattern generated on the PFTE surface by the evaporation of the microdroplets with different size (V is the average Kelvin potential). Figure with permission [11], copyright © National Academy of Sciences

charge of DI water drop in Fig. 2.8g is 17.88pC. If there is no electric field between the two electrodes, the droplet will stay in the center of the sound field, as shown in Fig. 2.8. The deionized water droplets were pre-charged with negative charge and + 15 kV voltage was applied to the right electrode, as shown in Fig. 2.8i. The droplets were attracted to the right electrode. Using the same method to measure the motion of a negatively charged water drop of 0.7 mm, the negative charge of the DI water drop in Fig. 2.8i is calculated to be − 10.72 pC. The camera used in the experiment has a resolution of 0.1 mm, so the sensitivity of the method can be as low as 1.55 pC, and can be further optimized by enhancing the electric field or changing the sound wave [52].

2.6 AFM-KPFM

2.6.1 Kelvin Probe Force Microscopy

In KPFM experiments for measuring the surface charge of an insulator, the insulator is usually a thin film deposited on a smooth conducting electrode. When the conductive probe is scanned on the sample surface, the probe, the insulator film and the conductive electrode can be regarded as a parallel flat plate capacitance model, as shown in Fig. 2.9a and b [53]. Let the charge density on the dielectric surface be σ, and the induced charge densities on the electrode tip surface and the electrode back surface be σ_1 and σ_2, respectively. the electric field at the gap between the tip of the needle and the dielectric surface is E_1, and that inside the dielectric is E_2.

According to Gauss's law, we have

$$E_1 = \frac{\sigma_1}{\varepsilon_0} \tag{2.11}$$

$$E_2 = \frac{\sigma_2}{\varepsilon_0 \varepsilon_d} \tag{2.12}$$

$$\sigma_1 + \sigma_2 + \sigma = 0 \tag{2.13}$$

where ε_0 is the vacuum dielectric constant and ε_d is the relative dielectric constant of the dielectric layer.

In KPFM mode, an AC bias ($V_{AC} \sin(\omega t)$) is applied between the tip and the conductive substrate to drive the vibration of the tip, and a DC bias (V_{DC}) is applied between the tip and the sample to compensate for the CPD. Thus, the potential difference between the tip and the conductive substrate can be written as

$$V = V_{DC} + V_{AC} \sin(\omega t) + V_{CPD} \tag{2.14}$$

where V is the potential difference between the tip and the conductive substrate.

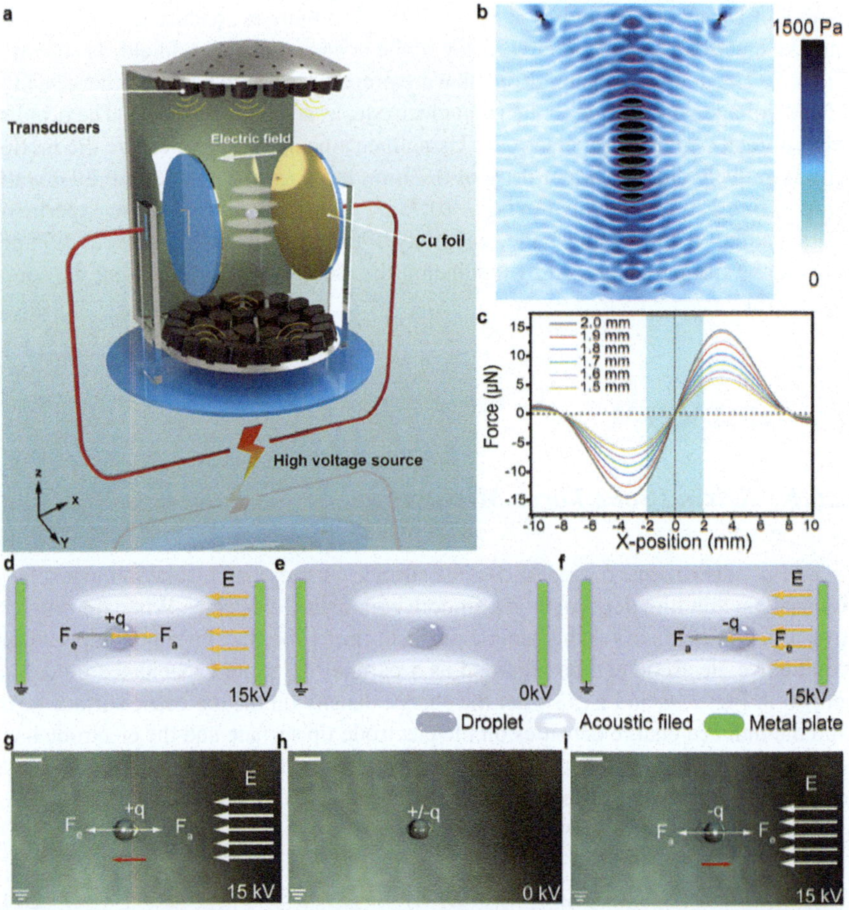

Fig. 2.8 Schematic illustration about the measurement method of charges on a freely suspended liquid droplet. **a** Schematic diagram showing experimental setup for directly measuring the quantity of charges carried by the droplet. **b** Simulated distribution of acoustic pressure in the acoustic field. **c** The acoustic field force on the liquid droplet in *x*-direction. **d–f** schematic diagram of the working principle. **e** Charged droplet stably suspends in position due to ambient acoustic force, when electric field is off. Attracted by the electrostatic force **d** droplet with negative charges moves toward to the positive plate and **f** droplet with positive charges repelled toward the negative plate, when electric field is on. **g–i** Photographs of the charge measurement process (The red arrow represents the movement direction of droplets; the image of the shadow represents the original position of the droplet). Scale bar: 2 mm. Figure with permission [52], copyright © 2021, Wiley-VCH GmbH

Furthermore, the potential difference between the tip and the conductive substrate can be described as

$$V = E_1 d_1 - E_2 d_2 \tag{2.15}$$

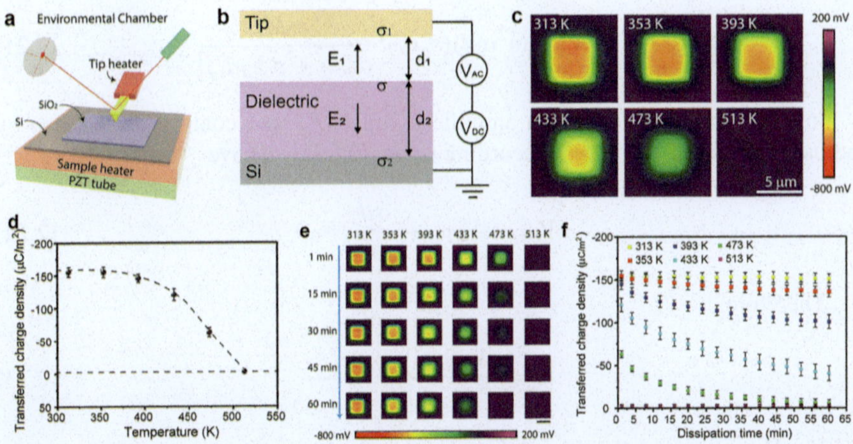

Fig. 2.9 The contact electrification studied by using KPFM. **a** The experiment setup for studying temperature effect on the contact electrification. **b** The capacitor model for the KPFM. **c–f** The charge decay under different temperature. Figure with permission [53], copyright © 2020, The Author(s). Figure with permission [54], copyright © 2019, WILEY-VCH Verlag GmbH & Co. KGaA, Weinheim

where d_1 is the distance from the tip to the dielectric surface and d_2 is the thickness of the dielectric layer.

Combined with the supplementary Eqs. (2.11–2.15), the electric field between the tip and the dielectric surface can be expressed as:

$$E_1 = \frac{V\varepsilon_o\varepsilon_d - \sigma d_2}{\varepsilon_o d_2 + \varepsilon_o\varepsilon_d d_1} \tag{2.16}$$

The electric field force at the tip is

$$F_{tip} = E_1\sigma_1 M = E_1^2\varepsilon_0 M \tag{2.17}$$

where M is the effective area of the tip.

Combining Eqs. (2.14, 2.16, 2.17), we have

$$F_{tip} = F_c + F_\omega + F_{2\omega} \tag{2.18}$$

where

$$F_c = -2(V_{DC} + V_{CPD})\sigma d_2\varepsilon_o\varepsilon_d + (\sigma d_2)^2 + (V_{DC} + V_{CPD})^2(\varepsilon_o\varepsilon_d)^2 \cdot \frac{\varepsilon_0 M}{(\varepsilon_o d_2 + \varepsilon_o\varepsilon_d d_1)^2} \tag{2.19}$$

$$F_\omega = 2V_{AC}\sin(\omega t)\left(V_{DC} + V_{CPD} - \frac{\sigma d_2}{\varepsilon_o\varepsilon_d}\right)(\varepsilon_o\varepsilon_d)^2\frac{\varepsilon_0 M}{(\varepsilon_o d_2 + \varepsilon_o\varepsilon_d d_1)^2} \tag{2.20}$$

$$F_{2\omega} = V_{AC}^2 \sin^2(\omega t)(\varepsilon_o \varepsilon_d)^2 \frac{\varepsilon_0 M}{(\varepsilon_o d_2 + \varepsilon_o \varepsilon_d d_1)^2} \tag{2.21}$$

In KPFM mode, the lock-in amplifier extracts F_ω and controls F_ω to zero by adjusting the DC bias (V_{DC}). According to Eq. (2.20), we have

$$V_{DC} + V_{CPD} - \frac{\sigma d_2}{\varepsilon_0 \varepsilon_d} = 0 \tag{2.22}$$

Therefore.

$$\sigma = \frac{(V_{DC} + V_{CPD})\varepsilon_0 \varepsilon_d}{d_2} \tag{2.23}$$

The V_{CPD} between the tip and the conducting substrate can be measured by scanning the substrate directly in KPFM mode before the substrate is covered by an insulator. The surface charge density can be calculated by knowing d_2, ε_0, ε_d and the obtained V_{DC}.

Figure 2.9c–f show the triboelectric charge on the silica surface measured by KPFM, and it can be seen that its resolution can reach the nanometre scale, and the charge decay can be tested by using KPFM [54]. KPFM is the most widely used probe technique in CE research, and it has played a very crucial role in the history of CE research. One of the most important discoveries based on KPFM is the mosaic effect in CE, which has led to the realization that there may be both positive and negative charges in the microscopic region of the contact, rather than uniform charges of the same polarity. This finding has important implications for the study of CE mechanisms [18].

Based on the KPFM method, the ion transfer and electron transfer can be quantified at the liquid and solid interface. The KPFM-based thermal electron emission method was applied to distinguish the transfer of electrons from ions in liquid–solid CE by Lin et al. They assumed that both O-ions and electrons are present in the CE on the surface of SiO_2 as shown in the inset of Fig. 2.10a. When a SiO_2 sample is heated by a sample heater, electrons are thermally excited and emitted from the surface, as shown in Fig. 2.10b, while O-ions may remain on the surface because they form covalent bonds with Si atoms on the SiO_2 surface. This means that if heating can cause a significant decay of the CE charge on the SiO_2 surface, it may be mainly caused by the thermal emission of electrons. In the experiment, the CE between SiO_2 and deionized water was carried out first, and the results of the effect of temperature on the decay of CE charge on the surface of SiO_2 are given in Fig. 2.10c. The results show that the charge density on the surface of SiO_2 in contact with deionized water is about $-810\ \mu Cm^{-2}$ (negative sign means negatively charged). In Fig. 2.10c, the temperature has a significant effect on the decay of negative charges on the surface of SiO_2. At 313 K, the surface charge density of SiO_2 remains essentially constant, while at 343 K, the surface charge density decays slightly. The decay rate of the surface charge increases as the temperature of the sample increases. However, there are some charges (~-180 μCm^{-2}) that cannot be removed even at temperatures as high as 434 and 473 K (these charges can be called "sticky" charges that remain on the surface even at elevated temperatures). For mobile charges, the decay behaviour is consistent with the theory of thermionic emission, i.e., electrons gain more energy at higher temperatures and the electron density decays faster. And the charge density was found to decay exponentially and the decay curve was consistent with the electron thermionic emission model. Therefore, the mobile charges in the CE between SiO_2 and DI water can be recognized as electrons. For the "sticky" charges, charging and heating cycle tests were performed to observe their behaviour, as shown in Fig. 2.10d. In each cycle of the test, the SiO_2 sample was first contacted with deionized water, then heated to 513 K and held for 10 min to remove the electrons from the surface. In the first cycle, SiO_2 in contact with deionized water was negatively charged with a "sticky" charge density of $-$ 180 μCm^{-2}. It was found that in the second cycle, the density of the "sticky" charge increased to $-$ 300 μCm^{-2}. It was found that in the second cycle, the "sticky" charge density increased to $-$ 300 μCm^{-2}, and increased with the increase of the number of cycles. After five cycles, the density of "sticky" charges reaches a saturation value and no mobile charges exist on the SiO_2 surface. These behaviours suggest that the "sticky" charges should be ions, such as O-ions, rather than electrons. By means of thermionic emission experiments and KPFM, Lin et al. demonstrated that electrons and ions can be distinguished in CE. The mobile and "sticky" charges in the experiment were identified as electrons and ions, respectively. The results show that there is a transfer of electrons and ions between SiO_2 and DI water in CE. The electron transfer density is -630 μCm^{-2} and the ion transfer density is about $-$ 180 μCm^{-2}, which means that the electron transfer accounts for 77% of the total charge in the first contact between SiO_2 and DI water [53].

2.6.2 *Electrostatic Force Microscopy*

In addition to KPFM, electrostatic force microscopy (EFM) is another relatively widely used method for detecting the surface charge of solids [55–57]. EFM sensing the charges on the surface by sensing the electrostatic force on the tip. In addition to the atomic interactions between the tip and the sample, the electrostatic interactions between the tip and the sample also affect the resonant frequency of the tip [55, 56]. If the tip is driven to vibrate at a fixed resonance frequency above a solid surface, the electrostatic attraction or repulsion between the charge on the solid surface and the tip results in a change in the tip amplitude and phase, both of which are related to the electrostatic force applied to the tip. If the amplitude of the control tip remains constant, the frequency driving the tip vibration will be displaced, and this displacement reflects the strength of the electrostatic force applied to the tip. EFM plays an important role in CE studies. In Fig. 2.11–e, sample preparation methods and simulation methods were elaborated. In Fig. 2.11f, PDMS and PET generate CE charge transfer during the demoulding process, and the triboelectric charge on the

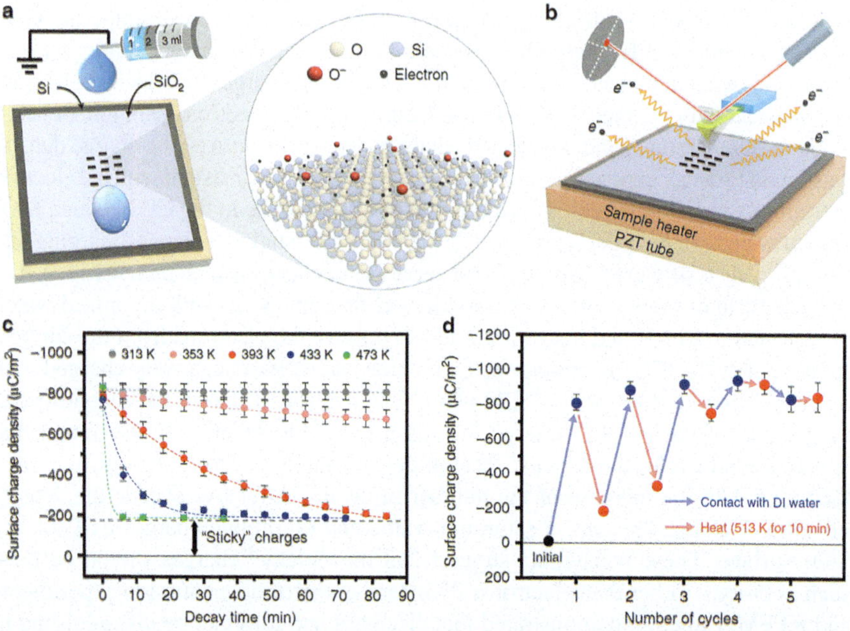

Fig. 2.10 Quantifying the ion and electron transfer at liquid–solid interface by using KPFM. **a–b** The experiment setup for quantifying the charge transfer at liquid–solid interface. **c–d** The charge decay and accumulation at liquid–solid interface. Figure with permission [53], copyright © 2020, The Author(s)

PET surface can be detected by EFM [57]. As shown in Fig. 2.11g and h, a bias voltage is applied to the tip of the needle to enhance the electrostatic force between the tip of the needle and the surface charge of the sample. The electrostatic force shifts the vibration frequency of the needle tip, and the frequency shift of the needle tip reflects the surface charge density of the solid surface.

2.6.3 *Scanning TENG*

KPFM is currently the mode of choice for measuring localized surface charge density. However, in the KPFM mode, an AC bias (AC bias) is required to generate electrostatic force on the conducting tip and induce tip vibration. It is difficult to ensure that the applied AC bias does not affect the accuracy of the surface charge density measurement. On the other hand, the output signal in the KPFM mode is an applied voltage bias (DC bias), which compensates for the potential difference between the conducting tip and the sample surface. The relationship between the potential difference and the surface charge density is not intuitively obvious. Therefore, there is a

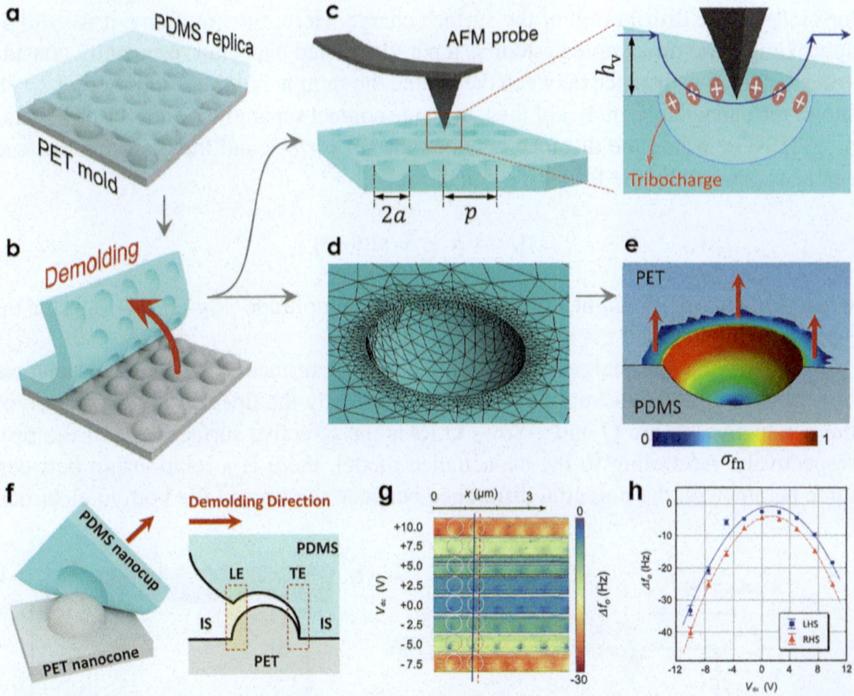

Fig. 2.11 The contact electrification studied by using EFM. **a–d** The sample fabrication and the experiment-calculation setup for the surface charge measurement by using EFM and simulation method. **f** Schematic of Demolding between PDMS nanocup and nanocone. **g** The mapping of static charges on the PET surface taking by EFM. **f** Frequency shift of the tip vibration extract from **h**. Figure with permission [57], copyright © 2019, American Chemical Society

need to develop a scanning probe microscopy technique for localized surface charge density measurements.

The contact separation mode TENG is one of the most common types of TENG, which usually consists of a conductive electrode and a dielectric layer with a conductive bottom electrode. When the conductive electrode is in contact with the dielectric layer, an electrostatic charge is generated on the dielectric surface due to the contact charge. If the conductive electrode is further vibrated near the charged dielectric surface, an AC displacement current will be generated due to electrostatic induction.

Inspired by the contact separation mode TENG, Lin et al. designed an AFM-based localized surface charge density measurement mode and named it as scanning TENG mode. As shown in Fig. 2.12a, a nanoscale contact separation mode TENG can be considered when a conductive tip with a bottom electrode taps the dielectric surface. When the dielectric surface is charged, an AC current between the conductive tip and the bottom electrode will be generated by electrostatic interactions, as occurs in a normal-size contact separation TENG. The amplitude of the AC current depends on the density of the surface charge, so the mapping of the AC current amplitude in turn reflects the distribution of the surface charge. Here, for simplicity, it is valid to approximate the tip-sample system as a parallel plated capacitor, especially considering that the gap distance between the tip and the sample is quite small. Figure 2.12b shows the capacitance model of the nanoscale contact separation mode TENG (scanning TENG), where the distance between the tip surface and the dielectric surface can be expressed by the following equation:

$$x(t) = h + A_0 \sin(\omega t) \tag{2.24}$$

where h is the tip lift height, A_0 is the tip tapping amplitude (Fig. 2.12a), ω is the tip tapping frequency, and t is the time.

Assuming that the charge density of the dielectric surface is σ, the induced charges on the bottom electrode surface (the area covered by the tip) and on the surface of the conductive tip are Q and $-S\sigma - Q$ (S is the effective surface area of the tip), respectively. According to the capacitance model, there is a relationship between three parameters: the potential difference between the tip and the bottom electrode

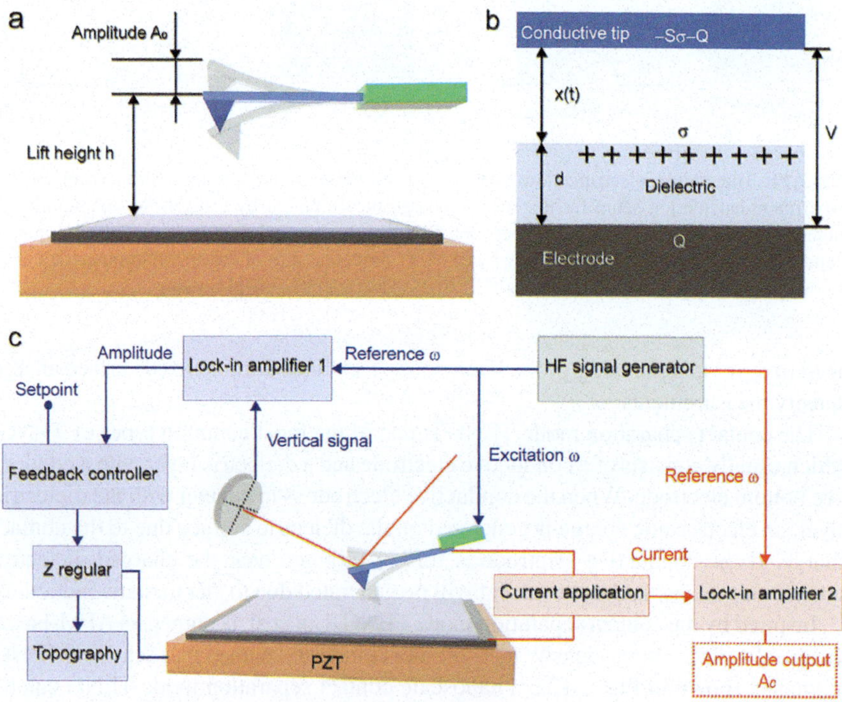

Fig. 2.12 The setup of the scanning triboelectric nanogenerator (scanning TENG). **a** A conductive tip tapping above a dielectric surface with lift height h and tapping amplitude A_0, which can be considered as a nanoscale contact-separation TENG or scanning TENG. **b** The capacitance model for scanning TENG. **c** Block diagram of scanning TENG mode. Figure with permission [58], copyright © AIP Publishing

(V), the amount of charge transferred between the two (Q), and the separation distance between the tip and the surface of the medium (x(t)) as follows:

$$V = -\frac{Q}{S\varepsilon_0}\left(\frac{d}{\varepsilon} + x(t)\right) - \frac{\sigma x(t)}{\varepsilon_0} \tag{2.25}$$

where d is the thickness of the dielectric film, ε_0 is the vacuum dielectric constant, and ε is the relative dielectric constant of the dielectric film.

Combined with Eqs. (2.24), (2.25) can also be expressed as:

$$Q(t) = \frac{S(-\sigma(h + A_0\sin(\omega t)) - \varepsilon_0 V)}{\frac{d}{\varepsilon} + h + A_0\sin(\omega t)} \tag{2.26}$$

By deriving Eq. (2.26), the expression for the current between the tip and the bottom electrode during tapping can be obtained as:

$$I(t) = Q'(t) = \frac{S\left(\varepsilon_0 V - \frac{\sigma d}{\varepsilon}\right)A_0\omega\cos(\omega t)}{\left(\frac{d}{\varepsilon} + h + A_0\sin(\omega t)\right)^2} \tag{2.27}$$

From Eq. (2.27), the AC current between the top and bottom electrodes has different frequency components. Here, we are concerned with the first harmonic of the AC current, which is extracted as follows.

$$I(t)_\omega = \left(\frac{d + \varepsilon h}{\sqrt{(d+\varepsilon h)^2 - (\varepsilon A_0)^2}} - 1\right)\frac{2\omega S(\varepsilon\varepsilon_0 V - \sigma d)}{\varepsilon A_0}\cos(\omega t) \tag{2.28}$$

In Eq. (2.28), the first harmonic amplitude of the current is linearly related to the surface charge density (σ), which means that if all other parameters in Eq. (2.28) are known and the first harmonic amplitude of the current is measured, the charge density on the dielectric surface can be obtained.

The block diagram of scanning TENG is given in Fig. 2.12c, where the amplitude of the first harmonic of the induced current is measured. In scanning TENG mode, a two-pass technique is used. The blue block diagram shows the first pass of the topographic mapping, where the AFM is operated in standard tapping mode. It should be noted that other modes of topographic mapping, such as peak force tapping mode, can also be used in the first pass. The scanning TENG mode is operated in the second pass (red box diagram in Fig. 2.12c), in which the tip is raised above the sample surface at a certain elevation (h) to scan the surface along the topographic profile. At the same time, the tip is mechanically excited by tapping the piezoelectric ceramic and vibrates at a resonant frequency (ω) with a certain tap amplitude (A_0). An AC current signal between the conductive tip and the bottom electrode is collected using a current application program. Further, the AC current signal is fed to a lock-in amplifier (lock-in 2), and the reference frequency of the lock-in 2 is set to ω so that the amplitude of the ω frequency component current can be extracted from the AC current signal. It should be clarified that the AC current signal needs to be converted to a voltage signal before it is fed into the lock-in amplifier. There is signal attenuation in the coupling between the current application and the lock-in amplifier. However, in any case, the output amplitude of the lock-in amplifier is linearly related to the amplitude of the ω-frequency component of the AC current between the top and bottom electrodes by a certain linearity factor (α). This means that if the output amplitude of the lock-in amplifier is Ac, the first-harmonic amplitude (ω-frequency component) of the induced AC current should be αAc. From Eq. (2.28), the surface charge density (σ) can be expressed as:

$$\sigma = \frac{\varepsilon\varepsilon_0 V}{d} - \frac{\alpha\varepsilon A_0 A_c}{2\omega Sd\left(\frac{d+\varepsilon h}{\sqrt{(d+\varepsilon h)^2-(\varepsilon A_0)^2}} - 1\right)} \tag{2.29}$$

Several parameters need to be measured in Eq. (2.29), among which the effective area of the leaf tip, S, is difficult to measure accurately. To ensure the accuracy of scanning TENG, we propose a self-calibration method. We let the constant

$$k = \frac{\alpha\varepsilon A_0}{2\omega Sd\left(\frac{d+\varepsilon h}{\sqrt{(d+\varepsilon h)^2-(\varepsilon A_0)^2}} - 1\right)} \tag{2.30}$$

Then Eq. (2.29) can be expressed as

$$A_c = \frac{\varepsilon\varepsilon_0 V}{kd} - \frac{\sigma}{k} \tag{2.31}$$

and

$$\sigma = -kA_c + \frac{\varepsilon\varepsilon_0 V}{d} \tag{2.32}$$

According to Eq. (2.31), the constant k can be calibrated by applying different biases between the top and bottom electrodes. The scanning TENG output amplitude (Ac) varies linearly with the applied bias, and the slope indicates the parameter ($\frac{\varepsilon\varepsilon_0}{kd}$), where ε and ε_0 are related to the dielectric material and the dielectric film thickness, d, can be precisely measured. After calibrating the constant k, the surface charge density can be calculated from Eq. (2.32) [58].

2.6.4 *Electrostatic Force Curve*

Essentially, the AFM tip is a force sensor that detects the interaction force between the tip and the sample by detecting the bending of the tip cantilever beam. Thus, as the tip approaches the sample, the electrostatic force between the tip and the sample can be obtained by directly detecting the deflection of the tip cantilever. This measurement is known as the electrostatic force curve (EFC) method [59] and provides insight into the electrostatic charge on the sample surface. Two typical force curves are given in Fig. 2.13a. As the tip approaches the sample, the positive deflection indicates an upward force on the tip, indicating a repulsive electrostatic force between the tip and the sample. This indicates that the sample has the same polarity as the charge on the surface of sthe tip. Conversely, a downward force on the tip indicates an attractive electrostatic force between the tip and the sample, suggesting that the sample has the opposite charge to the tip surface [59]. This method obtains the exact value of the force on the tip by calibrating the mechanical properties of the tip cantilever beam. However, the problem is that the charge on the tip surface cannot be precisely measured. Even if the exact electrostatic force is obtained, there is no way to get the charge density on the surface of the sample. Therefore, the force curve method is only a qualitative method. In addition, as shown in Fig. 2.13b, c, the force curve method has very limited resolution in measuring the sample surface topography and the surface electrostatic force strength. Therefore, the force curve method has not been widely used in CE research, but its simplicity and directness still have some advantages.

2.6.5 *Conducting Atomic Force Microscopy*

In recent years, the CE phenomenon involving semiconductor materials has received a great deal of attention, and it has been found that two semiconductors produce a continuous direct current during friction, called the tribovoltaic effect [60, 61]. The microscale methods described above are not applicable to the study of CE involving semiconductors because semiconductors are electrically conductive and electrostatic charges are carried away from their surfaces. Here we introduce a method to study the tribovoltaic effect involving semiconductors, the conducting atomic force microscopy (CAFM) method [62]. The principle of CAFM is very simple. Based on the contact method of AFM, a high-precision ammeter is connected to the loop connecting the tip of the needle to the sample, and the current generated when the tip of the needle rubs against the sample can be measured. The current measured by the CAFM method can reach a sensitivity of fA. As shown in Fig. 2.14a, different doped tips and samples are usually used in experiments using CAFM to study the friction volt effect. For example, the tip is an n-type semiconductor and the sample is a p-type semiconductor and vice versa, forming a heterojunction [62]. Prior to CAFM current measurements, the rectification effect at the tip-sample contact interface can be measured to determine if the tip and sample indeed form a heterojunction (Fig. 2.14b). The tribovoltaic current can then be studied under different loading conditions (or other different experimental conditions). Figures 2.14e–g show the current mapping measured using CAFM, which indicates that the tribovoltaic current increases with increasing load. The tribovoltaic effect is a very interesting phenomenon that has

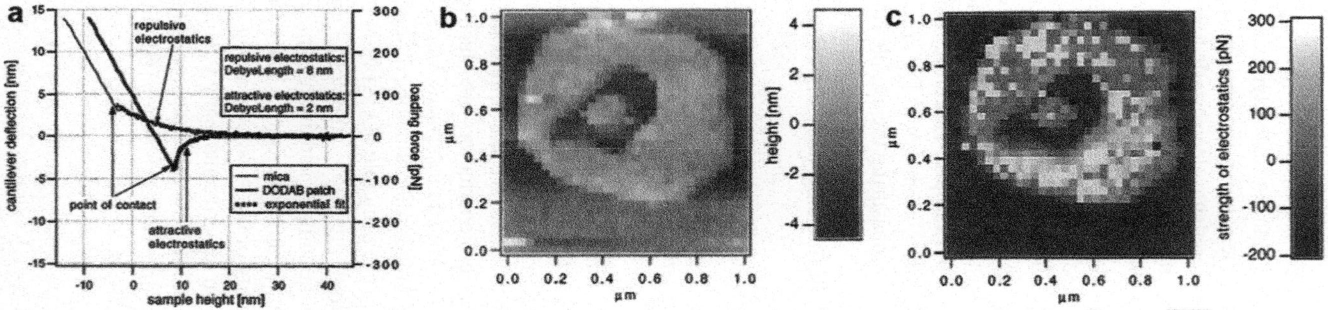

Fig. 2.13 EFC for charge measurement. **a** Two force curves when the tip receives attractive or repulsive force when it approaching charged surfaces. **b** Topography of the sample. **c** Strength of the electrostatic between tip and sample taking by electrostatic force mode. Figure with permission [59], copyright © 1997, American Chemical Society

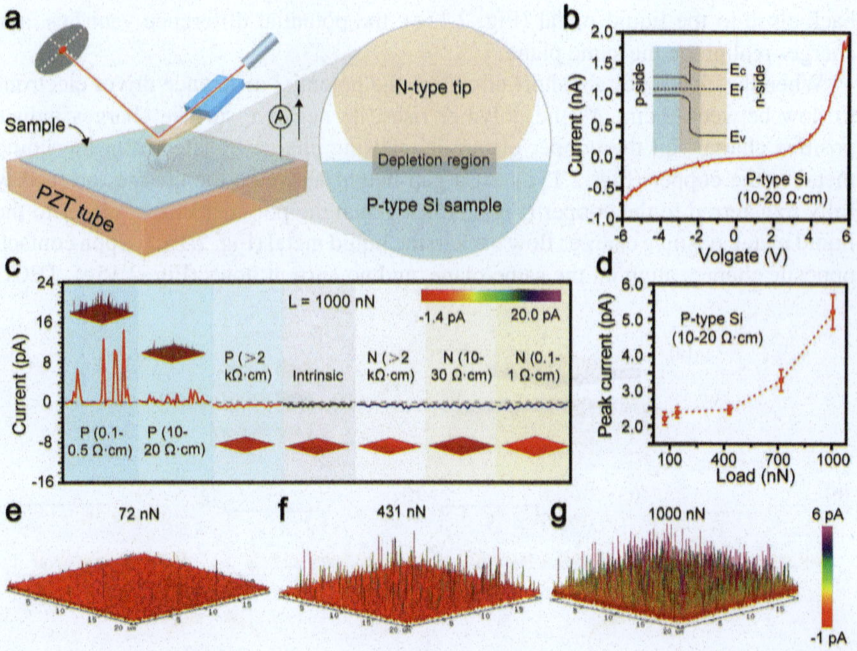

Fig. 2.14 Measuring of the tribovoltaic current between semiconductors by using CAFM. **a** Schematic of CAFM setup. **b** I-V curve between semiconductor tip and Si sample. **c** Current between tip and sample during friction takin by CAFM. **d** Relation between tribo-current and load. **e–f** Current maps taking by CAFM when the tip scanning on the surface. Figure with permission [62], copyright © 2020, WILEY-VCH Verlag GmbH & Co. KGaA, Weinheim

attracted the attention of many researchers in recent years. CAFM has been widely used in the study of tribovoltaic effect and plays an important role in the development of tribovoltaic theory.

2.7 Triboelectric Nanogenerator

2.7.1 Mechanism of Quantification for Solids by TENG

The triboelectric charge density (TECD) measurement by TENG is proposed by Zou et al. Figure 2.15 illustrates the process under open-circuit (Fig. 2.15b–e) and short-circuit (Fig. 2.15f–i) conditions. When two materials contact, charges transfer, leaving one material with excess electrons and the other with a deficit (Fig. 2.15b). Separation of the polymer and liquid metal (Fig. 2.15c) creates a potential difference due to the separation of opposite charges. At a gap distance of ~ 10 times the material thickness, the open-circuit voltage peaks (Fig. 2.15d). When the polymer is pushed back close to the liquid metal (Fig. 2.15e), the potential difference vanishes, and charges realign on the same plane.

When the electrodes are short-circuited, the potential difference drives electrons to flow between them. As the polymer rises, its negative surface charges induce positive charges on the copper electrode, causing charge transfer from the liquid metal to the copper (Fig. 2.15g). At a gap height above L, the charges are nearly fully transferred to the copper (Fig. 2.15h). When the polymer moves closer to the liquid metal, positive charges flow back to the liquid metal (Fig. 2.15i). Upon contact, opposite charges align on the same plane, and no current flows (Fig. 2.15g). TECD

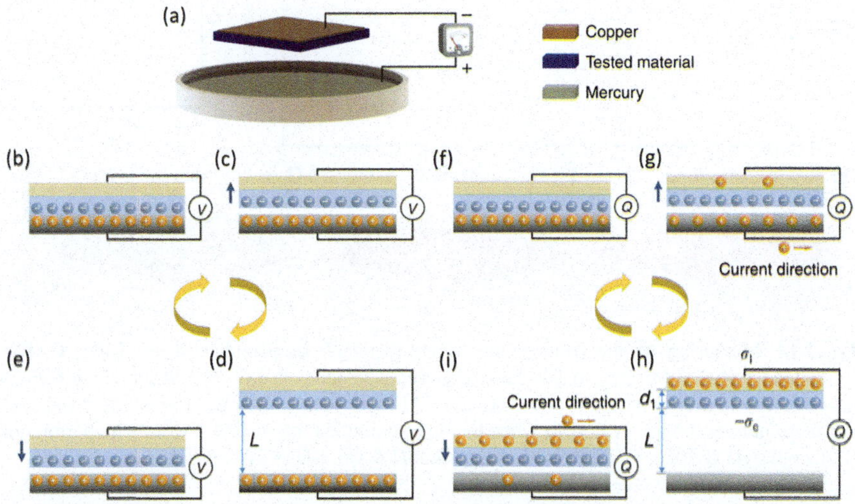

Fig. 2.15 Principle for measuring the TECD. **a** Simplified model of the measurement method. The tested material contacts the liquid metal of mercury, and then separates periodically. The positive electrode of the electric meter is connected to the mercury, and the negative is connected to the copper electrode. **b–e** The Theoretical model under open-circuit condition. **b** The contact electrification causes charge transfer between materials, the charges coincide at the same plane. The system has no net charge, there is no potential difference. **c** Voltage is generated between the two electrodes. Suppose the polymer has a strong capacity to absorb electrons, when the materials are separated, the polymer has negative charges, mercury has positive charges. Therefore, the potential difference is created between the two parts. **d** The potential reaches the maximum when the gap reaches certain distance *L*. The copper electrode is only influenced by the electric fields of the charges on the surface of the polymer. **e** When the polymer approaches the mercury, voltage drops due to the combined influence of the two electric fields. Finally, they are fully contacted, there is no voltage between the two electrodes (back to **b**). **f–i** The theoretical model under a short-circuit condition. **f** The two materials fully contact each other, there is no potential difference. **g** When the materials are separated, the negative charges on the surface of the polymer induce positive charges in copper, as copper and mercury are electronically connected, the positive charges in mercury flow into the copper side. **h** Approximately all charges flow into copper side to equalize the potential difference when the gap reaches certain distance *L*. **i** When the sample approaches the mercury, the negative charges on the surface of the polymer induces positive charges in mercury, the positive charges flow from copper to mercury until the charges are neutralized finally at the same plane when they are fully contacted (back to **f**). Figure from [65] with permission, copyright © 2019, The Author(s)

is determined by the total charge transfer between the copper electrode and liquid metal.

A conductor-to-dielectric parallel-plate model is used for simplicity. Upon contact, opposite static charges form on the surfaces at a charge density σ_c[66]. While the dielectric's surface charge density remains constant, the system's capacitance varies with mechanical motion [67]. The induced charge density ($\sigma_I(L, t)$) depends on the gap distance $L(t)$, especially for small gaps ($L \ll 10\, d_1$). From Gauss's theorem, the electric field in the media and gap can be approximated, neglecting edge effects:

$$E_1 = \frac{\sigma_I(L,t)}{\varepsilon_1} \tag{2.33}$$

$$E_{air} = \frac{\sigma_I(L,t)-\sigma_c}{\varepsilon_0} \tag{2.34}$$

where the dielectric permittivity of the material is ε_1, and its thickness is d_1. The voltage between the two electrodes is

$$V = \frac{\sigma_I(L,t)}{\varepsilon_1} d_1 + \frac{\sigma_I(L,t)-\sigma_c}{\varepsilon_0} L \tag{2.35}$$

Under short-circuit condition, the two electrodes are electrically connected, so $V = 0$, then,

$$\sigma_I(L,t) = \frac{L\sigma_c}{\frac{d_1\varepsilon_0}{\varepsilon_1} + L} \tag{2.36}$$

From Eq. (2.36), when the separation distance L is much larger than the materials thickness ($L \gg d_1$), the free electron charge density $\sigma_I(L, t)$ closely approximates the surface charge density σ_c. With $d_1 = 0.5$ mm, $L = 75$ mm and when $\frac{\varepsilon_0}{\varepsilon_1}$ ~ 2, the measured charge density σ_I is 99.67% of the surface charge density σ_c. Thus, the measured value of the charge density reflects the TECD on the dielectric surface.

Configuration of the triboelectric series measurement for solids

A standardized TECD measurement system (Fig. 2.16) includes a support, Faraday Cage, static, and motion parts. A linear motor moves samples vertically (up to 75 mm) to contact and separate from mercury. Test materials (38.1 mm × 38.1 mm) were cleaned, coated with 15 nm Ti and 300 + nm Cu (leaving a 2 mm margin), and adhered to the acrylic substrate with epoxy to prevent noise. Magnets enable easy and consistent sample replacement.

Noise from the AC, light, linear motor, and electrified objects was minimized by using a grounded Faraday cage and shielded wires. All electrified components, including the motor, were grounded to reduce interference. The system was placed in a glove box with controlled conditions: ultra-high purity nitrogen (99.999%) to prevent arc-over (per Paschen's law), a temperature of 20 ± 1 °C, pressure ~ 1 atm, and a dew point ≤ -46 °C (0.43% RH). Calcium chloride and phosphorus pentoxide absorbed moisture, and samples were stored overnight to remove surface water vapor

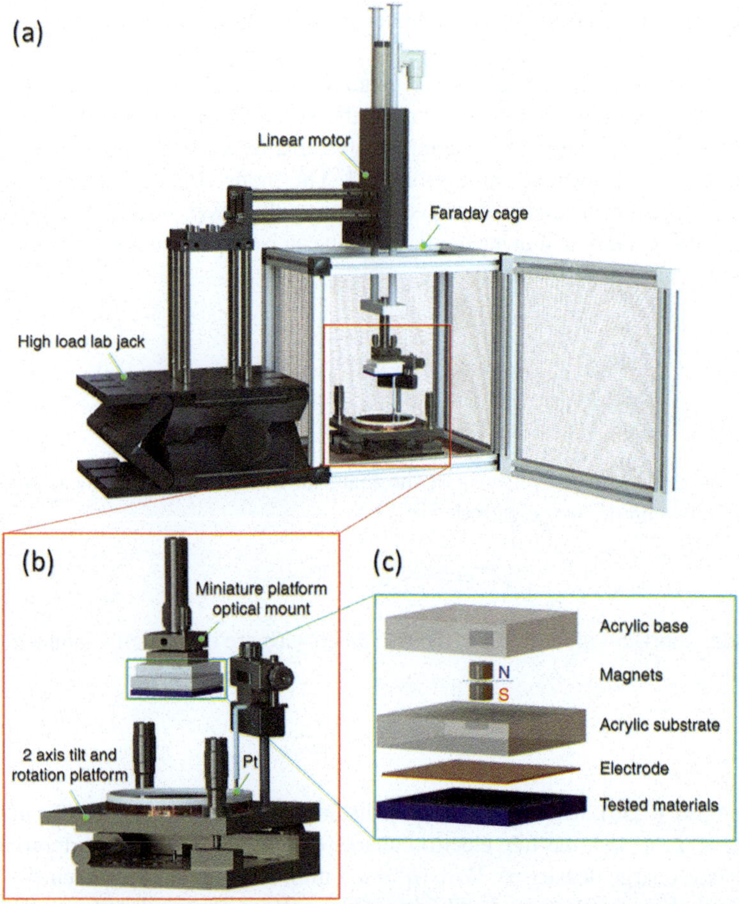

Fig. 2.16 Experimental set-up for the triboelectric series measurement. The whole measurement was set in a glove box filled with ultra-high purity of nitrogen gas at fixed temperature, pressure, and humidity. The linear motor was settled on a high-load lab jack **a**. The height of the sample to contact the liquid metal can be finely adjusted by both the high-load lab jack and a linear motor. The static part has the liquid metal as the electrode. The motion part consists of the tested sample, and it is controlled by the linear motor **b**. An acrylic base is attached to the end of the linear motor. A magnet is engraved into the acrylic base to attract another magnet engraved in the sample substrate. Each sample consists of an acrylic substrate with the magnet, electrode, and the tested materials **c**. Both the miniature platform optical mount and the two-axis tilt and rotation platform can adjust the orientations of the sample and liquid metal level. Figure from [65] with permission, copyright © 2019, The Author(s)

before measurement. Triboelectric charge was measured during contact electrification, and TECDs were calculated as the charge divided by the contact area. Variations in measured charge densities reflect the materials' ability to gain electrons. Initial surface charges, influenced by the cleaning process, were eliminated by running each sample until saturation. Steady-state values were recorded after no significant change was observed over 100 cycles.

2.7.2 *Mechanism of Quantification for Liquids by TENG*

Liquid frictional electrification poses both challenges and opportunities due to its diverse sources and the substantial energy it involves [68]. Its implications span fundamental physics [69–73] and practical applications, such as power generation [10], medical uses [74], agriculture [75], transportation [76], microfluidics [77], and environmental protection [78]. This broad relevance has garnered significant global research interest. Understanding contact electrification in liquids is crucial across scientific and engineering fields. The triboelectric series, developed through experimental observations, is a reliable tool for assessing material triboelectric properties. However, beyond material characteristics, factors like friction dynamics (e.g., contact pressure and relative movement velocity) and experimental conditions (e.g., temperature, pressure, humidity, and atmosphere) [79, 80] also influence the triboelectric effect. These complexities can hinder accurate quantification and potentially distort measurement results.

A liquid-tube triboelectric nanogenerator (TENG) in single-electrode mode was selected as the basic unit for triboelectric sensing, as illustrated in Fig. 2.17a. This setup features a liquid mercury electrode layer wrapped around the tube's outer surface and connected to an electrometer for non-contact electrostatic induction-based charge measurement. A pneumatic pump regulates the liquid's motion within the tube, maintaining a stable liquid–solid contact interface along the tube's hydrophobic inner surface, which ensures complete separation during operation. This in-situ characterization method reduces uncertainties in charge measurements. Additionally, the grounded front electrode neutralizes any initial charges in the test liquid before triboelectrification, enhancing measurement accuracy.

The principle for measuring triboelectric charge for liquids is based on the mechanism of TENG, which involves contact electrification and electrostatic induction. As negative charges accumulate on the tube's inner surface, they eventually reach saturation, as shown in Fig. 2.17b(i). These negative charges induce positive charges on the liquid's surface, which also reach saturation (Fig. 2.17b(ii–iii)). When the liquid moves into the electrode coverage area, the positive charges on the liquid and the negative charges on the tube's surface align on the same plane, resulting in a neutral local potential. Electrons then flow from the ground to the electrode to maintain balance (Fig. 2.17b(iv–vi)), with no net charge induced on the electrode in the contact area. When the liquid exits the electrode coverage area, the tube surface becomes positively charged again due to contact electrification. Consequently, the

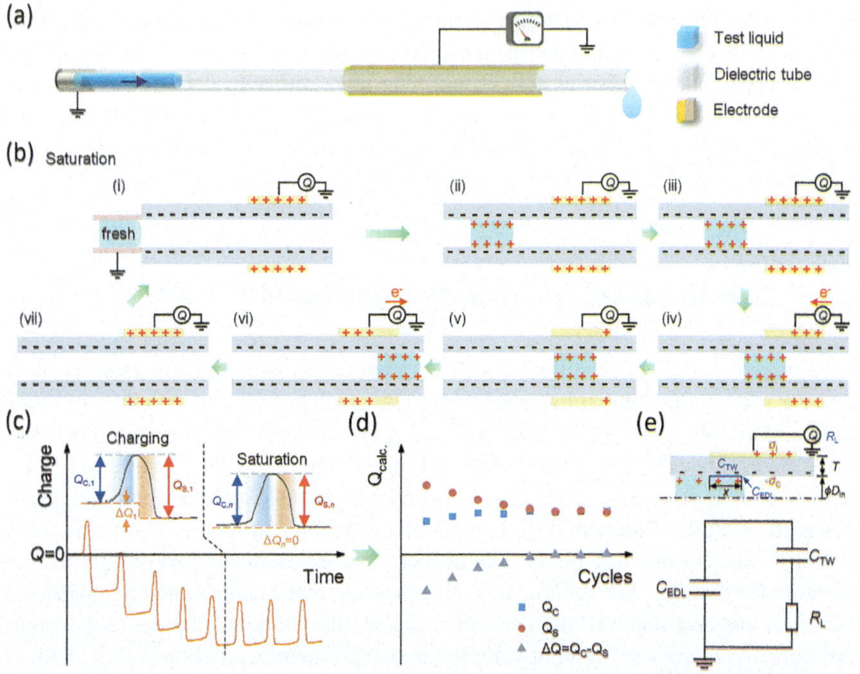

Fig. 2.17 Principle for measuring the triboelectric property in liquids. **a** Schematic diagram of a typical triboelectric sensing unit developed from the liquid-tube type TENG in single electrode mode. **b–c** Charge transfer process at charging stage **b** and saturation stage **c**. **d** Measured typical time-variable charge signal. **e** Triboelectric charges during the contact and separation, denoted as Q_C and Q_S, with their difference ΔQ, $\Delta Q = Q_C - Q_S$. **f** Equivalent-circuit model

net charge on the electrode remains unchanged before and after the liquid passes through the tube (Figs. 2.17b(vii) and 2.17b(i)).

The typical time-variable output signal is depicted in Fig. 2.17c. The transferred charge during contact and separation is denoted as Q_C and Q_S, respectively, with subscripts indicating the triboelectric cycle order. The difference $\Delta Q = Q_C - Q_S$ indicates charge saturation. During the charging stage, $\Delta Q_1 < 0$, reflecting the accumulation of net positive charges on the electrode. After several cycles, ΔQ_n approaches zero, signifying that saturation has been achieved (Fig. 2.17d). Subsequently, Q_S or Q_C is measured as the triboelectric charge for each liquid.

To simplify analysis, an equivalent circuit model for the liquid-tube TENG in single-electrode mode is presented, as shown in Fig. 2.17e. After multiple triboelectric cycles reach saturation, the two tribo materials exhibit opposite static charges with a fixed surface charge density, σ_c. An electrical double layer (EDL) forms between the test liquid and the tube's inner surface, which is surrounded by the outer electrode. The capacitance of this EDL, denoted as C_{EDL}, changes dynamically as the liquid flows through the tube. As a result, the induced static charge density, σ_I, transferred between the electrodes depends on time (t) and the overlap length (x) between the liquid column and the electrode coverage area. The open-circuit voltage between the electrode and the ground is then expressed by the following equation [81]:

$$V_{OC} = \frac{\sigma_I(x,t)}{\varepsilon} T \tag{2.37}$$

where ε is the dielectric permittivity of the tube material, and T is its wall thickness. Under short-circuit conditions, the transferred charges Q_{SC} are determined by surface charge density σ_C and effective contact area S, described as:

$$Q_{SC} = \frac{\sigma_C S}{1 + C_{EDL}/C_{TW}(\ll 1)} \approx \sigma_C \pi D_{in} \cdot x \tag{2.38}$$

where C_{EDL} and C_{TW} are the capacities of the electrical double-layer and tube wall, respectively. A mathematic approximation is used, as C_{EDL} is generally much smaller than C_{TW}. D_{in} represents the inner diameter of the tube.

Configuration of the triboelectric series measurement for liquid

A standardized experimental setup for measuring liquid triboelectric charge, shown in Fig. 2.18a, was constructed within a nitrogen-filled glove box (N_2, 99.99%) to ensure a stable testing environment. The conditions were maintained at 1 atm pressure with an additional ~ 0.2 kPa, 20% relative humidity, and a temperature of 20 ± 1°C. A low-pass filter reduced noise to ± 0.1 nC within a range of 0–200 nC. The system includes a liquid supply segment, a gas supply component, and a triboelectric sensing unit, connected via a grounded 316L stainless tee joint to neutralize the liquid's initial charge. Liquid columns were formed using a syringe pump and pneumatic pump, controlled by a computer program and signal generator, as depicted in Fig. 2.18b and simulated in Fig. 2.18c.

Before starting, the test liquid is positioned at the tee joint intersection (Fig. 2.18c(i)), with both pumps off. The syringe pump is first activated to inject the liquid into the dielectric tube (Fig. 2.18c(ii)), then turned off as the pneumatic pump propels the liquid into a bubble column using gas flow (Fig. 2.18c(iii)). The liquid column moves forward and exits into a container (Fig. 2.18c(iv)). The column length is controlled by a stepper motor, and its velocity is adjusted via a throttle valve. This periodic process, defined by time parameters and gas flow rates, is automated for efficiency.

To standardize liquid triboelectric property measurements, the test liquid must be electrically neutral, achieve thorough frictional contact with the tribo-material, and fully separate without significant charge loss. Reliable results require controlling variables like material, geometry, dynamics, and environmental factors. A novel approach integrates a liquid metal electrode into the liquid-tube TENG. A section of the dielectric tube is placed in a 3D-printed container filled with liquid metal, fully immersing the tube surface to prevent air exposure and minimize noise. A platinum wire (99.9%) serves as the measurement terminal (Fig. 2.18a).

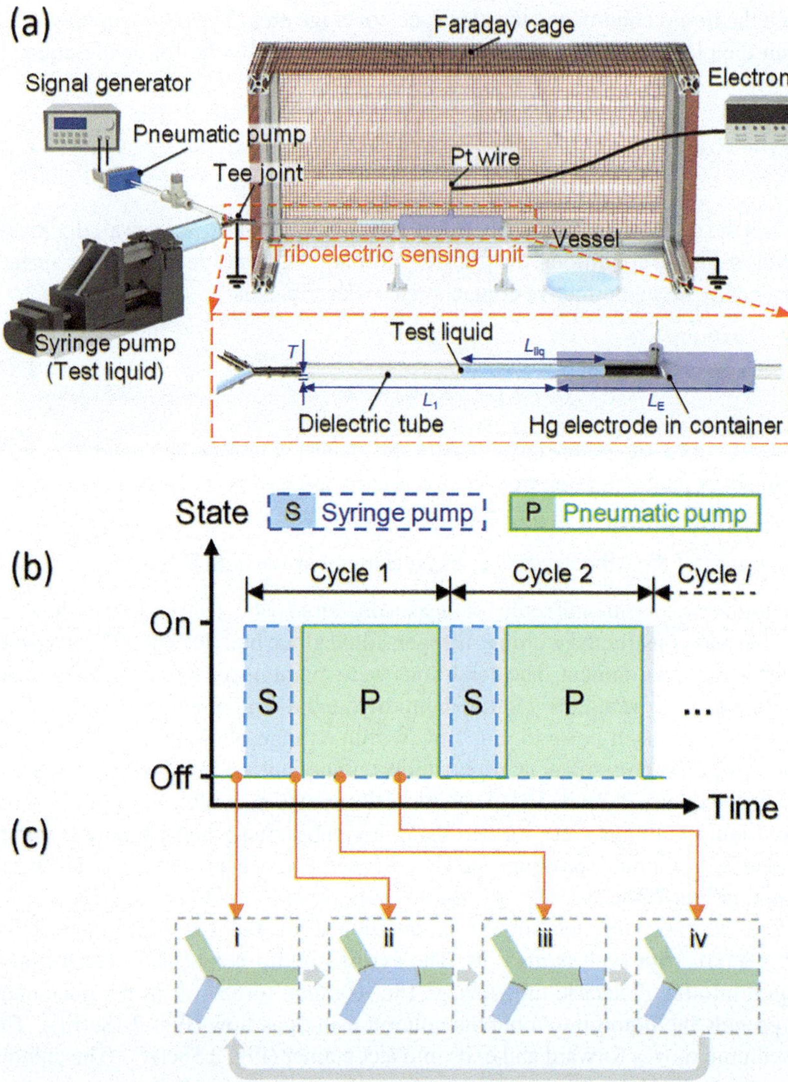

Fig. 2.18 Experimental setup for quantifying the triboelectric series of liquid materials. **a** The setup consists of a liquid supply unit, a driving-gas supply unit, and a triboelectrific sensing unit. **b** The working process of the operating system, demonstrating the formation and movement of the liquid column **c**

To prevent physical or chemical reactions from gas contact, a nitrogen atmosphere with 20% RH was maintained. The length of the inlet section for fully developed laminar flow was estimated using the Boussinesq formula $L^* = 0.065d{\cdot}\mathrm{Re}$ [82], where d is the feature size (specifically, $d = D_{\mathrm{in}}$ for in-tube flow). Reynolds number Re is a dimensionless parameter used to characterize fluid flow conditions, defined as $\mathrm{Re} = \rho ud/\eta$, where ρ and η denote the density (in $\mathrm{kg{\cdot}m^{-3}}$) and dynamic viscosity (in Pa·s) of liquids, and u represents the velocity (in $\mathrm{m{\cdot}s^{-1}}$). Accordingly, the distance between the grounded electrode and Hg electrode L_1 was set at 300–700 mm for analysis. Experiments showed that larger L_1 led to faster saturation but did not affect the triboelectric charge after sufficient cycles.

Accurately measuring liquids with low charge transfer (<0.5 nC) is challenging under fixed noise conditions. Increasing both the liquid column length (L_{liq}) and electrode length (L_{E}), with $L_{\mathrm{liq}} < L_{\mathrm{E}}$, amplifies measured charges. Results in Fig. 2.18i show a linear relationship between saturated charges and L_{liq} (contact area$S = \pi D_{\mathrm{in}} L_{\mathrm{liq}}$), with $R^2 > 0.99$, consistent with the circuit model. Improved singal-to noise ratio allows conversion of Q_{sat} into charge density σ_{sat}, extending the measurable lower limit. So far, the geometry parameters are standardized, as outlined in Fig. 2.18j.

Fluid dynamics play a crucial role in liquid triboelectrification, as increased contact force enhances electron transfer between materials [83]. To study its effect on Q_{sat}, flow velocity was precisely controlled using a pneumatic pump under programmable settings. Tests with DI water, glycerol, and ethanol were conducted at calibrated gas flow rates measured by a high-precision mass flowmeter. Experimental results verified that Q_{sat} increases with velocity initially due to faster contact-separation rates in laminar flow, then levels off as the maximum triboelectric potential is reached. This behavior aligns with prior studies showing maximum streaming current is velocity-independent [84]. In this study, velocity was adjusted until Q_{sat} plateaued, and the resulting value was reported.

2.8 Quantification of Triboelectric Series

2.8.1 *Measurement of Polymers*

By summarizing the measurement results of over 50 materials, the triboelectric order is shown in Table 2.2 (Fig. 2.19). The data represent the average values of tests of at least three samples of the same materials. When the tested material is negatively charged after contact with mercury, the measured charge density is recorded as a negative value; when the material is positively charged after contact with mercury, it is recorded as a positive value. PTFE is most widely used in TENG devices, therefore, we can consider the TECD of PTFE with a thickness of 20 mil (~0.5 mm) as the reference TECD. Then the normalized triboelectric charge densities $\alpha = \frac{\sigma}{|\sigma_{PTFE}|}$ of measured materials are listed in Table 2.1. For a material with $\alpha < 0$, then the material

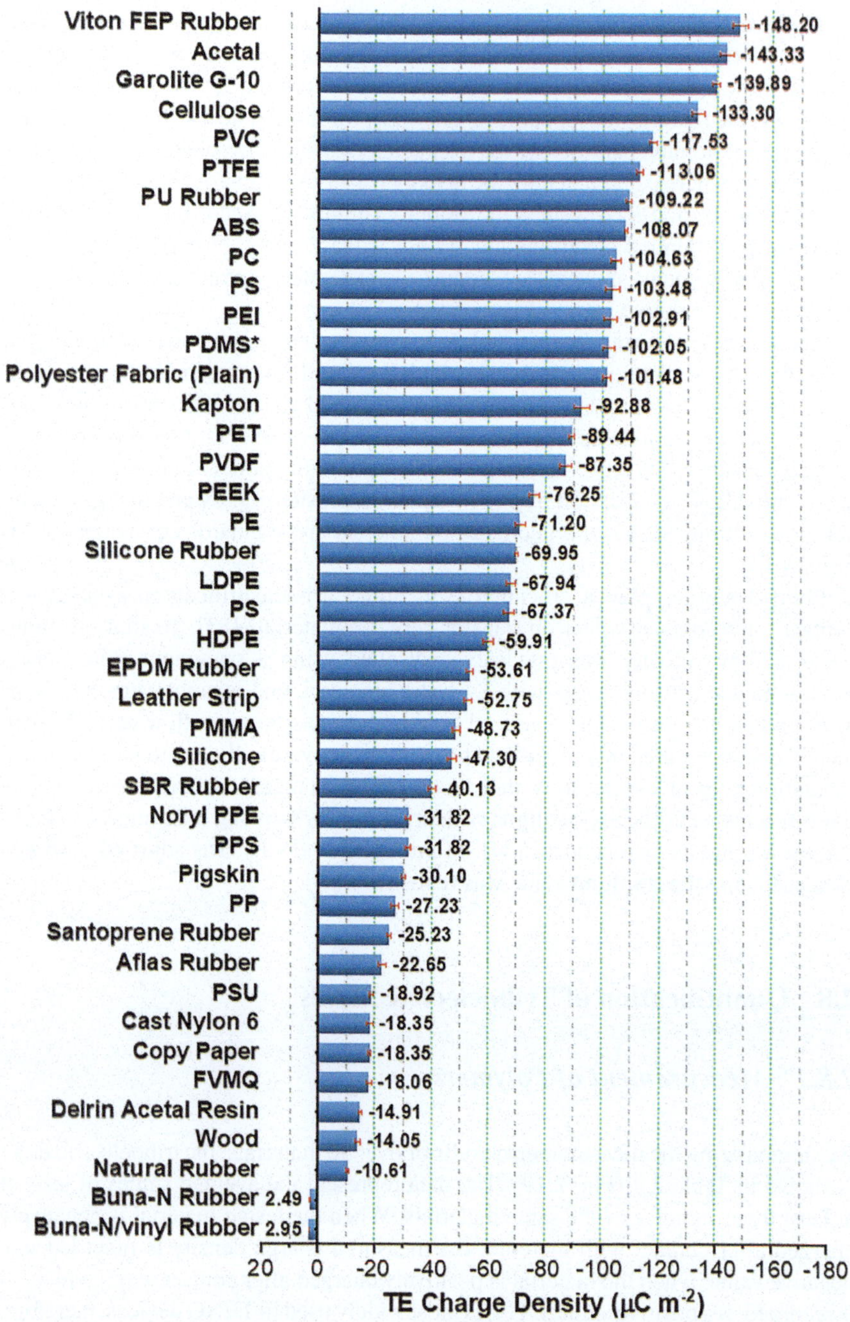

Fig. 2.19 Triboelectric series of common polymers. Figure from [65] with permission, copyright © 2020, The Author(s)

Table 2.1 The settled parameters for the following measurement

Main parameters	
Item	Value
Tube materials	FEP
Electrode materials	Hg
Driven-gas material	N2
Inner diameter, Din	3.0 mm
Wall thickness, T	0.5 mm
Specific length, L1	300 mm
Electrode length	100 mm
Liquid length	75 mm

Table 2.2 Triboelectric series of materials and their TECD [65]

Materials	Average TECD (mC m^{-2})	STDEV	α
Chemical-resistant viton® Fluoroelastomer rubber	– 148.2	2.63	– 1.31
Acetal	– 143.33	2.48	– 1.27
Garolite G-10	– 139.89	1.31	– 1.24
Clear cellulose	– 133.3	2.28	– 1.18
Clear polyvinyl chloride (PVC)	– 117.53	1.31	– 1.04
Polytetrafluoroethylene (PTFE)	**– 113.06**	**1.14**	**– 1**
Abrasion-resistant polyurethane rubber (PU Rubber)	– 109.22	0.86	– 0.97
Acrylonitrile butadiene styrene (ABS)	– 108.07	0.5	– 0.96
Clear glossy polycarbonate (PC)	– 104.63	1.79	– 0.93
Polystyrene (PS)	– 103.48	2.48	– 0.92
Ultem PEI	– 102.91	2.16	– 0.91
Polyester (PE)	– 101.48	1.49	– 0.9
Silicone rubber	– 94.03	0.99	– 0.83
DuraLar polyester film	– 89.44	0.86	– 0.79
Polyvinylidene fluoride (PVDF)	– 87.35	2.06	– 0.77
Polyetheretherketone	– 76.25	1.99	– 0.67
Polyethylene	– 71.2	1.71	– 0.63
Silicone rubber	– 69.95	0.5	– 0.62
Wear-resistant garolite	– 68.51	1.99	– 0.61
Low-density polyethylene (LDPE)	– 67.94	1.49	– 0.6
High impact polystyrene (PS)	– 67.37	1.79	– 0.6
Polyethylene (PE)	– 64.61	1.51	– 0.57

(continued)

Table 2.2 (continued)

Materials	Average TECD (mC m^{-2})	STDEV	α
High-density polyethylene (HDPE)	– 59.91	1.79	– 0.53
Weather-resistant EPDM rubber	– 53.61	0.99	– 0.47
Leather strip (Smooth)	– 52.75	1.31	– 0.47
Oil-filled cast nylon 6	– 49.59	0.99	– 0.44
Clear cast acrylic	– 48.73	1.31	– 0.43
Silicone	– 47.3	1.49	– 0.42
Fluorinated ethylene propylene	– 42.71	1.79	– 0.38
Abrasion-resistant SBR rubber	– 40.13	1.31	– 0.35
Flexible leather strip (Smooth)	– 34.4	0.86	– 0.3
Poly (phenylene Sulfide)	– 31.82	0.86	– 0.28
Noryl polyphenyl ether (PPS)	– 31.82	0.86	– 0.28
Pigskin	– 30.1	0.86	– 0.27
Polypropylene (PP)	– 27.23	1.31	– 0.24
Santoprene rubber	– 25.23	0.5	– 0.22
Aflas rubber	– 22.65	1.31	– 0.2
Polysulfone (PSU)	– 18.92	0.86	– 0.17
Copy paper	– 18.35	0.5	– 0.16
Cast nylon 6	– 18.35	0.99	– 0.16
Chemical-resistant and low-temperature Fluorosilicone (FVMQ)	– 18.06	0.86	– 0.16
Delrin® Acetal resin	– 14.91	0.5	– 0.13
Wood (Marine-grade plywood)	– 14.05	0.99	– 0.12
Natural rubber	– 10.61	0.5	– 0.09
Oil-resistant buna-N rubber	2.49	0.23	0.02
Buna-N/vinyl rubber	2.95	0.13	0.03

Note STDEV refers to the standard deviation. The α refers to the measured TECD of tested materials over the absolute value of the measured TECD of the reference material

is more negative than the reference liquid mercury; if $0 < \alpha < 1$, then this material is more negative than the reference liquid mercury and more positive than PTFE; if $\alpha > 1$, then this material is more negative than PTFE. Materials that are farther apart on the list create a stronger charge when rubbed together. Materials near to each other on the series may exchange little charge.

2.8.2 *Measurement of Inorganic Non-metallic Materials*

By using the universal method discussed above for standardized evaluations of the TECDs of various materials in a strictly controlled environment. It provides a new "gene" of materials. The solid–liquid contact between tested materials and shape-adaptive liquid metal can lead to more accurate measurements. A quantitative triboelectric series of non-metallic materials was also constructed based on the measured TECD (Fig. 2.20, Table 2.3). The normalized TECD reveals the materials' capability to obtain or lose electrons after contact electrification. The quantified triboelectric series and the standards evaluation methods provided here develop the standards and techniques enabling the acquisition, representation, and discovery of the very

Table 2.3 Triboelectric series of materials and their TECD [85]

Materials	Average TECD ($\mu C\ m^{-2}$)	STDEV	α
Mica	61.8	1.63	0.547
Float glass	40.2	0.85	0.356
Borosilicate glass	38.63	1.18	0.342
BeO	9.06	0.21	0.08
PZT-5	8.82	0.16	0.078
$MgSiO_3$	2.72	0.07	0.024
$CaSiO_3$	2.38	0.15	0.021
$Bi_4Ti_3O_{12}$	2.02	0.21	0.018
$Bi_{0.5}Na_{0.5}TiO_3$	1.76	0.05	0.016
$NiFe_2O_4$	1.75	0.07	0.0155
$Ba_{0.65}Sr_{0.35}TiO_3$	1.28	0.11	0.011
$BaTiO_3$	1.27	0.08	0.0112
PZT-4	1.24	0.12	0.011
ZnO	0.86	0.04	0.008
NiO	0.53	0.05	0.005
SnO_2	0.46	0.02	0.004
SiC	0.31	0.07	0.003
$CaTiO_3$	0.24	0.02	0.002
ZrO_2	0.09	0.07	0.001
Cr_2O_3	0.02	0.01	0.00013
Fe_2O_3	0	0.02	0
Al_2O_3	– 1.58	0.14	– 0.014
TiO_2	– 6.41	0.18	– 0.057
AlN	– 13.24	1.35	– 0.117
BN	– 16.9	0.97	– 0.149
Clear very high-temperature glass ceramic	– 39.95	2.04	– 0.353
Ultra-high-temperature quartz glass	– 62.66	0.47	– 0.554

Note STDEV refers to the standard deviation. The *α* refers to the measured TECD of tested materials over the absolute value of the measured TECD of the reference material (PTFE)

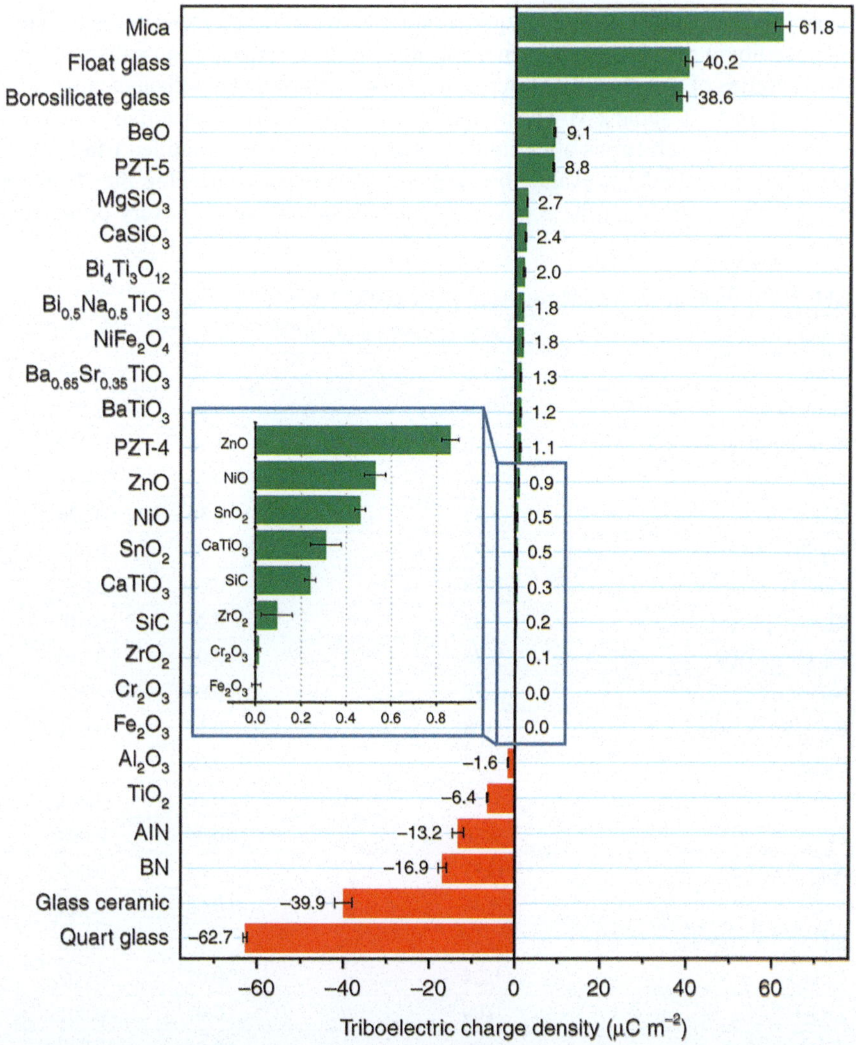

Fig. 2.20 Quantified triboelectric series of some common inorganic non-metallic materials. Figure from [85] with permission, copyright © 2020, The Author(s)

basic properties of materials, which could set the foundation for further applications and related fundamental research (i.e., the assessment of materials data, models, and simulations), which will be a strong support to the Materials Genome Project for applications in energy harvesting, sensors and human–machine interfacing.

2.8.3 Measurement of Liquid Materials

About 50 different liquids were assessed and organized into a liquid triboelectric series in order of triboelectric charge densities (TECD: $\sigma_{sat} = Q_{sat}/S$, S is the effective contact area), displayed in Fig. 2.21. In this study, the criterion for evaluating the test liquids involved ensuring that Q_{sat} varied non-monotonously within 10% of its value for a minimum of ten periods, which was statistically considered to be no significance (N.S.). The Q_{sat} was determined as the average of these ten values. To mitigate potential irregularities, each liquid underwent measurements at least three times using new tubes, with the mean value of these three experiments representing the final result for each test liquid (Fig. 2.21).

This series encompasses organic liquids and inorganic salt solutions, with data derived from the arithmetic mean of at least three independent experiments. It's worth noting that when the original charge is less than 0.5 nC for a particular liquid, its TECDs were determined by proportionately amplifying the length of the liquid column and electrode.

When the test liquid interacts with FEP and carries a positive charge, the measured charge density is recorded as a positive value; otherwise, it's negative. Although no negative value was observed within our testing scopes, they have been reported in HCl solution with pH < 3 [86]. Given that water is the most common and readily available pure liquid, the DI water is chosen as the reference liquid. The normalized TECD of a liquid is defined as a dimensionless symbol: $\beta = \frac{\sigma_{sat}(\text{test liquid})}{\sigma_{sat}(\text{DI water})}$. The $\beta > 1$ indicates that its triboelectric polarity is more positive than DI water, while the $\beta <$ 1 means more negative.

2.8.4 Measurement of 2-Dimensional Layered Materials

Minsu Seol et al. [87] have studied the triboelectric series of 2D layered material (Fig. 2.22). To characterize the triboelectric charge of 2D materials in the series, the 2D materials and nylon were worked as the two tribo layers, as nylon is known as a positive material in the triboelectric series and all 2D materials used exhibited negative triboelectric charges. These 2D films were prepared with a similar thickness of 200 nm, by controlling the concentration of the dispersion before filtering.

The triboelectric charging behaviors of various 2D layered materials, for example, MoS_2, $MoSe_2$, WS_2, WSe_2, graphene, and graphene oxide have been investigated based on the concept of triboelectric nanogenerator. 2D materials could be listed in the order of $MoSe_2$, GR, GO, WSe_2, and WS_2 according to the maximum output voltage and current of the corresponding TENGs.

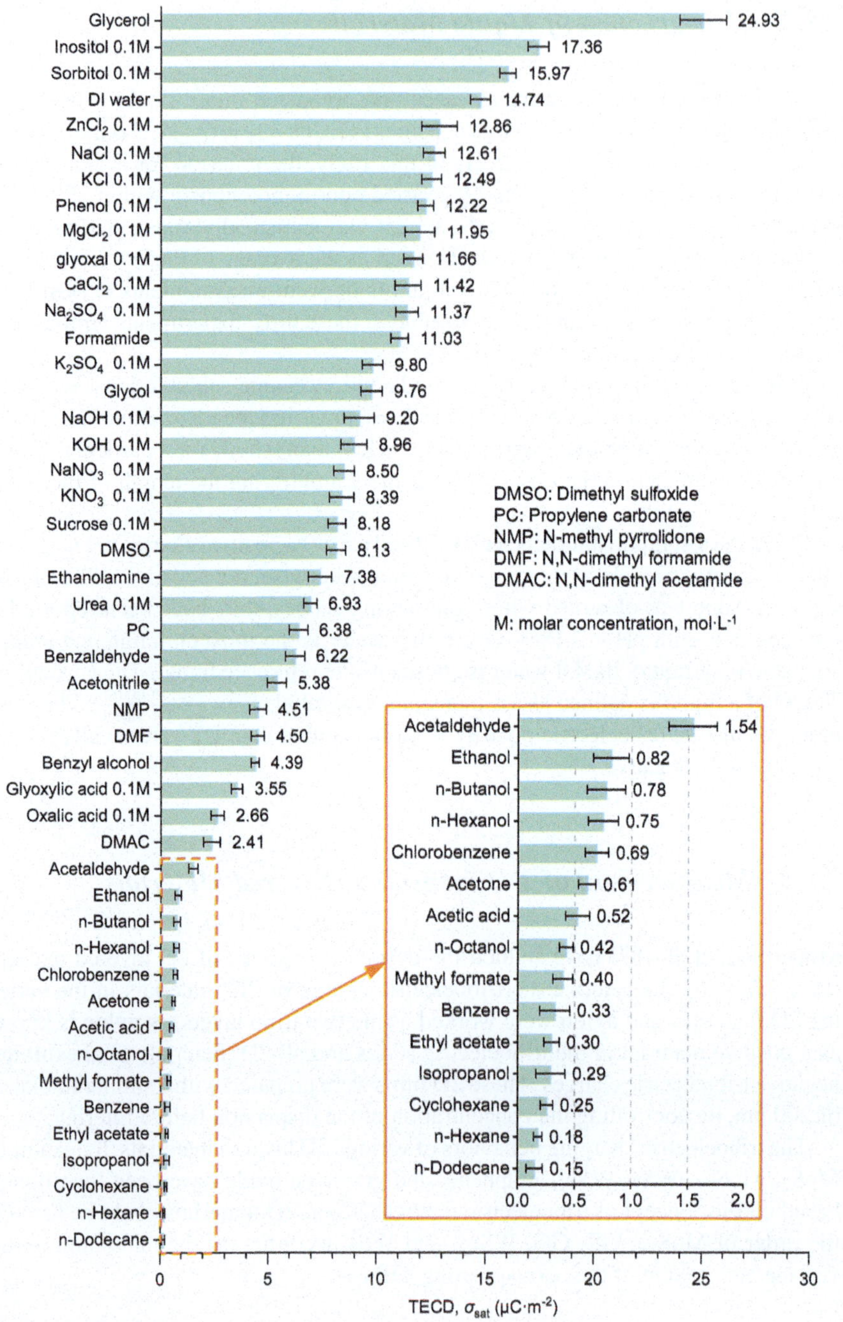

Fig. 2.21 The quantified triboelectric series of liquid materials. The error bar indicates the range within a standard deviation

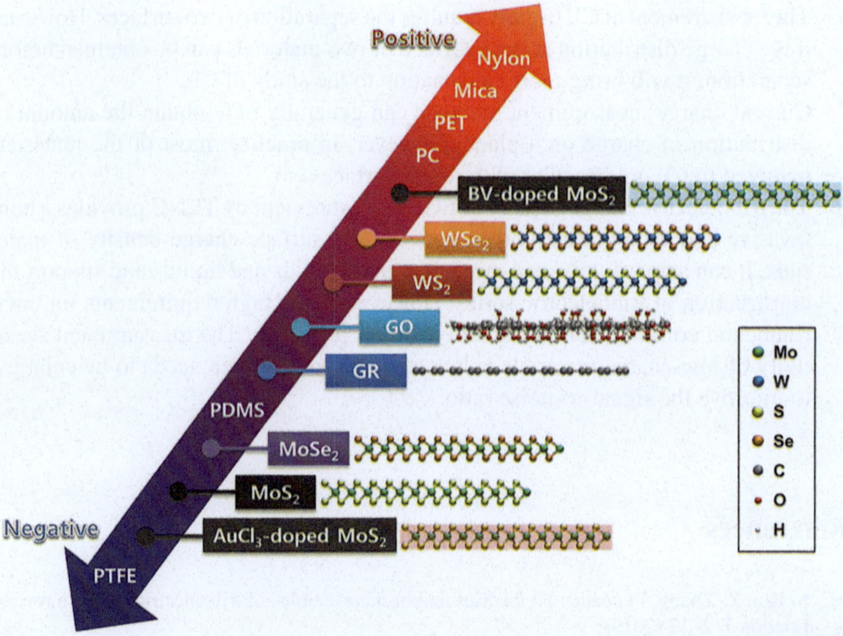

Fig. 2.22 Triboelectric series including 2D materials. The molecular structures are presented beside the corresponding materials. Figure from [87] with permission, copyright © 2018, WILEY-VCH Verlag GmbH & Co. KGaA, Weinheim

2.9 Summary

In this Chapter, we present six commonly used experimental methods to study CE. The experimental designs for CE studies are varied and very flexible. Based on these methods and the development of environmental performance research, we offer the following perspectives on the design of environmental performance methods.

1. The invention of the KPFM enabled surface charge detection at nanometre resolution, leading to the important discovery of the mosaic effect. It is natural to think that if the resolution continues to increase, new effects may be discovered at the atomic scale, revealing the mechanism of CE from the atomic scale.
2. The invention of KPFM solved the problem of experimental detection on tiny spatial scales, but how to realize charge transfer measurements on ultrafast time scales is still an important issue.
3. Existing CE measurement methods are mainly for solid–solid surfaces. Currently, researchers are very interested in the study of liquid–solid coupled charge, and how to realize the real-time measurement of charge transfer at the liquid–solid interface is one of the hotspots in the study of triboelectric charge.

4. The measurement of CE usually requires the separation of two surfaces. However, if the charge distribution at the interface of two materials can be obtained before separation, it will bring more information to the study of CE.
5. Current charge measurement methods can generally only obtain the amount or distribution of charge on a plane. However, in practice, most of the materials involved in CE are irregular with rough surfaces.
6. The triboelectric charge density (TECD) measurement by TENG provides a non-invasive and accurate method to quantify the surface charge density of materials. It can achieve universal measurement of solids and liquids and support the construction of triboelectric series. However, it has high requirements for environmental control (such as humidity and air pressure). The measurement sensitivity of low-charge materials is low, and the contact area needs to be enlarged to improve the signal-to-noise ratio.

References

1. S. Pan, Z. Zhang, Fundamental theories and basic principles of triboelectric effect: a review. Friction **7**, 2–17 (2019)
2. R. Elsdon, F.R.G. Mitchell, Contact electrification of polymers. J. Phys. D Appl. Phys. **9**(10), 1445–1460 (1976)
3. J. Lowell, Contact electrification of metals. J. Phys. D Appl. Phys. **8**(1), 53–63 (1975)
4. M. Sow, R. Widenor, A. Kumar, S. Lee, D. Lacks, R. Sankaran, Strain-induced reversal of charge transfer in contact electrification. Angew. Chem. Int. Ed. **51**(11), 2695–2697 (2012)
5. W.R. Harper, Contact electrification of semiconductors. Brit. J. Appl. Phys. **11**(8), 324–331 (1960)
6. D.A. Hays, Contact electrification between mercury and polyethylene: effect of surface oxidation. J. Chem. Phys. **61**(4), 1455–1462 (1974)
7. M.D. Sosa, M.L.M. Ricci, L.L. Missoni, D.H. Murgida, A. Canneva, N.B. D'Accorso, R.M. Negri, Liquid-polymer triboelectricity: chemical mechanisms in the contact electrification process. Soft Mater. **16**(30), 7040–7051 (2020)
8. T. Burgo, F. Galembeck, G.H. Pollack, Where is water in the triboelectric series? J. Electrostat. **80**, 30–33 (2016)
9. Y. Chen, X. Li, C. Xu, D. Wang, J. Huang, Z. Guo, W. Liu, Electron transfer dominated triboelectrification at the hydrophobic/slippery substrate-water interfaces. Friction **11**(6), 1040–1056 (2023)
10. J. Nie, Z. Wang, Z. Ren, S. Li, X. Chen, Z.L. Wang, Power Generation from the Interaction of a Liquid Droplet and a Liquid Membrane. Nat. Commun. **10**(1), 2264 (2019)
11. S. Lin, L. Cao, Z. Tang, Z.L. Wang, Size-dependent charge transfer between water microdroplets. P. Natl. Acad. Sci. USA **120**(31), e2307977120 (2023)
12. H. Gibson, Linear free energy relations. V. triboelectric charging of organic solids. J. Am. Chem. Soc. **97**(31), 3832–3833 (1975)
13. M. Sakaguchi, M. Makino, T. Ohura, T. Iwata, Contact electrification of polymers due to electron transfer among mechano anions, mechano cations and mechano radicals. J. Electrostat. **72**(5), 412–416 (2014)
14. C. Liu, A. Bard, Chemical redox reactions induced by cryptoelectrons on a PMMA surface. J. Am. Chem. Soc. **131**(18), 6397–6401 (2009)
15. C. Liu, A. Bard, Electrostatic electrochemistry at insulators. Nat. Mater. **7**(6), 505–509 (2008)

16. L. McCarty, G. Whitesides, Electrostatic charging due to separation of ions at interfaces: contact electrification of ionic electrets. Angew. Chem. Int. Ed. **47**(12), 2188–2207 (2008)
17. S. Thomas, S. Vella, M. Dickey, G. Kaufman, G. Whitesides, Controlling the kinetics of contact electrification with patterned surfaces. J. Am. Chem. Soc. **131**(25), 8746 (2009)
18. H. Baytekin, A. Patashinski, M. Branicki, B. Baytekin, S. Soh, B. Grzybowski, The mosaic of surface charge in contact electrification. Science **333**(6040), 308–312 (2011)
19. H. Baytekin, B. Baytekin, J. Incorvati, B. Grzybowski, Material transfer and polarity reversal in contact charging. Angew. Chem. Int. Ed. **51**(20), 4843–4847 (2012)
20. K. Francisco, T. Burgo, F. Galembeck, Tribocharged polymer surfaces: solvent effect on pattern formation and modification. Chem. Lett. **41**(10), 1256–1258 (2012)
21. T. Burgo, C. Silva, L. Balestrin, F. Galembeck, Friction coefficient dependence on electrostatic tribocharging. Sci. Rep. **3**(1), 2384 (2013)
22. T. Burgo, T. Ducati, K. Francisco, K. Clinckspoor, F. Galembeck, S. Galembeck, Triboelectricity: macroscopic charge patterns formed by self-arraying ions on polymer surfaces. Langmuir **28**(19), 7407–7416 (2012)
23. H. Qin, L. Xu, F. Zhan, Z.L. Wang, Electron transfer induced contact-electrification at oil and oleophobic dielectric interface. Nano Energy **116**, 108762 (2023)
24. Z. Tang, S. Lin, Z.L. Wang, Effect of surface pre-charging and electric field on the contact electrification between liquid and solid. J. Phys. Chem. C **126**(20), 8897–8905 (2022)
25. P. Ireland, Contact charge accumulation and separation discharge. J. Electrostat. **67**(2–3), 462–467 (2009)
26. Y. Zhang, T. Shao, A method of charge measurement for contact electrification. J. Electrostat. **71**(4), 712–716 (2013)
27. Y. Zhang, T. Shao, Contact electrification between polymers and steel. J. Electrostat. **71**(5), 862–866 (2013)
28. D. Davies, The examination of the electrical properties of insulators by surface charge measurement. J. Sci. Instrum. **44**(7), 521–524 (1967)
29. D. Davies, Charge generation on dielectric surfaces. Brit. J. Appl. Phys. **2**(11), 1533–1537 (1969)
30. S. Singh, Y. Kusano, P. Morgen, P. Michelsen, Surface charging, discharging and chemical modification at a sliding contact. J. Appl. Phys. **111**(8), 083501 (2012)
31. A. Kazzaz, Roseinnes, the use of liquid metals for contact charging experiments. J. Phys. D Appl. Phys. **20**(12), 1616–1622 (1987)
32. Z. Yu, P. Watson, J. Facci, The contact charging of PTFE by mercury: the effect of a thiophene monolayer on charge exchange. J. Phys. D Appl. Phys. **23**(9), 1207–1211 (1990)
33. K. Nakayama, Tribocharging and friction in insulators in ambient air. Wear **194**, 185–189 (1996)
34. R. Nevshupa, Effect of gas pressure on the triboluminescence and contact electrification under mutual sliding of insulating materials. J. Phys. D: Appl. Phys. **46**(18), 185501 (2013)
35. D. Debeau, The effect of adsorbed gases on contact electrification. Phys. Rev. **66**(1–2), 9–16 (1944)
36. P. Wagner, Electrostatic charge separation at metal-insulator contacts. J. Appl. Phys. **27**(11), 1300–1310 (1956)
37. S. Vella, X. Chen, S. Thomas, X. Zhao, Z. Suo, G. Whitesides, The determination of the location of contact electrification induced discharge events. J. Phys. Chem. C **114**(48), 20885–20895 (2010)
38. S. Soh, X. Chen, S. Vella, W. Choi, J. Gong, G. Whitesides, Layer-by-layer for tunable and rewritable control of contact electrification. Soft Matter. **9**(43), 10233 (2013)
39. J. Wiles, B. Grzybowski, A. Winkleman, G. Whitesides, A tool for studying contact electrification in systems comprising metals and insulating polymers. Anal. Chem. **75**(18), 4859–4867 (2003)
40. J. Wiles, M. Fialkowski, M. Radowski, G. Whitesides, B. Grzybowski, Effects of surface modification and moisture on the rates of charge transfer between metals and organic materials. J. Phys. Chem. B **108**(52), 20296–20302 (2004)

41. Schella, S. Herminghaus, M. Schroter, Influence of humidity on trobo-electric charging and segregation in shaken granular media. Soft Matter. **13**(2), 394 (2017)
42. Harris, M. Lim, H. Jaeger, Temperature dependence of nylon and PTFE triboelectrificaitoin. Phys. Rev. Mater. **3**(8), 085603 (2019)
43. L. Wang, J. Li, J. Tao, M. Hu, Z. Dai, Sensitive self-powered particles detection based on cumulative triboelectric charging. Nano Energy **89**, 106393 (2021)
44. S. Soh, S. Kwok, H. Liu, G. Whitesides, Contact de-electrification of electrostatically charged polymers. J. Am. Chem. Soc. **134**(49), 20151–20159 (2012)
45. B. Kwetkus, K. Sattler, H. Siegnmann, Gas breakdown in contact electrification. J. Phys. D Appl. Phys. **25**(2), 139–146 (1992)
46. W. Hu, L. Xie, X. Zheng, Contact charging of silica glass particles in a single collision. Appl. Phys. Lett. **101**(11), 114107 (2012)
47. Y. Jiang, S. Zhang, W. Lu, C. Ao, K. Lim, K. Zeng, S. Soh, Nanocavities stabilize charge: surface topology is a general strategy for controlling charge dissipation. Mater. Today Phys. **35**, 101105 (2023)
48. S. Cezan, A. Nalbant, M. Buyuktemiz, Y. Dede, H. Baytekin, B. Baytekin, Control of triboelectric charges on common polymers by photoexcitation of organic dyes. Nat. Commun. **10**(1), 276 (2019)
49. S. Matsusaka, M. Ghadiri, H. Masuda, Electrification of an elastic sphere by repeated impacts on a metal plate. J. Phys. D Appl. Phys. **33**(18), 2311–2319 (2000)
50. Q. Chen, B. Cheng, T. Wang, H. Shang, T. Shao, Method for the measurement of triboelectric charge transfer at solid-liquid interface. Friction **11**(8), 1544–1556 (2023)
51. F. Wang, P. Yang, X. Tao, Y. Shi, S. Li, Z. Liu, Z.L. Wang, Study of contact electrification at liquid-gas interface. ACS Nano **15**(11), 18206–18213 (2021)
52. Z. Tang, S. Lin, Z.L. Wang, Z.L. Quantifying contact-electrification induced charge transfer on a liquid droplet after contacting with a liquid or solid. Adv. Mater. **33**(42), 2102886 (2021)
53. S. Lin, L. Xu, A. Wang, Z. L. Wang, Quantifying electron transfer in liquid-solid contact electrification and the formation of electric double layer. Nat. Commun. **11**(1), 399(2020)
54. S. Lin, L. Xu, C. Xu, X. Chen, A. Wang, B. Zhang, P. Lin, Y. Yang, H. Zhao, Z. L. Wang, Electron transfer in nanoscale contact electrification: effect of temperature in the metal-dielectric case. Adv. Mater. **31**(17), 1808197 (2019)
55. F. Chien, C. Hsu, Local photo-assisted poling of azo copolymer films by scanning probe microscopy. J. Phys. D: Appl. Phys. **41**(23), 235502 (2008)
56. R. Sanikop, N. Arya, V. Balakrishnan, C. Sudakar, Charge pumping by contact electrification using electrostatic force microscopy in Bi Trilayered MoS2 nanosheets. J. Phys. Chem. C **125**(22), 12155–12165 (2021)
57. Q. Li, I. Cho, R. Biswas, J. Kim, Nanoscale modulation of friction and triboelectrification via surface nanotexturing. Nano Lett. **19**(2), 850–856 (2019)
58. S. Lin, Z.L. Wang, Scanning triboelectric nanogenerator as a nanoscale probe for measuring local surface charge density on a dielectric film. Appl. Phys. Lett. **118**(19), 193901 (2021)
59. C. Rotsch, M. Radmacher, Mapping local electrostatic forces with the atomic force microscope. Langmuir **13**(10), 2825–2832 (1997)
60. Z.L. Wang, A. Wang, On the origin of contact electrification. Mater. Today **30**, 34–51 (2019)
61. J. Liu, A. Goswami, K. Jiang, F. Khan, S. Kim, R. McGee, Z. Li, Z. Hu, J. Lee, T. Thundat, Direct current triboelectricity generation by a sliding Scottky nanocontact on MoS_2 multilayers. Nat. Nanotech. **13**(2), 112 (2018)
62. M. Zheng, S. Lin, L. Xu, L. Zhu, Z.L. Wang, Scanning probing of the tribovoltaic effect at the sliding interface of two semiconductors. Adv. Mater. **32**(21), 2000928 (2020)
63. S.M. Niu, S. Wang, L. Lin, Y. Liu, Y.S. Zhou, Y. Hu, Z.L. Wang, Theoretical study of contact-mode triboelectric nanogenerators as an effective power source. Energy Environ. Sci. **6**(12), 3576–3583 (2013)
64. G. Zhu, C. Pan, W. Guo, C.Y. Chen, Y. Zhou, R. Yu, Z.L. Wang, Triboelectric-generator-driven pulse electrodeposition for micropatterning. Nano Lett. **12**(9), 4960–4965 (2012)

65. H. Zou, Y. Zhang, L. Guo, P. Wang, X. He, G. Dai, Z.L. Wang, Quantifying the triboelectric series. Nat. Commun. **10**(1), 1427 (2019)
66. Z.L. Wang, T. Jiang, L. Xu, Toward the blue energy dream by triboelectric nanogenerator networks. Nano Energy **39**, 9–23 (2017)
67. T. Sun, R. Wang, R. Liu, C. Wu, Y. Zhong, Y. Liu, Y. Wang, Y. Han, Z. Xia, Y. Zou, T. Song Investigation of MoOx/n-Si strong inversion layer interfaces via dopant-free heterocontact. physica status solidi. Rapid Res. Lett. (RRL) **11**(7), 1700107 (2017)
68. L. Jiang, X. Liu, J. Lv, G. Li, P. Yang, Y. Ma, H. Zou, Z.L. Wang, Fluid-based triboelectric nanogenerators: unveiling the prolific landscape of renewable energy harvesting and beyond. Energy and Environ. Sci. (2024)
69. S. Lin, X. Chen, Z.L. Wang, The tribovoltaic effect and electron transfer at a liquid-semiconductor interface. Nano Energy **76**, 105070 (2020)
70. C. Xu, Y. Zi, A.C. Wang, H. Zou, Y. Dai, X. He, P. Wang, Y.C. Wang, P. Feng, D. Li, Z.L. Wang, On the electron-transfer mechanism in the contact-electrification effect. Adv. Mater. **30**(15), 1706790 (2018)
71. M. Sun, Q. Lu, Z.L. Wang, B. Huang, Understanding contact electrification at liquid–solid interfaces from surface electronic structure. Nature Commun. **12**(1), 1752 (2021)
72. M.D. Sosa, M.L.M. Ricci, L.L. Missoni, D.H. Murgida, A. Cánneva, N.B. D'Accorso, R.M. Negri, Liquid–polymer triboelectricity: chemical mechanisms in the contact electrification process. Soft Matter **16**(30), 7040–7051 (2020)
73. J. Nauruzbayeva, Z. Sun, A. Gallo Jr, M. Ibrahim, J.C. Santamarina, H. Mishra, Electrification at water–hydrophobe interfaces. Nature Commun. **11**(1), 5285 (2020)
74. T. Zhan, H. Zou, H. Zhang, P. He, Z. Liu, J. Chen, M. He, Y. Zhang, Z.L. Wang, Smart liquid-piston based triboelectric nanogenerator sensor for real-time monitoring of fluid status. Nano Energy **111**, 108419 (2023)
75. Q. Zhang, C. Jiang, X. Li, S. Dai, Y. Ying, J. Ping, Highly efficient raindrop energy-based triboelectric nanogenerator for self-powered intelligent greenhouse. ACS Nano **15**(7), 12314–12323 (2021)
76. S. Wang, Y. Wang, D. Liu, Z. Zhang, W. Li, C. Liu, T. Du, X. Xiao, L. Song, H. Pang, M. Xu, A robust and self-powered tilt sensor based on annular liquid-solid interfacing triboelectric nanogenerator for ship attitude sensing. Sens. Actuators, A **317**, 112459 (2021)
77. X. Ge, Z. Gao, L. Zhang, H. Ji, J. Yi, P. Jiang, Z. Li, L. Shen, X. Sun, Z. Wen, Flexible microfluidic triboelectric sensor for gesture recognition and information encoding. Nano Energy **113**, 108541 (2023)
78. K. Han, J. Luo, Y. Feng, Q. Lai, Y. Bai, W. Tang, Z.L. Wang, Wind-driven radial-engine-shaped triboelectric nanogenerators for self-powered absorption and degradation of NOX. ACS Nano **14**(3), 2751–2759 (2020)
79. D.J. Lacks, T. Shinbrot, Long-standing and unresolved issues in triboelectric charging. Nat. Rev. Chem. **3**(8), 465–476 (2019)
80. D. Liu, L. Zhou, S. Cui, Y. Gao, S. Li, Z. Zhao, Z. Yi, H. Zou, Y. Fan, J. Wang, Z.L. Wang, Standardized measurement of dielectric materials' intrinsic triboelectric charge density through the suppression of air breakdown. Nat. Commun. **13**(1), 6019 (2022)
81. S. Niu, Y. Liu, S. Wang, L. Lin, Y.S. Zhou, Y. Hu, Z.L. Wang, Theoretical investigation and structural optimization of single-electrode triboelectric nanogenerators. Adv. Func. Mater. **24**(22), 3332–3340 (2014)
82. L.D. Landau, *Lifschitz* (Pergamon, Fluid mechanics, 1959)
83. Z.L. Wang, A.C. Wang, On the origin of contact-electrification. Mater. Today **30**, 34–51 (2019)
84. M.S. Vihacencu, P.V. Notingher, T. Paillat, S. Jarny, Flow electrification phenomenon for Newtonian and non-Newtonian liquids: influence of liquid conductivity, viscosity and shear stress. IEEE Trans. Dielectr. Electr. Insul. **21**(2), 693–703 (2014)
85. H. Zou, L. Guo, H. Xue, Y. Zhang, X. Shen, X. Liu, P. Wang, X. He, G. Dai, P. Jiang, H. Zheng, Quantifying and understanding the triboelectric series of inorganic non-metallic materials. Nature Commun. **11**(1), 2093 (2020)

86. J. Nie, Z. Ren, L. Xu, S. Lin, F. Zhan, X. Chen, Z.L. Wang, Probing contact-electrification-induced electron and ion transfers at a liquid–solid interface. Adv. Mater. **32**(2), 1905696 (2020)
87. M. Seol, S. Kim, Y. Cho, K. E. Byun, H. Kim, J. Kim, S. Park, Triboelectric series of 2D layered materials. Adv. Mater. **30**(39), 1801210 (2018)

Chapter 3
Quantum Theory of Triboelectricity and Related Phenomena

Abstract The current article provides a comprehensive survey of quantitative models and calculations of a variety of closely-related contact electrification phenomena that focus on the microscopic origins of the phenomena. Beyond a few *ab initio* calculations, the most systematic and comprehensive study was carried out by our group using a Hamiltonian formalism. The main focus of this chapter is on three problems we have applied our microscopic theories to: (1) a calculation of electron transfer due solely to quantum-mechanical tunneling, (2) the study of electron transfer coupled with light emission, and (3) a model of tribovoltaics.

3.1 Introduction

Contact electrification (CE, also known as triboelectricity), the generation of electrostatic charge via contact or friction, is a natural phenomenon known since the early days and that has traditionally been viewed as firmly grounded in classical electrostatics [1]. It has been applied in technologies such as laser printing [2, 3] though still disruptive in natural and industrial environments. More recently, it has been proposed as a clean form of energy via triboelectric nanogenerators [4–12] and tribovoltaic nanogenerators [13–26].

Even to date, CE contains fundamental surprises such as the observed generation of X-rays by peeling adhesive tape [27, 28]. Yet, triboelectricity has escaped much quantitative analysis and it has been rather challenging to propose a unified physical model for an effect that starts at the microscopic level and is typically observed macroscopically, while covering a broad range of materials [29]. This is primarily due to two factors: (1) it involves multiple and multiphysics processes (e.g., mechanical deformation, electric charge transport, carrier heating, etc.), and (2) the macroscopic observation goes beyond the intrinsic material property and depends on many external conditions such as surface roughness, environmental conditions, …It is, therefore, clear that any quantitative analysis needs to remove as many of the variables as possible. An example of this approach is the recent standardization of the triboelectric series by using the liquid phase for the reference material since this allows the almost complete removal of the surface morphology variable [30].

Z. L. Wang, *Contact-Electrification of Matter*, Springer Series in Surface Sciences 75,
https://doi.org/10.1007/978-3-031-88504-4_3

The early experimental investigation of CE has been well documented by Harper [1] and more recent reviews [29, 31, 32]. It has advanced somewhat in recent years, since the introduction of Kelvin probe force microscopy [33]. On the theory side, one still relies predominantly on the qualitative picture of CE provided by the Volta-Helmholtz hypothesis for understanding electron transfer [1, 34, 35], that contact is required for electrification, and that a double charge layer is formed as a result, leading to a contact potential difference. This has been refined by the scheme in terms of band energies introduced by Vick [36] and Montgomery [37] and we refer to this state of knowledge as the Volta-Helmholtz-Montgomery hypothesis, and this is described pictorially in Fig. 3.1. They sought to establish the why of CE and led to the widely accepted band picture for different types of solid contacts.

For example, for metal-metal contacts, charge transfer is postulated to occur due to Fermi-level differences whereas, for metal-insulator contacts, empty surface states on the insulator side are needed. A modern scheme in terms of the electron-cloud overlap has been proposed but is also illustrative only [29]. Neither scheme provided for a quantitative calculation of electron transfer. To date, the only two direct calculations of electron transfer have been based on *ab initio* calculations [38–46] and our related calculations based on using empirical Hamiltonians [47–51] in order to provide for a more physical understanding. Recently, a quantum theory of tribovoltaics was developed using both an atomistic and a continuum approach [52]. They both demonstrated that the effect is quadratic in the speed to the lowest order.

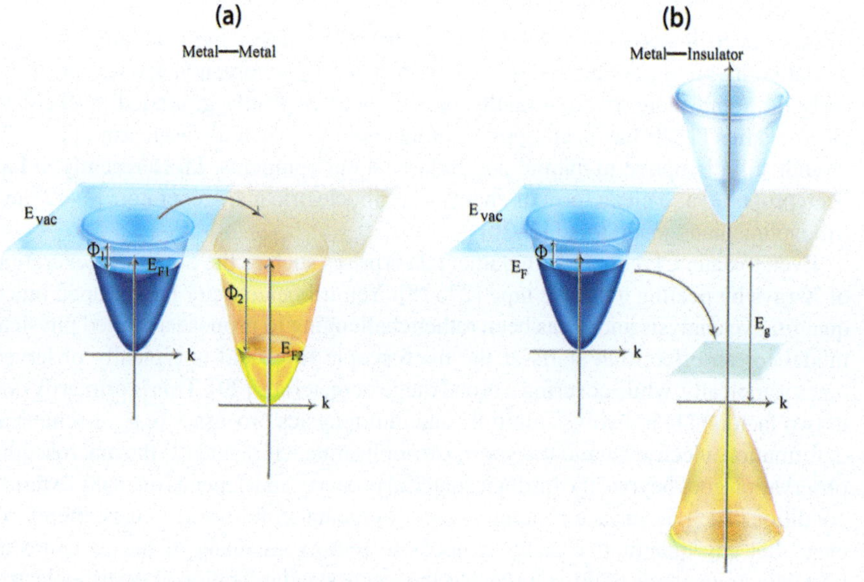

Fig. 3.1 Volta-Helmholtz-Montgomery hypothesis. Shown for **a** metal-metal, and **b** metal-insulator. Reproduced with permission from Ref. [49]

3.2 Overview of Past Theories

We present our recent microscopic models of CE starting with the next section. In the current section, we summarize other distinct types of modeling of CE to date: classical, phenomenological or density-functional calculations. We do not include the older theories discussed in detail in older reviews and books [1, 37].

3.2.1 Phenomenological Models

Two of the more recent models sought to provide a driving mechanism for CE. In 2008, Lacks et al. [53] attempted to explain charging of the same materials via a rate equation model where the driving mechanism was artificially introduced through constraints on which electron transfers are allowed without any microscopic basis and the model parameters were randomly chosen.

In 2019, flexoelectricity was proposed as a mechanism to explain charging due to indentations [54] yet the model ignored potentially more significant effects such as piezoelectricity, had electric potential estimates that were rather large, and is not applicable for situations when flexoelectricity is not present (e.g., for CE in general). We note that both of the above models lacked predictions.

3.2.2 Alicki-Jenkins

In 2020, the related problem of charge transfer when two materials are in relative motion was studied theoretically using a phenomenological quantum-mechanical theory of two cylindrical materials spinning with respect to each other [55]. It was argued that a motion-induced population inversion was responsible for the generation of an electromotive force in an irreversible process which does not rely on classical electrodynamics (i.e., does not arise from the Maxwell-Faraday law) and is independent of the sign of the rotation but dependent on the relative velocity. No explicit calculation of the effect was included.

3.2.3 Density-Functional Calculations

Some early *ab initio* calculations were performed by Duke and Fabish, though more direct calculations of charge transfer are more recent. The work of Duke and Fabish [38, 39]followed the Montgomery scheme by using (1) ab initio calculations to identify the existence of discrete levels on the insulator side of a metal-insulator junction (as intrinsic localized molecular-ion levels to explain charge transfer for metal-polymer [38]) and (2) a fitted energy structure and density-of-states to calculate charge transfer for polymer-polymer contacts [39].

First-principles calculations [and mostly density-functional theory (DFT)] have been carried out for quantitative studies of charge transfer but do not provide explanations for all observed effects and are currently restricted to periodic solids and elastic electron transfer [40–46]. Indeed, it is even questionable as to whether they all computed charge transfer. For instance, Zhang et al. [42] did not since they relaxed a computational system consisting of both materials—this amounts to bond creation between the two materials and the "charge transfer" they calculated is simply a bond polarity. On the other hand, Wu et al. [45] did a very careful analysis and uncovered a possible regime where the driving force is electrostatic. In general, while DFT is a great semi-predictive tool, it is much very like an experimental tool, requiring other models and theories in order to analyze its results.

3.3 Hamiltonian Model

In our program of work since 2019, we have proposed a unifying physical and microscopic model of electron transfer based on a quantum-mechanical Hamiltonian [47–51]. We focussed on electron transfer (as opposed to ion and material transfer), contact electrification (instead of frictional), and did not consider environmental factors such as humidity. Beyond that, the model is very general in being applicable to a variety of materials and transfer mechanisms. As many CE researchers have observed [53, 54], a "driver" for electron transfer is needed. The most fundamental one for electrons is the Hamiltonian quantum dynamics. Our group has started a quantum-mechanical study of CE by computing two possible mechanisms of electron transfer: (1) quantum tunneling across a gap [48] and, (2) the photon-assisted transition rate [47]. The lossless quantum-tunneling mechanism between two materials did allow for the explicit calculation of the charge transfer. The models were developed using both localized orbitals and Bloch states. This allowed us to study practically any kind of contact and display an ultrafast time dynamics not previously studied.

3.3.1 Scheme

Our scheme of contact electrification consists of the following elements:

- Only electron transfer occurs.
  Our goal was to demonstrate quantitatively the phenomenon of electron transfer when two materials are brought together and separated. Thus, we are automatically excluding systems whereby ions might be involved (e.g., in the presence of an ionic fluid); it is now indeed believed that most CE involves electron transfer [29, 56, 57].

- A microscopic quantum-mechanical Hamiltonian model is introduced. most fundamental approach to studying electron dynamics is by starting with an appropriate Hamiltonian and solving the time-dependent Schrödinger equation:

$$H|\psi(t)\rangle = i\hbar\frac{\partial|\psi(t)\rangle}{\partial t}. \tag{3.1}$$

- Our initial work was implemented using one-dimensional (1D) models. This apparent restriction was to simplify the model computationally while retaining much of the physics. Indeed, it is known that "any quantum-mechanical model can be transformed into one or more chain models" [58]. Hence, even the differing surface physics associated with crystal faces of different orientations can be modeled by using chains with differing hopping parameters. One-dimensional models are often used in physics and have been introduced to study related phenomena, such as in a model of frictional electrification via electron-phonon interaction [59]. It could also be relevant to modeling 1D arrays of electrostatically-charged spheres [60].

Our model contains the most fundamental physics and should, therefore, display the key properties of CE without the need to introduce thermal or flexoelectric driving forces to account for CE. As P. W. Anderson said in his 1977 Nobel lecture, "Very often such, a simplified model throws more light on the real workings of nature than any number of "ab initio" calculations."

3.3.2 *Tight-Binding Hamiltonian*

Our Hamiltonian models a system consisting of two materials with a coupling between them in order to drive the electronic charge transfer. Using an atomic-like, tight-binding (TB) basis set, the coupling is described by an effective hopping parameter t_C which can be chosen to describe various electron transfer mechanisms such as quantum tunneling, photon-assisted, phonon-assisted, ... Additionally, it could account for the imperfect contact between solids as well as the separation of contacts after charging. One advantage of the localized basis set is that it can be chosen to model finite materials, and both solid-solid and fluid-solid contacts.

A 1D model is illustrated in Fig. 3.2. The system of two materials in proximity to each other is described by a Hamiltonian $H = H_L + H_R + H_C$ which consists of three TB-like components. Each of the component Hamiltonians can be written down in terms of, for example, a nearest-neighbor TB approximation [61–63]. The TB Hamiltonian can be written down as [49–51]

$$H = \sum_{l=0}^{N}\left[t_l\left(a_l^\dagger a_{l+1} + a_{l+1}^\dagger a_l\right) - e\phi_l a_l^\dagger a_l\right], \tag{3.2}$$

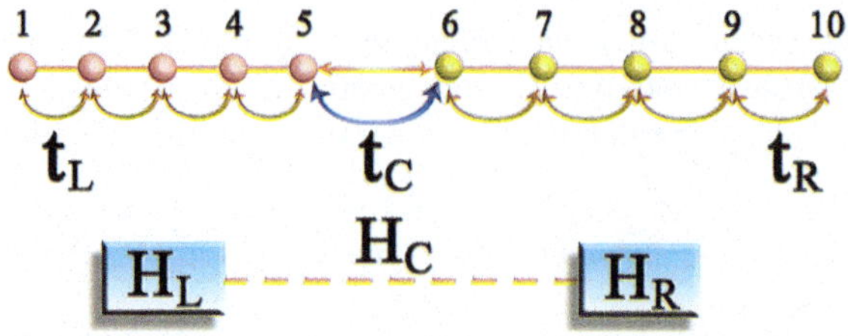

Fig. 3.2 Top: 1D model of two materials in contact (shown with a total of 10 atoms as an example). Here, atoms 1–5 form the left (L) material and atoms 6–10 form the right (R) material. Bottom: a schematic of the Hamiltonian of the system, consisting of the tight-binding Hamiltonian of the left and right materials, coupled via a hopping parameter t_C. Reproduced with permission from Ref. [49]

where l labels the atomic sites, ϕ_l is a scalar electric potential at site l, and e is the positive elementary charge. While a TB basis is most efficient for narrow-band materials, in its parameterized formulation [61], it can be used to reproduce the electronic structure of metal, semiconductors and insulators alike.

In addition to modeling the initial contact charging, one can also model the charge transfer as the two materials are moved with respect to each other if the coupling mechanism is distance $d(t)$ dependent. For example, there will be a reduced quantum tunneling as the separation increases. This can be represented by a distance-dependent hopping parameter t_C; one can use the Harrison rule [62] to simulate the latter:

$$t_C(t) = t_C \left(\frac{d}{d(t)} \right)^2 . \tag{3.3}$$

Charge Transfer

Charge transfer is computed by finding the time dependence of the electronic wave function. In the TB picture, a general electronic quantum state can be written as

$$|\psi(t)\rangle = \sum_{l=1}^{N} c_l(t)|l\rangle, \tag{3.4}$$

where $c_l(t)$ are time-dependent coefficients determined by the Schrödinger equation, Eq. (3.1). One can then define charging to be a measure of charge transfer in units of electron charge:

$$\text{charging} = -\sum_{m=1}^{N_L} (|c_m(t)|^2 - |c_m(0)|^2). \tag{3.5}$$

In order to find the $c_m(t)$'s, one first insert Eq. (3.4) into the Schrödinger equation, Eq. (3.1) giving

$$i\hbar\dot{c}_n = t_n c_{n+1} + t_{n-1}c_{n-1} - e\phi_n c_n, \quad 1 \le n \le N, \tag{3.6}$$

This is a system of linear first-order equations in time; for example, for a system of two chains, it can be rewritten as

$$\begin{aligned}
i\hbar\dot{c}_1 &= t_L c_2 - e\phi_1 c_1, \\
i\hbar\dot{c}_2 &= t_L c_3 + t_L c_1 - e\phi_2 c_2, \\
&\;\;\vdots \\
i\hbar\dot{c}_M &= t_C c_{M+1} + t_L c_{M-1} - e\phi_M c_M, \\
i\hbar\dot{c}_{M+1} &= t_R c_{M+2} + t_C c_M - e\phi_{M+1} c_{M+1}, \\
&\;\;\vdots \\
i\hbar\dot{c}_{N-1} &= t_R c_N + t_R c_{N-2} - e\phi_{N-1} c_{N-1}, \\
i\hbar\dot{c}_N &= t_R c_{N-1} - e\phi_N c_N.
\end{aligned} \tag{3.7}$$

This system can be easily solved numerically (e.g., using MATLAB) subject to specified initial conditions

$$c_l(0) = c_l^0, \quad 1 \le l \le N. \tag{3.8}$$

Example Results

As an example, an exact solution is obtained for the case of only two sites in the absence of the Coulomb interaction; this simulates the insulator-insulator case [49]. Then,

$$c_1(t) = c_1(0)e^{-it_C t/\hbar}, \; c_2(t) = c_2(0)e^{-i(t_C t/\hbar + \eta)}, \tag{3.9}$$

where c_1 and c_2 are the wave-function coefficients on the two sites and η is an arbitrary phase factor. This solution shows that there is a natural charge oscillation with a frequency proportional to the hopping parameter; for $t_C = -1\,\text{eV}$, $f = h/\text{eV} = 0.24\,\text{fs}^{-1}$. It is analogous to the Rabi frequency in optics.

Numerical calculations showed that the oscillation is preserved but modified for longer chains and inclusive of the Coulomb interaction. An example calculation for a total of 10 atoms is reproduced in Fig. 3.3.

The two insulators are taken to be different materials and simulated by choosing different starting charge population on the two sites (0.01 and 0, respectively), even though both materials are electrically neutral to start with. The main features of these results are as follows:

1. Charge oscillates back and forth due to the hopping coupling, and the transfer rate is enhanced by the presence of the Coulomb interaction. It remains in the femtosecond regime and should be observable in time-resolved experiments.
2. At low carrier concentration, Coulomb effects are unimportant and the exact results given in Eq. (3.9) are valid. For example, the period of charge oscillation is proportional to t_C and maximum charge transfer is 100 % irrespective of t_C.

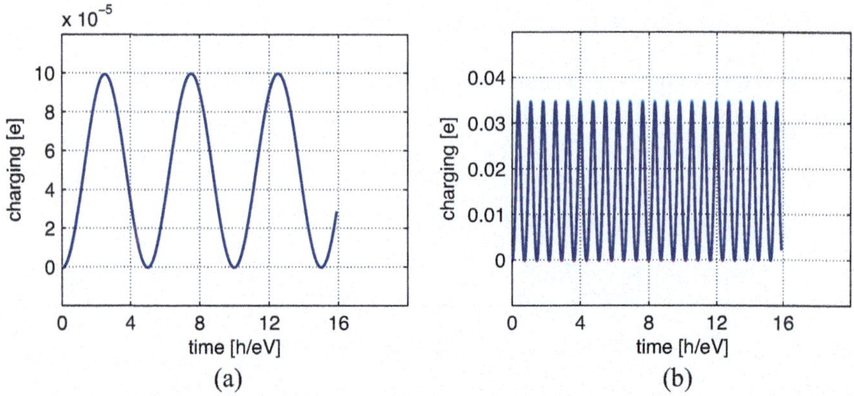

Fig. 3.3 Insulator-insulator contact ($N_L = N_R = 5$): **a** solid-solid ($c_C = 0.01$), **b** high carrier concentration, solid-solid ($c_C = 1.0$). Reproduced with permission from Ref. [49]

3. At high concentration, maximum charge transfer went down from 100% to about 3.5% and the period of oscillation decreased almost 10-fold.

The above results demonstrate "isoenergetic" charge transfer between two similar insulators provided there is a coupling between the two systems.

3.3.3 Continuum Model

Complementary to the atomistic model, a continuum model has also been developed [48] and the electrical current calculated via the Esaki-Tsu formalism.

Effective Schrödinger Equation

A canonical system is the structure depicted in Fig. 3.4 consisting of two materials denoted by subscripts 1 (left) and 2 (right) with conduction-band edges E_{c1} and E_{c2}, respectively.

A gap of thickness $2a$ is present between them and acts as a tunneling barrier of potential energy V. Using the effective mass approximation for the electrons and the effective mass to be the same in both materials, the effective Schrödinger equation for the electrons is

$$-\frac{\hbar^2}{2m^*}\frac{d^2\psi}{dx^2} + V_{\text{pot}}(x)\psi(x) = E\psi(x), \tag{3.10}$$

where m^* is the effective mass, $V_{\text{pot}}(x)$ is the potential energy profile, and E is the electron energy.

Charge Transfer

The charge transfer is obtained by solving the following equations

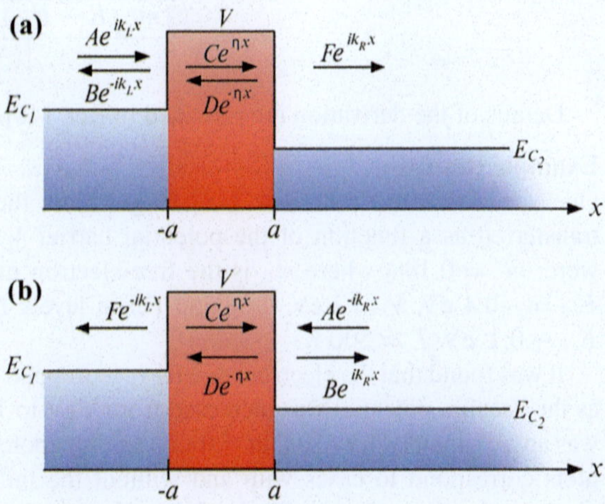

Fig. 3.4 Band diagram for tunneling of electrons through a vacuum gap. **a** Left-to-right tunneling and **b** right-to-left tunneling. Reproduced with permission from Ref. [48]

$$\frac{dN_1/\mathcal{A}}{dt} = -J, \tag{3.11}$$

$$\frac{dN_2/\mathcal{A}}{dt} = J, \tag{3.12}$$

where N_1 (N_2) is the number of electrons in the left (right) material, $\mathcal{A}$ the cross-sectional area, and the particle current density per area J is given by [48]

$$J = (2)2\pi \frac{m^*}{h^3} \int_{E_{\min}}^{V} dE_x T(E_x) D(E_x), \tag{3.13}$$

where

$$D(E_x) = k_B T \ln\left(\frac{1 + \exp\left(-\frac{E_x - E_{F1}}{k_B T}\right)}{1 + \exp\left(-\frac{E_x - E_{F2}}{k_B T}\right)}\right), \tag{3.14}$$

$$T = k_L k_R \left|\frac{4i\eta}{(-ik_L - \eta)(-ik_R - \eta)e^{-2\eta a} e^{i(k_L + k_R)a} - (-ik_L + \eta)(-ik_R + \eta)e^{2\eta a} e^{i(k_L + k_R)a}}\right|^2, \tag{3.15}$$

where $T(E(k_{x,1}))$ is the tunneling transmission coefficient assumed to only depend on the momentum perpendicular to the interface, $\frac{1}{L_x}$ is the electron density (plane wave normalized over L_x), $E = E(k_{x,1})$ and

$$k_L = \sqrt{\frac{2\,m^*\,(E - E_{c1})}{\hbar^2}}, \tag{3.16}$$

$$\eta = \sqrt{\frac{2\,m^*\,(V - E)}{\hbar^2}}, \tag{3.17}$$

$$k_R = \sqrt{\frac{2\,m^*\,(E - E_{c2})}{\hbar^2}}. \tag{3.18}$$

Details of the derivation are provided in Ref. [48].

Example Results

One set of results from Ref. [48] (Fig. 3.5) is the dependence of the electron transferred as a function of the potential barrier V. Additional parameter values were: $m^* = 0.1 m_0$ where m_0 is the free-electron mass, $L = 100$ Å, $E_{c1} = 0$ eV, $E_{c2} = -0.4$ eV, $V = 2$ eV, the inital Fermi levels $E_{F1} = E_{c1} + 0.01$ eV, $E_{F2} = E_{c1} - 0.1$ eV, $T = 300$ K.

It was found that the electron transfer rate dropped by several orders of magnitude as the vacuum potential was increased from 3 eV to 7 eV. The initial carrier density was approximately $4.9 \cdot 10^{16}$ m^{-2} for all vacuum potential values. The upper (lower) plots correspond to cases with and without the influence of Coulomb repulsion, respectively. The effect of Coulomb repulsion is most pronounced for the smallest potential barrier $V = 3$ eV. The carrier density N_1 is higher in the case with Coulomb repulsion as compared to the case without Coulomb repulsion and most noticeably at times above approximately 30 ps. Calculations with respect to separation was also

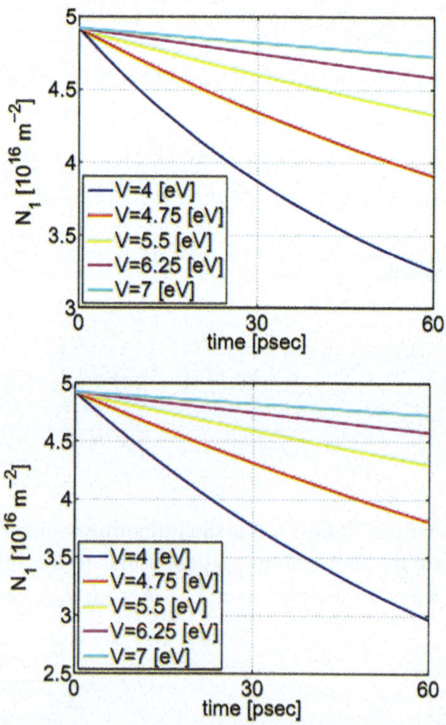

Fig. 3.5 Time dependence of the electron number in the left material N_1 as a function of time t for a set of potential values V. The upper (lower) plot is with (without) the effect of Coulomb repulsion due to electron transfer and effective charging of the two materials. The distance between the materials $2a$ was fixed to 10 Å. Other parameters are given in the main text. Reproduced with permission from Ref. [47]

carried out. It is important to notice that our model demonstrates that the tunneling time may vary by several orders of magnitude (from picosecs to microsecs or millisecs) depending on model parameters. Therefore, for large separations of contacting materials, tunneling times can be of the same order of magnitude as typical triboelectric time constants. The general trends and influence of distance and Fermi levels obtained were captured by experimental results [29]. All of the above results are consistent with the ones obtained using the TB model.

3.4 Optical Coupling Mechanisms

The previous model involved the isoenergetic electron transfer via quantum-mechanical tunneling. In the case when energy states ae not aligned for two materials, the charge transfer can occur with energy change. In particular, energy loss to either phonons or photons can be expected. The latter process has been computed by us using both a continuum and an atomistic formalism; hence, both situations are discussed next.

3.4.1 Photon Emission in a Two-Level System

The first demonstration of the emission of light upon charge transfer was carried out for a system of two atoms, each assumed to have a single energy level [47]. All atoms were considered independent and non-interacting and transitions between the two levels are possible via spontaneous emission, absorption, and stimulated emission.

Einstein Rate Equations
For a system of N two-level atoms of the same material with N_b (N_a) electrons occupying the upper (lower) level, the rate equations are

$$\frac{dN_b}{dt} = -N_b A_{ba} - N_b B_{ba} \rho(\omega) + N_a B_{ab} \rho(\omega), \tag{3.19}$$

$$\frac{dN_a}{dt} = N_b A_{ba} + N_b B_{ba} \rho(\omega) - N_a B_{ab} \rho(\omega), \tag{3.20}$$

where A_{ba} denotes the spontaneous emission rate, B_{ba} (B_{ab}) is the rate of stimulated emission (absorption) of photons, and $\rho(\omega)$ is the energy density in the field evaluated at the transition frequency $\omega = \frac{E_b - E_a}{\hbar}$ (assuming Boltzmann statistics):

$$\rho(\omega) = \frac{A_{ba}}{\frac{N_a}{N_b} B_{ab} - B_{ba}} = \frac{\hbar}{\pi^2 c^3} \frac{\omega^3}{e^{\frac{\hbar\omega}{k_B T}} - 1}, \tag{3.21}$$

where E_m, k_B, T is the energy of level m, Boltzmann's constant, and the absolute temperature, respectively. Additionally, one finds that [47]

$$B_{ab} = B_{ba}, \tag{3.22}$$

$$B_{ba} = \frac{\pi^2 c^3}{\omega^3 \hbar} A_{ba}, \tag{3.23}$$

$$A_{ba} = \frac{\omega^3}{3\pi c^3 \epsilon_0 \hbar} |\langle f | e\mathbf{r} | i \rangle|^2. \tag{3.24}$$

The system dynamics is then found upon assuming initial conditions, say,

$$N_b(t = 0) = N, \tag{3.25}$$

$$N_a(t = 0) = 0, \tag{3.26}$$

once the dipole matrix elements are obtained. The above Einstein rate equations apply to a system of identical atoms in the presence of light interaction where transitions take place between two states on the *same* atom. The generalization of these rate equations to CE between two different atom species is achieved by using transition matrix elements between two states on *different* atoms (assuming one state per atom for simplicity).

Example Results

An investigation of how the electron transfer depends on the atomic separation was carried out in Ref. [47]. The distance dependence was simulated by changing the optical coupling strength between the two atoms.

In Fig. 3.6, the absolute squared momentum matrix element $|\langle \psi^2_{100} | \mathbf{p} | \psi^1_{100} \rangle|^2$ is plotted as a function of the distance x_0 between two (identical) atoms for which $Z_1 = Z_2 = 1$. As expected, the matrix element vanishes when the two atoms overlap completely. This occurs since the two wave functions are the same (and, therefore, have the same parity) and the fact that $\mathbf{p}$ has negative parity. There is a maximum of

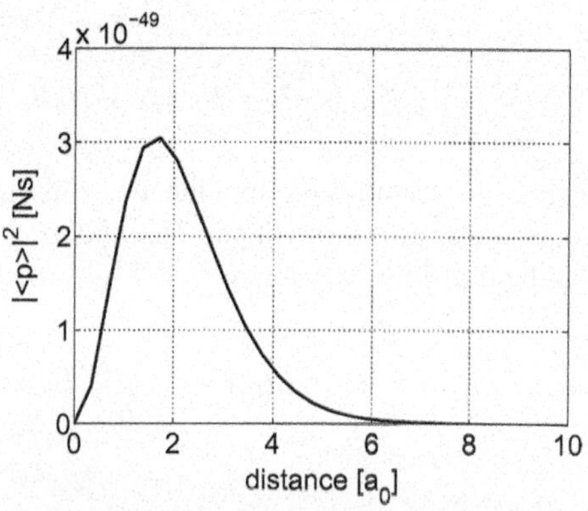

Fig. 3.6 The absolute momentum matrix element, $|\langle \psi^2_{100} | \mathbf{p} | \psi^1_{100} \rangle|^2$, as a function of the distance between two atoms. Parameters are $Z_1 = Z_2 = 1$ and the distance is measured in units of the Bohr radius a_0. Reproduced with permission from Ref. [47]

the matrix element at a distance when the two atoms are separated by approximately 1.5 Å. As the atoms separate, the matrix element decreases rapidly due to the small overlap of the wave functions at large distances.

3.4.2 Triboelectroluminescence

The phenomenon of triboelectrification with the emission of light has been referred to by others [29]; we coined it triboelectroluminescence [50]. This is a different phenomenon from the more familiar triboluminescence [64, 65]. The latter refers to the emission of light by stressing or fracturing a single material.

Hamiltonian

The model we developed is based on our standard TB Hamiltonian for the electrons and an electron-photon interaction term that is responsible for the light emission. The spontaneous emission process was modeled in Ref. [50] by generalizing the Weisskopf-Wigner model of spontaneous decay for a two-level atom [66]. The Weisskopf-Wigner model treats the transition of an electron from an excited atomic state to a ground state with the spontaneous emission of photons and was initially applied to the hydrogen atom. It led to an exponential decay of the electron from the excited state and to a Lorentzian spectrum of emitted photons. We generalized the above model to study the transfer of electrons from one material (left) to another (right) with the emission of photons. The neglect of stimulated emission is justified if there is no photon bath available. Experimentally, this would occur if the system is not enclosed in a cavity, allowing the emitted photons to radiate away and not reach equilibrium in the surrounding of the system. Since the operation of a TENG involves the relative motion of two materials, calculations of charge transfer as a function of the separation of the two materials was also carried out. Furthermore, as experiments have been carried out as a function of temperature in order to elucidate the mechanism of charge transfer, we had investigated the impact of temperature dependence via the Fermi-Dirac function.

The total Hamiltonian of the problem represents electrons in the materials interacting with the electromagnetic field:

$$H = H_{\text{elec}} + H_F + H_{\text{int}}, \tag{3.27}$$

where H_{elec} is a one-electron Hamiltonian [Eq. (3.2)], H_F is the Hamiltonian of the transverse electromagnetic field, and H_{int} will be chosen to be a dipole interaction. The field Hamiltonian H_F is

$$H_F = \sum_{\mathbf{k},s} \hbar\omega_{\mathbf{k}} \widehat{n}_{\mathbf{k},s}, \tag{3.28}$$

where $\hbar\omega_{\mathbf{k}}$ is the photon energy, $\widehat{n}_{\mathbf{k},s}$ is the number operator, and $\mathbf{k}$ and s denote the wave vector and polarization, respectively. The electron-photon interaction representing spontaneous emission between two electronic eigenstates (m denoting a higher energy ("excited") state on the left and n a lower energy state ("ground") on the right) is given by

$$H_{\text{int}} = -\sum_{m,n}\sum_{\mathbf{k},s} \hbar g_{\mathbf{k},s}^{m,n} \widehat{\sigma}_{e_m g_n} \widehat{a}_{\mathbf{k},s} + \text{h.c.}, \tag{3.29}$$

where $\widehat{\sigma}_{e_m g_n} = e_m^\dagger g_n$, with $e_m^\dagger$ the creation operator for an electronic excited state m on the left, g_n the annihilation operator for an electronic ground state n on the right, $\widehat{a}_{\mathbf{k},s}$ the annihilation operator for a photon, and the coupling constant is

$$g_{\mathbf{k},s}^{m,n} = i\sqrt{\frac{\omega_{\mathbf{k}}}{2\hbar\epsilon_0 V}}\mathbf{d}_{m,n} \cdot \varepsilon_{\mathbf{k},s}, \quad \mathbf{d}_{m,n} = \langle g_n | e\mathbf{r} | e_m \rangle, \tag{3.30}$$

with V the field normalization volume, $\mathbf{d}_{m,n}$ the electric dipole matrix element coupling states e_m and g_n, and $\varepsilon_{\mathbf{k},s}$ the photon polarization.

Charge Transfer

The most general first-order differential equation governing the amplitude of the electrons on the left is

$$\frac{da_m(t)}{dt} = -\sum_{m'} a_{m'}(t) e^{i\omega_{mm'}t} \sum_n \frac{\Gamma_{mm'}^n}{2}, \tag{3.31}$$

if a general state of an electron at a time t is given by the wave function

$$|\psi(t)\rangle = \sum_m a_m(t) e^{-i\omega_m t} |e_m, 0\rangle + \sum_n b_n(t) e^{-i\omega_{\mathbf{k},s} t} |g_n, 1_{\mathbf{k},s}\rangle. \tag{3.32}$$

Here, $\hbar\omega_m$ is the energy of the electron in the excited state with respect to the ground-state energy. In Ref. [50], we carried out calculations for 1D solids represented by finite linear chains of identical atoms. In order to allow for multiple bands and for band gaps, we used a two-band model for H_{elec}. The decoupling of the left and right surface states gives an indication that the chains are long enough to simulate bulk properties; 30 atoms were found as the minimum chain length.

We chose electrons in the left material to have a higher energy than those on the right and they can then transfer to the right material to occupy empty states there with the emission of photons. A model of atomic dipoles was introduced in Ref. [47]. We presented results of calculations of the charge transfer between two finite chains with the emission of photons. With M electrons on the left at time $t = 0$, the fractional charge transfered was calculated as

$$f_c(t) = 1 - \frac{1}{M} \sum_{m=1}^{M} |a_m(t)|^2. \tag{3.33}$$

Example Results

We have pointed out the importance of studying the temperature dependence as a way of getting insights into the charge transfer mechanism. We assumed that the effect of temperature is to change the electron distribution in the materials according to the Fermi-Dirac distribution:

$$f(E_m) = \frac{1}{1 + e^{\beta(E_m - E_F)}}, \beta = 1/(k_B T), \tag{3.34}$$

where E_m is the electron energy as before, E_F is the Fermi energy, and T the temperature. We assumed that, at time $t = 0$ s and $T = 0$ K, electrons occupy all levels up to the Fermi energy. However, the temperature dependence is again expected to be negligible if the material is an insulator or even a semiconductor since the amount of electrons excited across a gap is very small even at room temperature. Indeed, we found this to be the case in our model when the uppermost electrons were located in the surface states. Recent experimental data show an approximately linear increase of charge transfer of the ordeo of 50% from 313 K to 433 K between a Au-coated AFM tip (metal) and a SiO_2 layer (insulator); that mechanism was attributed to thermionic emission and a lowering of the potential barrier for electron transport with increasing temperature. We considered in our paper whether temperature can impact spontaneous emission. Indeed, for electrons originating from a metal, i.e., when electrons do not completely fill a band, we did find some small changes in the transition rate. We calculated the fractional charge transfer associated with these initial electron distributions (Fig. 3.7, left panel). The different curves at different temperatures can be seen to be slightly changed, showing that there is some temperature dependence.

The charge transfer was also studied as a function of the separation of the two materials. The results obtained at $T = 0$ K for separations from 0 Å (i.e., touching) to 10 Å for an insulator-insulator contact is given in Fig. 3.7 (right panel).

We illustrate with an example of the initial electron population filling the surface states on the left and available states on the right being the surface states. It can be seen that the charge transfer is strongly dependent on the separation. Indeed, there is no charge transfer for our model system for separations of the order of 5 Å. Hence, spontaneous emission is equally short-ranged as quantum tunneling is [48] in facilitating charge transfer between two materials.

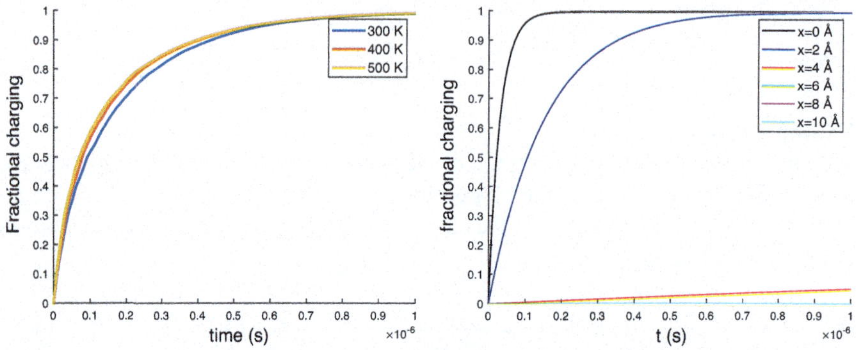

Fig. 3.7 (Left)Fractional charging as a function of time and temperature (from $T = 300 - 500$ K). Assuming electrons are in eigenstates up to #32 on left (at $T = 0$ K) and transfering to eigenstate # 30 (a surface state) on right. (Right) Charge transfer as a function of time for separations of 2Å, 4Å, 6Å, 8Å and 10Å beyond distance of closest approach ($T = 0$ K). Reproduced with permission from Ref. [50]

3.5 Tribovoltaics

Tribovoltaics is the transfer of charges leading to a perpendicular DC current between two materials generated by their relative motion [13–16, 18, 19, 21–24]. The detailed mechanisms behind tribovoltaics are still not well understood [15, 16, 19].

3.5.1 *Semiconductor* $p - n$ *Equations*

Recently [26], we proposed a 1D theory for the current generation due to tribovoltaics based on the semiconductor equations applied to $p - n$ structures (Fig. 3.8). The three differential equations are the Poisson equation and the two current continuity equations,

$$\frac{\partial^2 \phi}{\partial x^2} = -\frac{q}{\epsilon}\left(N_d - N_a + p - n\right), \tag{3.35}$$

$$\frac{\partial n}{\partial t} = \frac{1}{q}\frac{\partial J_n}{\partial x} + G_n - R_n, \tag{3.36}$$

$$\frac{\partial p}{\partial t} = -\frac{1}{q}\frac{\partial J_p}{\partial x} + G_p - R_p, \tag{3.37}$$

where n, p are the electron and hole concentrations, ϕ is the electrostatic potential, J_n (J_p) is the electron (hole) current density, G_n (G_p) is the electron (hole) generation rate, R_n (R_p) is the electron (hole) recombination rate, N_a (N_D) is the acceptor (donor) concentration, and ϵ is the material permittivity. Variables x and t are the

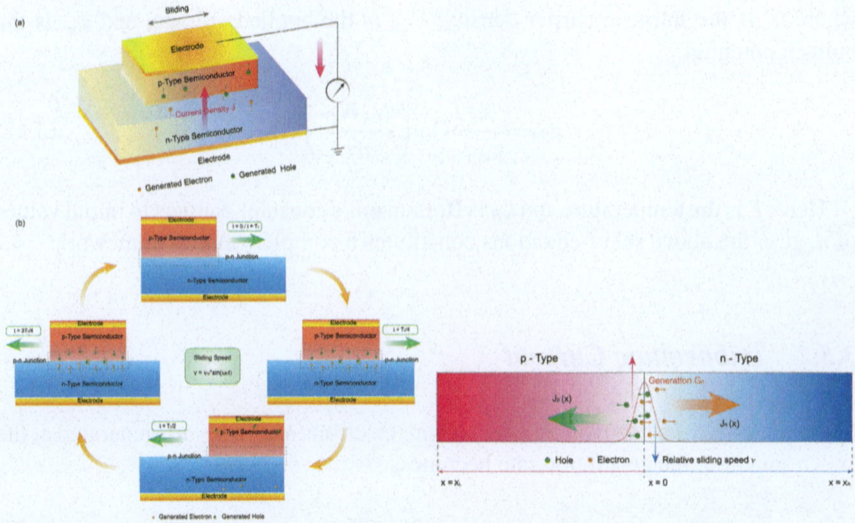

Fig. 3.8 Left. **a** A 3D schematic of a tribovoltaic $p-n$ semiconductor interface device. **b** Schematic of a sinusoidally-driven tribovoltaic cycle. As the p layer moves relative to the n layer, electron-hole pairs are generated at the semiconductor interface. The generation rate is a function of the relative speed between the two layers. Once an electron-hole pair is generated, the electron (hole) moves toward the electrode on the n-side (p-side). The tribovoltaic current can be harvested in an external circuit. Right. Electron-hole pair generation at the semiconductor interface due to the sliding of a p-doped semiconductor on a n-doped semiconductor. Reproduced with permission from Ref. [26]

spatial and time coordinates, respectively, and q is the positive elementary charge. Using a drift-diffusion current density model assumption,

$$J_n = qD_n\frac{\partial n}{\partial x} + q\mu_n n\mathcal{E}, \tag{3.38}$$

$$J_p = -qD_p\frac{\partial p}{\partial x} + q\mu_p p\mathcal{E}, \tag{3.39}$$

where $\mathcal{E} = -\frac{\partial \phi}{\partial x}$ is the electric field, μ_n (μ_p) is the electron (hole) mobility, and D_n (D_p) is the electron (hole) diffusivity.

The above set of equations includes three second-order differential equations in space. Hence, for *each* dependent variable n, p, and ϕ, two boundary conditions are required [67],

$$p\text{-side:} \qquad p(x = x_L) = N_A, \;\; n(x = x_L) = \frac{n_i^2}{N_A}, \;\; \phi(x = x_L) = 0, \tag{3.40}$$

$$n\text{-side:} \qquad n(x = x_R) = N_D, \;\; p(x = x_R) = \frac{n_i^2}{N_D}, \;\; \phi(x = x_R) = \phi_{bi} - V_{\text{apl}}, \tag{3.41}$$

where n_i is the intrinsic carrier density, V_{apl} is the applied voltage, and ϕ_{bi} is the built-in potential,

$$\phi_{\mathrm{bi}} = \frac{k_B T}{q} \ln\left(\frac{N_A N_D}{n_i^2}\right). \tag{3.42}$$

Here, T is the temperature and k_B is Boltzmann's constant. Subject to initial values of n, p, ϕ the above set of equations constitutes a complete model framework.

3.5.2 *Tribovoltaic Current*

In the presence of a tribovoltaic effect where electrons and holes are generated at the $p - n$ junction, the generation rate becomes,

$$G_n(x) = G_p(x) = \tilde{G}(t)\delta(x), \tag{3.43}$$

where δ is a Dirac delta function. The tribovoltaic current density can be determined analytically as a function of the time dependence of the relative absolute velocity of the p and n sides, $v(t)$:

$$J^T(t) = -q\tilde{G}(t) = -q\gamma\,\frac{\kappa}{a^2}\,\tilde{\eta}_{\mathrm{eh-pair}}\,v(t). \tag{3.44}$$

3.5.3 *Example Results*

Using the parameter values: $\gamma = 1 \cdot 10^5\ \mathrm{m}^{-1}$, $\kappa = 2$ is a crystal structure factor equal to 2 for GaAs (FCC lattice), $a = 5.64$ Å is the lattice constant of GaAs, and $\tilde{\eta}_{\mathrm{eh-pair}} = 1$ is the electron-hole pair conversion efficiency [26], one gets [26]

$$J^T(t) = -1.007 \cdot 10^5 \mathrm{C/m^3} \times v(t), \tag{3.45}$$

which is in perfect agreement with the COMSOL numerical result in Fig. 3.9.

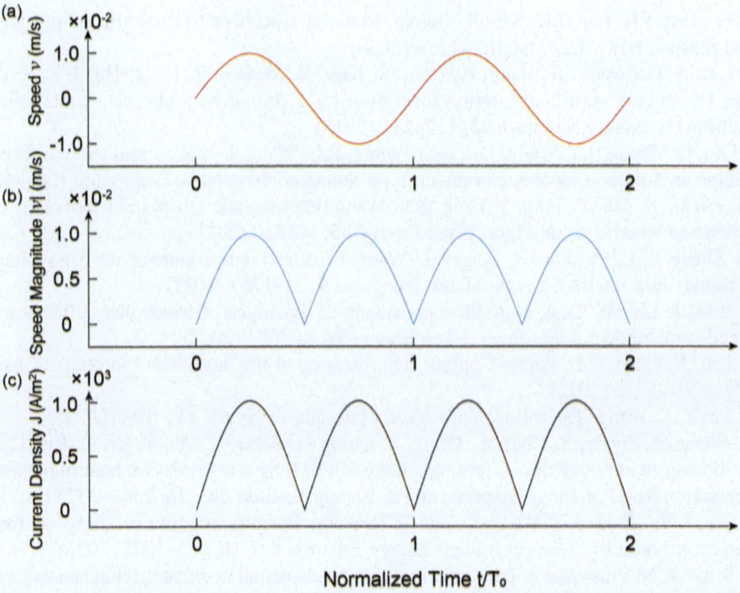

Fig. 3.9 Calculated time-dependent plot of the positive current density for GaAs [lower panel (**c**)], for a specified sinusoidal time dependence of the sliding velocity [upper panel (**a**)]. The middle panel (**b**) shows the absolute value of the sliding speed. Results are obtained numerically in COMSOL by solving the full set of drift-diffusion equations, Eqs. (3.35)–(3.37), and boundary conditions, Eqs. (3.40)-(3.41). Reproduced with permission from Ref. [26]

References

1. W.R. Harper, *Contact and Frictional Electrification* (Clarendon Press, Oxford, 1967)
2. P.E. Shaw, The electrical charges from like solids. Nature **118**, 659–660 (1926)
3. L.B. Schein, Recent progress and continuing puzzles in electrostatics. Science **316**, 1572–1573 (2007)
4. F.R. Fan, Z.Q. Tian, Z.L. Wang, Flexible triboelectric generator. Nano Energy **1**, 328334 (2012)
5. S. Niu, Y. Liu, S. Wang, L. Lin, Y.S. Zhou, Y. Hu, Z.L. Wang, Theory of sliding-mode triboelectric nanogenerators. Adv. Mater. **25**, 61846193 (2013)
6. S. Niu, Z.L. Wang, Theoretical systems of triboelectric nanogenerators. Nano Energy **14**, 161192 (2015)
7. Z.L. Wang, L. Lin, J. Chen, S. Niu, Y. Zi, *Triboelectric Nanogenerators* (Green Energy and Technology. Springer, Berlin, 2016)
8. A. Li, Y. Zi, H. Guo, Z.L. Wang, F.M. Fernandez, Triboelectric nanogenerators for sensitive nano-coulomb molecular mass spectrometry. Nat. Nanotechnol. **12**, 481487 (2017)
9. C. Yao, X. Yin, Y. Yu, Z. Cai, X. Wang, Chemically functionalized natural cellulose materials for effective triboelectric nanogenerator development. Adv. Funct. Mater. **27**, 1700794 (2017)
10. J.W. Lee, H.J. Cho, J. Chun, K.N. Kim, S. Kim, C.W. Ahn, I.W. Kim, J.Y. Kim, S.W. Kim, C. Yang, J.M. Baik, Robust nanogenerators based on graft copolymers via control of dielectrics for remarkable output power enhancement. Sci. Adv. **3**, e1602902 (2017)
11. R. Hinchet, H.-J. Yoon, H. Ryu, M.-K. Kim, E.-K. Choi, D.-S. Kim, S.-W. Kim, Transcutaneous ultrasound energy harvesting using capacitive triboelectric technology. Science **365**, 491 (2019)

12. D.W. Kim, J.H. Lee, J.K. Kim, U. Jeong, Material aspects of triboelectric energy generation and sensors. NPG Asia Materials **12**, 6 (2020)
13. J. Liu, A. Goswami, K. Jiang, F. Khan, S. Kim, R. McGee, Z. Li, Z. Hu, J. Lee, T. Thundat, Direct-current triboelectricity generation by a sliding Schottky nanocontact on MoS_2 multilayers. Nature Nanotech **13**, 112–116 (2018)
14. R. Xu, Q. Zhang, J.Y. Wang, D. Liu, J. Wang, Z.L. Wang, Direct current triboelectric cell by sliding an n-type semiconductor on a p-type semiconductor. Nano Energy **66**, 104185 (2019)
15. M. Zheng, S. Lin, Z. Tang, Y. Feng, Z.L. Wang, Photovoltaic effect and tribovoltaic effect at liquid-semiconductor interface. Nano Energy **83**, 105810 (2021)
16. M. Zheng, S. Lin, L. Zhu, Z. Tang, Z.L. Wang, Effects of temperature on the tribovoltaic effect at liquid-solid interfaces. Adv. Mater. Interfaces **9**, 2101757 (2022)
17. G. Liu, J. Liu, W. Dou, Non-adiabatic quantum dynamics of tribovoltaic effects at sliding metal-semiconductor interfaces. Nano Energy **96**, 107034 (2022)
18. S. Lin, X. Chen, Z.L. Wang, Contact electrification at the liquid-solid interface. Chem. Rev. **122**, 5209–5232 (2022)
19. S. Lin Z. L. Wang, The tribovoltaic effect, Materials Today **62**, 111–128 (2023)
20. Z. Wang, Z. Zhang, Y. Chen, L. Gong, S. Dong, H. Zhou, Y. Lin, Y. Lv, G. Liu, C. Zhang, Achieving an ultrahigh direct-current voltage of 130 V by semiconductor heterojunction power generation based on the tribovoltaic effect. Energy Environ. Sci. **15**, 2366–2373 (2023)
21. C. Xu, J. Yu, Z. Huo, Y. Wang, Q. Sun, Z.L. Wang, Pursuing the tribovoltaic effect for direct-current triboelectric nanogenerators. Energy Environ. Sci. **16**, 983–1006 (2023)
22. S. Sriphan, N. Vittayakorn, Tribovoltaic effect: fundamental working mechanism and emerging applications. Mater. Today Nano **22**, 100318 (2023)
23. X. Guo, J. You, D. Wei, J. Shao, Z. L. Wang, A generalized model for tribovoltaic nanogenerator. Appl. Phys. Rev. **1** 11 (2) 021415 (2024)
24. Z. Zhang, L. Gong, R. Luan, Y. Feng, J. Cao, C. Zhang, Tribovoltaic effect: Origin, interface, characteristic, mechanism & application. Adv. Sci. **11**, 2305460 (2024)
25. J. Meng, C. Lan, C. Pan, G. Liu, X. Pu, C. Zhang, Z. L. Wang, Coupling of tribovoltaic effect and tribo-electrostatic effect at dynamic semiconductor heterojunction interfaces. Nano Energy, 110395 (2024)
26. M. Willatzen, Z. Zhang, Z.L. Wang, Theory of tribovoltaics: Direct current generation at a *p-n* semiconductor interface. PRX Energy **3**, 013009 (2024)
27. C.G. Camara, J.V. Escobar, J.R. Hird, S.J. Putterman, Correlation between nanosecond X-ray flashes and stick-slip friction in peeling tape. Nature **455**, 1089 (2008)
28. E. Constable, J. Horvat, R.A. Lewis, Mechanisms of x-ray emission from peeling adhesive tape. Appl. Phys. Lett. **97**(13), 131502 (2010)
29. Z.L. Wang, A.C. Wang, On the origin of contact electrification. Materials Today **30**, 34 (2019)
30. H. Zou, Y. Zhang, L. Guo, P. Wang, X. He, G. Dai, H. Zheng, C. Chen, A.C. Wang, C. Xu, Z.L. Wang, Quantifying the triboelectric series. Nature Commun. **10**, 1427 (2019)
31. S. Pan, Z. Zhang, Fundamental theories and basic principles of triboelectric effect: A review. Friction **7**, 2–16 (2019)
32. D.J. Lacks, T. Shinbrot, Long-standing and unresolved issues in triboelectric charging. Nature Reviews **3**, 465–476 (2019)
33. W. Melitz, J. Shen, A.C. Kummel, S. Lee, Kelvin probe force microscopy and its application. Surf. Sci. Rep. **66**, 1 (2011)
34. J. Henniker, Triboelectricity in polymers. Nature **196**, 474 (1962)
35. J. Lowell, A.C. Rose-Innes, Contact electrification. Adv. Phys. **29**, 947 (1980)
36. F.A. Vick, Theory of contact electrification. Br. J. Appl. Phys. **4**, S1 (1953)
37. D. J. Montgomery, Static electrification of solids. In *Solid State Physics* (eds: Seitz and Turnbull), vol. 9, 139–157 (1959)
38. C.B. Duke, T.J. Fabish, Charge-induced relaxation in polymers. Phys. Rev. Lett. **37**, 1075 (1976)
39. C.B. Duke, T.J. Fabish, Molecular charge states and contact charge exchange in polymers. J. Appl. Phys. **49**, 315 (1978)

40. M. Yoshida, N. Ii, A. Shimosaka, Y. Shirakawa J. Hidaka, Experimental and theoretical approaches to charging behavior of polymer particles. Chem. Eng. Sci. **61**, 2229–2248 (2006)
41. Y. Shirakawa, N. Ii, M. Yoshida, R. Takashima, A. Shimosaka, J. Hidaka, Quantum chemical calculation of electron transfer at metal/polymer interfaces. Adv. Powder Tech. **21**, 500–505 (2010)
42. Y. Zhang, T. Shao, Effect of contact deformation on contact electrification: a first-principles calculation. J. Phys. D **46**, 235304 (2013)
43. X. Shen, A.E. Wang, R.M. Sankaran, D.J. Lacks, First-principles calculation of contact electrification and validation by experiment. J. Electrostatics **82**, 11–16 (2016)
44. S. Lin, T. Shao, Bipolar charge transfer induced by water: experimental and first-principles studies. Phys. Chem. Chem. Phys. **19**, 29418–29423 (2017)
45. J. Wu, X. Wang, H. Li, F. Wang, W. Yang, Y. Hu, Insights into the mechanism of metal-polymer contact electrification for triboelectric nanogenerator via first-principles investigations. Nano Energy **48**, 607–616 (2018)
46. J. Wu, X. Wang, H. Li, F. Wang, Y. Hu, First-principles investigations on the contact electrification mechanism between metal and amorphous polymers for triboelectric nanogenerators. Nano Energy **63**, 103864 (2019)
47. M. Willatzen, Z.L. Wang, Theory of contact electrification: Optical transitions in two-level systems. Nano Energy **52**, 517–523 (2018)
48. M. Willatzen and Z. L. Wang, Contact electrification by quantum-mechanical tunneling. Research, Article ID 6528689 (2019)
49. M. Willatzen, L. C. Lew Yan Voon, and Z.-L. Wang, Quantum theory of contact electrification of solids and fluids. Adv. Func. Mat. 1910461 (2020)
50. L. C. Lew Yan Voon, J. Hasbun, M. Willatzen and Z.-L. Wang, Generalized Weisskopf-Wigner model of triboelectroluminescence. EcoMat 1-11 (2021)
51. J. Hasbun, L. C. Lew Yan Voon, M. Willatzen, On chain models for contact electrification. J. Phys. Condens. Matter **34**, 135501 (2022)
52. W. Morten, L.C. Lew Yan Voon, Quantum theory of tribovoltaics, APL Electron. Devices **1**, 026101 (2025)
53. D.J. Lacks, N. Duff, S.K. Kumar, Nonequilibrium accumulation of surface species and triboelectric charging in single component particulate systems. Phys. Rev. Lett. **100**, 188305 (2008)
54. C.A. Mizzi, A.W. Lin, L.D. Marks, Does flexoelectricity drive triboelectricity? Phys. Rev. Lett. **123**, 116103 (2019)
55. R. Alicki, A. Jenkins, Quantum theory of triboelectricity. Physical Review Letters **125**, 186101 (2020)
56. C.-Y. Liu, A.J. Baird, Electrons on dielectrics and contact electrification. Chem. Phys. Lett. **480**, 145–156 (2008)
57. C. Xu, Y. Zi, A.C. Wang, H. Zou, Y. Dai, X. He, P. Wang, Y.C. Wang, P. Feng, D. Li, Z.L. Wang, On the electron-transfer mechanism in the contact-electrification effect. Adv. Mater. **30**, 1706790 (2018)
58. R. Haydock, in *Solid State Physics* (eds: Ehrenreich, Seitz and Turnbull), vol. **35**, 215 (1980)
59. S. Pan, Z. Zhang, Triboelectric effect: A new perspective on electron transfer process. J. Appl. Phys. **122**, 144302 (2017)
60. B.A. Grzybowski, A. Winkleman, J.A. Wiles, Y. Brumer, G.M. Whitesides, Electrostatic self-assembly of macroscopic crystals using contact electrification. Nature Materials **2**, 241 (2003)
61. J.C. Slater, G.F. Koster, Simplified LCAO method for the periodic potential problem. Phys. Rev. **94**, 1498 (1954)
62. W.A. Harrison, *Electronic Structure and the Properties of Solids* (Dover, Mineola, New York, 1989)
63. L. C. Lew Yan Voon, Electronic Properties of Semiconductor Nanostructures. PhD, Worcester Polytechnic Institute (1993)
64. D.O. Olawale, O.I. Okoli, R.S. Fontenot, W.A. Hollerman (eds.), *Triboluminescence* (Springer, Switzerland, 2016)

65. Y. Xie, Z. Li, Triboluminescence: Recalling interest and new aspects. Chem. **4**, 943–971 (2018)
66. V. Weisskopf, E. Wigner, Berechnung der natürlichen Linienbreite auf Grund der Diracschen Lichttheorie. Z. Physik **63**, 54–73 (1930)
67. J. Mørk, M. Willatzen, J. Mark, M. Preisel, C.P. Seltzer, Characterization and modelling of ultrafast carrier dynamics in quantum well optical amplifiers. SPIE **2146**, 52 (1994)

Chapter 4
Contact-Electrification at Metal-Insulator Interfaces

Abstract Contact electrification (CE) at metal-insulator interfaces is a complex phenomenon that has intrigued researchers for centuries, yet it remains inadequately understood. This chapter provides a comprehensive overview of the mechanisms underlying CE at metal-insulator interface, emphasizing the interaction between metals and insulators leads to the formation of potential barriers that govern charge flow, with significant implications for electronic device performance. The theoretical model discussed in this chapter includes the Schottky contact effect and work function difference, to elucidate the principles of charge transfer and the dynamics of electron movement at these interfaces. Advanced techniques such as KPFM are highlighted for their ability to provide nanoscale insights into charge distribution and transfer processes, while the influence from CE to the measurements of AFM is also discussed. By synthesizing recent advancements in the field, this chapter aims to enhance our understanding of CE at metal-insulator interfaces, paving the way for the development of innovative materials and devices that leverage these fundamental principles for practical applications in energy harvesting and electronic technologies.

4.1 Introduction

Contact-electrification (CE), also known as triboelectrification (TE), is a phenomenon that has been documented for over 2600 years, yet its scientific understanding remains inconclusive, unclear, and ununified. This phenomenon widely occurs between all known materials, including solids, liquids, and gases [1, 2]. Further, when it comes to solids, they include metals, semiconductors, and insulators. Rubbing a metal against a hard rock surface leads to the generation of sparks and discharge, which is the result of CE in a metal-insulator. The study of CE at metal-insulator interfaces is complex due to the multitude of factors that can influence the charge transfer process. These factors include the nature of the materials in contact, the surface conditions, the presence of environmental contaminants, and the specific conditions under which contact and separation occur. Therefore, there are also many debates about the CE principle of metal insulator interfaces. Despite

Z. L. Wang, *Contact-Electrification of Matter*, Springer Series in Surface Sciences 75,
https://doi.org/10.1007/978-3-031-88504-4_4

these complexities, significant progress has been made in recent years, particularly with the development of more sophisticated experimental techniques and theoretical models.

The study of CE at the metal-insulator interface involves a detailed examination of the mechanisms that lead to charge transfer. This includes understanding the possess of Schottky contact, the role of the work function, the property of surface states, and the process of electron transfer at the interface. Experimental research about electron transfer process at the metal-insulator interface requires precise control of contact conditions, strict measurement techniques, and the ability to isolate the effects of environmental variables such as humidity and temperature. With the development of Atomic Force Microscope (AFM) technology, especially Kelvin Probe Force Microscopy (KPFM), it can provide nanoscale charge transfer information for the CE process [3, 4]. KPFM compensates for the contact potential difference between the sample and the probe by controlling the oscillation amplitude of the probe, thereby achieving high-resolution imaging of the surface potential of the sample [5]. It works by detecting the contact potential difference between the sample surface and the probe, providing information on charge distribution at the nanoscale. In the study of charge electrification and transfer process, KPFM technology is particularly useful because it can directly observe and quantify changes in surface charge density caused by TE [6, 7].

This chapter will delve into various theoretical models that can explain the process of CE, including Schottky contact model, work function model and electron transfer model of CE process at metal-insulator interface. Moreover, due to the differences in energy bands among different types of materials, the problem of charge leakage cannot be ignored for special metal-insulator interfaces. This chapter will also present the principles of triboelectric nanogenerator (TENG) as a probe for measuring CE charges. This includes elucidating the use of TENG as a probe to measure its macroscopic output and quantify the charge on insulators, including evaluating contact area and charge density, as well as manipulating environmental conditions to achieve research objectives, and specifically using KPFM technology to accurately measure contact charge. This chapter will also explore the challenges and limitations of current understanding of metal-insulator CE. These factors include the difficulty of measuring real contact charges, the uncertainty of determining charge penetration depth, and the complexity of unraveling the contributions of electron transfer, ion transfer, and material transfer to the entire charging process. In summary, the study of CE at metal-insulator interface holds significant scientific and engineering implications, encompassing the development of fundamental theories in material science and electronic engineering, and it directly impacts the practical application of modern technology and industrial progress. By investigating CE at metal-insulator interfaces, we can enhance our comprehension of charge transfer, electronic structure, and interface states. Simultaneously, a profound understanding of the mechanism behind contact electrogenesis can facilitate advancements in new electronic devices and materials such as high-performance semiconductors, insulators, and conductive materials while promoting the progression of electronic technology.

4.2 Schottky Contact

The Schottky contact refers to a rectifying junction that is established between a metal and a semiconductor. Upon the metal contact with the semiconductor, an interface potential barrier arises due to the disparity in work function between the two materials. The presence of this barrier regulates the flow of charge across the junction, facilitating current conduction in one direction more effectively than in the opposite direction, thereby demonstrating diode characteristics. On the contrary, Ohmic contact refers to a conductive path with low resistance, where there is no or only a very small potential barrier at the contact interface, allowing charge carriers to flow freely. The concept of Schottky contact serves as the fundamental basis for comprehending metal-semiconductor interfaces.

In principle, the barrier height of a Schottky contact is a critical parameter that fundamentally determines the electrical characteristics of the metal-semiconductor junction. According to the Schottky-Mott rule [8, 9], the barrier height (Φ_{Bn} for an n-type semiconductor) primarily relies on the work function of the metal (Φ_m) and the electron affinity (X) of the semiconductor, with a smaller influence from the semiconductor's band gap (E_g). The relationship can be expressed as:

$$\Phi_{Bn} = \Phi_m - X - \Delta\Phi_n \tag{4.1}$$

where $\Delta\Phi_n$ accounts for the image force barrier lowering, which is a quantum mechanical effect that reduces the barrier height.

The classical Schottky contact model assumes that the barrier height remains unaffected by the doping concentration of the semiconductor. However, the presence of surface states can significantly influence the barrier height. These surface states may arise from either termination of the semiconductor's crystal lattice at its surface or from adsorbed impurities, and they have the ability to fixate the Fermi level at the surface, resulting in a barrier height that is more reliant on the metal work function. The model proposed by Cowley and Sze [10] enhances the comprehension of Schottky contacts through the incorporation of surface states and the interfacial layer. The proposed model suggests that the barrier height can be expressed as a linear combination of the metal work function and a quantity Φ_0, which represents the energy level below which surface states must be occupied to maintain charge neutrality at the semiconductor surface. The model also introduces a weighting factor γ that depends on the surface state density and the thickness of the interfacial layer. The model has been instrumental in explaining the variation in barrier heights observed experimentally for different metal-semiconductor systems. For instance, the model has been applied to various systems including Si, GaP, GaAs, and CdS systems. It offers a more precise description of the barrier height compared to the simplistic Schottky model, particularly in cases where surface states play a significant role. More specifically, the formula can be expressed as follows:

$$\Phi_{Bn} = \gamma(\Phi_m - X) + (1 - \gamma)(E_g - \Phi_0) - \Delta\Phi_n \tag{4.2}$$

where X and E_g are electron affinity and the band gap of the semiconductor, respectively, $\Delta\Phi_n$ is the image force barrier lowering, and γ is a weighting factor which depends mainly on the surface state density and the thickness of the interfacial layer. Figure 4.1a illustrates the relationship between these parameters very well.

Furthermore, within the realm of TENG, Schottky contacts also possess diverse applications. The sliding Schottky diode, in contrast to conventional TENG that converts relative friction between insulators into alternating current, has been experimentally proven to generate direct current (DC) output without the need for external bias voltage. Additionally, it can efficiently power electronic circuits without requiring electrical rectification. The TENG structure depicted in Fig. 4.1b exemplifies a typical configuration based on the Schottky junction of the AFM probe. The AFM probe is usually made of Pt. When it slides on the semiconductor surface, the friction process induces direct electronic excitation at the metal-semiconductor interface, leading to diffusion drift conduction in the whole material system, and consequently generates carrier movement in the system and outputs current to the outside.

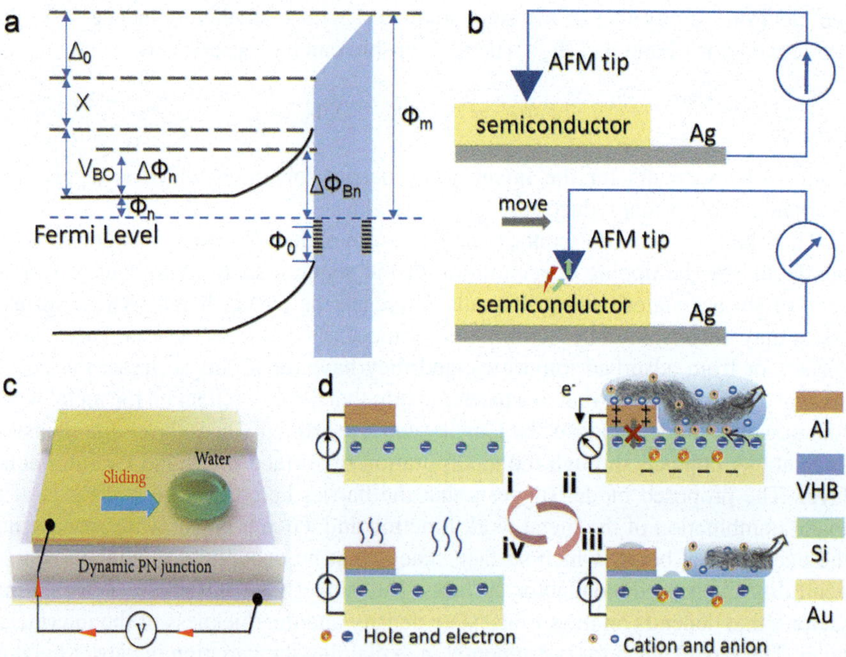

Fig. 4.1 **a** Energy band diagram of a metal-n-type-semiconductor contact with an interfacial layer. **b** Schematic of the working principle of a metal-insulator system with an AFM tip. **c** The diagram of the dynamic liquid-semiconductor junction structure generator, which consists of two semiconductors, Ti/Au electrodes, liquid, and PVC layer [13]. Reproduced with permission from AAAS. **d** The working mechanism of the wiping generator (the ions inside the water bridge undergo polarization between Al and Si, leading to the generation of a current loop from the top surface of the semiconductor to the bottom electrode) [14]. Reproduced with permission from Royal Society of Chemistry

TENG can be fabricated based on Schottky contacts, continuous direct-current can be directly generated by a sliding Schottky nanocontact without the application of an external voltage. Thundat [11] et al. slide a platinum-coated silicon AFM tip at a MoS_2 to generate a triboelectric current. This method was widely used to fabricate TENGs based on Schottky contacts. The range of metals has expanded from the initial AFM probes to metal sheets or blocks; semiconductor materials also include various inorganic semiconductors such as Si, GaN, MoS_2, AgS_2, etc. In addition to the aforementioned applications, the concept of Schottky junction generator has also been extended to encompass the interface between liquids and semiconductors. Under the influence of dynamic Schottky junction, the periodic formation and disappearance of the Schottky junction occur when there is relative motion between metal and semiconductor. This dynamic process disrupts the original equilibrium state of drift current and diffusion current in the static Schottky junction region [12]. By substituting metals with liquids like water, we can harness external mechanical energy to induce a dynamic polarization process at the interface between water molecules and semiconductors. The resulting rebound polarized charge at this dynamic PN junction has the capability to generate electrical energy, thereby presenting a novel perspective: the polarization and depolarization process of water molecules have significant potential for macroscopic electrical energy generation. As shown in Fig. 4.1c, by utilizing the polarization and depolarization processes of water molecules and the principle of Schottky junction, Lu et al. have proposed an interactive power generation device [13] based on water and all semiconductor system. The water droplet is located between a N-type silicon and a P-type silicon, and it can move in this channel in a forward or backward direction. The unique isotropy of the dynamic silicon-water-silicon generator arises from the intrinsic Fermi level difference of two semiconductors, leading to the directional water-semiconductor polarization, which can be used to harvest those low-frequency unordered water flow mechanical energy. Liu [14] et al. have extended to the interaction between water-bearing solid materials or hydrated biological tissues and semiconductors, resulting in the development of a wiping generator that utilizes water-containing solids to wipe across the surface of a semiconductor for charge generation. The principle of this device is shown in Fig. 4.1d, this innovative approach achieves an impressive transfer charge density of up to 186 mC m^{-2}, providing valuable insights for advancing new energy conversion technologies.

In summary, Fig. 4.2 shows a schematic diagram of the band structure of the metal-semiconductor contact (Schottky contact), which is the foundation for us to study the CE on metal-insulator interface due to the strict band structure of metals and semiconductors. The Schottky contact model has evolved to include the complex interplay between the metal work function, semiconductor properties, and surface states. This comprehensive understanding is crucial for the design and optimization of metal-semiconductor devices, such as diodes, transistors, and sensors, where the barrier height at the interface plays a pivotal role in determining device performance.

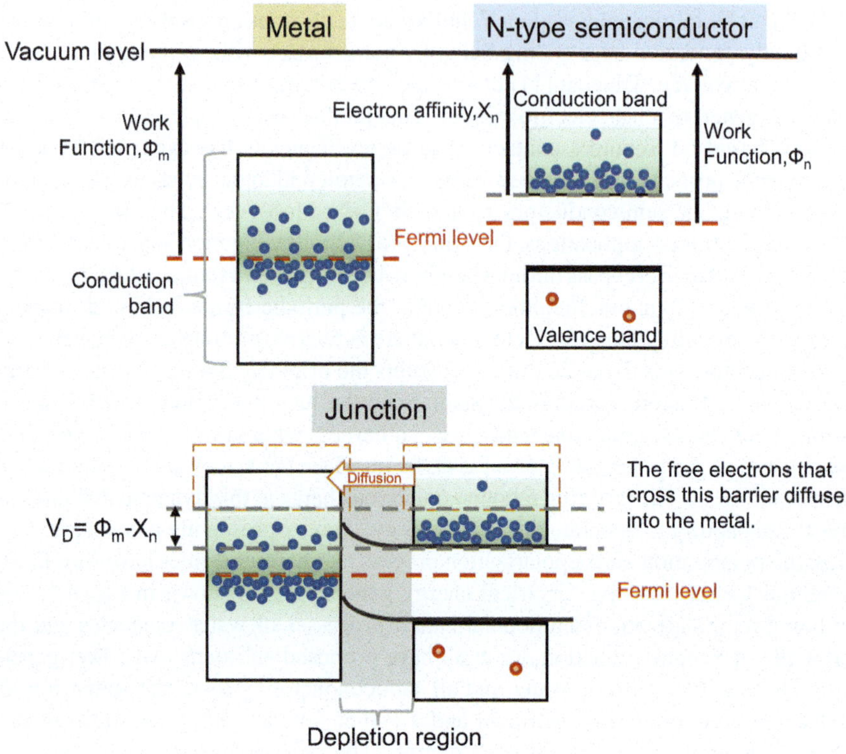

Fig. 4.2 Schematic diagram of electron transfer during Schottky contact process

4.3 Electron Transfer at Insulator-Metal Interface

The phenomenon of CE, in which objects acquire an electric charge upon contact, has perplexed scientists and engineers for centuries. This intriguing phenomenon not only plays a fundamental role in enhancing our comprehension of TE but also holds practical implications across various fields, ranging from electronics to materials science. The latest research indicates that the fundamental principle of CE lies in the transfer of electrons, ions, or material species between contact surfaces. In fact, there is ongoing debate about the main mechanism of CE, with electron transfer emerging as a prominent contender, especially in solid-solid interactions. The insulator-metal interface is a critical area of investigation within the broader context of CE, as it represents a boundary where two materials with distinct electronic properties meet. The behavior of electrons at this interface is governed by the differences in their work functions, electron affinities, and the potential barriers that arise due to the contact between dissimilar materials. Understanding the dynamics of electron transfer at this interface is essential for developing technologies that rely on controlled charge generation and management, especially as TENGs. Therefore, this section mainly introduces electron transfer at insulator metal interface.

The electron transfer process at insulator-metal interface has been investigated using KPFM over these years. By utilizing the capabilities of the KPFM system to regulate contact force, area, velocity, and cycle count during triboelectric interactions, a comprehensive analysis of CE phenomena can be conducted. This includes examining the distribution of triboelectric charges, exploring the multi-friction effect, and examining the subsequent diffusion process of charges on metal or insulator surfaces. The process of scanning a platinum-coated AFM tip across a SiO_2 surface using a tapping mode can be used to analyze electron transfer at insulator metal interface [15]. By repeatedly scanning an area approximately 5 μm in size, electrostatic charges gradually accumulate on the SiO_2 surface, reaching a saturation point after approximately eight scans (as shown in Fig. 4.3a–e). The line profile of the scanned area reveals the potential distribution, signifying the presence of negative electrostatic charges on the surface. The experimental results have demonstrated that the quantity of charge transferred to the surface is influenced by the duration of contact. Using the tapping mode, where the tip makes point-by-point contact, due to the fact that a slower contact speed can make the contact more sufficient, a slower scanning velocity is more efficient for transferring charges. Conversely, in the sliding mode, where the tip maintains continuous contact, charges are effectively transferred even at higher velocities, as indicated in Fig. 4.3b. The research suggests that continuous contact through sliding or rubbing with the tip is more effective for surface charging compared to the intermittent contact in the tapping mode. Nevertheless, the maximum surface charge density remains consistent regardless of whether charges are transferred through tapping or continuous contact mode, as long as the contact duration is sufficient to achieve saturation, thereby validating the process of electron transfer between metals and insulators.

The study by Lin et al. as shown in Fig. 4.4a has used a sliding mode Au-SiO_2 TENG [16] to investigate the surface charge decay behavior during operation at high temperatures, providing further evidence that electrons are the dominating transferred charges in CE, and the temperature dependence of TENG electric output contributes to understanding the characteristic metal-insulator and insulator-insulator processes. As the temperature increases, the thermionic emission of electrons from the surface of the insulator becomes more prominent, thereby potentially overshadowing the charge generation and transfer effect of the TENG. This thermionic emission is the key phenomenon explaining electron dominated CE, as when the material is heated during the CE process, electrons can obtain sufficient thermal energy to overcome the work function and emit from the material. In the sliding mode Ti-SiO_2 TENG proposed by Wang et al., the surface charge output is rationalized as a direct consequence of the coupling of the rate of electron thermionic emission, the charge transfer rate of CE, and the changing rate of the contacted area between the two materials. When the contacting area is larger than the displaced area, CE between the two materials is the major contributor to the measured surface charge. Conversely, when the contacting area is smaller, the thermionic emission of the exposed surfaces is the dominant factor. Based on experimental results, a modified electron cloud-potential

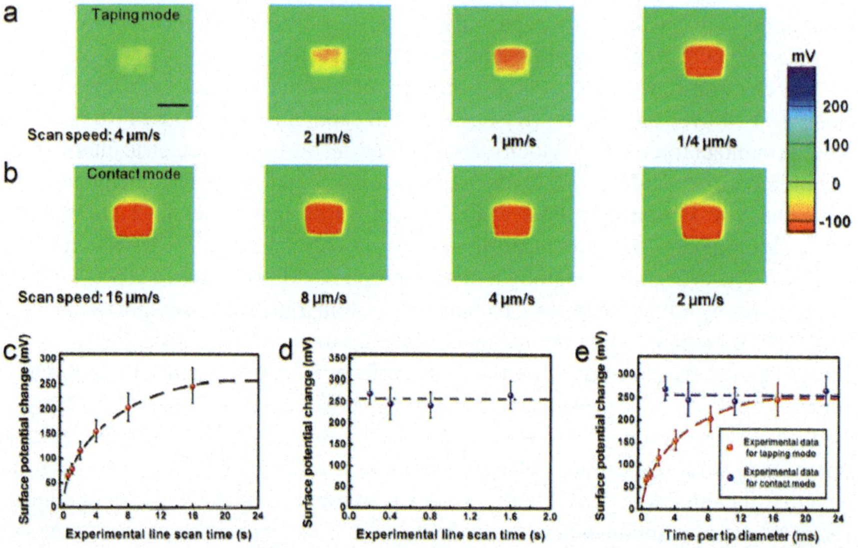

Fig. 4.3 Comparison of charge transfer dynamics between CE and TE. Surface potential mapping of parylene sample after the AFM probe scanned the central area at different scan speeds in **a** the tapping mode and **b** the contact mode. **c** and **d** Corresponding plots of surface potential change as functions of line scan time. **e** Comparison of the charge transfer density triggered by CE and TE as a function of the tip-sample interaction time [15]. Reproduced with permission from Springer Nature

well model has been proposed to explain the charge transfer and release between two materials especially under high temperature in the sliding-mode TENG as shown in Fig. 4.4c [17]. The model demonstrates that at high temperatures, the negatively pre-charged insulator exhibits thermionic emission of surface charges when not in contact with the metal. When the two materials come into contact, electrons may transfer from the metal to the insulator due to CE, resulting in equal and opposite charges on both surfaces. As the materials begin to separate, both thermionic emission from the exposed areas of the insulator and charge transfer from the areas in contact affect the measured total charges in the sliding mode TENG. The study by Wang et al. provides a comprehensive analysis of electron transfer at the insulator-metal interface, highlighting the significance of temperature on the CE process. The sliding mode TENG serves as a model system to understand the complex interplay between CE, thermionic emission, and the changing contacted area. The modified electron cloud-potential well model offers a unified framework to explain the observed phenomena, emphasizing the role of electrons in the charge transfer process at the insulator-metal interface. This understanding is crucial for the development of more efficient and reliable TENGs for energy harvesting and sensing applications, especially under varying temperature conditions.

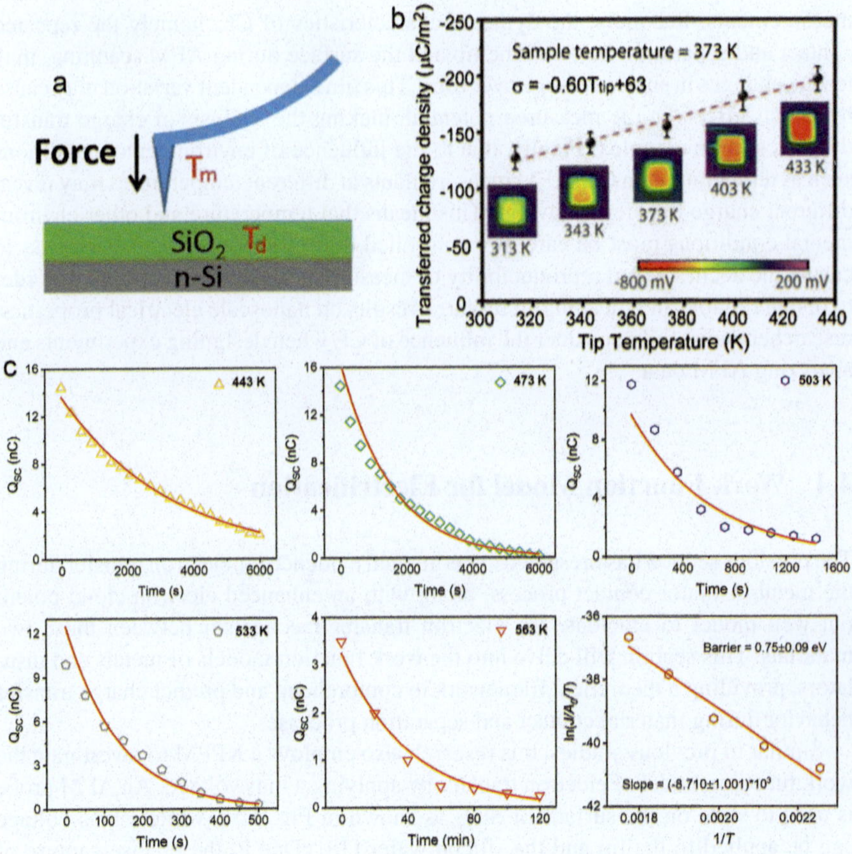

Fig. 4.4 Proving the effect of temperature differences on CE between the Au-coated tip and the SiO_2 sample by KPFM at nano-scale. **a, b** The sample temperature was set at 373 K, while the tip temperature varied from 313 to 433 K. **c** The measured (data points) and simulated curve based on electron thermionic emission model (solid) Q_{SC} as a function of the time at various temperatures [17]. Reproduced with permission from Wiley-VCH

The present section presents a comprehensive and systematic investigation of the electron transfer process at the metal-insulator interface, highlighting the pivotal role played by electron transfer in establishing electrical contact at this interface through the utilization of KPFM technology and hot electron emission. It is worth noting that the phenomenon of CE at the metal-insulator interface also has a certain impact on the measurement results of AFM and KPFM. Firstly, the charge transfer during the CE process will generate a potential difference between the metal tip and the insulator surface, which will affect the contact force and area in AFM measurements, thereby affecting the measurement results of surface potential. Specifically, contact forces may enhance charge transfer, resulting in an increase in surface charge density, which may lead to deviations in the measurement results of surface potential in KPFM measurements. Secondly, the dynamic characteristics of CE, namely the repeated contact and separation between the tip and the surface during AFM scanning, may lead to changes in surface charge over time. This time-dependent variation may cause hysteresis effects in the measured potential, making the analysis of charge transfer dynamics more complex. Finally, due to the influence of environmental conditions such as temperature on CE, AFM measurements at different temperatures may reveal different charge transfer behaviors. This means that temperature and other environmental conditions must be carefully controlled during the experimental process to ensure the accuracy and reproducibility of measurement results. Therefore, in order to obtain reliable and meaningful research results on nanoscale electrical properties, researchers must fully consider the influence of CE when designing experiments and analyzing AFM data.

4.4 Work Function Model for Electrification

The previous section has presented experimental evidence for electron transfer during the metal-insulator contact process, along with an enhanced electron cloud potential well model to elucidate the electron transfer mechanism between these two materials. This section will delve into the work function models of metals and insulators, providing a theoretical framework to comprehend and predict charge transfer behavior during material contact and separation processes.

Similar to previous studies, this research also employs a KPFM to investigate the work function model of electron transfer by applying a bias voltage. An AFM probe is used to slide on the surface of SiO_2 as shown in Fig. 4.5a, while a bias voltage can be applied to the tip and the silicon wafer [18]. Due to the precise capture of interface charges by the AFM probe, the metal insulator interface can be studied. The amount and sign of the charges to be delivered on the SiO_2 surface can be manipulated and switched by the applied bias voltage (Fig. 4.5b). The mechanism of CE in metal-dielectric systems can be comprehensively understood through an energy band diagram as shown in Fig. 4.5c, which illustrates the Fermi level of the metal. The energy band diagram of the insulator surface includes its conduction and valence bands, with surface and defect states assumed to be present at the surface (Fig. 4.5c-i). When the valence band edge of the insulator is situated below the metal's Fermi level (E_f), electrons from the metal can fill the surface states within the bandgap (E_g) that are below E_f, leading to a negative charge build-up on the insulator surface (Fig. 4.5c-ii). By applying a negative bias to the metal tip (Fig. 4.5c-iii), the energy levels of the surface states on the insulator side are lowered relative to the metal's Fermi level, facilitating further electron transfer from the metal to the insulator. Conversely, a positive bias (Fig. 4.5c-iv) raises the energy levels of the surface states on the insulator side, allowing electrons to transfer back to the metal and resulting in a positive charge on the insulator side. Recent AFM studies support the notion that CE in metal-dielectric systems is fundamentally an electron transfer phenomenon. KPFM has been instrumental in providing detailed nano-scale charge transfer data for these systems. This explanation holds true if the insulator's electronic structure can be characterized by a band diagram as described; otherwise, alternative models must be considered.

Xu et al. measured the surface charge density of materials in detail using various TENGs, and studied the long-term evolution of TENG surface charge density under different conditions [19]. Using TENG as a detect tool, surface states model and work function mode have been proposed to illustrate the charge transfer process in CE between metals and insulators. Work function, defined as the minimum energy required for an electron to escape from the surface of material into vacuum, directly influences the ease with which electrons can transfer from one material to another. In the study, various high-temperature-resistant TENGs composed of different metals with Al_2O_3 have been prepared to investigate the influence of metal work function on charge transfer in CE. Figure 4.6a depicts the energy band alignment prior to contact between a metal and a dielectric. Notably, even though dielectrics have a relatively wide bandgap, their highest occupied surface energy states typically lie below the Fermi levels (E_F) of metals, due to the lower electron density in these surface states.

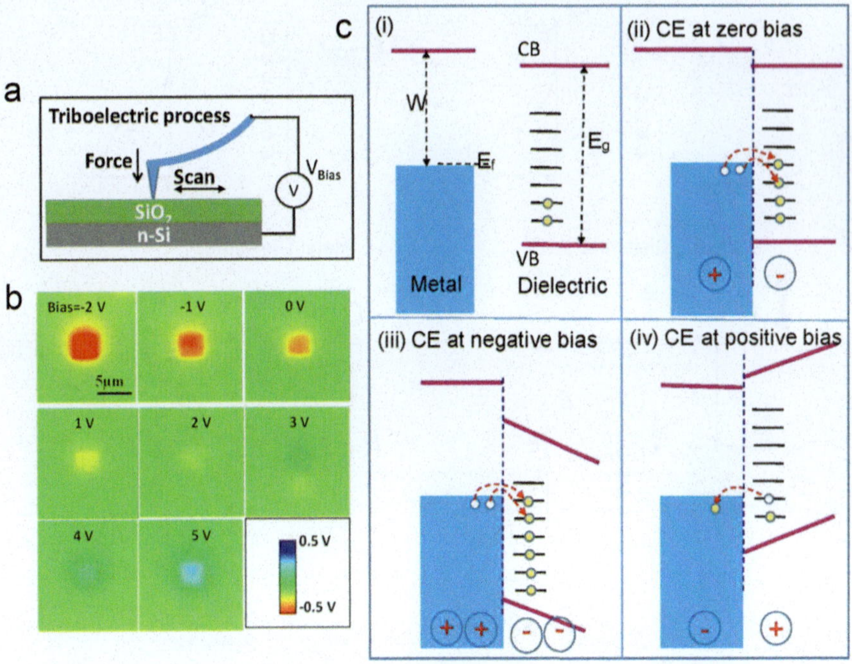

Fig. 4.5 **a** Schematic of experimental setup of the AFM tip and the SKPM to map the surface potential distribution. **b** Surface potential distributions of the parylene film including the areas that were rubbed by the Pt-coated AFM tip at different bias from − 2 to 5 V. **c** Schematic energy band diagrams for the metal and dielectric materials in the situations of precontact (i), in contact with no bias (ii), and in equilibrium with negative (iii) and positive (iv) bias [18]. Reproduced with permission from American Chemical Society

In this revised model, electron distribution adheres to the Fermi-Dirac distribution function given by:

$$f = \frac{1}{\exp^{(E-E_F)/kT} + 1} \quad (4.3)$$

where f represents the probability of an electron having energy E, E_F is the Fermi level energy, k is the Boltzmann constant, and T is the temperature. At T_0 (absolute zero), electrons occupy all energy states below E_F. At a higher temperature T_1 (where $T_1 > T_0$), some electrons gain enough energy to transition above the Fermi level. Upon contact with the dielectric, high-energy electrons from the metal can transfer to the surface states of the dielectric. In the separation process after contact the resulting charge disparity creates an electric field that facilitates the return flow

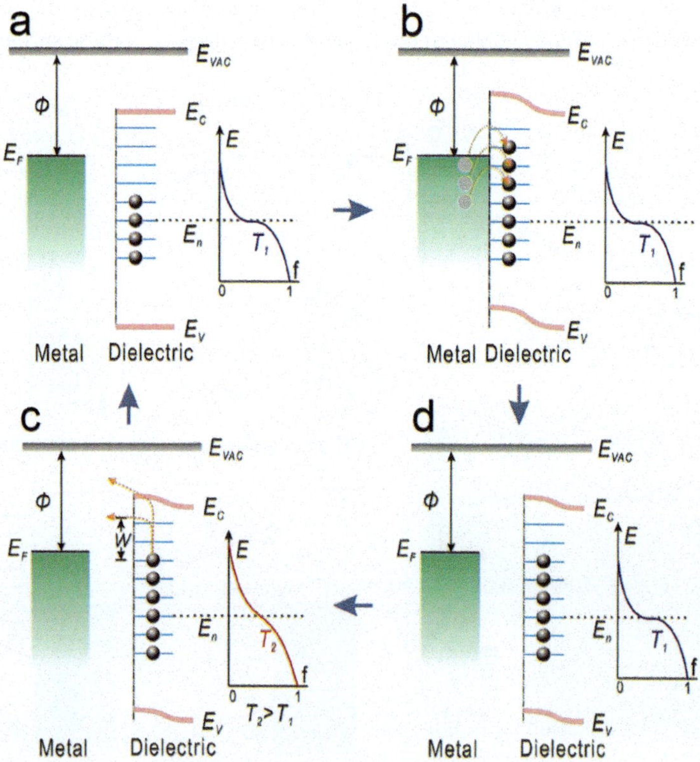

Fig. 4.6 Modified surface states model for explaining the charge transfer during and after CE for the case of metal-dielectric (semiconductor). The CE for a metal and a dielectric **a** before contact, **b** in contact, and **c** after contact. **d** Charge release. Φ, metal work function; E_F, Fermi level; E_{VAC}, vacuum level; E_C, conduction band; E_V, valence band; E_n, neutral level of surface states; f(E), Fermi-Dirac distribution probability; W, potential barrier; T_1 and T_2, temperature [19]. Reproduced with permission from Wiley-VCH

of electrons from the dielectric's surface states back to the metal (as illustrated in Fig. 4.6c). At an even higher temperature T_2 (with $T_2 > T_1$), electron hopping across the potential barrier W becomes more feasible (as shown in Fig. 4.6d). This indicates that the rate of "charge release" diminishes as the temperature decreases, explaining the residual surface charges after the separation of the two materials. However, it is worth knowing that, although the previously mentioned surface states model effectively elucidates the CE mechanisms for metal-semiconductor and metal-insulator interfaces, it encounters challenges when applied to metal-polymer and polymer-polymer CE systems. This limitation arises because the surface states model is rooted in the band structure theory, which is less applicable to polymers and amorphous materials. This includes two primary aspects: Firstly, once charges are excited into the conduction band and become free electrons, their energy states can no longer be adequately measured by band models [20, 21]. In contrast, the electron cloud model depicts the spatial distribution of electrons outside the nucleus using statistical methods, and it presents the likelihood of electron presence around the atomic nucleus in terms of probability density, offering a more intuitive and vivid description than band models. Secondly, band models can only explain the charge transfer between two materials from the perspective of electrons. They fail to account for atomic-level factors, such as atomic electronegativity [22, 23], especially the changes in band structure caused by strong electron-withdrawing groups, which may be reasons for increased charge density. The electron cloud model, on the other hand, is capable of describing the dynamic behavior of electrons outside the atomic nucleus, including their response to changes at the atomic level. Therefore, when studying the CE process of various material types, including polymers and amorphous materials, electron cloud/potential models [24] will be used to explain, which will be introduced in subsequent chapters.

4.5 Charge Leakage and Its Related Phenomena

Charge leakage is a key factor affecting the performance and efficiency of CE processes, especially in applications such as TENG that rely on charge accumulation and retention. Specifically, due to significant differences in the band structure between metallic and insulating materials. The conduction band of metals is partially filled with a large number of free electrons, while insulators have a wide bandgap and generally do not have observable conductivity. When a metal comes into contact with an insulator, free electrons in the metal may cross the interface and enter the conduction band of the insulator, causing charge leakage. Understanding the mechanism of charge leakage and its related phenomena is crucial for optimizing the design and operation of TENG and other energy harvesting devices.

The presence of humidity, temperature, and conductive pollutants in the environment can significantly affect charge leakage. For example, increasing humidity [25] can enhance the conductivity of the surrounding air and promote charge leakage. In TENG dynamic leakage, charge leakage can occur during operation (dynamic leakage) or when the system is in a static state (static leakage) [26, 27]. Dynamic leakage is usually more significant because continuous motion and contact separation cycles can promote charge transfer. Charge leakage in CE process can occur through several pathways, including environmental medium conduction, surface recombination, hot electron emission, and material defects and impurities. In the presence of conductive media such as humid air or conductive liquids, charges leak through these media, which is called environmental medium conduction. Surface recombination refers to the recombination of electrons and ions at the interface [28], which neutralizes the charge separation caused by the CE process. This leakage path is particularly important at the liquid-solid interface, as ions can move to the surface and recombine with electrons. Thermionic emission refers to that at elevated temperatures, electrons may gain sufficient thermal energy to overcome the work function barrier of the material, resulting in charge leakage. Defects in materials or impurities at interfaces can serve as channels for charge leakage, promoting the movement of electrons or ions and reducing the effectiveness of charge separation.

Chen et al. designed an electric-field-induced second harmonic generation (EFI-SHG system [29] to visualize three types of relaxation induced by friction on the surface of polyimide films and studied different charge diffusion pathways on the polymer surface. The EFI-SHG measurements reveal that the relaxation of tribo-induced charges on the surface of polyimide is a complex process involving briefly three processes from the saturated surface state. Firstly, a fast relaxation of charges occurs within about 180 s, possibly due to diffusion into the open air. Secondly, the net discharging along the surface happens with a time constant on the order of 10^3 s. After this relaxation, the SHG signal returned to 0.3 and showed a very slow decrease afterward, indicating an extremely long relaxation with a time constant on the order of 10^4 s. When a water droplet contacts the polyimide surface, the SHG signal decreases dramatically, indicating that tribo-induced surface charge diffuses due to the water, which means environmental medium conduction. As shown in Fig. 4.7, Nie et al. have studied the CE and charge leakage process between different liquids and a polytetrafluoroethylene (PTFE) film was systematically studied to clarify the electrification mechanism of the solid-liquid interface [30]. The experimental results and theoretical calculations show that the charge transfer between deionized water and PTFE is dominated by electron transfer, which accounts for more than 90% of CE charges. As the concentration of ions in the aqueous solution increases, ions accumulate at the liquid-solid interface, hindering the electron transfer process due to the screening effect. Furthermore, the CE between ion-free transformer oil and PTFE also confirms the presence of electron transfer between liquid and solid, as there are no ions in oil droplets.

The phenomenon of charge leakage is detrimental to TENG in the majority of cases, as it significantly diminishes the surface charge of TENG, thereby resulting in a reduction in device output. However, under specific circumstances, a substantial amount of charge can be rapidly discharged through charge leakage, enabling certain ionization applications. The TENG designed by Li et al. in Fig. 4.8 exhibits a high output voltage, enabling the generation of charge pulses with single or alternating polarity [31]. This feature makes it well-suited for inducing nano-electrospray

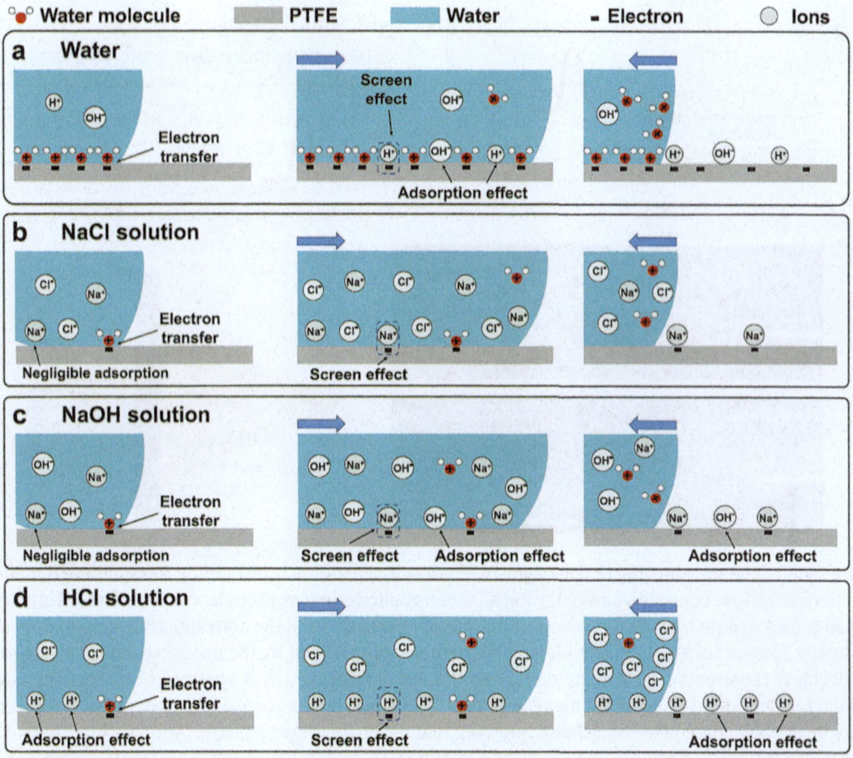

Fig. 4.7 Mechanism of CE between water and PTFE film. **a** The mechanism of CE between deionized water and PTFE. **b** The mechanism of CE between high concentration of NaCl solution and PTFE. The high concentration of the ionic layer on the PTFE surface suppresses the electron transfer of H_2O molecules to PTFE. **c** The mechanism of CE between high concentration of NaOH solution and PTFE. **d** The mechanism of CE between high concentration of HCl solution and PTFE. The adsorption of hydrogen ions on the surface of PTFE plays a leading role, and a small amount of electron transfer occurs [30]. Reproduced with permission from Wiley-VCH

ionization (nanoESI) and plasma discharge ionization. Precisely controlled charge pulses ranging from 1.0 to 5.5 nC can be delivered to a given nanoESI emitter, with an onset charge of 1.0 nC. Spray pulses can be generated at a frequency as high as 17 Hz (with a period of 60 ms), and the duration of each pulse is adjustable between 60 ms and 5.5 s upon request. By utilizing minimal sample volume (18 pl per pulse), highly sensitive mass spectrometry analysis ($\approx$ 0.6 zeptomole) has been successfully achieved using a cocaine sample concentration of 10 pg ml^{-1}.

In this section, we have discussed that charge leakage poses both advantages and disadvantages for TENG devices. To prevent charge leakage, it is crucial to control environmental conditions and optimize the materials used in TENGs. However, there are also certain application scenarios where charge leakage can be utilized effectively. Specifically, reducing environmental humidity, maintaining appropriate temperature levels, and avoiding contact with conductive pollutants can significantly mitigate the

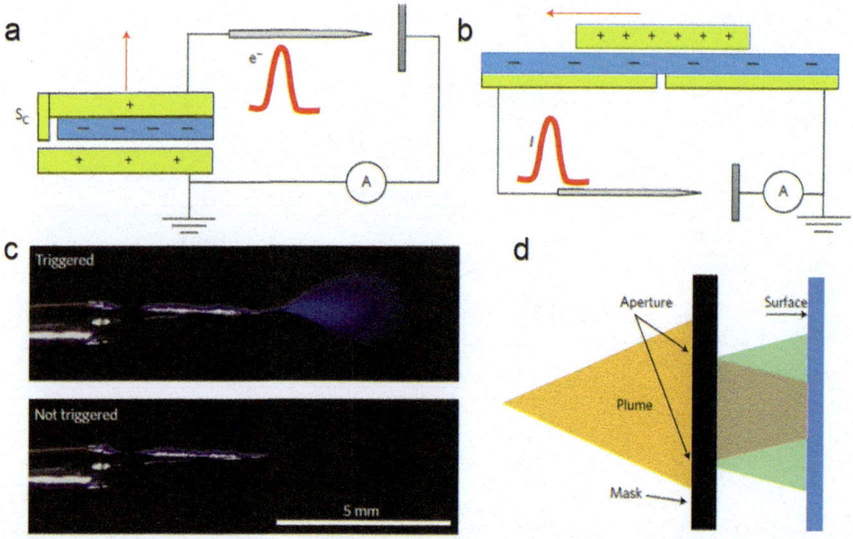

Fig. 4.8 Schemes showing the mechanism of contact-separation **a** and sliding freestanding **b** mode TENGs. Yellow, copper electrode layers; blue, fluorinated ethylene propylene layers. The red arrows and pulses denote moving directions of the TENG electrodes and the corresponding charge flow to the ion source, such as the nano electrospray emitter represented by the needle shape. The vertical rectangle represents a steel plate collecting the ion current, which is measured by a picoampere meter. **c** Dark-field images of a nano electrospray emitter showing an electrospray plume triggered by the TENG charge flow. **d** Scheme showing that the charged spray plume (yellow), after passing through the mask aperture, was focused (brown) with grounded conducting target surfaces, or defocused (green) with insulating target surfaces [31]. Reproduced with permission from Springer Nature

negative impact of charge leakage on TENG performance. Furthermore, selecting insulation materials with wider bandgaps, improving interface design, and utilizing materials with stronger leakage resistance can all contribute to suppressing charge leakage to a certain extent. When intentional charge leakage is required for specific ionization applications such as nano electric spray ionization, it can be achieved by designing a tailored TENG structure and controlling the operating conditions accordingly.

4.6 Summary

This chapter offers a comprehensive exploration of CE phenomenon at the metal-insulator interface, elucidating crucial scientific inquiries and recent advancements in this field. CE, also known as TE, is a multifaceted process influenced by various factors such as material properties, surface conditions, environmental contaminants, and specific circumstances during contact and separation. The metal-insulator interface, which represents the boundary where two materials with distinct electronic properties converge, constitutes a pivotal research focus in the field of CE. Investigating electron transfer at this interface is crucial for gaining a comprehensive understanding of CE. At the same time, in-depth research on the band model of metal semiconductor contact has laid a solid foundation for understanding and exploring the electron transfer mechanism at the metal insulator interface. It highlighted the role of work function, electron affinity, and surface states in determining the barrier height at the metal-semiconductor junction, ultimately affecting the performance of the device. Moreover, KPFM technology can provide nanoscale charge transfer information during the CE process. It detects the contact potential difference between the sample surface and the probe, offering insights into charge distribution at the nanoscale, and it can directly observe and quantify changes in surface charge density caused by triboelectrification. This is crucial for studying the charge transfer process at metal-insulator interfaces. This chapter also provides a detailed introduction to the use of KPFM to study Schottky contacts, electron transfer, and band models, emphasizing the role of charge transfer in the CE process. The work function model provides a theoretical framework for understanding and predicting charge transfer during material contact and separation processes. It focuses on the difference in work function between metals and insulators, which affects the ease of electron transfer between them. Based on experimental results, an improved electron cloud potential well model is proposed to explain the charge transfer and release between two materials. This model emphasizes the influence of temperature on the electron transfer process in contact electrification and provides a unified framework to explain the observed phenomena, highlighting the role of electrons in the charge transfer process at the insulator metal interface. These models are crucial for deepening the understanding of metal insulator interface contact charging and have practical significance for the development of electronic devices and energy harvesting technologies.

References

1. M.M. Apodaca, P.J. Wesson, K.J. Bishop, M.A. Ratner, B.A. Grzybowski, Contact electrification between identical materials. Angew. Chem. Int. Ed. **49**(5), 946–949 (2010)
2. D.J. Lacks, R.M. Sankaran, Contact electrification of insulating materials. J. Phys. D Appl. Phys. **44**(45), 453001 (2011)
3. B.D. Terris, J.E. Stern, D. Rugar, H.J. Mamin, Contact electrification using force microscopy. Phys. Rev. Lett. **63**(24), 2669–2672 (1989)
4. Y.S. Zhou, Y. Liu, G. Zhu, Z.-H. Lin, C. Pan, Q. Jing, Z.L. Wang, In situ quantitative study of nanoscale triboelectrification and patterning. Nano Lett. **13**(6), 2771–2776 (2013)
5. S. Lin, L. Zhu, Z. Tang, Z.L. Wang, Spin-selected electron transfer in liquid–solid contact electrification. Nat. Commun. **13**(1), 5230 (2022)
6. Y. Li, Y. Luo, S. Xiao, C. Zhang, C. Pan, F. Zeng, Z. Cui, B. Huang, J. Tang, T. Shao, X. Zhang, J. Xiong, Z.L. Wang, Visualization and standardized quantification of surface charge density for triboelectric materials. Nat. Commun. **15**(1), 6004 (2024)

7. K. Shi, B. Chai, H. Zou, D. Min, S. Li, P. Jiang, X. Huang, Dielectric manipulated charge dynamics in contact electrification. Research (2022)
8. W. Schottky, Zur Halbleitertheorie der Sperrschicht- und Spitzengleichrichter. Z. Phys. **113**(5), 367–414 (1939)
9. N. F. Mott, The theory of crystal rectifiers. Proc. Royal Soc. Lond. Ser. A. Math. Phys. Sci. **171**(944), 27–38 (1939)
10. A.M. Cowley, S.M. Sze, Surface states and barrier height of metal-semiconductor systems. J. Appl. Phys. **36**(10), 3212–3220 (1965)
11. J. Liu, A. Goswami, K. Jiang, F. Khan, S. Kim, R. McGee, Z. Li, Z. Hu, J. Lee, T. Thundat, Direct-current triboelectricity generation by a sliding Schottky nanocontact on MoS2 multilayers. Nat. Nanotechnol. **13**(2), 112–116 (2018)
12. S. Lin, Y. Lu, S. Feng, Z. Hao, Y. Yan, A High Current Density direct-current generator based on a moving van der Waals Schottky diode. Adv. Mater. **31**(7), 1804398 (2019)
13. Y. Lu, Y. Yan, X. Yu, X. Zhou, S. Feng, C. Xu, H. Zheng, Z. Yang, L. Li, K. Liu, S. Lin, Polarized water driven dynamic PN junction-based direct-current generator. Research (2021)
14. Z. Liu, S. Lin, P. Yang, S. Qin, J. Hu, X. Chen, A wiping-type semiconductor–liquid generator utilizing water-bearing solid materials and hydrated biological tissues. Energy Environ. Sci. **17**(18), 6582–6593 (2024)
15. Y.S. Zhou, S. Li, S. Niu, Z.L. Wang, Effect of contact- and sliding-mode electrification on nanoscale charge transfer for energy harvesting. Nano Res. **9**(12), 3705–3713 (2016)
16. S. Lin, L. Xu, C. Xu, X. Chen, A.C. Wang, B. Zhang, P. Lin, Y. Yang, H. Zhao, Z.L. Wang, Electron transfer in nanoscale contact electrification: effect of temperature in the metal-dielectric case. Adv. Mater. **31**(17), 1808197 (2019)
17. A.C. Wang, B. Zhang, C. Xu, H. Zou, Z. Lin, Z.L. Wang, Unraveling temperature-dependent contact electrification between sliding-mode triboelectric pairs. Adv. Func. Mater. **30**(12), 1909384 (2020)
18. Y.S. Zhou, S. Wang, Y. Yang, G. Zhu, S. Niu, Z.-H. Lin, Y. Liu, Z.L. Wang, Manipulating nanoscale contact electrification by an applied electric field. Nano Lett. **14**(3), 1567–1572 (2014)
19. C. Xu, Y. Zi, A.C. Wang, H. Zou, Y. Dai, X. He, P. Wang, Y.-C. Wang, P. Feng, D. Li, Z.L. Wang, On the electron-transfer mechanism in the contact-electrification effect. Adv. Mater. **30**(15), 1706790 (2018)
20. J. Singleton, *Band theory and electronic properties of solids*, Vol. 2 (OUP Oxford, 2001)
21. L.E. Brus, Electron–electron and electron-hole interactions in small semiconductor crystallites: the size dependence of the lowest excited electronic state. J. Chem. Phys. **80**(9), 4403–4409 (1984)
22. X. Tao, X. Chen, Z.L. Wang, Design and synthesis of triboelectric polymers for high performance triboelectric nanogenerators. Energy Environ. Sci. **16**(9), 3654–3678 (2023)
23. S. Li, J. Nie, Y. Shi, X. Tao, F. Wang, J. Tian, S. Lin, X. Chen, Z.L. Wang, Contributions of different functional groups to contact electrification of polymers. Adv. Mater. **32**(25), 2001307 (2020)
24. Z.L. Wang, A.C. Wang, On the origin of contact-electrification. Mater. Today **30**, 34–51 (2019)
25. S. Hoffmann, M. Koehl, Effect of humidity and temperature on the potential-induced degradation. Prog. Photovoltaics Res. Appl. **22**(2), 173–179 (2014)
26. J. Wang, S. Xu, C. Hu, Charge generation and enhancement of key components of triboelectric nanogenerators: a review. Adv. Mater. 2409833 (2024)
27. Y. Liang, X. Xu, L. Zhao, C. Lei, K. Dai, R. Zhuo, B. Fan, E. Cheng, M.A. Hassan, L. Gao, Advances of strategies to increase the surface charge density of triboelectric nanogenerators: a review. Small **20**(16), 2308469 (2024)
28. Z. Liu, Y.Z. Huang, Y. Shi, X. Tao, P. Yang, X. Dong, J. Hu, Z.X. Huang, X. Chen, J.P. Qu, Creating ultrahigh and long-persistent triboelectric charge density on weak polar polymer via quenching polarization. Adv. Funct. Mater. **33**(34), 2302164 (2023)
29. X. Chen, D. Taguchi, T. Manaka, M. Iwamoto, Z.L. Wang, Direct probing of contact electrification by using optical second harmonic generation technique. Sci. Rep. **5**(1), 13019 (2015)

30. J. Nie, Z. Ren, L. Xu, S. Lin, F. Zhan, X. Chen, Z.L. Wang, Probing contact-electrification-induced electron and ion transfers at a liquid-solid interface. Adv. Mater. **32**(2), 1905696 (2020)
31. A. Li, Y. Zi, H. Guo, Z.L. Wang, F.M. Fernández, Triboelectric nanogenerators for sensitive nano-coulomb molecular mass spectrometry. Nat. Nanotechnol.Nanotechnol. **12**(5), 481–487 (2017)

Chapter 5
Contact-Electrification at Insulator-Insulator Interface

Abstract Contact-electrification (CE) mechanism between insulators is the last puzzle to be solved between solid and solid, and it was ultimately proven to be dominated by electron transfer. The introduction of the energy band model incorporating surface states provides a possibility for the aforementioned electron transfer-dominated mechanism. In 2016, Wang's team made a groundbreaking discovery: the transfer of electrons occurs exclusively when the distance between the atoms of two substances is less than the standard bonding length, approximately 0.2 nm, which is within the range of repulsive forces. This finding laid the groundwork for the electron cloud model that was later proposed. By 2018, Wang's team, utilizing a newly engineered high-temperature resistant TENG, conducted research on charge transfer at elevated temperatures. Through these studies, they identified the dominant mechanisms of electron transfer and, in a ground-breaking move, proposed the innovative electron cloud model. This model can be broadened to encompass not only solid–solid interactions but also liquid–solid, liquid–gas, and even liquid–liquid scenarios, solving a scientific problem that has puzzled humankind for 2600 years. Interface electron transfer-transition induced photon emission was predicted by the electron cloud model and observed by Wang's team, achieving a unified explanation for the triboluminescence phenomenon previously discovered at the atomic and electronic levels. There are also self-discharging processes in the CE process between insulators, including thermal ion emission and photon excitation. Through researches on these two processes, the electron transfer-dominated CE mechanism was experimentally confirmed. In addition, there are some factors that also affect CE between insulators, including curvature effect, atomic electronegativity, material transfer and heterolytic bond, dipole polarization, and deep trapping. Through in-depth research on these influencing factors, CE processes can be further regulated, thereby enhancing the power generation and energy collection efficiency of the prepared TENG.

Z. L. Wang, *Contact-Electrification of Matter*, Springer Series in Surface Sciences 75,
https://doi.org/10.1007/978-3-031-88504-4_5

5.1 Introduction

Contact-electrification (CE) between insulators is very common, for example, as mentioned in textbooks, fur and glass rods will each carry an equal amount of different charges due to friction, and the human body often carries static electricity due to friction with clothing in daily life. However, the specific properties of these generated charges have not been fully understood. Harper first proposed that the CE between insulators belongs to ion transfer [1]. Lee et al. proposed that electrons and ions work together [2] McCarty et al. proposed an ion transfer model based on polymer experiments [3]. These reports all use electrical methods, covering studies on charge and bandgap, work function, and ion concentration, but it is difficult to prove whether it is electron or ion transfer. There are also numerous discussions on the use of nonelectrical methods, such as Bard et al. introducing charges on the surface of polytetrafluoroethylene (PTFE) using electrochemical methods and discovering that chemical reactions can occur in solution, leading to the inference of electrons [4]. However, Piperno et al. also used electrochemical methods and found that the rubbed PTFE did not undergo a reduction reaction with metal ions, indicating that it was not electron transfer [5]. Baytekin et al. proposed that the charge during the electrification process of polymers originates from material transfer [6]. The above studies were conducted at room temperature. Xu et al. proposed a high-temperature CE study and observed a strange phenomenon of rapid charge decay [7]. They found that the evolution of charges at high temperatures follows an exponential law of the hot electron emission equation, revealing that the main source is electron transfer. This chapter mainly introduces the energy band model and electron cloud model in the currently recognized electron transfer mechanisms, as well as some new phenomena that have emerged in the insulator-insulator CE, and other effects at the insulator-insulator interface.

5.2 Energy Band Model

Due to the large bandgap of insulators, electrons cannot be directly transferred from the valence band to the conduction band, hence there are no “free electrons”. How is the transfer of charge achieved when two insulators come into contact and generate electricity?

Gutman and Hartmann posited that the electron’s accessible energy levels are confined to the surface rather than the bulk, and these accessible levels are referred to as “surface states” [8]. Figure 5.1 illustrates the energy level diagram for an insulator-insulator interface (the low-density condition represents the “before contact” state, whereas the high-density condition signifies the “after contact” state [9]. Upon contact between the insulators, electrons migrate from the occupied surface states of dielectric 1 to the vacant surface states of dielectric 2. The impetus for this charge

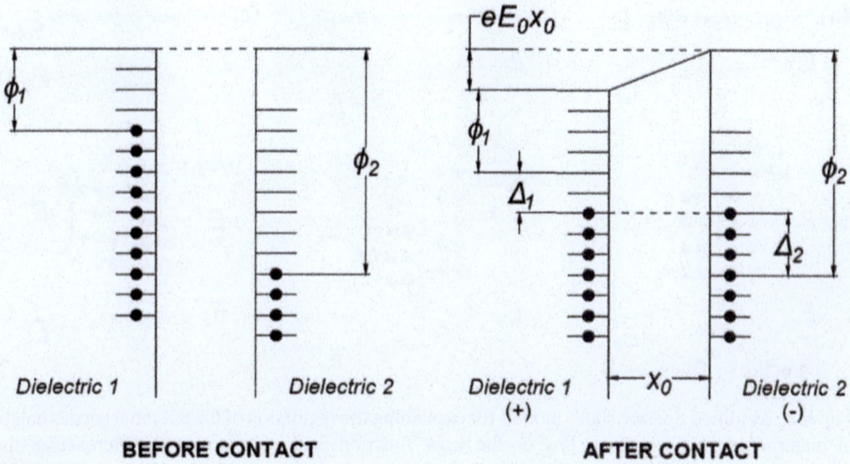

Fig. 5.1 Energy level diagram for insulator-insulator contact. Reproduced with permission from Elsevier [9]

transfer across the surfaces is rooted in the disparity of the effective work functions between the two surfaces. The transfer of charge will terminate once the Fermi levels of both materials align. This charge transfer results in alterations to the Fermi energy of dielectric 1 and dielectric 2 by $\Delta 1$ and $\Delta 2$, respectively. Additionally, the charge transfer engenders a potential difference across the surfaces denoted as E_0 x_0. Consequently, the Fermi levels of the surfaces become equal upon contact, and the expression for the energy level is denoted as

$$\Phi_1 + \Delta_1 + eE_0x_0 = \Phi - \Delta_2 \tag{5.1}$$

where Φ_1 and Φ_2 are the effective work functions of dielectric 1 and dielectric 2, respectively; Δ_1 and Δ_2 are changes in the fermi energy of dielectric 1 and dielectric 2, respectively; E_0 is the electric field at the contact point between the dielectrics; x_0 is the critical separation distance between the two dielectrics when charge transfer ceases; e is the elementary charge.

Xu et al. further refined the surface states models for CE [10]. They discovered that CE originates from electron transfer and that the barrier height of a material's surface is crucial for CE. The barrier heights of materials are influenced by their contacting materials. Changing the contacting materials, whether to metals or dielectrics, can alter the barrier height of the two materials in contact, a process termed "regulation". This regulation is attributed to the work function or contact potential difference of the two materials, affecting the direction and tendency of electron transfer, as explained through surface states models.

Figure 5.2a shows that when the neutral level (E_n) of dielectric A (initially negatively charged) and dielectric B are at the same level, the barrier for electron transfer is W when they contact. If the E_n of dielectric B is ΔE lower than that of dielectric

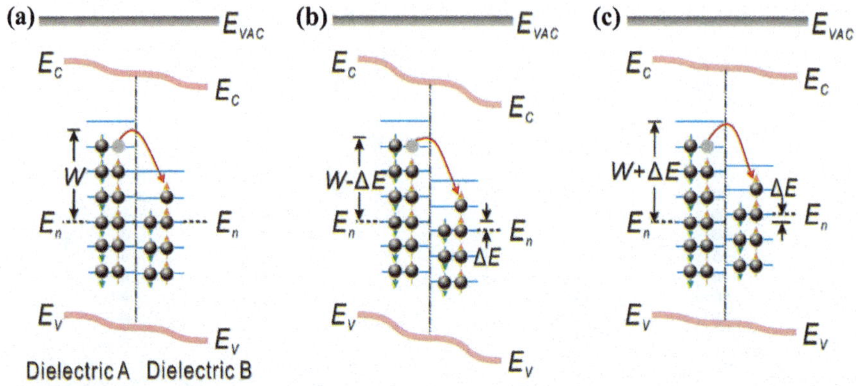

Fig. 5.2 Modified Surface states models for explaining the regulation of the potential barrier height of materials during CE depending on the work function or contact potential difference of the contacted materials. Charge transfer between dielectric A (with initial negative charges) and dielectric B when E_n of the former is **a** as high as **b** higher than and **c** lower than that of the later. E_{VAC}, vacuum level; E_C, conduction band; E_V, valence band; E_n, neutral level of surface states; W, potential barrier; Φ, work function; E_F, Fermi level. Reproduced with permission from Wiley [10]

A, the barrier for electron transfer from dielectric A to dielectric B reduces to W-ΔE (Fig. 5.2b). Conversely, if the En of dielectric B is ΔE higher than that of dielectric A, the barrier for electron transfer from A to B increases to $W + \Delta E$ (Fig. 5.2c). These models help in understanding the regulation effect on the potential barrier height of materials and the electron transfer mechanism in CE.

Additionally, Xu et al. extended the surface states model to explain the CE mechanism between two different types of insulator dielectrics at high temperatures [11]. Assuming that the E_n of dielectric A is higher than that of dielectric B (Fig. 5.3a), when the two dielectrics contact (Fig. 5.3b), electrons from the high energy states in dielectric A transition to the low energy states of dielectric B, as shown in process (1). This is similar to the metal-dielectric case in Fig. 5.2b. After separation (Fig. 5.3c), the transferred electrons in dielectric B escape from the surface due to thermionic emission (process 2). Meanwhile, dielectric A, now with more positive charges, captures electrons from the environment (process 3). Ultimately, dielectric A becomes positively charged, and dielectric B becomes negatively charged. The sign of the transferred charges is not expected to reverse with increasing temperature.

The energy band model based on surface states provides a reliable explanation for the CE between insulators and reveals that the CE may also be dominated by electron transfer. This provides an important foundation for further in-depth research on CE in the future, and for clarifying whether electrons or ions/materials are involved.

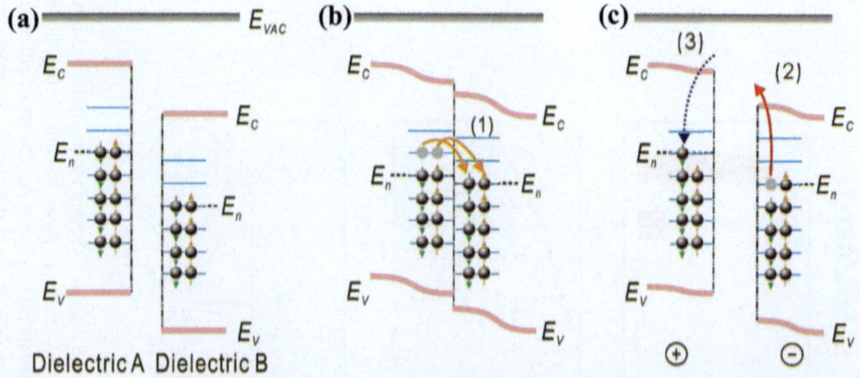

Fig. 5.3 CE mechanism explained by surface states model at high temperatures. Charge transfer before contact **a**, in contact **b**, and after contact **c** between two different insulator dielectrics for a case that E_n of the former is higher than that of the later. Φ, metal work function; E_F, Fermi level; E_{VAC}, vacuum level; E_C, conduction band; E_n, neutral level of surface states, E_V, valence band. Reproduced with permission from Wiley [11]

5.3 Electron Cloud Model

It is worth to note that, although the surface states model can be used to explain the mechanism of CE of insulators, there are still some difficulties. The reason is that the surface states model originates from the band structure theory of semiconductors, which is however not the real case for polymers and noncrystalline materials.

Wang's team has confirmed for the first time the occurrence of CE in KPFM insulation material surface potential measurement [12]. It has been established that a significant change in surface potential due to CE between the tip and the sample occurs only when the probe tip is within the repulsive region of the tip−sample interaction force. Figure 5.4a and b depict the schematic of the tip−sample interaction force during tapping scans with varying parameters. To apply the relationship between ΔV and $\Delta\varphi$ to forecast contact electrification during the tapping process, it is essential to verify its applicability across different KPFM probes and samples. Consequently, experiments were conducted on SiO_2 samples (approximately 300 nm thick) and with a KPFM probe featuring a reduced cantilever spring constant (k = 0.3 N/m). In the SiO_2 experiment, A_0 was set at 300 nm with varying A_{sp} values. The $\Delta V - Asp$ relationship (Fig. 5.4c, d) indicates CE when A_{sp} is below 155 nm. This aligns with the transition between $\Delta\varphi > 0$ and $\Delta\varphi < 0$ in the $\Delta\varphi - A_{sp}$ relationship (Fig. 5.4e). Furthermore, the highest occupied surface energy state of SiO_2 (sample) is positioned above the Fermi level of platinum (tip), prompting electron transfer from the sample to the tip, resulting in a positively charged contact area during the initial KPFM scan.

The study also examined the impact of different A_f values (50, 100, 150, and 200 nm) on SiO_2. In all tests, A_{sp} was set to 60% of A_0. For both materials, an increase in A_0 leads to greater vibrational energy of the cantilever. Consequently, even if the

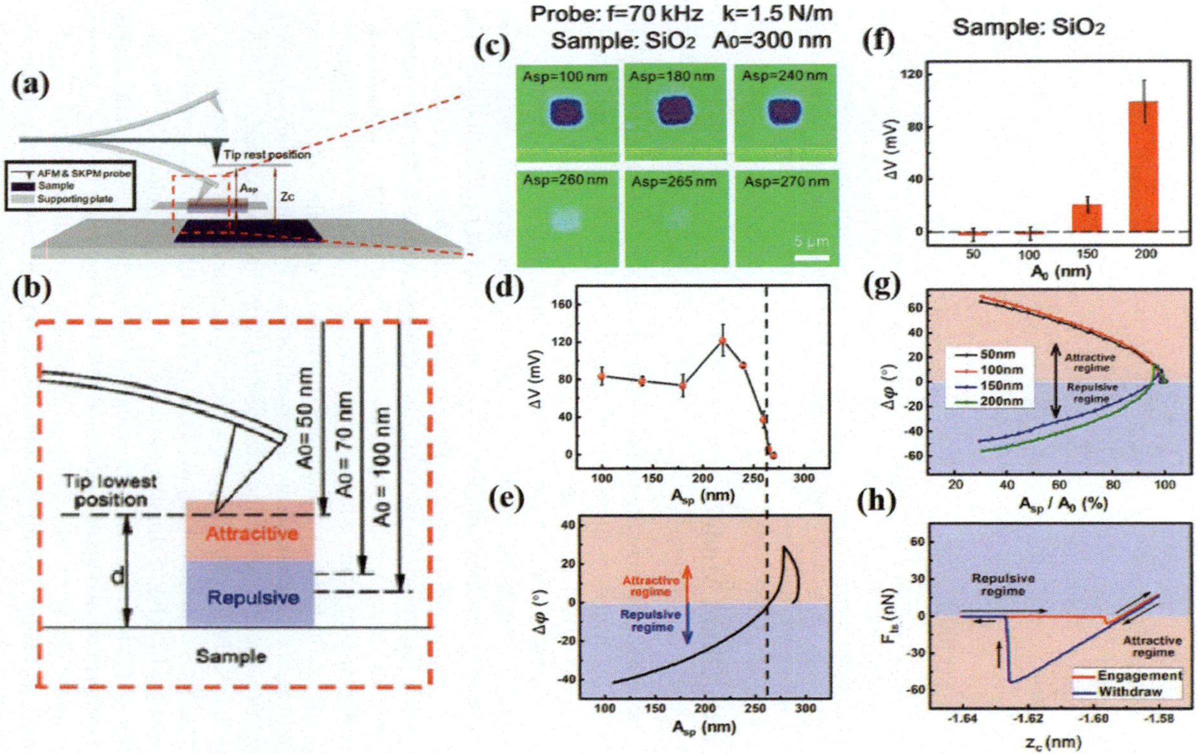

Fig. 5.4 (**a** Schematic of the tip—sample interaction force of tapping scans with different scanning parameters (A_0 and A_{sp}). **b** Enlarged image of the red box area in **a**. Relationships between CE and phase shift of different samples and KPFM probes **c–e** $\Delta V - A_{sp}$ and $\Delta\varphi - A_{sp}$ curves of $A_0 = 300$ nm on the silicon oxide sample. **f** Results of surface potential change caused by tapping scans with different A_0 (50, 100, 150, and 200 nm) on SiO_2. **g** $\Delta\varphi - A_{sp}$ curves of different A_0 (50, 100, 150, and 200 nm) on SiO_2. **h** $F_{ts} - z_c$ curves during the engagement and withdrawal of the probe tip on SiO_2 in the contact mode. Reproduced with permission from American Chemical Society [12]

A_{sp} to A_0 ratio remains constant, a larger A_0 results in a more pronounced phase shift decrease, with a reduced tip–sample distance at the lowest point above the surface during oscillation. The SiO_2 scanning experiment revealed no CE until A_0 reached 150 nm (Fig. 5.4f). Accordingly, phase shift increases were observed for $A_0 = 50$ nm and $A_0 = 100$ nm, with decreases emerging at $A_0 = 150$ nm and $A_0 = 200$ nm (Fig. 5.4g). This can be attributed to the adhesion force between the tip and the sample (Fig. 5.4h). As the KPFM probe tip approaches the sample surface, a "jump-to-contact" occurs in the force curve due to the tip—sample force gradient exceeding the cantilever's elastic constant. Conversely, a "jump-off-contact" occurs during the probe tip's retraction because the cantilever's elastic constant surpasses the tip—sample adhesive forces (F_{adh}). Generally, the jump-off-contact distance exceeds the jump-to-contact distance, leading to non-coinciding engagement and withdrawal traces and the formation of a hysteresis loop. This research suggests that for electron transfer to take place, the distance between the tip and the sample must be less than the equilibrium interatomic distance, where the long-range attractive force is balanced by the short-range repulsive force.

The concepts of repulsion and attraction between atomic entities can be readily grasped through the potential of interatomic interactions [13]. Figure 5.5a illustrates the typical depiction of these interactions. When two atoms engage in bonding, which involves some degree of electron cloud or wave function overlap, an equilibrium distance, denoted as a, is established, known as the bond length or interatomic separation. If the interatomic separation, denoted by x, is less than a, the atoms repel each other due to the enhanced electron cloud overlap, as depicted in Fig. 5.5b. A critical aspect of CE is the necessity of an external force to bring the two surfaces into contact. This force is crucial for generating localized high pressure at specific points of contact, even at the atomic and nanoscale, where the interatomic separation x is compressed to be less than a, leading to localized repulsion. Conversely, if the interatomic separation x exceeds a, the atoms are drawn towards each other due to diminished electron cloud overlap, nearly severing their bond, as shown in Fig. 5.5c.

The data presented in Fig. 5.4 suggest that for the AFM tip and the silica surface to undergo CE, their proximity must be less than the bond length. In the repulsion zone, the heightened superposition of electron clouds locally leads to the transfer of electrons between the two materials. Consequently, an atomic-scale charge transfer mechanism can be hypothesized for CE. To address this, Wang's team proposed an electron cloud/potential model based on fundamental electron cloud interactions to explain various CE phenomena in general materials [7]. As shown in Fig. 5.6, electron clouds are formed by electrons localized within specific atoms or molecules, occupying specific atomic or molecular orbitals. An atom can be represented by a potential well where the outer shell electrons are loosely bound, forming an electron cloud. In Fig. 5.6a, d is the distance between electron clouds, (E_A) and (E_B) are the occupied energy levels of electrons in materials A and B, and (E_1) and (E_2) are the potential energies required for electrons to escape from the surfaces of materials A and B, respectively. (E_A) and (E_B) are smaller than (E_1) and (E_2). Before the materials contact, electrons cannot transfer due to the local trapping effect of the potential wells. When material A contacts material B, the electron clouds overlap

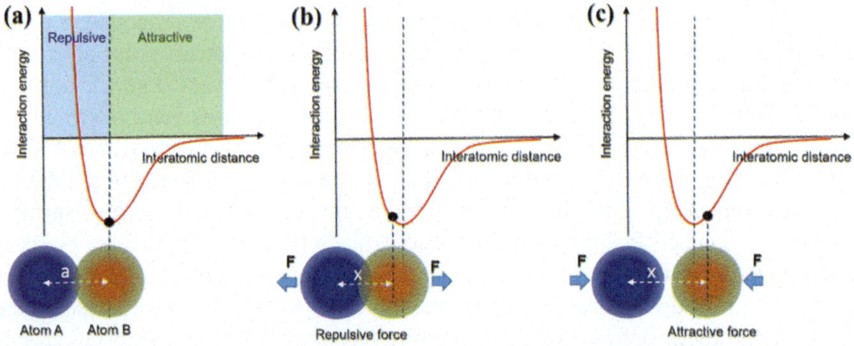

Fig. 5.5 Interatomic interaction potential between two atoms and the force between the two when they are **a** at equilibrium position, **b** with strong electron cloud/wave-function overlap in the repulsive region, and **c** with little electron cloud/wave-function overlap in the attractive region. Reproduced with permission from Elsevier [13]

due to the "screening" effect introduced by physical contact, transforming the initial single potential well into an asymmetric double-well potential, allowing electrons to hop from material A to material B (Fig. 5.6b). This asymmetric double-well potential is somewhat similar to the potential curve for the initial formation of a hydrogen bond between OH and H [14] or the proton motion in the NHN^+ system [15]. After separating materials A and B, most transferred electrons remain in material B due to the energy barrier (E_2) if the temperature is not too high (Fig. 5.6c). Thus, CE results in positively charged material A and negatively charged material B. Figure 5.6d illustrates the charge release process at an elevated temperature (T). As temperature increases, electron energy fluctuations grow with (kT), making it easier for electrons to escape the potential well and either return to their original material or emit into the air. This model clarifies why electrons generated from CE can be maintained due to the presence of potential barriers in all materials. Generally, electron transfer dominates the CE process, with ion or material transfer occurring as minor processes.

The electron cloud-potential well model explains the CE mechanism for any two solid materials at high temperatures, such as polymers, cellular, amorphous, or materials without a well-defined crystal structure. For these systems, neither band structure nor surface states can represent the electronic structure. Xu et al. suggested starting from the atomic or molecular orbits model, which applies to general materials [11]. Figure 5.7a shows that before atomic contact, the electron clouds of the two materials remain separate. The potential well binds the electrons, preventing them from escaping, typical for non-conducting materials. When atoms from the two materials contact, their electron clouds overlap, forming ionic or covalent bonds. The initial single potential wells become an asymmetric double-well potential, allowing electrons to transfer from one atom to another, i.e., triboelectrification, as shown in process (1) in Fig. 5.7b. Mechanical contact shortens the distance between atoms, causing electron cloud overlap at the contact area. This explains why more charges

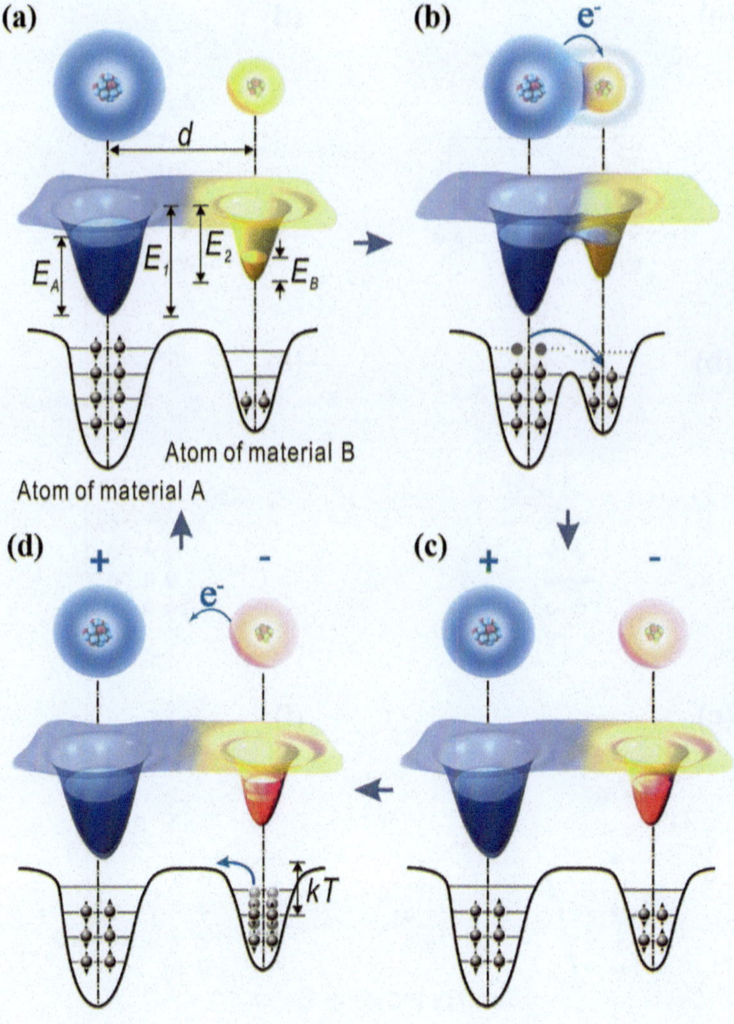

Fig. 5.6 An electron cloud-potential well model proposed for explaining CE and charge release. The status of the electron cloud and potential energy of the two materials before contact **a**, in contact **b** and after contact **c**. **d** Charge release at an elevated temperature T. d, distance between electron clouds; E_A and E_B, occupied energy levels of electrons; E_1 and E_2, potential energies for electrons to escape; k, Boltzmann constant; T, temperature. Reproduced with permission from Wiley [7]

transfer if one material scrubs the other harder. After separation (Fig. 5.7c), transferred electrons remain, but thermionic emission of triboelectric charges occurs, partially releasing electrons from material B (process (2) in Fig. 5.7c). Simultaneously, material A may capture electrons from the environment (process (3) in Fig. 5.7c), reducing its negative charge.

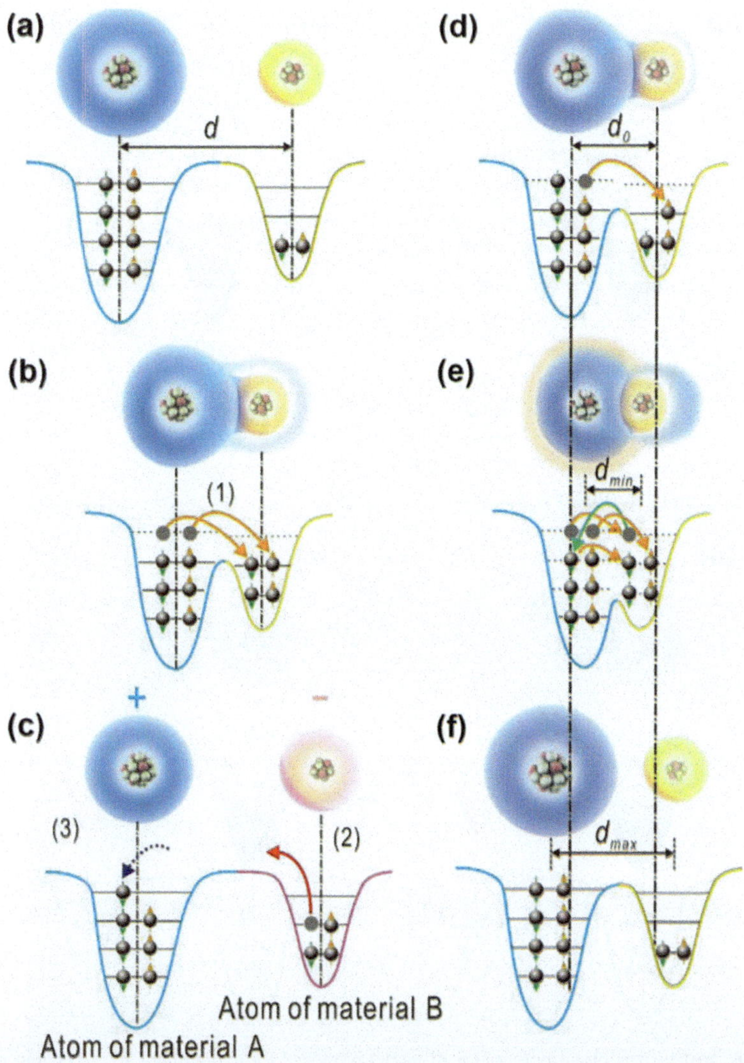

Fig. 5.7 CE mechanism explained by electron cloud-potential well model at high temperatures. **a–c** Charge transfer before contact, in contact, and after contact between two atoms of any two solid materials that cannot be simply described by a band structure model. d is the distance between electron clouds of the two atoms belonging to materials A and B, respectively. **d–f** Effect of atomic thermal vibration on charge transfer at the distance of d_0, d_{min} and d_{max} between the two atoms when they are in contact. Reproduced with permission from Wiley [11]

In practice, thermionically emitted electrons in process (2) can be significantly accelerated by the discharging field of triboelectrification in a vacuum, reaching energies of a few keV. This can generate X-ray emissions by exciting a metal target [15–18]. Alternatively, visible light, microwave, and even THz waves can be emitted during electron transition from material A to material B due to energy release. Light emission is also possible from thermionically emitted electrons accelerated by the discharging triboelectric field. Based on observed surface tribo-charge density in air (~250 $\mu C/m^2$) [19], the charge coverage on a surface is roughly one electron per 10,000 atoms. For a spherical particle 20 nm in diameter, the surface has about three triboelectric-induced electrons.

In the models discussed above, temperature not only influences electron distribution and thermionic emission but also affects the positions of atoms within the material. For crystalline materials, atomic vibrations are well described by phonon models with established dispersion relationships. One such model is the Einstein model, where all atoms vibrate randomly. This model helps illustrate the effect of atomic thermal vibrations on CE when two different atoms come into contact, using the electron cloud-potential well model. Figure 5.7d depicts the scenario where two atoms are in contact. This is a non-equilibrium dynamic process due to atomic thermal vibrations around their equilibrium positions at high temperatures. If the distance between the two atomic nuclei is (d), then:

$$d = d_0 + \Delta(T, t) \tag{5.2}$$

where (d_0) is the equilibrium distance between the atoms, and $\Delta(T, t)$ is the vector difference representing the deviation of the atoms from their equilibrium positions, dependent on temperature (T) and time (t). When the atoms momentarily separate by (d_0), their electron clouds overlap, allowing high-energy electrons from material A to transfer to material B, as electron transition speeds are much faster than atomic vibrations. When the distance between the atoms decreases to the minimum d_{min} (Fig. 5.7e), the electron clouds overlap maximally, and the potential well connection height decreases further. In this case, high-energy electrons from material A can transfer to material B, and vice versa. When the atoms separate to the maximum distance d_{max} (Fig. 5.7f), the electron clouds overlap minimally, preventing electron transfer. Thus, atomic thermal vibrations significantly influence CE, especially as temperature increases.

Additionally, based on new data from high-temperature sliding mode triboelectric nanogenerators (TENGs), Wang et al. proposed a modified electron cloud-potential well model for CE in sliding operational modes at elevated temperatures [20]. As shown in Fig. 5.6, this model applies to general triboelectric pairs in sliding contact, including metal-dielectric and dielectric-dielectric pairs. Since electrons in general dielectrics do not have discrete energy levels like those in metals and semiconductors [21, 22], the electron cloud and potential well model are presented as universal characteristics of all materials. Despite factors like surface impurities and the chemical/atomic structure of materials complicating dielectric CE studies, this model remains applicable, though these factors will also impact the results, especially at high temperatures. This model explains the prolonged charge retention and improved charge generation efficiency of sliding mode TENGs at high temperatures.

Initially, the model shows two materials completely separated, with their electron clouds filled with electrons localized to specific atomic or molecular orbitals (Fig. 5.8a). In this state, material A is negatively charged from initially induced charges, and no electron transfer occurs since it has not contacted material B. The charge difference between the materials is less apparent without precharging unless they are far apart on the triboelectric series. At high temperatures, the negatively precharged material A exhibits thermionic emission of surface charges, indicated by red arrows. Material B does not show significant thermionic emission as its electrons are in equilibrium, requiring much higher temperatures to induce emission. When the materials slide against each other (Fig. 5.8b), electrons transfer from material B to material A due to CE, creating equal and opposite charges on both surfaces. In this state, electron thermionic emission from SiO_2's surface is theoretically absent due to close contact. The electron clouds overlap, allowing charge transfer, represented by orange arrows.

As the materials begin to separate while sliding (Fig. 5.8c), both thermionic emission from the exposed areas of negative material A and charge transfer from contact areas affect the total measured charges in sliding mode TENG. Emission from material A is shown by red arrows, charge transfer by orange arrows, and electron flow through the circuit to balance positive material B by blue arrows. When the materials are fully separated again (Fig. 5.8d), residual charges on negative material A dissipate rapidly due to thermionic emission and lack of shielding by material B. As the displaced area exceeds the contact area, thermionic emission dominates over charge generation in sliding mode TENG. Emitted electrons are shown by red arrows, and electrons traveling through the circuit due to potential imbalance are shown by blue arrows. This model can describe the behavior of nearly all metal-dielectric and dielectric-dielectric pairs in sliding mode at high temperatures, based on electron thermionic emission rates, CE charge transfer rates, and the changing contact area between materials. When the contact area is dominant, CE is the major contributor. When the displaced area is larger, thermionic emission is the major contributor.

Hence, the process of electron transfer stands as the primary mechanism for CE in insulating materials. This transfer of electrons takes place exclusively when the distance between the atoms of the two substances is less than the standard bonding length (approximately 0.2 nm), which is within the range of repulsive forces. A pronounced overlap of electron clouds (or wave functions in physics terms, or chemical bonding in chemistry) between the two atoms/molecules in this repulsive zone results in the transfer of electrons, facilitated by the lowered potential barrier between atoms. The function of contact or frictional force is to elicit a robust overlap of electron clouds (or wave functions/bonding). The electron cloud model can be broadened to encompass not only solid–solid interactions but also liquid–solid, liquid–gas, and even liquid–liquid scenarios. Overall, the electron transfer triggered by the overlapping electron clouds under mechanical stress is posited as the principal mechanism for initiating CE across various states of matter, including solids, liquids, and gases. This concept not only offers a comprehensive initial comprehension of the physics

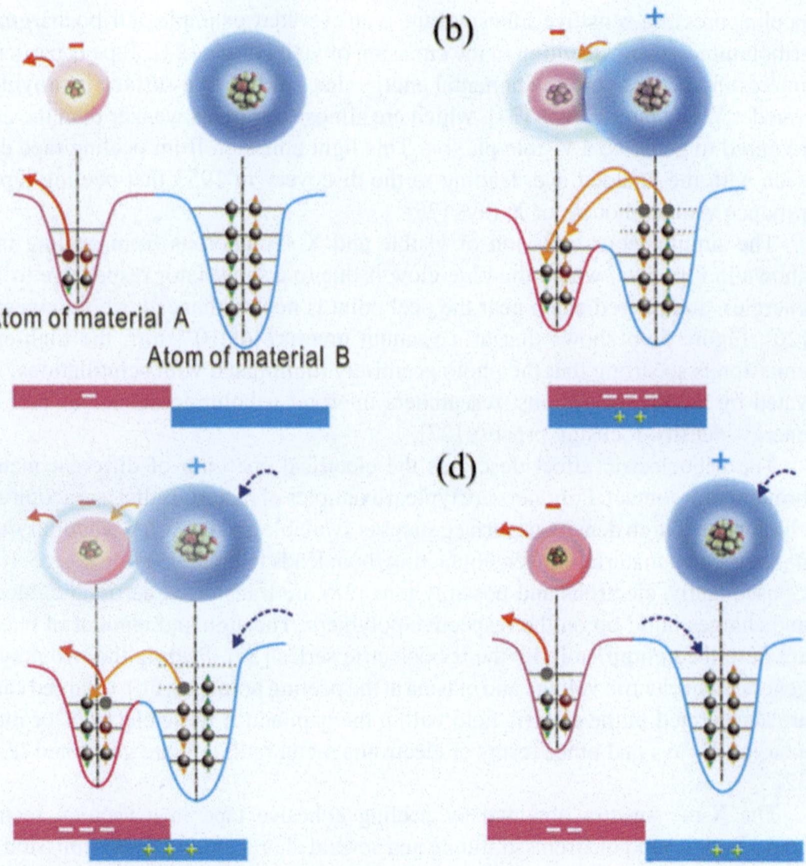

Fig. 5.8 An electron cloud-potential well model proposed for explaining CE (this is called Wang transition) and charge transfer and release between two materials in the sliding-mode TENG. **a** Initial state. **b** Contact state. **c** Semiseparated state. **d** Separated state. Reproduced with permission from Wiley [20]

underlying CE but also illustrates that TENGs are an effective tool for investigating the essence of CE across any materials.

5.4 Interface Electron Transfer-Transition Induced Photon Emission

Certain solids emit light when subjected to mechanical forces, a phenomenon known as triboluminescence. For example, sucrose (rock sugar), tartaric acid, and other crystals can produce a flash of light when squeezed. Walton et al. reported that peeling pressure-sensitive adhesive tape is an everyday example of tribocharging and triboluminescence, resulting in the emission of visible light [23]. Tape is particularly interesting because the fundamental energy that holds it to a surface is provided by van der Waals interactions [24], which are almost 100 times weaker than the energy required to generate a visible photon. This light emission from peeling tape can be seen with the unaided eye, leading to the discovery in 1953 that peeling tape can produce visible photons or X-rays [25].

The simultaneous emission of visible and X-ray photons from peeling tape is shown in Fig. 5.9a, where the blue glow is due to a scintillator responsive to X-ray energies, and the red patch near the peel point is neon-enhanced triboluminescence [26]. Figure 5.9b shows that at a vacuum pressure of 10^{-3} torr, the high-energy emission is so strong that the photo is entirely illuminated with scintillations. Motivated by these observations, researchers interpret triboluminescence as part of an energy–density-focusing process [27].

The triboelectric effect describes the electrical charging of different materials brought into contact. Polymers are typical examples of materials that can acquire such charging. If a high density of surface states is available and the work functions Φ_A and Φ_B of the two materials differ, contacting them leads to band bending (Fig. 5.10a-b). Consequently, electrons and possibly ions [28] are transferred across the interface, and charges build up on the respective polymers. The sign and amount of charging are described empirically by the triboelectric series [29]. Peeling the two polymers generates an electric voltage and plasma at the peeling point [18, 30]. Charged carriers are accelerated in the electric field within the gap, and if these electrons or ions hit a target, X-rays and other forms of electromagnetic radiation are generated [23, 31, 32].

The X-ray spectra obtained by peeling adhesive tape in a vacuum feature a broad background of Bremsstrahlung and several characteristic X-ray emission lines (Fig. 5.10c). The most intense line is related to the titanium foil fixed as an electron target underneath the adhesive tape (Fig. 5.10d). Further lines are related to the components of the vacuum chamber (stainless steel, O-rings, etc.) that are also excited to produce X-ray fluorescence. This was verified by measurements without the Ti foil.

The two previous works done by peeling off ordinary tapes allowed observation of phenomena such as visible light or X-ray emission. Wang's team predicted that virtually any two objects that come into contact with each other and separate may produce interface electron transfer-transition induced photon emission, and that it is possible to observe this light emission through TENG [11].

In 2021, Li et al. observed photon emission spectra of atomic features during CE at solid–solid interfaces by contacting fluorinated ethylene-propylene (FEP) with acrylic acid or FEP with quartz [33]. They designed the experimental setup shown in Fig. 5.11. The core components consisted of a hollow cylinder sandwiched between a metal cover and a metal base. Inside this cylinder, four metal fans driven by a motor were placed. Materials for CE could be attached to either the metal fans or the cylinder. When the fans rotated, CE occurred at the interface. To minimize air discharge effects, the core components were placed in a vacuum chamber (31 cm in diameter and 50

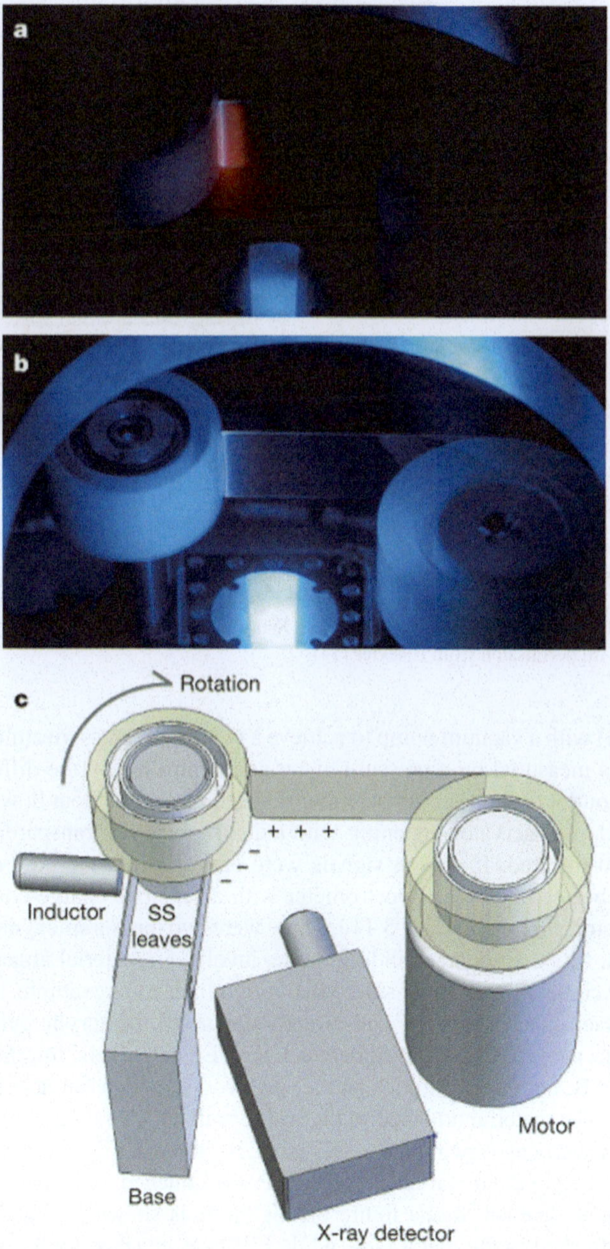

Fig. 5.9 Apparatus for studying high-energy emission from peeling tape. **a** Photograph of the simultaneous emission of triboluminescence (red line) and scintillations of a phosphor screen sensitive to electron impacts with energies in excess of 500 eV (under neon at a pressure of 150 mtorr). **b** Photograph of the apparatus (under a pressure of 1023 torr) illuminated entirely by scintillations. **c** Diagram of the apparatus used to measure peeling force; SS, spring steel. Reproduced with permission from Nature Publishing Group [27]

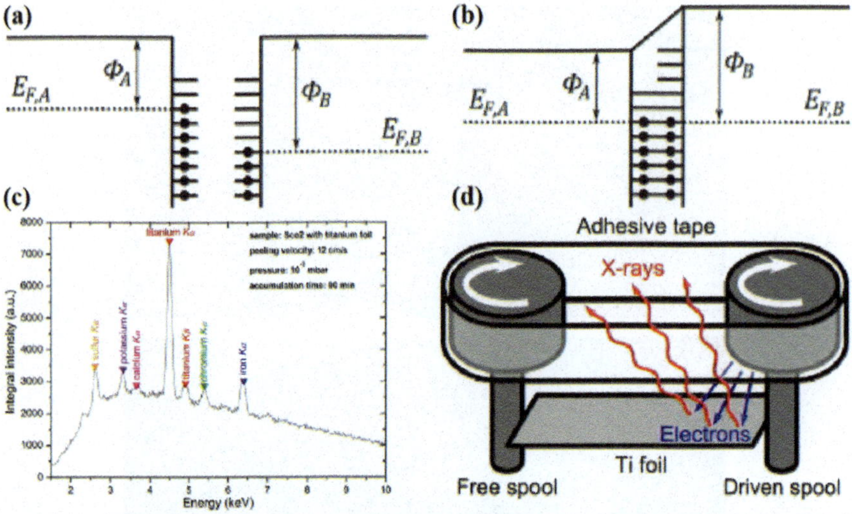

Fig. 5.10 Electronic band scheme at the interface of two polymers A and B before **a** and after contact **b**. **c** Assignment of characteristic X-ray emission lines in the generated spectrum for sample Sco2 (Scotch, cellulose acetate) at a velocity of 12 cm/s and a pressure of 10^{-3} mbar accumulated for 90 min. **d** Schematic setup for triboelectric X-ray generation by peeling of adhesive tape. Reproduced with permission from Elsevier [17]

cm in height) with a vacuum pump to achieve a controlled pressure atmosphere. The pressure was measured by a pressure meter and controlled by the differential flow of inlet and outlet of the vacuum chamber through glass tube float flow meters. The chamber had an observatory opening with a quartz window, transparent from deep ultraviolet to infrared. If photon signals were emitted from the core components, they would be recorded by a spectrometer with a sensitive charge-coupled device (CCD) detector, as shown in Fig. 5.11a. There were two measurement modes in these experiments. Mode A, contact-sliding mode, involved a material attached to metal fans directly contacting the inner side wall of the cylinder. For example, in Fig. 5.11b, FEP was attached to metal fans and directly contacted the acrylic cylinder. When the metal fans rotated, CE occurred between the FEP and acrylic interface in sliding mode. Mode B, contact-separation mode, involved one material attached to metal fans and another material attached to the inside wall of the cylinder with a tiny gap (<1 mm). The vertical-view image of this mode is shown in Fig. 5.11c, where copper foil was attached to the metal fans, while FEP was attached to the inside wall of the quartz cylinder. The schematic figure of this mode is shown in Fig. 5.11e, where nylon was attached to the metal fans, while FEP was attached to the inside wall of the quartz cylinder. The working principle of Mode B is demonstrated in Fig. 5.11d. When a metal fan rotated, it alternately contacted and released the surface of FEP. When in contact, the FEP was compressed to contact the inside wall of the quartz cylinder, causing electron transition at the interface due to CE. When released, the

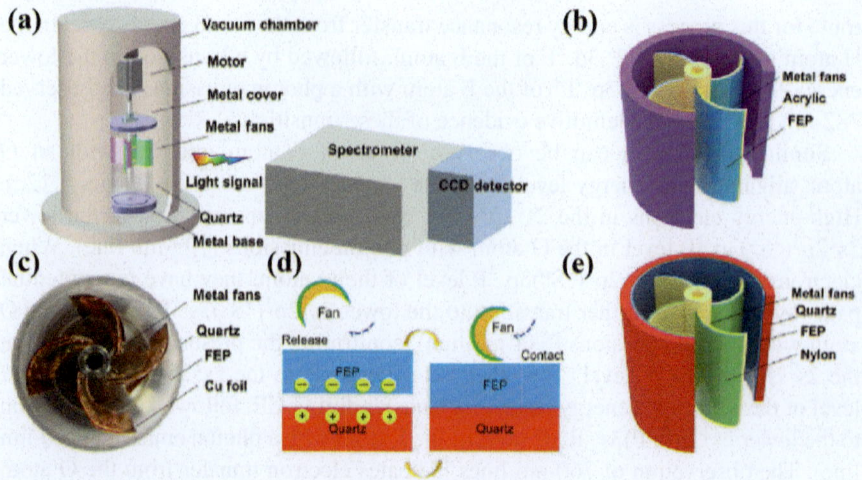

Fig. 5.11 The schematic diagram of experiments. **a** CE parts in vacuum chamber and spectrometer. **b** Measurement mode A: FEP on fans and directly contact with quartz or acrylic. **c** Optical photograph of the CE parts. **d** The working principle of the measurement mode B. **e** Measurement mode B: FEP attached to quartz or acrylic with nylon, etc. on fans. Reproduced with permission from American Association for the Advancement of Science [33]

FEP separated from the quartz cylinder, leaving negative charges on FEP and positive charges on quartz.

The atomic spectrum's characteristic photon emission lines, which are indicative of the underlying physical processes involving energy levels and electron transitions, are detailed in Fig. 5.12. Figure 5.12a provides an atomic-level interface diagram of FEP and quartz, where H, O, Si, F, and C atoms are represented by yellow, red, gray, blue, and brown spheres, respectively. The quartz is depicted in projection along the a-axis. Upon contact between FEP and quartz, electron transitions are anticipated. For instance, electron transfer might occur between F and O atoms, F and H atoms, and even H and O atoms, with accompanying photon emissions. When F and H atoms are in close proximity within the repulsive force region, indicating significant electron cloud overlap, electrons may transfer between these atoms via energy resonance transfer. The 656.279-nm line signifies the transition from the $n = 3$ level to the $n = 2$ level in the H atom, while the 782.3-nm line signifies the transition from $2s^22p^4(^3P)3d\ ^2F$ to $2s^22p^4(^3P)3p\ ^4P°$ in the F atom. With the vacuum level set at 0 eV and the first ionization energy defined as the energy difference between the vacuum level and the ground state, these can be plotted together in an energy diagram as shown in Fig. 5.12b. It is evident that the $n = 3$ energy level of the H atom is nearly identical to the $2s^22p^4(3P)3d\ ^2F$ of the F atom, allowing for facile electron transfer between these levels through energy resonance transfer. The observation of the 656-nm line of the H atom during CE suggests that electrons occupy the $n = 3$ energy level. Given that FEP becomes negatively charged after contact with quartz, electron transfer from quartz to FEP is expected. One plausible route for this process is energy resonance transfer from the $n = 3$ energy level in the H atom to the $2s^22p^4(^3P)3d\ ^2F$ of the F atom, followed by a transition to the lower energy level $2s^22p^4(^3P)3p\ ^4P°$ of the F atom with a photon emission. The observed 782-nm lines serve as definitive evidence of these transitions.

Similar phenomena can be observed when an F atom interacts with an O atom, aligning their energy levels with the vacuum level set at 0 eV (Fig. 5.12c). High-energy electrons in the $2s^22p^3(^2D°)3s\ ^3D°$ level can transition to the lower $2s^22p^3(^4S°)3p\ ^3P$ level in the O atom with photon emissions (799-nm line). When electrons are at the $2s^22p^3(^4S°)3p\ ^3P$ level of the O atom, they have two potential pathways. One is to further transition to the lower $2s^22p^3(^4S°)3s\ ^3S°$ level in the O atom with photon emissions (844-nm line), confirming the presence of electrons at the $2s^22p^3(^4S°)3p\ ^3P$ level. The other is to transition to the $2s^22p^4(^3P)3p\ ^2D°\ 3/2$ level of the F atom via energy resonance transfer during CE, followed by a transition to the lower $2s^22p^4(^3P)3s\ ^2P\ 3/2$ level of the F atom with a photon emission (760-nm line). The observation of 760-nm lines indicates electron transfer from the O atom to the F atom.

During CE between FEP and quartz, electron transitions between H and O atoms on the quartz surface are also evident, as shown in Fig. 5.12d. Electrons at the higher energy level [$2s^22p^3(^2D°)3p^1D$ of the O atom] transition to the lower [$2s^22p^3(^2D°)3s\ ^1D°$ in the O atom] with photon emissions (715-nm line). When electrons are at the $2s^22p^3(^2D°)3s\ ^1D°$ level of the O atom, they can transfer to another energy level through energy resonance transfer, with two options. One is to the $2s^22p^3(^2D°)3s\ ^3D°$ energy level, from which they can further transition to the lower $2s^22p^3(^4S°)3p\ ^3P$ in the O atom with photon emissions (799-nm line). The other is to transition to the $n = 4$ energy level of the H atom, followed by a transition to the lower $n = 2$ level of the H atom with a photon emission (486-nm line). The observed lines confirm these physical processes.

The potential physical process routes for electron transitions among different atoms during CE are summarized schematically in Fig. 5.12e–h, using atom A (yellow) and atom B (blue) as examples. When atom A and atom B are in close proximity within the repulsive force region, the energy potential barrier between them is reduced due to significant electron wave function overlap. Within the energy potential wells of each atom, there are energy levels available for electron occupation, ranging from the ground state to the vacuum level. The closer to the vacuum level, the higher the concentration of energy levels. Dotted lines represent this high concentration of energy levels near the vacuum level, with some denoted in the schematic figures. During CE, some electrons are temporarily transferred to excited states. There are two primary ways for electrons to transition to excited states (Fig. 5.12e): (i) The electron transitions from a molecular orbit to an excited state of an atom, potentially experiencing nonradiative or radiative decay; (ii) The electron is excited from a lower to a higher energy level within an atom. When an electron is in an excited state, it can transition to a lower energy level by emitting a photon. It can also transition to the excited state of another atom (from atom A to atom B) through energy resonance transfer if the excited state energy level in atom A is close to that in atom B (Fig. 5.12f). Subsequently, the electron, transferred from atom A, can

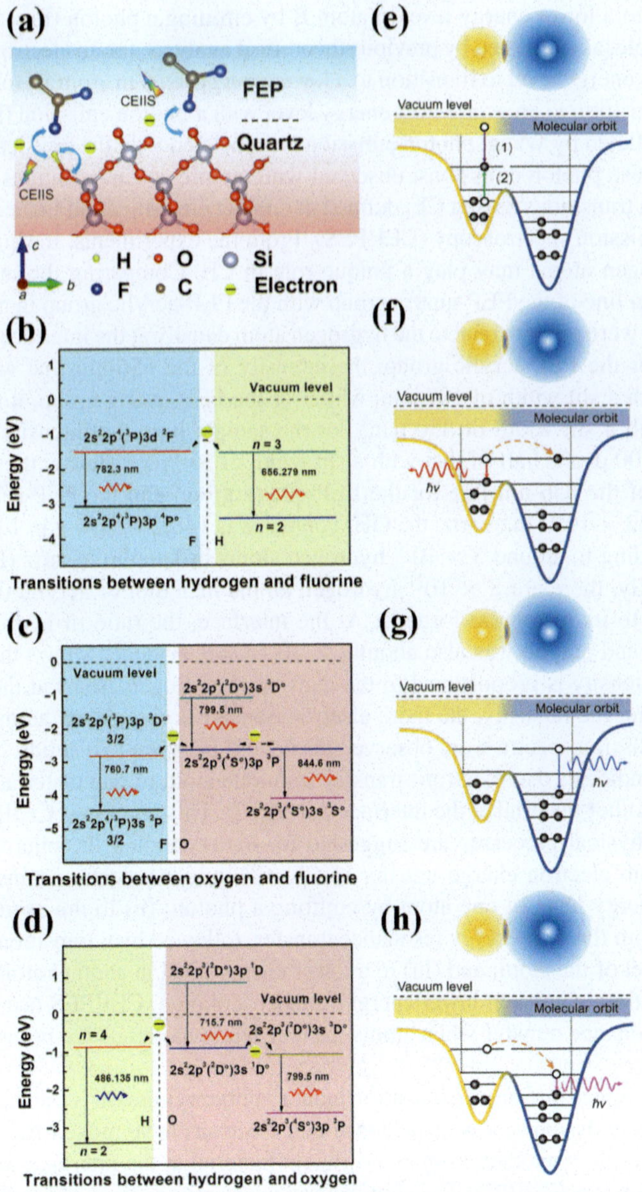

Fig. 5.12 Energy diagram for interface electron transition induced photo emission. **a** The schematic diagram of FEP and quartz interface at atomic level. **b** Energy diagram of electron transition between hydrogen and fluorine. **c** Energy diagram of electron transition between oxygen and fluorine. **d** Energy diagram of electron transition between hydrogen and oxygen. In addition, the schematic diagram of possible physical processes of electrons transitions and the associated photon emission, also known as Wang transition, when two atoms are close to each other (**e**–**h**). Reproduced with permission from American Association for the Advancement of Science [33]

transition to a lower energy level in atom B by emitting a photon (Fig. 5.12g). It is also possible, as suggested by previous theoretical analysis, for an electron in atom A at a higher energy level to transition to a lower energy level in atom B, followed by a further transition to an even lower energy level with a photon emission (Fig. 5.12h).

As predicted by Wang, photon emission is associated with the physical processes of CE. These photon emissions, observed with atomic spectral features, are linked to electron transitions during CE, defined as contact electrification induced interface photon emission spectroscopy (CEIIPES). From the experiments, it can be inferred that hydrogen atoms may play a unique role in CE. Comparing the intensities of the 656-nm line of the FEP-quartz group with the FEP-acrylic group in mode A, the intensities were proportional to the hydrogen atom density at the interface (Fig. 5.13a and c). For the FEP-acrylic group, the intensity of the 656-nm line was 451,355 counts with a slit width of 100 μm, while for the FEP-quartz group, it was 19,374 counts with a slit width of 350 μm. For the sample line, the intensity with a slit width of 100 μm is half of that with a slit width of 350 μm. Hence, the ratio of the intensity of the 656-nm line for the FEP-quartz group and the FEP-acrylic group was about 2×10^{-2}. In quartz, the OH- content was 148.2×10^{-6} g in 1 g of quartz, corresponding to around 5×10^{19} hydrogen atoms in 1 mol of quartz (Fig. 5.13b). Additionally, there are 8×10^{21} hydrogen atoms in 1 mol of acrylic (Fig. 5.13d) according to its molecular formula. At the interface, the ratio of hydrogen atoms for quartz and acrylic was also about 2×10^{-2}. This evidence shows that the 656-nm line intensity is proportional to the density of hydrogen atoms at the interface. The stronger the intensity, the more electron transfer occurs. Here, atomic-featured photon emission spectra were observed during CE between two solids. The photon emission indicates that electrons transfer from one atom in one material to another atom in another material at the interface during CE. This process is CEIIPES. Three possible physical processes are suggested for understanding the photon emission arising from electron charge transfer in CE: CE-induced electron transition (i) to a lower energy level in one atom by emitting a photon; (ii) to the excited state of another atom through energy resonance transfer, followed by a transition to a lower energy level of the atom; and (iii) to a lower energy level in another atom, followed by a transition to an even lower energy level by a photon. CEIIPES occurs through energy resonance transfer when atoms from different materials are brought close to each other.

Interface electron transfer-transition induced photon emission is a prevalent occurrence. This emission serves as evidence that electrons have moved from one atom in one material to another atom in a different material at the interface during CE, a process known as CEIIPES. The photon emission stemming from electron transfer in CE can be understood through three potential physical mechanisms: (i) CE prompting an electron to transition to a lower energy level within the same atom, emitting a photon in the process; (ii) CE facilitating an electron's transfer to the excited state of a different atom via energy resonance transfer, followed by a descent to a lower energy level of that atom; and (iii) CE causing an electron to move to a lower energy level in a different atom, which then transitions to an even lower energy state through photon emission. CEIIPES is triggered by energy resonance transfer when

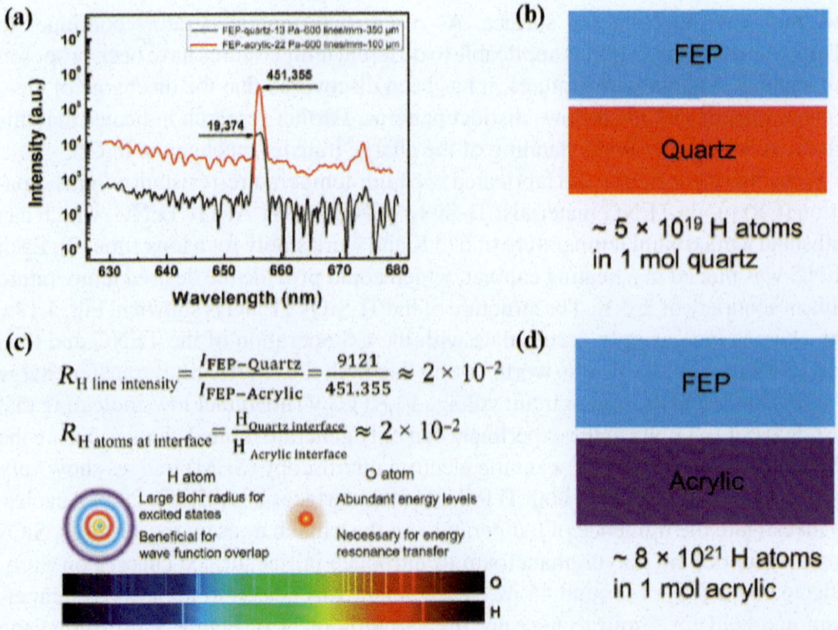

Fig. 5.13 Interface electron transition induced photo emission intensity is comparable to the H atoms at the interfaces for FEP-acrylic group and FEP-quartz group. **a** Take the H 656.2-nm line for example and the corresponding illustrations in **b** and **d**. **c** Color spectra of elements H and O in the range of 400 to 700 nm, showing different functions of them for electron transfer at CE. Reproduced with permission from American Association for the Advancement of Science [33]

atoms from distinct materials come into close proximity. This naturally opens up the possibility for spectroscopic analysis corresponding to CE at interfaces, potentially revolutionizing our comprehension of interactions between solids, liquids, and gases. While this research concentrated on photon emission in solid–solid CE scenarios, it could be broadened to encompass Auger electron excitation, X-ray emission, and electron emission in CE for a broader range of cases, which are yet to be investigated. This field could evolve into a comprehensive discipline known as contact electrification induced interface spectroscopy.

5.5 Self-discharging Processes

5.5.1 Thermal Ion Emission

Thermionic ion emission involves the energy distribution and thermal motion of particles in materials. In both scenarios mentioned above, the emitted particles acquire sufficient energy through thermal excitation to overcome the surface potential barrier and thus leave the material surface. As research on nanogenerators continues to advance and expand, TENGs applicable to different temperatures have been proposed and studied. At high temperatures, it has been discovered that the discharge of electrons or ions in TENGs follows distinct patterns. Further research indicates that this can provide a deeper understanding of the charge transfer mechanism in CE.

Wang's team designed and fabricated two high-temperature-resistant contact separation (CS) mode TENG materials, Ti-SiO_2 TENG and Ti-Al_2O_3 TENG, which can withstand a maximum temperature of 673 K and work stably for a long time [7]. Each TENG was placed in a heating cabinet, which could provide the desired temperature with an accuracy of $\pm$ 5 K. The structure of the Ti-SiO_2 TENG is shown in Fig. 5.14a. The charges increasingly accumulate with the CS operation of the TENG and then tend to balance at 293 K. It is worth noting that both the short-circuit transfer charge Q_{SC} (0.45 nC) and the open-circuit voltage V_{OC} (1.3 V) are rather low, indicating that the CS Ti-SiO_2 TENG in the experiment can only generate limited charges during the CS operation. Additionally, scanning electron microscopy (SEM) images show very slight structural changes for both Ti foil and SiO_2 surfaces after CE for 20,000 cycles. To investigate the influence of temperature on the tribocharges on the surfaces, SiO_2 was first rubbed by polyurethanefoam to introduce initial surface charges on SiO_2, reflected by a Q_{SC} of around 45 nC. The cabinet was heated to the desired temperature and held for 5 min to measure the variation of Q_{SC}. Figure 5.14b shows the change of Q_{SC} at temperatures of 353, 533, and 583 K, demonstrating that the charge density decreased more rapidly at higher temperatures. When the temperature rose to 583 K, the charges quickly disappeared, and the total Q_{SC} was less than 1 nC. Figure 5.14c shows the residual charges on the TENG after 5 min of measurement at different temperatures, with the inset showing the working model of the CS Ti-SiO_2 TENG. The residual charges decreased more rapidly with increasing temperature, notably starting to decrease more rapidly once the temperature exceeded 533 K.

It is important to note that the aforementioned experiments reflect results after holding the TENG at different temperatures for 5 min. To further systematically explore the effect of temperature on Q_{SC}, long-term measurements were conducted on the TENG under various temperatures. Figure 5.14d shows long-term charge decay at high temperatures, indicating that increasing temperature facilitates charge decay. It also reveals that the Q_{SC} response follows an exponential decay pattern at high temperatures. Additionally, at 503 K, Q_{SC}, V_{OC}, and short-circuit current I_{SC} share the same decay characteristics, featuring a slow-fast-slow trend in decay speed.

To determine whether charge decays are consistent with the electron thermionic emission model, Q_{SC} values are fitted according to the thermionic emission equation [34, 35]

$$J = \lambda A_0 T^2 e^{-\frac{W}{kT}}\left[e^{\frac{\Delta W}{kT}} - 1\right] \tag{5.3}$$

where J is the current density, λ is the material-specific correction factor, A_0 is Richardson constant of a free electron, T is the temperature, W is the height of the potential barrier, k is Boltzmann constant, and ΔW is the potential barrier height

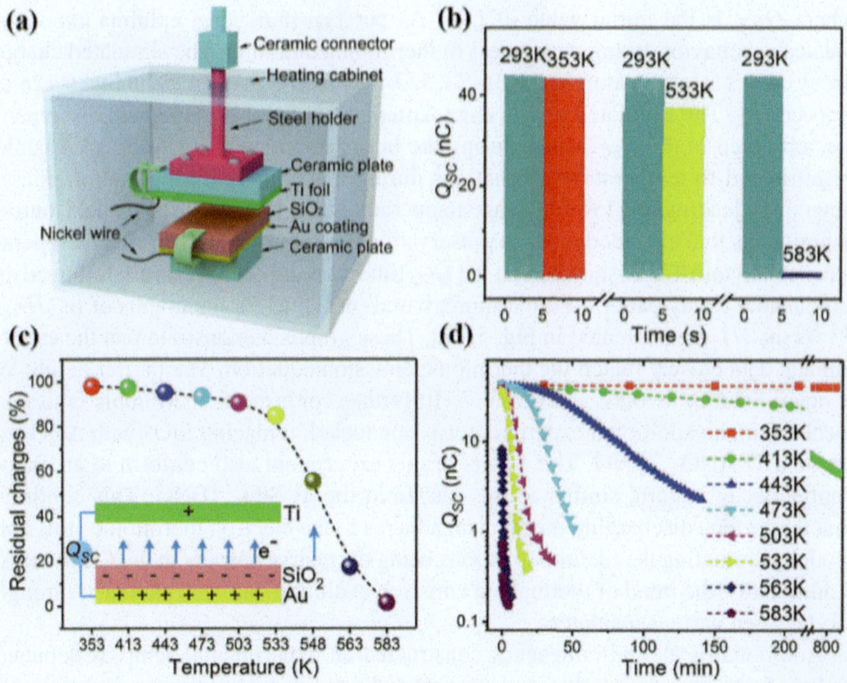

Fig. 5.14 Performance of the Ti-SiO_2 TENG at different temperatures. **a** Setup of the measurement platform. **b** The total transferred changes Q_{SC} at room temperature and various high temperatures for three groups of experiments, respectively. **c** The percentage of residual charges of the TENG at different temperatures. The residual charges are the Q_{SC} of the TENG after 5 min heat preservation at different temperatures. Inset is the diagram of the working model of the TENG. **d** Q_{SC} evolution with time under high temperatures. Reproduced with permission from Wiley [7]

variation due to the surface electric field E. Since $\propto \frac{\sigma}{\varepsilon 0} \propto Qsc$, we may assume that $\Delta W = \lambda_1 Q_{SC}/\lambda$ (λ_1 is a constant). When $\Delta W \ll$ kT $e^{\frac{\Delta W}{kT}} - 1 \approx \frac{\Delta W}{kT}$, *then*

$$J = \frac{\lambda_1 A_0}{k} T e^{-\frac{qV}{kT}} Q_{SC} \tag{5.4}$$

or equivalently

$$\ln\left(\frac{J}{A_0 T}\right) = -\frac{qv}{kT} + \ln\left(\frac{\lambda_1}{K} Q_{sc}\right) \tag{5.5}$$

where q is the electronic charge. By assuming $J = \frac{1}{A}\frac{dQ_{sc}}{dt} = SQ_{sc}$, where A is the surface area and S is used to replace $\frac{\lambda_1 A_0}{k} T e^{\frac{qV}{kT}}$, then

$$Q_{sc} = e^{-SAt} Q_{SC0} \tag{5.6}$$

where Q_{SC0} is the initial value of Q_{SC}. As per Eq. (5.6), Q_{SC} exhibits exponential decay behavior during the process of thermionic emission. The simulated charge decay curves at temperatures of 413, 473, 533, and 563 K are depicted in Fig. 5.12a-c, respectively. The simulated results align with the experimental data, with discrepancies observed in the Q_{SC} values during the initial minutes. This discrepancy could be attributed to temperature fluctuations during the brief heat preservation phase, potentially leading to a lower average temperature than the measured value. Control experiments that extended the heat preservation period from 5 to 25 min at temperatures of 443 and 473 K showed that the Q_{SC} time-varying curves closely followed an exponential decay pattern. Furthermore, based on Eq. (5.5), the graphs of ln ($J/A_0/T$) versus $1/T$ are presented in Fig. 5.15d. These graphs demonstrate that the experimental data closely match the thermionic emission equation. The barrier height W is determined to be 0.81 ± 0.06 eV 7. To further confirm the thermionic emission mechanism, an additional experiment was conducted, replacing SiO_2 with Al_2O_3 to create a Ti-Al_2O_3 TENG. The results of this experiment also conform to an exponential decay pattern, similar to the data from the Ti-SiO_2 TENG. This confirms that the surface discharging mechanism adheres to the electron thermionic emission model, eliminating the likelihood of ions being the charge carriers in the CE process. Additionally, the trend of thermal ion emission evolution has been validated through CE between various insulators.

Additionally, Zi and colleagues constructed an experimental setup as depicted in Fig. 5.16a, where a contact-separation (CS) mode triboelectric nanogenerator (TENG) was positioned within an oven and propelled by a linear actuator [36]. The short-circuit charge transfer (Q_{SC}) during the CS operation was quantified by a coulometer, representing the triboelectric surface charge. Photographs of the setup and the CS mode TENG are presented in Fig. 5.16b and c. Figure 5.16d illustrates two potential mechanisms for CE, which can be elucidated by the experiments, encompassing electron transfer and material/ion transfer processes. Initial studies focused on metal/polymer combinations, considering metals generally exhibit lower wear resistance compared to polymers, with the two possible mechanisms illustrated in Fig. 5.16d. Metals such as aluminum (Al), titanium (Ti), and stainless steel (SLS) were employed. The initial Q_{SC} at ambient temperature for the Al/Kapton pair was regulated to approximately 60 nC through the friction of nitrile butadiene rubber (NBR) against the Kapton film. Figure 5.16e summarizes the residual charge immediately after reaching the target temperature compared to the initial Q_{SC}, demonstrating a progressive charge reduction with increasing temperature. At 463 K, there was a negligible amount of charge left on the Kapton film, restricting the operational temperature to below 473 K. The charge residual ratio was introduced to evaluate the charge dissipation during the temperature increase, defined as the ratio of the residual charge at the target temperature to the initial Q_{SC} at room temperature. Subsequent experiments were designed to ascertain that thermionic emission during the separation phase was the sole cause of charge decay, excluding other factors such as charge dissipation through bulk and surface conductivity.

The exponential decay curves at 423 K, the evolution of Q_{SC} over time at various temperatures, and the plots of ln($J/A_0/T$) versus $1/T$ for the Kapton/Kapton pair are

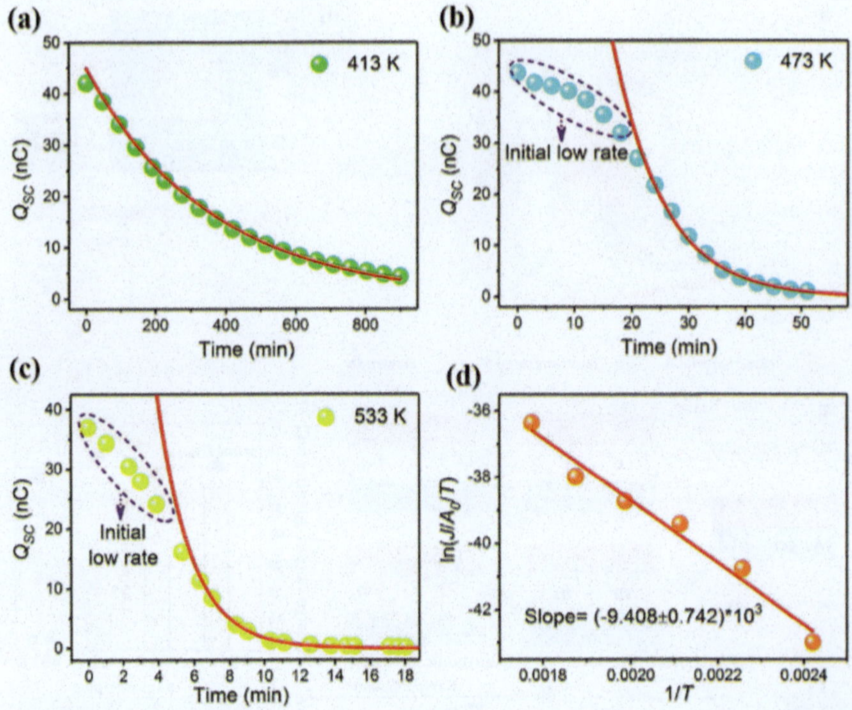

Fig. 5.15 The measured (dots) and simulated (line) data of the Ti-SiO_2 TENG. Q_{SC} as a function of the time at various temperatures: **a** 413 K, **b** 473 K, and **c** 533 K. **d** Plots of ln $(J/A_0/T)$ against $1/T$. Reproduced with permission from Wiley [7]

displayed in Fig. 5.13f–h. The slope in Fig. 5.16h, similar to that of metal/Kapton pairs, confirms Kapton as the sole material responsible for thermionic emission in metal/Kapton combinations. The work function (W) of 0.82745 ± 0.0173 eV derived from these tribo-pairs is attributed to the Kapton surface. As per the discussions in Fig. 5.16d, thermionic emission electrons from Kapton as material B were initially transferred from metals as material A, establishing the electron transfer as the dominant mechanism. This indicates that even with some NBR transfer, the conclusion remains unaffected. To further validate the role of NBR in preloading surface charge, an experiment with Ti/Kapton was conducted where the initial Q_{SC} was preloaded using an electron gun. The W extracted from this experiment was nearly identical to that with NBR, demonstrating the accuracy of using NBR to preload surface charge on polymers. This research establishes that the electrification in insulator-to-insulator contact is primarily due to electron transfer, and processes involving polymers may also involve a minor amount of material transfer. Furthermore, the work function in polymer/polymer pairs may be associated with multiple material surfaces and/or other factors such as chemical reactions, suggesting that both electron transfer and

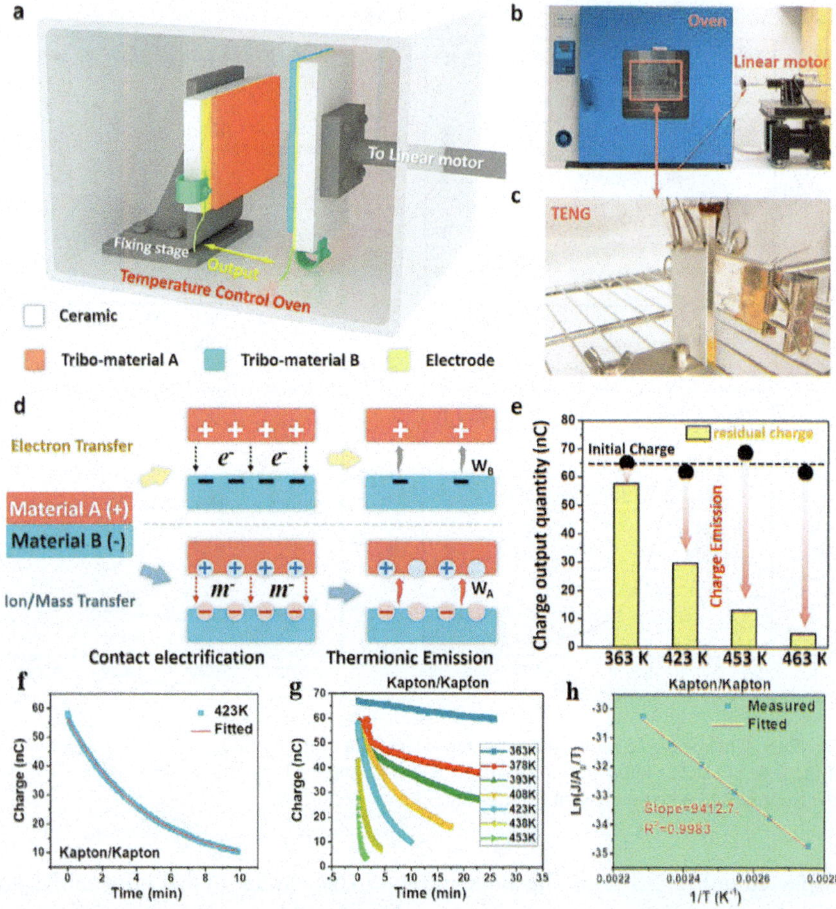

Fig. 5.16 Introduction of the experiment. **a** Setup of the experimental platform. **b** Photograph of the experiment system. **c** Photograph of the TENG device. **d** Illustrations of the assumptions on the two major CE mechanisms. **e** Column diagram of Al/Kapton for the comparison between initial charge at room temperature and residual charge at target temperature. **f–h** The measured (dots) and fitted (line) Q_{SC} as a function of the time at various temperatures, evolution of Q_{SC} with time at different temperatures, plots of $\ln(J/A_0/T)$ against $1/T$ with measured results in dots and linear fitted results in line for Kapton/Kapton. Reproduced with permission from Elsevier [36]

material transfer could coexist in polymer/polymer CE, with the specific mechanism varying significantly between different pairs.

To further demonstrate the thermionic emission of surface charges and corroborate the role of electrons in charge transfer, Wang engineered and fabricated a sliding-mode Ti-SiO_2 triboelectric TENG [20]. The Q_{SC} data from the pre-charged TENG experiments conducted at temperatures exceeding 443 K were compared with the previously established exponential decay model for contact-separation TENGs. While the data at 443 K (Fig. 5.17a) showed a close match to the model, the residual charges at 473 K (Fig. 5.17b) and 533 K (Fig. 5.17d) exhibited a faster-fast-slow decay pattern that deviated from the simulation, characterized by a slower initial drop and a less abrupt shift to the slower decay phase. Conversely, at 503 K (Fig. 5.17c) and 563 K (Fig. 5.17e), the initial decline was more rapid, and the transition into the slow decay phase occurred more swiftly compared to the simulated curve. These inconsistencies, also noted in prior research, might be resolved by increasing the number of tests averaged, potentially aligning the data points more closely with the exponential decay model. Additionally, the graph of $\ln(J/A_0/T)$ versus $1/T$ in Fig. 5.17f revealed a strong linear relationship, indicating a significant correlation between surface charge data and the thermionic emission equation. The good fit of the sliding-mode TENG data to the simulated thermionic emission curves further supports the notion that electron transfer is the predominant mechanism in CE. Comparing the sliding-mode data with prior contact-separation mode findings [7], it is plausible to attribute the extended timescale for near-complete emission of surface electrons to the sliding-mode TENG's capacity to retain and generate charge at elevated temperatures, owing to its protective nature against discharge. This operational mode, permitting prolonged contact between triboelectric surfaces, mitigates the effects of thermionic emission until a certain temperature threshold is reached. Beyond Q_{SC}, surface charge density is also a valid parameter for examining temperature effects on TENGs, and recent research utilizing tapping-mode AFM has determined that triboelectric charge decay adheres to the thermionic emission model at the nanoscale [37]. It would be intriguing to extend this research by employing contact scanning mode AFM to emulate a sliding-mode TENG.

Thermal ion emission provides an important method for a deeper understanding of charge transfer in TENG, solving the problem of charge properties that has plagued people for over two thousand years in CE. It points out that the CE between insulators is dominated by electron transfer and has significant value.

5.5.2 Photon Excitation

The electrons can be excited and emitted out of the insulator surface under the excitation of ultraviolet (UV) photons, which is called as photon excitation or photoelectron emission. Therefore, UV illumination is an effective approach for releasing the surface triboelectric charges. Lin and colleagues explored the CE between metals and various insulators, such as polymers, and studied the decay of friction-induced charges on insulator surfaces under UV light at the nanoscale using AFM and KPFM [40]. In their studies, the metal contact was a gold-coated silicon probe, while ceramic (SiO_2) and polymer films (poly(vinyl chloride) (PVC) and poly(methyl methacrylate) (PMMA)) on heavily doped silicon substrates served as the counter insulator samples. The charged region on the insulator was exposed to UV light, and the residual triboelectric charge density was measured at intervals to track the decay. To enhance the UV light intensity, a condensing lens was used to focus the light on the charge area, with the intensity being quantified using an optical power meter.

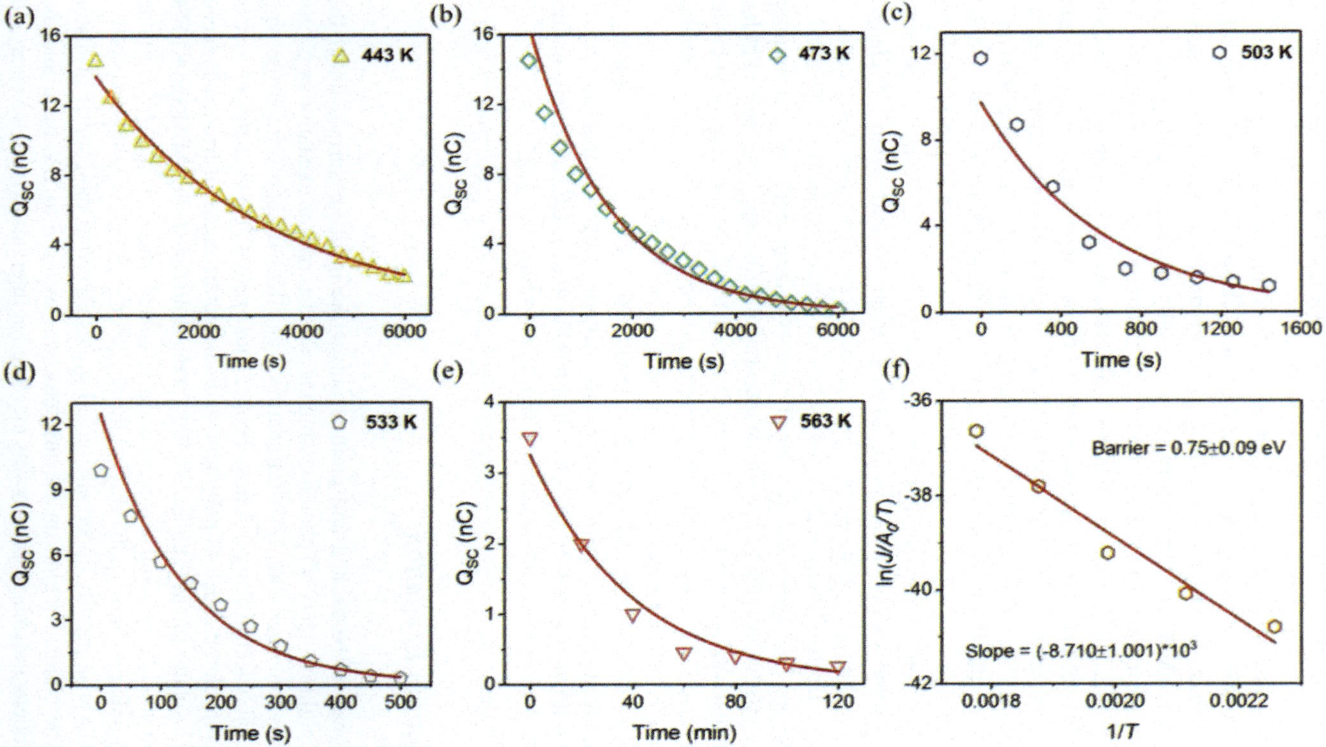

Fig. 5.17 The measured (data points) and simulated curve based on electron thermionic emission model (solid) Q_{SC} as a function of the time at various temperatures. **a** 443 K, **b** 473 K, **c** 503 K, **d** 533 K. **e** 563 K. **f** Plots of $\ln(J/A_0/T)$ against $1/T$. Reproduced with permission from Wiley [20]

Figure 5.18a illustrates the variation in triboelectric charge density on the SiO_2 surface under direct xenon light exposure at an intensity of 1200 W m^{-2} (with wavelengths ranging from 200 to 2500 nm). The triboelectric charge density on the SiO_2 surface dropped from $-160\,\mu C\,m^{-2}$ to below $-10\,\mu C\,m^{-2}$ within 45 min under xenon light. Subsequently, an optical filter was used to remove wavelengths below 400 nm from the xenon light, resulting in filtered light with wavelengths between 400 and 2500 nm. Figure 5.18b displays the triboelectric charge density on the SiO_2 surface under filtered xenon light, indicating that the filtered light did not significantly reduce the triboelectric charge, suggesting that UV light photons (wavelengths from 200 to 400 nm) were crucial for charge decay under xenon irradiation. To assess the thermal impact on charge decay, the temperature increase of the SiO_2 sample was monitored. Figure 5.18c indicates the sample's position during the experiments, while Fig. 5.15d and e depict the temperature profiles of the SiO_2 sample under xenon and filtered xenon light, respectively. The findings indicate that the sample's peak temperature after 10 min of irradiation was below 313 K, which is insufficient for thermionic emission as per recent research. Additionally, the temperature of the SiO_2 sample under monochromatic UV light irradiation was measured, revealing no temperature increase.

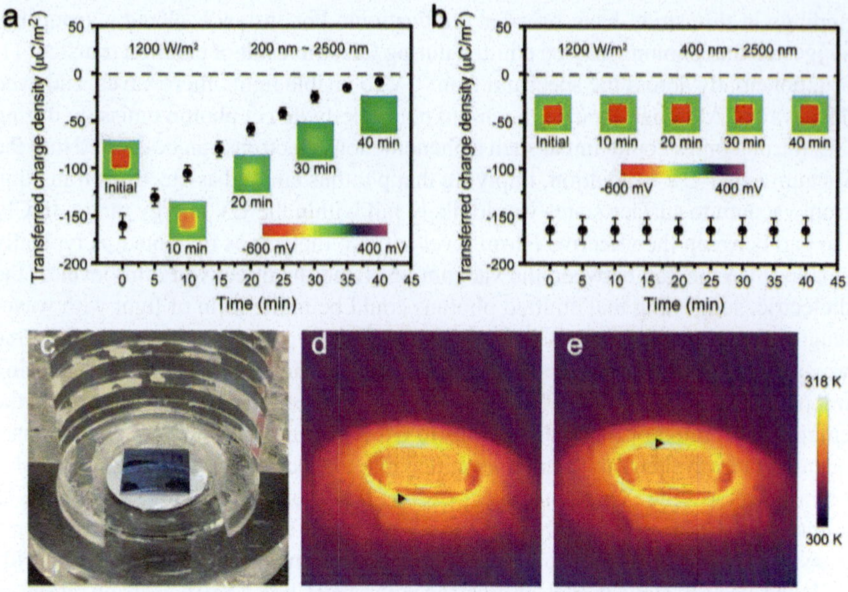

Fig. 5.18 The effect of light irradiation on the decay of triboelectric charges and the temperature rise of the samples. **a** and **b** The decay of triboelectric charges on the SiO_2 surface under the irradiation of the xenon light with wavelengths from 200 to 2500 nm **a**, and under the irradiation of the filtered xenon light with wavelengths from 400 to 2500 nm **b**. **c** The position of the SiO_2 sample in the light irradiation experiments. **d** and **e** The temperature distribution of the SiO_2 sample surface after 10 min irradiation of light with wavelengths from 200 to 2500 nm **d** and light with wavelengths from 400 to 2500 nm **e**. Reproduced with permission from Wiley [38]

The findings indicate that the decay of triboelectric charges observed in the experiments was due to photoelectron emission rather than thermionic emission, as ions cannot be ejected from the insulator surface by UV photons. Consequently, the charge carriers involved in the CE between SiO_2 and the Au-coated probe are electrons. In these experiments, light with wavelengths exceeding 400 nm did not contribute to the decay of triboelectric charges. Based on Einstein's equation for photovoltaics, as referenced in Eq. (5.7), photons with shorter wavelengths and higher frequencies possess greater energy. This suggests that the energy of the incident photons must surpass a specific threshold to release trapped electrons in the CE process. This aligns with the photoelectron emission theory, which posits that photons can only release electrons if their energy exceeds the electron's surface charge barrier energy

$$E = hv = \frac{hc}{\lambda} \tag{5.7}$$

where E denotes the energy of the photon, h denotes the Planck constant, v denotes the frequency of the light, c denotes the light speed, and λ denotes the wavelength of the light.

Photoelectron emission experiments strongly support the electron transfer mechanism in CE between metals and insulators, including polymers. Moreover, the findings in this study have broader implications. For instance, Wang's group has suggested that photons may be emitted during CE as a result of electron transfer [11, 39], potentially across the spectrum from UV to visible light, microwaves, and even THz waves. Although their research did not directly detect photon emission during CE, the outcomes could hint at such a phenomenon. Electrons can be excited into the vacuum under UV irradiation, implying that photons emitted as electrons transition from vacuum to surface states would likely fall within the UV energy range. In CE, the gap between the effective Fermi levels of two dielectrics in contact is typically smaller than the gap between the vacuum level and the effective Fermi level of the dielectric, suggesting that emitted photons could be in the form of light with wavelengths longer than UV. This is in line with Wang's prediction [11] regarding the wavelength of emitted photons. Importantly, the threshold wavelengths for releasing trapped electrons on SiO_2, PVC, and PMMA surfaces varied, indicating that the energy distribution of trapped electrons in CE is dependent on the material. Consequently, the wavelength of photons emitted due to electron transfer in CE may also be material-specific. Therefore, spectroscopy based on photon emission in CE, as mentioned in a previous study [11], could serve not only to investigate electronic transitions during CE but also as a novel method for material characterization [38].

In the electron cloud model, an electron with a weak bond can move from one atom to another if the interatomic distance is shorter than the typical bonding length, due to the reduced potential barrier. This model was later supported by quantum–mechanical calculations [39] and experimentally shown to occur only when two materials interact in the repulsive region [12, 40], aligning with the electron cloud model. The electron cloud model, also known as the Wang transition, is a significant theory for electron transfer in CE. However, the current model only describes intrinsic electron transfer between materials without considering external factors like temperature and electric fields, which are often used to control CE. For example, a temperature difference [37] and/or an electric field [41, 42] at the interface of two materials can significantly affect CE, both by altering the energy of electrons trapped in surface states and, in the case of electric fields, potentially leading to electron tunneling from one surface to another [43]. This suggests that strong overlap of electron clouds between atoms may not be necessary for electron transfer if an electric field is present at the interface. However, experiments on CE at varying atomic interaction distances have not yet been conducted.

Lin et al. extended their investigation to CE between Pt-coated silicon tips and ceramic films (Si_3N_4 and AlN), with 100 nm thick samples on heavily doped silicon wafers, using AFM and KPFM [44]. Figure 5.19a illustrates a typical interatomic interaction energy curve as a tip approaches an insulator sample surface. In tapping mode (as shown in Fig. 5.19b), the tip vibrates at a certain amplitude (free amplitude, A_f) before contact with the sample surface. When a force acts on the vibrating tip, its amplitude decreases, and the vibration phase shifts, reflecting an interference with its harmonic oscillation, with the phase shift direction depending on the force direction. When the amplitude of the tip is set to a value (set amplitude, A_0) less than the free amplitude, the tip interacts with the sample surface at a force that reduces the amplitude to the set value. Thus, the tip-sample interaction force is set-amplitude dependent when the free amplitude is constant. The phase shift of the vibrating tip also varies with the set amplitude, with the interaction regions of the tip-sample force reflected in the phase shift direction.

Figure 5.19c–e show the phase shift of the tip tapping on the Si_3N_4 surface at different set amplitudes, with free amplitudes of 50, 70, and 100 nm, respectively. It is observed that the phase shift was negative, indicating the tip did not enter the repulsive region when the free amplitude was 50 nm. When the free amplitude increased to 70 and 100 nm, the phase shift of the tip turned positive if the set amplitude was below a threshold value. Here, the scanning amplitude was set to 60% of the free amplitude in tapping mode CE. Under these parameters, the tip operated in the attractive region when the free amplitude was 50 nm (Fig. 5.19c) and in the repulsive region when the free amplitude was 70 and 100 nm (Fig. 5.19d, e). The inset in Figs. 5.16c-e shows the change in surface potentials after tapping CE with different free amplitudes. The results indicate that the Si_3N_4 surface acquired positive charges when the tip tapped in the repulsive region (Fig. 5.19d, e), while no charge transfer occurred between the tip and Si_3N_4 when the tip tapped in the attractive region (Fig. 5.19c). A similar phenomenon was observed in tapping CE between the tip and AlN. Furthermore, when the scanning amplitude of the tip was altered with a free amplitude of 100 nm, the sample surface potential changed when the tip was in the repulsive region; when the set amplitude increased to the threshold value and the tip was in the attractive region, the amount of charge transfer between the tip and the sample decreased significantly (close to zero). These findings suggest that CE between the tip and sample only occurs if the surfaces interact in the repulsive-force region, implying that electron cloud overlap is necessary for CE in the absence of external stimuli at the interface, consistent with the electron cloud model.

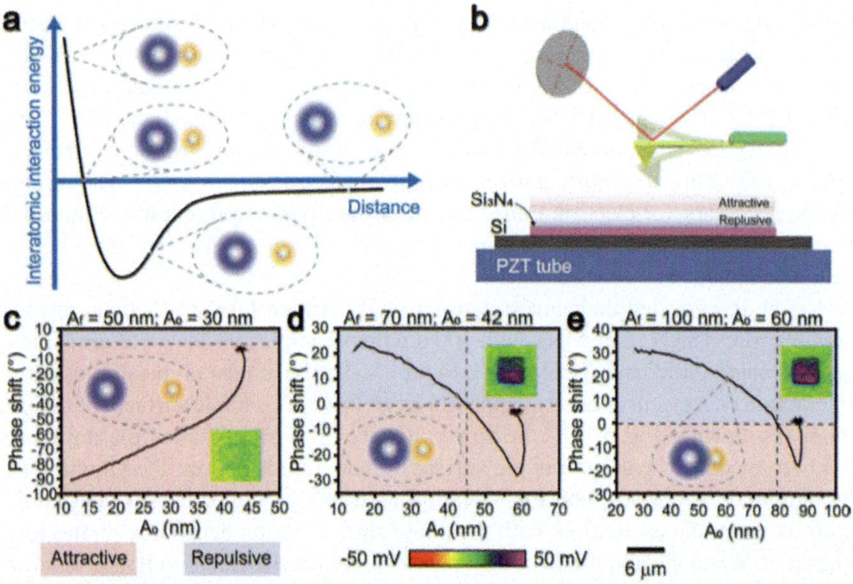

Fig. 5.19 Schematic illustration of the experiments and the interatomic interaction energy curves. **a** lustration of the tip-sample interatomic interaction energy curve and the interaction of the atoms on the tip and sample. **b** Tapping mode AFM and the illustration of attractive region and repulsive region on the sample surface. The phase shift at different set tapping amplitudes and the charge transfer between tip and Si_3N_4 sample when the **c** free amplitude (A_f) is 50 nm and set tapping amplitude (A_0)is 30 nm; **d** free amplitude is 70 nm and set tapping amplitude is 42 nm; **e** free amplitude is 100 nm and set tapping amplitude is 60 nm (tip temperature = sample temperature = 313 K). Reproduced with permission from Wiley [44]

In CE, including thermionic emission and photon excitation, the former confirms that the CE between insulators is dominated by electron transfer, while the latter implies that there must be sufficient external stimulation at the interface to allow the two to interact in the repulsive region, indicating that two electron clouds overlap and further verifying the electron cloud model.

5.6 Other Effects at Insulator-Insulator Interface

5.6.1 Curvature Effect

It is well-known that CE typically occurs between two different materials, which can be explained by several models for different material systems. However, contact between two pieces of the same chemically identical material can also result in electrostatic charges, although the charge density is relatively low, making it difficult to understand from a physics perspective.

Xu et al. designed and prepared a TENG using two pieces of a chemically identical material but with surfaces of different macroscale shapes or curvatures. They investigated the direction of charge transfer in real-time via the TENG operated in contact-separation mode [45]. Figure 5.20 shows the performances of PTFE TENGs with surfaces of various shapes. Figure 5.20a and c display the typical V_{OC} and Q_{SC} of the TENG with two flat surfaces. Eight sets of samples were assessed five times each, as shown in Fig. 5.20b and d. No apparent tendency in the direction of charge transfer was observed with two flat films. It is important to note that since it is necessary to remove the samples, rinse them with anhydrous ethanol to discharge after each test, and continue a new test after reinstallation, the surface contact position of the samples likely differs each time. This may lead to a reversal in the direction of charge transfer in CE at different times using the same set of samples. Therefore, they confirmed the consistency in surface contact position each time and reassessed sample 5 in Fig. 5.20b and d ten times. It was shown that the direction of charge transfer remained the same even though the transferred amount of charges differed each time. This is due to the so-called "memory effect." Therefore, it was believed that there is a certain regularity in the direction of charge transfer for CE between identical materials, which may be related to the surface status.

Figure 5.20e–h show the V_{OC} and Q_{SC} of the TENG with a convex surface of positive curvature and a flat surface. The convex PTFE film was negatively charged in most cases, indicating that it gains electrons more readily than the flat film. Figure 5.20i–l show the V_{OC} and Q_{SC} of the TENG with a concave surface of negative curvature and a flat surface. The concave PTFE films were positively charged in most cases, meaning they are less likely to gain electrons than flat films. Figure 5.20m–p show the V_{OC} and Q_{SC} of the TENG with a convex surface and a concave surface. To improve the quality of contact-separation operation and avoid unintentional contact in other modes, such as sliding, the curvature of the convex PTFE film was set to 0.20 cm^{-1} and that of the concave PTFE film to −0.07 cm^{-1}. Additionally, the V_{OC} and Q_{SC} of the TENG with PTFE films of opposite but the same magnitude of curvatures (0.20 cm^{-1} and −0.20 cm^{-1}) were also measured. These convex-concave pairings reveal that the convex PTFE film is more prone to acquiring electrons than the concave film. Figure 5.20e–p indicate that a surface with larger curvature gains negative charges more easily, consistent with the V_{OC} and Q_{SC} results of the contact-separation mode FEP TENG, Kapton TENG, polyester TENG, and nylon TENG with convex-flat, concave-flat, and convex-concave surfaces.

Based on the experimental results, a surface states model was proposed to explain the mechanism of CE between identical materials, as shown in Fig. 5.21. When two ideal flat surfaces composed of chemically identical materials A and B come into contact, no electron transfer occurs because the neutral levels of surface states (E_n) are equivalent and horizontally aligned (Fig. 5.21a). However, CE still occurs when two flat surfaces contact each other in the experiment. This is likely because flat surfaces are rarely perfectly smooth at the nano-microscale, and curvature differences influence the direction of charge transfer through microstructures. Figure 5.18b–d illustrate charge transfer before contact, in contact, and after contact between two identical materials A and B with different surface curvatures, where A has a convex

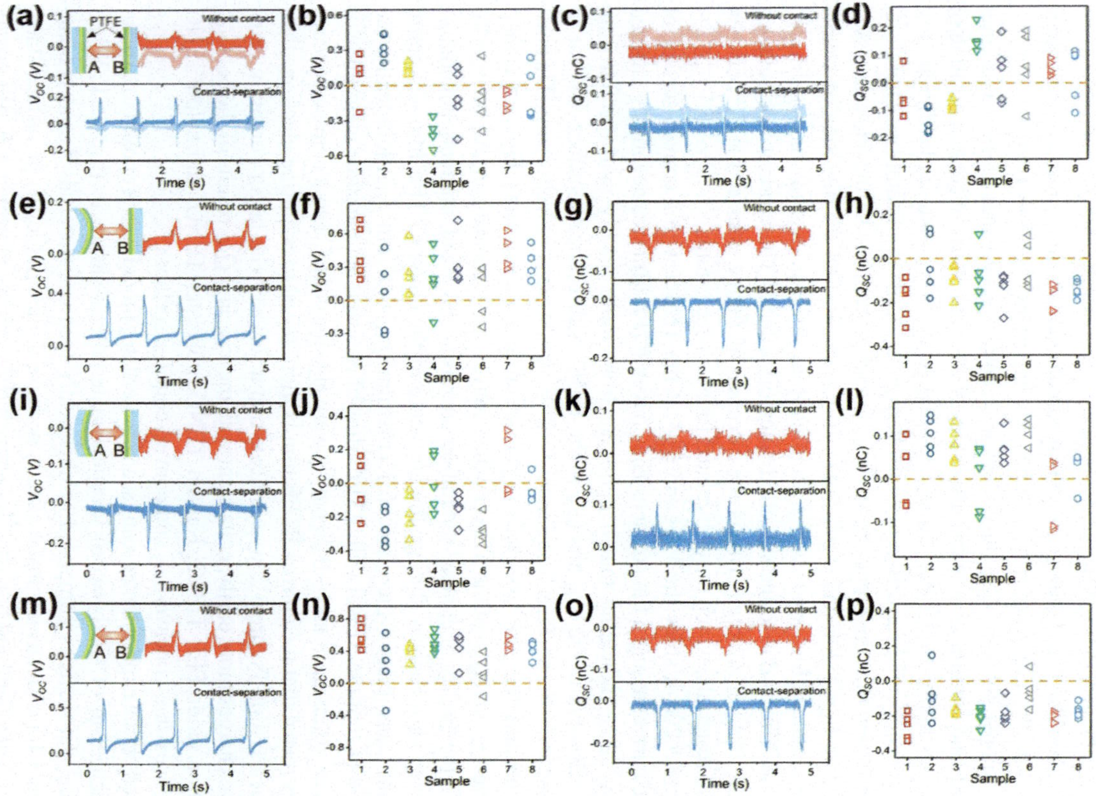

Fig. 5.20 Properties of the PTFE TENG with surfaces of different shapes. **a–d** V_{OC} and Q_{SC} of the TENG with two flat surfaces. **e–h** V_{OC} and Q_{SC} of the TENG with a convex surface (curvature of 0.20 cm^{-1}) and a flat surface. **i–l** V_{OC} and Q_{SC} of the TENG with a concave surface (curvature of −0.07 cm^{-1}) and a flat surface. **m–p** V_{OC} and Q_{SC} of the TENG with a convex surface (curvature of 0.20 cm^{-1}) and a concave surface (curvature of −0.07 cm^{-1}). The applied force during the contact-separation CE process is 2 ± 0.5 N. Reproduced with permission from American Chemical Society [42]

surface (i.e., larger curvature) and B has a concave surface (i.e., smaller curvature). The curvature difference may result in different surface energies, leading to a shift in surface states E_n. As shown in Fig. 5.21b, the E_n shifts to a lower level for the convex surface, while it shifts to a higher level for the concave surface compared to flat surfaces in Fig. 5.21a. When the convex and concave surfaces contact each other (Fig. 5.21c), electrons located at high-energy states in B transfer to the low-energy states of the surface states of A. In other words, the contact position on surfaces with larger curvature can accommodate more electrons than those with smaller curvature when in contact, resulting in the flow of electrons from the latter to the former. Figure 5.21d shows that the separation between the two identical materials results in negative charges on A and an equivalent amount of positive charges on B. Although the research findings are obtained on macrosurfaces, they may also apply to microsurfaces, considering the relationship between CE and micro/nanostructures of surfaces, which certainly needs further validation using other techniques such as AFM. It is worth noting that the prerequisite for the above mechanism is the utilization of a contact-separation mode with a gently applied force that is not expected to cause surface damage. Additionally, the micro/nanosurface roughness may change, and the local strain/stress generated in the films may vary according to the curvature difference. Both of these factors are likely to influence the direction of charge transfer, which still requires further study.

Roughness and curvature effects may have similar impacts on CE to some extent. Currently, increasing roughness is primarily used to enhance the starting power of TENGs. For example, one of the main methods to increase the amount of charge generated by contact electrification is to increase the surface roughness, as proposed by Li et al. [46]. Increasing surface roughness can enlarge the contact area, thereby generating a larger amount of charge. Various mechanical methods for roughening surfaces are possible, such as sandblasting. Micro-lithography or nano-lithography techniques (e.g., photolithography, electron beam lithography, or laser interference lithography) are commonly used to fabricate large-scale features on surfaces to increase roughness. A common technique for fabricating TENG devices involves first creating micrometer-scale patterns on a silicon wafer (i.e., the template) using photolithography to protect specific regions of the surface, followed by selective etching of the unprotected regions. After fabricating the template, the patterns are transferred onto a polymer (e.g., poly(dimethylsiloxane), PDMS) by polymerizing a layer of liquid monomer on the template (i.e., soft lithography; Fig. 5.22a). Different types of micrometer-sized features on the surfaces of PDMS were obtained after peeling the polymer off the templates. Examples of the fabricated microstructures include lines, cubes, and pyramids. The 3D feature of the pyramids was created by anisotropic etching of the (100) silicon wafer. These microstructured PDMS films were used in a TENG device to generate electrical outputs. Results indicated that these micrometer-sized features (especially the pyramids) enabled the TENG device to perform several times better than the surface without any patterned features (Fig. 5.22b). Selective etching can also be performed by first covering a polymeric film with a layer of metal nanoparticles, which serve as the "mask." By etching

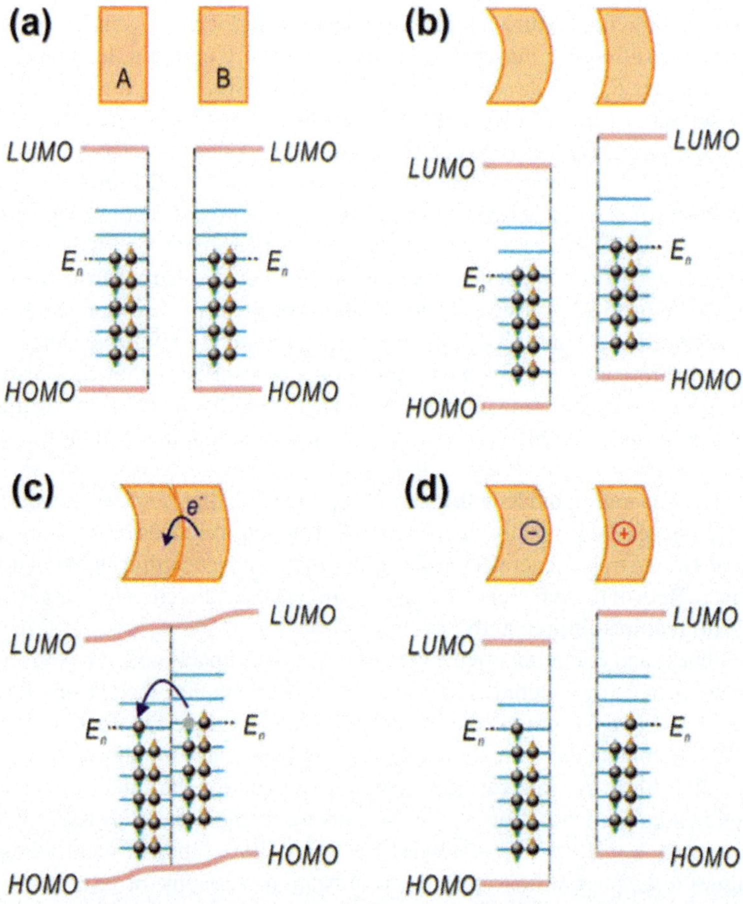

Fig. 5.21 Mechanism of CE between identical materials. **a** Surface states model of the identical materials A and B with idealflat surfaces. **b–d** Charge transfer before contact, in contact, andafter contact between the identical materials A and B with different surface curvatures (depicted by orange shapes) explained by thesurface states models. LUMO, lowest unoccupied molecularorbital; E_n, neutral level of surface states; HOMO, highest occupied molecular orbital. Reproduced with permission from American Chemical Society [42]

the regions of the polymeric film not protected by the mask, an array of polymeric nanowires can be fabricated.

A creative method for fabricating surfaces with patterned features is through the self-assembly of materials. For example, a layer of polymeric spheres (e.g., polystyrene, PS) was self-assembled onto a surface via force-directed assembly into a hexagonal close-packed colloidal array. After assembling the spheres, the liquid PDMS monomer was poured over the spheres, polymerized, and peeled off to obtain the ordered hemispherical hollow microstructures on its surface left behind by the

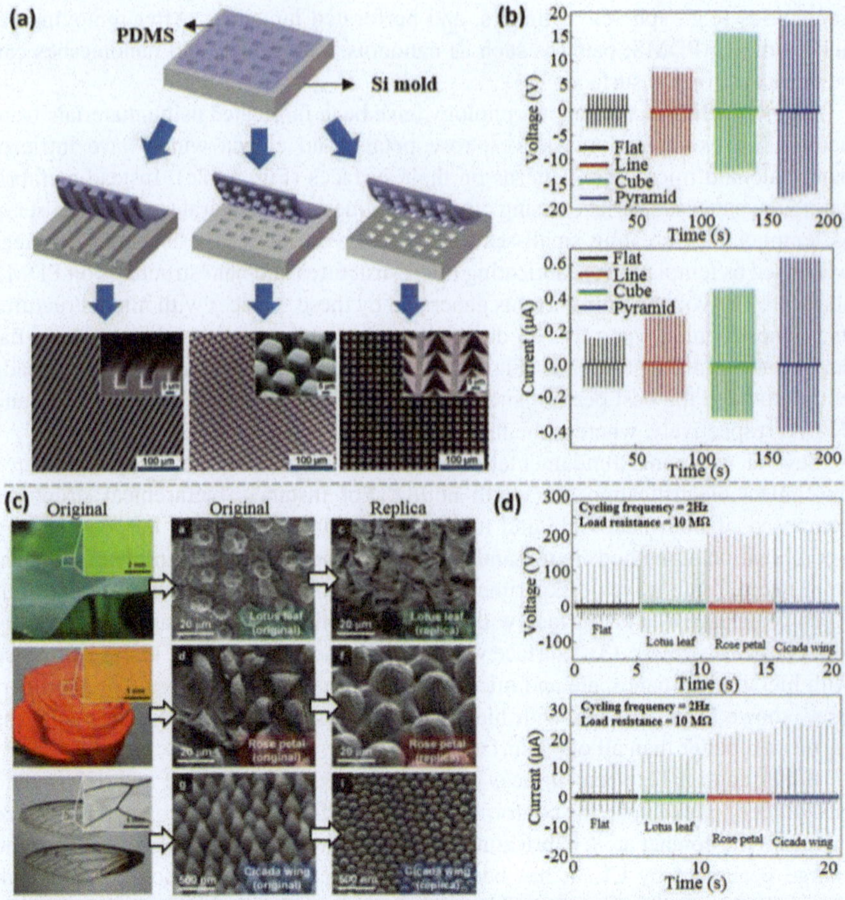

Fig. 5.22 General methods for increasing the amount of surface charge generated by contact electrification. **a** Surface charge can be increased by patterning the surface of PDMS with micrometer-sized features (e.g., lines, cubes, and pyramids). **b** Electrical outputs (i.e., voltage and current) generated by a TENG device that operated based on the piece of PDMS with or without the micrometer-sized features. **c** Surface charge can be increased by patterning the surface with biomimicking features derived from materials from nature, including lotus leaf, rose petal, and cicada wing. **d** Electrical performance of TENG using thin PDMS films fabricated from natural surfaces. Reproduced with permission from Wiley [46]

spheres. By using this surface as a template, a piece of PDMS with embossed hemispherical microstructures was also fabricated by casting the liquid PDMS monomer over the template modified with hydrophobic molecules (for easier separation of the polymers), polymerizing the monomer, and peeling the polymer off the template. Another type of self-assembly involves the microscopic phase separation of covalently linked block copolymers (i.e., block copolymer of PS and PDMS). This block copolymer can self-assemble into spatially well-defined microdomains of different geometries (e.g., spheres, cylinders, and perforated lamellae). After removing PS and oxidizing PDMS, patterns such as nanodots, nanogrates, and nanomeshes can be fabricated on the surface.

Surfaces with fascinating morphology have been fabricated using materials from nature. For example, lotus leaves, rose petals, and cicada wings have intricate nanoscale and microscale patterns on their surfaces (Fig. 5.22c). Instead of fabricating the patterned surfaces using other techniques, these natural surfaces can serve as templates for creating small-scale patterns on surfaces. These natural surfaces were used as templates for fabricating microstructures and nanostructures on PDMS surfaces. The voltages and currents generated by these surfaces with microstructures and nanostructures via a TENG device were larger than those generated by a flat surface without the structures (Fig. 5.22d). The surface fabricated from the cicada wing provided the best performance, with a peak voltage and current of 210 V and 23 μA, respectively, whereas the flat surface generated only 110 V and 11 μA.

Several additional fundamental strategies for increasing the charge generated by contact electrification are worth noting. For instance, hierarchical structures composed of both nanoscale and microscale features have been fabricated using a combination of methods (e.g., sandblasting and anodization) or templates obtained from nature (e.g., leaves). Experimental comparisons have been conducted on four types of surfaces: (1) a surface without any modification, (2) a surface with only microscale roughness, (3) a surface with only nanoscale roughness, and (4) a surface with hierarchical nanoscale and microscale roughness. Results from these comparisons showed that surfaces with hierarchical nanoscale and microscale roughness performed better than all other surfaces.

Better contact between surfaces can be achieved using flexible and deformable elastomers; for example, the conformal contact of PDMS with the contacting surface increases the contact area. Fabricating surfaces with nanowires can also enhance the charge generated by CE. It has been reported that when a surface covered with nanowires comes into contact with another surface, the nanowires bend, causing frictional contacts with the other surface, thus generating more charge than simple contacts. When both surfaces are covered with nanowires, contacting the surfaces allows the nanowires to interlock, increasing the amount of charge generated. Another common method involves using highly chargeable materials (e.g., PTFE and PDMS) to increase the amount of charge generated by contact electrification (e.g., via coating).

In summary, the curvature effect indicates that convex surfaces with positive curvature are more likely to become negatively charged, while concave surfaces with negative curvature tend to become positively charged. Additionally, an increase in applied pressure may cause a reversal in the direction of charge transfer between the two surfaces. Moreover, the effect of roughness is somewhat related to the curvature effect. However, the primary method currently used to enhance the power generation of TENGs is by increasing surface roughness.

5.6.2 Atomic Electronegativity

Electronegativity is a measure of an element's ability to attract electrons in a compound, also known as relative electronegativity. This concept, introduced by Linus Pauling in 1932, represents the relative strength of an atom's ability to attract electrons when forming chemical bonds with different atoms. Elements with high electronegativity strongly attract electrons, while those with low electronegativity do not.

The polarity of a triboelectric material is determined by its chemical properties and is closely related to the functional groups present on its surface. As the electronegativity of these functional groups increases, the material's ability to accept electrons also increases, allowing it to generate more charge. Fluorine, being the most electronegative element with a small radius and high ionization energy, is often used in polymers such as FEP, PVDF, and PTFE as negative triboelectric materials. While a wide variety of triboelectric materials are available, chemical modification can adjust surface polarity to expand the range of materials and improve the output performance of existing materials.

Shin et al. reported a straightforward surface functionalization method that effectively changes the triboelectric sequence of materials in the triboelectric series, significantly enhancing the output performance of TENGs [47]. They prepared four different TENG combinations to systematically investigate and experimentally verify whether the generated output performance is indeed induced by surface functionalization. As shown in the schematic representations (Fig. 5.23a), the contact-mode TENGs with four different functionalization pairs were prepared: (i) PET:PET, (ii) PET:P-PET, (iii) PET:F-PET, and (iv) P-PET:F-PET. Using these devices, the open-circuit voltage (V_{OC}) and short-circuit current density (J_{SC}) were measured (Figs. 5.20b and c). Figure 5.23b presents the Voc of the four TENGs during cyclic pushing and releasing using a push machine. The experimental conditions and dimensions of the TENGs were identical, with a device area of 4 cm^2 and an applied force normal to the surface of 0.5 MPa. Under the forward connection mode, the smallest V_{OC} (~4 V) was measured in the device with the (i) PET:PET contact pair (enlarged view in Fig. 5.23b). Weak output voltages were still generated between the same PET pair, likely due to capacitance changes between the top and bottom electrodes during cyclic contact motions. After functionalizing one of the TENG surfaces with poly-L-lysine [(ii) PET:P-PET pair], the V_{OC} immediately increased to ~ 110 V. In the TENG fabricated with the (iii) PET:F-PET pair, where only one surface was functionalized with FOTS, the V_{OC} reached up to ~ 205 V. Lastly, the maximum output V_{OC} of ~ 330 V was obtained using the TENG with the (iv) P-PET:F-PET contact pair, which is ~ 80 times higher compared to the V_{OC} measured in the (i) PET:PET pair. The measured short-circuit current densities showed a similar trend as the voltage outputs, increasing from the lowest ~ 2.6 mA/m^2 in the TENG with the (i) PET:PET pair to the highest ~ 270 mA/m^2 in the TENG with the (iv) P-PET:F-PET pair (Fig. 5.23c). Consequently, the output power density increased by approximately four orders of magnitude after functionalizing both surfaces with opposite polarity. The validity of the output signals was also confirmed using a switching polarity test, where V_{OC} and J_{SC} were measured by reversely connecting the TENG. The enhancement in output performance in surface-functionalized TENGs can be explained as follows: In the TENG with the (ii) PET:P-PET pair, when the PET surface contacts the positively charged P-PET surface, the positive charges on the P-PET surface increase the potential difference between the two ITO electrodes, enhancing output values. Similarly, in the case of the (iii) PET:F-PET paired TENG, the F-PET surface is partially negatively charged, induced by the fluorine (F) in the (-CF_3) end groups of FOTS. This causes a potential difference between the electrodes as the two surfaces (PET:F-PET) contact each other. The fluorinated polymer is well-known as an excellent electron-accepting material due to its high electronegativity. Thus, the triboelectric sequence difference between the PET and F-PET surfaces is relatively larger compared to that between the PET and P-PET pair, resulting in higher peak-to-peak output pulses. As expected, the maximum output pulses were obtained in the TENG with the (iv) P-PET:F-PET combination, originating from the highest potential difference between the two surfaces during cyclic contact motions.

Additionally, Song et al. presented significantly enhanced triboelectric properties by engineering a surface triboelectric nanogenerator using self-assembled monolayers (SAMs) [48]. They used PDMS as a substrate and applied PFOTES through liquid deposition and molecular self-assembly to graft the PDMS surface. The open-circuit voltage and short-circuit current of the modified TENG reached 105 V and 27 μA, respectively, which is more than 60 times that of the unmodified device.

Triboelectric properties were significantly altered depending on the type of SAMs, as shown in Fig. 5.24. All monolayers (METS) were measured at a frequency of 1.25 Hz with a vertical pressure of 3 N, and the data were collected after the properties stabilized over 5000 cycles. While an unmodified PDMS/Al device showed a maximum open-circuit voltage (V_{OC}) and maximum short-circuit current (I_{SC}) of 17.3 V and 6.23 μA, respectively, the 1H,1H,2H,2H-perfluorooctyltrichlorosilane (FOTS)-METS exhibited a maximum V_{OC} of 105 V and I_{SC} of 27.2 μA, as shown in Fig. 5.24a and b. The significant enhancement of more than 6 and 4 times in V_{OC} and I_{SC}, respectively, indicates the effectiveness of SAMs modification. When trichloro (3,3,3-trifluoropropyl) silane (TFPS) was used, both the maximum V_{OC} and I_{SC} decreased compared to the values with FOTS, due to its lower surface density of fluorine atoms, as confirmed by water contact angle measurement. When 3-aminopropyltriethoxysilane (APTES) and 3-glycidoxypropyltriethoxysilane (GPTES) were used, the maximum values of V_{OC} (-25.8 V) and I_{SC} (-6.43 μA) for APTES, and V_{OC} (-9.9 V) and I_{SC} (-2.51 μA) for GPTES were obtained, as shown in Fig. 5.24a and b. Both negative V_{OC} and I_{SC} values for APTES and GPTES clearly suggest that the surface of PDMS with these SAMs is positively charged, while the surface of bare PDMS and those treated with either TFPS or FOTS are negatively charged with respect to the Al electrode. To further elucidate the surface characteristics of SAM-modified PDMS substrates, they measured the zeta potential of the samples to quantify the magnitude of the surface charges. The values varied significantly with the types of SAMs, as shown in Fig. 5.24c. The bare PDMS displayed −11.04 mV, and GPTES, consisting of epoxy

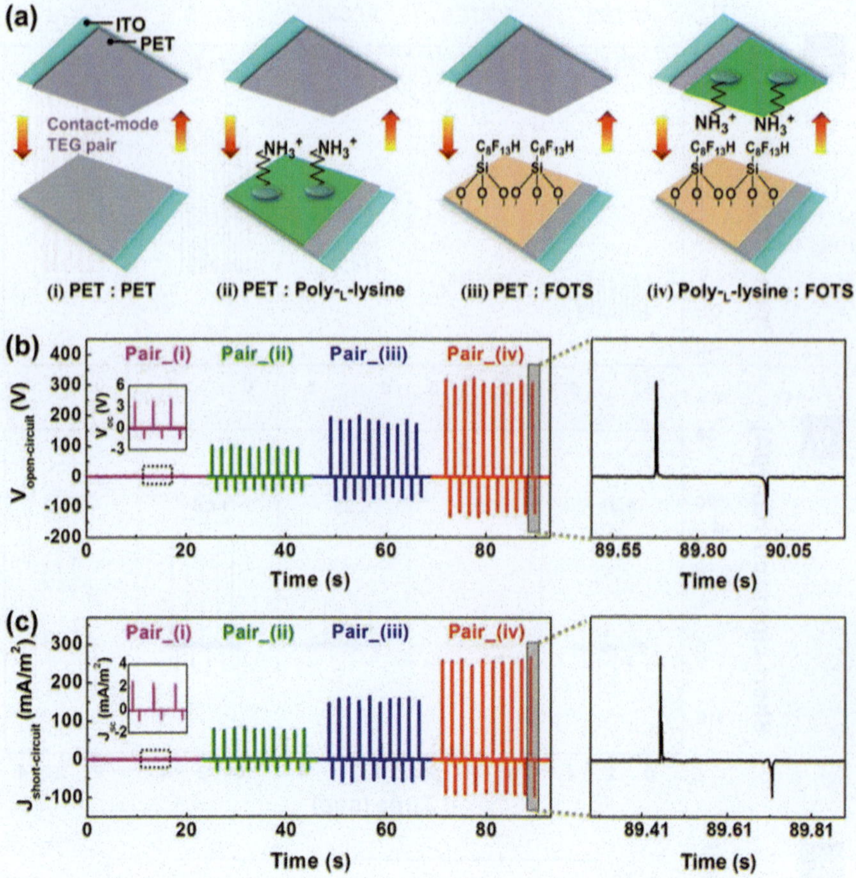

Fig. 5.23 **a** The prepared contact pairs of the TENGs. (i) PET:PET, (ii) PET:P-PET, (iii) PET:F-PET, and (iv) P-PET:F-PET. **b** Open-circuit voltages of the TENGs with different contact pairs. The generated output voltages increase from ~4 V in the device with PET:PET contact pair to ~330 V in the device with (iv) P-PET:F-PET pair. **c** Short-circuit current densities of the same TENGs measured in **b** demonstrates a similar trend as in **b**, where the current densities increase from ~2.6 to ~270 mA/m^2 as both surfaces are functionalized. The size of TENGs used was 4 cm^2. Reproduced with permission from American Chemical Society [47]

functional groups, gave a value of −11.74 mV, similar to the reference PDMS. While FOTS and TFPS modified films showed more negative zeta potentials of −34.72 and −31.79 mV, respectively, the APTES surface with amine molecules was nearly neutral with a potential of −0.45 mV, which is qualitatively consistent with the surface properties of the triboelectrified METS in Fig. 5.24a and b.

In summary, atomic electronegativity is crucial for the charge transfer efficiency of triboelectric nanogenerators. When two materials with different electronegativities come into contact, electrons spontaneously migrate from the material with lower

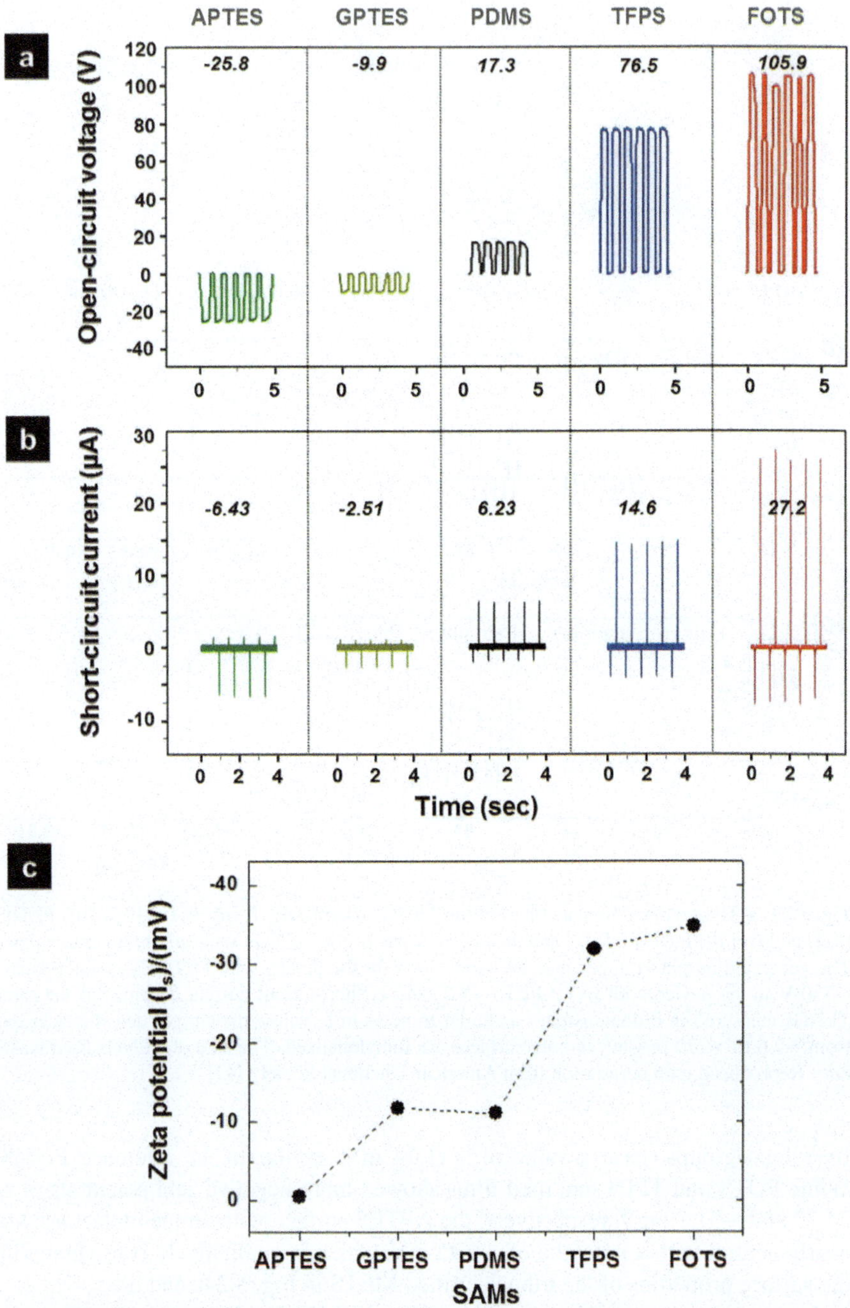

Fig. 5.24 Triboelectric properties of **a** the maximum open circuit voltages and **b** the maximum short circuit currents of METS measured at the frequency of 1.25 Hz with the vertical contact force of 3 N. **c** Zeta-potential values obtained, dependent upon SAMs in aqueous NaCl. Reproduced with permission from American Chemical Society [48]

electronegativity (weaker electron affinity) to the material with higher electronegativity (stronger electron affinity) due to the significant difference in their electronegativities. This process results in effective charge separation on the material surfaces, providing the necessary charge source for triboelectric nanogenerators and enabling the conversion of mechanical energy into electrical energy. It is important to note that the magnitude of the electronegativity difference directly determines the driving force for electron migration. The larger the electronegativity difference, the more pronounced the electron migration trend, thereby enhancing charge transfer efficiency. Consequently, under the same mechanical energy input, material combinations with larger electronegativity differences can generate electrical energy more effectively, thus improving the overall performance of TENGs.

5.6.3 *Material Transfer and Heterolytic Bond*

Previous studies have shown that most charges generated by the CE on insulator-insulator interfaces result from electron transfer. Additionally, chemical bond heterolysis, which involves material transfer, also influences CE due to electron transitions between atoms when chemical bonds break.

In their research, Tao et al. enhanced the charge retention capability of positive materials by incorporating materials and ions during the contact process [49]. They introduced an acid-ion chimerism strategy to create positive triboelectric materials with exceptional charge retention, beyond just electron-donating capabilities. They synthesized a series of ion-doped pyridine-based polybenzimidazoles (PBIs) by doping H_3PO_4, H_2SO_4, and HCl into PBI films, making them excellent positive partners for corona-polarized FEP. Thermal stimulation and ion mass spectrometry revealed that the formation of acid-ion chimerism, specifically polybenzimidazoliums, has high activation energy for stored charges. Selective anion migration helps compensate for the escape of polarized charges. The charge density of the H_3PO_4-doped PBI film was measured, showing an increase with higher H_3PO_4 solution concentrations, indicating a direct relationship with the amount of ion chimera in the 2,4-PBI film.

The charge density of the H_3PO_4-doped PBI film is measured according to a previous method. As shown in Fig. 5.25a, the charge density increases with the increasing concentration of the H_3PO_4 solution, indicating a direct relationship between the amount of ion chimera in the 2,4-PBI film. The charge density of the doped 2,4-PBI film is close to the absorbing phosphate concentration during treatment in different concentrations of H_3PO_4 solution and the surface properties. The ion chimerism of H_3PO_4 significantly increases the charge density and charge retention rate, showing a significant suppression of charge loss and escape (Fig. 5.25b). The charge retention rate of corona-polarized FEP coupled with 2,4-PBI treated with 5 mol L^{-1} H_3PO_4 solution is improved to 49.7%, which is significantly higher than that of the original PBI (30%). The maximum charge density can reach up to 596 $\mu C\ m^{-2}$, 1.67 times that of undoped 2,4-PBI (Fig. 5.25b). The charge density of the reported strategies such as low-energy ion irradiation/implantation, repeated rheological forging, ultrathin dielectric film chemical structure design (Fig. 5.25c). For other corona-polarized negative polymers such as PTFE, Kapton, and PET, the charge retention rate is similarly increased during contact with H_3PO_4-doped 2,4-PBI (Fig. 5.25d). The retention charge density after contact with the corona-polarized PTFE surface (814 μC m^{-2}) is increased to 335 μC m^{-2}, and the charge retention rate (41.2%) is also higher than that of the original 2,4-PBI (257 μC m^{-2}, 31.6%). It is worth noting that the charge retention capability provided by the ion chimerism strategy not only is applicable to negative electrets such as FEP and PTFE but also has a stabilizing effect on the injected charge of polar materials such as Kapton and PET. The charge retention rate is increased to over 30% during contact with Kapton and PET (over 200 μC m^{-2}). This suggests that the charge retention capability of 2,4-PBI increased by the ion chimerism strategy is universal for corona-polarized negative materials.

In addition, a TENG in contact separate mode based on 1 cm^2 of 2,4-PBI film treated with 5 mol L^{-1} H_3PO_4 solution can easily illuminate [47] LEDs (Fig. 5.25e). The maximum output power and matching resistance of the TENG are also measured under different resistance loads (Fig. 5.25f). The peak power density is realized up to 8.97 W m^{-2} at an external circuit load of 1.8 MΩ. It is worth noting that the low matching resistance here indicates a low internal resistance of the TENG, due to the conductivity of the ion-doped PBI compared to that of the insulating dielectric material. These results indicate that the H_3PO_4-doped 2,4-PBI film promises application as an excellent positive tribo-layer for various TENG devices.the anionic ions PO_3^- appear on the FEP surface after contact while the cations $H_2PO_3^+$ appear to a lesser extent and are almost absent (Fig. 5.25g).

Moreover, Yang et al. proposed a method to achieve ultra-high frictional charge density in the open air by using radical ion transfer as charge escape compensation [50]. Ionized-PTFE (I-PTFE) was prepared by injecting strongly reducing naphthalene radical anions into the surface of the PTFE, while the sufficient delocalization and heteroatoms of the I-PTEF surface can maintain the stability of the radical anion. After ionized treatment, the triboelectric polarity of I-PTFE changes to positive during contact electrification, while the ranking of I-PTFE in the triboelectric series is between natural rubber and copy paper (Fig. 5.26a). This phenomenon is related to the breaking of the C-F bond and the introduction of other functional groups mentioned above. Based on contact-separation mode TENG, the generated surface charge densities between I-PTFE and other dielectric materials are measured (Fig. 5.26b). It is worth noting that a record-high charge density of 525 μC m^{-2} is realized between I-PTFE and fully charged FEP (thickness: 15 μm). Meanwhile, the triboelectric charge densities between I-PTFE and fully charged FEP with other thicknesses (12.5, 20, and 30 μm) are 402, 451, and 414 μC m^{-2}, respectively. Then, the charge density of PTFE and I-PTFE are respectively compared and the enhancement of charge density reaches 1733% after radical anion injection. Here, they emphasized that the electrification process of I-PTFE and fully charged FEP is quite different from the previous case of common positive triboelectric materials (Fig. 5.26c). For the fully charged FEP, the charge escaping is inevitable, and a sharp decrease of surface

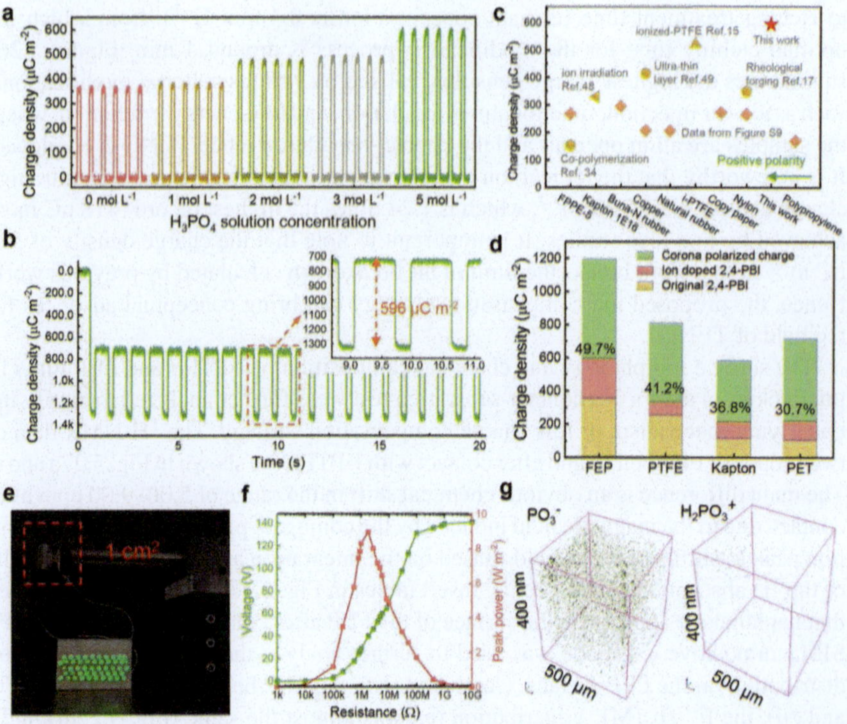

Fig. 5.25 **a** Charge density of 2,4-PBI films treated by soaking in different concentrations of H_3PO_4 solution, contacting with corona-polarized FEP. **b** Charge density of 5 mol L^{-1} H_3PO_4 solution treated 2,4-PBI, maximum stable charge density up to 596 μC m^{-2} after charge loss and leakage in contact separation. **c** Comparison of the charge density by various material design approaches in previous work with that in this work. **d** Charge density of negative polymers after corona polarization (CP) and that during coupling with 2,4-PBI film and H_3PO_4-doped 2,4-PBI after continuous contact. **e** Physical image of the 1 cm^2 contact-separate mode TENG lighting up 50 LEDs. **f** The output voltage and peak power curves of the contact-separate mode TENG based on 5 mol L^{-1} H_3PO_4 solution treated 2,4-PBI film. **g** The three-dimensional distribution of anion PO_3^- and cation $H_2PO_3^+$ on the FEP surface after a contact-separation test with 5 mol L^{-1} H_3PO_4 solution treated 2,4-PBI. Reproduced with permission from American Chemical Society [49]

charge density can always be observed. Accordingly, when the fully charged FEP continuously contacts with common positive dielectrics, the charge density usually shows decreased tendency before finally stable state, indicating continuous charge loss from the fully charged FEP. However, the electrification behavior of I-PTFE is quite different from previous cases. For the initial contact, the charge density sharply decreases from 1,830 μC m^{-2} to 358 μC m^{-2} due to the charge escaping caused mainly by air breakdown, which is in line with expected behavior. Then, with the increase of contact cycles, charge density gradually increases until it reaches 525 μC m^{-2} after 200 s. This is strong proof of the compensation effect of I-PTFE to the charge loss of fully charge FEP. Figure 5.26d shows the response of charge density to etching treatment time (etchant concentration is 0.4 mol L^{-1}), from which the optimal etching time for the modification process is around 4 min. Figure 5.26e summarizes the highest charge densities realized by various material modifications, such as ion-air injection, functionalities regulation, and low-energy irradiation, where the samples are all in open air and the contact-separation mode TENG is employed. It is noteworthy that the radical ion compensation method has realized an ultrahigh charge density of 525 $\mu C\ m^{-2}$, which is 1.24 times the highest record (418 $\mu C\ m^{-2}$) achieved by previous studies. It is important to note that the charge density of 525 $\mu C\ m^{-2}$ in this work breaks the limit of air breakdown calculated by previous work. Hence, the proposed ion compensation strategy can bring conceptual advances for the field of TENG.

The surface morphology and chemical composition of I-PTFE and FEP film (15 μm thickness) after a 3-h contact-separation test were further analyzed to clarify the underlying mechanism of this charge compensation strategy. The ^1H-NMR data of two groups of FEP before and after contact with I-PTFE are shown in Fig. 5.27a and b. The main difference is an obvious chemical shift in the range of 5.00–9.00 ppm after contact, due to the magnetic field induced by the conjugate plane, causing the absorption peak to shift to a lower field. Based on the integration and normalization results of the ^1H absorption peaks (see the insert tables in Fig. 5.27a and b), it is confirmed that naphthalene appears on the surface of the FEP after contact with I-PTFE. TOF–SIMS in positive ion mode was used to further analyze the chemical composition distribution on the FEP surface. Compared with the FEP before contact (Fig. 5.27ci and cii), the F^- (F, $[M]^-$) distribution remains almost the same (Fig. 5.27di), indicating that the main chemical composition of FEP has not changed significantly. However, there is a significant increase of $C_{10}H_8^+$ ($C_{10}H_8$, [M]) on the surface of FEP after contact with I-PTFE (Fig. 5.27dii), suggesting the possible transfer of naphthalene radical anion during contact. To rule out the influence of material fragment transfer during contact, if substantial material transfer occurred between I-PTFE and FEP, the TOF–SIMS result of FEP after contact should be similar to that of I-PTFE. According to the results in Fig. 5.27f, most naphthalene elements exist as complex ions ($C_{10}H_8Na$) on the surface of I-PTFE, with the naphthalene monomer or radical anion present at extremely low levels. However, no complex ions ($C_{10}H_8Na$) were observed on the surface of FEP after contact, and the naphthalene elements identified by TOF–SIMS are mainly in the form of ($C_{10}H_8^+$), as shown in Fig. 5.27dii. This indicates that selective ion transfer or ionic reaction occurs during contact due to interfacial chemisorption. Figure 5.27e reveals that $C_{10}H_8^+$ in FEP after contact is mainly located in the region with a depth of 0–30 nm near the surface, which aligns with typical contact electrification results, and the depth distributions of F^- in FEP are similar before and after contact.

The above work suppresses charge escape by introducing a material (radical anion) transfer process between interfaces, pioneering a new approach to achieve ultrahigh triboelectric charge density in open air. It also suggests that in some CE cases, material transfer and electron transfer have a relationship of both competition and cooperation. The observed coexistence and complementary behavior of electrons and materials at the contact interface can help further elucidate the mechanism of CE.

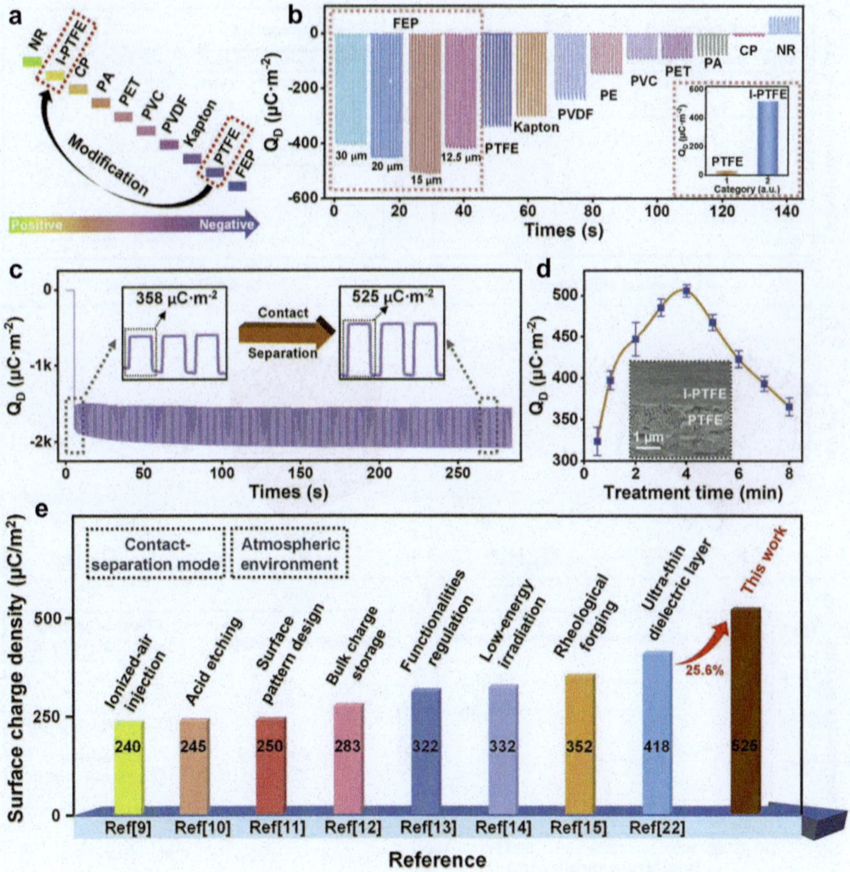

Fig. 5.26 Output performance of TENG with I-PTFE as positive triboelectric layer **a** The position of I-PTFE in the triboelectric series. **b** Triboelectric charge density (Q_D) between I-PTFE and different polymers including fully charged FEP with four different thicknesses in open air and contact-separation mode. **c** The charge density changes (Q_D) of the contact-separation TENG composed of I-PTFE and fully charged FEP. **d** Influence of etching treatment time on the charge density (Q_D). Inset: SEM image of I-PTFE existing on PTFE surface. Scale bar, 1 μm. **e** Comparison of various material design approaches in atmospheric environment and contact-separation mode, with the highest charge density of 525 μC m^{-2} obtained by radical ion modification, and a higher 1237 μC m^{-2} is realized in sliding mode. Reproduced with permission from Cell Press [50]

5.6.4 Dipole Polarization

A key determinant of the output power in TENGs is the concentration of triboelectric surface charges, which act as the impetus for the movement of free electrons. It is widely accepted that the transfer of charge between two different materials is due

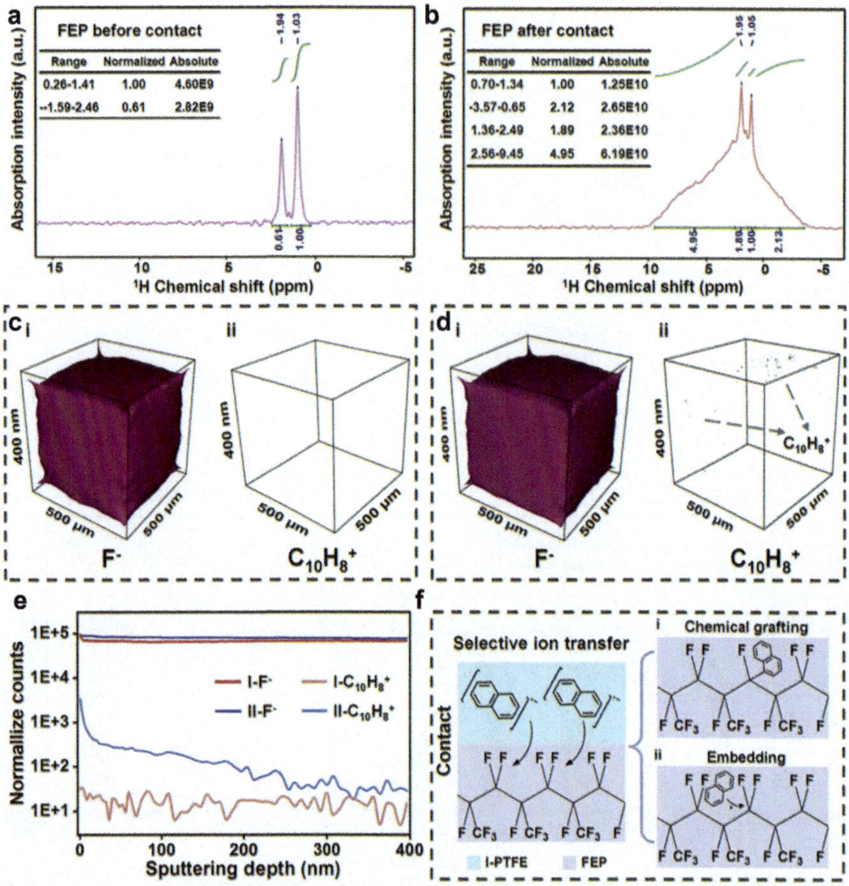

Fig. 5.27 Mechanism of high charge density between I-PTFE and fully charged FEP (**a** and **b**) The ^{1}H solid-state NMR spectra of FEP before **a** and after **b** contact with I-PTFE. Insert tables: Integration and normalization results of ^{1}H absorption peak. **c** The three-dimensional distribution of (i) F^- (F, $[M]^-$) and (ii) $C_{10}H_8{}^+$ ($C_{10}H_8$, [M]) in the FEP bulk before contact-separation test. **d** The spatial distribution of (i) F^- (F, $[M]^-$) and (ii) $C_{10}H_8{}^+$ ($C_{10}H_8$, [M]) in the surface of FEP after contact with I-PTFE. **e** The content changes along the depth direction, where I and II represent the FEP before and after the contact with I-PTFE, respectively. **f** Chemical mechanism of radical anion transfer between I-PTFE and FEP interface. Reproduced with permission from Cell Press [50]

to their varying capacities to acquire electrons. The concentration of these surface charges is largely influenced by the inherent characteristics of the materials involved, with environmental factors playing a secondary role. Consequently, there is a scarcity of effective strategies to boost the output power of TENGs by altering the intrinsic material properties involved in charge transfer [51].

Here, Wang et al. examined the impact of intrinsic dipole moments on the electrical performance of TENGs. They employed PVDF thin films to construct a TENG, eliminating the need for additional fabrication steps. The output voltage of a TENG made with polarized PVDF was found to be either significantly increased to 240% or substantially decreased to 70%, contingent upon the orientation of the polarization. It is suggested that the bond charges resulting from the alignment of dipoles can alter the surface potential energy of PVDF thin films. Consequently, the potential energy disparity between the polarized PVDF and aluminum is altered, leading to a variable amount of charge transfer upon contact. This research represents an initial attempt to modulate the electrical output of TENGs across a broad spectrum by altering the intrinsic electrical properties of materials. By introducing a mechanism that includes polarization-induced alteration of surface potential, this study also provides novel insights into contact electrification, indicating that charge transfer could potentially be controlled through surface potential engineering.

As depicted in Fig. 5.28a, a short-circuit current density (J_{SC}) of 8.34 $\mu A/cm^2$ was recorded for a forward-polarized TENG, marking a 36.3% increase compared to the 6.13 $\mu A/cm^2$ produced by a nonpolarized TENG. To further illustrate the effect of the dipole moment on TENG output, the J_{SC} of a reverse-polarized TENG was also measured under identical conditions, yielding a J_{SC} of 4.83 $\mu A/cm^2$, which represents a roughly 21% reduction compared to the nonpolarized TENG. Similarly, the V_{OC} values obtained from the three types of TENGs showed comparable outcomes, as shown in Fig. 5.28b. The highest V_{OC} of 215 V was associated with the forward-polarized TENG, while the lowest V_{OC} of 72 V was linked to the reverse-polarized TENG. The differentiated electrical outputs from the three categories of TENGs can also be substantiated by the charge output, as presented in Fig. 5.28c. Alternating electrons flowing between electrodes are rectified through a diode bridge, resulting in accumulative induced charges (accumulative Q). The forward-polarized TENG produced a charge output of 9.05 μC within 5 s, equating to induced charges of 0.51 μC per cycle. In contrast, the nonpolarized and reverse-polarized TENGs generated charge outputs of 4.72 and 2.50 μC over 5 s, corresponding to 0.25 and 0.14 μC per cycle, respectively. To mitigate random errors that might arise from fabrication and measurement, each category of TENG was tested in five batches under identical conditions. As shown in Fig. 5.28d, the average J_{SC} values obtained were 7.55 $\mu A/cm^2$, 5.40 $\mu A/cm^2$, and 4.28 $\mu A/cm^2$ for forward-polarized, nonpolarized, and reverse-polarized TENGs, respectively. The same trend was observed in the average amplitude of V_{OC} and the equivalent direct current density (equivalent J_{SC}), which is defined as the average amount of positive charges passing through the circuit per second post-rectification (refer to the ESM), as demonstrated in Fig. 5.25e and f, respectively. Taking the average amplitude of V_{OC} as an example, it exhibited over a three-fold enhancement between the forward-polarized and reverse-polarized TENGs. Resistors were used as external loads to assess the output power of the different TENGs under a compressive force of approximately 50 N. As shown in Fig. 5.28g, a peak power of 3.74 mW was achieved at a load resistance of 40 MΩ for the forward-polarized TENG, whereas a peak power of 1.19 mW was recorded at 40 MΩ for the nonpolarized TENG (Fig. 5.28h), and a lower peak power of 0.91 mW was achieved at 20 MΩ for the reverse-polarized TENG (Fig. 5.28i).

Despite multiple factors potentially impacting the electrical performance of TENGs, with other parameters such as material selection, fabrication techniques,

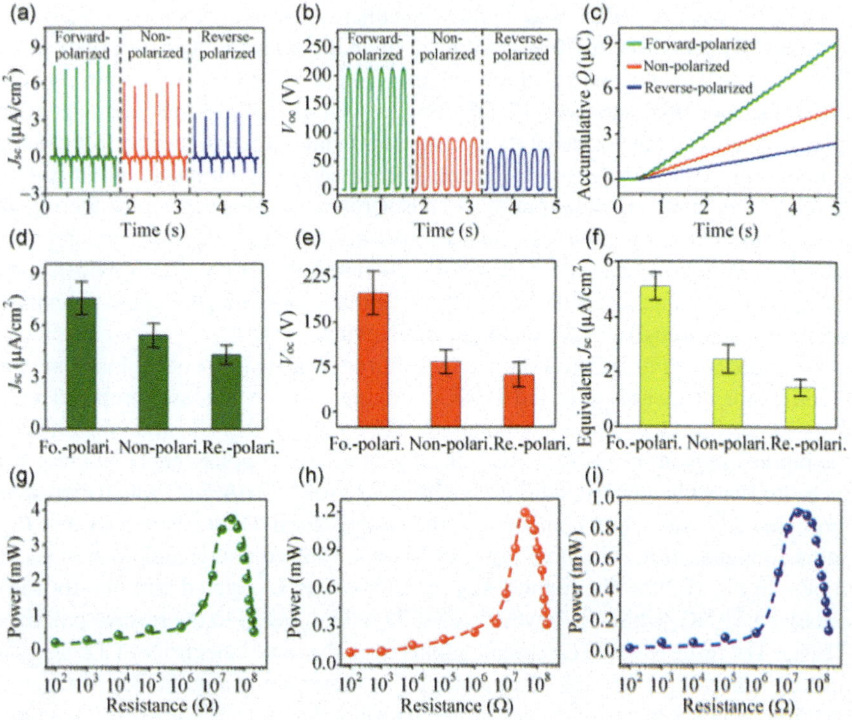

Fig. 5.28 **a** Jsc of TENGs fabricated using different types of PVDF films under a periodic compressive force around 50 N applied by an electric shaker at a frequency of 4 Hz. **b** Voc of TENGs fabricated using different types of PVDF films. **c** Accumulative Q generated by TENGs fabricated using different types of PVDF films. **d** Average Jsc of TENGs fabricated using different types of PVDF films. **e** Average Voc of TENGs fabricated using different types of PVDF films. **f** Average equivalent Jsc of TENGs fabricated using different types of PVDF films. **g** Dependence of the peak power output of the forward-polarized TENG on the resistance of the external load. **h** Dependence of the peak power output of the nonpolarized TENG on the resistance of the external load. **i** Dependence of the peak power output of the reverse-polarized TENG on the resistance of the external load. The curve is a fitted result. Reproduced with permission from Springer [51]

device dimensions, and measurement conditions held constant, the variation in electrical output among the three types of TENGs is likely due to differences in surface charge density. It is hypothesized that the intrinsic dipole moment modifies the surface potential of the PVDF film, thereby regulating the transfer of surface charges between the PVDF and the aluminum electrode. Extensive experimental data has demonstrated that charge transfer between metals and dielectrics occurs through tunneling, a process stemming from the disparate effective work functions (Φ) of the two materials. As depicted in the simplified band diagram in Fig. 5.29a, the surface energy state of nonpolarized PVDF is characterized by an energy level with an effective work function of approximately 5.0 eV, which exceeds that of aluminum at 4.2 eV. Consequently, upon contact between aluminum and PVDF, electrons tunnel from the Fermi level of aluminum (E_F) to the characteristic energy level of PVDF (E_0), resulting in a negatively charged PVDF surface. Electron transfer continues until the energy levels are aligned. It is clear that the surface charge density is directly proportional to the potential difference between the two energy levels, which is primarily a function of the materials' properties, such as their chemical composition. Depending on the relative sizes of E_F and E_0, electrons tend to move from the filled Fermi level of aluminum to the vacant surface states of PVDF. The likelihood of charge transfer is determined by the potential difference between the two adjacent materials. As shown in Fig. 5.29b, the electric field generated by charge transfer elevates the energy of the electrons in PVDF's surface states by $\delta E(Q)$ relative to E_F,, reducing the potential difference between aluminum and PVDF. When the charge reaches saturation, charge transfer halts. In the polar β-phase of PVDF, the centers of positive and negative charges do not coincide in every molecule, but the bond charge is zero in the naturally occurring β-phase due to the random orientation of dipoles. After polarization along the thickness, all dipole moments align in the same direction, creating bond charges on the surfaces of the PVDF thin film (refer to the ESM). Although molecules or space charges may adsorb onto the PVDF surfaces, bond charges are not always fully compensated, and partially compensated surfaces are likely the norm in air. If the film is forward-polarized with positive bond charges on the surface in contact with aluminum, while the other surface is covered by the back electrode, the characteristic energy level of PVDF will be lowered to E_0' on the surface, as shown in Fig. 5.29c. This energy level shift results in an increased potential difference with the Fermi level of aluminum, thereby increasing the likelihood of charge transfer by tunneling upon contact. Consequently, an increased density of surface triboelectric charge is anticipated. Conversely, the negative bond charges on the surface of reverse-polarized PVDF will elevate the characteristic energy level to E_0'', as shown in Fig. 5.29d, reducing the density of triboelectric charge compared to the nonpolarized PVDF film. According to previous research, surface triboelectric charges are the immediate driving force for the transport of induced electrons between electrodes in a TENG. The surface charge density, therefore, dictates the quantity of induced electrons that flow to neutralize the electric field from the triboelectric charges. As a result, forward-polarized PVDF films result in enhanced electrical output, while reverse-polarized PVDF has the opposite effect.

In summary, the electrical output of TENGs can be either enhanced or reduced by using forward-polarized or reverse-polarized PVDF as the contact material. This is due to the influence of intrinsic dipole moments on the characteristic potential energy on the surface of the films. This demonstrates the potential to adjust the electrical output of TENGs over a wide range by altering the dipole polarization of a material, offering new insights into CE.

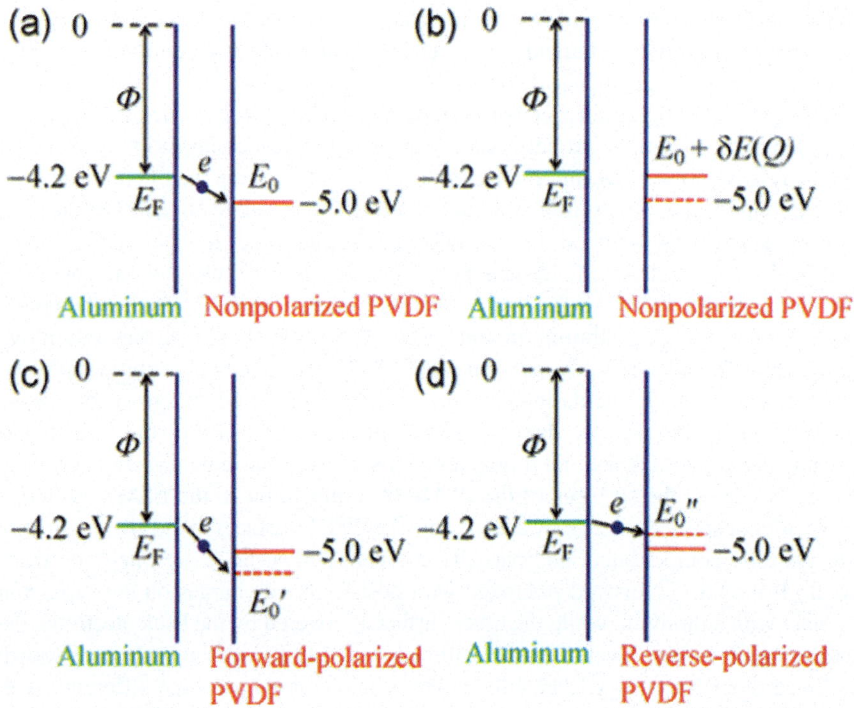

Fig. 5.29 **a** Schematic energy band diagram illustrating the process when electrons tunnel between E_F and E_0. **b** The electric field developed by charge transfer raises the energy of the electros in the PVDF's surface states by $\delta E(Q)$. **c** The positive bond charges of the forward-polarized PVDF reduce the characteristic energy level of PVDF to E_0'. **d** The negative bond charges of the reverse-polarized PVDF increase the characteristic energy level of PVDF to E_0''. Reproduced with permission from Springer [51]

5.6.5 Deep Trapping

In recent years, many researchers have focused on developing various types of dielectric layers [52–54], such as metallic conductive materials [55], inorganic nonmetallic materials [56–59], and polymer materials [60–62], to effectively enhance the output performance of TENGs. It has been shown that this improvement is related to the increase in space charge regions and electron trapping sites. However, factors affecting space charge density and charge storage depth, which are related to trap distribution, remain a concern. Additionally, it is still unclear whether the design of a multilayer structure to improve TENG output is universally applicable to all friction materials.

Lv et al. explored the integration of various dielectric materials, such as polyimide (PI), polypropylene (PP), nylon (polyamides; PA), polyethylene (PE), polytetrafluoroethylene (PTFE), silicone rubber, and polyoxymethylene (POM), between the triboelectric layer and the electrode to refine the trap distribution within triboelectric materials, aiming to enhance their electrical output capabilities [63]. Figure 5.30a illustrates the output voltage of porous PTFE functioning solo and in conjunction with several dielectric materials (e.g., PI, PP, PA, PE, dense PTFE film, POM, and silicone rubber). It is evident that incorporating any of the evaluated dielectric layers beneath the triboelectric layer (porous PTFE) of the TENGs improved their electrical output compared to a device with solely a porous PTFE layer. However, the extent to which each dielectric layer amplified the output varied. The most significant enhancement was observed with the addition of the PI layer, which increased the output voltage and transferred charges by 380% (to 1200 V) and 430% (to 160 nC), respectively, compared to the single-layer porous PTFE (Fig. 5.30a and b).

The generation of triboelectric charges is inherently linked to the electronegativity of the triboelectric layers involved. Thus, it is plausible to deduce that the initial quantity of charges generated in all cases should be uniform since the friction materials are identical across all TENGs. However, when an electrostatic potential is induced between the contact surface and the electrode, surface charges migrate through the electric field and neutralize with the induced opposite charges, leading to a substantial disparity in potential. These processes, which are tied to the charge trapping and retention capabilities of the triboelectric materials, are markedly altered with the addition of a dielectric layer.

To decipher the impact of charge trapping on the output performance of TENGs, an isothermal surface potential decay (ISPD) model was utilized to determine the trap distribution traits of each dielectric layer. This method involves measuring the decay behavior of the dielectric layer's surface charges under isothermal conditions, as shown in Fig. 5.30c. Each dielectric layer exhibits a distinct initial surface potential and decay rate. Notably, the PI layer possesses the most negative initial surface potential and the slowest decay, while the POM layer has the least negative initial potential and the quickest decay. These attributes align with the output performance of the respective TENGs.

The ISPD theory, as proposed by Simmons and Tam, posits that (i) the trapping process ceases once the charge is trapped during the surface potential decay and (ii) disregards the composite behavior of holes and electrons. The continuous distribution function of traps, $N_t(E)$, is given by Eq. (5.8),

$$N_t(E) = \frac{4\varepsilon_0\varepsilon_r}{qL^2K_BT}\left|t\frac{dv_{s(t)}}{dt}\right| \tag{5.8}$$

where ε_0 is the dielectric constant of the vacuum, ε_r is the relative dielectric constant of the sample, q is the number of hole charges, L is the thickness of the sample, k_B is the Boltzmann constant, T is the absolute temperature, and $V_s(t)$ is the surface potential of the sample at time, t. The trap energy, E, is defined by Eq. (5.9),

$$E = E_m - E_V = K_BT\ln(vt) \tag{5.9}$$

where v is the escape frequency factor of the trap charge, E_m defines a local state hole trap in the forbidden band, and E_V is the highest energy in the valence band.

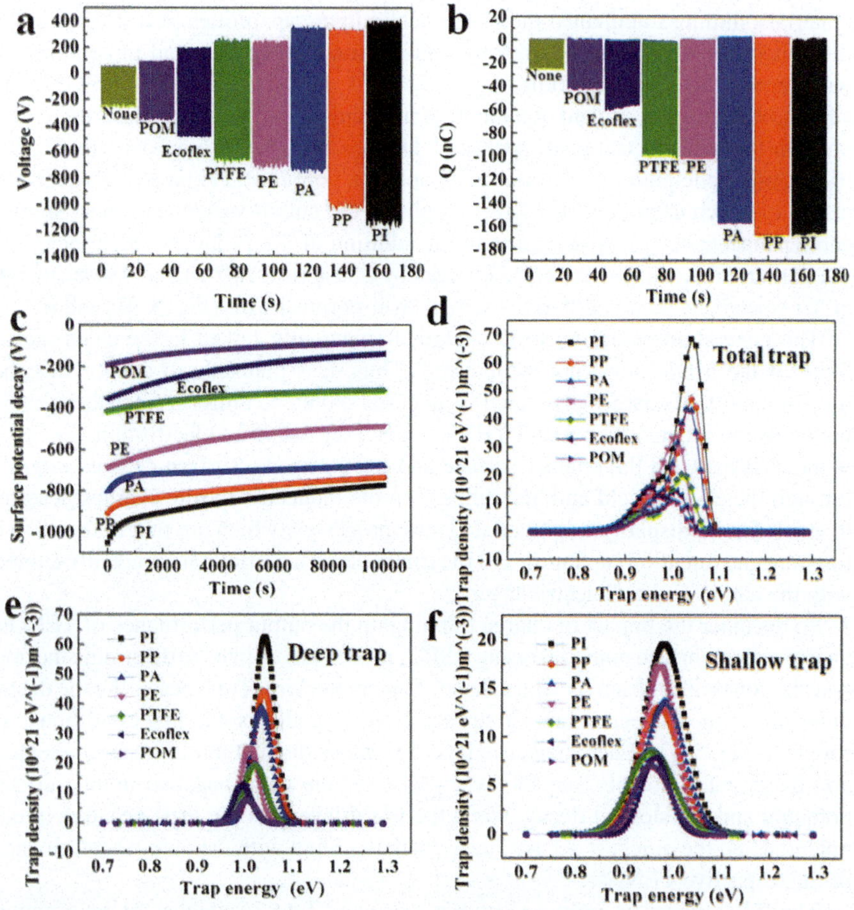

Fig. 5.30 **a**, **b** Comparison of output voltage and transferred charge, respectively, of the porous PTFE single layer and porous PTFE layer combined with PI, PP, PA, PE, PTFE, silicone rubber, and POM dielectric layers. **c** Surface potential decay curves of different dielectric layers. **d** Calculated total charge trap energy distribution of different dielectric layers. **e** Deep trap and **f** shallow trap energy distribution curves of different dielectric layers determined by peak fitting of the total trap distribution. Reproduced with permission from American Chemical Society [63]

Then, the surface potential decay curve is fitted to determine the $V_s(t)$ function expression. The ISPD curves of the porous PTFE dielectric layers can be fitted using the double-exponential function in Eq. (5.9), where A_1, A_2, t_1, and t_2 are the fitting parameters.

$$V_s(t) = A_1 e^{-t/\tau_1} + A_2 e^{-t/\tau_2} \tag{5.10}$$

Equations (5.8) and (5.9) can be used to calculate the trap energy distributions corresponding to the two terms, A_1 $e^{-t/t1}$ and A_2 $e^{-t/t2}$, and obtain the energy distribution characteristics corresponding to the two trap centers.

Consequently, the surface potential decay curves of the dielectric layers are fitted using the double-exponential function in Eq. (5.10) to extract the parameters A_1, A_2, t_1, and t_2 parameters. The total trap distribution curves displayed in Fig. 5.30d were derived from Eqs. (5.8) and (5.9). It is apparent that the PI layer has the highest total trap density among all tested dielectric layers, indicating that a greater number of charges generated at the surface of porous PTFE could be retained in the trap sites of the PI layer. A Gaussian fitting-based deconvolution process was also conducted to distinctly identify individual deep and shallow trap distributions, with the results presented in Fig. 5.30e and f, respectively.

In summary, both the deep and shallow trap energy distributions of the dielectric layers increase in correlation with the output performance of the associated TENGs, from the lowest to the highest. These findings suggest that charge trapping within the dielectric layers accounts for the high initial surface potential and slow surface potential decay. It implies that charges trapped in the deep traps of the dielectric layers escape slowly, thus being retained for a more extended period compared to those in shallow traps, significantly contributing to the enhancement of TENG output.

The implication is that the inclusion of a dielectric layer with substantial trap energy results in a higher retention of charges in deep traps, leading to superior electrical performance in TENGs. Furthermore, the improved output of TENGs when a dielectric layer is integrated with the triboelectric material is associated with the volume conductivity of the triboelectric material, rather than its electronegativity or surface morphology. Consequently, the creation of deep traps could be a viable strategy to enhance the power and energy harvesting efficiency of TENGs.

5.7 Summary

CE is a phenomenon that has puzzled scientists for centuries, particularly in the context of interactions between insulators. Recent advancements have shed light on the underlying mechanisms, revealing that electron transfer is the dominant process in CE.

The energy band model based on surface states has provided a reliable explanation for CE between insulators, suggesting that electron transfer is the dominant factor. The electron cloud-potential well model models contributed to the general CE mechanism with temperature effects for all types of materials, potentially broadening the range of TENG applications to diverse and extreme environments, including those on other planets and in outer space.

Interface electron transfer-transition induced photon emission are common phenomena that pave the way for a new spectroscopy corresponding to CE at interfaces. This could fundamentally impact our understanding of interactions among solids, liquids, and gases. While current studies have focused on photon emission in solid–solid CE, future research could expand to include Auger electron excitation, X-ray emission, and electron emission in more general cases, thus opening a new field of contact electrification-induced interface spectroscopy.

The self-discharging processes in CE including thermal ion emission and photon excitation. Thermal ion emission has provided a deeper understanding of charge transfer in TENGs, solving a long-standing problem in CE regarding charge properties. It has been established that CE between insulators is dominated by electron transfer, which is a significant finding in the field. Photon excitation suggests that there exists a threshold photon energy for releasing the triboelectric charges from the surface, and photons with energy smaller than this cannot effectively excite the surface electrostatic charges. This process is attributed to the photoelectron emission of the charges trapped in the surface states of the insulator material.

The curvature effect in CE suggests that convex surfaces with positive curvature tend to become negatively charged, while concave surfaces with negative curvature tend to become positively charged. An increase in applied pressure can lead to a reversal of charge transfer direction. The effect of roughness is related to the curvature effect and is currently used to increase the power generation of TENGs by increasing surface roughness.

Atomic electronegativity plays a crucial role in the charge transfer efficiency of TENGs. When two materials with different electronegativities come into contact, electrons spontaneously migrate from the material with lower electronegativity to the one with higher electronegativity, resulting in effective charge separation.

Suppressing charge escaping by introducing material transfer processes between interfaces can lead to ultrahigh triboelectric charge density in open air. This suggests a competitive and cooperative relationship between material transfer and electron transfer in CE. The coexistence and complementary phenomenon of electrons and materials on the contact interface can help further elucidate the mechanism of CE.

The electrical output of TENGs can be adjusted by using forward-polarized or reverse-polarized PVDF as the contact material. This is due to the influence of intrinsic dipole moments on the characteristic potential energy on the surface of films. This finding demonstrates the possibility of adjusting the electrical output of TENGs over a wide range by changing the dipole polarization of a material, offering new insights into CE.

Incorporating a dielectric layer with high trap energy can improve TENG performance by holding more charges in deep traps. The enhanced output is related to the volume conductivity of the triboelectric material, rather than its electronegativity or surface structure. Introducing deep traps could be an effective way to improve TENG power and energy collection efficiency.

In conclusion, the understanding of CE mechanisms, particularly the dominance of electron transfer, has progressed significantly. This knowledge is crucial for the optimization and development of TENGs, which rely on effective charge separation for energy conversion.

References

1. W.R. Harper, The volta effect as a cause of static electrification. Proc. R. Soc. A. **205**, 83–103 (1951)
2. L.H. Lee, Dual mechanism for metal-polymer contact electrification. J. Electrostat. **32**, 1–29 (1994)
3. L.S. McCarty, G.M. Whitesides, Electrostatic charging due to separation of ions at interfaces: CE of ionic electrets. Angew. Chem. Int. Ed. **47**, 2188–2207 (2008)
4. C. Liu, A.J. Bard, Electrons on dielectrics and contact electrification. Chem. Phys. Lett. **480**, 145–156 (2009)
5. S. Piperno, H. Cohen, T. Bendikov, M. Lahav, I. Lubomirsky, The absence of redox reactions for palladium (II) and copper (II) on electrostatically charged Teflon: relevance to the concept of "cryptoelectrons." Angew. Chem. Int. Ed. **50**, 5654–5657 (2011)
6. H. Baytekin, A. Patashinski, M. Branicki, B. Baytekin, S. Soh, B.A. Grzybowski, The mosaic of surface charge in contact electrification. Science **333**, 308–312 (2011)
7. C. Xu, Y. Zi, A.C. Wang, H. Zou, Y. Dai, X. He, P. Wang, Y. Wang, Z.L. Wang, On the electron transfer mechanism in the contact-electrification effect. Adv. Mater. **30**, 1706790 (2018)
8. E.J. Gutman, G.C. Hartmann, Triboelectric properties of two-component developers for xerography. J. Imaging Sci. Technol. **36**, 335–349 (1992)
9. P. Vasandan, Z.H. Mao, W. Jia, M. Sun, Relationship between triboelectric charge and contact force for two triboelectric layers. J. Electrostat. **90**, 147–52 (2017)
10. C. Xu, B. Zhang, A.C. Wang, W. Cai, Y. Zi, P. Feng, Z.L. Wang, Effects of metal work function and contact potential difference on electron thermionic emission in contact electrification. Adv. Funct. Mater. **29**(29), 1903142 (2019)
11. C. Xu, A.C. Wang, H. Zou, B. Zhang, C. Zhang, Y. Zi, L. Pan, P. Wang, P. Feng, Z. Lin, Z.L. Wang, Raising the working temperature of a triboelectric nanogenerator by quenching down electron thermionic emission in contact-electrification. Adv. Mater. **30**(38), 1803968 (2018)
12. S. Li, Y. Zhou, Y. Zi, G. Zhang, Z.L. Wang, Excluding Contact electrification in surface potential measurement using Kelvin probe force microscopy. ACS Nano **10**, 2528–2535 (2016)
13. Z.L. Wang, A.C. Wang, On the origin of contact-electrification. Mater. Today **30**, 34–51 (2019)
14. P.C. Mckinney, G.M. Barrow, A one-dimensional theory of the energetics and ionic character of the hydrogen bond. J. Chem. Phys. **31**, 294–9 (1959)
15. I. Majerzw, I. Olovsson, The shape of the potential energy curves for NHN+ hydrogen bonds and the influence of non-linearity. Phys. Chem. Chem. Phys. **10**, 3043–51 (2008)
16. C. G. Camara, J. V. Escobar, J. R. Hird, S. J. Putterman, Correlation between nanosecond X-ray flashes and stick–slip friction in peeling tape. Nature **455**, 1089-1092 (2008)
17. H. Stöcker, M. Rühl, A. Heinrich, E. Mehner, D.C. Meyer, Generation of hard X-ray radiation using the triboelectric effect by peeling adhesive tape. J. Electrostat. **71**, 905–909 (2013)
18. K. Nakayama, Triboplasma generation and triboluminescence: influence of stationary sliding partner. Tribol. Lett. **37**, 215–228 (2010)
19. J. Wang, S. Li, F. Yi, Y. Zi, J. Lin, X. Wang, Y. Xu, Z.L. Wang, Sustainably powering wearable electronics solely by biomechanical energy. Nat. Commun. **7**, 12744 (2016)
20. A.C. Wang, B. Zhang, C. Xu, H. Zou, Z. Lin, Z.L. Wang, Unraveling temperature-dependent contact electrification between sliding-mode triboelectric pairs. Adv. Funct. Mater. **30**(12), 1909384 (2020)
21. J.F. Hughes, Electrostatics: principles, problems and applications. Phys. Bull. **38**, 424 (1987)
22. F.A. Vick, Theory of contact electrification. Br. J. Appl. Phys. **4**, S1 (1953)
23. A.J. Walton, Triboluminescence. Adv. Phys. **26**, 887–948 (1977)
24. C. Gay, L. Leibler, Theory of tackiness. Phys. Rev. Lett. **82**, 936–939 (1999)
25. V.V. Karasev, N.A. Krotova, B.W. Deryagin, Study of electronic emission during the stripping of a layer of high polymer from glass in a vacuum. Dokl. Akad. Nauk. SSR. **88**, 777–80 (1953)
26. N.E. Harvey, The luminescence of adhesive tape. Science **89**, 460–461 (1939)
27. C.G. Camara, J.V. Escobar, J.R. Hird, S.J. Putterman, Correlation between nanosecond X-ray flashes and stick-slip friction in peeling tape. Nature **455**, 1089 (2008)

28. L. S. McCarty, G. M. Whitesides, Electrostatic charging due to separation of ions at interfaces: contact electrification of ionic electrets. Angew. Chem. Int. Ed. **47**, 2188-2207 (2018)
29. J. Henniker, Triboelectricity in polymers. Nature **196**(4853), 474 (1962)
30. K. Nakayama, Microplasma generated in a gap of sliding contact. J. Vac. Soc. Japan **49**, 618–623 (2007)
31. Z. Ma, J. Fan, J.T. Dickinson, Properties of the photon emission accompanying the peeling of a pressure-sensitive adhesive. J. Adhes. **25**, 63–77 (1988)
32. J. Horvat, R.A. Lewis, Peeling adhesive tape emits electromagnetic radiation at terahertz frequencies. Opt. Lett. **34**, 2195–2197 (2009)
33. D. Li, C. Xu, Y. Liao, W. Cai, Y. Zhu, Z. L. Wang, Interface inter-atomic electron-transition induced photon emission in contact-electrification. Sci. Adv. **7**(39), eabj0349 (2021)
34. C.R. Crowell, The Richardson constant for thermionic emission in Schottky barrier diodes. Solid-State Electron. **8**(4), 395–9 (1965)
35. J. Racko, A. Grmanová, J. Breza, Extended thermionic emission-diffusion theory of charge transport through a Schottky diode. Solid-State Electron. **39**(3), 391–7 (1996)
36. X. Xia, H.Y. Wang, H.Y. Guo, C. Xu, Y.L. Zi, On the material-dependent charge transfer mechanism of the contact electrification. Nano Energy **78**, 105343 (2020)
37. S. Lin, L. Xu, C. Xu, X. Y. Chen, Aurelia C. Wang, B. B. Zhang, P. Lin, Y. Yang, H. B. Zhao, Z. L. Wang, Electron transfer in nanoscale contact electrification: effect of temperature in the metal–dielectric case. Adv. Mater. **31**(17), 1808197 (2019)
38. S. Lin, L. Xu, L. Zhu, X.Y. Chen, Z.L. Wang, Electron transfer in nanoscale contact electrification: photon excitation effect. Adv. Mater. **31**(27), 1901418 (2019)
39. M. Willatzen, Z.L. Wang, Theory of contact electrification: Optical transitions in two-level systems. Nano Energy **52**, 517 (2018)
40. Y. Zhou, S. Li, S. Niu, Z.L. Wang, Effect of contact-and sliding-mode electrification on nanoscale charge transfer for energy harvesting. Nano Res. **9**, 3705 (2016)
41. A. Jaskiewics, M. Debska, The influence of an electric field on the contact electrification between a metal and a polymer. J. Phys. D Appl. Phys. **11**, 2577 (1978)
42. J. Lowell, The effect of an electric field on contact electrification. J. Phys. D Appl. Phys. **14**, 1513 (1981)
43. J. Lowell, Tunnelling between metals and insulators and its role in contact electrification. J. Phys. D Appl. Phys. **12**, 1541 (1979)
44. S. Lin, C. Xu, L. Xu, Z.L. Wang, The overlapped electron-cloud model for electron transfer in contact electrification. Adv. Funct. Mater. **30**(11), 1909724 (2020)
45. C. Xu, B. Zhang, A.C. Wang, H. Zou, G. Liu, W. Ding, C. Wu, M. Ma, P. Feng, Z. Lin, Z.L. Wang, Contact-electrification between two identical materials: curvature effect. ACS Nano **13**(2), 2034–2041 (2019)
46. L.F. Chen, Q.F. Shi, Y.J. Sun, T. Nguyen, C. Lee, S. Soh, Controlling surface charge generated by contact electrification: strategies and applications. Adv. Mater. **30**, e1802405 (2018)
47. S.H. Sung-Ho, Y.K. Hyeog, K. Young-Hwan et al., Triboelectric charging sequence induced by surface functionalization as a method to fabricate high performance triboelectric generators. ACS Nano **9**, 4621–4627 (2015)
48. G. Song, K. Giyoung, Kim et al., Molecularly engineered surface triboelectric nanogenerator by self-assembled monolayers (METS). Chem. Mater. **27**, 4749–4755 (2015)
49. X. Tao, P. Yang, Z. Liu et al., Acid-doped pyridine-based polybenzimidazole as a positive triboelectric material with superior charge retention capability. ACS Nano **18**(5), 4467–4477 (2024)
50. P. Yang, Y. Shi, X. Tao, Z. Liu, X. Dong, Z.L. Wang, X. Chen, Radical anion transfer during contact electrification and its compensation for charge loss in triboelectric nanogenerator. Matter **6**, 1295–1311 (2023)
51. P. Bai, G. Zhu, Y.S. Zhou, S. Wang, J. Ma, G. Zhang, Z.L. Wang, Dipole-moment-induced effect on contact electrification for triboelectric nanogenerators. Nano Res. **7**, 990–997 (2014)
52. J. Zhang, Y. Zheng, L. Xu, D. Wang, Oleic-acid enhanced triboelectric nanogenerator with high output performance and wear resistance. Nano Energy **69**, 104435 (2020)

53. G. Min, L. Manjakkal, D.M. Mulvihill, R.S. Dahiya, Triboelectric nanogenerator with enhanced performance via an optimized low permittivity substrate. IEEE Sens. J. **20**, 6856–6862 (2020)
54. D.W. Kim, J.H. Lee, J.K. Kim, U. Jeong, Material aspects of triboelectric energy generation and sensors. NPG Asia Mater. **12**, 1600665 (2020)
55. M. Lai, B. Du, H. Guo, Y. Xi, H. Yang, C. Hu, J. Wang, Z.L. Wang, Enhancing the output charge density of TENG via building longitudinal paths of electrostatic charges in the contacting layers. ACS Appl. Mater. Interfaces **10**, 2158–2165 (2018)
56. J. Du, X. Yang, J. Duan, Y. Wang, Q. Tang, Tailoring all-inorganic cesium lead halide perovskites for robust triboelectric nanogenerators. Nano Energy **69**, 104514 (2020)
57. R. Mohammadpour, Flexible triboelectric nanogenerator based on high surface area TiO_2 nanotube arrays. Adv. Eng. Mater. **20**, 1700767 (2018)
58. L. Su, Z. Zhao, H. Li, Y. Wang, S. Kuang, G. Cao, Z. Wang, G. Zhu, Photoinduced enhancement of a triboelectric nanogenerator based on an organolead halide perovskite. J. Mater. Chem. C **4**, 10395–10399 (2016)
59. L. Su, Z. Zhao, H. Li, J. Yuan, Z.L. Wang, G. Cao, G. Zhu, High-performance organolead halide perovskite-based self-powered triboelectric photodetector. ACS Nano **9**, 11310–11316 (2015)
60. C. Wu, T.W. Kim, H.Y. Choi, Reduced graphene-oxide acting as electron-trapping sites in the friction layer for giant triboelectric enhancement. Nano Energy **32**, 542–550 (2017)
61. C. Wu, T.W. Kim, J.H. Park, H. An, J. Shao, X. Chen, Z.L. Wang, Enhanced triboelectric nanogenerators based on MoS2 monolayer nanocomposites acting as electron-acceptor layers. ACS Nano **11**, 8356–8363 (2017)
62. H. Jiang, H. Lei, Z. Wen, J. Shi, D. Bao, C. Chen, J. Jiang, Q. Guan, X. Sun, S.-T. Lee, Charge-trapping-blocking layer for enhanced triboelectric nanogenerators. Nano Energy **69**, 105011 (2020)
63. S. Lv, X. Zhang, T. Huang, H. Yu, Q. Zhang, M. Zhu, Trap distribution and conductivity synergic optimization of high-performance triboelectric nanogenerators for self-powered devices. ACS Appl. Mater. Interfaces **13**, 2566–2575 (2021)

Chapter 6
Contact Electrification at Semiconductor Interface—The Tribovoltaic Effect

Abstract Contact-electrification (CE) is a common phenomenon between any two materials or two phases of a material. But contact between two different materials may produce different outputs. When a p-type semiconductor slides on the surface of an n-type semiconductor, the current flowing between the two electrodes at the top of the p-type and the bottom of the n-type is a direct current. This phenomenon is called the "tribovoltaic effect" and was discovered in recent years. The mechanism of the tribovoltaic effect is the release of energy due to the formation of new chemical bonds at the interface by mechanical sliding; the energy released by which can excite electron-hole pairs at the PN junction, and the internal field established at the PN junction separates the electrons from the holes, resulting in a DC output. The energy released by the formation of chemical bonds is called the "bindington", which acts as an internal exciton to excite the electron-hole pairs, similar to the photovoltaic effect. Here, we first review recent studies of the contact electrification involving semiconductors. Then, the mechanism of the tribovoltaic effect is introduced. Surface chemical methods to modulate the tribovoltaic effect are discussed. Finally, a hybrid tribovoltaic effect and its potential applications are described.

6.1 Introduction

As the world enters the era of the Internet of Things (IoT), there is an urgent need for high-entropy distributed energy sources to power a large number of miniature sensors [1, 2]. Triboelectric nanogenerators (TENGs) are a typical distributed energy source capable of harvesting mechanical energy from the environment to power sensors [3–6]. Conventional TENGs usually consist of insulators with conductive electrodes [7–10]. When two insulators are in contact with each other, contact energization (CE) occurs and the insulator surface is charged. Then, driven by mechanical energy, the insulators vibrate on the conductive electrodes, such as human body movement, wind [11, 12], wind [13, 14], and water waves [15, 16], generating induced alternating current (AC). Since most microsensors require direct current (DC), conventional TENG are usually equipped with rectifiers, which increases the size of the device

Z. L. Wang, *Contact-Electrification of Matter*, Springer Series in Surface Sciences 75,
https://doi.org/10.1007/978-3-031-88504-4_6

and reduces the energy harvesting efficiency. Recently, [17, 18] investigated the CE and triboelectric interaction between semiconductors and found that when a p-type semiconductor rubs against an n-type semiconductor it generates a direct DC current, which can be considered as a DC-TENG without a rectifier [19–22].

Due to the advantages of DC output, semiconductor DC- TENG has received much attention. However, the mechanism of CE or triboelectricity involving semiconductors has been unclear until the tribovoltaic effect was first proposed by Wang [21], pointing out that the generation of DC current at the interface of a sliding semiconductor is similar to the photovoltaic effect. Once a p-type semiconductor slides over an n-type semiconductor, the newly formed atomic bonds at the interface release quanta of energy, called "bindington", which excite electron-hole pairs at the PN junction. The electrons and holes are separated by an electric field built into the interior of the PN junction, producing a direct current. This is the tribovoltaic effect, which occurs at the PN-semiconductor interface [23, 24] or at the Schottky junction between the metal and semiconductor interfaces [25, 26]. Photon-like quanta of energy called "bindington" are introduced here [27–29]. In recent years, the tribovoltaic effect at different interfaces has been widely discussed, and the output of DC-TENG based on tribovoltaic effect has been continuously optimized based on different methods such as surface modification and material selection, suggesting that the tribovoltaic effect has a wide range of potential applications in the field of nano-energy.

This paper summarizes the research progress on the tribovoltaic effect in recent years. First, the photovoltaic effect and the energy released during friction are systematically introduced to elucidate the mechanism of the tribovoltaic effect at semiconductor interfaces. Second, the tribovoltaic effect at different interfaces such as metal-semiconductor interface, n-type and p-type semiconductor interfaces is summarized. Third, the surface modification and material optimization methods to improve the TENG output of tribovoltaic effect are systematically summarized. Finally, the contribution of CE to the tribovoltaic effect and the potential applications of the tribovoltaic effect are further discussed.

6.2 Contact Electrification at Semiconductor-Metal Interface

Recent studies have demonstrated that the tribovoltaic effect can occur at various interfaces, including metal-semiconductor interface (Schottky junction) [20, 30–35], P-N junction [33, 36–38], and even liquid-semiconductor interface [27, 28, 39–43].

In 2016, Shao et al. used a metal electrode to contact a conductive polymer without lateral friction, such as polypyrrole (PPy) [30]. They noticed that electrons excited by mechanical forces can cross the Schottky junction to generate DC electricity. Figure 6.1a gives the structure of the device designed by shao et al. A conducting polymer layer sandwiched between two metals is capable of outputting current under compression at both ends (Fig. 6.1b). In order to investigate the mechanism of this novel mechanical-electrical energy conversion, shao et al. measured the I–V curves and electrical impedance spectra (EIS) of the Au/PPy/Al device. As shown in Fig. 6.1c, nonlinear I–V curves characteristic of Schottky diodes were observed on the devices under mild contact (1% strain level to ensure electrical connection) and compression, with stronger rectification of the devices under compressive strain. For the Al/PPy/Al devices, nonlinear I–V relationships were observed with resistances up to 12 kΩ. this suggests that a Schottky contact should be formed between PPy and Al. A Schottky contact between PPy and Al was previously reported and an ohmic contact between PPy and Au [16]. For comparison, shao et al. also tested the mechanical-electrical transition response of Au/PPy/Au and Al/PPy/Al devices. As shown in Fig. 6.1d, the Au/PPy/Au device produced no current under compression, but the Al/PPy/Al device produced uneven positive and negative outputs under compression, with peak output voltages averaging around 0.1 V. However, the output voltage of the Al/PPy/Al device is not only significantly lower than that of the Al/PPy/Au device, but also in the AC mode. These results suggest that a single Schottky contact in the Al/PPy/Au device plays a key role in the conversion of mechanical energy into a DC power source. PPy is a p-type semiconductor with LUMO and HOMO energy levels of − 2.7 and − 5.6 eV, respectively. The work functions of the metals Al and Au are 4.06–4.26 eV and 5.1–5.47 eV, respectively. Figure 6.1e shows the output voltage of the Al/PPy/Au device energy band diagram. The Al-PPy Schottky contact leads to electron flow to the metal side.

Further, the contact electrification between semiconductor-metal interface was observed by atomic force microscopy (AFM). A semiconductor sample was rubbed using a conductive or semiconductor tip, the tribo-current was recorded by the conductive atomic force microscopy (CAFM) mode, and the tribo-voltage was obtained by applying a compensating voltage to eliminate the tribo-current. In addition, the rectification characteristics of the interface can be measured by current-voltage (I-V) experiments using an atomic force microscope (AFM) device, as shown in Fig. 6.2a and b. Conducting AFM tips have lower resistance compared to semiconductor AFM tips, and conducting tips are also more readily available than specifically doped semiconductor tips. Therefore, microscale investigations of the contact electrification involving semiconductor started at the metal-semiconductor interface (Schottky junction, Fig. 6.4a). Liu et al. used a Pt-coated silicon AFM tip to rub molybdenum disulphide (MoS2), which produced a continuous DC current with a maximum density of 106 A/m^{-2}, which is much larger than that of most conventional TENGs. I-V experiments by [20] showed the presence of Schottky barriers at the interface between the Pt-coated tip and MoS_2. Liu et al. combined finite element simulations to propose that the tribo-current under friction is generated by electronic excitation, with the rectifying Schottky barriers at the interface playing a key role. In order to avoid the influence of wear of metal-coated tips on the triboelectrification at the metal-semiconductor interface, Zheng et al. used pure Pt tips on p-type and n-type silicon samples and the direction of the tribo-currents in the external circuit was from the p-type silicon side to the Pt-tip side and from the Pt-tip side to the n-type silicon side [33]. Based on the contact electrification between semiconductor

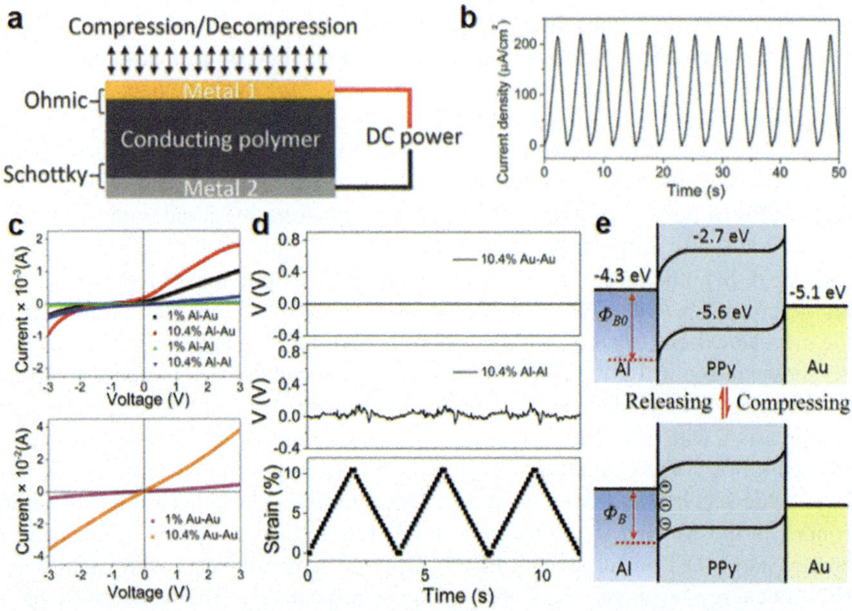

Fig. 6.1 Contact electrification at metal-semiconductor interface. **a** The DC power generator made by conducting polymer and metal. **b** The current density of the DC power generator. **c** The I-V curves at the interface of conducting polymer and metal. **d** The output of the DC power generator. **e** The band diagram of the interface between metal and conducting polymer. Figure with permission [30], copyright © 2015, WILEY-VCH Verlag GmbH & Co. KGaA, Weinheim

and metal, researchers designed a number of DC-TENG [35]. An advantage of these DC-TENG over the contact electrification based on the P-N junction is that the metal has a lower resistance, which corresponds to a lower internal resistance.

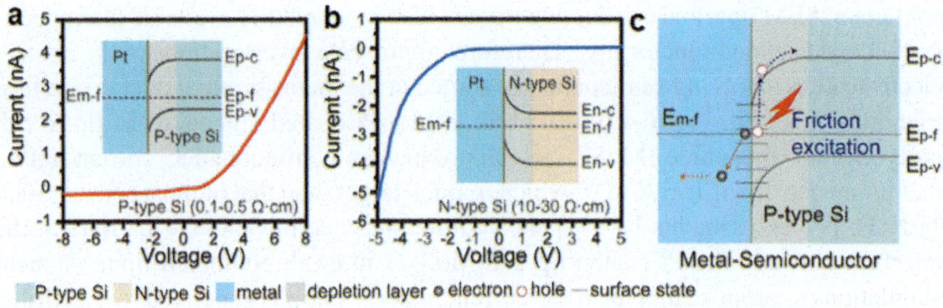

Fig. 6.2 The band diagram of the metal-semiconductor interface. **a** The I-V curve between Pt and p-type silicon. **b** The I-V curve between Pt and n-type silicon. **c** band diagram of the metal-semiconductor interface and generation of electron-hole pairs. Figure with permission [33], copyright © 2020 WILEY-VCH Verlag GmbH & Co. KGaA, Weinheim

6.3 Contact Electrification at Semiconductor-Semiconductor Interface

Similar to the contact electrification at the metal-Semiconductor interface, a DC current can be generated at the P-N junction (Fig. 6.3). Zhang et al. used n-type silicon in frictionless contact with p-type silicon and recorded the current between the two silicon samples [38]. Experimentally, it was found that the direction of the contact current is related to the chemical potential difference between the two silicon samples. n-type silicon usually has a higher Fermi energy level than p-type silicon. When they are in contact, electrons will be transferred from the n-type silicon surface to the p-type silicon surface. When they separate, the transferred electrons will return from the p-type silicon to the n-type silicon in the external circuit, thus generating a current. As shown in Fig. 6.3a–c, Xu et al. further utilized the n-type silicon to rub against the p-type silicon on a macroscopic scale, generating a DC current between the two sliding silicon. In this case, the direction of the DC current is from the p-type silicon in the external circuit to the n-type silicon, which is consistent with the direction of the electric field built into the P-N junction [37]. As shown in Fig. 6.3a, Xu et al. used a pair of semiconductors and metals being used directly as electrodes. One of the substrates is cut to a smaller size and placed on the other substrate so that the two planes face each other. The smaller electrode can be slid or rotated on the larger electrode. On top of the moving electrode, a weight of 100 g is applied to obtain a normal force of 1 N. The electrode is then moved to the top of the moving electrode. Figure 6.1b shows an arbitrary cycle of sliding at a constant speed of 50 mm/s. It was found that a continuous DC current of about 50 nA was generated during the movement regardless of the sliding direction. In stage 1, the top electrode was slid forward (backward) from the rest to a constant velocity with an acceleration of 5 m/s^2. In contrast, in stages 2 and 4, the top electrode decelerates to a complete stop with a velocity of 5 m/s^2. Xu et al. proposed a theoretical model for their experiments. Assuming the semiconductor electrode surfaces to be ideal, when the electrodes are in contact (Fig. 6.3c), electrons diffuse from the n-type (E_2) to the p-type semiconductor electrode (E_1) due to the difference in their chemical potentials (holes diffuse from E_1 to E_2), leading to the formation of a PN junction on the two contact surfaces. As a result, positively charged space charges remain in the depletion region of the n-type semiconductor, while negatively charged space charges reside on the p-type side, creating a built-in electric field in the junction pointing from the n-type to the p-type semiconductor electrode. In thermal equilibrium, a common flat Fermi energy level is established across the two entire electrodes, resulting in a PN junction built-in potential. During frictional sliding motion, energy is released through bond breaking at the contact surfaces to produce electrons and holes, which are expelled from the PN junction by the built-in electric field, forming a DC current. Due to the overall neutrality of the charge, the space charge in the p-n junction can be regarded as an effective dipole with a dipole moment pointing from the negatively charged space charge in the p-type electrodes to the positively charged space charge in the n-type electrodes perpendicular to the contact surfaces under thermal equilibrium.

As a result, the effective dipole moment deviates slightly from the dipole moment under thermal equilibrium, causing a change in the dipole potential energy. The energy released by the newly formed dipole can excite electron-hole pairs. On the microscopic scale, Zheng et al. performed CAFM scans of p-type and n-type silicon with different doping concentrations using n-type diamond-coated silicon tips. The results showed that tribo-currents could be generated at the interface of n-type tip and p-type silicon, but not at the interface of n-type tip and n-type silicon, and that the direction of tribo-currents was from p-type silicon to the diamond-coated tip in the external circuit. In addition, I-V tests showed that the current between the diamond-coated tip and the p-type silicon sample was highly dependent on the rectification characteristics of the tip-sample interface. The stronger the interfacial rectification characteristics, the higher the tribo-current.

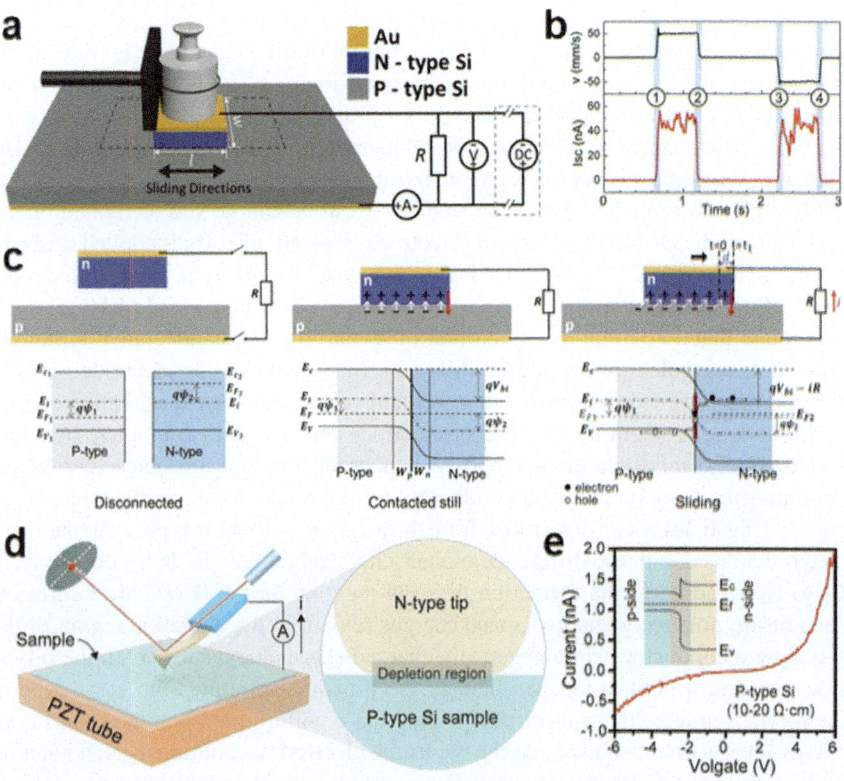

Fig. 6.3 Contact electrification at semiconductor-semiconductor interface. **a** The experiment setup for studying contact electrification between semiconductor-semiconductor interface. **b** The current generation between sliding p silicon and n silicon interface. **c** The band diagram for the generation of current between p-type and n-type silicon interface. **d** The use of AFM for the studying of contact electrification between p-type and n-type semiconductor interface. **e** The I-V curve at the p-type and n-type semiconductor interface tested by using AFM. Figure with permission [33], copyright © 2020, WILEY-VCH Verlag GmbH & Co. KGaA, Weinheim

6.4 Contact Electrification at Semiconductor-Liquid Interface

In recent years, contact electrification between liquid and solid has attracted much attention due to the invention of the liquid-solid TENG and its potential effects on chemical reactions [16, 29, 44–54]. Similar to the contact electrification effect at insulator and metal interface, the contact electrification occurs not only at the solid semiconductor-solid semiconductor interface but also at the liquid-semiconductor interface. Lin et al. designed an experiment to study the contact electrification at the interface between aqueous solution and silicon by dragging a DI water droplet on the surface of a silicon wafer with a syringe-conducting needle to slide, and recorded the tribo-currents and tribo-voltages, as shown in Fig. 6.4a [43]. The results show that a DC tribo-current is generated during the sliding process, as shown in Fig. 6.4b. In the external circuit, the tribo-current is directed from p-type silicon to aqueous solution or from aqueous solution to n-type silicon, which implies that electrons move from the p-type silicon side to the aqueous solution side or from the n-type silicon side to the aqueous solution side at the interface. Combined with I-V characterization, it was found that the direction of the tribo-current at the sliding aqueous solution-silicon interface is consistent with the direction of the built-in electric field at the interface. During the sliding process, the droplet contacts the new surface of the solid semiconductor, forming a new chemical bond and releasing a quantum of energy (called a "bindington"), as shown in Fig. 6.4c. An energy band diagram model was proposed, the whole process of the contact electrification at the liquid-solid interface is shown in Fig. 6.4d. First, a built-in electric field exists at the liquid-solid interface due to electron diffusion driven by Fermi energy level differences. During the sliding process, the "bindington" will be released, and electron-hole pairs will be excited at the interface, which will then be separated by the built-in electric field, generating a DC current.

Further liquid-solid sliding experiments under light irradiation were performed [28], which showed that light irradiation enhances the friction current between the aqueous solution and the silicon wafer, which is due to the increase of the electric hole pairs at the interface. Zheng et al. mounted a silicon wafer and a quartz tube on a moving platform and a stationary stand, respectively, as shown in Fig. 6.5a. The quartz tube dragged the DI water droplet to slide on the surface of the silicon wafer. Figure 6.5b gives a cycle of the relative position and relative velocity between the DI water droplet and the silicon wafer. In the experiment, a copper wire was fixed in the tube as an electrode in contact with deionized water, and the short-circuit current and open-circuit voltage were measured with an electrostatic meter (Keithley 6514). During the sliding process, the distance between the quartz tube and the silicon wafer was about 0.3 mm. these ensured that the connection between the quartz tube and the deionized water was transparent, and that the light emitted by the xenon lamp could reach the interface between the deionized water and the silicon wafer. Figure 6.5c and d give the tribo- voltages and tribo-currents of DI water

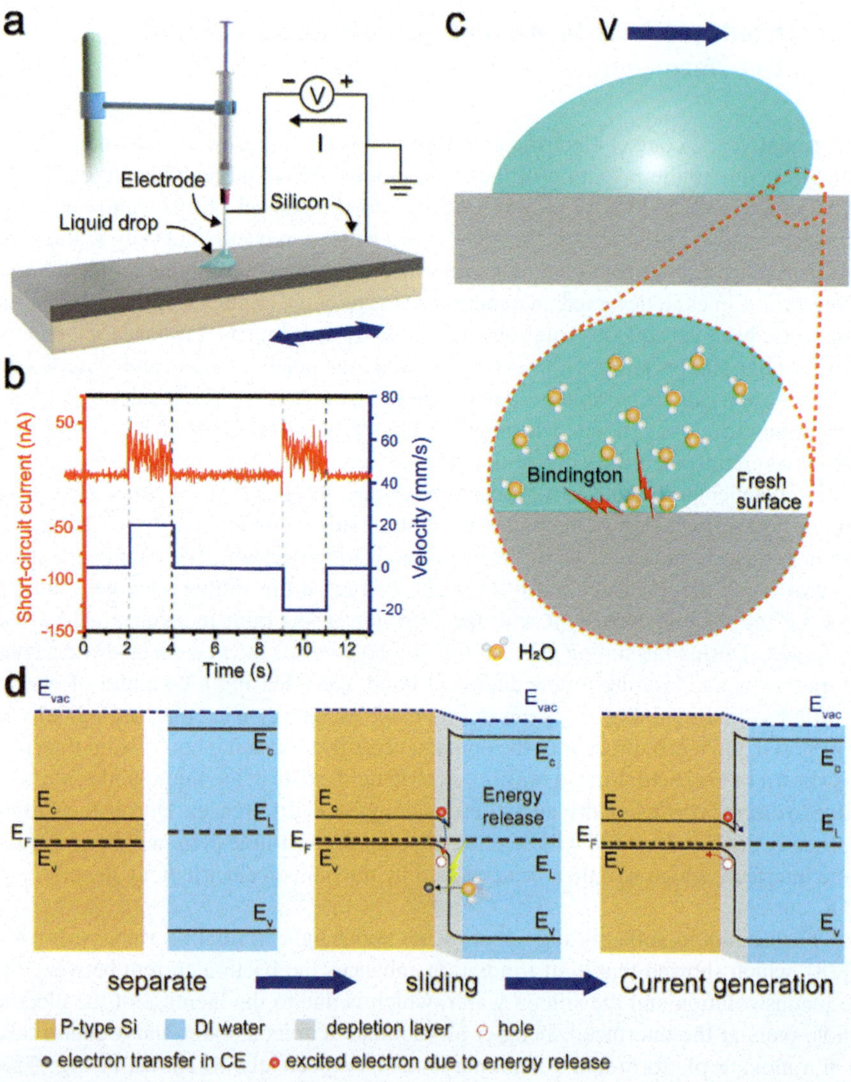

Fig. 6.4 Contact electrification between liquid-semiconductor interface. **a** The experiment setup for the studying of contact electrification between liquid-solid interface. **b** The output of current at sliding liquid-semiconductor interface. **c** The generation of bindington at liquid-solid interface. **d** The band diagram for the generation of current at sliding liquid-semiconductor interface. Figure with permission [43], copyright © 2020, Elsevier Ltd.

droplets sliding on a silicon wafer (n-type, Ωcm resistivity of 1) under dark conditions and light (1000 nm, 28.3 mW/cm^2 light intensity). Under dark conditions, the tribo-current and tribo-voltage are about − 0.3 μA and − 0.3 V, respectively, indicating that the direction of the tribo-current is moving from the n-type Si side to the DI water side at the interface. The current-voltage curves under dark conditions show the rectification performance of the junction. It indicates that a built-in electric field is formed at the interface when DI water is in contact with the n-type Si wafer, and the direction of the built-in electric field at the interface also points from the n-type Si side to the deionized water side. When the light was turned on, the light irradiated at the interface of Si and DI water, and the tribo-voltage and tribo-current were affected by the light irradiation (Fig. 6.5c, d). The voltage and current signals were extracted for 6 cycles before and after the light was turned on, as shown in Fig. 6.5e and f. The photovoltage and photocurrent are about − 0.08 V and − 0.3 μA (red region), respectively, which are the same as in the absence of sliding. For the friction signal component, the tribo-voltage decreases slightly while the tribo-current increases significantly. These results cannot be explained by ion desorption/adsorption on the silicon surface. This is because both the voltage and current generated by ion desorption/adsorption depend on the rate of desorption/adsorption. A decrease in the ion transfer induced voltage should result in a decrease in the ion transfer induced current. Therefore, the tribo-voltage and tribo-current should be caused by electron transfer. Notice that the photovoltage and photocurrent are in the same direction as the tribo-voltage and tribo-current, which is consistent with the physical mechanism of the tribovoltaic effect. The energy released by electron leaps during friction is hardly able to excite more electron-hole pairs because its concentration is already high under light irradiation. As a result, the tribo-voltage decreases when the photovoltage is generated. And the increase in tribo-current is due to the high concentration of electron-hole pairs at the interface, which can be considered as a decrease in internal resistance. According to the observation and explanation, the generation of tribo-voltage and tribo-current is caused by electron transfer rather than desorption/adsorption of ions.

A key point in understanding the contact electrification at the aqueous solution-semiconductor interface is that the aqueous solution can be considered as a liquid semiconductor [55–58]. When an aqueous solution is in contact with a solid semiconductor, electrons will diffuse from one surface to the other to establish a built-in electric field, as occurs in a P-N junction. The effect of temperature on the contact electrification between liquid-semiconductor supports the production of "bindington". Zheng et al. placed silicon wafers on a fixed heating stage and mounted quartz tubes with Cu electrodes on a moving stand as shown in Fig. 6.8a. In order to eliminate the noise generated by the heating stage during the testing process, the heating stage was first raised to 95 °C, and the tribo-current outputs at different temperatures were tested during the cooling process after the power was turned off. The Si/DI water interface temperature was measured with an infrared camera. In the experiments, the copper wire was in contact with deionized water. Tribo-current and tribo-voltage (or open circuit voltage and short circuit current) were measured with an electrostatic meter (Keithley 6514). The external tribo-voltage and tribo-current

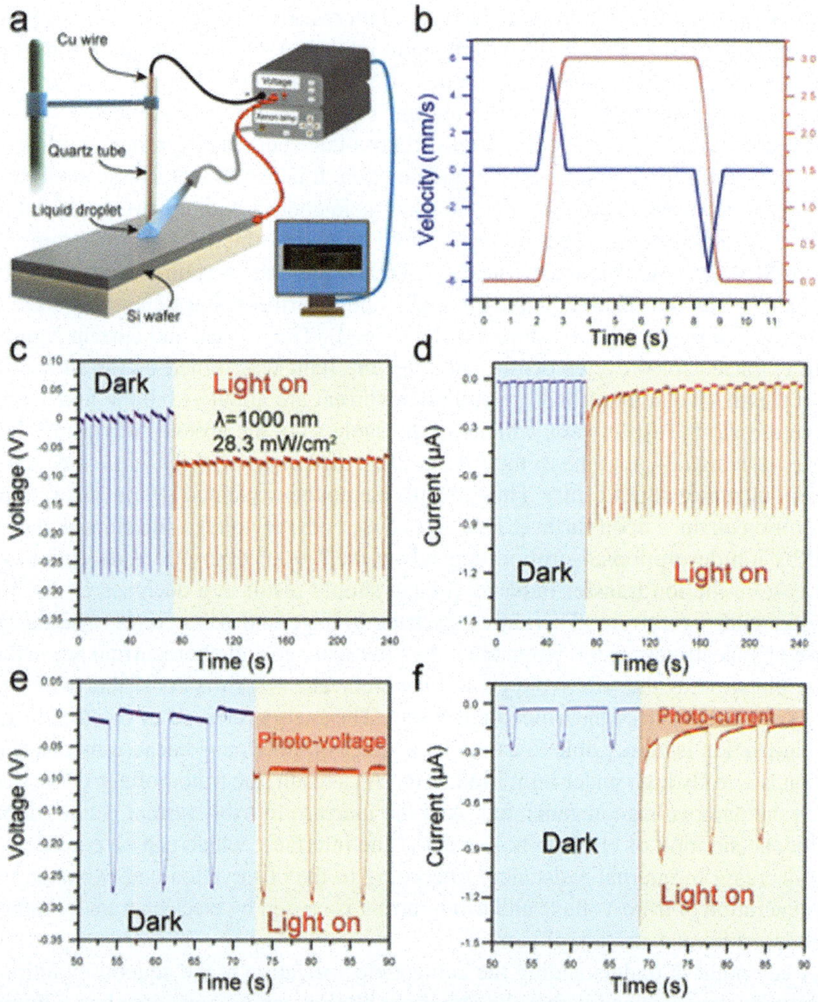

Fig. 6.5 The light irradiation effect on the contact electrification between liquid-solid interface. **a** The experiment setup for the light irradiation effect on the liquid-solid contact electrification. **b** The movement of the liquid droplet. **c, d** The current generated at the liquid-semiconductor interface with and without light irradiation. Figure with permission [28], copyright © 2021 Elsevier Ltd.

from the wafer side to the DI water droplet side were defined as positive. A schematic of the tribo-current generated by the tribovoltaic effect is given in Fig. 6.8b. When a water molecule collides with an atom belonging to the Si surface, energy will be released due to the bonds formed at the interface, called "binding". Then, the electron-hole pair will be excited by "binding" and separated by the built-in electric field at the DI water-Si interface. The electron and hole are transferred to the n-type Si side and the water side, respectively. In this case, a hole may be shared by several water molecules or water clusters instead of belonging to one molecule. Figure 6.8c gives the velocity and displacement profiles of a water droplet sliding on a silicon wafer. The contact angle between water and N-type silicon wafer (resistivity of 0.1 Ω cm) at different temperatures is shown in Fig. 6.8d. It can be clearly seen that the contact angle decreases from 86.72° to 60.36° as the temperature increases from 30 to 90 °C. In the experiment, a column of water with a length of about 1.0 cm was sealed inside the tube under the balanced effect of gravity and capillary action, and one end of the column was closed with silica gel. This structure has the ability to replenish evaporated water in a timely manner, resulting in a stable output of DI water droplets sliding on the wafer surface at high temperatures. It was excluded that the contribution of water evaporation to power generation because the external circuit is connected differently in the hydroelectric effect and the tribovoltaic effect, where two electrodes are connected to the solid surface in the hydroelectric effect, one electrode is connected to the liquid side and the other electrode is connected to the solid side for collecting the tribovoltaic current. The tribo-voltage and tribo-current for the tribovoltaic effect of P-type Si (resistivity of 0.01 Ω cm) with DI water at 90 °C are given in Fig. 6.8e and f. The tribo-voltage and tribo-current are stabilized over nine output cycles to 0.15 V and 70 nA, respectively. With the increase of Si resistivity and temperature, the reverse current of the I-V curves increases, indicating the existence of a depletion zone at the water droplet/Si interface, and the width of the depletion zone and the intensity of the built-in electric field are closely related to the resistivity and temperature of n-type Si. The results show that the tribo-current between the aqueous solution and the solid semiconductor increases as the temperature increases due to the more intense collisions of the water molecules with the solid surface, which results in the release of more "bindington" (Fig. 6.6).

6.5 Tribovoltaic Effect

6.5.1 *Mechanisms of Tribovoltaic Effect*

6.5.1.1 A Brief Introduction to Photovoltaic Effect

The mechanism of the contact electrification or triboelectrification involving semiconductors has not been clear until the tribovoltaic effect was first proposed by Wang [21]. The tribovoltaic effect is usually analogous to the photovoltaic effect except the source for exciting the electron-hole pair is different. Understanding the photovoltaic effect is important for studying the tribovoltaic effect. The photovoltaic effect is the physical basis for solar cells to convert light energy into electricity, which is an important clean energy source [59–61]. After decades of efforts, the mechanism of the photovoltaic effect has been relatively clearer compared to the friction voltammetry effect [62–65]. As shown in Fig. 6.7, PV cells are usually composed of different semiconductors. The energy band structures of p-type and n-type semiconductors

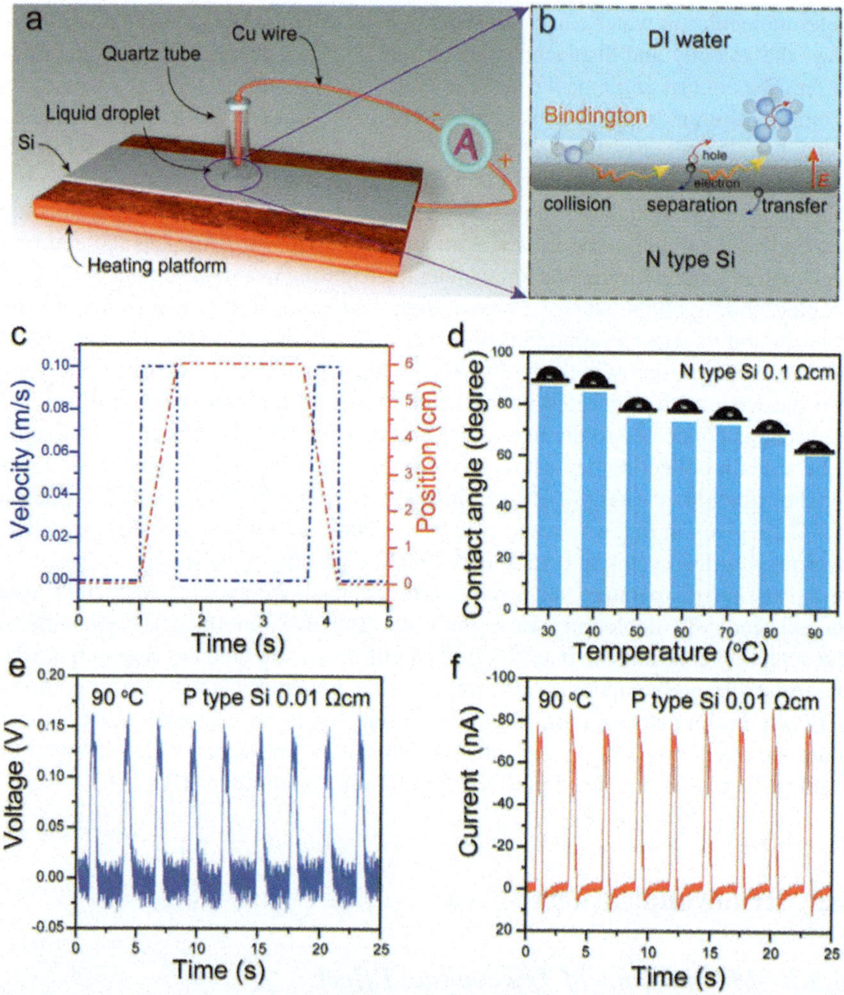

Fig. 6.6 The temperature effect on the contact electrification between liquid-solid interface. **a** The experiment setup for the temperature effect on the liquid-solid contact electrification. **b** The generation of "bindington" at liquid-semiconductor interface. **c** The movement of the liquid droplet. **d** The contact angle of the liquid on the semiconductor surface. **e, f** The current generated at the liquid-semiconductor interface under 90 degree temperature. Figure with permission [27], copyright © 2021, Wiley-VCH GmbH

before contact are given in Fig. 6.7a. When an n-type semiconductor is in contact with a p-type semiconductor, the majority carriers (electrons) in the n-type semiconductor diffuse toward the p-type semiconductor side, and holes diffuse from the p-type semiconductor side to the n-type semiconductor side until the Fermi energy levels of the two semiconductors are in equilibrium and a built-in electric field is established. Then, the incident photons provide energy to excite the electron-hole

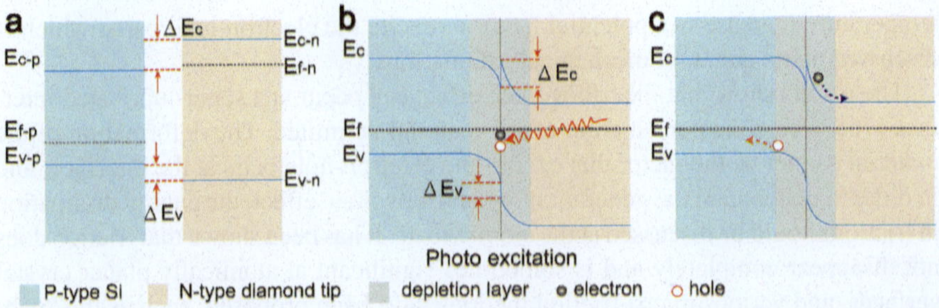

Fig. 6.7 Energy band diagram of the photovoltaic effect. **a** Energy band diagram of the p-type and n-type semiconductors before contact. **b** Energy band diagram of the PN junction and photoexcitatatoin of electron-hole pairs. **c** Seperation of electron-hole pairs under built-in electric field. Figure with permission [33], copyright © 2020, WILEY-VCH Verlag GmbH & Co. KGaA, Weinheim

pairs (Fig. 6.7b), which are further separated by the built-in electric field (Fig. 6.7c), realizing the conversion of light energy to electrical energy.

Photovoltaic cells have been successful in the field of clean energy and are now an important means of collecting clean energy. The friction voltaic effect is similar to the photovoltaic effect, the only difference is that the friction voltaic cell converts mechanical energy into electrical energy instead of light energy into electrical energy. And friction volt batteries, as a complement to photovoltaic batteries, are expected to be another important way to obtain clean energy.

6.5.1.2 Bonding and Energy Release During Friction

Unlike the photovoltaic effect, in the tribovoltaic effect, the energy to excite electron-hole pairs at the interface comes from friction. How friction excites electron-hole pairs at an interface is key to understanding the tribovoltaic effect. Friction is the resistance caused by the interaction between two sliding contact surfaces, and energy is lost at almost any sliding interface [66–69]. The tribovoltaic effect can be thought of as converting some of the energy dissipated during friction into electrical energy. Energy dissipation in friction is a complex problem that has been discussed for a long time. On the macroscopic scale, friction is proportional to the normal load, which is often referred to as Coulomb's law [70]. Different friction theories have been proposed to explain Coulomb's law, such as the mechanical engagement theory, where friction originates from surface roughness and frictional energy is lost in roughness engagement, collisions, and plastic deformations, such as embedding in soft surfaces through hard bumps. Adhesion effects in friction were also considered by [72, 73], indicating that friction may arise from a sharp increase in temperature at the microscopic scale leading to adhesion of two projections at the interface [74, 75]. It can be concluded from the macroscopic friction theory that the energy lost to friction is converted into thermal and deformation energy. The deformation of the material is unlikely to excite the electron-hole pairs at the interface, but the temperature increase is a potential factor to excite the electron-hole pairs, which is discussed using quantum mechanical calculations.

The results show that the tribovoltaic effect can occur at a super-lubricated interface [76], where the deformation of the material is limited. The deformation of the material cannot be the factor that excites the electron-hole pairs at the interface, and in order to understand the generation of the tribovoltaic effect, the energy dissipation in friction should be discussed at the atomic scale. It has been shown that friction does not disappear completely and is sometimes significant at atomically planar crystal surfaces, and various micro-friction theories have been proposed, such as the "pebble" model [77, 78], the oscillator model [79, 80], the phonon friction model [81] and the bonding model [82, 86]. In these models, energy is dissipated in the form of atomic vibrations and chemical bond formation, where the energy released by chemical bond formation may be large enough to excite electron-hole pairs. As shown in Fig. 6.8, some silicon atoms have a −OH group at the end of the atom during sliding, the −OH group forms a siloxane bond (Si–O–Si) with the −OH group on the other surface through a silanol condensation reaction, releasing an H_2O molecule [87, 88]. On the other hand, some of the formed siloxane bonds are torn, absorbing energy and causing a loss of mechanical energy and friction [89]. As the sliding continues, siloxane bonds are formed and broken, and the formation of chemical bonds releases a quantum of energy that is named "bindington". The energy value of the "bindington" depends on the energy difference between the electron orbitals before and after the bonding, and can be as high as a few electron volts. The energy of a few electron volts is sufficient to excite electric hole pairs at least at the semiconductor interface [27, 28]. In this process, the formation and breaking of chemical bonds can be viewed as a pump that converts mechanical energy into electrical energy. In fact, the friction process is very complex, and the current experimental studies on the tribovoltaic effect do not discuss the friction process in depth, which is a key issue for future research on the tribovoltaic effect. But in any case, the formation of chemical bonds may be an important source of energy for excited electrons.

In general, phonons are excited locally at interfaces due to local elastic and possibly plastic deformation during material sliding. However, we believe that phonon energies in the range of 10–50 meV are not sufficient to excite electron-hole pairs in PN junctions, whereas silicon PN junctions typically require energies above 1.1 eV for excitation. Localized heating may not be a problem at low sliding speeds, but may be more severe at high sliding speeds. In such cases, the situation may be more complicated.

6.5.1.3 Friction Produced DC Current During Sliding—The Tribovoltaic Effect

Based on the above discussion, the tribovoltaic effect can be clearly described as analogous to the photovoltaic effect, where we emphasize the role of the formation of chemical bonds at the interface to release energy. The energy band diagram of a p-type semiconductor in contact with an n-type conductor is given in Fig. 6.9a.

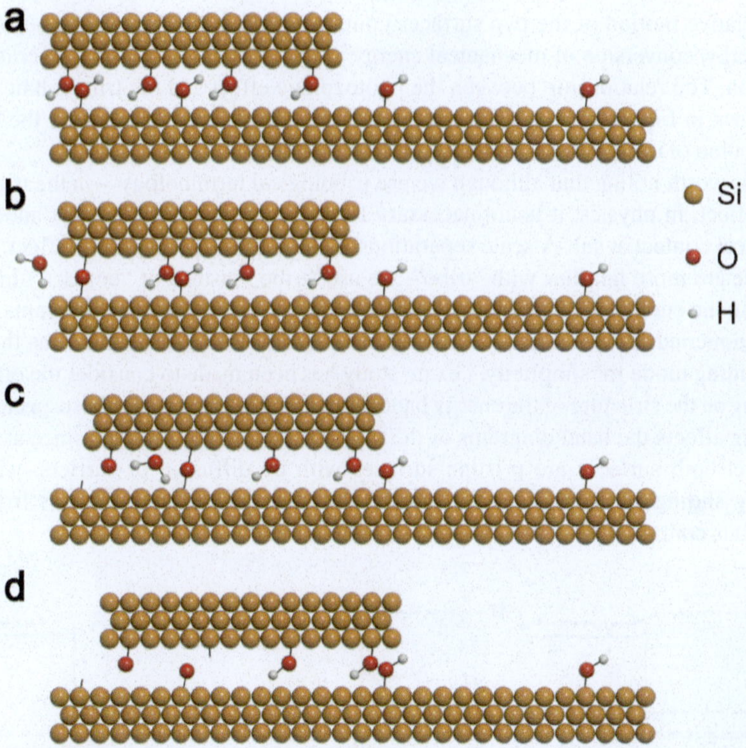

Fig. 6.8 Schematic of the chemical reactions at a sliding silicon interface in which the formation of chemical bonds occurs. Figure with permission [125], copyright © 2022 Elsevier Ltd.

When two semiconductors are in contact with each other, electrons diffuse from the n-type semiconductor side to the p-type semiconductor side, while holes diffuse from the p-type semiconductor side to the n-type semiconductor side, resulting in a built-in electric field to compensate for the difference in Fermi energy levels. At the semiconductor surface, the periodic arrangement of atoms is terminated by dangling bonds on the surface atoms. When there is relative motion between the upper and lower surfaces, the distance between the atoms belonging to the two surfaces changes. If the distance between the atoms is close enough, the dangling bonds interact, further forming a new chemical bond and releasing a quantum of energy ("bindington") (Fig. 6.9b), since in general the formation of bonds is energetically favourable. The "bindington" plays the role of an incident photon in the photovoltaic effect and excites an electron-hole pair at the interface (Fig. 6.9c). Then, under the action of the built-in electric field, the electron-hole pairs are separated, and the electrons are transferred to the n-type semiconductor side and the holes to the p-type semiconductor side (Fig. 6.9c), so that the tribo-current in the external circuit is directed from the p-type semiconductor side to the n-type semiconductor side. And it should be mentioned that the chemical bond formed between atoms is torn apart by mechanical forces as the relative motion of the two surfaces continues, which is a process of adsorption of energy, conversion of mechanical energy into chemical energy, and generation of friction. The relationship between the photovoltaic effect and the tribovoltaic effect is shown in Fig. 6.9d. The results show that the only difference between the two is the source of energy to excite the electron-hole pair.

It is worth noting that although we use tribological terminology—in the tribological effect, in physics, it is not necessary for one material to rub against another—physical contact at sub-Å scale separations is sufficient to produce this effect. Since people are more familiar with "tribo", we use "tribo" instead of "contact". In addition, in the current theory of the tribovoltaic effect, the energy band diagrams of the two semiconductors in the contact mode are considered to be the same as those in the sliding mode for simplicity, i.e., no study has been made to consider the effect of sliding on the structure of the energy band diagrams. In fact, it is highly suspected that sliding affects the band diagrams of the two semiconductor surfaces, since at initial contact both surfaces are pristine surfaces with no diffusion of carriers, whereas during sliding one surface is pristine and the other has diffused carriers from the previous contact. This effect is subject to further discussion.

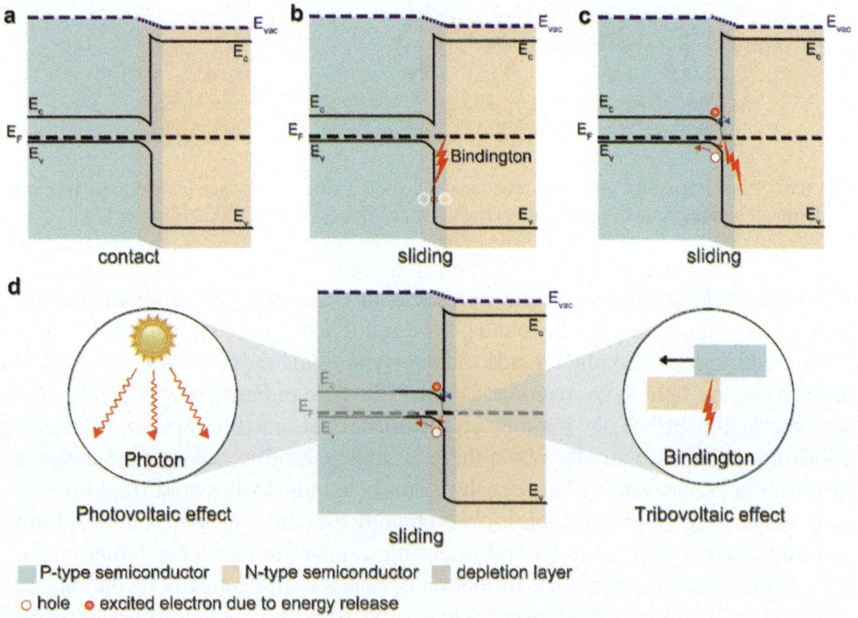

Fig. 6.9 Energy band diagram of the tribovoltaic effect. Energy band diagram of the p-type semiconductor and the n-type semiconductor **a** in contact and **b** in sliding. **c** The movement of the electrons and the holes at the p-type semiconductor and p-type semiconductor interface in sliding. **d** The illustration of the relation between photovoltaic effect and tribovoltaic effect. (E_c is the bottom of the conduction band, E_f is the Fermi level of the semiconductors, and Ev is the top of the valence band. Figure with permission [125], copyright © 2022 Elsevier Ltd.

6.5.2 *Interfacial Engineering for Tribovoltaic Effect*

The development of high-performance DC nanogenerators based on the tribovoltaic effect has been a hot topic of academic interest. The output tribovoltaic current and voltage are highly dependent on the interfacial properties of the tribovoltaic nanogenerator. The actual contact area, loading and surface state density of semiconductors have been reported to have a significant effect on the tribovoltaic effect. Therefore, changing the surface properties of semiconductors is an important way to improve the output power of tribovoltaic nanogenerators.

6.5.2.1 Surface States Enhancement

A surface state can be understood as the position of an electron on a surface due to the termination of the periodic arrangement of surface atoms. Theoretically, the density of semiconductor surface states can affect the tribovoltaic effect in three ways. First, electrons in the surface states are more easily excited than those in the valence band because less energy is required. The higher the density of surface states, the more electrons the surface can hold, and the more electrons that are excited, resulting in more electron-hole pairs and thus an increase in the tribovoltaic current. Second, a higher surface density of states corresponds to more surface contact bonds, leading to the formation of more chemical bonds and the release of more "bindingtons" during sliding. Third, the semiconductor surface state density can affect the interfacial band structure and the intensity of the built-in electric field, which in turn affects the tribovoltaic current and voltage.

Zheng et al. used inductively coupled plasma reactive ion etching (ICP-RIE) to enhance the surface density of states of silicon and tested the tribovoltaic effect between two semiconductors before and after ICP-RIE treatment, as shown in Fig. 6.10a [37]. The results showed that the tribovoltaic current between the tip and the treated semiconductor increased dramatically, as shown in Fig. 6.10b and c. I-V characterization showed that the ICP-RIE treatment of the semiconductor did not enhance the built-in electric field at the semiconductor interface. Whereas, x-ray photoelectron spectroscopy (XPS) indicates that fluorine and sulphur elements adsorb on the surface of the Si sample, which may introduce new surface states. On the other hand, the ICP-RIE treatment may produce defects on the surface, which may also increase the density of surface states in the Si sample. From these results, it can be concluded that increasing the density of semiconductor surface states can enhance the tribovoltaic effect, which is supported by some studies on dynamic Schottky generators [38].

At the solid-solid interface, the actual contact area is much smaller than the nominal contact area due to the presence of rough bumps. Increasing the actual contact area and involving more semiconductor surface states in friction is another implication of increasing the density of surface states. An effective way to increase the actual contact area is to choose flexible contact materials. For example, the

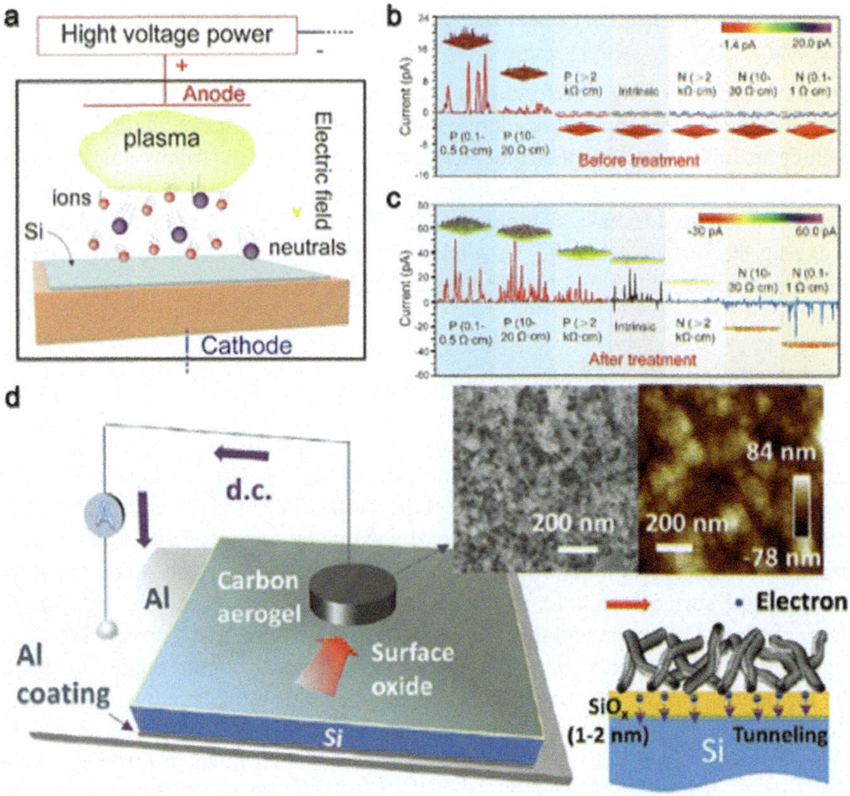

Fig. 6.10 Effect of the surface states density of the semiconductor and the contact area on the tribovoltaic effect. **a** The setup of the ICP-RIE treatment for enhancing the surface states density of the silicon. The tribovoltaic current between a N-type tip and the silicon samples **b** before ICP-RIE treatment, **c** after ICP-RIE treatment. **d** The tribovoltaic effect at the SiO_2 (with oxide layer) and carbon aerogel. Figure with permission [91], copyright © 2019 WILEY-VCH Verlag GmbH & Co. KGaA, Weinheim

presence of a large number of nanoscale protrusions on the surface of a flexible carbon aerogel increases the actual contact area at the interface. As a result, the carbon aerogel-Sio_2–Si tribovoltaic nanogenerator has a high output performance (Fig. 6.10d) [90].

6.5.2.2 Interfacial Electric Field Enhancement

In the tribovoltaic effect, excited electron-hole pairs are separated by a built-in electric field. A stronger built-in electric field separates the electron-hole pairs more efficiently. The Fermi energy level difference between two semiconductors before contact is an important factor affecting the strength of the built-in electric field at the interface. It has been reported that the tribovoltaic-current and voltage between the diamond tip and the silicon sample are highly dependent on the doping concentration of the silicon sample, which affects the Fermi energy level of silicon [37]. The doping concentration of the silicon sample also affects the tribovoltaic effect at the liquid-semiconductor interface [47]. Experiments have shown that the tribovoltaic currents and voltages of aqueous silicon solutions also increase with increasing silicon doping concentration within a certain range. Therefore, choosing a suitable semiconductor doping concentration to establish a relatively strong built-in electric field at the interface is an important method to optimize the tribovoltaic nanogenerator. Introducing an insulator layer at the semiconductor interface, is another method to enhance the built-in electric field and the interfacial barrier, thus further improving the output performance of the tribovoltaic nanogenerator [92].

Using GaN materials in the tribovoltaic effect is a way to increase the interfacial electric field, since GaN has a relatively low resistance. Based on the ball-and-disk structure, Zhang et al. proposed a TVNG consisting of a uGaN thin film (Fig. 6.11a-i) and a steel ball (Fig. 6.11a-ii), which maintain good physical contact to form a metal-semiconductor heterojunction [93]. It exhibits typical Schottky characteristics in the static state under a pressure of 3N (Fig. 6.1b). When a steel ball is pressed on the uGaN surface at 120 rpm (rotation radius $R = 16$ mm), the TVNG produces a maximum open-circuit voltage of 130 v at steady state (Fig. 6.11c) and an average short-circuit current of 11.8 μA (Fig. 6.11d). The amplified waveforms show periodic fluctuations in the open-circuit voltage and short-circuit current, which may be due to the load pressure or surface difference. The Hertzian theory is commonly used to estimate the contact area of a spherical disk model. The contact radius (Rc) of the disk can be estimated by Hertzian contact, and based on the Hertzian contact theory, the contact area of the ball-on-disk can be calculated, and the average short-circuit current density of the ball-on-disk TVNG can be up to 4.5 kAm^{-2}. After rotating for 3 cycles, the TVNG generates a short-circuit charge of 20.6 μC. The transferred charge density per cycle can be up to 2.6 kCm^{-2} (Fig. 6.11e). The friction coefficient of the TVNG was monitored and was found to fluctuate between 0.18 and 0.35 with an average value of about 0.253 (Fig. 6.11f). The power density versus short-circuit current variation curve was found to be a function of resistance in the range of 104–108 Ω (Fig. 6.11g). The peak power density reaches up to 153.1 kWm^{-2} at an impedance of 2.5 MΩ. At the optimal impedance, an APD of 49.2 kWm^{-2} is obtained by integration. Zhang et al. listed some normalized APDs and structures for high-performance tvng and teng (Fig. 6.11h). The normalized APD of the ball-and-disk TVNG is significantly increased by a factor of 282 (24.6 kWm^{-2} Hz^{-1}) compared to the previous results (87.26 Wm^{-2} Hz^{-1}).

6.5.2.3 Surface Chemistry for Tribovoltaic Effect

Surface functional groups play an important role in many effects, especially those related to friction, such as contact electrification and tribovoltaic effect. The results suggest that surface functionalization is an effective way to modulate the output

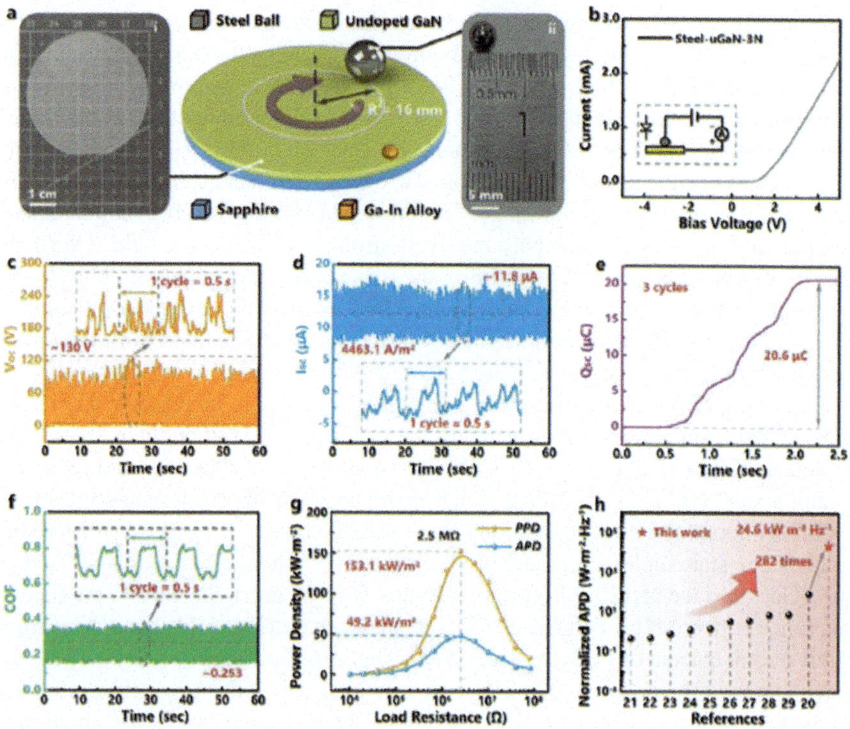

Fig. 6.11 The control of the interfacial electric field at semiconductor interface. **a** The experiment setup for measuring the current between sliding semiconductor interface. **b** The I-V curve at the semiconductor interface. **c–e** The output of the sliding semiconductor interface. **f** The coefficient of the friction. **g** power density generated at the semiconductor interface. **h** The references for the output of the semiconductor sliding interface. Figure with permission [93], copyright © 2023 Wiley-VCH GmbH

performance of tribovoltaic nanogenerators [94, 95]. The surface work function of semiconductors is very sensitive to the surface functional groups, which can be modulated by surface modification, which in turn controls the height of the potential barriers and the intensity of the built-in electric field at the semiconductor interfaces, further affecting the tribovoltaic effect. In addition, the energy of excited electron-hole pairs is released through the formation of chemical bonds during friction. The energy of the released "bindington" depends on the functional groups on the two sliding surfaces and the chemical bonds formed at the interface. The selection of functional groups with high chemical energy helps to release high-energy "bindingtons" during friction and improves the output performance of tribovoltaic nanogenerators.

Ferrie et al. grafted a layer of organic molecules with a range of terminal functional groups onto a silicon surface via silicon hydrogenation chemistry (Fig. 6.12a). [96, 97] investigated the tribovoltaic effect between functionalized silicon and Pt tips using CAFM (Fig. 6.11b–f). The results showed that the tribovoltaic current between the Pt tip and the monolayer functionalized silicon was significantly affected by the functional groups at the end of the monolayer. The influence of the functionalized surface's work function, adhesion, and wettability on the tribovoltaic effect is discussed. This work connects the fields of surface chemistry and molecular electronics, demonstrating for the first time the influence of surface chemical functional groups on the tribovoltaic effect. However, the formation of chemical bonds at the interface is not mentioned in this work. According to the mechanism of the tribovoltaic, controlling the bond formation at the interface is one of the ideas for surface chemistry studies in the tribovoltaic effect.

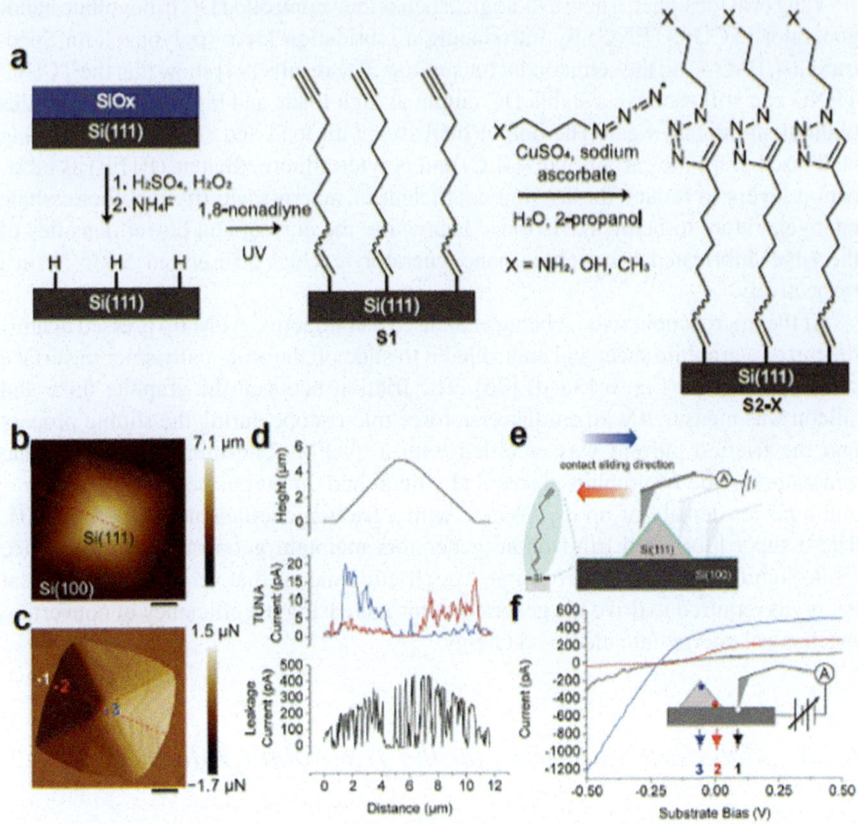

Fig. 6.12 Effect of the surface chemistry on the tribovoltaic effect. **a** The modification of the surface function group of the silicon sample. **b–d** The topography of the sample. **e** The experiment setup for the current measuring by using AFM. **f** The I-V tested by using AFM. Figure with permission [126], copyright © 2020 Elsevier Ltd.

6.5.2.4 Super-Lubricant Interface

Tribovoltaic nanogenerators convert mechanical energy into electrical energy through friction. Unfortunately, mechanical energy is also converted into other energies during friction, such as deformation energy and heat energy, and even only a small portion of it is converted into electrical energy, which makes tribovoltaic nanogenerators inefficient. On the other hand, semiconductor wear also leads to poor stability of tribovoltaic nanogenerators. Therefore, reducing the friction due to ploughing effect and plastic deformation, and further reducing the wear and friction coefficient can improve the energy conversion efficiency of tribovoltaic nanogenerators.

Yang et al. designed a new tribological behaviour-controlled DC tribovoltaic nanogenerator (TCDC-TENG) by introducing a lubrication layer (polyphaolefin SpectraSyn 4, PAO 4) on the semiconductor surface. The results [94] show that the TCDC-TENG can still produce a stable DC output at high loads and high sliding velocities without significant wear. The output increases with load and sliding speed. Zhang et al. used graphitic carbon-like (GLC) and polytetrafluoroethylene (PTFE) as lubrication layers to reduce the friction coefficient of macro-scale friction photovoltaic nanogenerators to below 0.01 [95]. Meanwhile, the current and power densities of the super-lubricated tribovoltaic nanogenerator reached 60 nA and 5.815 W/m^2, respectively.

At the microscopic scale, Huang et al. used a conductive AFM tip pressed against the top of a graphite sheet and controlled it to slide on the atomically smooth surface of n-type silicon (Fig. 6.13a–d) [76]. The friction between the graphite flake and silicon was measured with a transverse force microscope during the sliding process and the friction current was recorded with a CAFM. The miniature tribovoltaic generators made of graphite flakes and silicon had an output power of 210 Am^{-2} and a power density of up to 7 Wm^{-2} with a friction coefficient of less than 0.01. These super-lubricated tribovoltaic generators maintain good output performance while significantly reducing friction. Less friction means that very little mechanical energy is required to drive the generator, thus increasing the efficiency of converting mechanical energy into electrical energy.

6.5.3 Material Engineering for the Tribovoltaic Effect

Special materials have been designed for the production of tribovoltaic nanogenerators, which improves the output performance of tribovoltaic nanogenerators and also allows the nanogenerators to be used in some special occasions.

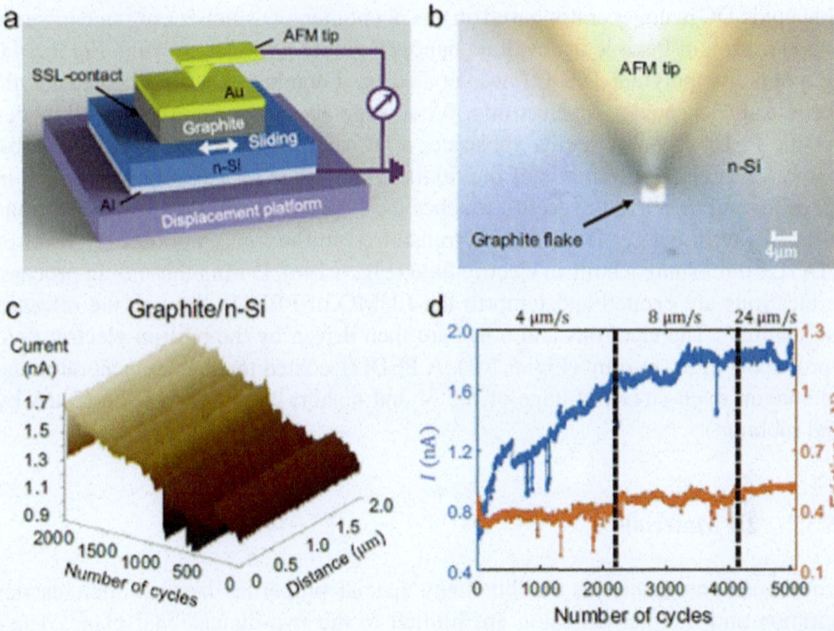

Fig. 6.13 The super-lubric interface tribovoltaic effect. **a** The experiment setup for the studying of tribovoltaic effect at super-lubric interface. **b** The imaging of the sliding interface. **c** The current generated at the sliding Go-Si interface. **d** The current and friction force at the sliding Go-Si interface. Figure with permission [76], copyright © 2021, The Author(s)

6.5.3.1 Flexible and Organic Materials

Flexible wearable devices have developed rapidly in recent years [96–99]. Directly harvesting mechanical energy from human motion to power wearable devices is a perfect strategy. Therefore, it is important to study the tribovoltaic effect of flexible materials and design flexible tribovoltaic nanogenerators. Yang et al. investigated the tribovoltaic effect at sliding metal-conducting polymer interfaces. In the experiment, different metals such as Al, Cu, Fe, Au and Pt were used [100]. PEDOT:PSS was chosen as the conducting polymer on flexible PDMS. The results showed that the tribovoltaic voltages were highly dependent on the difference in the figure of merit between the metal and the conducting polymer, with a maximum output current of 20 A m^{-2} and an open-circuit voltage of 0.8 V obtained at the Al-PEDOT:PSS interface. In addition, You et al. used aluminium alloys, Si, ITO, Cu, and graphene sliding on the surface of PEDOT:PSS, and discussed tribo-voltages at the interface of the flexible materials [101]. It was also found that tribovoltaic nanogenerators consisting of PEDOT:PSS and aluminium alloys are optimal outputs with short-circuit currents up to 309 uA and open-circuit voltages up to 1 V.

The above two works demonstrated that the tribovoltaic effect occurs at the interface of sliding metals and flexible conducting polymers. Meng et al. designed a textile tribovoltaic DC nanogenerator based on this phenomenon, which is a big step towards the application of flexible tribovoltaic nanogenerators [98]. As shown in Fig. 6.14a, the conductive polymer PEDOT was coated on a conductive nickel-coated textile substrate and selected as an electrode. When a metal, such as aluminium foil, slides over the PEDOT-coated textile substrate, a DC current is generated (Fig. 6.14b). The power generation process of the textile tribovoltaic nanogenerator is given in Fig. 6.14c and d. When the Al foil touches the PEDOT-coated textile, the electrons in the Al Fermi energy level will be transferred to the lower null energy level of PEDOT, establishing a built-in electric field (Fig. 6.14e). During the sliding process, the electrons are excited and jump to the LUMO of PEDOT through the released "bindington". The electrons and holes are then driven by the built-in electric field to produce a direct current (Fig. 6.14f). A PEDOT-coated textile nanogenerator can generate an open-circuit voltage of 0.5 V and a short-circuit current of 12 uA by hand rubbing.

6.5.3.2 2D Materials

Two-dimensional materials exhibit many special properties because their carrier migration and thermal diffusion are limited to the two-dimensional plane. Their bandgap tunable properties are widely used in field effect tubes, optoelectronic devices, and thermoelectric devices. The electronic structure of two-dimensional materials is simpler and clearer compared to the surface of solid materials, and it is of scientific significance to study the tribovoltaic effect in two-dimensional materials.

Molybdenum disulfide and graphene are the two most commonly used 2D materials to study the tribovoltaic effect. Liu et al. synthesized a MoS_2 thin film by PLD, and used a conducting tip to slide on the MoS_2 film to generate tribo-currents, which can reach up to 106 Am^{-2} [20] In addition, they discussed the output of a variety of sliding MoS_2 multilayered heterojunctions as shown in Fig. 6.15a. The results of reference [102] showed that the band bending of MoS_2 plays a dominant role in the direction of the tribo-current, supporting the mechanism of the tribovoltaic effect. In addition to MoS_2, the tribovoltaic effect involving graphene has also been investigated. However, due to the lack of relevant studies, graphene constitutes a tribovoltaic nanogenerator that does not find an advantage in terms of output performance. For example, according to You's report, the output performance of the PEDOT:PSS/graphene tribovoltaic nanogenerator is much inferior to that of the PEDOT:PSS/Al tribovoltaic nanogenerator, as shown in Fig. 6.15b [101]. But in any case, graphene-based generators should have the advantage of low friction because graphene is a good solid lubricant material.

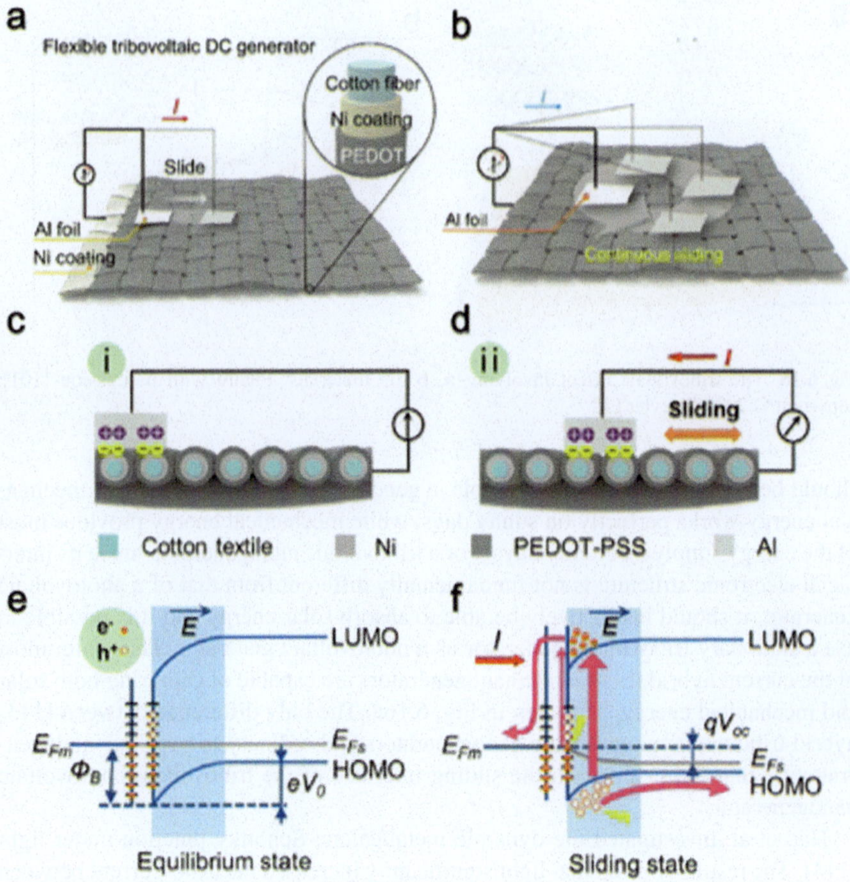

Fig. 6.14 The textile tribovoltaic DC nanogenerator. **a** Schematic diagram of a textile tribovoltaic DC nanogenerator and **b** the experiment set up of generating of DC current. **c, d** The electricity generation process of the textile tribovoltaic nanogenerator. **e, f** Energy band diagram of the Al–PEDOT interface in the equilibrium state and sliding state, respectively. Figure with permission [98], copyright © 2021, American Chemical Society

6.5.3.3 Hybrid Tribovoltaic Effect

As a generator, how to improve its output performance is a key issue. In the study of conventional TENGs, hybrid TENGs can significantly improve the output performance by harvesting triboelectric charges, induced currents, and other forms of energy, such as currents generated by electromagnetic induction [104, 105] and solar energy [106, 107]. Studies have also shown that designing hybrid tribovoltaic nanogenerators is a good strategy to improve the output performance of tribovoltaic nanogenerators [108–112]. For hybrid generators, ideally, the target energies collected

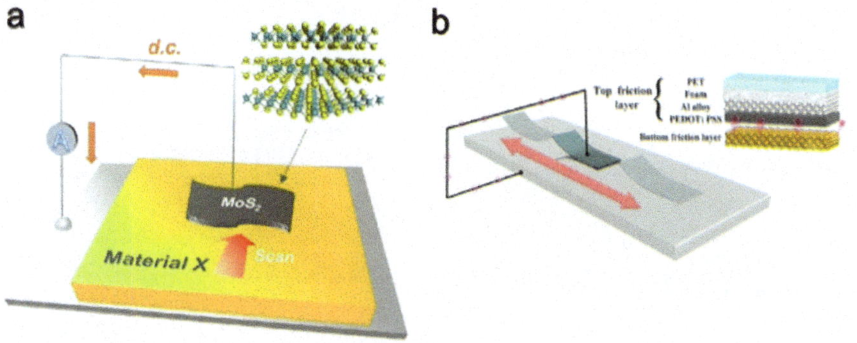

Fig. 6.15 The tribovoltaic effect involving **a, b** 2D materials. Figure with permission [101], copyright © 2021 Elsevier Ltd.

should be complementary. For example, a generator that collects solar and mechanical energy works perfectly on sunny days, while mechanical energy provides most of the energy supply on cloudy days. For a tribovoltaic nanogenerator, since its interfacial electronic structure is not fundamentally different from that of a photovoltaic generator, it should instinctively be able to absorb solar energy and it is possible to use a stationary tribovoltaic generator as a photovoltaic generator. Therefore, most of the current hybrid tribovoltaic nanogenerators are capable of capturing both solar and mechanical energy, as shown in Fig. 6.16a. The only difference between [113] hybrid tribovoltaic-photovoltaic nanogenerators and ordinary tribovoltaic nanogenerators is that light can reach the sliding interface of the tribovoltaic-photovoltaic nanogenerator.

Hao et al. investigated the dynamic metal/calcite Schottky junction under light [114]. The results showed that light significantly increased the tribo-current between metal and chalcogenide. The tribo-voltage and tribo-current density between Al and chalcogenide were 0.7 V and 41.1 Am^{-2}, respectively, in the dark environment. The output current density was increased by a factor of 3 in the light compared to that in the dark environment. Ren et al. irradiated sliding Si-GaN interfaces using UV light and found that the tribo-current was increased by a factor of 13 and the tribo-voltage was increased by a factor of 4 in the UV light irradiation [113]. The mechanism of tribovoltaic-photovoltaic hybrid nanogenerators seems to be easy to understand. It is noted that light irradiation excites photogenerated carriers, which increases the concentration of electron-hole pairs at the interface, leading to an increase in output current and voltage under light irradiation. However, it was noted that at some semiconductor interfaces, such as between DI-water and silicon, illumination can produce a constant background photocurrent without sliding [28]. The result is like a simple linear superposition of the photovoltaic effect and the tribovoltaic effect. At some interfaces, such as silicon and gallium nitride, light illumination increases the tribo-current without generating a background current, as shown in Fig. 6.16b [113]. The difference between these two phenomena remains to be further explained.

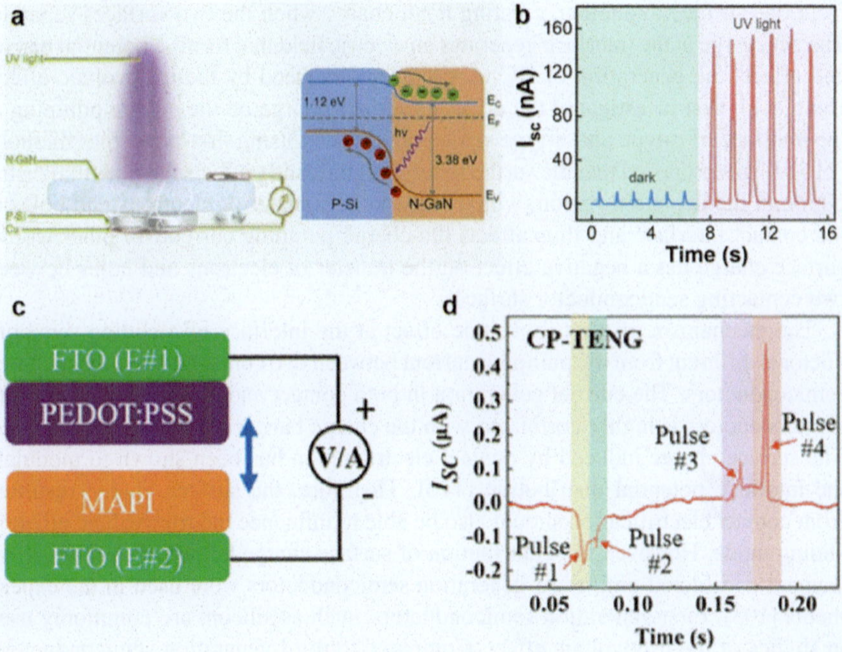

Fig. 6.16 Hybrid tribovoltaic effect. **a, b** Increasing the output of the tribovoltaic nanogenerator by UV irradiation. **c, d** The semiconductor contact-separation TENG and its output current. Figure with permission [115], copyright © 2022, American Chemical Society

TENG and tribovoltaic nanogenerators share the common feature that they generally have sliding or contact-separated interfaces, which offers the possibility of collecting both triboelectric induced and tribovotaic currents simultaneously. Wang et al. investigated the output currents between two contact-separated semiconductors, as shown in Fig. 6.16c and d [115]. The results showed that the current signal from the semiconductor contact separation TENG has two sources. One part is generated by electron transfer in the CE and electrostatic induction in the conventional TENG, and the other part is generated by electron transfer pumped by the PN junction. Similarly, if a semiconductor material is used in a slip-mode TENG, both electrostatic induction and tribovoltaic effects generate tribo-currents, which can be viewed as a hybrid generator consisting of a TENG and a tribovoltaic nanogenerator.

6.6 Effect of Contact Electrification to Tribovoltaic Effect

The tribovoltaic effect occurs at sliding interfaces. A highly related phenomenon, contact electrification, occurs at almost any sliding interface, including those involving semiconductors. During contact electrification, electrons are transferred from one surface to another, creating a net charge when the two surfaces separate. The net charge at the interface generates an electric field that forms a potential barrier and affects the generation of DC tribo-currents induced by the tribovoltaic effect. Deng et al. first investigated the effect of surface charge on the charge pumping at the interface of p-type and n-type semiconductors by using first-principles methods [116]. It was proposed that the surface charge significantly affects the potential distribution in the depletion region, which hinders the transfer of electrons and holes at the contact interface and thus affects the charge pumping current. In other words, surface charge has a negative effect on the transfer of electrons and holes between two contacting semiconductor surfaces.

The mechanism of the tribovoltaic effect at the interface of a sliding semiconductor is different from the pumping current between two contacting and independent semiconductors. The current generation in both contact and sliding modes for both semiconductors is highly correlated with the charge barrier formed at the interface. The surface charge induced by contact electrification has been shown to modulate the interface potential distribution [116]. Therefore, the surface charge resulting from contact electrification should also be able to influence the tribovoltaic effect in sliding mode. However, the contribution of surface charge to the tribovoltaic effect was not considered until third-generation semiconductors were used in the experiments [103]. First-generation semiconductors, such as silicon, are commonly used in studies of the tribovoltaic effect. Compared to third-generation semiconductors, first-generation semiconductors have narrower bandgap widths and higher electrical conductivity. The high conductivity of first-generation semiconductors makes it difficult for tribo-charges to be trapped on the surface, thus making it impossible to detect the effect of tribo-charges on the tribovoltaic effect.

Third-generation semiconductors have been introduced to the study of the tribovoltaic effect due to their wider bandgap, which is expected to produce higher tribo-voltages. Chen et al. investigated a series of gallium nitride (GaN)-based tribovoltaic nanogenerators, including GaN- si and GaN- Al. The maximum open-circuit voltage of tribovoltaic nanogenerators involving GaN reaches 25 V, which is four times the maximum voltage of conventional tribovoltaic nanogenerators [103]. The output tribo-voltage and tribo-current of GaN-based tribovoltaic nanogenerators were further increased to 130 V and 340 uA, respectively [117]. Such tribo-voltages are much higher than the strength of the built-in electric field. Zhang et al. Investigated the tribovoltaic effect at semiconductor interfaces, a heterojunction SDC-TENG employing p-type Bi_2Te_3 and n-type GaN friction substituents is proposed, as shown in Fig. 6.17a. The planar GaN film (Fig. 6.17a-i) and Bi_2Te_3 slider (Fig. 6.17a-ii) are attached to a flexible conducting strip to form a good physical contact. The voltametric characteristic curves show the Schottky properties of the pBi_2Te_3-nGaN heterojunction, as shown in Fig. 6.17b. When a normal pressure of 4 N was applied to the GaN slider and reciprocated over a range of 3 cm with a maximum speed of 50 cm s^{-1} (Fig. 6.17c), a DC pulsed electrical output from the motion was observed. The peak open-circuit voltage was approximately 40 V and the peak short-circuit current was 0.34 mA, as shown in Fig. 6.17d, e. It is worth noting that the directions of the output voltages and currents are opposite to the direction of the built-in electric field and are much higher than the typical output values of the previous tribovoltaic effect-based SDC-TENG. Figure 6.17f shows the peak currents and voltages at different external load resistances from 10 Ω to 10 MΩ. As shown in Fig. 6.17g, with a matching impedance of 160 kΩ, the PPD of the SDC-TENG is about 11.85 Wm^{-2} (1.58 mW at a working area of 1.33 cm^2), and the APD is about 9.23 Wm^{-2}. The short-circuit transfer charge for this time period was 18.13 μC, and the charge transfer density was up to 136.3 m cm^{-2}. In order to exclude the piezoelectric effect of gallium nitride, pBi_2Te_3 was pressed onto nGaN to measure the open-circuit voltage and the short-circuit current. The maximum output values of the piezoelectricity are about 30 mV and 1 nA, respectively, when the normal force is varied from 0 to 20 N. Both values are much smaller than those observed in SDC-TENG. Therefore, the contribution of the verified piezoelectric effect can be excluded. This pBi_2Te_3-nGaN heterojunction SDC-TENG is significantly different from previously reported silicon-based SDC-TENGs. The directions of the generated voltages and currents are not governed by the built-in electric field. Similar phenomena were observed in experiments on the tribovoltaic effect of zinc oxide and tungsten oxide. In addition, the open-circuit output voltage of SDC-TENG far exceeds the semiconductor bandgap limit. Previous mechanisms based on built-in electric field and energy band theories are difficult to explain. Therefore, a new mechanism of dominant tribovoltaic effect should exist at the interface between Bi_2Te_3 and GaN. Unlike the conventional tribovoltaic effect, the external tribo-current always flows from the GaN side to the Si/Al side rather than relying on the built-in electric field caused by charge carrier diffusion. This suggests that there are other driving forces separating the electron-hole pairs during friction, most likely the surface charge induced by contact electrification, since the electric field generated by electrostatic charge can reach several thousand volts [118, 119].

In conjunction with contact electrification, Chen et al. proposed a model to explain the tribovoltaic effect involving third-generation semiconductors [103]. As shown in Fig. 6.18a, electrons belonging to Si and GaN are trapped in an electron cloud potential well. When Si slides on the GaN surface, due to the high energy level of Si, the potential barrier for electron leaps decreases and electrons are transferred from the Si surface to the GaN surface (Fig. 6.18b). Increasing the friction intensity, more electrons are transferred from the Si side to the GaN side, as shown in Fig. 6.18c. Assuming that the Fermi energy level of n-type GaN is higher than that of Si (Fig. 6.18d), the electrons will diffuse from the GaN side to the Si side at the contact (Fig. 6.18e) and establish a built-in electric field point from the GaN side to the Si side. The higher the friction, the stronger the built-in electric field at the interface (Fig. 6.18f). Notice that the direction of electron transfer at the contact charge is opposite to the direction of diffusion caused by the electron concentration difference. In the third-generation semiconductor tribovoltaic effect, the electric field generated by contact electrification plays a dominant role, and the direction of the external friction current flows from the GaN side to the Si side. The introduction of third-generation triboelectric surface charge in the tribovoltaic effect provides a new idea for the development of high-performance tribovoltaic nanogenerators.

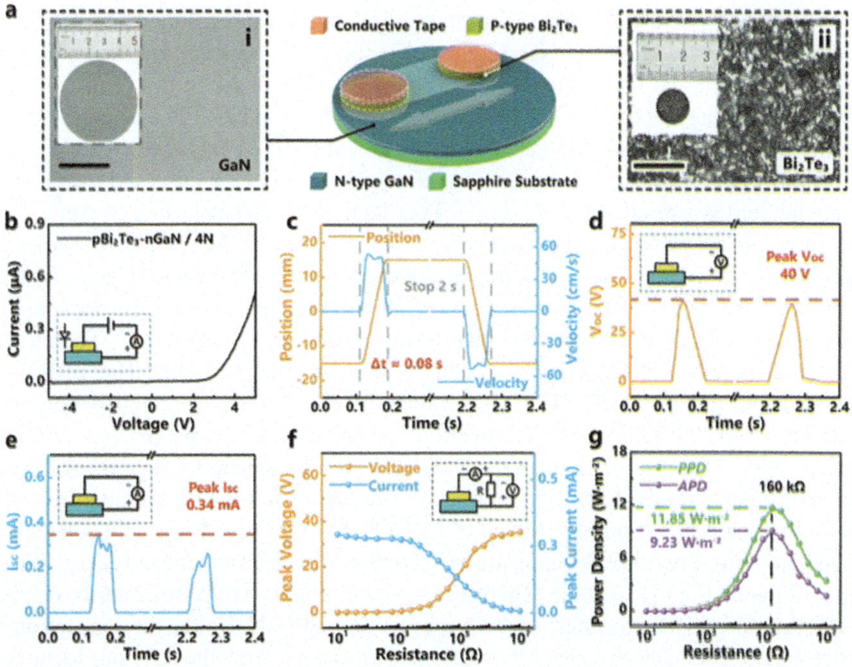

Fig. 6.17 The tribovoltaic effect involving GaN. **a** The experiment setup for the studying of tribovoltaic effect involving GaN. **b** The I-V curves at the pBi$_2$Te$_3$-nGaN interface. **c** The movement of the samples. **d–g** The output of the tribovoltaic effect at pBi$_2$Te$_3$-nGaN interface. Figure with permission [127], copyright © 2022, Wiley-VCH GmbH

6.7 Applications of the Tribovoltaic Effect

The tribovoltaic effect is a newly discovered phenomenon. Although the tribovoltaic effect has been applied to energy harvesting and sensor design, the research on the application of tribovoltaic effect is still in its infancy when compared to the conventional TENG, as elaborated in details in Chap. 11. In fact, the use of tribovoltaic nanogenerators as energy harvesting devices has the advantage of direct DC output. They are usually designed by introducing a semiconductor material at a sliding interface, which causes sliding at the semiconductor interface to generate a tribovoltaic current, as shown in Fig. 6.19a. The force driving the sliding of the interface can be various mechanical forces such as wind, water and human power. For example, Yu et al. designed a tribovoltaic nanogenerator that can be driven by wind and can generate a tribovoltaic current of 4.4 mA [122]. The tribovoltaic effect was shown to be sensitive to many friction parameters, such as sliding velocity and sliding load [94]. Depending on the relationship between the tribovoltaic output current and the friction parameters, different sensors can be designed. Siliconization is necessary to produce the tribovoltaic effect. The tribovoltaic current can only be generated when two semiconductors have relative silicification and can be used to detect whether

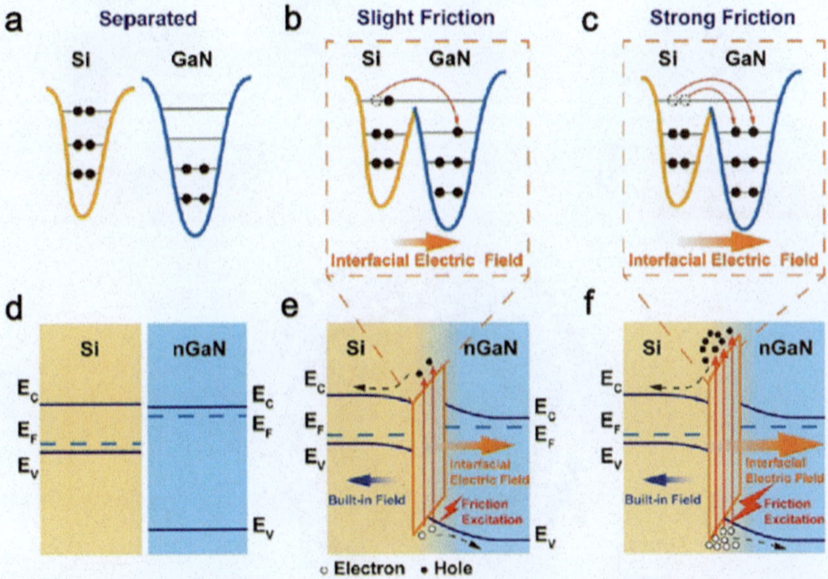

Fig. 6.18 Contribution of charge transfer induced by contact electrification to the tribovoltaic effect. Electron-cloud-potential-well model for contact electrification between GaN and Si **a** in separated state, **b** slight friction state, and **c** strong friction state. Band diagram of Si and nGaN **d** in separated state, **e** slight friction state, and **f** strong friction state. Figure with permission [103], copyright © 2022, American Chemical Society

two objects have relative motion. By integrating the tribovoltaic current, the relative displacement of the objects can be obtained as a position sensor, as shown in Fig. 6.19b. In addition, the tribovoltaic current between water and fluorinated graphite has been reported to be proportional to the relative velocity between them. Reference [40] and designed a graphite fluoride-based tribovoltaic nanogenerator to detect the water velocity.

In addition to serving as energy harvesters and smart sensors, tribovoltaic nanogenerators can also be used as a new data storage technology. As we know, electron transfer at the interface of ferromagnets is related to spin [121, 122]. When the magnetization directions of adjacent ferromagnetic films are the same, the whole interface presents a low-resistance state; on the contrary, when the magnetization directions of adjacent ferromagnetic films are opposite, the whole interface presents a high-resistance state, which is called the giant magnetoresistance effect [123, 124]. If magnetic semiconductor materials are used as sliding pairs, the tribovoltaic current is likely to be affected by the magnetization direction of both sliding materials. A reasonable assumption is that the current generated at the sliding interface is larger when both have the same direction of magnetization than when both have opposite directions of magnetization. If this assumption is confirmed, the data can be stored by

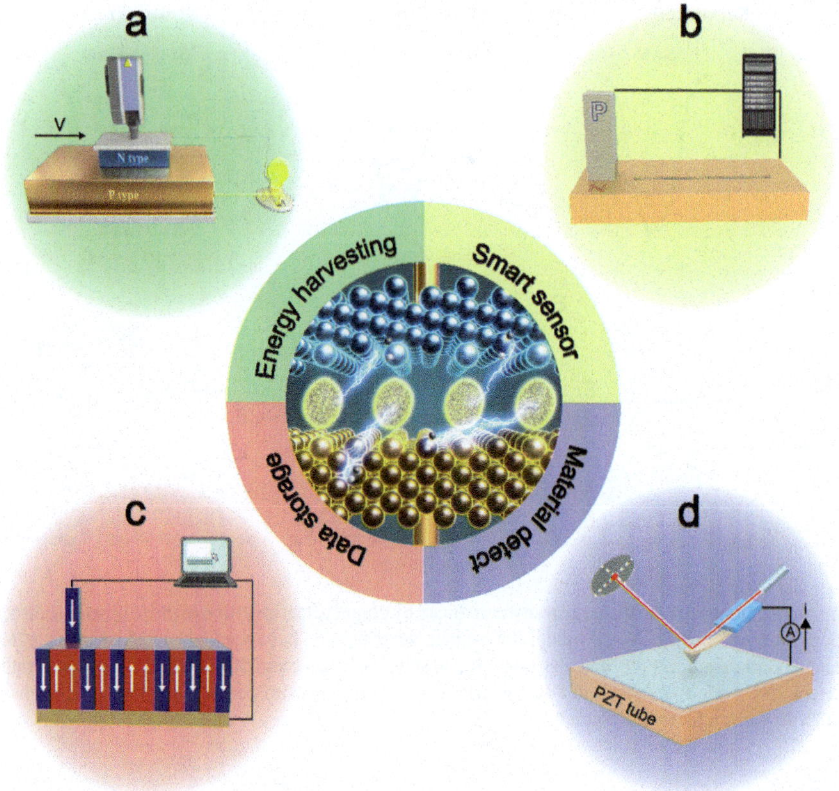

Fig. 6.19 Potential applications of tribovoltaic effect. Tribovoltaic nanogenerator for **a** energy harvesting, **b** smart sensor, **c** data storage and **d** material characterization. Figure with permission [125], copyright © 2022 Elsevier Ltd.

magnetizing a semiconductor ferromagnet, as shown in Fig. 6.19c. Since the tribovoltaic current also depends on semiconductor properties such as doping concentration and electrical resistance [33, 43], the tribovoltaic effect can also be used to characterize material properties at the microscopic scale, as shown in Fig. 6.19d.

6.8 Conclusion

In this paper, we summarize the fundamental studies that have carried out related to the tribovoltaic effect and provide a brief introduction to the mechanism of the tribovoltaic effect. At a sliding semiconductor interface, the formation of a new bond releases a quantum of energy, called "bindington", which excites electron-hole pairs

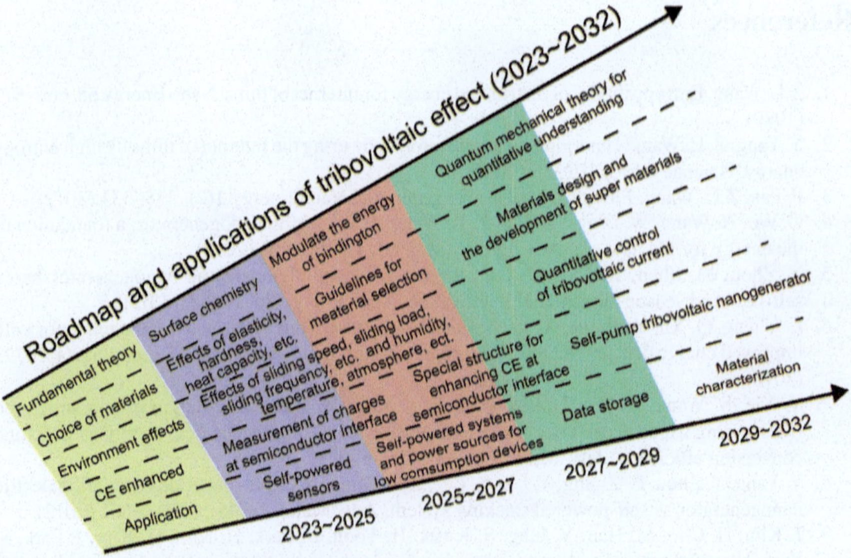

Fig. 6.20 Proposed technology roadmap about the fundamental studies and applications of the tribovoltaic effect. Figure with permission [125], copyright © 2022 Elsevier Ltd.

at the interface. The electron-hole pairs are further separated by a built-in electric field, generating a DC tribo-current.

Tribovoltaic effects at metal-semiconductor interfaces, P-N junctions, and liquid-semiconductor interfaces are discussed. It is demonstrated that the tribovoltaic effect is influenced by the semiconductor surface properties. The results show that the tribovoltaic effect can be modulated by surface modifications that can change the semiconductor's surface state density, built-in electric field strength, and the interaction between the two surfaces. In addition, by selecting materials with special features, such as flexible materials, the tribovoltaic effect can be applied to the appropriate applications. In particular, if third-generation semiconductors involve sliding, the contribution of the electric field induced by the tribo-charge at the interface to the tribovoltaic effect is significant.

In order to increase the output power of friction volt nanogenerators, hybrid friction volt nanogenerators are proposed to harvest friction volt current and other forms of energy, such as solar and thermal energy. In addition to harvesting energy, friction volt-beat nanogenerators have also been shown to be used as self-powered sensors.

In contrast to the photovoltaic effect, the tribovoltaic effect has only just been discovered. Although the tribovoltaic effect has been shown to occur widely at various semiconductor material interfaces, it has not yet been discussed in depth. The following are the challenges that need to be investigated (a roadmap of the tribovoltaic effect is shown in Fig. 6.20):

References

1. Z.L. Wang, Entropy theory of distributed energy for internet of thins. Nano Energy **58**, 669–672 (1989)
2. Y. Yang, Z. L. Wang, Emerging nanogenerators: powering the internet of thins by high entropy energy. Iscience **24**(5), 102358 (2021)
3. F. Fan, Z.L. Wang, Flexible triboelectric generator! Nano Energy **1**(2), 328–334 (2012)
4. C. Wu, A. Wang, W. Ding, H. Guo, Z. L. Wang, Triboelectric nanogenerator: a foundation of the energy for the new era. Adv. Energy Mater. **9**(1), 1802906 (2019)
5. Y. Zhou, M. Shen, X. Cui, Y. Shao, L. Li, Y. Zhang, Triboelectric nanogenerator based self-powered sensor for artificial intelligence. Nano Energy **84**, 105887 (2021)
6. L. Cheng, Q. Xu, Y. Zheng, X. Jia, Y. Qin, A self-improving triboelectric nanogenerator with improved charge density and increased charge accumulation speed. Nat. Commun. **9**(1), 3773 (2018)
7. Y. Xie, S. Wang, S. Niu, L. Lin, Q. Jing, J. Yang, Z. Wu, Z.L. Wang, Grating-structured freestanding triboelectric layer nanogenerator for harvesting mechanical energy at 85% total conversion efficiency. Adv. Mater. **26**(38), 6599–6607 (2014)
8. Y. Yang, Y. Zhou, H. Zhang, Y. Liu, S. Lee, Z.L. Wang, A single-electrode based triboelectric nanogenerator as self-powered tracking system. Adv. Mater. **25**(45), 6594–6601 (2013)
9. J. Kim, H. Cho, M. Han, Y. Jung, S. Kwak, H. Yoon, B. Park, H. Kim, H. Kim, J. Park, S. Kim, Ultrahigh power output from triboelectric nanogenerator based on serrated electrode via spark discharge. Adv. Energy Mater. **10**(44), 2002312 (2020)
10. W. Cao, H. Ouyang, W. Xin, S. Chao, C. Ma, Z. Li, F. Chen, M. Ma, A stretchable high output triboelectric nanogenerator improved by MXene liquid electrode with high electronegativity. Adv. Funct. Mater. **30**(50), 2004181 (2020)
11. F. Xing, Y. Jie, X. Cao, Y. Li, N. Wang, Natural triboelectric nanogenerator based on soles for harvesting low-frequency walking energy. Nano Energy **42**, 138–142 (2017)
12. N. Cui, J. Liu, L. Gu, S. Bai, X. Chen, Y. Qin, Wearable triboelectric generator for powering the portable electronic devices. ACS Appl. Mater. Interfaces **7**(33), 18225–18230 (2015)
13. L. He, C. Zhang, B. Zhang, O. Yang, W. Yuan, L. Zhou, Z. Zhao, Z. Wu, J. Wang, Z.L. Wang, A dual-mode triboelectric nanogenerator for wind energy harvesting and self-powered wind speed monitoring. ACS Nano **16**(4), 6244–6254 (2022)
14. B. Chen, Y. Yang, Z.L. Wang, Scavenging wind energy by triboelectric nanogenerators. Adv. Energy Mater. **8**(10), 1702649 (2018)
15. Z. Lin, G. Cheng, L. Lin, S. Lee, Z.L. Wang, Water solid surface contact electrification and its use for harvesting liquid-wave energy. Angew. Chem. Int. Edit. **52**(48), 12545–12549 (2013)
16. X. Yang, S. Chan, L. Wang, W. Daoud, Water tank triboelectric nanogenerator for efficient harvesting of water wave energy over a broad frequency range. Nano Energy **44**, 388–398 (2018)
17. G. Qiao, J. Wang, X. Yu, R. Jia, T. Cheng, Z.L. Wang, A bidirectional direct current triboelectric nanogenerator with the mechanical rectifier. Nano Energy **79**, 105408 (2021)
18. Y. Zi, J. Wang, S. Wang, S. Li, Z. Wen, H. Guo, Z.L. Wang, Effective energy storage from a triboelectric nanogenerator. Nat. Commun. **7**(1), 10987 (2016)
19. Y. Song, N. Wang, Y. Wang, R. Zhang, H. Olin, Y. Yang, Direct current triboelectric nanogenerators. Adv. Energy Mater. **10**(45), 2002756 (2020)
20. J. Liu, A. Goswami, K. Jiang, F. Khan, S. Kim, R. McGee, Z. Li, Z. Hu, J. Lee, T. Thundat, Direct current triboelectricity generation by a sliding Schottky nanocontact on MoS2 multilayers. Nat. Nanotechnol. **13**(2), 112–116 (2018)
21. Z.L. Wang, A.C. Wang, On the origin of contact electrification. Mater. Today **30**, 34–51 (2019)
22. R. Yang, R. Xu, W. Dou, M. Benner, Q. Zhang, J. Liu, Semiconductor based dynamic heterojunctions as an emerging strategy for high direct current mechanical energy harvesting. Nano Energy **83**, 105849 (2021)
23. W. Chen, A. Andreev, L. Glazman, Quantum criticality in a mott PN junction in an armchair carbon nanotube. Phys. Rev. Lett. **106**(21), 216801 (2011)

24. M. Buscema, D. Groenendijk, G. Steele, H. van der Zant, A. Catellanos, Photovoltaic effect in few layer black phosphorus PN junctions defined by local electrostatic gating. Nat. Commun. **5**(1), 4651 (2014)
25. L. Su, W. Yang, J. cai, H. Chen, X. Feng, Self-powered ultraviolet photodetectors driven by built-in electric field. Small **13**(45), 1701687 (2017)
26. Y. Dou, Y. Liang, H. Li, Y. Xue, H. Ye, Y. Han, Integration of $H_2V_3O_8$ nanowires and GaN thin film for self-powered UV photodetectors. Chem. Commun. **58**(61), 8548–8551 (2022)
27. M. Zheng, S. Lin, L. Zhu, Z. Tang, Z. L. Wang, Effects of temperature on the tribovoltaic effect at liquid-solid interfaces. Adv. Mater. Interfaces **9**(3), 2101757 (2022)
28. M. Zheng, S. Lin, Z. Tang, Y. Feng, Z.L. Wang, Photovoltaic effect and tribovoltaic effect at liquid-semiconductor interface. Nano Energy **83**, 105810 (2021)
29. S. Lin, X. Chen, Z.L. Wang, Contact electrification at the liquid-solid interface. Chem. Rev. **122**(5), 5209–5232 (2022)
30. H. Shao, J. Fang, H. Wang, L. Dai, T. Lin, Polymer-metal Schottky contact with direct-current outputs. Adv. Mater. **28**(7), 1461–1466 (2016)
31. H. Zheng, R. Shen, H. Zhong, Y. Lu, X. Yu, S. Lin, Dynamic Schottky diode direct-current generator under extremely low temperature. Adv. Funct. Mater. **31**(40), 2105325 (2021)
32. J. Han, X. Yang, L. Liao, G. Zhou, G. Wang, C. Xu, W. Hu, M. Debora, Q. Song, Photoinduced triboelectric polarity reversal and enhancement of a new metal/semiconductor triboelectric nanogenerator. Nano Energy **58**, 331–337 (2019)
33. M. Zheng, S. Lin, L. Xu, L. Zhu, Z. L. Wang, Scanning probing of the tribovoltaic effect at the sliding interface of two semiconductors. Adv. Mater. **32**(21), 2000928 (2020)
34. S. Lin, R. Shen, T. Yao, Y. Lu, S. Feng, Z. Hao, H. Zheng, Y. Yan, E. Li, Surface states enhanced dynamic Schottky diode generator with extremely high power density over 1000 Wm^{-2}. Adv. Sci. **6**(24), 1901925 (2019)
35. Z. Zhang, T. He, J. Zhao, G. Liu, Z.L. Wang, C. Zhang, Tribo-thermoelectric and tribovoltaic coupling effect at metal-semiconductor interface. Mater. Today Phys. **16**, 100295 (2021)
36. Y. Lu, Z. Hao, S. Feng, R. Shen, Y. Yan, S. Lin, Direct-current generator based on dynamic PN junctions with the designed voltage output. iScience **22**, 58–69 (2019)
37. R. Xu, Q. Zhang, J. Wang, D. Liu, J. Wang, Z.L. Wang, Direct current triboelectric cell by sliding an n-type semiconductor on a p-type semiconductor. Nano Energy **66**, 104185 (2019)
38. Q. Zhang, R. Xu, W. Cai, Pumping electrons from chemical potential difference. Nano Energy **51**, 698–703 (2018)
39. Y. Yan, X. Zhou, S. Feng, Y. Lu, J. Qian, P. Zhang, X. Yu, Y. Zheng, F. Wang, K. Liu, S. Lin, Direct current electricity generation from dynamic polarized water semiconductor interface. J. Phys. Chem. C **125**(26), 14180–14187 (2021)
40. H. Huang, R. Wang, Y. Chen, Fluorinated graphite paper used for self-powered water speed sensors by immersion type tribovoltaic effect dominated triboelectric nanogenerators. Nano Energy **93**, 106887 (2022)
41. H. Zhong, J. Xia, F. Wang, H. Chen, H. Wu, S. Lin, Graphene-piezoelectric material heterostructure for harvesting energy from water flow. Adv. Funct. Mater. **27**(5), 1604226 (2017)
42. Y. Lu, Y. Yan, X. Yu, X. Zhou, S. Feng, C. Xu, H. Zheng, Z. Yang, L. Li, K. Liu, S. Lin, Polarized water driven dynamic PN junction-based direct current generator. Research **2021**, 7505638 (2021)
43. S. Lin, X. Chen, Z.L. Wang, The tribovoltaic effect and electron transfer at a liquid-semiconductor interface. Nano Energy **76**, 105070 (2020)
44. Z. Lin, G. Cheng, S. Lee, K. Pradel, Z.L. Wang, Harvesting water drop energy by a sequential contact-electrification and electrostatic-induction process. Adv. Mater. **26**(47), 4690–4696 (2014)
45. X. Liu, K. Cheng, P. Cui, H. Qi, H. Qin, G. Gu, W. Shang, S. Wang, Cheng, G., Du, Z, Hybrid energy harvester with bi-functional nano-wrinkled anti-reflective pdms film for enhancing energies conversion from sunlight and raindrops. Nano Energy **66**, 1041889 (2019)

46. G. Cheng, Z. Lin, Z. Du, Z.L. Wang, Simultaneously harvesting electrostatic and mechanical energies from flowing water by a hybridized triboelectric nanogenerator. ACS Nano **8**(2), 1932–1939 (2014)
47. G. Zhu, Y. Su, P. Bai. J. Chen, Q. Jing, W. Yang, Z.L. Wang, Harvesting water wave energy by asymmetric screening of electrostatic charges on a nanostructured hydrophobic thin-film surface. ACS Nano **8**(6), 6031–6037 (2014)
48. L. Pan, J. Wang, P. Wang, R. Gao, Y. Wang, X. Zhang, J. Zou, Z.L. Wang, Liquid-FEP-based U-tube triboelectric nanogenerator for harvesting water-wave energy. Nano Res. **11**, 4062–4073 (2018)
49. X. Li, J. Tao, X. Wang, J. Zhu, C. Pan, Z.L. Wang, Networks of high performance triboelectric nanogenerators based on liquid-solid interface contact electrification for harvesting low-frequency blue energy. Adv. Energy Mater. **8**(21), 1800705 (2018)
50. C. Le, C. Vo, T. Nguyen, D. Vu, K. Ahn, Liquid-solid contact electrification based on discontinuous-conduction triboelectric nanogenerator induced by radially symmetrical structure. Nano Energy **80**, 105571 (2021)
51. M. Xu, S. Wang, S. Zhang, W. Ding, P. Kien, C. Wang, Z. Li, X. Pan, Z.L. Wang, A highly-sensitive wave sensor based on liquid-solid interfacing triboelectric nanogenerator for smart marine equipment. Nano Energy **57**, 574–580 (2019)
52. H. Cho, J. Chung, G. Shin, J. Sim, D. Kim, S. Lee, W. Hwang, Toward sustainable output generation of liquid-solid contact triboelectric nanogenerators: the role of hierarchical structures. Nano Energy **56**, 56–64 (2019)
53. Z. Lin, G. Cheng, L. Lin, S. Lee, Z. L. Wang, Water-solid surface contact electrification and its use for harvesting liquid-wave energy. Angew. Chem. Int. Ed. **52**(48), 12545−125499 (2013)
54. J. Ha, J. Chung, S. Kim, J. Kim, S. Shin, J. Park, S. Lee, J. Kim, Transfer-printable micropatterned fluoropolymer-based triboelectric nanogenerator. Nano Energy **36**, 126–133 (2017)
55. A.W. Copeland, O.D. Black, A.B. Garrett, The photovoltaic effect. Chem. Rev. **31**(1), 177–226 (1942)
56. F. Williams, A.J. Nozik, Solid-state perspectives of the photoelectrochemistry of semiconductor-electrolyte junctions. Nature **312**, 21–27 (1984)
57. N.S. Lewis, Progress in Understanding electron-transfer reactions at semiconductor/liquid interfaces. J. Phys. Chem. B **102**(25), 4843–4855 (1998)
58. A. Iqbal, M.S. Hossain, K.H. Bevan, The role of relative rate constants in determining surface state phenomena at semiconductor-liquid interfaces. Phys. Chem. Chem. Phys. **18**(42), 29466–29477 (2016)
59. Q. Dong, Y. Fang, Y. Shao, P. Mulligan, J. Qiu, L. Cao, J. Huang, Electron-hole diffusion lengths > 175 mu m in solution-grown CH3NH3PbI3 single crystals. Science **347**(6225), 967–970 (2015)
60. G. Dennler, M. Scharber, C. Brabec, Polymer fullerene bulk heterojunction solar cells. Adv. Mater. **21**(13), 1323–1338 (2009)
61. Y. Li, Molecular design of photovoltaic materials for polymer solar cells: toward suitable electronic energy levels and broad absorption. Accounts Chem. Res. **45**(5), 723–733 (2012)
62. P. Landsberg, An introduction to the theory of photovoltaic cells. Solid State Electron. **18**(12), 1043–1052 (1975)
63. M. Archer, Photovoltaic and photoelectrochemistry similarities and differences. Phys. E **14**, 61–64 (2022)
64. B. Parida, S. Iniyan, R. Goic, A review of solar photovoltaic technologies. Renew. Sust. Energy Rev. **15**(3), 1652–1636 (2011)
65. L. Kronik, Y. Shapira, Surface photovoltage phenomena: theory, experiment, and applications. Surf. Sci. Rep. **37**, 1–206 (1999)
66. K. Holmberg, A. Erdemir, Influence of tribology on global energy consumption, costs and emissions. Friction **5**, 263–284 (2017)
67. H. Sakuma, K. Kawai, I. Katayama, S. Suehara, What is the origin of macroscopic friction. Sci. Adv. **4**(12), eaav2268 (2018)

68. J. Krim, Friction at macroscopic and microscopic length scales. Am. J. Phys. **70**(9), 890 (2002)
69. J. Luo, M. Liu, L. Ma, Origin of friction and the new frictionless technology-superlubricity: advancements and future outlook. Nano Energy **86**, 106092 (2021)
70. M. Jean, The non-smooth contact dynamic method. Comput. Method Appl. M. **177**, 235–257 (1999)
71. M. Raous, L. Cangemi, M. Cocu, A consistent model coupling adhesion, friction, and unilateral contact. Comput. Method Appl. M. **177**, 383–399 (1999)
72. Z. Zhang, X. Shi, D. Guo, Dynamic temperature rises mechanism and some controlling factors of wet clutch engagement. Math. Probl. Eng. **2016**(1), 6530213 (2016)
73. F. Bowden, D. Tabor, *The friction and lubrication of solid* (Clarendon Press, Oxford, 1964)
74. G. Sutter, N. Ranc, Flash temperature measurement during dry friction process at high sliding speed. Wear **268**, 1237–1242 (2010)
75. B. Perssion, Rubber friction: role of the flash temperature. J. Phys. Condens. Matter **18**(32), 7789–7823 (2006)
76. X. Huang, X. Xiang, J. Nie, D. Peng, F. Yang, Z. Wu, H. Jiang, Z. Xu, Q. Zheng, Microscale Schottky superlubric generator with high direct current density and ultralong life. Nat. Commun. **12**(1), 2268 (2021)
77. I. Sivebaek, B. Persson, Shearing nanometer-thick confined hydrocarbon films: friction and adhesion. Tribol. Lett. **62**, 5 (2016)
78. K. Kristiansen, H. Zeng, P. Wang, J, Israelachvili. Microtribology of aqueous carbon nanotube dispersions. Adv. Funct. Mater. **21**(23), 4555–4564 (2011)
79. Z. Xu, P. Huang, Composite oscillator model for the energy dissipation mechanism of friction. Acta Phys. Sin-ch Ed. **55**, 2427–24329 (2006)
80. Z. Xu, P, Huang, Study on the energy dissipation mechanism of atomic scale friction with composite oscillator model. Wear **262**, 972–977 (2007)
81. K. Nakayama, H. Hashimoto, Triboemission, tribochemical reaction, and friction and wear in ceramics under various n-butane gas pressures. Tribol. Int. **29**(5), 385–393 (1996)
82. M. Zheng, Y. Su, L. Chen, T. Zhang, Frictional effects of interface bonding energy in blister testing. Theor. Appl. Fract. Mec. **32**(2), 75–80 (1999)
83. K. Tian, Z. Li, N. Gosvami, D. Goldsby, I. Szlufarska, R. Carpick, Memory distance for interfacial chemical bond induced friction at the nanoscale. ACS Nano **13**(7), 7425–7434 (2019)
84. M. Srinivasan, Binding site models of friction due to the formation and rupture of bonds: state function formalism, force velocity relations, response to slip velocity transients and slip stability. Phys. Rev. E **80**(4), 0461249 (2009)
85. A. Filippov, J. Klafter, M. Urbakh, Friction through dynamical formation and rupture of molecular bonds. Phys. Rev. Lett. **92**(13), 135503 (2004)
86. K. Tian, Z. Li, N. Gosvami, D. Goldsby, Y. Liu, I. Szlufarska, R. Carpick, Load and time dependence of interfacial chemical bond induced friction at the nanoscale. Phys. Rev. Lett. **118(7)**, 07610399 (2017)
87. J. Batteas, X. Quan, M. Weldon, Adhesion and wear of colloidal silica probe by force microscopy. Tribol. Lett. **7**, 121–128 (1999)
88. J. Batteas, M. Weldon, K. Raghavachari, Bonding and Inter-particle interactions of silica nanoparticles: probing adhesion at asperity-asperity contacts. Nanotribology Crit. Assess. Futur. Res. Needs, 387–398 (2003)
89. A. Li, Y. Liu, I. Szlufarska, Effects of interfacial bonding on friction and wear at silica/silica interfaces. Tribol. Lett. **56**, 481–490 (2014)
90. S. Ferrie, A. Brun, G. Krishnan, G. Andersson, N. Darwish, S. Ciampi, Sliding silicon based Schottky diodes: maximizing triboelectricity with surface chemistry. Nano Energy **93**, 106861 (2022)
91. J. Liu, M. Cheikh, R. Bao, H. Peng, F. Liu, Z. Li, K. Jiang, J. Chen, T. Thundat, Tribo-tunneling DC generator with carbon aerogel/silicon multi-nanocontacts. Adv. Electron. Mater. **5**(12), 1900464 (2019)

92. M. Benner, R. Yang, L. Li, M. Liu, H. Li, J. Liu, Mechanism of in-plane and out-of-plane tribovoltaic direct current transport with a metal/oxide/metal dynamic heterojunction. ACS Appl. Mater. Interfaces **14**(2), 2968–2978 (2022)
93. Z. Zhang, N. Wu, L. Gong, R. Luan, J. Cao, C. Zhang, An ultrahigh power density and ultralow wear GaN-based tribovoltaic nanogenerator for sliding ball bearing as self-powered wireless sensor node. Adv. Mater. **36**(6), 2310098 (2024)
94. D. Yang, L. Zhang, N. Luo, Y. Liu, W. Sun, J. Peng, M. Feng, Y. Feng, H. Wang, D. Wang, Tribological behaviour controlled direct current triboelectric nanogenerator based on the tribovoltaic effect under high contact pressure. Nano Energy **99**, 107370 (2022)
95. L. Zhang, H. Cai, L. Xu, L. Ji, D. Wang, Y. Zheng, Y. Feng, X. Sui, Y. Guo, W. Guo, F. Zhou, W. Liu, Z.L. Wang, Macro superlubric triboelectric nanogenerator based on tribovoltaic effect. Matter **5**(5), 1532–25469 (2022)
96. J. Gao, K. Shang, Y. Ding, Z. Wen, Material and configuration design strategies towards flexible and wearable power supply devices: a review. J. Mater. Chem. A **9**(14), 8950–8965 (2021)
97. Q. Huang, D. Wang, Z. Zheng, Textile based electrochemical energy storage devices. Adv. Energy Mater. **6**(22), 1600783 (2016)
98. J. Meng, Z. Guo, C. Pan, L. Wang, C. Chang, L. Li, X. Pu, Z.L. Wang, Flexible textile direct current generator based on the tribovoltaic effect at dynamic metal-semiconducting polymer interfaces. ACS Energy Lett. **6**(7), 2442–2450 (2021)
99. K. Keum, J. Kim, S. Hong, J. Son, S. Lee, J. Ha, Flexible/stretchable supercapacitors with novel functionality for wearable electronics. Adv. Mater. **32**(51), 2002180 (2020)
100. R. Yang, M. Benner, Z. Guo, C. Zhou, J. Liu, High performance flexible Schottky DC Generator via metal conducting polymer sliding contacts. Adv. Funct. Mater. **31**(43), 2103132 (2021)
101. Z. You, S. Wang, Z. Li, Y. Zou, T. Lu, F. Wang, B. Hu, X. Wang, L. Li, W. Fang, Y. Liu, High current output direct current triboelectric nanogenerator based on organic semiconductor heterojunction. Nano Energy **91**, 106667 (2022)
102. J. Liu, F. Liu, R. Bao, K. Jiang, F. Khan, Z. Li, H. Peng, J. Chen, A. Alodhayb, T. Tundat, Scaled up direct current generation in MoS2 multilayer based moving heterojunctions. ACS Appl. Mater. Interfaces **11**(38), 35404–35409 (2019)
103. Y. Chen, Z. Zhang, Z. Wang, T. Bu, S. Dong, W. Wei, Z. Chen, Y. Lin, Y. Li, H. Zhou, W. Sun, C. Zhang, Friction dominated carrier excitation and transport mechanism for GaN based direct current triboelectric nanogenerators. ACS Appl. Mater. Interfaces **14**(20), 24020–24027 (2022)
104. H. Guo, Z. Wen, Y. Zi, M. Yeh, J. Wang, L. Zhu, C. Hu, Z.L. Wang, A water proof triboelectric electromagnetic hybrid generator for energy harvesting in harsh environments. Adv. Energy Mater. **6**(6), 1501593 (2016)
105. X. Cui, S. Cao, H. Zhang, Building self-powered emergency electronics based on hybrid nanogenerators for field survival/rescue. Energy Sci. Eng. **8**(3), 574–581 (2020)
106. S. Jeon, D. Kim, Y. Choi, Self cleaning hybrid energy harvester to generate power from raindrop and sunlight. Nano Energy **12**, 636–645 (2015)
107. D. Yoo, S. Park, S. Lee, J. Sim, I. Song, D. Choi, H. Lim, D. Kim, Biomimetic anti reflective triboelectric nanogenerator for concurrent harvesting of solar and raindrop energies. Nano Energy **57**, 424–431 (2019)
108. C. Ma, B. Kim, S. Kim, N. Park, Dynamic halide perovskite heterojunction generates direct current. Energy Environ. Sci. **14**(1), 374 (2021)
109. H. Zhong, Z. Wu, X. Li, W. Xu, S. Xu, S. Zhang, Z. Xu, H. Chen, S. Lin, Graphene based two dimensional hybrid nanogenerator for concurrently harvesting energy from sunlight and water flow. Carbon **105**, 199–204 (2016)
110. X. Xu, Y. Hu, Z. Guo, S. Liu, Q. Zhang, X. Zhang, Y. Xiang, High energy conversion efficiency direct alternating current hybrid generator with piezoelectric polyactide and dynamic Schottky diode. J. Mater. Sci. **55**, 9014–9026 (2020)

111. V. Sharov, P. Alekseev, B. Borodin, M. Dunaevskiy, R. Reznik, G. Cirlin, InP/Si heterostructure for high current hybrid triboelectric/ Photovoltaic generation. ACS Appl. Energy Mater. **2**(6), 4395–4401 (2019)
112. J. Liu, Y. Zhang, J. Chen, R. Bao, K. Jiang, F. Khan, A. Goswami, Z. Li, F. Liu, K. Feng, J. Luo, T. Thundat, Separation and quantum tunneling of photogenerated carries using a tribo-induced field. Matter **1**(3), 650–660 (2019)
113. L. Ren, A. Yu, W. Wang, D. Guo, M. Jia, P. Guo, Y. Zhang, Z.L. Wang, J. Zhai, P-N junction based direct current triboelectric nanogenerator by conjunction of tribovoltaic effect and photovoltaic effect. Nano Lett. **21**(23), 10099–10106 (2021)
114. Z. Hao, T. Jiang, Y. Lu, S. Feng, R. Shen, T. Yao, Y. Yan, Y. Yang, Y. Lu, S. Lin, Co-harvesting light and mechanical energy based on dynamic metal/perovskite Schottky junction. Matter **1**(3), 639–649 (2019)
115. H. Wang, S. Huang, H. Kuang, T. Zou, P. Rajagopalan, X. Wang, Y. Li, H. Jin, S. Dong, H. Zhou, T. Hasan, L. Occhipinti, J. Kim, Coexistence of contact electrification and dynamic p-n junction modulation effects in triboelectrification. ACS Appl. Mater. Interfaces **14**(26), 30410–30419 (2022)
116. S. Deng, R. Xu, M. Li, L. Li, Z.L. Wang, Q. Zhang, Influences of surface charges and gap width between p-type and n-type semiconductors on charge pumping. Nano Energy **78**, 105287 (2020)
117. Z. Wang, Z. Zhang, Y. Chen, L. Gong, S. Dong, H. Zhou, Y. Lin, Y. Lv, G. Liu, C. Zhang, Achieving an ultrahigh direct current voltage of 130 V by semiconductor heteojunction power generation based on the tribovoltaic effect. Energy Environ. Sci. **15**(6), 2366 (2022)
118. F. Liu, Y. Liu, Y. Lu, Z. Wang, Y. Shi, L. Ji, J. Cheng, Electrical analysis of triboelectric nanogenerator for high voltage applications exampled by DBD microplasma. Nano Energy **56**, 482–493 (2019)
119. Q. Zheng, L. Fang, X. Tang, L. Zheng, H. Li, Indoor air dust removal system based on high voltage direct current triboelectric nanogenerator. Nano Energy **97**, 107183 (2022)
120. X. Yu, H. Zheng, Y. Lu, R. Shen, Y. Yan, Z. Hao, Y. Yang, S. Lin, Wind driven semiconductor electricity generator with high direct current output based on a dynamic Schottky junction. RSC Adv. **11**(31), 19106 (2021)
121. H. Reittu, Analysis of spin-dependent tunneling of electrons in solid state structures using the transfer Hamiltonian method. J. Phys. Condensed Matter **9**(48), 10651–10660 (1997)
122. F. Blobner, R. Han, A. Kim, W. Wurth, P. Feulner, Spin dependent electron transfer dynamics probed by resonant photoemission spectroscopy. Phys. Rev. Lett. **112**(8), 086801 (2014)
123. H. Saarikpski, W. Wetzels, G. Bauer, Charge and spin transport in spin valves with anisotropic spin relaxation. Phys. Rev. B **75**(7), 075313 (2007)
124. F. Mancoff, N. Rizzo, B. Engel, S. Tehrani, Area dependence of high frequency spin transfer resonance in giant magnetoresistance contacts up to 300 nm diameter. Appl. Phys. Lett. **88**(11), 112507 (2006)
125. S. Lin, Z.L. Wang, The tribovoltaic effect. Mater. Today **62**, 111–128 (2023)
126. S. Ferrie, N. Darwish, J.J. Gooding, S. Ciampi, Harnessing silicon facet-dependent conductivity to enhance the direct-current produced by a sliding Schottky diode triboelectric nanogenerator. Nano Energy **78**, 105210 (2020)
127. Z. Zhang, Z. Wang, Y. Chen, Y. Feng, S. Dong, H. Zhou, Z. L. Wang, C. Zhang, Semiconductor contact-electrification-dominated tribovoltaic effect for ultrahigh power generation. Adv. Mater. **34**(20), 2200146 (2022)

Chapter 7
Contact Electrification at Liquid-Solid Interfaces

Abstract The liquid-solid (L-S) interface, ubiquitous in nature, serves as a critical boundary for addressing fundamental scientific problems. Its prevalence and intricate dynamics underpin a wide range of essential physical and chemical phenomena essential to fields such as catalysis, energy, sensors and environmental science, offering profound opportunities for exploration in various disciplines. The functionality of the L-S interface is rooted in charge accumulation and release driven by potential gradients or disparities in material properties across the phases. This chapter delves into the mechanisms of contact electrification (CE) at L-S interfaces, emphasizing charge transfer processes across interfaces involving conductors, dielectrics, and semiconductors. It emphasizes the formation and dynamics of the electrical double layer (EDL), a fundamental structure regulating charge behavior at interfaces. Wang's two-step model is introduced to elucidate the role of CE in EDL formation on dielectric surfaces. Key discussions include ion and electron transfer quantification, advanced electron transfer modeling, and the hybrid characteristics of the EDL, shedding light on the physics of L-S CE. The chapter explores techniques for dynamic EDL regulation through electric fields or mechanical methods to control charge carrier behavior. By modulating ion migration and optimizing charge concentration at interfaces, these strategies directly influence energy scavenging, storage, and processes of information flow etc. Practical applications of L-S CE are demonstrated across various devices. At liquid-conductor interfaces, EDL regulation underpins efficient energy storage in supercapacitors and the functionality of electrochemical sensors. For liquid-dielectric interfaces, the dynamic regulation of the EDL has catalyzed innovations in high-efficiency energy harvesting and information flow, such as triboelectric nanogenerators (TENGs) and triboiontronic nanogenerators (TINGs). At liquid-semiconductor interfaces, the tribovoltaic effect harnesses L-S CE for effective direct current (DC) energy generation. These advancements highlight the vast and new application potential of EDL regulation at L-S interfaces, laying the foundation for breakthroughs in energy and information technologies.

Z. L. Wang, *Contact-Electrification of Matter*, Springer Series in Surface Sciences 75,
https://doi.org/10.1007/978-3-031-88504-4_7

7.1 Introduction

The phenomenon of contact electrification (CE) at liquid-solid interfaces has become a cornerstone of modern interfacial science, with the electrical double layer (EDL) theory serving as its foundational framework. Over nearly two centuries, the EDL model has evolved from basic conceptualizations to sophisticated formulations that underpin a myriad of technological applications. These advancements have reshaped our understanding of interfacial charge dynamics across conductors, dielectrics, and semiconductors in liquid CE systems. The origins of EDL theory can be traced to Helmholtz in the nineteenth century [1], who proposed a model likening interfacial charge distribution to a parallel-plate capacitor. While this provided an initial understanding, it failed to account for ion thermal motion and diffusion, limiting its applicability to dynamic systems. In the early twentieth century, Gouy and Chapman independently introduced the concept of a diffuse layer [2, 3], refining the understanding of charge distribution. This insight highlighted the dynamic nature of interfacial phenomena but overlooked finite ion sizes and specific adsorption effects. The Gouy-Chapman-Stern (Stern) model [4], introduced in 1924, addressed these limitations by combining the strengths of prior models. It divided the EDL into the Stern layer and the diffuse layer, with the Stern layer further subdivided into the Inner Helmholtz Plane (IHP) and Outer Helmholtz Plane (OHP). Beyond these regions lay the diffuse layer, where free ions moved under the influence of electrostatic forces and thermal energy. This comprehensive model provided a robust framework for understanding charge distribution at liquid-conductor interfaces and has remained invaluable for applications in energy storage, chemical sensing, and other advanced technologies.

Traditional EDL models primarily focused on liquid-conductor interfaces, leaving the complexities of liquid-dielectric systems less explored. Wang et al. bridged this gap by performing thermionic emission experiments across various liquid-dielectric systems [5], revealing critical nanoscale insights into charge transfer mechanisms. These studies demonstrated that interfacial charge exchange involves both electron transfer and ionization reactions, with their relative contributions shaped by the surface chemistry of the dielectric and the liquid's properties. Wang et al. proposed a two-step EDL model for liquid-dielectric interfaces [6]. In the initial step, electron transfer occurred due to the overlap of atomic electron clouds between the solid and liquid molecules. Concurrently, surface ionization reactions generated electrons and ions depending on the material's chemical properties and the liquid's ionization potential. In the subsequent step, electrostatic forces drove counter-ions in the liquid to form a structured EDL similar to that at conductor interfaces. This refined model elucidated the interplay between electron transfer and ionization, advancing our understanding of interfacial behaviors between dielectric materials and liquids. The EDL serves as an ideal ionic-electronic coupling interface, enabling dynamic modulation of ion and electron behaviors in iontronics. Unlike traditional electronics, which rely on electrons or holes as charge carriers, iontronics facilitates the simultaneous manipulation of ionic and electronic charges, significantly enhancing energy efficiency. For instance, the human brain, a highly integrated iontronic system, operates on approximately 12 W of power, whereas electronic systems replicating similar integration would require megawatts, underscoring the efficiency advantage of iontronics. Wei et al. advanced this field by dynamically regulating EDL properties using the triboelectric charge from CE [7, 8]. This innovation led to the development of triboiontronics, which leveraged the interaction between triboelectricity and iontronics to precisely control ionic-electronic coupling interfaces. The dynamic modulation of charge carriers within the EDL enabled highly efficient energy harvesting and information flow, paving the way for groundbreaking applications in interface control, biosensing, and human-computer interaction.

From Helmholtz's initial parallel-plate capacitor model to Wang's two-step EDL model, the evolution of EDL theory at liquid-solid interfaces reflects a journey of profound scientific advancement. These developments have not only deepened our understanding of interfacial charge dynamics but also unlocked transformative applications across energy and information technologies. By continuing to refine these models and integrate emerging insights, the field of CE at liquid-solid interfaces stands poised to address critical challenges in sustainable energy and advanced information systems for the post-Moore era.

7.2 Electrical Double Layer

The L-S contact interface is a critical boundary governing interfacial dynamics between liquids and solids, forming the basis for diverse physical, chemical, and electrochemical processes in catalysis, corrosion, energy scavenging and sensors etc. Upon contact, differences in electrochemical potential and surface properties, such as electronegativity, surface energy, and roughness, drive charge redistribution, facilitating L-S CE. This charge redistribution governs the formation of the EDL, which in turn affects ionic transport kinetics, reaction rates, and colloidal stability, while driving key phenomena such as ion adsorption, surface reactivity, and electrostatic interactions. These functions are vital for applications in capacitive energy storage, electrochemical sensing, and advanced separation technologies. Dynamic regulation of the EDL via electric fields or mechanical methods enables control over charge carrier behavior, optimizing ion migration and charge concentrations at the interface. This capability directly enhances energy scavenging, storage, sensors, and information flow processes, underscoring the transformative potential of EDL engineering in advancing interfacial science and practical applications.

At the interface of L-S CE, EDL theory has been refined over nearly two centuries, advancing from foundational models to sophisticated frameworks with broad technological applications [9]. It originated with Helmholtz in the nineteenth century, who conceptualized interfacial charge distribution as a parallel-plate capacitor [1] (Fig. 7.1a). In this model, a rigid layer of counter-ions formed near the charged solid surface, balanced by an opposing charge on the solid. While groundbreaking, the Helmholtz model failed to account for ion thermal motion and diffusion, limiting its ability to describe the EDL's dynamic behavior. In the early twentieth century, Gouy and Chapman independently extended the understanding of interfacial charge distribution by introducing the concept of the diffuse layer [2, 3] (Fig. 7.1b). They proposed that, in addition to ions accumulating near the contact surface, some ions diffuse into the bulk solution, forming a region where ion concentration follows a Boltzmann distribution. It resulted in an exponential decay of charge density with increasing distance from the interface. This dynamic representation marked a significant improvement but overlooked factors such as finite ion sizes and specific adsorption effects, which were critical to accurately describing interfacial phenomena. To address these limitations, Stern introduced the Gouy-Chapman-Stern (Stern) model in 1924 [4] (Fig. 7.1c), which integrated the strengths of both Helmholtz and Gouy-Chapman approaches. This model divided the EDL into two regions: the Stern layer and the diffuse layer. The Stern layer is further subdivided into the IHP and the OHP. The IHP was composed of specifically adsorbed ions and solvent molecules, influenced by electrostatic and chemical interactions, while the OHP contained solvated ions interacting with the surface primarily through long-range electrostatic forces. Beyond the Stern layer lies the diffuse layer, where free ions are governed by electrostatic forces and thermal motion. Together, these layers define the EDL's spatial structure, with their combined thickness approximating the Debye length, a key parameter determining ionic interaction strength in solution. The Debye length varied with electrolyte concentration and the dielectric constant of the medium, ranging from 0.2 to 20 nm under typical conditions. The Stern model has proven invaluable for understanding charge distribution at L-S contact interfaces. By synergizing the interactions among electrostatic forces, ion adsorption, and thermal dynamics, this model has continued to guide the development of advanced technologies, including energy storage, chemical sensing, and beyond, reaffirming its enduring relevance in both foundational research and practical innovation.

The classical models of EDL primarily emphasized charge distribution at liquid-conductor interfaces, often neglecting the complex interactions at liquid-nonconductor interfaces. To bridge this gap, Wang et al. performed thermionic emission experiments across various liquid-dielectric systems, offering critical nanoscale insights into charge transfer mechanisms [5]. These experiments revealed that charge exchange at the liquid-dielectric interface involves both electron transfer and ionization reactions, with their relative contributions influenced by the material's surface properties and the liquid's characteristics. Specifically, at liquid-hydrophobic dielectric interfaces, electron transfer dominated due to the limited occurrence of ionization reactions, as the solid surface lacked sufficient affinity for dissociating liquid molecules. In contrast, at liquid-hydrophilic dielectric interfaces, ionization reactions played a more significant role, enabling ions to mediate charge transfer processes more effectively. Based on these findings, Wang et al. proposed a two-step EDL model that extended traditional EDL concepts to liquid-dielectric interfaces [6] (Fig. 7.2). In the first step, when a liquid came into contact with a pristine dielectric surface, electron transfer occurred due to the overlap of atomic electron clouds between the solid and water molecules. Concurrently, surface ionization reactions might generate both electrons and ions, depending on the material's chemical properties and the

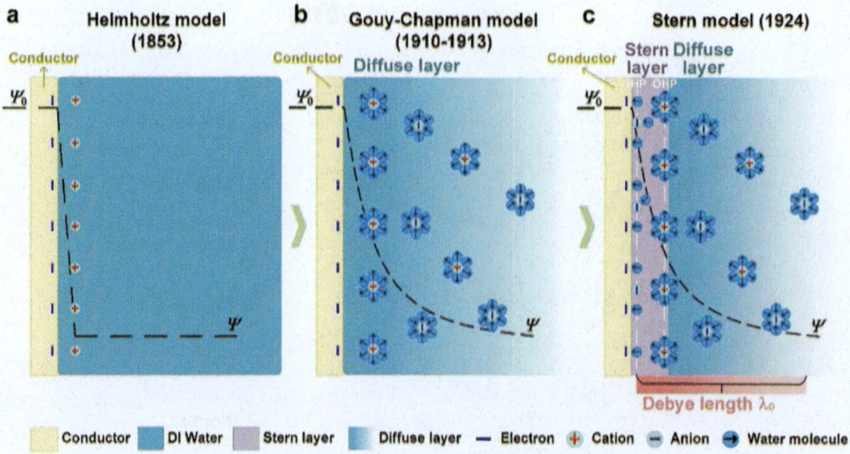

Fig. 7.1 The evolution of the EDL model at liquid-conductor interfaces. **a** The Helmholtz model. **b** The Gouy-Chapman model. **c** The Stern model. With permission from Wiley-VCH GmbH (2024) [9]

liquid's ionization potential. In the second step, electrostatic forces drive counter-ions in the liquid toward the charged solid surface, forming a structured EDL similar to that observed at conductor interfaces. This refined model enhanced the understanding of the EDL structure at non-conductor interfaces, particularly by elucidating the interplay between electron transfer and ionization in shaping interfacial charge distributions. The model also highlighted the critical role of surface chemistry and liquid properties in defining interfacial behaviors, with significant implications for applications in dielectric materials, energy storage devices, and environmental engineering.

The EDL served as an ideal ionic-electronic coupling interface in the field of iontronics, enabling the dynamic modulation of ion and electron behaviors. Unlike traditional electronics, which relied primarily on electrons and/or holes as charge carriers, iontronics facilitated the simultaneous manipulation of ionic and electronic charges, offering the potential for significantly enhanced energy efficiency [10]. For example, the human brain, a highly integrated iontronic system, operates on approximately 12 W of power. In contrast, replicating brain-like integration with electronic systems could demand tens of megawatts of energy [11], illustrating the efficiency advantage of iontronics. In this context, the EDL played a pivotal role by enabling the transport and reorganization of ions to modulate electronic properties, presenting a promising paradigm for achieving efficient energy and information flow, an essential feature for next-generation technologies in the post-Moore era [12]. Recently, the dynamic regulation of charge carriers within the EDL at liquid-dielectric interfaces by the triboelectric charge from CE was demonstrated by Wei et al. [7], as illustrated in Fig. 7.3. This breakthrough led to the development of triboiontronics, which leveraged the interplay between triboelectricity and

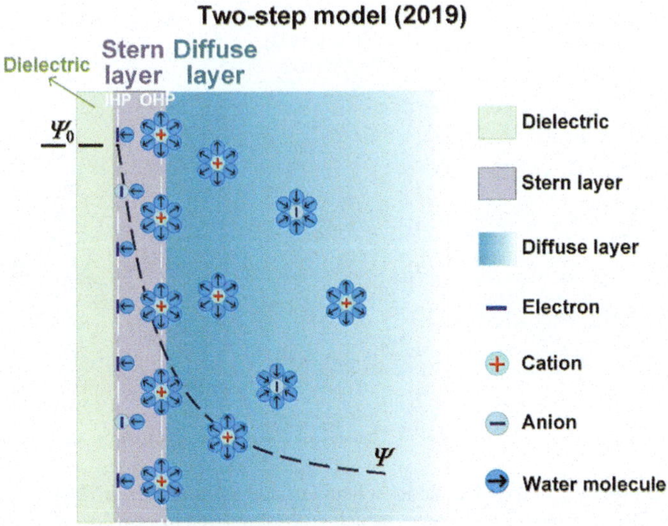

Fig. 7.2 The two-step EDL model at the liquid-dielectric interface was proposed by Wang et al. With permission from Wiley-VCH GmbH (2024) [9]

iontronics to precisely control the ionic-electronic coupling interface. This innovative approach enabled highly efficient energy scavenging and information flow by dynamically regulating EDL properties. The effective modulation of ionic and electronic charge carriers paved the way for controllable ionic-electronic information flow, unlocking new applications in interface control, biosensing, and human-computer interaction. These advancements underscored the transformative potential of triboiontronics across diverse fields, driving the seamless integration of ionic and electronic systems for next-generation energy and information technologies.

7.3 Contact Electrification at Liquid-Conductor Interfaces

Charge transfer at the liquid-conductor interface plays a crucial role in electrochemical processes, governing the movement of electrons between the conductive material and the surrounding electrolyte. This interaction is fundamental to processes such as electrolysis, corrosion, and energy storage. At the interface, the electrochemical potential difference between the liquid (electrolyte) and conductor (electrode) drives the transfer of charge, resulting in the formation of an EDL, which regulates ion and electron movements. In electrochemistry, the liquid-conductor interface is central to processes like oxidation and reduction reactions, where electrons are transferred between the solid electrode and the liquid phase. The rate of charge transfer depends on the nature of the electrode surface, its reactivity, and the ionic composition of the electrolyte. Additionally, the interface reaction is governed by the Nernst equation,

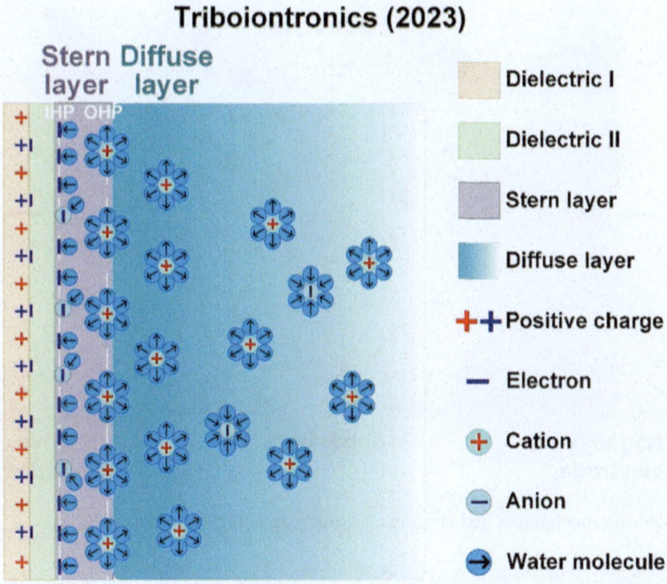

Fig. 7.3 Triboiontronics based on dynamically regulating the EDL at liquid-dielectric interfaces. With permission from Wiley-VCH GmbH (2024) [9]

which relates the electrode potential to the concentration of ions in the solution, and the Butler-Volmer equation, which describes the kinetics of electron transfer reactions. The efficiency of charge transfer at the liquid-conductor interface influences the overall performance of electrochemical devices such as batteries, capacitors, and sensors. Enhanced charge transfer can improve the power and energy densities of storage devices, while a better understanding of the interface dynamics can optimize electrochemical sensing for applications in chemical analysis, environmental monitoring, and medical diagnostics.

7.3.1 Supercapacitors

Since Helmholtz introduced the concept of the EDL, it has become a cornerstone in understanding electrochemical processes. The EDL effect facilitates the formation of a highly concentrated ion-electron layer at the electrode surface, providing a robust mechanism for energy storage. This principle is foundational to the design of EDL supercapacitors, a classical electrochemical energy storage device that leverages the EDL effect to deliver high power density and rapid charge-discharge capabilities as shown in Fig. 7.4. These characteristics make supercapacitors a critical component in modern energy systems. Unlike traditional capacitors, which rely primarily on dielectric materials for charge storage, supercapacitors accumulate a substantial

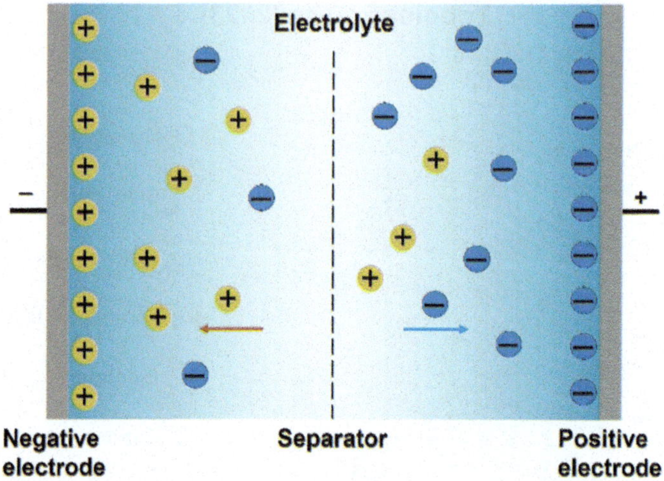

Fig. 7.4 Working mechanism and structure of supercapacitors based on EDL

amount of ionic charge directly within the interface layer. This physical process, free from chemical reactions, enables supercapacitors to sustain their performance over thousands of charge-discharge cycles. Such a unique mechanism grants supercapacitors significant advantages in power density, and cycle life, positioning them as indispensable in applications where rapid charging and long-term durability are paramount [13].

EDL supercapacitors arose in the mid-twentieth century, driven by advances in EDL theory and electrochemical materials, with prototypes employing activated carbon electrodes and aqueous electrolytes. [14]. By the 1970s, supercapacitors began to be used in industrial applications, but early devices were limited by low energy and power densities. Late in the twentieth century, advances in high-surface-area carbon materials and gel electrolytes significantly enhanced performance. The twenty-first century ushered in a new phase, with nanomaterials technology accelerating supercapacitor development. The introduction of novel electrode materials, such as graphene and carbon nanotubes, markedly improved device performance and expanded application potential. Today, supercapacitors are integral to energy-intensive sectors, from transportation to wearable electronics, cementing their role as key components in modern energy systems. In recent years, extensive modifications to electrode materials, electrolytes, and asymmetric electrode configurations have been pursued to enhance supercapacitor performance [15]. The efficiency of supercapacitors is highly dependent on the properties of electrode materials, including specific surface area, conductivity, and pore structure, which directly influence the EDL's effective surface. Notably, metal oxides like MnO_2 and RuO_2 function as pseudocapacitive materials, adding Faradaic reactions to conventional EDL charge storage, thereby significantly enhancing energy density [15]. Second, electrolytes are equally critical to achieving high energy and power densities in supercapacitors. Conventional aqueous electrolytes offer high ionic conductivity and low cost, but their narrow electrochemical window, limited by water decomposition at elevated voltages, restricts energy density. As a result, organic electrolytes and ionic liquids have emerged as key alternatives to extend the electrochemical window. Organic electrolytes, with electrochemical windows up to 2.5–3 V, significantly enhance energy density, while ionic liquids, recognized for their wide electrochemical window, low vapor pressure, and chemical stability, excel in high-voltage applications. They improve ion concentration at the electrode-electrolyte interface, thereby enhancing capacitance and charge-discharge efficiency. Optimizing the surface and structural characteristics of electrodes to enhance charge storage has become a critical area of focus. Functional groups, such as hydroxyl and carboxyl, improve electrode hydrophilicity, facilitating efficient ion penetration and adsorption. Nanostructural modifications, including porous or layered architectures, significantly increase the specific surface area and shorten ion diffusion pathways, thereby maximizing charge storage density. These structural enhancements directly influence EDL processes: the expansion of the electrode surface area generates additional double layers, thereby augmenting charge storage capacity, while the use of advanced electrolytes extends the voltage window, enabling higher energy density within the same volume. Moreover, modifications to the physicochemical properties of the electrode interface optimize ion mobility, thereby improving the kinetics of EDL charge and discharge processes and resulting in enhanced power density.

Supercapacitors hold significant promise across renewable energy, transportation, and consumer electronics sectors, where their high power density, rapid response, long cycle life, and environmental compatibility have positioned them as pivotal components in modern energy systems. Unlike traditional batteries, supercapacitors excel in delivering bursts of high power within short durations, making them ideal for applications such as starting mechanisms and energy recovery systems in electric vehicles. In renewable energy setups like TENG, photovoltaic, and wind power systems, supercapacitors serve a vital role in smoothing power fluctuations, thereby enhancing overall stability and reliability [16]. As a highly efficient energy storage solution, supercapacitors are poised to play a transformative role in future low-carbon and intelligent energy infrastructures, aligning with global shifts toward sustainable and resilient power systems.

7.3.2 *Various Types of Electrochemical Sensors*

Electrochemical sensors utilizing the EDL have emerged as a versatile and promising technology across various applications, from environmental monitoring to implantable biomedical devices. These sensors exploit the distinctive properties of the EDL at the interface between conductors and ionic materials to convert physical or chemical variations into electrical signals with remarkable sensitivity and stability. Based on their structural design, EDL-based sensors can be categorized into two primary types: EDL capacitive sensors and EDL electrode sensors, both of which have shown significant potential in overcoming challenges across diverse fields.

7.3.2.1 EDL Capacitive Sensors

The EDL capacitive sensor typically consists of two conductors and an ionic material arranged in a cavity or sandwich structure, as illustrated in Fig. 7.5. It exploits EDL capacitance effects to transduce external physical or chemical variations, such as pressure, strain, and temperature, into electrical signals [17]. These sensors rely on the EDL properties at the interface to detect changes in capacitance or impedance, with ionic materials rich in mobile ions serving as the critical components. Unlike traditional parallel-plate capacitors, EDL capacitive sensors utilize ion modulation mechanisms, offering enhanced sensitivity and resistance to interference. This makes them particularly suitable for high-resolution, wide-dynamic-range applications.

EDL capacitive sensors are increasingly pivotal in the sensing domain due to their exceptional sensitivity, resistance to interference, and rapid response times. In smart wearables and portable medical devices, these sensors can detect subtle pressure variations, facilitating health monitoring and motion analysis, and providing reliable support for real-time human signal monitoring. In environmental monitoring, the interference-resistant ion modulation mechanism of EDL capacitive sensors enables precise signal detection in complex electromagnetic environments. As advancements in ionic materials and structural designs continue, the application scope of EDL capacitive sensors expands, particularly in emerging areas such as smart materials, flexible electronics, and adaptive control systems. With ongoing progress in materials science and micro-nano manufacturing technologies, the performance and multi-functionality of EDL capacitive sensors will further improve, driving innovations in intelligent, portable, and multifunctional sensing devices.

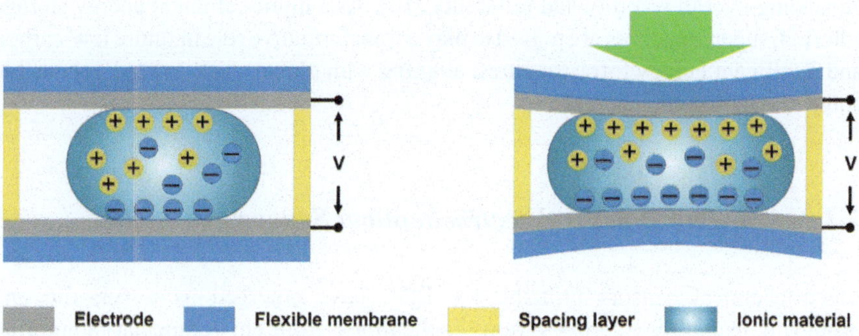

Fig. 7.5 Typical mechanism and structure of the EDL capacitive sensors

7.3.2.2 Implantable Sensors Based on EDL

The challenge of facilitating direct charge exchange between electronic electrodes and ion-regulated biological systems has highlighted the limitations of traditional conductive electrodes in implantable medical sensors and devices [18]. Such electrodes can cause electrochemical reactions at the electrode-tissue interface, leading to tissue damage through local pH shifts, heat generation, and hydrogen ion accumulation [19]. Implantable electrodes based on the EDL interface, typically consisting of a conductive electrode and an interfacial ionic gel as shown in Fig. 7.6, based on the EDL interfacial charge transfer mechanism have recently emerged as a promising solution. It leverages EDL effects to enable controlled ion transport and charge modulation at the interface between electrodes and biological tissues [20]. These ionic gels offer unique advantages in biocompatibility, stability, and signal transmission [21], making them an increasingly crucial technology in implantable devices, including pacemakers, neurostimulation electrodes, and brain-machine interfaces. Thus, it can establish an efficient charge transmission between the electrode and the biological tissue, enhance the biocompatibility of the electrode and make the signal stable and accurate.

Early explorations of EDL electrode interfaces in biology trace back to Luigi Galvani's seminal concept of "animal electricity" in the late eighteenth century [22], with implantable electrode technology achieving pivotal advancements by the mid-twentieth century. Milestones included the development of cardiac pacemakers and spinal stimulators by Medtronic in the 1960s, where the role of electrode capacitance in facilitating charge transfer across electrode-tissue interfaces was first realized [22]. In 1978, the cochlear implant, a platinum electrode array implanted in the cochlea, converted acoustic signals to electric pulses, marking a breakthrough in auditory prosthetics. In 2002, deep brain stimulation was approved for the treatment of Parkinson's disease. Implantable electrodes can also selectively identify and transmit electrical signals from specific biomolecules, such as glucose and lactic acid, and are widely used in the monitoring of diseases such as diabetes and Rheumatoid Arthritis. After that, with the rapid development of bioelectronics and ionic gel materials, EDL ionic

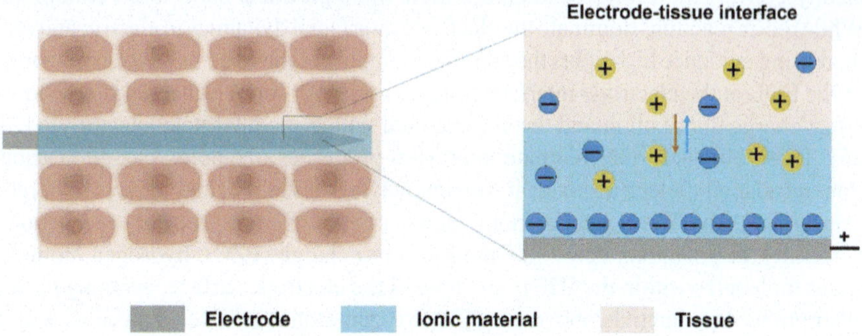

Fig. 7.6 Schematic of insertion of an EDL electrode sensor inside the tissue

gel technology has been widely used in electrical stimulation electrodes and brain-computer interfaces because of their biocompatibility and flexibility. Especially in the field of nerve repair, and muscle regeneration, by forming an EDL between the ionic gels and surrounding tissue, these gels establish efficient electrical signal pathways, enabling bioelectronic signal transduction while minimizing the risk of tissue damage. While in the field of brain-computer interface, the ionic gels can accurately convey brain signals to electronic systems, playing a critical role in intelligent prosthetic control and electroencephalogram [23]. To ensure optimal EDL performance in complex biological environments, researchers have incorporated a variety of modifications to enhance the biocompatibility, conductivity, mechanical durability, and surface properties of ionic gels. For example, natural polymers like alginate, gelatin, and chitosan, as well as bio-compatible conductive polymers such as polythiophene, polypyrrole, and PEDOT, have been utilized to optimize electrode safety. Mechanical stability has been improved by integrating cross-linking agents or layered structures within the gels to create support networks that withstand stretching and twisting forces. For applications requiring high conductivity, functional nanomaterials such as carbon nanotubes, graphene, and gold nanoparticles are added to enhance signal transmission and responsiveness. In short-term or transient implants, biodegradable polymers like poly(lactic-co-glycolic) acid (PLGA) reduce the need for removal surgery, minimizing patient discomfort and risk.

In summary, implantable electrodes based on EDL are positioned at the forefront of biomedical engineering, underpinning a new class of implantable devices that achieve higher signal fidelity, biocompatibility, and functional versatility. As material science and bioelectronics continue to advance, EDL technology is expected to play an increasingly vital role in personalized healthcare, providing precise and efficient solutions for diagnostics, therapeutic interventions, and human-machine interaction.

7.3.2.3 Electrocatalysis Based on EDL

Electrocatalysis plays a pivotal role in material recycling, environmental protection, and clean energy conversion, attracting significant attention for its capacity to efficiently convert molecules or ions into high-value products under mild conditions. While redox reactions dominate the electrode interfacial dynamics in electrocatalysis, increasing evidence highlights the significant impact of the structure and composition of the EDL at the electrode interface. Interactions with surface charges significantly affect charge distribution and the electric field within the interfacial electrolyte [24, 25]. These effects further regulate interfacial properties, including charge distribution, interfacial dielectric constant, solvent network structure, and local pH, thereby exerting effective kinetic and thermodynamic control over electrocatalytic reactions. Generally, at potentials below the point of zero charge (PZC), hydrogen atoms in water molecules within the IHP orient toward the electrode surface, enabling proton adsorption. At potentials above the PZC, oxygen atoms in water molecules within the IHP face the surface, facilitating the specific adsorption of anions. At even higher potentials, substantial adsorption of hydroxide ions and the formation of surface oxide species occur. On the other hand, electrolyte cations with large hydration shells are believed to reside in the outer Helmholtz plane (OHP). Hydrated cations can influence surrounding water molecules and adsorption structures through non-covalent interactions, such as ion-dipole, dipole-dipole, and hydrogen bonding interactions [26]. The interface structure mediated by cations, including the electrode surface, water, and adsorbed species, is crucial for understanding high-performance electrocatalytic systems [27]. Recent studies have demonstrated that the EDL effect plays a pivotal role in various electrocatalytic reactions. For instance, in the hydrogen evolution reaction (HER) and oxygen evolution reaction (OER), as shown in Fig. 7.7, the EDL influences the reaction rate by modulating local ion or molecule concentrations. In the oxygen reduction reaction (ORR), the EDL stabilizes intermediates, enhancing reaction efficiency. In carbon dioxide reduction reactions (CO_2RR), the EDL regulates the adsorption of CO_2 molecules and the selectivity of products. Similarly, in nitrogen reduction reactions (NRR), the EDL facilitates the activation of reactants. These findings underscore the critical role of the EDL in electrocatalysis, providing new insights and directions for optimizing catalytic performance.

Electrocatalysis, as a scientific discipline, has undergone significant development since its origins in the late 19th and early twentieth centuries [28]. The early development of electrocatalysis centered on quantitative evaluations of reaction rates at electrode interfaces, exemplified by Bowden and Rideal's 1928 studies, which analyzed electrocatalytic activity through current densities at defined potentials. Frumkin later advanced the field by introducing reaction rate constants influenced by the· EDL, providing a formalized framework [29]. Historically, the hydrogen evolution reaction (HER) served as a foundational model for electrocatalytic investigations [30–32]. By the early twentieth century, industrial applications, such as chlorine and hypochlorite production, began to utilize electrocatalysis, with early catalysts like

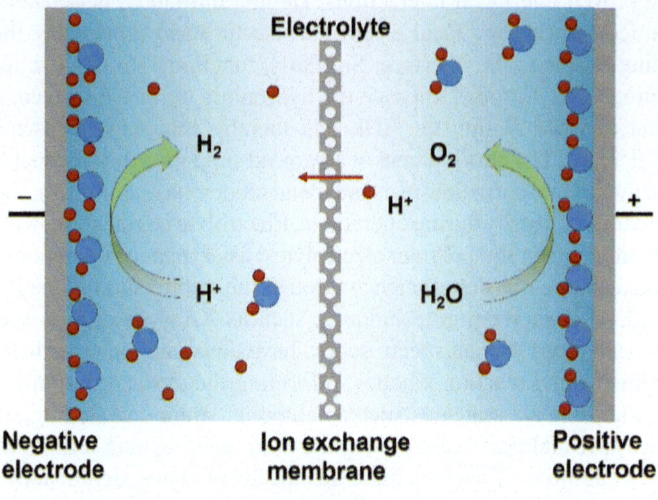

Fig. 7.7 Schematic of HER and OER electrocatalysis

platinized platinum offering enhanced surface area and activity. Groundbreaking work by scientists like Tafel and Haber during this period established key principles, including the relationship between overpotential and catalytic efficiency [33, 34]. The term "electrocatalysis" was formally introduced in the 1930s by researchers like Kobosew and Monblanowa, who investigated the correlation between hydrogen-metal bond energy and catalytic overpotentials [35]. This era also marked the introduction of model systems and the utilization of single crystals, enabling the investigation of catalytic properties under controlled conditions [36, 37]. By the mid-twentieth century, advancements in nanomaterials and surface science catalyzed further progress. Today, electrocatalysis encompasses diverse applications ranging from fuel cells and electrolyzers to carbon dioxide reduction and water splitting. The field has evolved to integrate complex catalysts, computational modeling, and advanced in situ characterization techniques, ensuring its continued relevance in addressing modern energy and environmental challenges.

Significant advancements in electrocatalysis have been achieved through innovations in materials, surface engineering, and system optimization, addressing the challenges of efficiency, selectivity, and durability. At the heart of many of these advancements is the EDL, whose structure and dynamics at the electrode-electrolyte interface play a critical role in determining catalytic performance. By tailoring materials and interfaces to modulate the EDL, researchers have achieved significant improvements in catalytic efficiency and stability. Material innovations, such as nanostructured catalysts and multi-element alloys, have enhanced catalytic activity by increasing active site density and tailoring electronic properties. For example, transition metal oxides, 2-dimensional (2D) materials like graphene and MXenes, and alloyed systems create enhanced electronic environments that influence the EDL by modulating surface charge density and local electric fields. Surface modifications, including doping and functionalization, have further optimized the binding strength of reaction intermediates and improved interfacial interactions. Doping introduces heteroatoms into the catalyst lattice, altering the local electronic structure and influencing the EDL by redistributing charge at the interface. Similarly, functional groups like hydroxyl (–OH) or amine ($-NH_2$) groups improve the hydrophilicity of the surface, promoting efficient ion transport within the EDL and thereby enhancing mass transfer and catalytic efficiency. The development of advanced support materials, such as carbon-based frameworks and corrosion-resistant metal oxides, has stabilized active sites and ensured structural integrity during operations. Electrolyte engineering, with the introduction of ionic liquids and polymer electrolytes, has expanded operational windows and improved charge transfer efficiency through the stabilization of the EDL. In situ and operando characterization techniques, such as X-ray absorption spectroscopy and surface-enhanced Raman spectroscopy, have elucidated intermediate stabilization mechanisms and reaction kinetics, informing the rational design of advanced catalysts. Additionally, strategies such as alloying, strain engineering, and protective coatings have addressed catalyst degradation, ensuring long-term stability under operational conditions. Together, these advancements have significantly improved the efficiency and resilience of electrocatalysts, positioning them as key enablers of sustainable energy technologies such as fuel cells, electrolyzers, and CO_2 reduction systems.

Overall, the EDL, as a fundamental mechanism connecting interfacial chemistry with electrocatalytic processes, offers distinct advantages in enhancing electrocatalytic performance by modulating ion distribution, electric field intensity, and solvent environments. It has become integral to electrocatalysis research, underpinning advancements in green energy technologies such as fuel cells and tribo-electrocatalysis [38]. With continued progress in advanced characterization methods and theoretical simulations, the precise control of EDL effects is poised to evolve further. Future endeavors will focus on leveraging these effects in increasingly complex and efficient electrocatalytic systems, propelling innovations in clean energy and sustainable technologies.

7.3.3 The Tribovoltaic Effect at Water-Metal Interfaces

When water interacts with a metal surface, beyond traditional electrochemical applications where EDLs are regulated by external electric fields, the tribovoltaic effect, driven by the Schottky junction, can also arise in the absence of external fields. The tribovoltaic effect at solid semiconductor-metal interfaces was first discovered by Wang et al. in 2020 [39]. The study demonstrated that a Schottky junction formed when a high-work-function metal was in static contact with n-type Si, characterized by a lower work function. This interaction facilitated electron transfer from the semiconductor to the metal, creating a built-in electric field. During sliding, frictional energy released through bond formation excited non-equilibrium carriers, such as electron-hole pairs in the semiconductor's space charge region and electrons in the surface states of both the metal and the semiconductor. Driven by the built-in electric field, these carriers drifted, producing a steady output in the external circuit.

In the tribovoltaic effect observed at water-metal interfaces explored by Wang et al. in 2020 [41], deionized (DI) water was considered analogous to a wide-band gap semiconductor. As illustrated in Fig. 7.8a, the Fermi level of the metal was lower than that of DI water. Upon contact, a built-in electric field was established to equalize the Fermi level difference, as shown in Fig. 7.8b. In solid-solid (S-S) Schottky junctions, the depletion region was always located on the semiconductor side. Similarly, at DI water-metal interfaces, the depletion region is formed on the DI water side. Recent experimental studies have confirmed the presence of the tribovoltaic effect at DI water-metal interfaces, highlighting its potential for energy scavenging and sensing applications. These findings suggested that interfacial charge dynamics in liquid-metal systems could play a significant role in developing next-generation technologies.

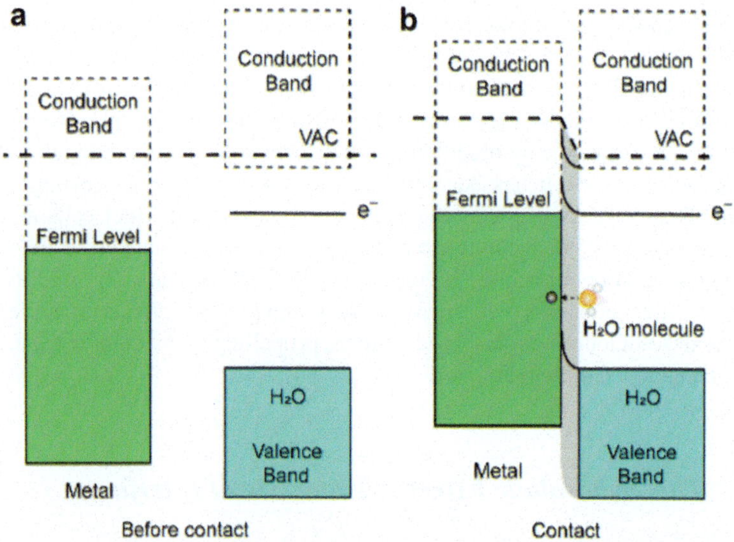

Fig. 7.8 The tribovoltaic effect observed at water-metal interfaces. **a** The Fermi level of the metal was lower than that of DI water. **b** Upon contact, a built-in electric field was established to equalize the Fermi level difference. With permission from American Chemical Society (2022) [40]

7.4 Contact Electrification at Liquid-Dielectric Interfaces

7.4.1 Quantifying Ion Transfer and Electron Transfer

In the context of the EDL at liquid-conductor interfaces, it was commonly assumed that the primary charge carriers were ions, as liquids typically contain ion species like H^+ and OH^- or other species. The EDL at the liquid-solid interface has traditionally been attributed to the ionization or dissociation of surface functional groups and the adsorption of ions from the liquid onto the solid surface. However, Wang et al. proposed an innovative "electron-cloud-potential-well" model in 2018 [6, 42], as illustrated in Fig. 7.9, which broadened the understanding of charge exchange processes beyond the conventional S-S CE framework. This model suggested that under mechanical stress, the electron clouds of two adjacent atoms could overlap significantly, reducing the potential barrier between them and facilitating electron transfer. Mechanical stress was critical for bringing the atoms into close proximity, thereby maximizing the overlap of their electron clouds and promoting electron transitions. This model has been extended to L-S interactions, where liquid molecules collide with atoms on a solid surface due to liquid pressure. Such collisions may enhance the overlap of electron clouds, enabling electron transfer at the interface. This perspective suggested that electron transfer, in addition to ion exchange, played a significant role in L-S CE. The ongoing debate surrounding the identity of charge carriers in L-S CE underscores a fundamental question in interfacial charge exchange

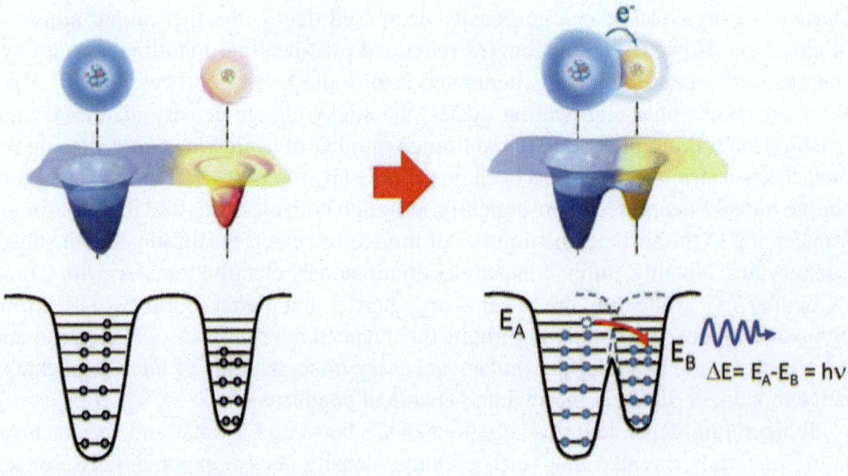

Fig. 7.9 The "electron-cloud-potential-well" model of L-S CE was proposed by Wang et al. With permission from Elsevier Ltd. (2019) [6]

and physical chemistry. Gaining deeper insights into the mechanisms governing charge transfer at these interfaces, including both ionic and electronic contributions, is crucial for advancing theoretical frameworks.

The attenuation of surface triboelectric charges via thermionic emission has been demonstrated as an effective approach to distinguish between electron and ion transfer in L-S CE. Lin et al. leveraged this principle by designing temperature-dependent charge decay experiments at the nanoscale in 2020 [5, 40, 43], employing Kelvin Probe Force Microscopy (KPFM) to quantitatively study CE processes between aqueous solutions and various insulating materials, such as SiO_2, Si_3N_4, and MgO (Fig. 7.10a). This methodology provided a comprehensive understanding of charge transfer dynamics at the L-S interface, shedding light on interfacial charge behavior. In the experiments, the initial surface charge density of insulating samples was measured using KPFM. Subsequently, a DI water droplet slid across the dielectric surface, facilitating charge transfer between the liquids and solid materials. The charges generated at the L-S contact interface might consist of either electrons or ions, with ions formed through ionization reactions on the oxide and nitride surfaces. After charge exchange, the dielectric was heated to a predetermined temperature (e.g., 513 K), enabling the thermionic emission of electrons as per thermionic emission theory [44, 45] (Fig. 7.10b). In contrast, ions chemically bonded to the dielectric's atoms underwent chemical adsorption, characterized by significantly higher detachment energy thresholds compared to electrons. For instance, detaching hydroxide ions (OH^-) from the SiO_2 surface requires approximately 8.5 eV, while removing protons (H^+) demands around 20 eV. These energy barriers indicate that chemically adsorbed ions are far more challenging to remove at moderate temperatures. The results revealed that SiO_2's initial surface charge density was nearly negligible. Upon water contact, negative charges were transferred from the liquid to the solid surface.

During heating, surface charge density decreased due to the thermionic emission of electrons. However, some charges remained post-heating, identified as "sticky" charges, corresponding to ions generated during the L-S CE process (Fig. 7.10c). With successive charging-heating cycles, the sticky charge density increased until reaching saturation, likely due to the limited number of available charge sites on the dielectric surface. As these sites became occupied by ions, the availability for further charge transfer decreased. Consequently, although both electrons and ions continued transferring to the surface, the number of transferred electrons diminished in subsequent cycles. Notably, initial contact was dominated by electron transfer, with a ratio exceeding 75%, reflecting the lower energy barrier and higher mobility of electrons compared to ions. This study highlights the nuanced interplay between electron and ion transfer at the L-S interface and advances the understanding of interfacial charge behaviors under different thermal and chemical conditions.

Temperature-dependent investigations of CE between DI water and various insulating materials revealed that surface charge density decays exponentially, consistent with thermionic emission principles. This decay pattern indicated the emission of electrons or holes from the material surface at elevated temperatures, helping clarify charge transfer mechanisms (Fig. 7.11a). For instance, materials such as MgO primarily exhibited an increase in positive surface charge density, attributed to the transfer of electrons to the solid surface during CE. Conversely, materials like Si_3N_4, HfO_2, Ta_2O_3, and AlN showed a decrease in positive charges due to the transfer of holes rather than electrons. This behavior highlighted a material-specific mechanism of charge reception during CE, which depended on their energy band structures and surface chemical properties. Negative surface charges arose from the transfer of electrons or negative ions, whereas positive charges originated from holes or positive ions. At elevated temperatures, thermionic emission could reduce surface charge density by releasing electrons and holes. However, exceptions such as MgO and Al_2O_3 display increased surface charge density over time. This phenomenon suggested that MgO absorbs both electrons and positive ions, while Al_2O_3 interacts with holes and negative ions. These differences stemmed from each material's specific surface adsorption properties and their ability to accommodate various charge species. The study also highlighted the role of surface hydrophilicity in charge transfer mechanisms (Fig. 7.11b). Hydrophilic surfaces, characterized by a water contact angle (CA) $< 90°$, form stronger interactions with water molecules, resulting in higher interfacial energy compared to hydrophobic surfaces (CA $> 90°$). These stronger interactions promoted the formation of covalent bonds between water molecules and surface atoms, facilitating ionization reactions at the interface and generating ions. In contrast, hydrophobic surfaces, with weaker water-solid interactions, were less likely to undergo ionization, leading to electron transfer dominating the CE process. One of the significant insights from this research was the apparent independence of electron transfer and ion transfer during L-S CE. While these processes may occur simultaneously, they are spatially segregated, with distinct regions of the surface facilitating different types of charge transfer. For hydrophilic materials, ion transfer was more pronounced due to stronger water-surface interactions, whereas hydrophobic materials favor electron transfer due to weaker molecular interactions.

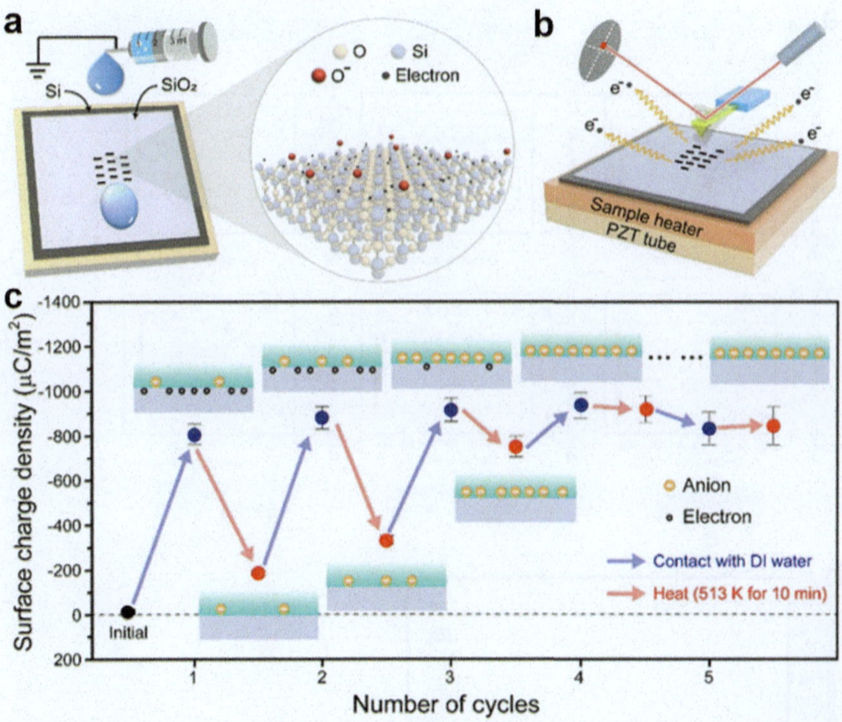

Fig. 7.10 The temperature-dependent charge decay experiments at the nanoscale were designed, employing KPFM to quantitatively study CE processes between aqueous solutions and various insulating materials. **a** A DI water droplet slid across the dielectric surface, facilitating charge transfer between the liquids and solid materials. **b** After charge exchange, the dielectric was heated to a predetermined temperature, enabling the thermionic emission of electrons as per thermionic emission theory. **c** During heating, surface charge density decreased due to the thermionic emission of electrons. However, some charges remained post-heating, identified as "sticky" charges, corresponding to ions generated during the L-S CE process. With successive charging-heating cycles, the sticky charge density increased until reaching saturation, likely due to the limited number of available charge sites on the dielectric surface. With permission from the authors (2020) [5]

These findings not only provided a deeper understanding of charge transfer dynamics but also offered practical guidance for designing and optimizing materials for electrochemical and energy-scavenging applications. By tailoring surface properties such as hydrophilicity, adsorption capacity, and thermal stability, it becomes possible to control charge transfer pathways, thereby improving the performance of CE-based systems across diverse environments and applications.

In addition to KPFM, Nie et al. developed a squeezing system in 2020 to investigate the electrification behavior at the L-S contact interface[46]. PTFE was selected as the insulating material due to its excellent chemical stability in both acidic and alkaline environments. In the experimental setup, a DI water droplet was positioned between two fluorine-doped tin oxide-coated PTFE (FTO-PTFE) substrates. The

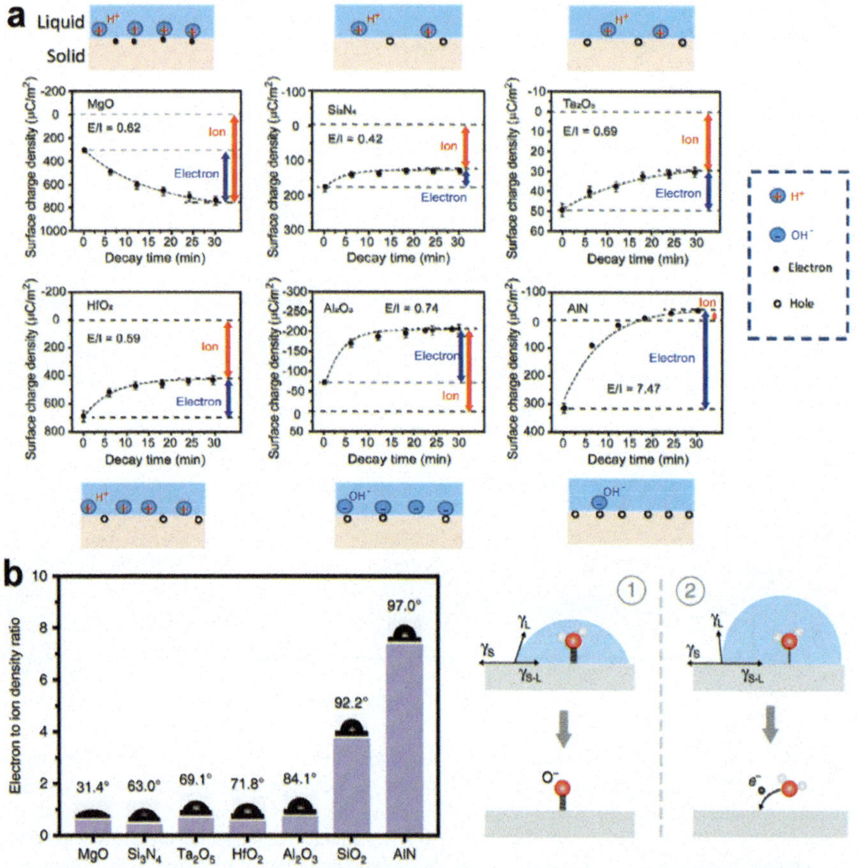

Fig. 7.11 The influence of dielectric materials properties on the L-S CE. **a** The comparative amounts of electron transfer and ion transfer during the CE process between DI water and various dielectric materials With permission from WILEY-VCH Verlag GmbH & Co. (2020) [43]. **b** The correlation between the ratio of electron transfer to ion transfer and the hydrophilicity of dielectric materials. With permission from the authors (2020) [5]

squeezing motion of the substrates was precisely controlled by a linear motor. As shown in Fig. 7.12a, the water droplet spread uniformly during compression, forming a thin liquid film and significantly increasing the contact area between the liquid and PTFE. After a contact-separation process, the triboelectric charge generated in the DI water droplet (50 μL) was measured, showing a positive correlation with the contact area, as depicted in Fig. 7.12b. To evaluate the contributions of electron transfer and ion adsorption to the total triboelectric charge, the authors performed a simplified calculation based on a pure ion adsorption model. Figure 7.12b also presented the calculated triboelectric charges under the assumption that the L-S CE was solely caused by ion adsorption. This calculation represented an ideal scenario where all OH^- ions near the PTFE surface were adsorbed. However, several real-world factors, such as the inhibitory effects of H^+ ions, may suppress the adsorption of OH^- during experiments. Comparing the experimental results with the calculated values revealed that the calculated ion adsorption charge was over 10 times smaller than the experimental total triboelectric charge. This discrepancy indicated that ion adsorption alone could not account for the observed electrification and that electron transfer played a dominant role in the L-S CE process.

To further explore ion adsorption on the PTFE surface, solid-state ^1H nuclear magnetic resonance (NMR) spectroscopy was used to analyze changes in hydrogen signals after CE. As shown in Fig. 7.12c, a consistent peak around 0.7 ppm corresponded to primary alcohol ethoxylate (AEO, $RO(CH_2CH_2O)_nH$) in the PTFE dispersion. The water signal appeared near 4.8 ppm, with similar peak values observed for PTFE samples contacted with DI water, NaCl solution, and NaOH solution. However, the charge transfer after contact with DI water was significantly higher than those observed with NaCl or NaOH solutions, suggesting that ion adsorption alone cannot fully explain the CE process. Conversely, when PTFE was exposed to an acidic HCl solution, a prominent peak near 6.2 ppm was observed, indicative of substantial H^+ ion adsorption. This behavior highlighted that in acidic conditions, both electron transfer and ion adsorption occur, but the high concentration of H^+ ions suppresses electron transfer between water molecules and PTFE. These findings, corroborated by the proposed model in Fig. 7.12d, confirmed that both electron transfer and ion adsorption contribute to the L-S CE process. Ion adsorption, in particular, played a critical role in modulating the electrification behavior under acidic conditions.

7.4.2 *The Wang's Two-Step EDL Model at Liquid-Dielectric Interfaces*

The process of electron transfer at the L-S contact interface played a crucial role in the formation and dynamics of the EDL. To understand this phenomenon in depth, Wang et al. proposed a two-step model of the EDL in 2019 [5, 40], which reinterpreted the conventional concept of the EDL at liquid-dielectric interfaces. This model provided a comprehensive framework for understanding the charge transfer mechanisms that occur when a dielectric surface comes into contact with a liquid. In the first step of the model (Fig. 7.13a), upon contact with a liquid, various particles, including water molecules, cations, and anions, underwent thermal motion and pressure-driven collisions. These interactions promoted electron transfer between the solid's atomic structure and the surrounding liquid molecules. More specifically, the overlap of electron clouds between the solid surface and liquid molecules facilitated electron exchange. Concurrently, ionization reactions could occur at the contact interface, resulting in the simultaneous generation of electrons and ions. As electron transfer occurs on the timescale of picoseconds to nanoseconds, significantly faster than ion diffusion, it precedes ion movement. The electrons attract ions, with a portion being

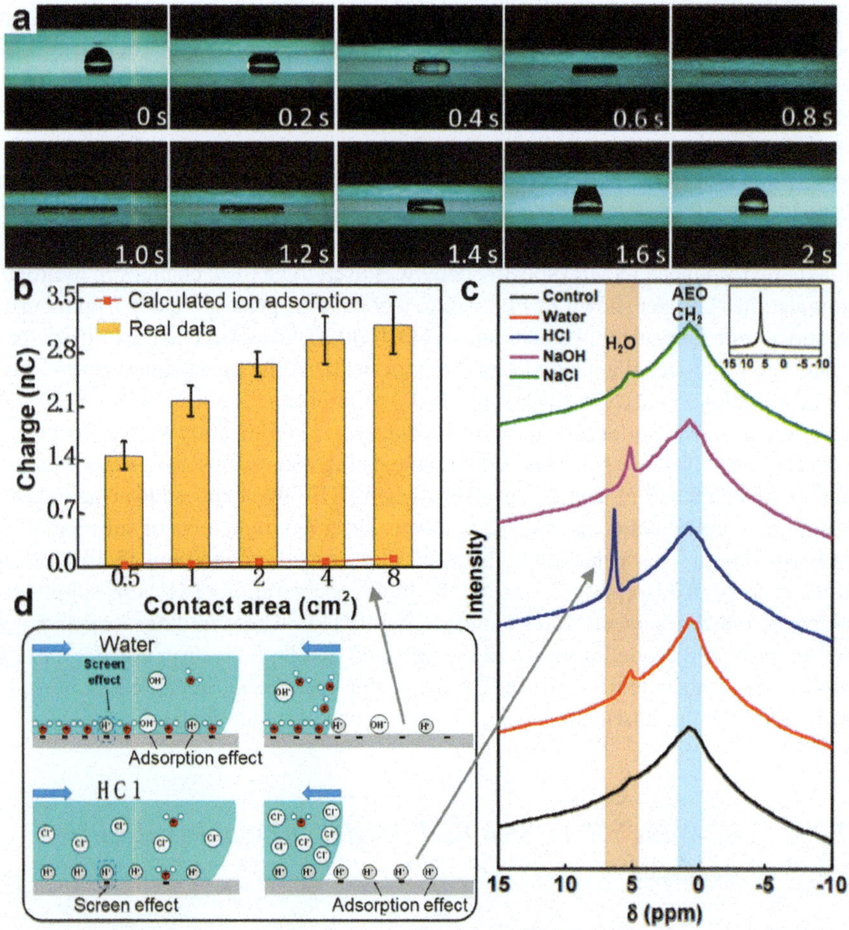

Fig. 7.12 The electrification behavior at the L-S contact interface was systematically investigated. **a** The water droplet spread uniformly during compression, forming a thin liquid film and significantly increasing the contact area between the liquid and PTFE. **b** Comparing the experimental results with the calculated values revealed that the calculated ion adsorption charge was over 10 times smaller than the experimental total triboelectric charge. **c** To further explore ion adsorption on the PTFE surface, solid-state ^{1}H NMR spectroscopy was used to analyze changes in hydrogen signals after CE. **d** These findings confirmed that both electron transfer and ion adsorption contribute to the L-S CE process. With permission from WILEY-VCH Verlag GmbH & Co. (2019) [46]

gradually adsorbed onto the solid surface, forming the IHP, a critical component of the EDL structure. In the second step (Fig. 7.13b), oppositely charged ions in the liquid are drawn to the charged solid surface through electrostatic interactions. This attraction drove the migration of these ions towards the surface, leading to the formation of the OHP. These two layers, IHP and OHP, together constituted the Stern layer, which was crucial for the overall structure of the EDL. In addition to these structured layers, further oppositely charged free ions in the surrounding liquid contributed to the diffuse layer, completing the EDL. Overall, the two-step model is also referred to as Wang's model, which suggests that the formation of EDL is a result of liquid-solid interfacial CE. Secondly, both electron transfer and ion adsorption are possible in the process, but electron transfer is much faster than ion adsorption. The surface transferred charges (both electrons and ions) would attract the ions in the liquid and thus form an EDL. This type of EDL can be called the dynamic EDL, in which the compositions of ions and electrons could vary depending on the experimental conditions.

Wang's two-step model of the EDL offered significant insights into the charge transfer dynamics that occur at liquid-dielectric interfaces. It provided a nuanced understanding of how both electron transfer and ion adsorption contribute to the formation of the EDL, which was essential for the development of new materials and technologies for energy scavenging and information transmission. Specifically, the model enhanced the understanding by recognizing that ionization at L-S contact interfaces resulted in increased surface charge density, leading to charge dispersion in the diffuse layer. Furthermore, the electron transfer process could occur alongside ion adsorption, which was further divided into chemisorption (strong chemical bonding)

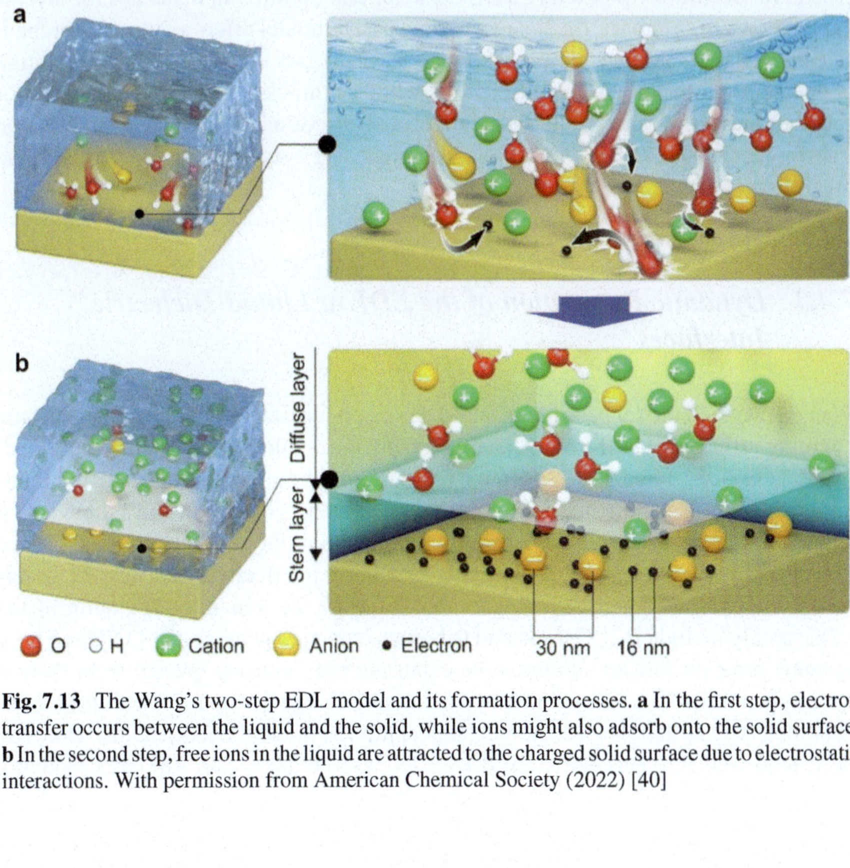

Fig. 7.13 The Wang's two-step EDL model and its formation processes. **a** In the first step, electron transfer occurs between the liquid and the solid, while ions might also adsorb onto the solid surface. **b** In the second step, free ions in the liquid are attracted to the charged solid surface due to electrostatic interactions. With permission from American Chemical Society (2022) [40]

and physisorption (weak physical bonding). For certain materials, such as SiO_2 and polytetrafluoroethylene (PTFE), electron transfer might dominate over ion transfer, leading to a more pronounced generation of surface charges. In these cases, the transferred electrons might become trapped in surface states, which were distinct from the atomic orbitals of the solid. The potential barrier for electrons in surface states was lower than that in atomic orbitals, making transferred electrons more mobile and less stable, a phenomenon that could be observed during heating experiments. In addition, the capacity of a material to either donate or accept electrons (electronegativity) could play a pivotal role in determining its behavior during charge exchange. Materials such as PTFE and fluorinated ethylene propylene (FEP), which were electron-withdrawing polymers (higher electronegative), exhibited stronger electron transfer during CE with water, further illustrating the importance of material properties in dictating charge transfer dynamics. It was also crucial to consider the spatial arrangement of charges in the EDL model. For example, in the case of CE between SiO_2 and DI water, calculations showed that the distance between neighboring electrons on the SiO_2 surface is approximately 16 nm, while the distance between adjacent ions is around 30 nm. These distances far exceed the thickness of the Stern layer, highlighting the necessity for any comprehensive EDL model to consider the spatial distribution of charges and their molecular-level interactions. This arrangement is crucial for accurately predicting EDL behavior and optimizing materials for diverse applications. In summary, the Wang's two-step EDL model offers a more comprehensive understanding of the complex interactions at the L-S interface. By incorporating both electron transfer and ion adsorption, while acknowledging the distinct roles of surface states and atomic orbitals, the model provides valuable insights into the charge transfer mechanisms fundamental to advancing energy-scavenging technologies and other applications.

7.4.3 Dynamic Regulation of the EDL at Liquid-Dielectric Interfaces

Wang's two-step model positioned the EDL as a pivotal interface for ionic-electronic coupling in the field of iontronics, enabling dynamic regulation of both ionic and electronic behaviors. Recent advancements underscore the potential of actively modulating the EDL to influence ionic migration, thereby enhancing the efficiency of electrical current generation. These developments lay the groundwork for innovations in energy scavenging and storage, particularly in effectively harnessing water-based energy. Emerging technologies that leverage the dynamic regulation of the EDL mainly include L-S TENGs and triboiontronic nanogenerators (TINGs). These systems offer promising solutions for capturing high-entropy energy from natural environments. L-S TENGs operate by modulating charge distribution within the EDL near dielectric materials through mechanical motion or electrostatic fields. They can be categorized into mechanically driven DC L-S TENGs [47, 48], mechanically driven alternating current (AC) L-S TENGs [49, 50], and electrostatic-field-driven AC L-S TENGs [51, 52]. These devices exploit the high internal resistance of dielectric materials to produce high-voltage outputs. TINGs, by contrast, generate effective pulsed DC outputs via two primary mechanisms: creating asymmetric EDLs at ultra-thin conductor-liquid interfaces on dielectric substrates[8] or regulating charge density within the diffuse layer of the EDL adjacent to the dielectric surface [7]. Additionally, the inherent sensitivity of the EDL to external stimuli has enabled the development of self-powered information flow systems, unlocking applications in human-machine interaction and advanced interface probes [53]. The dynamic regulation of the EDL presents significant opportunities to enhance energy and information flow, with profound implications for intelligent, autonomous, and efficient systems in the Internet of Things (IoT) and artificial intelligence (AI). Innovative EDL modulation strategies show immense potential across diverse domains, from energy harvesting to the creation of self-powered devices, positioning these advancements to transform sustainable energy technologies and smart systems.

7.4.3.1 Various Types of L-S TENGs

L-S TENGs were operated by modulating charge distribution and migration within the EDL near dielectric material surfaces via mechanical motion or electrostatic fields. One of the first L-S TENGs were invented by Wang's group[49, 54, 55]. After over a decade of advancements, this technology has evolved into a well-established system. These devices can be classified into three main types: mechanically driven DC L-S TENGs, mechanically driven AC L-S TENGs, and electrostatic-field-driven AC L-S TENGs[56]. In mechanically driven DC L-S TENGs, where power generation was based on the movement of the EDL boundary along the dielectric material surface, the process was intricately linked to the coupling of the CE effect and the dynamic behavior of the EDL. When the liquid contacted the dielectric surface with the higher electronegativity, charge transfer occurred at the L-S contact interface, forming a negatively net-charged Stern layer tightly bound to the dielectric surface (Fig. 7.14a). Simultaneously, cations remained distributed within the diffuse layer in the liquid, consistent with Wang's two-step EDL model. The liquid slid across the dielectric surface driven by mechanical forces, causing the EDL boundary to move along the surface. It could generate continuous ion migration within the liquid. As the liquid moved away from the dielectric surface, cations within the diffuse layer interacted with the charge-collecting layer (Fig. 7.14b). This interaction redistributed charge carriers, inducing a unidirectional electron flow through an external circuit and sustaining a continuous current output. Power generation persisted as long as the liquid remained in motion, dynamically adjusting charge distribution within the EDL until a stable state was established near the dielectric material's surface. The reliability and efficiency of this process underscored its potential for effective energy scavenging and precise surface charge regulation. These attributes positioned mechanically driven DC L-S TENGs as highly promising technologies for practical applications in energy generation and advanced sensor systems.

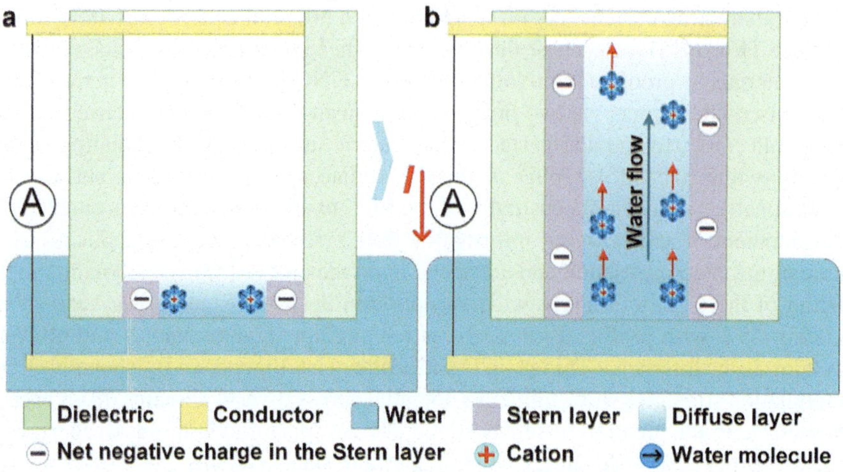

Fig. 7.14 The operating mechanism of mechanically driven DC L-S TENGs. **a** When the liquid contacted the dielectric surface with the higher electronegativity, charge transfer occurred at the L-S contact interface, forming a negatively net-charged Stern layer tightly bound to the dielectric surface. **b** As the liquid moved away from the dielectric surface, cations within the diffuse layer interacted with the charge-collecting layer, which redistributed charge carriers, driving a unidirectional flow of electrons through an external circuit and generating a continuous current output. With permission from the authors (2024) [56]

In 2015, Zhang et al. developed a microfluidic generator (MFG) as a representative mechanically driven DC L-S TENG[47] (Fig. 7.15a). This device was designed to achieve stable output performance by dynamically moving the EDL boundary at the contact interface between the KCl solution and the polydimethylsiloxane (PDMS) surface. The MFG was designed to produce a stable electrical current through L-S CE as the KCl solution flowed across the PDMS surface. Experimental results demonstrated that at a flow rate of 0.4 mL/min, the MFG produced a stable open-circuit voltage (V_{OC}) of approximately 0.16 V and a short-circuit current (I_{SC}) of around 1.75 nA (Fig. 7.15b and c). These outputs remained consistent across multiple on-off flow cycles, highlighting the device's operational stability. When the flow was stopped by turning off the syringe pump, the voltage output gradually decayed to zero, confirming the liquid-flow-dependent nature of the device. Further analysis demonstrated a nearly linear correlation between flow rate and current output, where an increase in flow rate proportionally enhanced the current, provided the flow direction and electrical connection were aligned in the forward configuration (Fig. 7.15d). Reversing the flow direction yielded a negative current output, demonstrating a predictable and reversible relationship between flow direction and electrical output (Fig. 7.15e). The effect of different ions, including Li^+, Na^+, and K^+, on current output was also evaluated (Fig. 7.15f). The MFG exhibited similar performance across these ions, indicating its compatibility with various ionic solutions. However, the KCl concentration significantly influenced the current output. At lower concentrations (< 10 μM), the current increased with concentration, peaking at approximately 2.45 nA at 10 μM. This enhancement was attributed to reduced internal resistance and improved solution conductivity. At concentrations exceeding 10 μM, the current output declined sharply. This reduction was likely caused by the formation of a denser EDL via the ion adsorption on the PDMS surface at higher ion concentrations, which induced a screening effect that hindered charge transfer at the contact interface, thereby decreasing the generator's performance (Fig. 7.15g).

In addition to external mechanical forces, liquid evaporation offered an autonomous and effective mechanism for driving the motion of the EDL boundary at the liquid-dielectric interface. In 2022, Chi et al. developed a mechanically driven DC L-S TENG that utilized both liquid evaporation and capillary forces to guide water through Al_2O_3 ceramic sheets[48] (Fig. 7.16a). This process initiated L-S CE and generated a consistent electrical output. By regulating the environmental seal, the liquid evaporation rate, and consequently the continuous flow through nanochannels, could be precisely controlled, ensuring a stable electrical output (Fig. 7.16b). Environmental parameters such as temperature and airflow significantly influenced the evaporation rate, leading to higher current outputs (Fig. 7.16c). These findings underscored the critical role of environmental factors in optimizing the performance of energy-scavenging devices and provided strategies to enhance operational efficiency. The optimized DC L-S TENG demonstrated notable performance metrics, achieving a V_{OC} of 0.7 V and an I_{SC} of 0.3 μA, highlighting its potential for diverse self-powered systems and energy-scavenging applications. In mechanically driven DC L-S TENGs, prolonged liquid flow over solid surfaces could lead to the formation of a dense EDL, which might impede subsequent charge transfer and reduce output stability. To address this limitation, Dong et al. in 2022 introduced a gas-liquid two-phase flow-based triboelectric nanogenerator (GL-TENG) [57] (Fig. 7.16d). This design leveraged dielectric breakdown to dissipate accumulated charges in the dense EDL, maintaining the solid surface's initial charge density and enabling efficient subsequent charge transfer at the L-S contact interface. The GL-TENG utilized gas-liquid flow dynamics to optimize the contact area and relative velocity between the water and the PTFE tube, enhancing both ion mobility and charge density within the diffuse layer. This configuration ensured a stable current output by dynamically facilitating charge exchange across the interface. During continuous operation, accumulated charges induced dielectric breakdown (Fig. 7.16e), producing high pulse voltages and effectively resetting the tube's surface charge state to sustain subsequent charging cycles. By incorporating dielectric breakdown, the GL-TENG achieved significantly enhanced operational stability and energy-scavenging performance. The device exhibited exceptional pulse output values, with a V_{OC} of 3789 V, an I_{SC} of 867 μA (Fig. 7.16f), and a volumetric peak power (P_R) density of 143.6 mW/cm^3. These metrics underscored the GL-TENG's potential for high-voltage output and reliable energy scavenging across diverse applications.

In addition to addressing the dense EDL's screening effect on charge transfer at L-S contact interfaces through the introduction of the dielectric breakdown effect in mechanically driven DC L-S TENGs, mechanically driven AC L-S TENGs offered a complementary approach by enabling effective control over ion migration within the

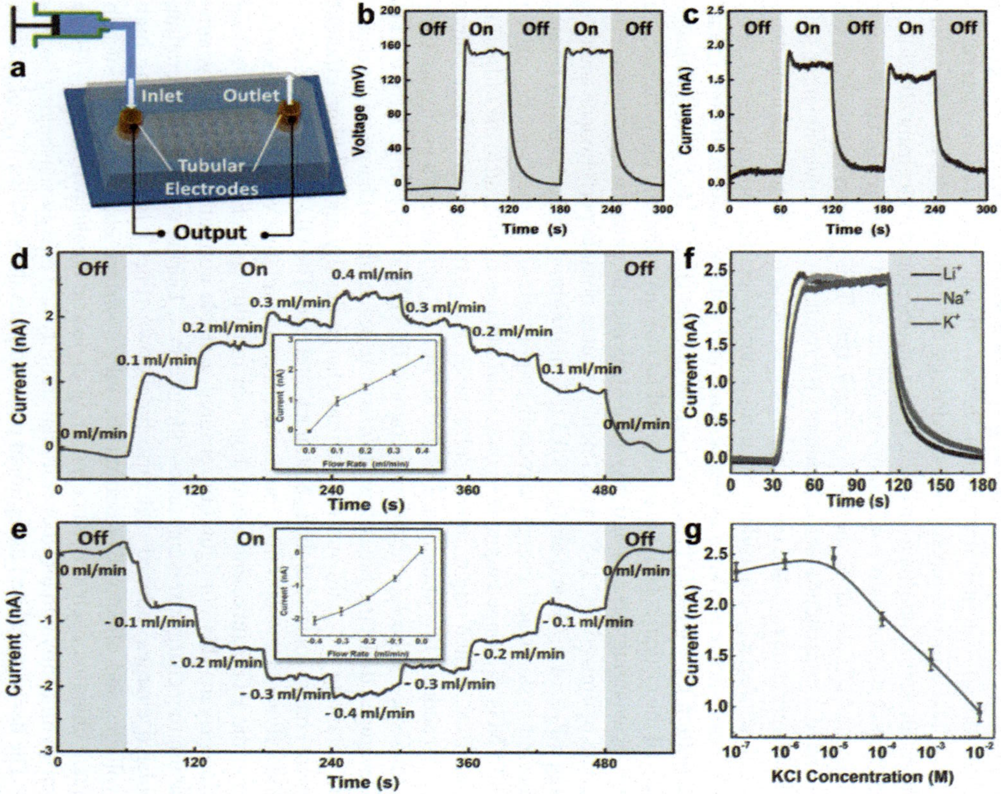

Fig. 7.15 An MFG as a typical mechanically driven DC L-S TENG and its output performance. **a** Schematic diagram of the MFG. **b, c** At a flow rate of 0.4 mL/min, the MFG generated a stable V_{OC} of approximately 0.16 V and an I_{SC} of around 1.75 nA. **d, e** The output performance of MFG was related to the direction and flow rate of the liquid. **f** The MFG exhibited similar performance across Li^+, Na^+, and K^+, indicating its compatibility with various ionic solutions. **g** Effect of ion concentration in solution on the output performance of the MFG. With permission from WILEY-VCH Verlag GmbH & Co. (2015) [47]

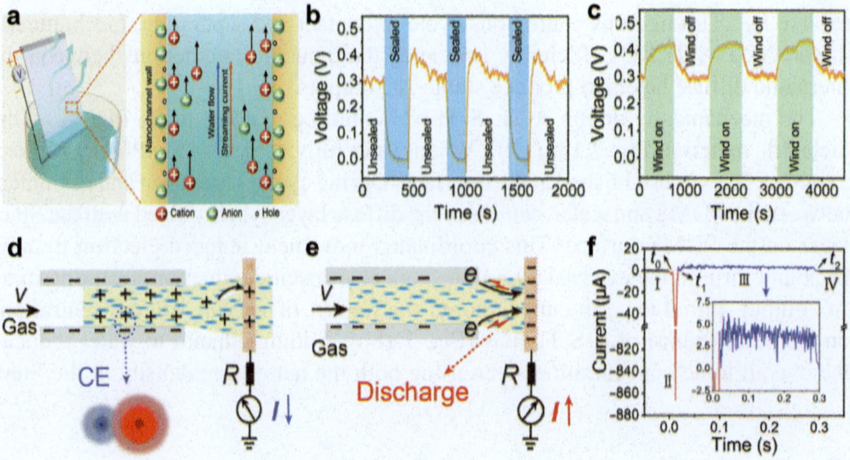

Fig. 7.16 A series of mechanically driven DC L-S TENG and their output characteristics. **a** A mechanically driven DC L-S TENG was introduced in 2022, which harnesses both liquid evaporation and capillary forces to guide the water through Al_2O_3 ceramic sheets. **b** By adjusting the environmental seal, the rate of liquid evaporation could be precisely controlled, resulting in a stable electrical output. **c** Environmental parameters such as temperature and airflow significantly amplified the evaporation rate, leading to increased current output. With permission from the authors (2022) [48]. **d** The GL-TENG was developed in 2022. **e** During continuous operation, the accumulated charges initiate dielectric breakdown, generating higher pulse voltages and effectively resetting the tube's surface charge state to support subsequent charging cycles. **f** The GL-TENG could generate a higher I_{SC} of 867 μA. With permission from the authors (2022) [57]

diffuse layer. This was achieved by manipulating the mechanical motion of the Stern layer adjacent to the dielectric surface by external mechanical forces. This strategy facilitated the flexible and consistent generation of stable AC outputs by dynamically harnessing the processes of charge accumulation and release at the dielectric-liquid interface. During the operation of mechanically driven AC L-S TENGs, when the liquid came into contact with a dielectric surface possessing higher electronegativity, an EDL formed stably at the contact interface (Fig. 7.17a). The Stern layer, characterized by a net negative charge, adhered closely to the dielectric, while cations from the liquid accumulated in the adjacent diffuse layer, forming a distributed charge region. At this stage, charge equilibrium was established between the dielectric and liquid. As mechanical motion separates the dielectric from the liquid, the interaction between the Stern and diffuse layers diminishes, weakening the electrostatic binding of cations to the charged dielectric surface. This reduction triggered electron transfer between the two metal charge-collecting layers via electrostatic induction, generating a forward I_{SC} in the external circuit (Fig. 7.17b). Once the dielectric and liquid were fully separated, electron transfer concluded, and a new equilibrium state was reached (Fig. 7.17c). When the dielectric re-approached the liquid, cations migrated back toward the Stern layer, re-establishing the diffuse layer within the EDL (Fig. 7.17d). This process induced a reverse electron flow in the external circuit, generating a reverse I_{SC}. Through this continuous cycle of contact and separation, mechanically driven AC L-S TENGs efficiently leveraged the dynamic interactions between the Stern and diffuse layers to produce stable AC outputs.

The mechanically driven AC L-S TENG could be realized using PDMS as the dielectric material (Fig. 7.18a) [49]. When water fully contacted the PDMS surface, a stable EDL was established at the interface. During cyclic separation and re-contact between the PDMS and water, cations in the diffuse layer synchronized with the Stern layer on the PDMS surface. This coordinated movement induced electron transfer back and forth in the external circuit through electrostatic induction, resulting in an AC output. Initial experiments explored the impact of varying ion concentrations on the performance of L-S TENGs (Fig. 7.18b). Adding ethanol to water reduced the overall ionic concentration, decreasing both the net charge density in the Stern

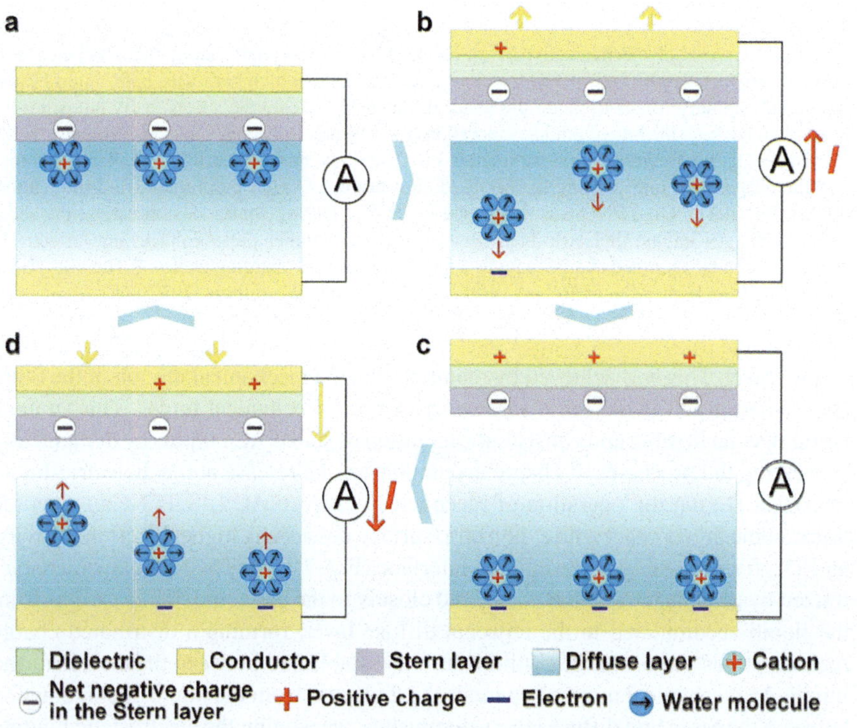

Fig. 7.17 The operating mechanism of mechanically driven AC L-S TENGs. **a** As the liquid came into contact with a dielectric surface with higher electronegativity, a stable EDL formed. **b** The net negative charge in the Stern layer induced electron transfer in the external circuit via electrostatic induction, generating a forward I_{SC}. **c** Upon full separation of the dielectric from the liquid, electron transfer was completed, and the system attained a new equilibrium. **d** When the dielectric re-approached the liquid, cations migrated back toward the Stern layer, re-forming the diffuse layer in the EDL. This process induced a reverse electron flow in the external circuit, generating a reverse I_{SC}. With permission from the authors (2024) [56]

layer and the cation density in the diffuse layer. This reduction impaired electron transfer efficiency, lowering the overall output. Modifications to the PDMS surface microstructure were also explored, increasing the L-S contact area and enhancing ion migration in the diffuse layer. These structural optimizations significantly improved output performance. Under optimal conditions, the L-S TENG achieved a V_{OC} of approximately 80 V and an I_{SC} density of 0.25 μA/cm^2 at an operating frequency of 5 Hz. With an 88 MΩ load resistance, the P_R density reached 5 μW/cm^2 (Fig. 7.18c), demonstrating its potential for efficient energy scavenging in self-powered applications. Building on the practical potential of mechanically driven AC L-S TENGs for ocean and wave energy scavenging, You et al. developed an advanced single-electrode AC L-S TENG in 2021 [50] (Fig. 7.18d). They systematically analyzed the effects of L-S contact area and relative sliding velocity on output performance. Experimental results showed that expanding the PTFE-water contact area increased ion density in both the Stern and diffuse layers, significantly enhancing electron transfer efficiency (Fig. 7.18e). Additionally, higher relative sliding velocities between the PTFE membrane and water improved output performance (Fig. 7.18f). This improvement was attributed to intensified interfacial collisions, which expanded the microscopic contact area and promoted controlled, directional ion migration. By optimizing contact area and sliding speed, the single-electrode L-S TENG achieved a remarkable V_{OC} of approximately 400 V, underscoring its promise for scavenging energy from ocean and wave environments. This innovation highlighted the potential of mechanically driven AC L-S TENGs for sustainable energy solutions in marine applications and offers a roadmap for further advancements in L-S energy scavenging technologies.

While ocean wave energy has garnered substantial attention for mechanically driven AC L-S TENG applications, the untapped potential of raindrop kinetic energy presents a promising alternative for sustainable power generation. In 2014, Lin et al. first introduced a mechanically droplet-driven AC L-S TENG specifically designed to harvest raindrop energy[54] (Fig. 7.19a). This device utilized the negatively charged Stern layer on the PTFE surface to drive ion migration within the diffuse layer of droplets, producing efficient AC output. By optimizing the droplet impact height, the contact area between droplets and the PTFE film was maximized, increasing the charge transfer per droplet and improving overall performance (Fig. 7.19b). At an optimized impact height of 90 cm, a single 30-μL droplet sliding across the PTFE surface generated a V_{OC} of 9.3 V and an I_{SC} of 17 μA (Fig. 7.19c), corresponding to a P_R density of 9.1 μW/cm^2. This work demonstrated the feasibility of utilizing raindrops as a renewable energy source, marking a significant breakthrough in sustainable energy scavenging. To further enhance energy-scavenging efficiency, Xu et al. developed an advanced droplet-driven AC L-S TENG in 2020 [58] (Fig. 7.19d). In this design, droplets striking the negatively charged PTFE surface spread across it while simultaneously contacting a top aluminum (Al) charge-collecting layer. This configuration exploited a "bulk effect", wherein rapid electron transfer occurred between the Al layer and the bottom indium tin oxide (ITO) via an external circuit, generating a robust positive current. As droplets moved across the Al layer, reverse electron flow produced a low-amplitude current. This bulk effect significantly improved

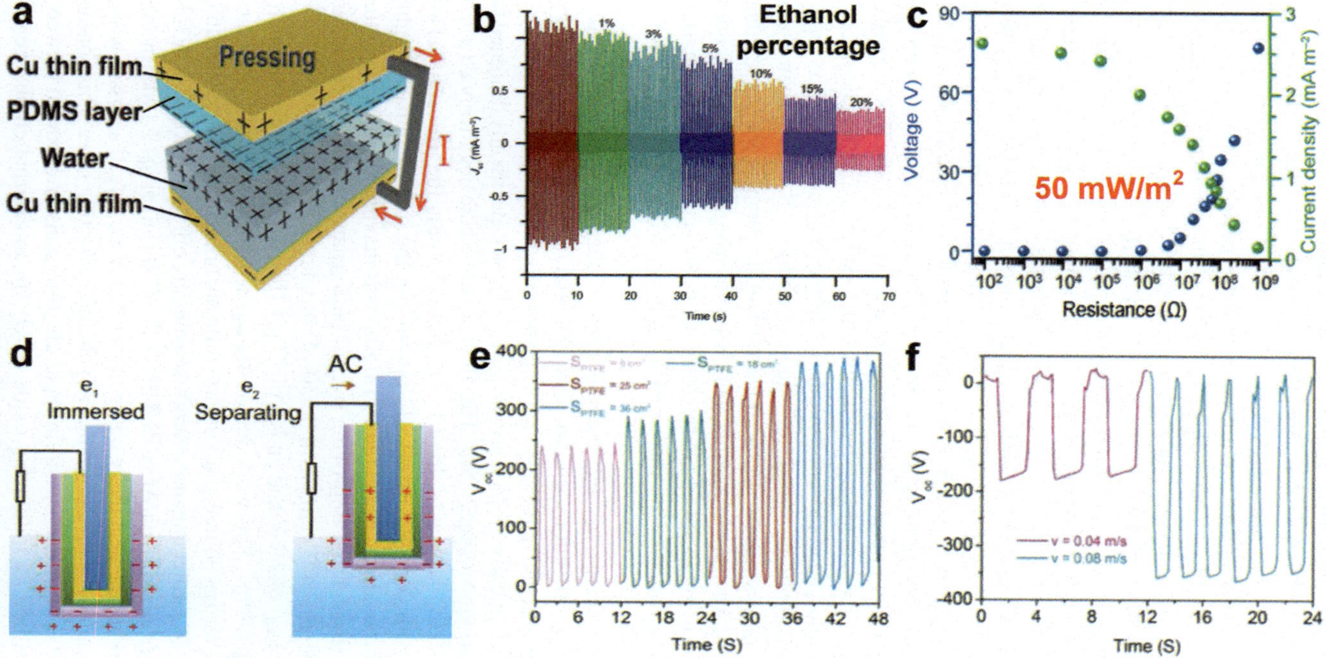

Fig. 7.18 A series of mechanically driven AC L-S TENG and their output characteristics. **a** The first mechanically driven AC L-S TENG was developed in 2013. **b** Increasing ethanol concentration in the water led to a reduction in electron transfer efficiency and the overall output. **c** With an 88 MΩ load resistance, this L-S TENG achieved a P_R density of 5 μW/cm^2. With permission from WILEY-VCH Verlag GmbH & Co (2013) [49]. **d** To harvest ocean and wave energy, a single-electrode AC L-S TENG was developed in 2021. **e** Expanding the PTFE-water contact area directly increased the ion density in both the Stern and diffuse layers, significantly increasing the output performance of the L-S TENG. **f** The higher relative sliding velocity between the PTFE membrane and water further improved the output performance. With permission from American Chemical Society (2021) [50]

performance, yielding a V_{OC} of 143.5 V and an I_{SC} of 270 μA (Fig. 7.19e). These results far exceeded those of single-electrode configurations without the Al layer. Under an optimal load resistance of 332 kΩ, the P_R density reached 5.01 mW/cm^2, representing a three-order-of-magnitude increase compared to conventional designs. Building on the bulk effect, Li et al. developed a high-voltage droplet-driven AC L-S TENG in 2023 [59] (Fig. 7.19f). Their work identified directional ion migration within the droplet's diffuse layer, guided by the electrostatic field of the Stern layer on the PTFE surface, as a key factor for power enhancement. When droplets struck the negatively charged PTFE and contacted the top charge-collecting layer, rapid anion migration caused the electron transfer to generate a high-magnitude positive current. As droplets moved away, cations migrated back more slowly, resulting in a low-magnitude reverse current. By optimizing droplet dynamics and refining circuit designs, this generator achieved an exceptional V_{OC} of 1200 V. These innovations highlight the transformative potential of droplet-driven AC L-S TENGs for renewable energy applications. By precisely controlling ion migration and L-S interactions, these devices offer novel strategies to maximize energy output and represent a promising pathway for sustainable energy solutions.

The various contact states between the droplet and the dielectric material significantly impact the output performance of mechanically droplet-driven AC L-S TENGs while also influencing the wettability characteristics of the dielectric surface. Wettability, a fundamental property dictating L-S interface interactions, is critical for numerous industrial and scientific applications. However, the dynamic interplay between surface charge and wettability, especially in the context of CE, remains intricate and incompletely understood. To address it, Tang et al., systematically investigated the relationship between triboelectric charge generation and the subsequent alteration in the wettability of dielectric surfaces, elucidating the CE induced wetting (CEW) phenomenon [60] (Fig. 7.20a). A custom-designed droplet generator was developed to simulate and monitor the behavior of water droplets sliding across an inclined dielectric surface, emphasizing the influence of triboelectric charge accumulation on the dynamic wettability of these materials. The apparatus facilitated precise, real-time measurements of CA variations and surface potential changes during droplet interaction. The spreading behavior of the droplets, driven by surface tension, governed the CA at the L-S contact interface (Fig. 7.20b). A smaller CA corresponds to enhanced wettability, while a larger CA indicates reduced wettability. When a water droplet is deposited, dielectric surfaces, especially those with strong electronegativity like PTFE, acquire residual surface charges via CE. The polarity of the charges generated adheres to the principle of charge opposition: the dielectric surface typically accumulates negative charges, while the water droplet carries positive charges. This triboelectric charge induces an electric field gradient at the interface, electrostatically attracting polar water molecules and ions, thereby modulating the surface's wetting behavior (Fig. 7.20c). In this study, the PTFE surface potential and CA of PTFE films were monitored as they interacted with water droplets. The results revealed a marked reduction in hydrophobicity, with the CA decreasing from 114° to 99° as the number of water droplets sliding across the surface increased from 0 to 1000 (Fig. 7.20d). Concurrently, as measured by KPFM, the PTFE surface

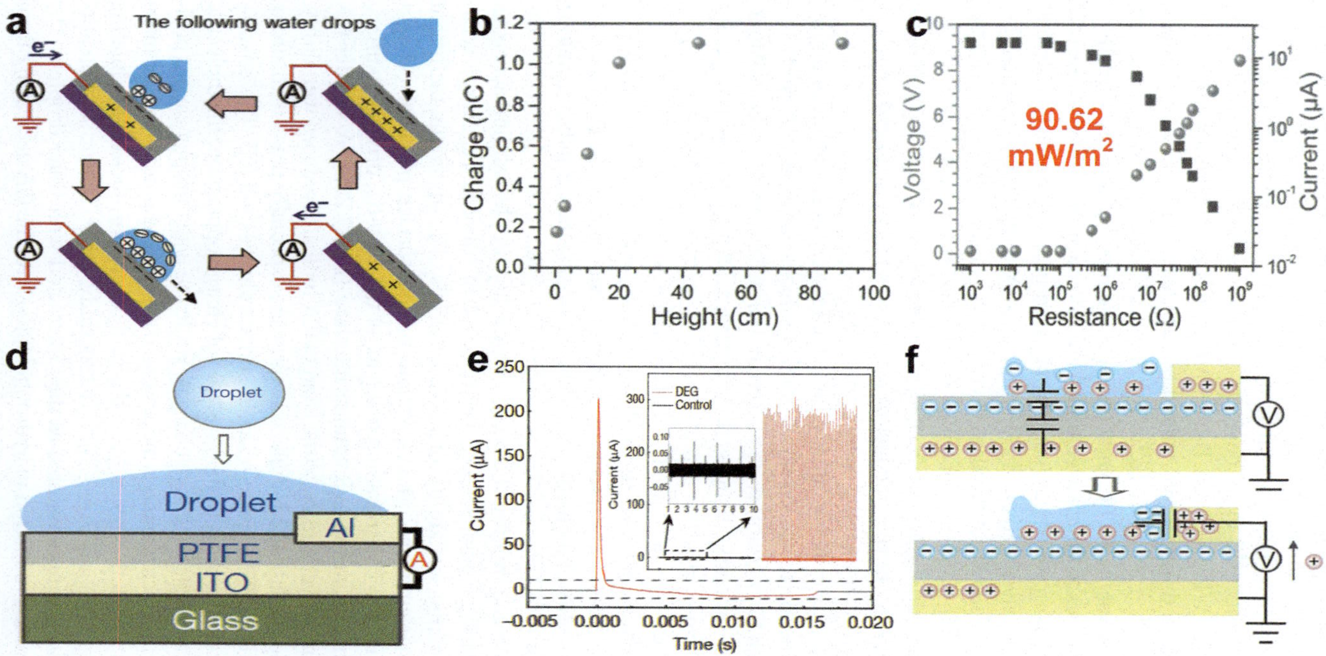

Fig. 7.19 A series of mechanically droplet-driven AC L-S TENG and their output characteristics. **a** A mechanically droplet-driven AC L-S TENG specifically designed to harness raindrop energy was pioneered in 2014. **b** By optimizing the droplet impact height, the contact area between droplets and the PTFE film was enhanced, thereby improving output performance. **c** After height optimization to 90 cm, this droplet-driven AC L-S TENG could achieve a P_R density of 9.1 μW/cm^2. With permission from WILEY-VCH Verlag GmbH & Co (2014) [54]. **d** An advanced droplet-driven AC L-S TENG was developed in 2020. **e** Its V_{OC} and I_{SC} could reach approximately 143.5 V and 270 μA, respectively. With permission from the Authors (2020) [58]. **f** A high-voltage droplet-driven AC L-S TENG was developed by Li et al. in 2023. With permission from the Authors (2023) [59]

potential shifted from a positive value of 11 mV to a negative potential of 663 mV, indicating the accumulation of negative charges on the surface (Fig. 7.20e). This shift in surface potential is indicative of the triboelectric charge transfer that occurred during the CE process.

The CEW phenomenon can be described as a two-step process: firstly, the generation of triboelectric charges at the contact interface, followed by the alteration in wettability due to the accumulation of these charges (Fig. 7.21a). The charge storage capacity of dielectric films enables sustained retention of surface charges, leading to prolonged modifications in wettability. To gain a deeper understanding of the mechanism, molecular dynamics (MD) simulations were employed to model the behavior of

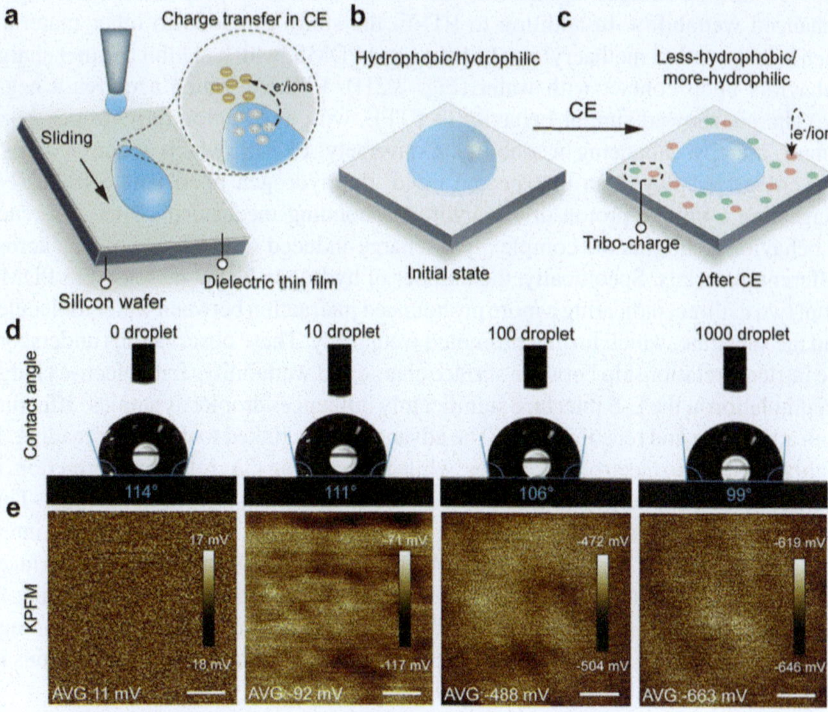

Fig. 7.20 The relationship between triboelectric charge generation and the subsequent alteration in the wettability of dielectric surfaces was systematically investigated in 2024, elucidating the CEW phenomenon. **a** A custom-made droplet generator was developed, which allowed for the simulation and monitoring of the behavior of water droplets sliding across an inclined dielectric surface. **b** The spreading behavior of the droplets, governed by surface tension, dictates the CA formed at the L-S contact interface. **c** Triboelectric charge creates an electric field gradient at the interface, which electrostatically attracts both polar water molecules and ions, thereby influencing the wetting behavior of the surface. **d** The results revealed a marked reduction in hydrophobicity, with the CA decreasing from 114° to 99° as the number of water droplets sliding across the surface increased from 0 to 1000. **e** The surface potential, as measured by KPFM, shifted from a positive value of 11 mV to a negative potential of 663 mV, indicating the accumulation of negative charges on the surface. With permission from Wiley-VCH GmbH (2024) [60]

water droplets on PTFE surfaces under different charge states (Fig. 7.21b). The simulations indicated a significant reduction in the CA, from 118° to 96°, as a result of the induced negative charges on the surface. Furthermore, ab initio calculations were used to examine the intermolecular interactions between water molecules and charged PTFE surfaces (Fig. 7.21c). These calculations revealed that the presence of triboelectric charges leads to stronger hydrogen bonding, particularly through O–H···F interactions, between water molecules and the fluorine atoms on the PTFE surface (Fig. 7.21d). The enhanced hydrogen bonding was observed to decrease the distance between the oxygen and hydrogen atoms in the O–H···F bond, thereby reducing the interfacial tension and promoting a more hydrophilic surface (Fig. 7.21e). These findings were supported by experimental data, which demonstrated that the presence of surface charges significantly alters the interfacial properties, resulting in enhanced wettability. In addition to PTFE, the study extended to other materials such as polymethyl methacrylate (PMMA) and PDMS, which exhibit distinct charge behaviors upon contact with water (Fig. 7.21f). PMMA, which acquires a negative charge, showed similar behavior to PTFE, with a reduction in hydrogen bond length, thereby enhancing wettability. Conversely, PDMS, which acquires a positive charge, displayed an unexpected trend: the hydrogen bond length increased, suggesting a unique proton-driven hydrogen bonding mechanism. This difference in behavior highlights the complexity of charge-induced wetting phenomena across different materials. Specifically, the number of hydrogen bonds increased in PDMS from two to three, indicating a more pronounced interaction between water molecules and the substrate, which further promoted wettability. These observations underscore the intricate relationship between surface charge and wettability. Triboelectric charge accumulation at the L-S interface significantly influences droplet dynamics, affecting both advancing and receding CAs. The advancing CA, linked to droplet spreading, is highly sensitive to electrostatic forces, while the receding CA, reflecting retraction, is less impacted due to the stabilized electrostatic environment after initial contact. This study highlights the role of triboelectric charge in modulating wettability and interface interactions. By leveraging the CEW effect, droplet movement on solid surfaces can be controlled for applications ranging from TENG-based energy harvesting to surface engineering for improved liquid management. Additionally, these findings enable the design of advanced materials with tailored wettability for applications in self-cleaning, droplet manipulation, and microfluidics.

Advances in mechanically driven AC L-S TENGs have been propelled by a deeper understanding of the dynamic modulation of diffuse layers at dielectric-liquid interfaces, with a focus on stabilizing the EDL. However, visualizing the complexities of EDL formation has remained challenging. The influence of the EDL during phase transitions is an intriguing area of study. A recent investigation explored how the phase transition from ice to water affects charge transfer dynamics with dielectric materials [53] (Fig. 7.22a). The stages of charge transfer and EDL formation during ice melting in contact with PTFE were systematically elucidated (Fig. 7.22b and c). Initially, during the S-S contact stage, electron transfer dominated. The limited contact area resulted in relatively weak charge transfer. As melting began, microdroplets formed on the ice surface, substantially increasing the contact area with

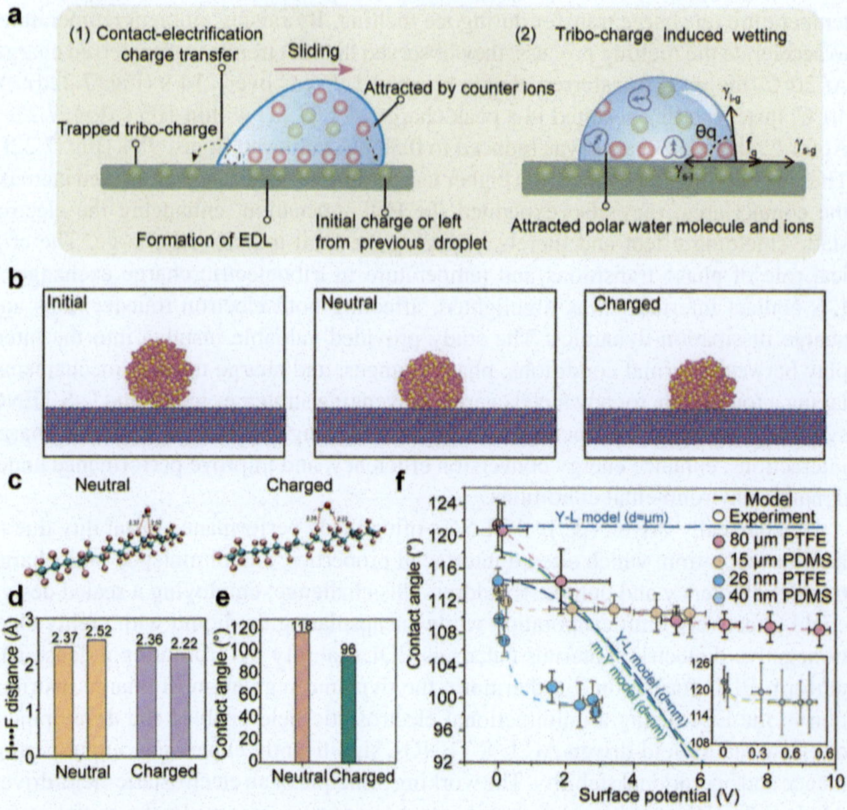

Fig. 7.21 The theoretical model and influencing factors of CEW were systematically investigated. **a** The CEW phenomenon can be described as a two-step process. **b** MD simulations were employed to model the behavior of water droplets on PTFE surfaces under different charge states. **c** Ab initio calculations were used to examine the intermolecular interactions between water molecules and charged PTFE surfaces. **d** The presence of triboelectric charges leads to stronger hydrogen bonding between water molecules and the fluorine atoms on the PTFE surface. **e** The enhanced hydrogen bonding was observed to decrease the distance between the oxygen and hydrogen atoms in the O–H···F bond, thereby reducing the interfacial tension and promoting a more hydrophilic surface. **f** In addition to PTFE, the study extended to other materials such as PMMA and PDMS, which exhibit distinct charge behaviors upon contact with water. With permission from Wiley-VCH GmbH (2024) [60]

PTFE and enhancing electron transfer, leading to a sixfold increase in transferred charge. With continued melting, larger droplets condensed, transitioning the S-S contact stage toward the L-S contact stage. At this stage, EDL formation began, progressively screening the dielectric surface charge. This shifted the primary mechanism from electron transfer to charge migration in the diffuse layer, reducing the total transferred charge. Upon full melting, a stable EDL formed at the contact interface, generating strong electrostatic shielding that minimizes further charge transfer. To further investigate these phenomena, Wei et al. [53] studied the influence of temperature on charge transfer during ice melting. By raising ambient temperatures to accelerate the melting process, they observed distinct trends in transferred charge. At 20°C, the peak transferred charge reached 1.75 nC over 214 s (Fig. 7.22d). At 40°C, faster melting resulted in a peak charge of 1.21 nC within 105 s (Fig. 7.22e). At 60°C, the peak charge was reduced to 0.58 nC, achieved in just 98 s (Fig. 7.22f). These results indicated that while higher temperatures accelerate melting and increase the contact area, they also expedited the EDL formation, enhancing the electrostatic shielding effect and thereby reducing the total transferred charge. The critical role of phase transitions and temperature in triboelectric charge exchange at L-S contact interfaces was highlighted, affecting both electron transfer rates and charge dissipation dynamics. The study provided valuable insights into the interplay between thermal conditions, phase changes, and charge transfer mechanisms, laying a foundation for advanced energy-scavenging strategies in S-S and L-S TENG systems. These insights opened avenues for designing devices that optimize charge interactions, enhance energy conversion efficiency, and improve performance under dynamic environmental conditions.

Mechanically driven AC L-S TENGs often faced performance instability due to liquid evaporation, which altered interfacial properties, and diminished both charge transfer efficiency and output. To address this challenge, employing a sealed design could effectively limit evaporation while encapsulating the liquid with highly electronegative dielectric materials enhances EDL stability by promoting a dense and uniform EDL formation. Furthermore, the dynamic regulation of charge distribution in the EDL using a unidirectional electrostatic field enabled the development of electrostatic-field-driven AC L-S TENGs, significantly enhancing output performance and operational stability. The working principle of an electrostatic-field-driven AC L-S TENG could be delineated into four distinct stages. In the initial stage, the liquid established contact with dielectric material I, triggering the formation of an EDL at the contact interface (Fig. 7.23a). Concurrently, dielectric I interacted with a more electronegative dielectric II on its opposite side, facilitating electron transfer from dielectric I to dielectric II and imparting a net positive charge to the surface of dielectric I. During the second stage, as dielectric I and dielectric II began to separate (Fig. 7.23b), the positive charge on dielectric I generated an electric field that drove cations in the EDL's diffuse layer away from the liquid-dielectric I interface, pushing them toward the charge-collecting layer. This movement induced electron flow from the ground to the charge-collecting layer, generating a positive current. In the third stage, as dielectric I and dielectric II reached full separation (Fig. 7.23c), the migration of ions ceased, halting electron transfer and stabilizing the system momentarily. Finally, during the fourth stage, as dielectric II returned toward dielectric I (Fig. 7.23d), the electric field associated with the positive charge on dielectric I was rebalanced through electrons distributed onto dielectric II. Simultaneously, cations migrated back to the liquid-dielectric I interface, re-establishing the diffuse layer and generating a reverse current. This cyclic process, driven by continuous mechanical stimulation, ensured stable AC output. The sealed design not only mitigated evaporation-induced disruptions but also strengthened the structural integrity of the dielectric-liquid interface, enhancing charge transfer efficiency and

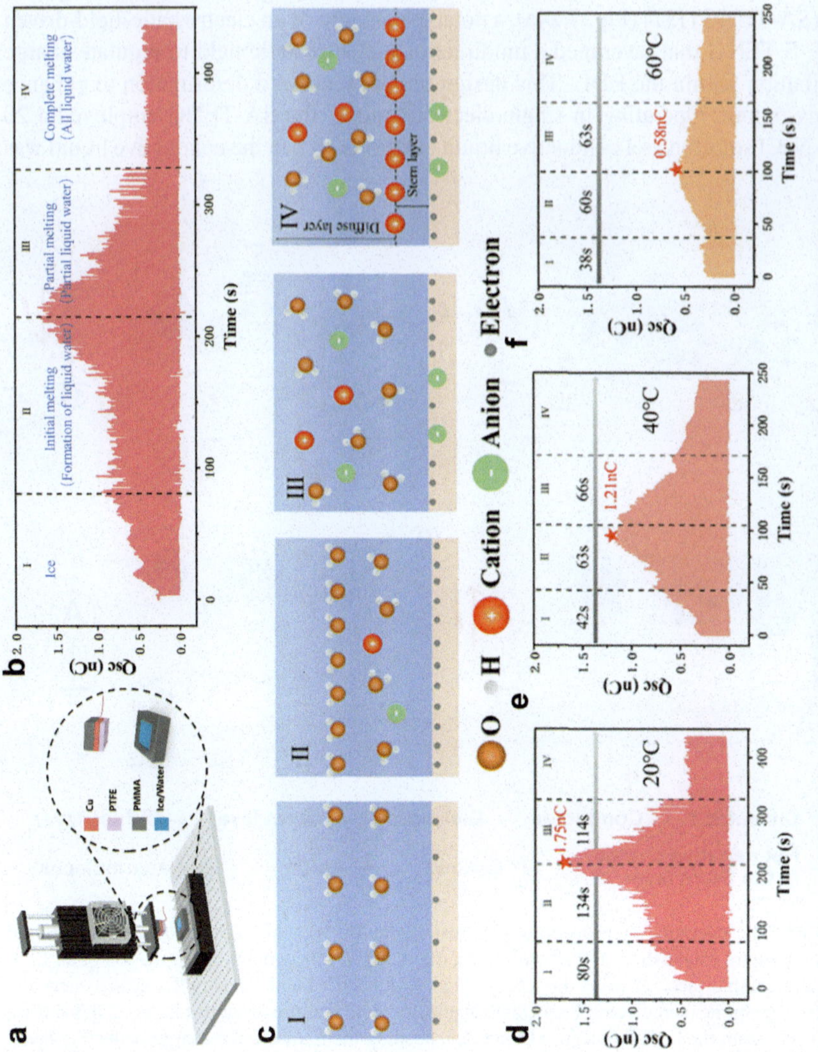

Fig. 7.22 A study examining how the phase transition from ice to water influences charge transfer dynamics with dielectric materials was conducted in 2024. **a** The experimental system. **b** The charge transfer stages between ice and PTFE at different melting stages. **c** The EDL formation stages between ice and PTFE at different melting stages. **d–f** The influence of temperature on charge transfer and EDL formation during ice melting was studied. With permission from Elsevier Ltd. (2024) [53]

energy conversion reliability. These innovations significantly expanded the adaptability of L-S TENGs to diverse environmental conditions, paving the way for efficient, stable energy-scavenging solutions in real-world applications. By combining structural enhancements with advanced charge modulation strategies, these devices held great promise for powering next-generation sustainable technologies.

In 2016, Yi et al. introduced an innovative shape-adaptive triboelectric nanogenerator (SA-TENG) [51] (Fig. 7.24a), a notable example of an electrostatic-field-driven AC L-S TENG that leveraged a unidirectional electrostatic field to regulate charge distribution within the EDL. This design uniquely utilized deformation to enhance energy output. Operating in single-electrode mode, the SA-TENG employed a 20 wt% NaCl solution as a conductive liquid electrode. When the conductive liquid was

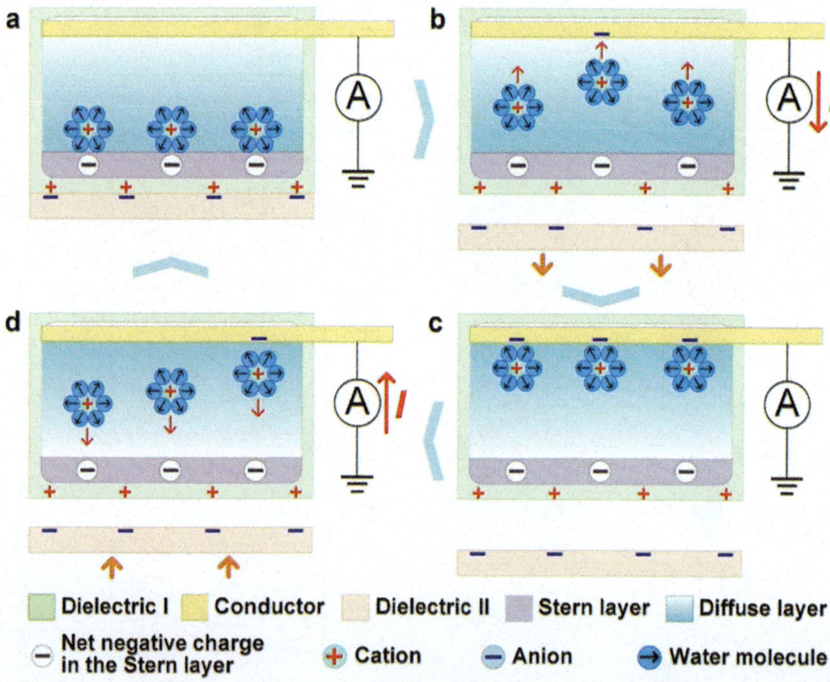

Fig. 7.23 The operating mechanism of electrostatic-field-driven AC L-S TENGs. **a** In the initial stage, the liquid established contact with dielectric material I, triggering the formation of an EDL at the contact interface. **b** During the second stage, as dielectric I and dielectric II began to separate, the positive charge on dielectric I generated an electric field that drove cations in the EDL's diffuse layer away from the liquid-dielectric I interface, pushing them toward the charge-collecting layer. **c** In the third stage, as dielectric I and dielectric II reached full separation, the migration of ions ceased, halting electron transfer and stabilizing the system momentarily. **d** During the fourth stage, as dielectric II returned toward dielectric I, the electric field associated with the positive charge on dielectric I was rebalanced through electrons distributed onto dielectric II. With permission from the authors (2024) [56]

grounded through an external load, the flexible rubber shell of the SA-TENG interacted with an external dielectric material, such as nylon (PA), to efficiently harvest mechanical energy. The performance of the SA-TENG was governed by two critical parameters: the deformation of the SA-TENG (denoted as d_0) and the distance (d_1) between the external PA film and the rubber shell. Both factors significantly influenced the transferred charge (Q_{SC}) and V_{OC} (Fig. 7.24b and c). Increased deformation d_0 enlarged the contact area between the PA film and the SA-TENG surface, thereby enhancing Q_{SC} and V_{OC}. Similarly, increasing d_1 while maintaining constant deformation strengthened electric field interactions, further boosting Q_{SC} and V_{OC}. Under fixed d_0 and d_1, Q_{SC} and V_{OC} stabilized, however, the I_{SC} increased with the frequency of contact between the rubber shell and the PA film, underscoring the role of contact rate in optimizing power output. Adjusting the dimensions of the SA-TENG also impacted performance. Lengthening or widening the device enhanced Q_{SC} and V_{OC} by increasing the contact area between the triboelectric surfaces. Conversely, a thicker rubber shell reduced Q_{SC} and V_{OC} under fixed d_0 and d_1 conditions due to the increased distance between the liquid electrode and the dielectric surface, which lowered capacitance and diminished electrical output. Experimental testing under periodic contact at ~ 0.45 Hz revealed that the SA-TENG achieved a V_{OC} of approximately 67.71 V (Fig. 7.24d), a Q_{SC} density of ~ 35.35 μC/m^2 (Fig. 7.24e), and an I_{SC} density of ~ 0.83 mA/m^2 (Fig. 7.24f). Encased in a flexible polymer, the SA-TENG was highly adaptable and capable of scavenging energy through various operational modes while enduring strains of up to 300%. Its seamless adaptability to complex three-dimensional and curved surfaces establishes it as an optimal candidate for wearable power sources and self-powered sensors in biomechanical motion monitoring. The scalable, flexible design of the SA-TENG marks a notable breakthrough in stretchable electronics, delivering a robust, adaptable energy solution and multifunctional sensing capabilities across diverse applications.

To further enhance the output performance of electrostatic-field-driven AC L-S TENGs, expanding the effective microscopic contact area of the electrostatic field has been demonstrated as a promising approach. Building on this principle, Chen et al. developed a hierarchical micro-nanostructured triboelectric nanogenerator (HM-TENG) in 2019 [52], designed to efficiently harvest both mechanical energy and raindrop energy (Fig. 7.24g). The elastomer surface of this device was engineered with a hierarchical micro-nanostructure fabricated through electrospinning, significantly increasing its surface roughness. This enhancement amplified the microscopic contact area between the elastomer and the dielectric film, thereby strengthening the electrostatic field. As a result, a greater number of free ions were driven to migrate directionally across the elastomer-ion conductor interface, increasing the current output. Moreover, applying pressure on the dielectric film further deformed the elastomer, enhancing its contact with the dielectric film (Fig. 7.24h). This enhancement amplified electrostatic interactions, significantly improving the device's output performance. The optimized HM-TENG achieved remarkable results, including a V_{OC} of approximately 170 V, an I_{SC} of 12.5 μA/cm^2, and a Q_{SC} density of 24 nC/cm^2. When paired with an external load resistance of 13 MΩ, the HM-TENG produced a power density of 0.72 mW/cm^2 (Fig. 7.24i). This study highlighted the crucial role of

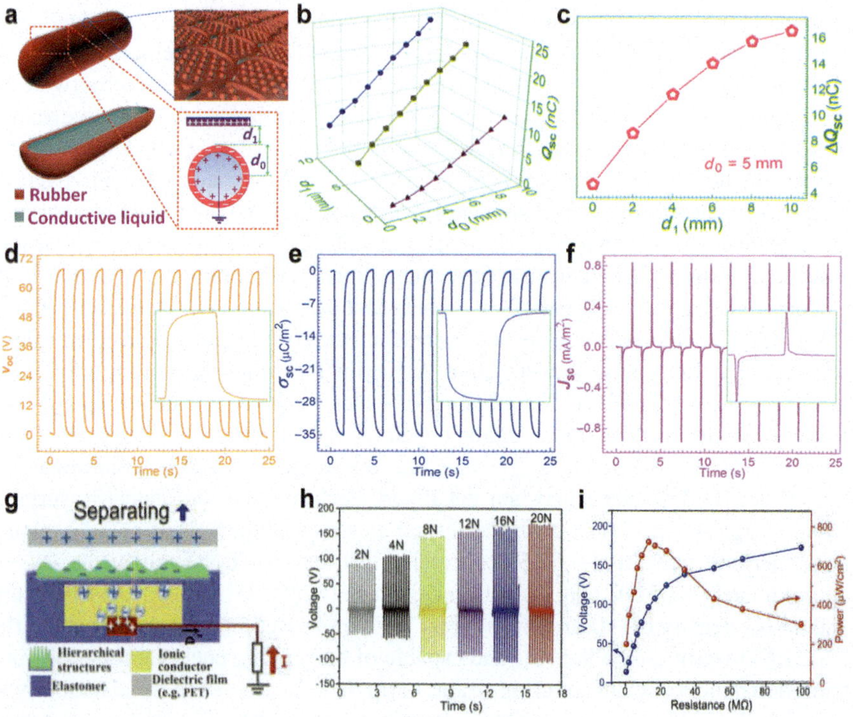

Fig. 7.24 A series of electrostatic-field-driven AC L-S TENGs and their output characteristics. **a** The SA-TENG as an exemplary electrostatic-field-driven AC L-S TENG was developed in 2016. **b**, **c** The output performance of the SA-TENG is strongly influenced by two main parameters including the deformation of the SA-TENG and the distance between the external PA film and the rubber shell. **d**–**d** At an operating frequency of 0.45 Hz, the SA-TENG achieved a V_{OC} of ~ 67.71 V, a Q_{SC} density of ~ 35.35 μC/m^2, and an I_{SC} density of ~ 0.83 mA/m^2. With permission from the authors (2016) [51]. **g** The HM-TENG was developed in 2019 to effectively harvest both mechanical and raindrop energy. **h** Applying pressure on the dielectric film increased elastomer deformation, which could intensify the electrostatic field and raise the output signals from the HM-TENG. **i** With an external load resistance of 13 MΩ, the HM-TENG generated a power density of 0.72 mW/cm^2. With permission from Elsevier Ltd. (2019) [52]

mechanical force in determining the output performance of electrostatic-field-driven AC L-S TENGs and underscores hierarchical micro-nanostructuring as an effective design strategy for optimizing energy scavenging technologies. The HM-TENG demonstrated the potential of surface engineering to enhance energy conversion efficiency, offering valuable insights for advancing the field of TENGs.

Electrostatic-field-driven AC L-S TENGs have surpassed traditional liquid systems by exploiting dynamic EDL regulation at the dielectric-hydrogel interface, achieving enhanced output performance. In 2020, Bao et al. introduced a single-electrode triboelectric nanogenerator featuring an anti-freezing hydrogel electrode (AH-TENG) [61], as illustrated in Fig. 7.25a. This design incorporated thin silicone rubber layers as the top and bottom dielectric surfaces, while a 1 M LiCl ionic hydrogel, selected for its exceptional mechanical stability and sub-zero temperature resilience, serves as the conductive electrode. This configuration rendered the AH-TENG highly efficient in freezing environments. During operation, human skin, serving as an electropositive dielectric, interacted with the electronegative silicone rubber. Contact induced negative charging on the silicone, while subsequent separation triggered ion redistribution within the EDL at the dielectric-hydrogel interface, facilitating electron flow through the grounding circuit. This cyclic contact-separation mechanism enabled continuous power generation. Performance tests demonstrated a V_{OC} of approximately 285 V and a Q_{SC} of ~ 90 nC. The I_{SC} increased from 6.7 to 15.5 μA as the frequency rose from 0.5 to 2.5 Hz (Fig. 7.25b), indicating a frequency-dependent enhancement in output. When connected to various load resistances, the AH-TENG exhibited P_R scaling with frequency, achieving a maximum P_R density of 626 mW/m^2 at 2.5 Hz (Fig. 7.25c). The AH-TENG demonstrated exceptional output performance even under sub-zero conditions, exhibiting remarkable adaptability to both low-temperature and high-stretch environments. These attributes position the AH-TENG as an outstanding candidate for energy scavenging in demanding conditions.

Additionally, differences in electronegativity between dielectric materials significantly influence charge transfer during S-S CE, thereby impacting the output performance of electrostatic-field-driven AC L-S TENGs. In 2021, Xia et al. introduced a fully transparent and stretchable triboelectric nanogenerator (EC-TENG) based on edible-grade silicone and crystal mud [62], as shown in Fig. 7.25d. By exploiting a substantial difference in electronegativity between materials, the EC-TENG achieved ultra-high voltage outputs. Systematic selection of dielectric materials with varying electronegativities interacting with silicone created electrostatic fields of different strengths, modulating the EDL at the silicone-crystal mud interface. This approach generated diverse electrical signals, directly correlating the electronegativity disparity to EC-TENG performance. Experimental results revealed that greater electronegativity differences led to stronger electrostatic fields, which enhanced EC-TENG efficiency (Fig. 7.25e). Under compression using nitrile gloves at 5 Hz, the EC-TENG achieved an I_{SC} of 25.96 μA, a V_{OC} of 1400 V, and a Q_{SC} of 150 nC. With an optimal load resistance of 40 MΩ, the device attained a P_R density of 0.17 mW/cm^2 (Fig. 7.25f). Furthermore, leveraging triboelectric sequence principles improved the device's electrical performance, significantly enhancing its energy-scavenging efficiency. These results underscore the potential of employing unidirectional electrostatic fields to regulate EDL charge distribution in energy-scavenging systems. Encapsulating liquid or gel components in electrostatic-field-driven AC L-S TENGs further enhanced their flexibility and adaptability, making them ideal for wearable applications, such as biomechanical energy scavenging or self-powered motion monitoring. Future research should focus on optimizing design parameters and broadening the applications of this technology in self-powered wearable systems, which show great promise for advanced biomechanically responsive electronics.

Overall, L-S TENGs effectively generated either constant current or AC outputs, especially in scenarios where traditional power sources are limited or impractical. However, their reliance on the insulating properties of dielectric materials typically

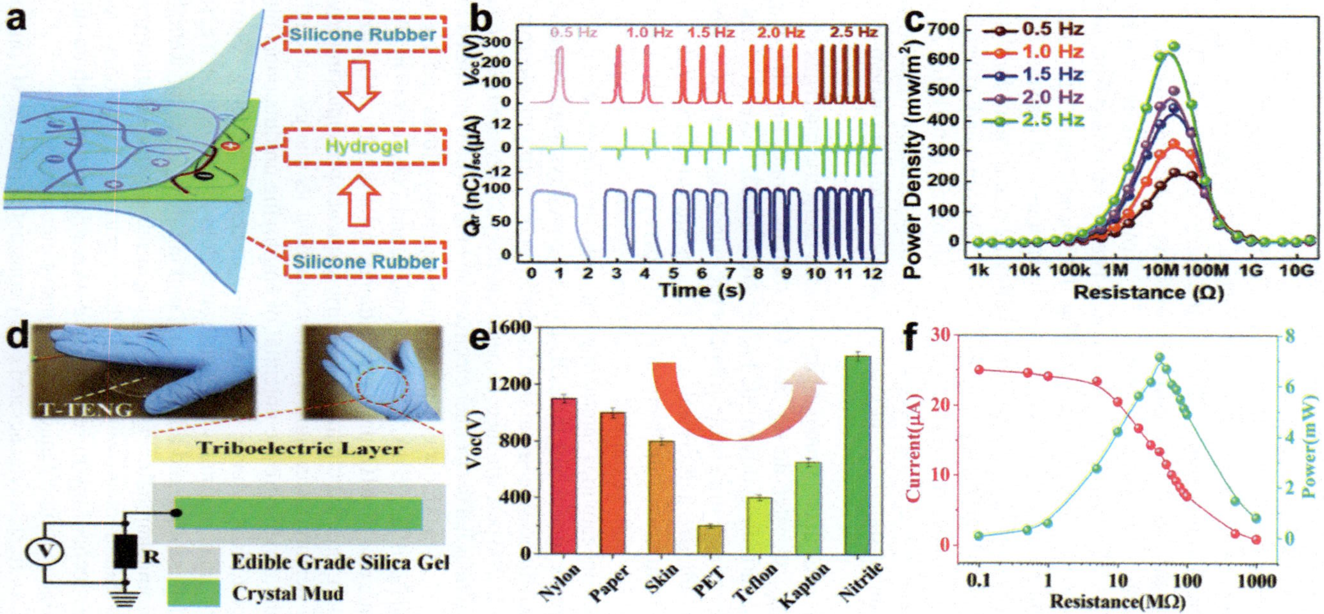

Fig. 7.25 A series of electrostatic-field-driven AC L-S TENGs effectively leveraging dynamic EDL regulation at the dielectric-hydrogel interface and their output characteristics. **a** The AH-TENG was developed in 2020. **b** The I_{SC} increased from 6.7 to 15.5 μA as the frequency rose from 0.5 to 2.5 Hz. **c** The P_R density could reach 626 mW/m^2 at 2.5 Hz. With permission from Royal Society of Chemistry (2020) [61]. **d** In 2021, an EC-TENG based on edible-grade silicone and crystal mud was developed. **e** Larger electronegativity differences allowed for a stronger electrostatic field, which, in turn, enhanced EC-TENG performance. **f** With an optimal load resistance of 40 MΩ, the EC-TENG achieved a P_R density of 0.17 mW/cm^2. With permission from Elsevier Ltd. (2021) [62]

resulted in higher internal resistance. While this elevated resistance facilitated the production of high-voltage outputs suitable for high-voltage applications, it concurrently constrained current output. Thus, optimizing L-S TENG performance for practical use necessitated a delicate balance between voltage and current outputs, particularly in scenarios demanding higher power density. Striking this balance is crucial to unlocking their potential in advanced technologies such as flexible electronics, wearable devices, and renewable energy systems.

7.4.3.2 Various Types of TINGs

Wang's two-step model demonstrated the establishment of a highly efficient ionic-electronic coupling interface in iontronics. This approach marked a pivotal advancement over traditional electronics by enabling optimized ionic-electronic charge exchange at the interface, thereby achieving energy efficiency far exceeding that of conventional electronic systems. A compelling analogy was the human brain, which functions as an integrated iontronic processor, consuming only about 12 W of power [11], whereas replicating its functions with an electronic system would require tens of megawatts. This pronounced contrast underscored the transformative potential of iontronics in advancing the next generation of energy-efficient technologies. In 2023, Wei et al. presented a triboiontronics system that harnessed bidirectional electrostatic fields generated by S-S CE to precisely regulate charge distribution within the EDL [7]. This system effectively controlled charge redistribution within the Stern layer and the polarity of charges in the diffuse layer. In Stage I, the water made contact with an electronegative dielectric material (Dielectric I), forming a stable EDL. The Stern layer carried a small net negative charge, while the diffuse layer contained a modest concentration of free cations (Fig. 7.26a). In Stage II, when Dielectric I contacted a pre-positively charged, electropositive dielectric material (Dielectric II), S-S CE occurred, resulting in additional charge transfer. The positive electrostatic field from Dielectric II intensified the net negative charge in the Stern layer and increased the concentration of free cations in the diffuse layer (Fig. 7.26b). In Stage III, the removal of Dielectric II caused electrons to be retained on Dielectric I, generating a negative electrostatic field that attracted cations to the L-S contact interface. This process resulted in a net positive charge in the Stern layer and an abundance of free anions in the diffuse layer (Fig. 7.26c). This triboiontronic system demonstrated how bidirectional electrostatic fields could dynamically regulate charge redistribution and ion migration within the Stern layer, providing precise control over the charge polarity in the diffuse layer. This approach laid the foundation for advanced iontronic systems capable of high-efficiency charge manipulation and information processing, driving innovation in energy-efficient applications.

Wei et al. made a notable advancement in triboiontronic research by developing a bionic neural circuit capable of dynamically regulating the EDL through bidirectional electrostatic fields [7]. This circuit could emulate the brain's coordination of body movements, such as walking, by efficiently transmitting signals from the central nervous system to control limb motion with minimal energy consumption.

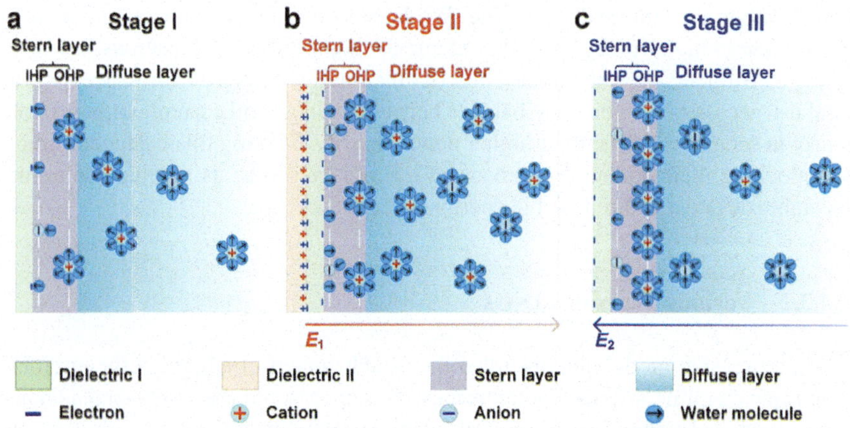

Fig. 7.26 A triboiontronics system was introduced in 2023, which utilized bidirectional electrostatic fields generated by S-S CE to regulate charge distributions in the EDL **a** In Stage I, the water contacted an electronegative Dielectric I, establishing a stable EDL. **b** In Stage II, Dielectric I contact with a pre-positively charged Dielectric II. The pre-positive charges on the Dielectric II generate a positive electrostatic field, intensifying the net negative charge in the Stern layer and increasing the concentration of free cations in the diffuse layer. **c** In Stage III, as dielectric II is removed, its electrons transferred to the surface of Dielectric I form a negative electrostatic field that attracts cations to the L-S interface, resulting in a net positively charged Stern layer and abundant free anions in the diffuse layer. With permission from Elsevier Ltd. (2023) [7]

Replicating similar functionality in traditional electronics would necessitate complex systems comprising multiple diodes and external power supplies. In contrast, the triboiontronic bionic neural circuit provided a more streamlined, energy-efficient alternative through a three-stage operation, as illustrated in Fig. 7.27a. In Stage I, the system remained in a resting state, simulating the central nervous system at rest. In this stage, the positively charged rabbit fur did not contact the initial polyethylene terephthalate (PET) film and a water droplet was added to the charge-collecting layer, with all external circuits remaining disconnected. In Stage II, contact between the positively charged rabbit fur and the PET film caused the water droplet, now enriched with free cations, to induce forward electron transfer in the external circuit, as indicated by the red arrows. Finally, in Stage III, when the rabbit fur was removed, the charge polarity of the water droplet shifted to free anions, inducing reverse electron transfer in the external circuit, highlighted in blue. This bionic neural circuit encoded physical signals as ionic charge polarities within the diffuse layer and converted them into electronic signals, demonstrating effective cross-media information transmission. Through this mechanism, the bionic circuit enabled the control of a virtual robot's coordinated walking movements, as shown in Fig. 7.27b. In Stage I, with the circuit at rest, the virtual robot remained stationary. By adjusting the contact states between the rabbit fur and the PET film in Stages II and III, the ionic polarities in the water droplet enabled direct control over the robot's multi-degree-of-freedom rhythmic and coordinated movements. Unlike conventional silicon-based circuits, triboiontronic systems leveraged ionic-electronic coupling to create streamlined, self-powered platforms. These platforms could revolutionize human-computer interactions by providing secure, energy-efficient interfaces. The integration of triboiontronic bionic neuromorphic circuits, which feature distributed processing and high biocompatibility, could advance information flows in fields such as in-sensor computing, implantable neural interfaces, and low-power neuromorphic devices. In 2023, Wang et al. provided an insightful preview of the "triboiontronics" concept introduced by Wei et al. [63] (Fig. 7.27c), emphasizing the importance of dynamic EDL control in applications like interface control, biosensing, and human-computer interactions. A key challenge in energy-scavenging systems that dynamically manage the EDL is optimizing ionic migration to maximize energy transfer efficiency and ensure accurate information flow. This necessitates precise control over ionic dynamics and flux. Enhancing system design to minimize energy loss and improve transmission efficiency, particularly through real-time sensing and feedback control of ion migration, holds significant promise. Addressing these challenges is crucial for advancing the next generation of energy-scavenging and information-processing technologies.

In addition to the influence of external electrostatic fields, the type and concentration of ions within a liquid also played a crucial role in modulating the charge density and distribution within the EDL. This regulation mechanism was explored in detail by Wei et al. in 2024, and the process could be broadly divided into three stages [64]. In Stage I (Fig. 7.28a), when the water droplet initially contacted the dielectric surface, the ion concentration was too low to support effective ion migration at the L-S contact interface. As a result, an initial EDL formed at the interface, but the charge density within the diffuse layer of the droplet remained minimal. In Stage II (Fig. 7.28b), as the ion concentration in the LiCl solution droplet increased slightly, the solution's conductivity improved, facilitating enhanced ion migration. This, in turn, led to an increase in charge density within the EDL, with the triboelectric charge in the diffuse layer of the droplet reaching its peak as the concentration rises to around 10^{-5} M. In Stage III (Fig. 7.28c), further increases in LiCl concentration led to a decrease in the contact charge within the diffuse layer. This reduction was likely due to the screening effect created by the dense EDL, where a higher ion concentration limited further charge transfer at the dielectric-liquid interface. As illustrated in Fig. 7.28d, droplets of LiCl solutions at varying concentrations slid across the PTFE surface, showing distinct patterns of triboelectric charge in the diffuse layer at each stage of EDL adjustment. The generated triboelectric charge levels varied according to the ion concentration, reflecting the influence of ion concentration on charge migration behavior. Across all tested concentrations, a positive triboelectric charge was produced. However, when droplets of [EMIM]Cl reach a concentration of 0.1 M or higher, the polarity of the triboelectric charge reversed to negative, likely due to the unique EDL structures induced by the asymmetry of anions and cations at higher concentrations (Fig. 7.28e). This demonstrated that by adjusting the ion type and concentration in the liquid, the charge density and distribution in the EDL could be controlled, offering a tunable approach to modulating the charge polarity in the droplet.

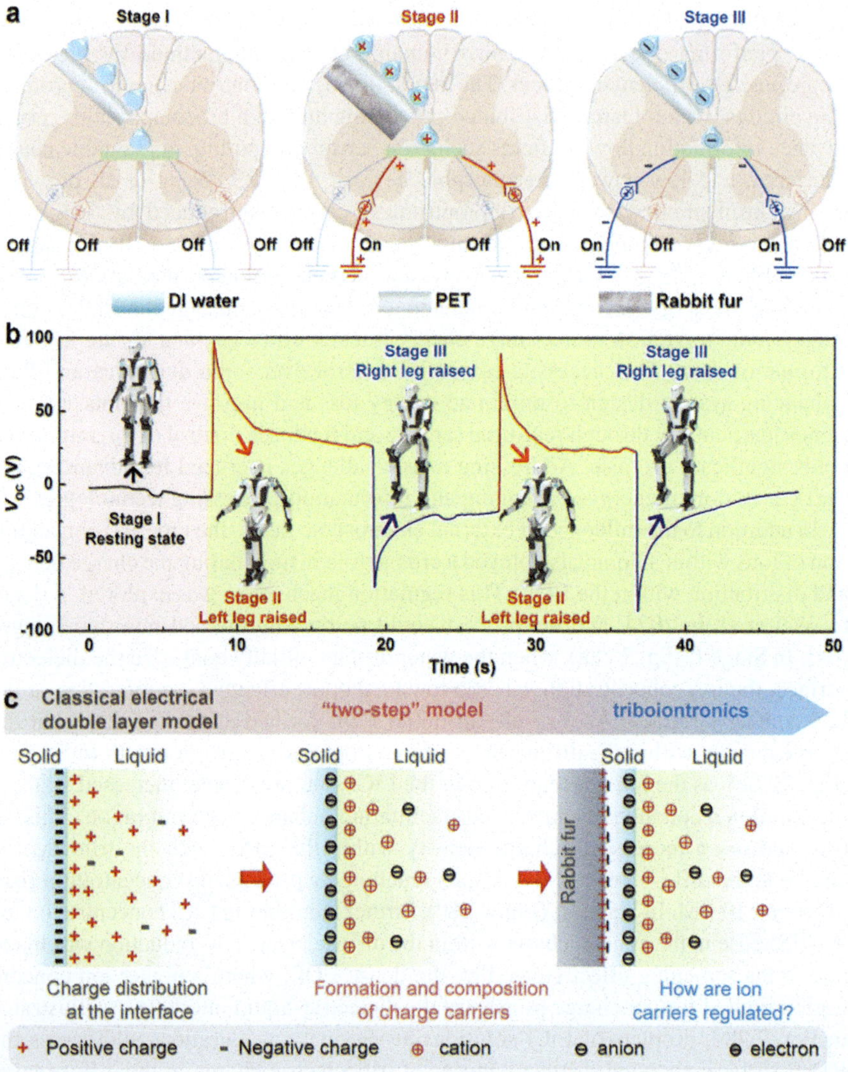

Fig. 7.27 A bionic neurologic circuit capable of dynamically regulating the EDL using bidirectional electrostatic fields was developed in 2023. **a** The three operation stages of the bionic neurologic circuit. **b** The bionic neurologic circuit can direct a virtual robot to perform coordinated walking motions. With permission from Elsevier Ltd. (2023) [7]. **c** In 2023, Wang et al. provided an insightful preview of the "triboiontronics" concept introduced by Wei et al. With permission from Elsevier Ltd. (2023) [63]

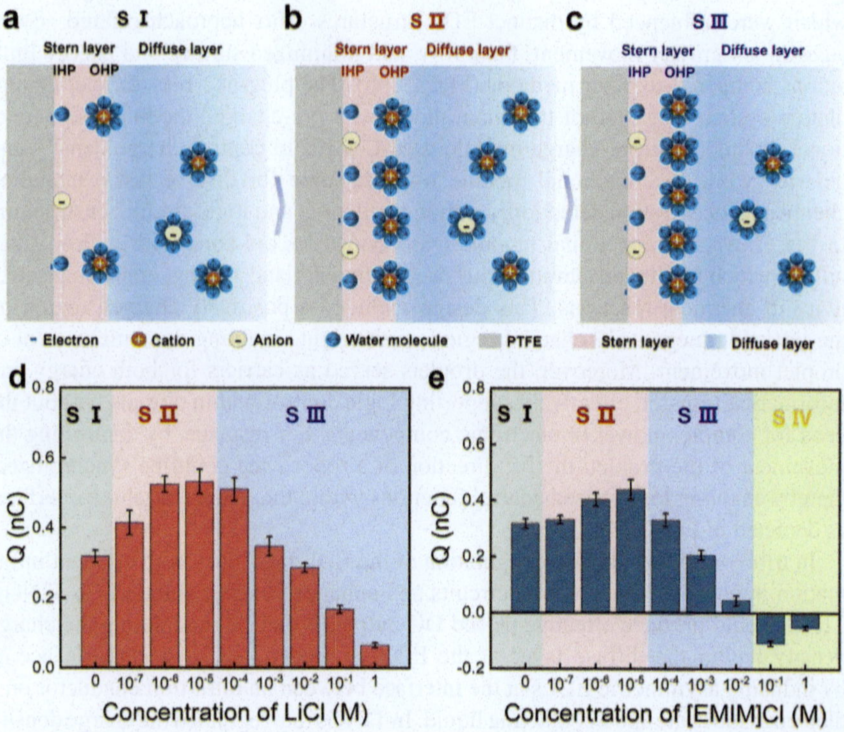

Fig. 7.28 The regulation mechanism of the ion type and concentration on modulating charge density and distribution within the EDL was explored in detail in 2024. **a** In Stage I, when the water droplets initially contact the dielectric surface, the ion concentration in the droplet is too low to fully support effective ion migration at the L-S contact interface. **b** In Stage II, as the ion concentration in the LiCl solution droplet slightly increases, the solution's conductivity improves, facilitating greater ion migration. **c** In Stage III, upon further increases in LiCl concentration, the contact charge within the diffuse layer begins to decrease. **d** Droplets of LiCl solutions at various concentrations slide across the PTFE surface, displaying distinct patterns of triboelectric charges within the diffuse layer at each stage of the EDL adjustment. **e** When droplets of [EMIM]Cl reach a concentration of 0.1 M or above, the triboelectric charge polarity reverses to negative. With permission from Wiley-VCH GmbH (2024) [64]

The flexible regulating of charge carried by droplets enabled effective control of their movement under the influence of an electrostatic field generated by S-S CE[64]. As shown in Fig. 7.29a, droplets with varying ion concentrations exhibited distinct motion behaviors when driven by a constant electrostatic field. Notably, the 10^{-5} M LiCl droplet traveled the longest distance and attained the highest average velocity, while droplets of 0.1 M and 1 M [EMIM]Cl showed a reversal in direction. This reversal was likely due to differences in triboelectric charge polarity, which were influenced by distinct EDL structures. This approach offered advantages in antigravity movement, flexibility, and minimized surface and energy limitations compared to other methods (Fig. 7.29b). The plug-and-play droplet manipulation system consisted of two main modules: a pre-charged module and a functional module. The pre-charge module used L-S CE to control charge density and polarity, while the functional module was adaptable for diverse tasks, including chemical concentration detection, reaction regulation, and logic circuit construction. In Fig. 7.29c, the functional module was divided by the zone with micron-sized pillars, which facilitated simultaneous droplet transfer and sorting, enhancing selectivity in chemical reactions. This design included a positively charged region on the left and a negatively charged region on the right, enabling dynamic control of droplet movement. Moreover, the droplets served as carriers for both energy and information transfer, effectively supporting logic control within circuits without the need for complex power or electronic components. For instance, by controlling the movement of the droplet, the lift direction of a robot's leg could be synchronized, thereby enabling logic-based control functions within the human-machine interface, as depicted in Fig. 7.29d.

In triboiontronics, dynamic regulation of the EDL facilitates both efficient information transfer in bionic neural circuits and enhanced energy solutions in TINGs. TINGs could produce effective pulsed DC outputs either by modulating the charge density within the diffuse layer of the EDL adjacent to the dielectric surface or by inducing asymmetric EDLs at the interface between an ultra-thin conductor on a dielectric substrate and a contacting liquid. In TINGs that regulated the charge density of the diffuse layer, an ion concentration gradient was established, promoting directed ionic migration to generate ionic current. This ionic current modulated the electrical displacement current, leading to high-power-density DC ionic-electronic coupling output[7], as shown in Fig. 7.30. In the initial state (Fig. 7.30a), the TING was in the separation phase. A mist of positively charged DI water, generated by L-S CE, was sprayed onto the dielectric layer, forming an EDL with a high density of cations in the nano-confined diffuse layer. As the bottom charge-collecting layer approached the dielectric layer, now covered by the positively charged water film, the positive charges in the diffuse layer rapidly transferred to the surface of the bottom charge-collecting layer, generating a larger ionic current (Fig. 7.30b). In the subsequent stage, when the dielectric layer fully contacted the bottom charge-collecting layer, any remaining water film was squeezed out of the TING (Fig. 7.30c). Electrons were transferred from the bottom charge-collecting layer to the dielectric layer surface through the S-S CE. As the dielectric layer began to separate from the bottom charge-collecting layer, electrostatic induction drove electrons from the top charge-collecting layer to the bottom charge-collecting layer, generating an electrical displacement current (Fig. 7.30d). The ionic current and electrical displacement current flow in the same direction, demonstrating that the TING could achieve high-power-density DC ionic-electronic coupling output. When the TING returned to the initial separation phase, the positively charged water mist was reapplied, reestablishing the EDL. This cyclic application of water mist illustrated how L-S CE dynamically modulated charge density within the diffuse layer, producing a stable and consistent DC ionic-electronic

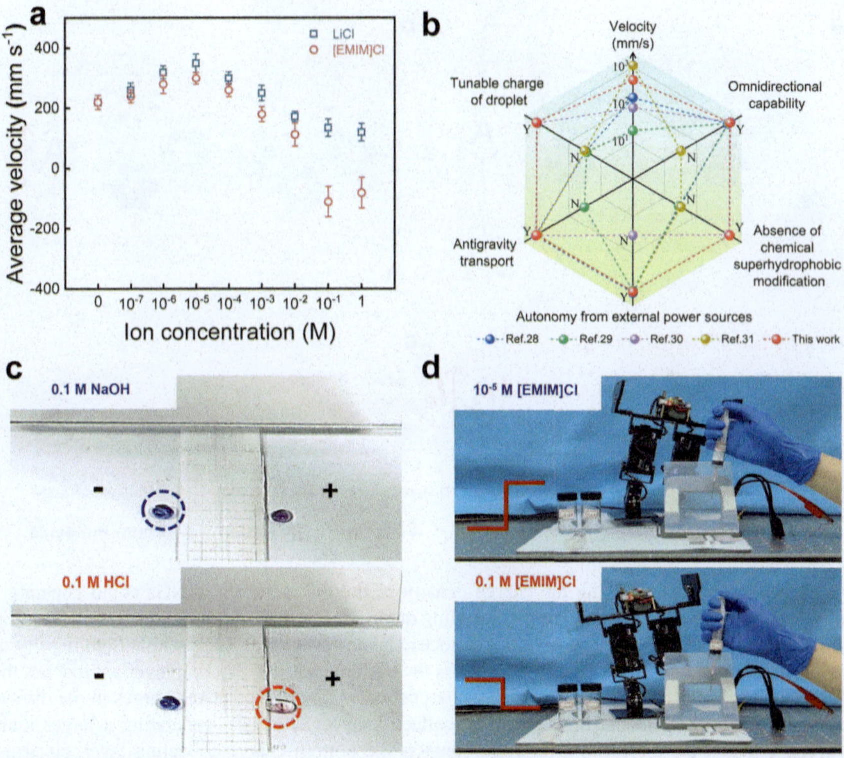

Fig. 7.29 The flexible tuning of charge carried by droplets enables effective control of their movement under the influence of an electrostatic field generated by S-S CE, which was studied in detail in 2024. **a** Droplets with varying ion concentrations exhibit distinct motion behaviors when driven by a constant electrostatic field. **b** This approach exhibits advantages in antigravity movement, flexibility, and minimized surface and energy limitations compared to other methods. **c** The functional module divided by micron-sized pillars supports simultaneous droplet transfer and sorting, enhancing selectivity in chemical reactions. **d** Directing droplet movement can synchronize the lift direction of a robot's leg, thus enabling logic-based control functions. With permission from Wiley-VCH GmbH (2024) [64]

coupling output. Such an approach underscored the potential of EDL regulation in enhancing energy scavenging efficiency, power density, and overall system stability, offering significant advancements in sustainable energy technology.

Based on the principle of regulating charge density in the diffuse layer, the DC-TING was introduced in 2023 (Fig. 7.31a) [7], achieving enhanced charge collection on the dielectric surface by incorporating a carbon nanotube film with a high specific surface area as the charge-collecting layer. Initially, in the absence of the positively charged DI water mist (i.e., in a dry state), the DC-TING operated as a conventional mechanically droplet-driven AC L-S TENG, generating an AC signal of approximately 0.8 μA through S-S CE and electrostatic induction at a frequency of 1 Hz (Fig. 7.31b). However, upon the application of positively charged DI water mist to

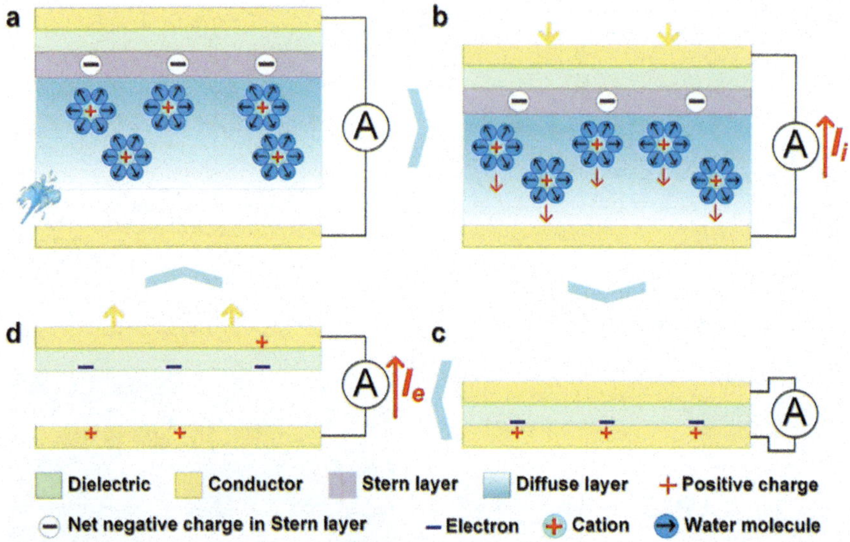

Fig. 7.30 Based on regulating the charge density of the diffuse layer, TINGs could generate a high-power-density DC ionic-electronic coupling output. **a** The mist of positively charged DI water generated by L-S CE is sprayed onto the dielectric layer, forming an EDL with a high density of cations in the nano-confined diffuse layer. **b** As the bottom charge-collecting layer approaches the dielectric layer covered with the positively charged water film, the positive charges in the diffuse layer rapidly transfer to the bottom charge-collecting layer's surface, generating a larger ionic current. **c** When the dielectric layer fully contacts the bottom charge-collecting layer, electrons are transferred from the bottom charge-collecting layer to the dielectric layer surface based on the CE effect. **d** As the dielectric layer begins to separate from the bottom charge-collecting layer, electrostatic induction drives electrons from the top charge-collecting layer to the bottom charge-collecting layer, resulting in an electrical displacement current. With permission from the authors (2024) [56]

the dielectric surface, the DC-TING quickly switched to a DC output mode. As ionic charges accumulated on the dielectric surface, the DC output stabilized at around 6.0 μA. This result confirmed that regulating charge density within the diffuse layer of the EDL could generate a stable DC ionic-electronic coupling output. Further analysis of the charge replenishment time revealed a crucial parameter affecting the DC-TING output (Fig. 7.31c). At lower operational frequencies, extended replenishment times facilitated greater charge accumulation in the diffuse layer, resulting in increased current output when the stored charge was transferred to the charge-collecting layer. Moreover, the hydrophilicity of the dielectric surface played a significant role in determining the output performance of the TING (Fig. 7.31d). Micro-engraving techniques increased the dielectric surface area but introduced air pockets within the microstructure, reducing L-S contact and, consequently, diminishing the surface's hydrophilicity. To address this issue, plasma treatment was employed to introduce oxygen-containing groups on the micro-engraved surface, enhancing hydrophilicity and promoting greater accumulation of positive charges in the diffuse layer, thereby improving device performance (Fig. 7.31e). Through these charge replenishment strategies, the dielectric surface's charge density was effectively sustained, allowing the TING to deliver consistent and efficient DC output. The device achieved an impressive I_{SC} density of 360 mA/m^2 and a Q_{SC} density of 5.6 mC/m^2 (Fig. 7.31f and g), while its P_R density reached 126.4 W/m^2 (Fig. 7.31h). These findings underscored the importance of optimizing charge density to maximize TING efficiency, offering significant potential for advancing energy scavenging technologies. However, a key challenge for energy scavenging devices that rely on dynamic regulation of the diffuse layer is the reduced charge transfer resulting from increased internal resistance of the dielectric material. Therefore, future research should focus on materials with lower internal resistance, such as those with enhanced conductivity or low resistivity. Additionally, optimizing dielectric materials and improving electrode materials could further enhance charge transfer efficiency.

In addition to generating high-power-density DC ionic-electronic coupling output through regulation of the diffuse layer charge density, TINGs based on asymmetric EDLs exhibited a distinct operational mechanism that significantly enhanced Q_{SC} density [8]. These TINGs employed a hybrid material, consisting of a dielectric substrate with an ultra-thin conductive layer closely adhered to it. This ultra-thin conductive layer functioned both as the charge-collecting layer and as a modulator of the liquid-substrate CE properties, which in turn regulated the charge distribution and migration in the EDL. In the first step, a thin gold (Au) layer was sputtered onto a PET dielectric substrate, forming the hybrid material (Au/FEP). Upon contact with water, the Au layer interacted with the PET surface through microscopic cracks, generating L-S CE and establishing a stable EDL on the substrate (Fig. 7.32a). Subsequently, as the top Au/PET layer made contact with the water, a second L-S CE event occurred, establishing a new EDL at the top contact surface and forming a dual EDL system with pronounced asymmetry. This asymmetry generated an ion concentration gradient, which drove ion migration within the asymmetric EDLs. The ion migration induced electron transfer in the external circuit, thereby generating an ionic current (Fig. 7.32b). When the top Au/PET layer detached, ion migration briefly ceased (Fig. 7.32c). Through repeated contact-separation cycles, the ion concentration gradient gradually diminished, but ion migration persisted until equilibrium was reached between the two EDLs (Fig. 7.32d). This innovative mechanism enhanced Q_{SC} efficiency and underscored the potential of TINGs for various energy-scavenging applications, effectively converting mechanical motion into electrical output.

A physically adsorptive direct-current triboiontronic generator (PDC-TING) based on asymmetric EDLs was also developed [8] (Fig. 7.33a). The device utilized the Au/PET hybrid layer that included a PET dielectric substrate with an ultra-thin Au layer deposited via magnetron sputtering. To optimize the PDC-TING's output performance, three key strategies were introduced to enhance the ion concentration gradient, thereby maximizing device efficiency. The first strategy involved electrostatic field adjustment (Fig. 7.33b). By applying a positive electrostatic field to the bottom Au/PET hybrid layer in the PDC-TING, the I_{SC} was significantly increased. This enhancement was attributed to the influence of the positive electrostatic field on the bottom EDL density, which strengthened the ion concentration

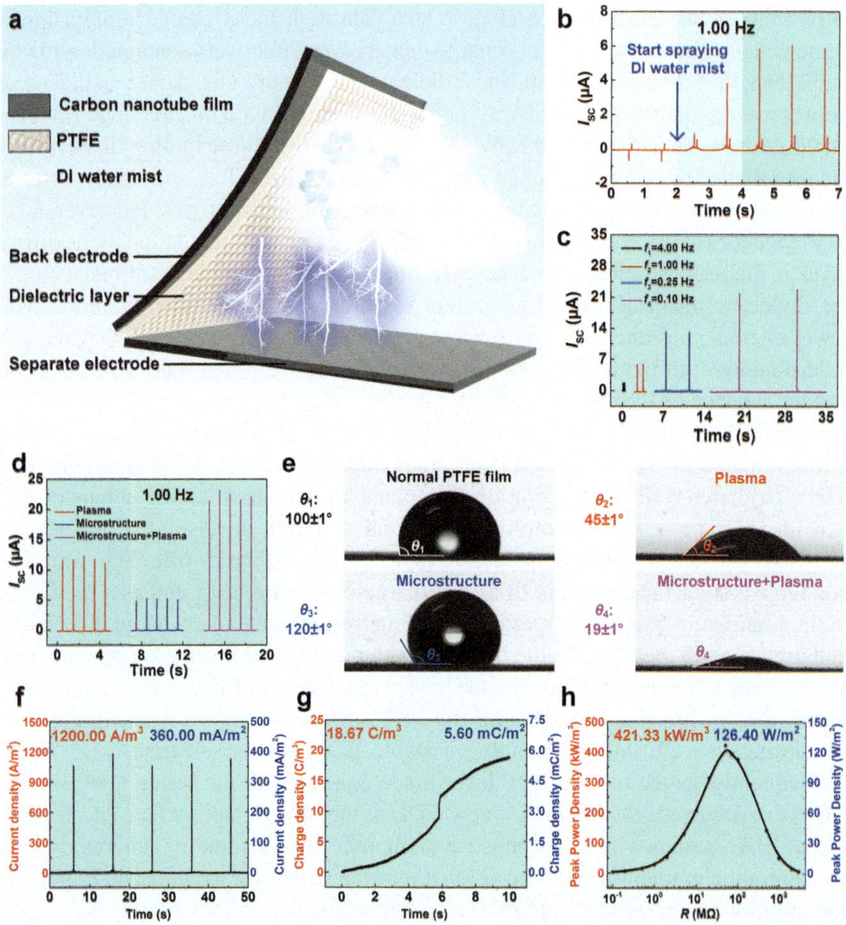

Fig. 7.31 Based on the above principle of regulating the charge density of the diffuse layer, a DC-TING was introduced. **a** The schematic diagram of the DC-TING. **b** Upon application of positively charged DI water mist to the dielectric surface, the DC-TING quickly shifted to a DC output mode from the AC output mode. **c** At lower operational frequencies, prolonged replenishment times enabled a greater buildup of charge in the diffuse layer, resulting in increased current output when this charge was subsequently transferred to the charge-collecting layer. **d** The hydrophilicity of the dielectric surface was found to impact DC-TING performance significantly. **e** Plasma treatment to introduce oxygen-containing groups on the micro-engraved surface successfully enhanced hydrophilicity, facilitating greater accumulation of positive charges in the diffuse layer and improving device performance. **f–h** The DC-TING achieved an impressive I_{SC} density of 360 mA/m^2 and Q_{SC} density of 5.6 mC/m^2 respectively, while the DC-TING's P_R density reached 126.4 W/m^2. With permission from Elsevier Ltd. (2023) [7]

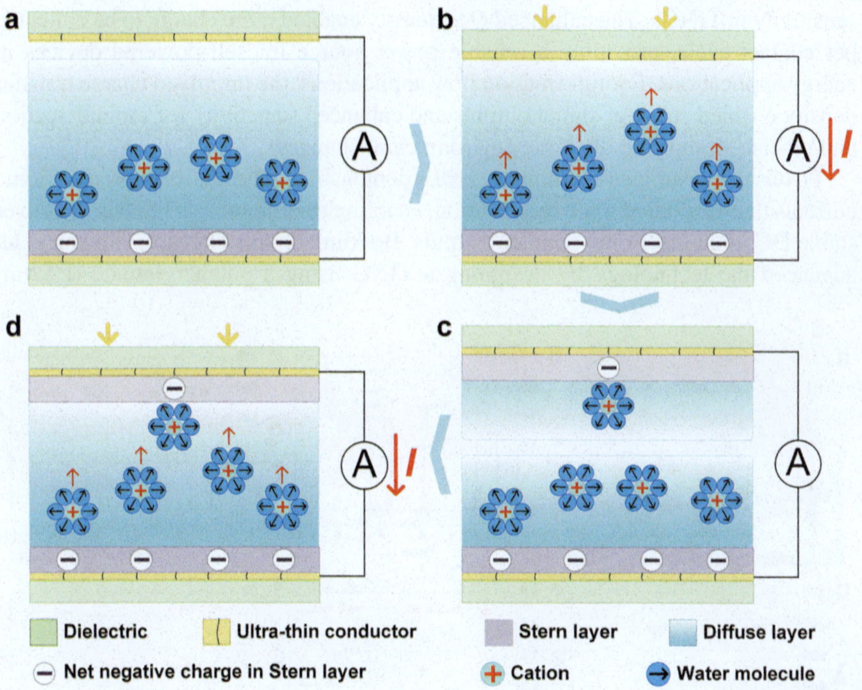

Fig. 7.32 The operating mechanism of TINGs based on asymmetric EDLs. **a** When the water contacts the bottom Au/PET surface, it interacts directly with the PET through microscopic cracks in the Au layer, generating L-S CE and establishing a stable EDL. **b** When the top Au/PET layer descends into contact with the water, asymmetric EDLs are created and an ion concentration gradient is established, driving electron transfer within the external circuit and producing an ionic current. **c** The subsequent detachment of the top Au/PET layer momentarily halts ion migration. **d** Through repeated contact-separation cycles, the ion concentration gradient gradually dissipates, although ion migration continues until equilibrium is achieved between the two EDLs. With permission from the authors (2024) [56]

gradient, promoting greater ion migration and thus improving device performance. The second strategy focused on enhancing the hydrophilicity of the bottom Au/PET hybrid layer (Fig. 7.33c). Plasma treatment was applied to improve the surface's hydrophilicity, which significantly boosted the EDL density at the bottom contact interface. This, in turn, amplified the ion concentration gradient, further improving output performance. The third strategy aimed at increasing the ion concentration in the liquid (Fig. 7.33d). By replacing water with a higher concentration LiCl solution, the bottom EDL density was increased, lowering the internal resistance and reducing energy losses. This modification enhanced both signal sensitivity and output strength. Collectively, these strategies resulted in a substantial improvement in PDC-TING performance, with an I_{SC} density of 26.00 A/m^2 (Fig. 7.33e), a Q_{SC} density of 412.54 mC/m^2 (Fig. 7.33f), and a P_R density of 8.45 W/m^2. The increased Q_{SC} density was pivotal in maximizing energy output efficiency, signal strength, and detection sensitivity in TINGs. The enhanced Q_{SC} density enabled more charge to be collected per contact cycle, providing a reliable power source for self-powered devices in energy applications. For information flow applications, the improved charge transfer density enabled stronger signal outputs and enhanced sensitivity for rapid detection and precise monitoring of subtle environmental changes.

TINGs that leveraged asymmetric EDLs demonstrated the ability to produce ionic currents that facilitated ionic rectification, enabling traditional AC TENGs to deliver stable DC ionic-electronic coupled outputs. Building on this concept, Ouyang et al. advanced the technology by designing a TING using a polyacrylamide (PAAm)

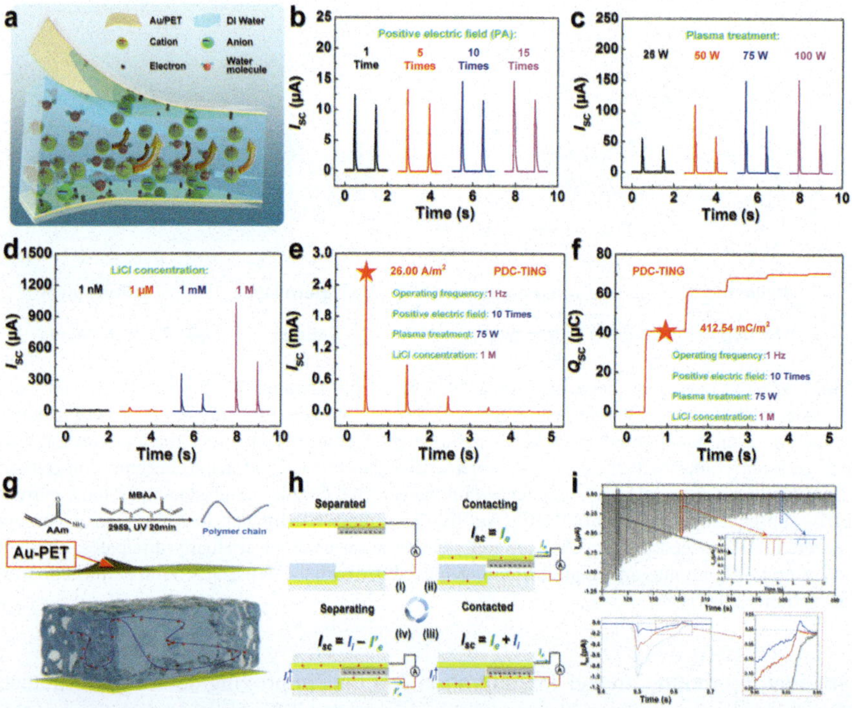

Fig. 7.33 A series of TINGs based on asymmetric EDLs and their output characteristics. **a** The PDC-TING based on asymmetric EDLs was developed in 2024 **b** Applying a positive electrostatic field on the bottom of the PDC-TING successfully increased I_{SC}. **c** Plasma treatment was applied to improve the hydrophilicity of the bottom Au/PET hybrid layer, significantly boosting the EDL density at the bottom contact layer. **d** Replacing water with a higher concentration of LiCl solution increased the bottom EDL density and lowered internal resistance, which minimized energy losses and enhanced signal sensitivity and strength. **e–f** The PDC-TING could achieve an optimal output with an I_{SC} density of 26.00 A/m^2 and a Q_{SC} density of 412.54 mC/m^2. With permission from the authors (2024) [8]. **g** In 2024, a TING via asymmetric EDLs was designed, which facilitated the rectification of AC TENG outputs. **h** The principle of ionic rectification. **i** The close correlation between ionic and electronic current variations highlights the crucial role of ionic current in enhancing both energy output and rectification efficiency. With permission from American Chemical Society (2024) [65]

hydrogel in asymmetric EDLs in 2024, which facilitated the rectification of AC TENG outputs[65] (Fig. 7.33g). The hydrogel was created by injecting a reagent solution into a custom mold, followed by UV-induced polymerization, and then integrated with a hybrid material consisting of a PET substrate coated with an ultrathin Au layer. As the hydrogel approached the top hybrid material, the ionic current, driven by the ion concentration gradient, flowed in the same direction as the electrical displacement current induced by electrostatic induction. Upon separation, the direction of the ionic current remained unchanged, while the electrical displacement current was modulated by the ionic current, thereby validating efficient ionic rectification and ultimately generating a DC ionic-electronic coupled output (Fig. 7.33h). Experimental data confirmed a strong correlation between the variations in ionic and electronic currents, underscoring the role of ionic current in enhancing both energy output and rectification efficiency (Fig. 7.33i). This flexible rectification approach overcame the limitations of conventional rigid rectifiers, offering advantages in adaptability, lightweight design, and energy efficiency. This advancement positioned it as a significant step forward for wearable and flexible electronics.

Overall, triboiontronic systems and various types of TINGs were expected to garner increasing research attention due to their remarkable potential in several key areas. These systems held promise for efficient energy scavenging, enabling sustainable power generation from environmental sources. Additionally, the precise control of interfacial EDLs within these systems offered new avenues for optimizing energy conversion efficiency and regulating ionic currents in real time. This regulation was pivotal for the development of biomimetic nervous systems, which were designed to replicate the dynamic response characteristics of biological neural circuits. As a result, triboiontronic systems and TINGs showed significant promise in a variety of applications, ranging from self-powered wearable electronics and flexible energy devices to advanced health-monitoring technologies and intelligent robotics, thereby showcasing their broad and diverse potential for future technological advancements.

7.5 Contact Electrification at Liquid-Semiconductor Interfaces

Beyond the conventional liquid-conductor or liquid-dielectric interfaces, the movement of liquids over semiconductor surfaces could also generate a constant current output, a phenomenon known as the tribovoltaic effect, in which the DI water was considered analogous to a wide-bandgap semiconductor. The conventional tribovoltaic effect at solid semiconductor-semiconductor interfaces was first discovered by Wang et al. in 2019 and was considered to be a DC output mechanism [6]. Specifically, when p-type and n-type semiconductors came into contact, the difference in their Fermi energy levels drove the transfer of electrons between them, with holes migrating in the opposite direction [66]. This electron-hole transfer established a built-in electric field at the contact interface, effectively forming a p-n junction.

Upon relative motion between the two semiconductor surfaces, interactions between surface atoms led to the formation of new chemical bonds, releasing what Wang et al. referred to as "Bindington", a form of frictional energy. In the tribovoltaic effect, this energy behaves similarly to excitons in the photovoltaic effect, as it excites electron-hole pairs at the interface, enabling them to move either along or against the built-in electric field. This movement generates a constant current within an external circuit, underscoring the potential of semiconductor-semiconductor interface interactions for sustainable DC power generation.

The tribovoltaic effect also could be effectively harnessed in liquid-semiconductor interfaces to generate a stable output current. Additionally, this approach offered a promising strategy for reducing friction and wear between solid semiconductor surfaces, potentially extending the device's lifespan. In 2020, Lin et al. developed a nanogenerator based on liquid-semiconductor interfaces using water droplets sliding across a silicon (Si) surface[67], as illustrated in Fig. 7.34a. Notably, when the water droplets moved over p-type Si, the nanogenerator generated consistent V_{OC} and I_{SC}, which remained stable regardless of the sliding direction, as shown in Fig. 7.34b and c. The operational mechanism relied on the difference in Fermi levels between the water and the Si surface. The Fermi level of water was higher than that of p-type Si, and upon separation, the band diagram in Fig. 7.34d showed the resulting difference. When the water droplet contacted the p-type Si, electrons transferred from the droplet to the Si surface, establishing a built-in electric field at the interface. Concurrently, new chemical bonds formed at the contact interface (Fig. 7.34e). As the droplet moved, the bond formation released frictional energy, which excited electron-hole pairs on the p-type Si surface (Fig. 7.34f). These pairs, driven by the built-in electric field, then migrated in alignment with or opposite to the electric field direction, generating a stable output in the external circuit (Fig. 7.34g). When the water droplet interacted with n-type Si, where the Fermi level was higher, electrons were transferred from the Si surface to the droplet, creating a reverse built-in electric field. This reverse field drove electron-hole pairs in a directional motion, resulting in an opposite constant output in the external circuit. The corresponding energy band diagrams for this process at various stages of operation are shown in Fig. 7.34h–k.

Further investigations by Zheng et al. explored the impact of temperature on the performance of nanogenerators based on liquid-semiconductor interfaces[68], focusing specifically on the interactions between the liquid and semiconductor surfaces (Fig. 7.35a). Their study revealed that as the temperature increased from 30 to 90 °C, the CA of water droplets on the Si surface decreased from 86.72° to 60.36° (Fig. 7.35b). This reduction in CA effectively expanded the droplet's contact area with the Si surface, thereby enhancing output performance by increasing the available interface for charge transfer. For example, when water droplets slid over an n-type Si surface, the elevated temperature resulted in higher-density electron-hole pair excitation. This was due to the increased frictional energy at the interface as the droplet moved, leading to a significant enhancement in DC output performance, as shown in Fig. 7.35c and d. Similarly, when the droplet contacted p-type Si, the reverse constant output also increased with temperature, as depicted in Fig. 7.35f–h.

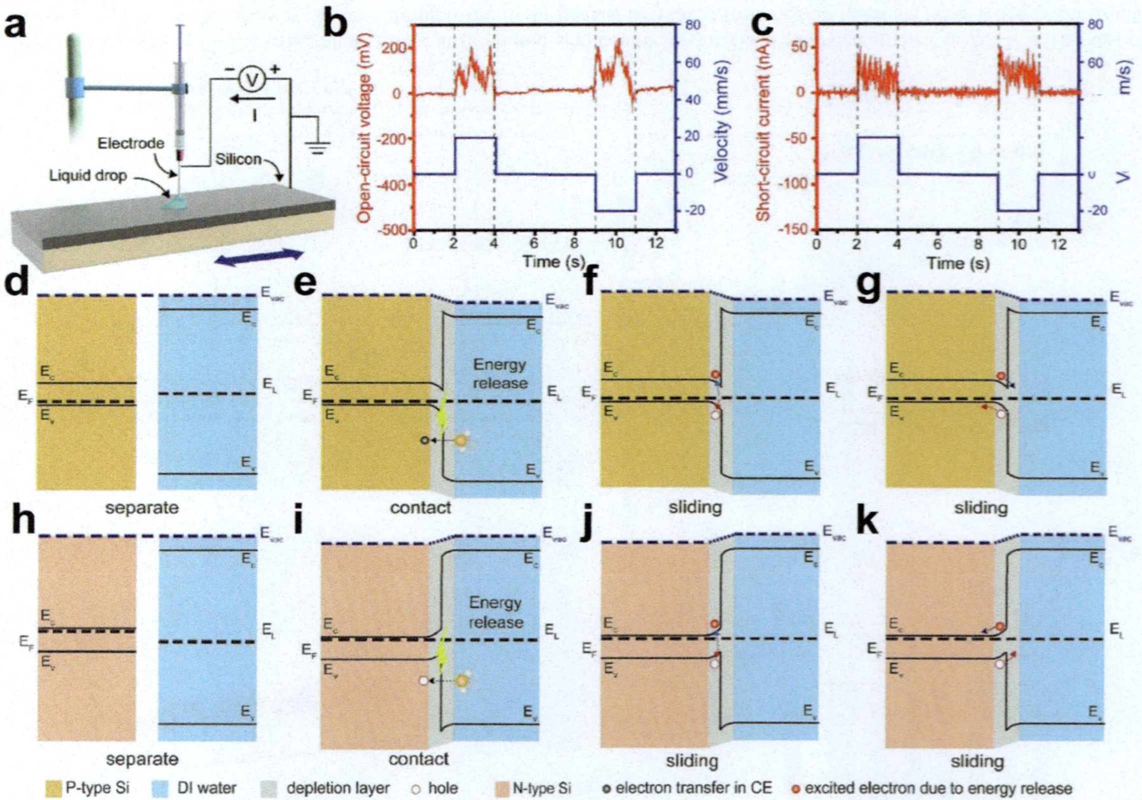

Fig. 7.34 Tribovoltaic effect at liquid-semiconductor interfaces. **a** In 2020, Lin et al. developed a nanogenerator based on semiconductor-liquid interfaces using droplets of water sliding across the Si surface. **b–c** When the water droplets moved over p-type Si, the nanogenerator generated a consistent V_{OC} and I_{SC} that remained stable regardless of the sliding direction. **d–g** The energy band diagram transitions when the droplet slides on the p-type Si surface. **h–k** The energy band diagram transitions when the droplet slides on the n-type Si surface. With permission from Elsevier Ltd. (2020) [67]

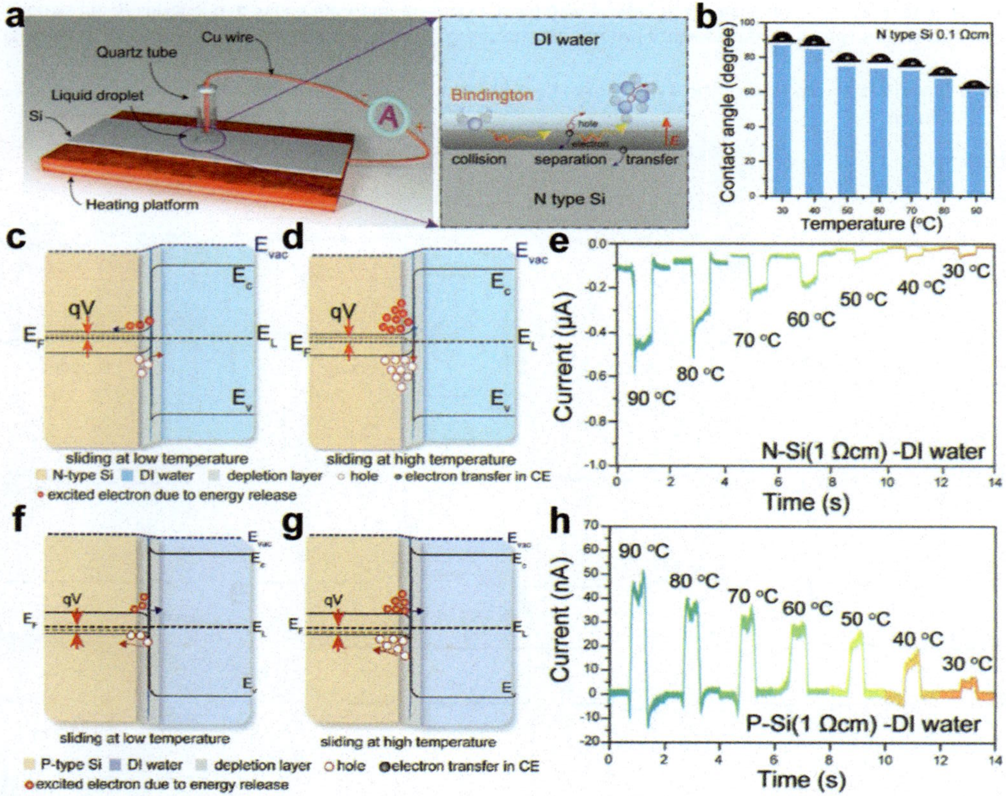

Fig. 7.35 The effect of temperature on the tribovoltaic effect at the liquid-semiconductor interface has been studied in 2021. **a** The test system. **b** As the temperature increased from 30 to 90 °C, the CA of water droplets on the Si surface decreased from 86.72° to 60.36°. **c–e** When water droplets slide over an n-type Si surface, the rising temperature results in higher-density electron–hole pair excitation due to increased friction energy at the interface during droplet sliding, leading to a notable enhancement in DC output performance. **f–h** Similarly, when the droplet contacts p-type Si, the reverse constant output increases with temperature. With permission from Wiley-VCH GmbH (2021) [68]

Overall, these findings underscored the potential of nanogenerators based on liquid-semiconductor interfaces to improve device performance by leveraging the inherent properties of semiconductors. The temperature-dependent changes in CA and frictional energy at the contact interface contributed to higher charge transfer efficiency, resulting in a more stable and enhanced DC output. Additionally, the intrinsic characteristics of semiconductors, such as their ability to reduce internal resistance, further improved current density. These results indicated that nanogenerators based on semiconductor-liquid interfaces can be optimized for greater efficiency and durability, making them suitable for applications that require consistent energy output and minimal wear.

7.6 Summary

In this chapter, a thorough analysis of the mechanisms behind CE at L-S contact interfaces was provided, with a focus on charge transfer processes across interfaces involving solid conductors, dielectrics, and semiconductors. The formation of the EDL, a fundamental structure that governed charge dynamics at these interfaces, was discussed in detail. Key topics included the quantification of ion and electron transfer, advanced modeling of electron transfer mechanisms, and an exploration of the hybrid characteristics of the EDL, offering deeper insights into the fundamental physics underlying L-S CE. Furthermore, this chapter explores techniques for dynamically regulating the EDL using electric fields or mechanical means to control charge carrier behavior, including polarity and ionic flux. By manipulating ion migration and optimizing charge carrier concentration at the contact interface, these methods enable precise regulation of energy scavenging, storage, and information flow. Devices such as supercapacitors, electrochemical sensors, L-S TENGs, and TINGs have been developed based on L-S contact interface regulation. Dynamic control of the EDL at liquid-conductor interfaces has enabled stable energy storage and sensitive sensors, while at liquid-dielectric interfaces, it has facilitated high-efficiency energy scavenging. At liquid-semiconductor interfaces, the tribovoltaic effect, driven by L-S CE, provides an effective energy generation method. These advancements highlighted the significant potential of dynamic EDL regulation across various contact interfaces for supporting efficient energy generation and information flow.

References

1. H. Helmholtz, Ueber einige gesetze der vertheilung elektrischer ströme in körperlichen leitern mit anwendung auf die thierisch-elektrischen versuche. Ann. Phys. **165**(6), 211–233 (1853)
2. M. Gouy, Sur la constitution de la charge électrique à la surface d'un électrolyte. J. Phys. Theor. Appl. **9**(1), 457–468 (1910)

3. D.L. Chapman, A contribution to the theory of electrocapillarity. Philos. Mag. **25**(148), 475–481 (1913)
4. D.C. Grahame, The electrical double layer and the theory of electrocapillarity. Chem. Rev. **41**(3), 441–501 (1947)
5. S. Lin, L. Xu, A. Chi Wang, Z.L. Wang, Quantifying electron-transfer in liquid-solid contact electrification and the formation of electric double-layer. Nat. Commun. **11**(1), 399 (2020)
6. Z.L. Wang, A.C. Wang, On the origin of contact-electrification. Mater. Today **30**, 34–51 (2019)
7. X. Li, S. Li, X. Guo, J. Shao, Z.L. Wang, D. Wei, Triboiontronics for efficient energy and information flow. Matter **6**(11), 3912–3926 (2023)
8. X. Li, R. Li, S. Li, Z.L. Wang, D. Wei, Triboiontronics with temporal control of electrical double layer formation. Nat. Commun. **15**(1), 6182 (2024)
9. X. Li, Z.L. Wang, D. Wei, Scavenging Energy and Information through dynamically regulating the electrical double layer. Adv. Funct. Mater. **34**(42), 2405520 (2024)
10. J. Maier, Nanoionics: ion transport and electrochemical storage in confined systems. Nat. Mater. **4**(11), 805–815 (2005)
11. R. Sarpeshkar, Analog versus digital: extrapolating from electronics to neurobiology. Neural Comput. **10**(7), 1601–1638 (1998)
12. S.Z. Bisri, S. Shimizu, M. Nakano, Y. Iwasa, Endeavor of iontronics: from fundamentals to applications of ion-controlled electronics. Adv. Mater. **29**(25), 1607054 (2017)
13. L.L. Zhang, X.S. Zhao, Carbon-based materials as supercapacitor electrodes. Chem. Soc. Rev. **38**(9), 2520–2531 (2009)
14. M. Şahin, F. Blaabjerg, A. Sangwongwanich, A comprehensive review on supercapacitor applications and developments. Energies **15**(3), 674 (2022)
15. A. González, E. Goikolea, J.A. Barrena, R. Mysyk, Review on supercapacitors: technologies and materials. Renew. Sustain. Energy Rev. **58**, 1189–1206 (2016)
16. R. Liu, Z.L. Wang, K. Fukuda, T. Someya, Flexible self-charging power sources. Nat. Rev. Mater. **7**(11), 870–886 (2022)
17. N. Liu, R. Chen, Q. Wan, Recent advances in electric-double-layer transistors for bio-chemical sensing applications. Sensors **19**(15), 3425 (2019)
18. R. Feiner, T. Dvir, Tissue–electronics interfaces: from implantable devices to engineered tissues. Nat. Rev. Mater. **3**(1), 17076 (2017)
19. S. Zhao, P. Tseng, J. Grasman, Y. Wang, W. Li, B. Napier, B. Yavuz, Y. Chen, L. Howell, J. Rincon, F.G. Omenetto, D.L. Kaplan, Programmable hydrogel ionic circuits for biologically matched electronic interfaces. Adv. Mater. **30**(25), e1800598 (2018)
20. G. Schiavone, S.P. Lacour, Conformable bioelectronic interfaces: mapping the road ahead. Sci. Transl. Med. **11**(503), eaaw5858 (2019)
21. G.G. Wallace, S.E. Moulton, G.M. Clark, Applied physics. Electrode-cellular interface. Science **324**(5924), 185–186 (2009)
22. R. Jalili, A. Kanneganti, M.I. Romero-Ortega, G.G. Wallace, Implantable electrodes. Curr. Opin. Electrochem. **3**(1), 68–74 (2017)
23. M. Hu, J. Ren, Y. Pan, L. Cheng, X. Xu, C.L. Tan, H. Sun, Y. Shi, S. Yan, Scaled elastic hydrogel interfaces for brain electrophysiology. Adv. Funct. Mater. **34**(46), 2407926 (2024)
24. M.T. Darby, C.S. Cucinotta, The role of water at electrified metal-water interfaces unravelled from first principles. Curr. Opin. Electrochem. **36**, 101118 (2022)
25. P. Sebastián-Pascual, Y. Shao-Horn, M. Escudero-Escribano, Toward understanding the role of the electric double layer structure and electrolyte effects on well-defined interfaces for electrocatalysis. Curr. Opin. Electrochem. **32**, 100918 (2022)
26. D. Strmcnik, K. Kodama, d.V.D. van, J. Greeley, V.R. Stamenkovic, N.M. Markovic, The role of non-covalent interactions in electrocatalytic fuel-cell reactions on platinum. Nat. Chem. **1**(6), 466–472 (2009)
27. H. Shi, Z. Cai, J. Patrow, B. Zhao, Y. Wang, Y. Wang, A. Benderskii, J. Dawlaty, S.B. Cronin, Monitoring local electric fields at electrode surfaces using surface enhanced raman scattering-based stark-shift spectroscopy during hydrogen evolution reactions. ACS Appl. Mater. Inter. **10**(39), 33678–33683 (2018)

28. T. Asset, A.G. Oshchepkov, F. Maillard, G.A. Tsirlina, Electrocatalysis, diverse and forever young. Electrochim. Acta **507**, 145174 (2024)
29. A. Frumkin, N. Polianovskaya, I. Bagotskaya, N. Grigoryev, Electrocatalysis and electrode surface properties. J. Electroanal. Chem. Interfacial Electrochem. **33**(2), 319–328 (1971)
30. S. Trasatti, Work function, electronegativity, and electrochemical behaviour of metals. J. Electroanal. Chem. Interfacial Electrochem. **39**(1), 163–184 (1972)
31. R. Parsons, J.O.M. Bockris, Calculation of the energy of activation of discharge of hydrogen ions at metal electrodes. Trans. Faraday Soc. **47**(8), 914–928 (1951)
32. B.E. Conway, B.V. Tilak, Interfacial processes involving electrocatalytic evolution and oxidation of H2, and the role of chemisorbed H. Electrochim. Acta **47**(22–23), 3571–3594 (2002)
33. J. Tafel, Über den Verlauf der elektrolytischen Reduktion schwer reduzierbarer Substanzen in schwefelsaurer Lösung. Z. Phys. Chem. **34U**(1), 187–228 (1900)
34. F. Haber, R. Russ, Über die elektrische Reduktion. Z. Phys. Chem. **47U**(1), 257–335 (1904)
35. N. Kobosew, N.I. Nekrassow, Bildung freier Wasserstoffatome bei Kathodenpolarisation der Metalle. Zeitschrift für Elektrochemie und angewandte physikalische Chemie **36**(8), 529–544 (2010)
36. A.T. Hubbard, Electrochemistry of well-defined surfaces. Acc. Chem. Res. **13**(6), 177–184 (2002)
37. J.C. Canullo, W.E. Triaca, A.J. Arvia, Electrochemical faceting of single crystal platinum electrodes. J. Electroanal. Chem. Interfacial Electrochem. **200**(1–2), 397–400 (1986)
38. J. Liu, Z. Yang, S. Li, Y. Du, Z. Zhang, J. Shao, M. Willatzen, Z.L. Wang, D. Wei, Nonaqueous contact-electro-chemistry via triboelectric charge. J. Am. Chem. Soc. **146**(46), 31574–31584 (2024)
39. Z. Zhang, D. Jiang, J. Zhao, G. Liu, T. Bu, C. Zhang, Z.L. Wang, Tribovoltaic effect on metal–semiconductor interface for direct-current low-impedance triboelectric nanogenerators. Adv. Energy Mater. **10**(9), 1903713 (2020)
40. S.Q. Lin, X.Y. Chen, Z.L. Wang, Contact electrification at the liquid-solid interface. Chem. Rev. **122**(5), 5209–5232 (2022)
41. M. Zheng, S. Lin, L. Xu, L. Zhu, Z.L. Wang, Scanning probing of the tribovoltaic effect at the sliding interface of two semiconductors. Adv. Mater. **32**(21), 2000928 (2020)
42. C. Xu, Y. Zi, A.C. Wang, H. Zou, Y. Dai, X. He, P. Wang, Y.-C. Wang, P. Feng, D. Li, Z.L. Wang, On the electron-transfer mechanism in the contact-electrification effect. Adv. Mater. **30**(15), 1706790 (2018)
43. Z.L. Wang, Triboelectric nanogenerator (TENG)—sparking an energy and sensor revolution. Adv. Energy Mater. **10**(17), 2000137 (2020)
44. K. von Burg, P. Delahay, Photoelectron emission spectroscopy of inorganic anions in aqueous solution. Chem. Phys. Lett. **78**(2), 287–290 (1981)
45. Y. Takakuwa, M. Niwano, M. Nogawa, H. Katakura, S. Matsuyoshi, H. Ishida, H. Kato, N. Miyamoto, Photon-stimulated desorption of H+ ions from oxidized Si(111) surfaces. Jpn. J. Appl. Phys. **28**(12R), 2581 (1989)
46. J. Nie, Z. Ren, L. Xu, S. Lin, F. Zhan, X. Chen, Z.L. Wang, Probing contact-electrification-induced electron and ion transfers at a liquid-solid interface. Adv. Mater. **32**(2), 1905696 (2020)
47. R. Zhang, S. Wang, M.-H. Yeh, C. Pan, L. Lin, R. Yu, Y. Zhang, L. Zheng, Z. Jiao, Z.L. Wang, A streaming potential/current-based microfluidic direct current generator for self-powered nanosystems. Adv. Mater. **27**(41), 6482–6487 (2015)
48. J. Chi, C. Liu, L. Che, D. Li, K. Fan, Q. Li, W. Yang, L. Dong, G. Wang, Z.L. Wang, Harvesting water-evaporation-induced electricity based on liquid-solid triboelectric nanogenerator. Adv. Sci. **9**(17), 2201586 (2022)
49. Z.-H. Lin, G. Cheng, L. Lin, S. Lee, Z.L. Wang, Water-solid surface contact electrification and its use for harvesting liquid-wave energy. Angew. Chem. Int. Ed. **52**(48), 1–6 (2013)
50. J. You, J. Shao, Y. He, F.F. Yun, K.W. See, Z.L. Wang, X. Wang, High-electrification performance and mechanism of a water-solid mode triboelectric nanogenerator. ACS Nano **15**(5), 8706–8714 (2021)

51. F. Yi, X. Wang, S. Niu, S. Li, Y. Yin, K. Dai, G. Zhang, L. Lin, Z. Wen, H. Guo, J. Wang, M.-H. Yeh, Y. Zi, Q. Liao, Z. You, Y. Zhang, Z.L. Wang, A highly shape-adaptive, stretchable design based on conductive liquid for energy harvesting and self-powered biomechanical monitoring. Sci. Adv. **2**(6), e1501624 (2016)
52. X. Chen, J. Xiong, K. Parida, M. Guo, C. Wang, C. Wang, X. Li, J. Shao, P.S. Lee, Transparent and stretchable bimodal triboelectric nanogenerators with hierarchical micro-nanostructures for mechanical and water energy harvesting. Nano Energy **64**, 103904 (2019)
53. Y. Wei, X. Li, Z. Yang, J.J. Shao, Z.L. Wang, D. Wei, Contact electrification at the solid-liquid transition interface. Mater. Today **74**, 2–11 (2024)
54. Z.-H. Lin, G. Cheng, S. Lee, K.C. Pradel, Z.L. Wang, Harvesting water drop energy by a sequential contact-electrification and electrostatic-induction process. Adv. Mater. **26**(27), 4690–4696 (2014)
55. G. Zhu, Y. Su, P. Bai, J. Chen, Q. Jing, W. Yang, Z.L. Wang, Harvesting water wave energy by asymmetric screening of electrostatic charges on a nanostructured hydrophobic thin-film surface. ACS Nano **8**(6), 6031–6037 (2014)
56. X. Li, Z.L. Wang, D. Wei, Nanogenerators via dynamic regulation of electrical double layer. Nano Trends **8**, 100062 (2024)
57. Y. Dong, S. Xu, C. Zhang, L. Zhang, D. Wang, Y. Xie, N. Luo, Y. Feng, N. Wang, M. Feng, X. Zhang, F. Zhou, Z.L. Wang, Gas-liquid two-phase flow-based triboelectric nanogenerator with ultrahigh output power. Sci. Adv. **8**(48), eadd0464 (2022)
58. W. Xu, H. Zheng, Y. Liu, X. Zhou, C. Zhang, Y. Song, X. Deng, M. Leung, Z. Yang, R.X. Xu, Z.L. Wang, X.C. Zeng, Z. Wang, A droplet-based electricity generator with high instantaneous power density. Nature **578**(7795), 392–396 (2020)
59. L. Li, X. Li, W. Deng, C. Shen, X. Chen, H. Sheng, X. Wang, J. Zhou, J. Li, Y. Zhu, Z. Zhang, J. Yin, W. Guo, Sparking potential over 1200 V by a falling water droplet. Sci. Adv. **9**(46), eadi2993 (2023)
60. Z. Tang, D. Yang, H. Guo, S. Lin, Z.L. Wang, Spontaneous wetting induced by contact-electrification at liquid-solid interface. Adv. Mater. **36**(25), 2400451 (2024)
61. D. Bao, Z. Wen, J. Shi, L. Xie, H. Jiang, J. Jiang, Y. Yang, W. Liao, X. Sun, An anti-freezing hydrogel based stretchable triboelectric nanogenerator for biomechanical energy harvesting at sub-zero temperature. J. Mater. Chem. A **8**(27), 13787–13794 (2020)
62. K. Xia, Y. Tian, J. Fu, Z. Zhu, J. lu, Z. Zhao, H. Tang, Z. Ye, Z. Xu, Transparent and stretchable high-output triboelectric nanogenerator for high-efficiency self-charging energy storage systems. Nano Energy **87**, 106210 (2021)
63. L. Zhang, D. Wang, Triboiontronics based on dynamic electric double layer regulation. Matter **6**(11), 3698–3699 (2023)
64. R. Li, X. Li, Z. Zhang, M. Willatzen, Z.L. Wang, D. Wei, Triboelectric programmed droplet manipulation for plug-and-play assembly. Adv. Funct. Mater. **35**(10), 2416457 (2024)
65. Y. Ouyang, X. Li, S. Li, Z.L. Wang, D. Wei, Ionic rectification by dynamic regulation of the electric double layer at the hydrogel interface. ACS Appl. Mater. Interfaces **16**(14), 18236–18244 (2024)
66. S. Lin, Z. Lin Wang, The tribovoltaic effect. Mater. Today **62**, 111–128 (2023)
67. S. Lin, X. Chen, Z.L. Wang, The tribovoltaic effect and electron transfer at a liquid-semiconductor interface. Nano Energy **76**, 105070 (2020)
68. M. Zheng, S. Lin, L. Zhu, Z. Tang, Z.L. Wang, Effects of temperature on the tribovoltaic effect at liquid-solid interfaces. Adv. Mater. Interfaces **9**(3), 2101757 (2021)

Chapter 8
Contact-Electro-Catalysis (CEC)

Abstract Catalysis has played a crucial role in the advancement of human society by facilitating reactions that would otherwise be kinetically sluggish or thermodynamically unfavorable. Considerable efforts have been dedicated to achieving a comprehensive understanding of the fundamental mechanisms underlying catalysis, leading to the development of novel catalytic strategies with exceptional performance. Contact-electro-catalysis (CEC) represents an emerging field that employs contact-electrification-driven electron transfer to facilitate redox reactions. Due to the widespread nature of the contact-electrification effect, nearly all organic and inorganic materials can potentially serve as CEC catalysts, with the energy source being universally applicable but often wasted mechanical energy. This paradigm-shifting technology unveils promising avenues for the devise of novel catalysts and catalytic processes. This chapter provides a comprehensive overview of the theoretical foundation of CEC, strategies for improving CEC, and the unique advantages of CEC. Additionally, a roadmap for future research and development of CEC has been proposed. We aspire for this chapter to motivate in-depth investigations within the chemistry and materials community to further investigate CEC, a catalytic process in nature.

8.1 Introduction

8.1.1 Brief History of Catalysis

Catalytic phenomena are ubiquitous, with enzymes acting as catalysts within the human body. Perhaps one of the earliest catalytic processes utilized by human beings is fermentation, which has played a significant role in the production of wine and vinegar for millennia. Although the phenomenon had been observed and utilized much earlier, the concept of catalysis has roots dating back to the early nineteenth century. Before the term "catalysis" was coined by Jöns Jacob Berzelius in 1835, Döbereiner [1] discovered in 1823 that hydrogen could burn at room temperature in the presence of platinum. This was, in fact, a form of catalysis. The presence

Z. L. Wang, *Contact-Electrification of Matter*, Springer Series in Surface Sciences 75,
https://doi.org/10.1007/978-3-031-88504-4_8

of the catalyst reduced the activation energy for the oxidation of hydrogen, making the reaction conditions milder, which was still miraculous for people at that time. Early research focused on the essence of catalysis for decades until Ostwald eventually made a definition close to our modern understanding that "Catalysis is the acceleration of a slow chemical process by the presence of a foreign material" [2].

Over centuries, advancements in catalysis have been pivotal in various fields, impacting industries ranging from pharmaceuticals [3, 4] to petrochemicals [5, 6]. Catalysts enable the production of essential chemicals, fuels, and materials with greater efficiency and lower environmental impact. The significance of catalysis has also been recognized through more than twenty Nobel Prizes. Representative examples include the 2001 Nobel Prize in Chemistry shared by Prof. William S. Knowles, Ryoji Noyori, and K. Barry Sharpless for their development of chirally catalyzed hydrogenation reactions, which have profound implications for the pharmaceutical industry. Another significant award was the 2021 Nobel Prize awarded to Prof. Benjamin List and David W. C. MacMillan for their groundbreaking research in the field of asymmetric organocatalysis, which has revolutionized the selective synthesis of complex organic molecules, significantly impacting the development of new pharmaceuticals and sustainable chemical processes.

8.1.2 General Processes of Catalysis

Catalysis generally involves three fundamental steps: adsorption, reaction, and desorption. The significance of adsorption lies in its ability to allow reactants to reach the catalyst's surface, thereby improving the effectiveness of their interactions. The reaction mechanism involves transforming initial substances into end products, generally through the rearrangement of electrons between reactants with the assistance of catalysts. Finally, the process of desorption guarantees that the products detach from the catalyst's surface, making the active sites accessible for future cycles. In the early 1900s, The formulation of the Sabatier principle provided a fundamental guideline for designing efficient catalysts, emphasizing that the ideal catalyst should exhibit moderate adsorption strength [7]. When adsorption is too weak, it fails to initiate the activation of reactants; conversely, when it is excessively strong, it impedes the desorption of products, leading to a decrease in catalytic performance. This principle is often visualized using a volcano plot [8], which graphically represents the relationship between catalyst activity and adsorption energy, highlighting the optimal adsorption for maximum catalytic efficiency.

Despite the commonality of adsorption and desorption steps across different catalytic processes, the driving forces for electron excitation or transfer during the reaction step can vary significantly. For example, in electrocatalysis, electrons originate from an external circuit and are driven by an applied voltage to facilitate electrochemical reactions; in photocatalysis, photoexcited electrons are promoted from the valence band to the conduction band under light irradiation. Optimizing catalyst design necessitates a thorough understanding of the specific electronic demands linked to each catalytic process. By elucidating these requirements, researchers have developed more effective catalysts that enhance both the rate and selectivity of the desired reactions [9, 10].

8.1.3 *Important Applications of Catalysis*

Catalysis is critical for contemporary chemistry and various industrial applications. A prominent example is the Haber-Bosch process (HBP). Utilizing an iron-based catalyst, this method synthesizes ammonia (NH_3) with a massive annual production from nitrogen and hydrogen present in the atmosphere [11]. They are integral to the synthesis of polymers [12], the refining of petroleum [13] and the production of fertilizers [14], thereby supporting industry and agriculture.

The twenty-first century has witnessed a growing awareness and emphasis on environmental protection, leading to an increased preference for non-contaminating and eco-friendly chemical reactions. Catalysis plays a critical role in this context by reducing the activation energy required for reactions, thereby making the conditions under which these reactions occur more moderate. This not only facilitates the achievement of energy efficiency but also aligns well with sustainable development goals. Additionally, catalytic processes can effectively degrade pollutants [15, 16], contributing to the mitigation of environmental issues. The use of catalysts thus represents a promising approach in advancing both industrial productivity and ecological sustainability.

To better accommodate a wide array of application scenarios, the methods of catalysis have become increasingly diverse. Various catalytic approaches operate under different reaction conditions, each with its own unique mechanism. For one specific chemical reaction, it is also possible to combine multiple catalytic methods to achieve optimal catalytic efficiency [17, 18]. This integration enhances the effectiveness and selectivity of the catalytic process. Moreover, by tailoring the catalytic strategy to specific requirements, the sustainability and economic viability of chemical processes can be further improved.

8.1.4 *Electrocatalysis*

Electrocatalysis, originating in the early nineteenth century, is among the most important catalytic methods for centuries. One of the earliest records of electrocatalysis dates back to Michael Faraday's experiments in the 1830s, where the water could be decomposed into hydrogen and oxygen gases through the application of an electric current. This foundational work laid the groundwork for understanding the interaction between electrical energy and chemical reactions, which later evolved into the field of electrocatalysis. Electrocatalysis has been pivotal in fulfilling global energy demands and confronting environmental challenges through reactions transforming readily available natural resources into valuable fuels and chemicals including the hydrogen evolution reaction (HER) [19], oxygen evolution reaction (OER) [20], oxygen reduction reaction (ORR) [21], carbon dioxide reduction reaction (CO_2RR) [22], and nitrogen reduction reaction (NRR) [23].

HER is a key process in water splitting [24], where water molecules are reduced to produce hydrogen gas. Hydrogen is a clean and efficient fuel, making HER essential for sustainable energy storage and conversion. ORR is fundamental in fuel cells and metal-air batteries, where oxygen is reduced to form water. Efficient ORR catalysts are vital for enhancing the performance and durability of these energy devices.

In the process of electrocatalysis, the transfer of electrons is driven by an external electrical potential. To be specific, when an external voltage is applied to the electrochemical cell, it creates a potential difference between the electrode and the electrolyte solution. This potential difference induces the movement of electrons from the electrode to the reactants or from the reactants to the electrode. At the electrode surface, the applied potential can cause the excitation of electrons within the catalyst material. These excited electrons then participate in redox reactions, transferring from the electrode to the reactants or from the reactants to the electrode.

The HER is a key example of electrocatalysis. In a typical HER setup, as is demonstrated in Fig. 8.1, the external circuit pumps electrons into the cathode, rendering it negatively charged. In an acidic solution, positively charged protons are attracted to and captured by the cathode due to the attraction between opposite charges. Some of these protons combine with electrons to form adsorbed hydrogen species, while others remain in the solution. When the concentration of adsorbed hydrogen on the cathode surface becomes sufficiently high, these species can combine with each other to form molecular hydrogen gas. However, when the concentration is low, one adsorbed hydrogen specie interacts with an additional electron and adsorb another proton from the solution, subsequently forming hydrogen gas. During the whole process, the external voltage that drives the electrons plays a critical role.

The ORR is another essential electrocatalytic process, particularly important for fuel cells and metal-air batteries. ORR involves the reduction of molecular oxygen to water. Similar to the HER, electrons from the external circuit accumulate at the cathode, attracting positively charged protons. Oxygen molecules, upon bond cleavage, form oxygen radicals which react with protons and electrons to produce hydroxyl radicals. These hydroxyl radicals further react with protons and electrons to form water. Alternatively, oxygen molecules may be activated to directly combine with protons and electrons, forming peroxide radicals. These peroxide radicals can then react with additional protons and electrons to yield hydrogen peroxide, a common byproduct in the ORR. In the whole reactions, the external circuit plays a critical role by providing the necessary electrons, which are essential for the reaction to proceed. The external supply of electrons in the reactions ensures precise control and optimization of the processes.

The process of electrocatalysis requires a continuous supply of external circuit power, and the reactions occur exclusively in the vicinity of the electrodes. This imposes high demands on the material and durability of the electrodes. For example, Pt-based catalysts exhibit unparalleled efficiency for HER due to their optimal

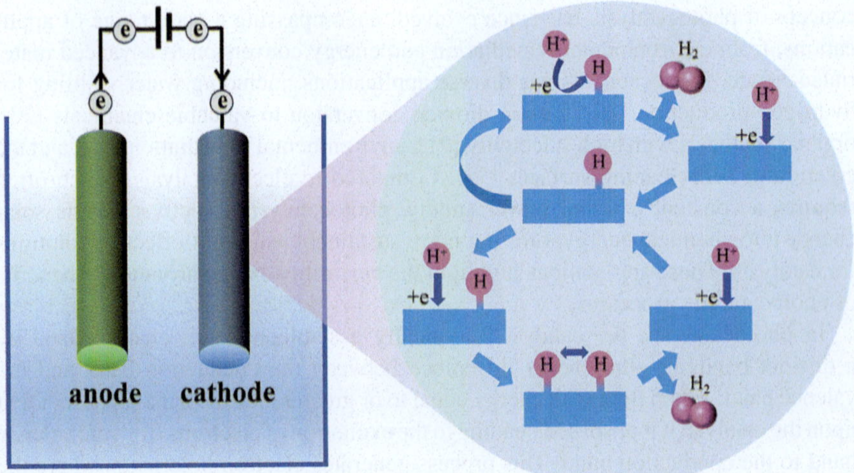

Fig. 8.1 Typical electrocatalytic hydrogen evolution reaction (HER) setup in an acidic solution

hydrogen adsorption free energy [25]. Similarly, Ir and Ru-based catalysts offer superior catalytic sites for OER by forming active surface oxide layers [26]. However, the high cost and scarcity of these noble metals pose a significant barrier to the application. Therefore, developing alternative catalysts composed of non-precious metals contributes to broader applications. Moreover, the intrinsic catalytic performance of materials is primarily governed by their microstructure and the electronic environment surrounding the active metal centers [27]. Therefore, deeper research into the structure—activity relationship can further improve the efficiency of electrocatalysis.

8.1.5 Photocatalysis

Solar energy is one of the most ubiquitous sources of power on earth. Producers, such as plants and certain microorganisms, harness solar energy to absorb carbon dioxide and produce oxygen through the process of photosynthesis. The Earth's surface also relies on solar energy to maintain its temperature, which is crucial for sustaining life and driving various ecological and atmospheric processes. Photocatalysis, a technology that harnesses light to drive chemical reactions, has its roots in seminal work conducted by Prof. Fujishima and his colleagues in the early 1970s [28]. In a groundbreaking experiment, they observed that titanium dioxide (TiO_2) could decompose water into hydrogen and oxygen when illuminated. This discovery not only laid the foundation for the field of photocatalysis but also highlighted the potential of semiconductors in converting light energy into chemical energy. The concept of photocatalysis has since evolved, encompassing a wide range of applications, from environmental remediation and energy conversion to advanced materials science. Photocatalysis has diverse applications, including water splitting for hydrogen production [29], carbon dioxide conversion to valuable chemicals [30], organic synthesis with high selectivity [31], environmental remediation [32], and the creation of self-cleaning surfaces [33]. Compared to electrocatalysis, which often requires a constant external power supply, photocatalysis directly converts solar energy into chemical energy, offering more sustainable and cost-effective solutions for catalysis. These applications highlight the versatility and environmental benefits of photocatalytic processes.

In photocatalysis, the catalyst is typically a semiconductor, characterized by a distinct bandgap—the energy difference between the conduction band and the valence band. When light with energy equal to or greater than the bandgap is incident upon the catalyst, it is absorbed, leading to the excitation of electrons from the valence band to the conduction band. This process generates electron-hole pairs, where the electrons are promoted to the conduction band by absorbing the photon energy, leaving behind positively charged holes in the valence band. These photo-excited electrons and holes are referred to as photogenerated electrons and photogenerated holes, collectively known as photogenerated carriers. These photogenerated carriers undergo diffusion within the semiconductor material, and only those that successfully migrate to the surface of the catalyst can participate in catalytic reactions. At the surface, these carriers interact with species adsorbed on the catalyst to undergo redox reactions. Photogenerated electrons possess reducing properties, while photogenerated holes exhibit oxidizing properties, enabling them to drive reduction and oxidation reactions.

Exemplified by the water splitting reaction in Fig. 8.2, the principle of photocatalysis can be illustrated as follows. First of all, a semiconductor material, such as titanium dioxide (TiO_2), is stimulated by incidental light with sufficient energy, leading to the generation of electron-hole pairs by exciting electrons from the valence band (VB) to the conduction band (CB). The excited electrons can then reduce hydrogen ions into hydrogen gas, while the holes can oxidize water to produce oxygen. This mechanism is fundamental to various photocatalytic applications, including carbon dioxide reduction to produce fuels and the degradation of organic pollutants for environmental remediation. By harnessing solar energy, photocatalysis provides an efficient and sustainable approach to driving these important chemical reactions. It is a green and environmentally friendly technology that reduces reliance on fossil fuels, thereby lowering carbon emissions and mitigating environmental contamination.

However, the recombination of photogenerated electrons and holes poses a significant challenge for the improvement of photocatalytic efficiency. Due to their opposite charges, these photogenerated electrons and holes have a natural tendency to recombine, which can occur during the diffusion process or even after reaching the surface but before participating in a reaction. Such recombination reduces the efficiency of the redox reactions and, consequently, the overall photocatalytic performance. Therefore, minimizing recombination is crucial for enhancing photocatalytic efficiency. Moreover, photocatalysis faces challenges in scaling up for industrial

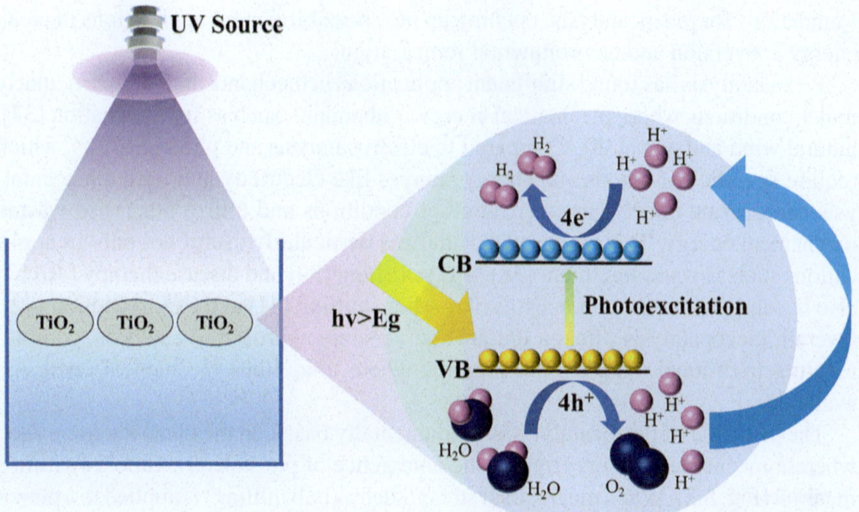

Fig. 8.2 Typical photocatalytic process in water splitting, using titanium dioxide (TiO_2) as a catalyst

applications. The efficiency and stability of photocatalytic materials often decline when transitioning from laboratory-scale experiments to large-scale operations, and the cost and complexity of fabricating high-performance photocatalysts can pose economic barriers to widespread adoption. More importantly, due to the limitations of light penetration through liquid, photocatalysis achieves satisfactory efficiency only in a narrow region close to the surface or interface. This hinders the scaled industrialization of photocatalytic processes.

8.1.6 Piezocatalysis

The piezoelectric effect, where mechanical stress generates an electric charge in certain materials, has been recognized since 1880 [34]. However, the application of this effect to catalyze chemical reactions is a more recent development. Pioneering works by Hong et al. [35] and Starr et al. [36] marked the beginning of piezocatalysis. Hong's work discovered that water can be catalyzed by piezoelectric materials to generate hydrogen under ultrasound, indicating that mechanical energy could be harnessed through the piezoelectric properties of materials to drive chemical reactions. Starr's work deepened the understanding of the piezocatalytic mechanism, explored how piezoelectric potentials affect the electronic energy levels of electrodes and drive electrochemical reactions. They further delved into the influencing factors for piezocatalytic efficiency, including capacitive currents and the formation of the electric double layer (EDL). These studies laid experimental and theoretical foundations for piezocatalysis, opening up new possibilities for applications in clean energy conversion and environmental remediation.

Piezocatalysis has found significant applications in mechanocatalysis, particularly under conditions where mechanical energy is abundant, such as from vibration [37], natural wind and tides [38]. Compared to electrocatalysis and photocatalysis, which require specific conditions and energy sources like electricity and light, piezocatalysis can operate under a broader range of conditions and utilize otherwise wasted mechanical energy. This makes piezocatalysis particularly useful not only in applications such as water treatment [38], SiC polishing [39] and disease therapy [40] but also in energy conversion processes like water splitting [41] and CO_2 reduction [42]. Overall, piezocatalysis offers a unique and versatile approach to catalysis, complementing traditional methods in scenarios where ubiquitous mechanical energy is available.

The principle of piezocatalysis is fundamentally based on the piezoelectric effect, wherein mechanical energy triggers the emergence of piezoelectric dipoles within a material (Fig. 8.3). When mechanical stress (such as ball milling) is applied to a piezoelectric material (such as barium titanate $BaTiO_3$) [43], it causes a displacement of atoms within the crystal lattice. This atomic displacement creates a disparity between the centers of cations and anions, thereby generating localized electric dipoles. The collective action of these dipoles establishes a macroscopic internal electric field, termed the piezopotential. The piezopotential facilitates the separation and migration of charge carriers within the material. Consequently, the polarized piezoelectric material is capable of functioning as both an oxidant and a reductant. The material can donate an electron to a reactant, which upon gaining an electron, transforms into a free radical, constituting a single-electron transfer (SET) reduction process [43]. Subsequent free radical reactions lead to the formation of additional free radical intermediates. These intermediates are then oxidized by the polarized piezoelectric material, yielding the final product and completing the SET oxidation process. This whole process is a SET cycle, during which the polarized piezoelectric material loses an electron and gains another. The external mechanical stimulation drives the polarization and thus promote the catalytic process.

In comparison with electrocatalysis and photocatalysis, piezocatalysis presents numerous advantages. The utilization of mechanical energy, the exhibition of higher efficiency, and the ability to achieve solvent-free conditions [43, 44] enables it to be applicable across a broader spectrum of scenarios. Nevertheless, the selection of catalysts for piezocatalysis is confined to materials that exhibit piezoelectric properties. This limitation in catalyst options emphasize the necessity for the development of a novel catalytic technology that can exploit phenomena more prevalent in nature.

8.1.7 Contact-Electro-Catalysis (CEC)

Contact electrification (CE) is an ancient phenomenon that could be dated back to over 2,600 years ago. However, the understanding of the physical essence of CE

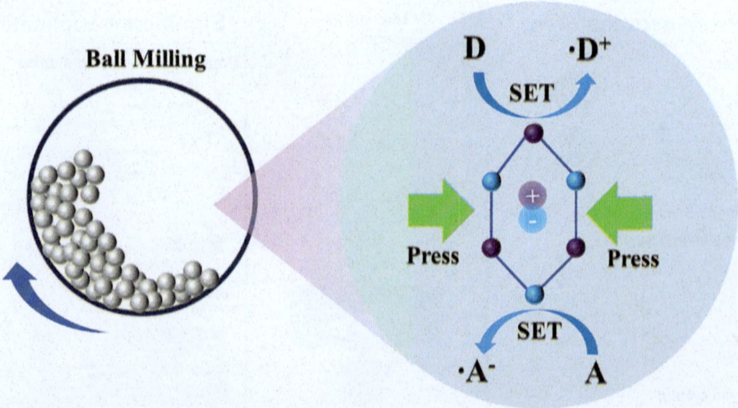

Fig. 8.3 Typical piezocatalysis under ball milling, with a single-electron transfer (SET) mechanism

remains vague for centuries because of lacking intuitive and precise characterization methods, and CE is usually recognized as a negative effect. In 2012, Prof. Zhong Lin Wang has proposed the concept of triboelectric nanogenerator (TENG) [45] by coupling the CE effect with electrostatic induction. The TENG technique not only provides strategies for how to positively exploiting the CE effect, but is also a powerful probe for *in-situ* investigate the charge transfer process during CE [46]. This marks a milestone for exploring the essence of CE. In 2020, Lin et al. [47] conducted a precise investigation into electron transfer during CE using Kelvin probe force microscopy (KPFM). By observing the different responses of electrons and ions to various external stimuli, multiple research revealed that electron transfer is the primary carrier of charge in most CE processes. In 2022, the research team led by Prof. Zhong Lin Wang has proposed the concept of using CE-driven electron transfer to promote chemical reactions (Fig. 8.4). The underly mechanism is summarized as contact-electro-catalysis (CEC) [48]. This innovative approach provides a sustainable and effective alternative to existing catalytic strategies.

CEC is an emerging field with various applications, as is depicted in Fig. 8.4. The current scenarios of CEC encompass the degradation of organic pollutants [48], the synthesis of crucial chemical raw materials such as hydrogen peroxide (H_2O_2) [49], the recycle of cathode materials from spent lithium-ion batteries [50], and the continuous ammonia synthesis from nitrogen [51] at room temperature. The ubiquity of CE phenomena broadens the selection range of catalytic materials for CEC and facilitates the integration of CEC with current catalytic mechanisms. In a typical CEC process, mechanical excitation leads to the frequent contact-separation cycles between the CEC catalyst to surrounding substrates. Such contact-separation cycles are accompanied by the CE effect that would drive electron transfer across the contact interface, rendering the production of highly reactive radicals for catalyzing subsequent redox reactions.

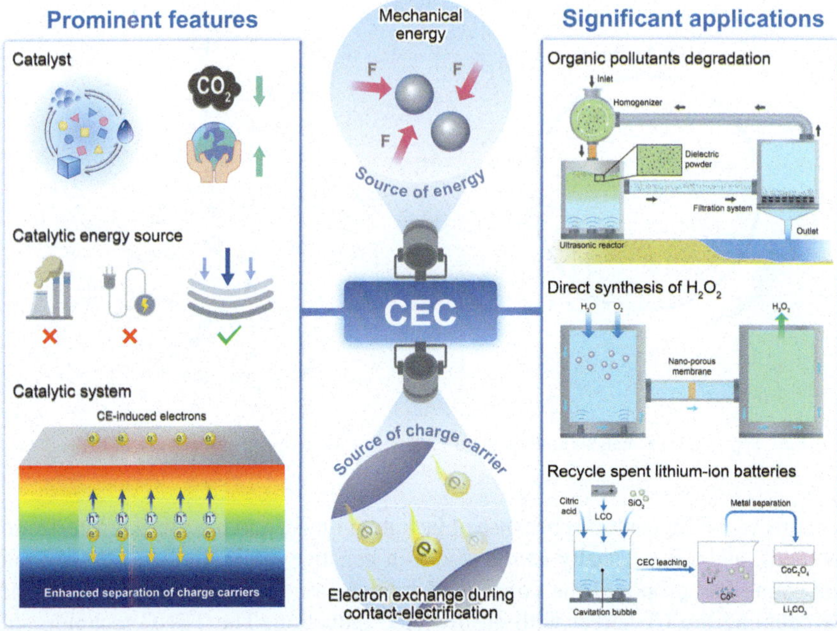

Fig. 8.4 Schematic of the fundamental principle, significant applications and future prospect of contact-electro-catalysis (CEC). Reproduced with permission from Refs. [48–50]. Copyright 2022, Springer Nature and Elsevier

CEC possesses several distinct advantages. First of all, owing to the ubiquity of CE among various interfaces, nearly all materials have the potential to be utilized in CEC. Moreover, the energy source of CEC is the universal but often-overlooked mechanical energy, thereby reducing reliance on fossil fuels, aligning with the principles of green chemistry. In addition, materials in any scale can potentially be CEC catalysts, and the mild reaction condition during CEC enable them to be facility recycled and reused with no apparent diminution in catalytic efficiency. The last but not the least, the CE effect not only occurs at inert materials, but also takes place at surfaces of functional materials. Thus, the CEC could be integrated with existing catalytic strategies for a much-improved overall efficiency.

8.2 Fundamentals of Contact-Electro-Catalysis

8.2.1 The Origin of Contact-Electro-Catalysis

8.2.1.1 Typical Process of Catalysis

Catalytic process is vital in modern society as it enables reactions that would otherwise be kinetically sluggish or thermodynamically unfavorable. As shown in Fig. 8.5, typical catalytic cycle encompasses three distinct stages: adsorption, reaction, and desorption [52, 53]. In the absorption step, reactants A and B are sequentially adsorbed onto the surface of the catalyst (C). Initially, reactant A forms bonds with the catalytic sites and transitions into an adsorbed state, denoted as C-A. This adsorption process is crucial as it brings reactant A into close proximity to the catalyst's active centers. Subsequently, reactant B is adsorbed onto the catalyst surface, forming C-B. The interaction between reactants and the catalyst is essential for the formation of these adsorbed species. Following adsorption, the reaction step occurs. During this phase, electrons are rearranged between A and B with the assistance of the catalyst. For instance, electrons may be transferred from C-A to C-B. This transfer results in C-A losing electrons and getting oxidized, forming oxidized-A, which can be referred to as A_{oxd}. Meanwhile, C-B gains electrons and gets reduced, forming reduced-B, or B_{red}. The catalytic cycle concludes with the desorption step, where the products A_{oxd} and B_{red} are released from the catalyst surface. This release signifies the end of one catalytic cycle, freeing the catalyst to engage in a new cycle with fresh reactants A and B. The composition and properties of the catalyst remain basically unchanged before and after the process.

The electron transfer step is the key process for catalysis as the driving force for electron exchanging is one of the fundamental differences among various catalytic strategies. For instance, in electrocatalysis [19, 20, 54], the driving force for electron transfer is provided by an external electrical circuit. In photocatalysis [29, 32, 55], the electron-hole pairs for subsequent redox reactions are produced by light irradiation. Piezocatalysis that activated by physical pressure [56–58] is yet another example, which utilizes piezo-charges for facilitating redox reactions.

8.2.1.2 The Dominant Role of Electron Transfer During CE

A series of recent studies have highlighted the dominant role of electron transfer in CE processes by using the different response of electrons and ions to certain stimuli. To elaborate, electrons and ions differ significantly in their physical and chemical properties (e.g., spin properties), which leads to distinct behaviors under thermal excitation, photoexcitation, and the influence of magnetic fields. Therefore, the observation of these behaviors can facilitate the identification of the dominant charge carrier in CE. For example, electrons are tend to be excited due to thermionic emissions while ions should remain on the surface under thermal excitation [47].

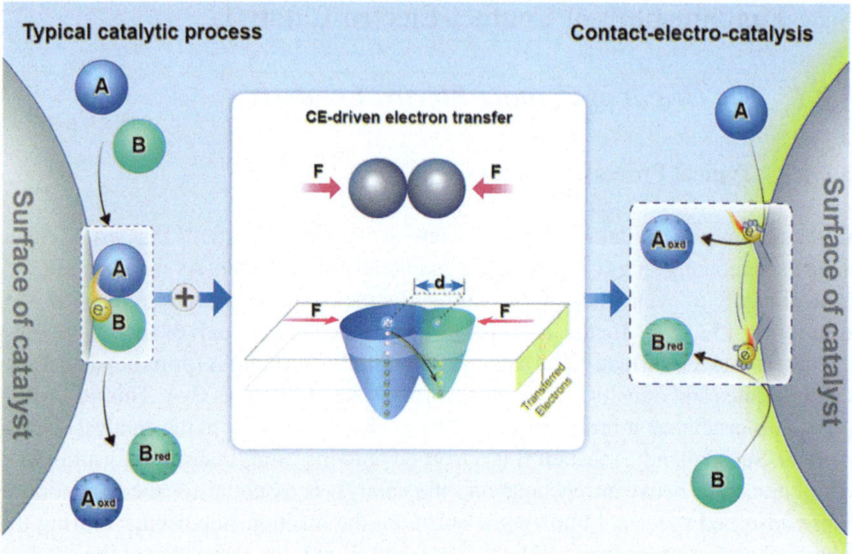

Fig. 8.5 From CE-driven electron transfer to contact-electro-catalysis. In typical catalysis, the two reacting molecules have to meet at the same time and at the same site in order for the catalysis to assist the transfer of electrons. But in the CEC, the two reacting molecules do not have to meet at the same time, and the dielectric surface is the site for electron transfer. Reproduced with permission from Ref. [59]. Copyright 2024, Royal Society of Chemistry

Therefore, an increase in temperature will result in a significant decrease in the solid's surface charge density. This assumption is verified by measuring the change in surface charge density of SiO_2 under different thermal utilizing Kelvin probe force microscopy (KPFM) to [47]. In addition, electrons and ions can also be distinguished by the photoelectric effect. Under photoexcitation, only electrons will be emitted from the original surface, resulting in corresponding decrease in surface charge density. Recent studies on the evolution of residual charges on FEP surface under different UV light irradiations [60] have also verified the dominant role of electron transfer during CE. The spin-conservative feature of electron is another vital difference between ions and electrons. Once a magnetic field is applied, the transfer of electrons that should follow the Pauli exclusion principle during transitions would be influenced. Ions, on the contrary, can hardly response to the magnetic excitation. Lin et al., have employed this principle to verify the dominant role of electrons during CE by observing the charge transfer behavior between an oxygen-containing water droplet and a ferrimagnetic Fe_3O_4 film under a 0.5 T magnetic field [61]. A significant increase in the surface charge density of the Fe_3O_4 film was observed when the magnetic field was applied.

A unified theory for illustrating the electron transfer during CE was proposed by Prof. Zhong Lin Wang in 2019. [62] This electron-cloud-potential-well model proposes that upon the contact of two materials induced by mechanical agitation, the electron clouds of atoms belonging to these two surfaces would overlap. This consequently reduce the energy barrier for electron to transfer, resulting in two surfaces being oppositely charged once sufficient energy is supplied. This model offers insights into understanding electron transfer in CE between various interfaces, specifically, in the context of liquid-solid contact-electrification (LS-CE), the electric double layer (EDL) model is widely accepted. Upon contacting with a liquid, the solid surface forms a tightly packed Stern layer due to adsorption and ionization. Outside of the Stern layer, a more loosely structured diffusion layer is formed due to the electric field and the gradient of ion concentration. This conventional model highlights the crucial role of ion distribution and migration in the overall structure and stability of the EDL, while the role of electron is ignored. However, further LS-CE studies revealed that electron transfer is prevalent during LS-CE, and it is the dominant mechanism when the solid surface is hydrophobic. Based on this, a hybrid EDL model [62] is proposed, which considers both electron transfer and ion adsorption responsible for the formation of EDL. Initially, molecules and ions in the liquid will be propelled by thermal motion or fluid pressure to collide with the solid surface. Such contact will cause interfacial electron transfer according to the electron-cloud-potential-well model. In the meantime, ions will be adsorbed on the surface, or some ionization reactions could take place under suitable conditions. Subsequently, ions within the liquid are drawn to the charged solid surface through electrostatic forces, resulting in the formation of the EDL. This model involving the contribution of both electrons and ions is called Wang's hybrid EDL model. It offers a more comprehensive interpretation of charge transfer phenomena at liquid-solid interfaces, which may give rise to the discovery of unexpected mechanisms.

8.2.1.3 From CE to CEC

One of the fundamental differences among various catalytic strategies is the driving forces for facilitating electron transfer in the reaction step. Recent progress on the underlying mechanisms of CE has proved the existence of electron transfer during contact-separation cycles. Moreover, electrons appear as the dominant charge carrier in a majority cases of CE. Hence, the CE effect should also be a feasible driving force for electron transfer in catalysis.

CEC is a catalytic strategy that utilizes the CE-driven electron transfer to promote chemical reactions. Under external mechanical stimuli, such as ultrasonication or ball milling, there are frequent contact-separation cycles between the reactants and the CEC catalyst. According to Wang's electron-cloud-potential-well model, each contact-separation cycle induces the overlap of electron clouds between the reactants and the CEC catalyst, thus facilitating electron transfer between them. The right panel in Fig. 8.5 demonstrates a general CEC process for facilitating redox reactions. Once the reactant A comes into contact with the CEC catalyst C, electrons are supposed to transfer from A to C due to the CE effect. During this, reactant A undergoes oxidation to form A_{oxd} as it loses electrons, while the catalyst C obtains these electrons and transitions into a charged state, denoted as C*. Subsequently, when another reactant

B encounters the charged catalyst C*, it can grab electrons from C*, leading to its reduction to B_{red}. In the meantime, the charged catalyst C* to release the accumulated electrons, reverting to its original uncharged state C. Cycles like this constantly promote the yield of the product with the input of external mechanical stimuli.

This CE-induced electron transfer is the cornerstone of the CEC mechanism, distinguishing it from other catalytic processes that rely on other driving forces. This unique nature offers several advantages for CEC. First and foremost, the ubiquity of CE enables a much broader selection range of catalysts in CEC. Secondly, the mild condition under which CE is able to occur enlightens multiple eco-friendly strategies for initiating CEC. Lastly, when the catalyst is dispersed in a solution, CE can occur at any contact interface, thus enabling each particle to act as a reaction chamber, enabling a promising future for large-scale production.

8.2.2 Two-Step Mechanism of CEC

The two-step mechanism of CEC for producing reactive oxygen species (ROS) was first proposed by Prof. Zhong Lin Wang's group, which involves the oxidation of water and reduction of oxygen molecules, as depicted in Eq. 8.1 and 8.2 [48]. At the first step, the contact between FEP and water causes the interfacial electron transfer. During this process, FEP seizes an electron from water, owing to the strong electronegativity of F groups in FEP. When a water molecule loses an electron, it transforms into a water radical cation, which subsequently engages in a swift proton transfer with another water molecule, resulting in the formation of hydronium cations and hydroxyl radicals [63, 64]. FEP* was applied to depict the charged state of FEP after gaining electrons from water molecules. In the second step, when dissolved oxygen in aqueous solutions interacts with FEP*, electrons from FEP* are exchanged with the oxygen molecules, leading to the formation of superoxide radicals. Simultaneously, the discharged FEP* will return to ground state FEP. This recovery of the catalyst material marks the completion of a CEC cycle.

$$2H_2O + FEP \rightarrow H_3O^+ + OH^{\cdot} + FEP^* \tag{8.1}$$

$$O_2 + FEP^* \rightarrow {}^{\cdot}O_2^- + FEP \tag{8.2}$$

In Eq. (8.1), H_2O^+ is generated through the electron transfer between FEP and H_2O. However, this radical cation is very unstable, with a lifetime for the briefest of instants - about 200 ps. H_2O^+ instantaneously merges with another H_2O molecule, experiencing the following reaction.

$$H_2O + H_2O^+ \rightarrow H_3O^+ + OH^{\cdot} \tag{8.3}$$

A further study on the understanding of the "two-step" mechanism is conducted by Prof. Jiang's group [65], and the obtained result suggests that the energy requirement for realizing these two steps is quite different. Images of experimental setup is available in Fig. 8.6a. Specifically, nitrogen gas was utilized to propel DI water droplets through a PTFE tube, ensuring adequate LS-CE while minimizing oxygen interference. The electron paramagnetic resonance (EPR) results shown in Fig. 8.6b indicates that hydroxyl radicals could be generated even in the absence of ultrasonic treatment, though ultrasonic treatment does promote the process. In contrast, the generation of superoxide radicals was only observed under ultrasonication. This could be attributed to the cavitation effect induced by ultrasound that could not only increases the contact frequency, but also inducing a high-pressure environment to facilitate the electron transfer process. As a result, electrons can transfer from the PTFE surface to the oxygen molecules under ultrasonication, leading to the formation of superoxide radicals.

To further reveal the underlying reasons for the different energy requirements in producing hydroxyl and superoxide radicals, Zhao et al. have [49] utilized density functional theory (DFT) calculations (Fig. 8.6c) to investigate the electron transfer process under different conditions. They found that the strong electronegativity of fluorine increases the electron-withdrawing capability of the PTFE surface. However, the fluorine-rich surface makes it more challenging for PTFE to donate electrons to oxygen, raising the energy barrier by approximately 1 eV. Due to this barrier, extra energy input, such as the ultrasound, is necessary. Here, ultrasound serves as both an external energy source and a means to reduce the energy barrier in the CEC process, making it a reliable approach to initiate various CEC reactions.

Exemplified by PTFE, Wang et al. [59] have presented a more comprehensive two-step mechanism for the generation of ROS through CEC in Fig. 8.6d. The first step is water oxidation reaction (WOR). Electrons can be transferred from water molecules to PTFE surface due to the CE effect once these two come into contact. Water is oxidized to form water radical cations after losing electrons. These cations then undergo a rapid proton transfer, converting into hydronium ions and hydroxyl radicals. In the following step (oxygen reduction reaction, ORR), additional energy is necessary to remove an electron from the electron-affinitive surface of PTFE* for producing superoxide radicals. The differential energy requirements of the WOR and ORR are a key feature of the CEC process. The WOR involves the spontaneous generation of hydroxyl radicals, while the ORR requires additional energy to reduce oxygen molecules to superoxide radicals.

In conclusion, a more comprehensive "two-step" mechanism for producing ROS by CEC has been established by taking the energy requirement into consideration. This mechanistic understanding is essential for the advancement of CEC technologies, as it provides insights into the energy dynamics and potential optimization strategies for enhancing the efficiency and control of radical generation in these systems. With ORR being the rate determining step, further research can boost the overall efficiency of CEC by optimizing ORR.

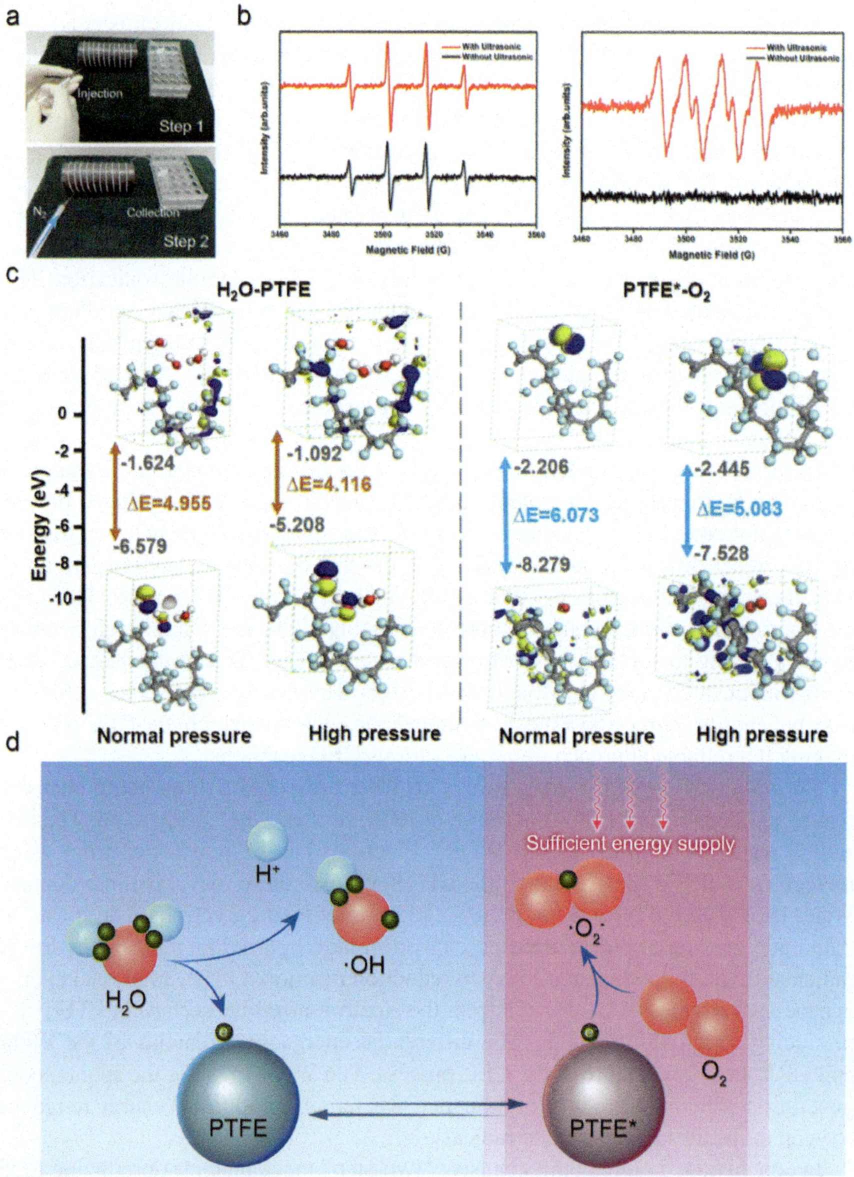

Fig. 8.6 The two-step mechanism of CEC. a Optical images for illustrating the process of CEC. **b** Measured EPR profiles in conditions of with and without ultrasonication. Reproduced with permission from Ref. [65]. Copyright 2022, Elsevier. **c** DFT calculations of the values of LUMO and HOMO levels for H_2O-PTFE and O_2-PTFE in various conditions. Reproduced with permission from Ref. [49]. Copyright 2022, Wiley. **d** Proposed two-step mechanism for producing radical oxygen species through CEC. Reproduced with permission from Ref. [59]. Copyright 2024, Royal Society of Chemistry

8.2.3 *Broad Selection Range of CEC Catalysts*

As one of the most ubiquitous phenomena, contact-electrification is prevalent across a variety of interfaces, such as solid-solid, solid-liquid, and even liquid-liquid or solid-gas. This widespread occurrence of CE significantly enhances the range of catalyst options available for CE-based catalytic energy conversion, i.e., CEC (Fig. 8.7). On the other hand, CE is available at various scales, including both micro and macro levels. Therefore, a variety of forms such as micrometer-scale powders, millimeter-sized spheres, centimeter-scale films or even bulk materials have been proven to be effective options for CEC catalysts. Consequently, the selection range of CEC catalysts can be significantly broadened, which can be now divided into three main categories: polymers, oxides and matrix composites. Furthermore, the CE at liquid-liquid interface has also been proved feasible to catalyze the production of ROS, which indicates that the liquid can also potentially be CEC catalysts.

8.2.3.1 From Powders to Films

Powders possess high specific surface areas compared to their peers, which is beneficial for enhancing the CEC efficiency as more area is available for CE. The first application of powder-based CEC [48] is degrading organic pollutants using FEP powders. An obvious discoloration of MO was observed after ultrasonication for 3 h. Further analysis using liquid chromatography-mass spectrometry (LC-MS) confirmed the chemical degradation of MO and ruled out the possibility of physical adsorption of MO onto the FEP powder. Powders with different size can also be employed as CEC catalysts. For example, Su et.al [70] utilized FEP powder with varying particle sizes (from 30 μm to 0.2 μm) to investigate its CEC efficiency in the reduction of noble metal ions such as $AuCl^{4-}$, Ag^{+}, Rh^{3+}, $[PtCl_4]^{2-}$, Pd^{2+}, and Ir^{3+} from aqueous solutions under ultrasonic treatment.

Balls are another representative configuration of CEC catalyst, and its first application was demonstrated by Wang et al. [71] in a CEC-based ball milling process. Specifically, 5-mm PTFE, PDMS, and PP balls and corresponding milling vials are employed for a liquid-assisted-grinding (LAG) setup. The electron paramagnetic resonance (EPR) spectra proved the generation of ROS during such process, and the produced ROS could degrade MO. Ball catalysts offer two main advantages comparing to powders. On the one hand, balls can withstand higher mechanical stress, allowing for its durability in CEC reactions. On the other hand, balls are easier to separate from the reaction system, simplifying the recycle and reuse process.

Films possess several advantages in CEC applications. One key benefit of films is their ease of separation from liquids, which greatly facilitates the process of recycling used CEC catalysts. For example, a study by Wei et al. [72] focused on the effectiveness of PTFE films as reusable CEC catalysts for inactivating Microcystis aeruginosa, a harmful cyanobacterium responsible for algal blooms. The PTFE films can be easily separated from the solution after reaction. The structure and properties

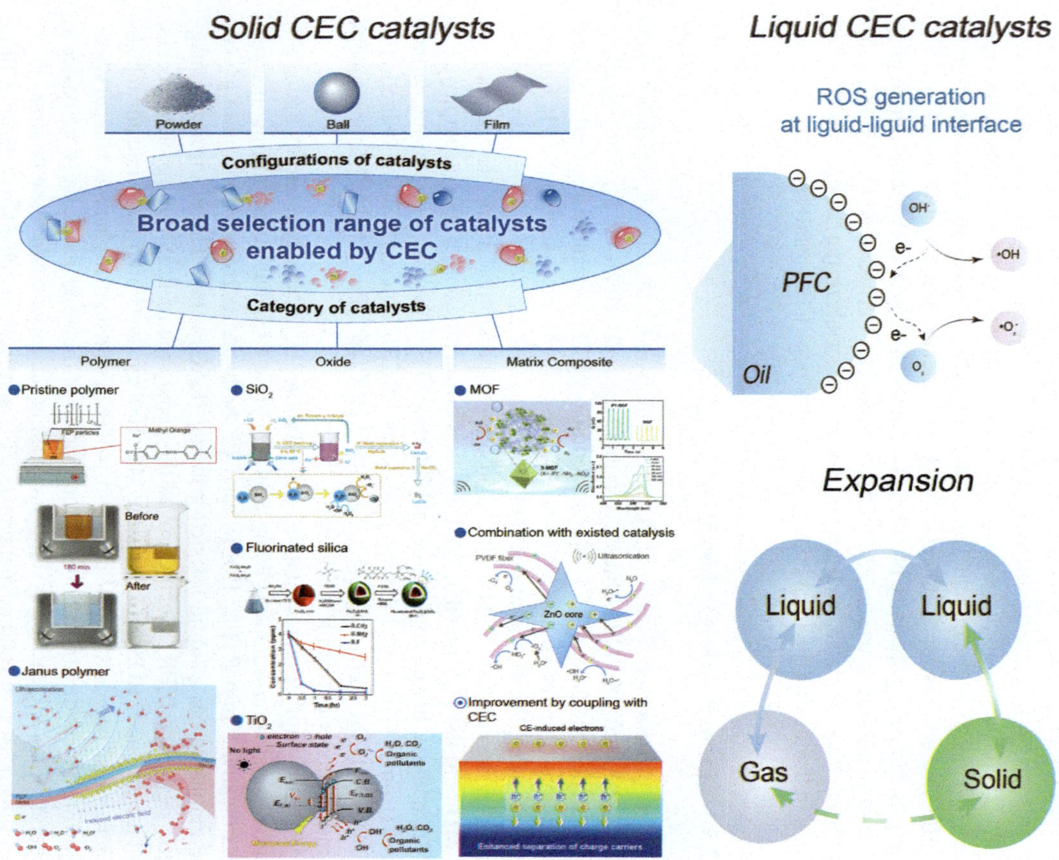

Fig. 8.7 Contact-electro-catalysis has enabled much broader choices for catalysts. Reproduced with permission from Ref. [48, 50, 66–69]. Copyright 2022, Springer Nature and Elsevier

of the PTFE film before and after the reaction have been systematically characterized, and obtained results indicated that there were no significant changes in the PTFE film before and after CEC.

Owing to the ubiquity of CE, several commonly available laboratory consumables with diverse shapes can also be used to achieve CEC. For instance, Wang et al. [73] utilized a PTFE stir bar and ultrasound to synthesize H_2O_2 from water with a production rate of 256.6 mM h^{-1} under ambient conditions, demonstrating its potential for cost-efficient applications.

Overall, powders are among the most commonly utilized catalysts in CEC processes. However, the recovery of powdered catalysts remains a significant challenge. In Chen's study [66], a novel approach was employed to address this issue by utilizing magnetite (Fe_3O_4), a magnetic material, as the core of the powdered catalyst. This innovative strategy leverages magnetic forces to facilitate the separation of the powdered catalyst from the liquid phase, thereby achieving efficient recovery. On the other hand, while balls, films and bars are more recyclable than powders, the limited surface area still hinders their further applications. Further studies utilizing surface modification techniques to introduce micro/nano-scale topographies may offer a viable solution. In summary, the expansion in catalyst morphologies not only broadens the selection range of CEC catalysts but also paves the way for more efficient and sustainable catalytic systems for various applications.

8.2.3.2 From Polymers to Composites

From the perspective of material, the CEC catalysts can be divided into polymers, oxides, and matrix composites. Polymers are the most widely used material for CEC. On the one hand, polymers, especially those with fluorinated surface functional groups, possess strong CE capacity. On the other hand, polymers also offer an ideal platform to investigate CEC due to their chemical inertness. Wang et al. have provided compelling evidence for the feasibility of CEC by using chemically inert and highly insulated FEP to degrade MO aqueous solutions under ultrasonication. In addition to FEP, other polymers such as PTFE and rubber are also feasible for CEC. For example, a study by Liu et al. [74] investigated the use of a PTFE membrane for fouling control in membrane separation systems through CEC under ultrasonic irradiation. The untreated PTFE membrane demonstrates significant effectiveness in degrading organic pollutants, with the degradation rate of crystal violet reaching 94.54% after 5 h of ultrasound treatment. Another study by Liu et al. [75] proposed a strategy to degrade antibiotics using a low-cost, modified ethylene-propylene-diene-monomer (EPDM) rubber to activate peroxymonosulfate (PMS). The study found that the optimized catalyst could degrade over 90% of tetracycline within 30 min under neutral pH conditions. These findings show the high efficiency of polymer catalysts, indicating promising future for broader applications.

Owing to the different rates of two reaction paths in CEC, charges tend to accumulate on the surface of polymers, especially those F-containing polymers. These charges can induce a high-intensity electric field, and this electric field could polarize ambient conductive materials. Dong et al. [76] have utilized this principle to fabricate a polymer/metal Janus catalyst for significantly enhancing the CEC efficiency. According to the "mirror charge" concept, the high-intensity electric field could induce electrons on the outer surface of the adjacent metal through the electrostatic induction effect. While the electrons on the polymer surface are bounded due to the strong electronegativity, the charges induced on the metal surface are easier to be removed. Thus, the efficiency of CEC can be significantly enhanced by facilitating the ORR process. Besides, obtained results suggest a synergistic effect in this polymer/metal Janus catalyst, further highlighting its potential application in high performance CEC.

Although polymers are widely utilized in CEC, they usually suffer from rapid decrease of CE ability under elevated temperatures. Consequently, polymers are not suitable for reactions that require high temperatures to take place. To address this limitation, thermally stable oxides have been proposed as an ideal candidate for CEC under high temperatures. Li et al. [50] have proposed a CEC-based approach for recycling cathode materials from spent lithium-ion batteries, with SiO_2 as the CEC catalysts. A high leaching efficiency (> 90%) is obtained for both LCO and NMC batteries within 6 h at 90 °C, outperforming conventional polymers such as PTFE and FEP. However, the intrinsic CE ability of SiO_2 is limited compared to fluorine-containing polymers, which hinders further improvement of the CEC efficiency. Therefore, efforts have been made to enhance the CE properties of oxides through proper surface modifications. For example, Chen et al. [66] have investigated the CEC performance of silica-coated magnetic nanoparticles with various terminal groups modification (–F, $-NH_2$, $-CH_3$). The fluorinated catalyst performed the best, with a MO removal efficiency of over 90% within 3 h. Since CE is mainly a surface effect, proper modification of the SiO_2 surface presents a promising strategy to enhance its CE performance without damaging its thermal stability. Besides nonmetal oxides, metal oxides are also applied to CEC. For example, Lin et al. [67] explored the degradation of Rhodamine B (RhB) using an aluminum (Al) and titanium dioxide (TiO_2) system under magnetic stirring and ultrasonication. They hypothesized that the synergistic degradation effect observed in the Al-TiO_2 system was due to the tribovoltaic effect at the dynamic metal/semiconductor contact interfaces, where mechanical energy excited electron-hole pairs. Those pairs reacted with water to generate hydroxyl radicals, thus degrading the RhB. This approach achieved ~ 95% degradation efficiency after 5 cycles with Al powder replenishment.

The CE effect not only exists at inert materials' surface, but also occurs on the surface of various functional materials. Thus, CEC can be combined with existing catalytic strategies by using matrix composites as the catalyst. One example is the use of a PTFE/ZSM-5 (PZ) catalyst, which combines the hydrophobicity of PTFE with the high surface area and water absorption capacity of ZSM-5, a representative oxide-based molecular sieve belonging to the aluminosilicate family. Li et al. [77] have introduced an innovative Fe^{III}-initiated self-cycling Fenton reaction system that enhances the catalytic efficiency of CEC. Obtained PTFE/ZSM-5 (PZ) catalyst could significantly enhance the H_2O_2 production yield through the CEC process, reaching

1978.8 μmol g^{-1} h^{-1}. This system could efficiently decompose H_2O_2 to ROS, realizing sixfold enhancement in dye degradation efficiency. Methyl orange is nearly completely degraded (98.7%) within 10 min. Additionally, the system dissolves precious metals in mildly acidic solutions at room temperature, achieving an 80% gold dissolution rate in 2 h without external oxidants.

MOFs are porous materials with high surface area and tunable properties, which provides plenty rooms for surface modifications to integrate with CEC. Zhang et al. [68] have investigated the CEC properties of a series of molecular-functionalized MIL-101(Cr) MOFs for the degradation of organic dyes. By modifying the MOF framework with pyridine, amino, and nitro groups respectively, the CE ability of MIL-101(Cr) MOF is highly tunable. Among the modified MOFs, pyridine-functionalized PY-MOF exhibited the highest CEC efficiency, achieving a degradation rate of 97% for methylene blue (MB) within 3 h.

Jiang et al. have [69] devised a contact-piezoelectric catalytic system using ZnO@PVDF composite membranes for efficient dye degradation. Piezoelectric ZnO nanoparticles were integrated into PVDF fibers through electrospinning. The composited membrane achieves a degradation efficiency of 55.29% for 5 ppm methyl orange after ultrasonication for 360 min, compared to only 10.60% for the pure PVDF membrane. The synergistic effect between CEC and piezocatalysis derives from the high CE performance of PVDF and piezoelectricity in both ZnO and the β-phase of PVDF. This study presents a promising strategy for promoting chemical reactions by combining CEC with piezocatalysis.

8.2.3.3 From Solid to Liquid

In addition to the CE at liquid-solid interface, the CE could also take place once a liquid contacts with another liquid, i.e., liquid-liquid contact-electrification (L-L CE). Therefore, the CEC should also be feasible if ultrasonication is introduced into a L-L system. This assumption has been verified by Li et al. [78] that utilize perfluorocarbon (PFC) as the CEC catalyst. Under ultrasound-induced high-frequency contact-separation, electrons will be transferred from water to the PFC, catalyzing the production of ROS for in-situ and highly-controllable cancer therapy.

The propose of liquid-liquid CEC offers several advantages over conventional solid catalysts. One of the key benefits lies in its unique ability to catalyze reactions inside human body. The source of the radicals during liquid-liquid CEC is water, which means they are not influenced by the properties of other endogenous substances in the tumor microenvironment. Additionally, the use of ultrasound to control reactions within the human body does no harm to visceral organs. Therefore, this CEC strategy is available for cancer treatment inside human body. In contrast, current researches provide no evidence in support of solid CEC catalysts with similar ability.

This pioneering study of liquid-liquid CEC has sparked interest in exploring other interfaces, such as gas-liquid and gas-solid, for CEC applications. For example, Wang et al. [79] utilized acoustic levitation to suspend droplets to eliminate the influence of solid-liquid or liquid-liquid contact. They found that ultrasound-induced droplet-air friction charges the droplets positively. This charge transfer can be enhanced with the presence of solid particles in the droplets. Li et al.'s study on acoustic levitated microdroplets [80] further showed that high electric field at the gas-liquid interface could generate hydroxyl radicals, indicating a possible gas-liquid CEC.

8.2.4 Strategies for Initiating CEC

8.2.4.1 Ultrasonication

Among the representative strategies illustrated in Fig. 8.8, ultrasonication was the earliest proposed method for initiating CEC [48] and remains the most widely used technique. This technique harnesses the unique properties of ultrasound waves to generate reactive species at the interface between dielectric materials and water, leading to efficient chemical reactions. The use of ultrasound could provide high-frequency contact-separation cycles and create a high-pressure environment to decrease the energy barrier for electrons to transfer.

During ultrasonication [82], high-frequency sound waves propagate through the liquid, creating alternating regions of high and low pressure. In the low-pressure regions, small bubbles known as cavitation bubbles (CBs) forms through a nucleation process when the local pressure falls below the vapor pressure. These CBs are more likely to form in areas where oxygen molecules are dissolved in the liquid and can encapsulate these oxygen molecules to facilitate ORR. These CBs rapidly grow due to the influx of gas from the surrounding liquid and reach a critical size. Upon reaching this size, the bubbles collapse due to the surrounding high-pressure environment, creating localized temperatures reaching up to 5000°C and pressures approaching 500 ATM [83]. This process, known as the implosion, generates powerful microjets that collide with the surface of dielectric materials, such as PTFE or FEP. This impact induces frequent contact and separation cycles between the dielectric material and the water molecules, creating opportunities for electron transfer. Moreover, theoretical calculations have proved that the localized high-pressure environment is beneficial for electron transfer as the energy difference between LUMO and HOMO is decreased [49].

The first ultrasonication-based CEC was proposed by Wang et al. [48] that used FEP particles and ultrasound to degrade organic pollutants. Moreover, in Zhao's study [49], H_2O_2 could be directly synthesized from water under ambient and even anerobic conditions by ultrasonication, with PTFE being the catalyst. Both studies proved ultrasonication as an effective approach to convert mechanical input to chemical energy without generating harmful byproducts. Therefore, this approach offers a sustainable and efficient alternative for environmental remediation and the synthesizing of valuable chemicals. Moreover, ultrasonication is also feasible for initiating CEC at L-L interface [78], and the produced ROS has been proved to be effective for cancer therapy. Given the non-harmful nature of ultrasound to human tissues,

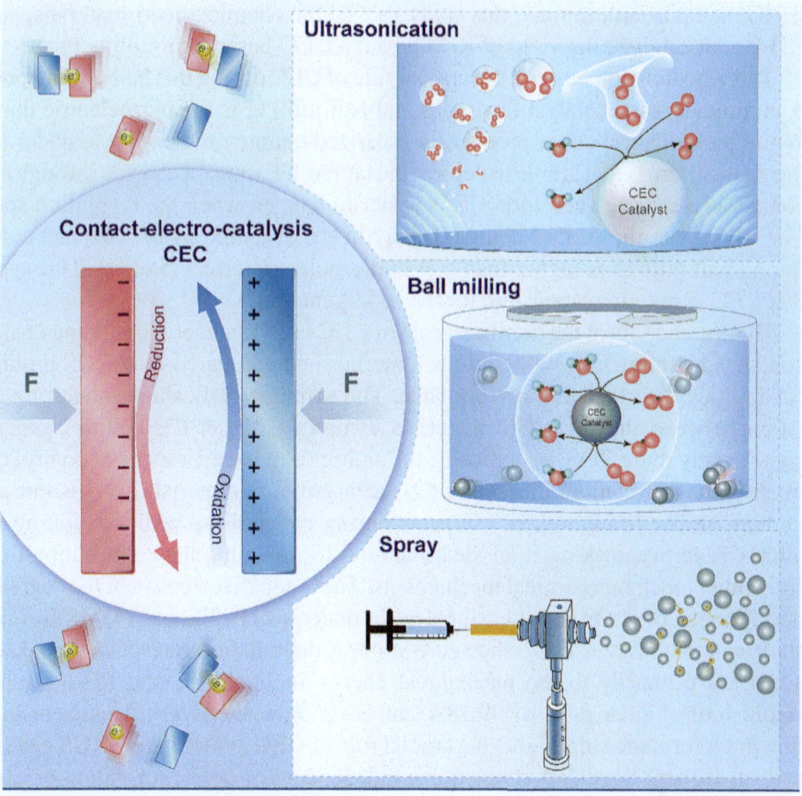

Fig. 8.8 Representative strategies for introducing contact-electro-catalysis at liquid-solid contact interfaces. Reproduced with permission from Refs. [48, 71, 81]. Copyright 2022, Springer Nature and Elsevier

it is viable to utilize ultrasonication-based CEC for in-situ and highly controllable anti-cancer therapy for human tissue.

8.2.4.2 Ball Milling

Ball milling is another effective strategy for initiating CEC (Fig. 8.8). As a widely employed mechanochemical technique, ball milling could naturally bring about frequent contact-separations even at low revolution speeds. Conventional ball milling utilizes mechanical energy for subsequent reactions by inducing extreme conditions, surface defects, or other effects, such as piezoelectric effects. This process of ball milling involves high-energy impact and friction during highspeed rotation. However, during relatively low speed ball-milling, the frequent contact-separation has long been dismissed. Due to this, Wang et al. have [71] first introduced a CEC-based ball milling process for promoting target reactions. Using a liquid-assisted grinding

(LAG) setup as an example, this study proved the chemical inert materials, such as PTFE, can catalyze the yield of ROS through CEC-based ball milling process.

This research emphasized the critical role of CEC during this ball milling process. A commonly used catalyst for traditional ball milling is the piezoelectric material. While piezoelectric effect produces a polarized charge on the surface under strong mechanical stress that is able to deform the lattice, CE transfer charges through milder contact-separation. Therefore, CE is easier to happen when the revolution speed is low. Moreover, the surface charge density in CE triggered by the frequent collision during ball milling is higher than that in piezoelectric effect [58, 71]. This suggests that CEC is mainly responsible for the ROS generation.

To further confirm the dominant role of CEC in ROS generation, Wang et al. have conducted experiments with various materials and compared their ROS production capabilities with their CE performance. The results clearly showed a strong correlation between the two, with materials exhibiting higher CE abilities generating significantly more ROS. Specifically, the authors conducted a series of control experiments using zirconium dioxide (ZrO_2) as a grinding material. ZrO_2 is known for its high ability to high-energy impacts during ball milling while exhibiting negligible CE ability, making it an ideal material for isolating the contribution of CEC from other mechanochemical mechanisms. The comparison between the degradation performance of ZrO_2 and the triboelectric materials (PTFE and PDMS) revealed a stark contrast. While ZrO_2 showed a modest degradation rate of around 12.04%, attributed primarily to the mechanical energy input, PTFE and PDMS achieved significantly higher rates of 99.18% and 32.05%, respectively. This clear distinction in performance highlights the crucial role of CEC in facilitating ROS generation and subsequent degradation reactions, independent of defect formation or extreme conditions.

Comparing to traditional ball milling, the CEC-based ball milling is independent on high energy impact, allowing for milder conditions. The relatively low revolution speed induces milder collision, minimizing the abrasion of CEC catalysts. Moreover, ball-milling also offers an energy-efficient alternative to ultrasonication in the context of initiating CEC, as it generates less heat without the implosion of cavitation bubbles.

8.2.4.3 Microdroplets

In 2022, Prof. Richard N. Zare from Stanford University have reported that hydrogen peroxide (H_2O_2) emerges spontaneously within water microdroplets formed by the condensation of water vapor on cold surfaces. This finding challenges the conventional understanding of water molecules as stable and inert, particularly in bulk liquid form. Since then, the field of microdroplet chemistry has garnered worldwide attention for the unique ability for microdroplets to promote chemical reactions under ambient conditions, including yielding ammonia from nitrogen [81], degrading hydrocarbon [84], and synthesizing urea from carbon dioxide and nitrogen [85]. This uniqueness of microdroplets is rooted in the interfacial electron transfer.

Although the fact that electron transfer happens when water is splitted into droplets was found in 1892 [86], deeper explorations into this phenomenon have long been overlooked. Lin et al. have [87] demonstrated a size-dependent charge transfer phenomenon in water microdroplets during ultrasonic atomization, where larger droplets tend to acquire positive charges while smaller ones carry negative charges. This suggests a transfer of negative charges from the larger to the smaller droplets. The authors proposed a theoretical model attributing this charge transfer to the curvature-induced surface potential/energy differences between droplets of varying sizes, which facilitate the movement of negative charges from areas of lower to higher surface potential. Further study by Xia [88] visualized the charging process using a high-speed camera. They found that the electric field strength between separating droplets can reach up to 3×10^9 V/m, sufficient to cause micro-lightning and ionize the surrounding gas molecules, with Open-air mass spectrometry revealing the presence of ions such as N_2^+, O_2^+, NO^+, and NO_2^+. The strong electric field at the interface is also present in CEC, and it is believed to facilitate electron transfer.

CE is believed to be a crucial effect in microdroplet chemistry. A well-accepted perspective [89–91] is that CE, which occurs during the separation of droplets, plays a pivotal role in driving the electron transfer for the formation of free radicals. The presence of radicals within microdroplets has been observed by Zhao et al. [92], employing pyridine to trap hydroxyl radicals within microdroplets. Qiu et al. have [93] reported the presence of both hydronium and hydroxide radicals within microdroplets, enabling the simultaneous occurrence of both oxidation and reduction reactions.

Given those commonalities between microdroplet chemistry and CEC, it is considered feasible to utilize microdroplets for achieving CEC (Fig. 8.8). Chen et al. have [94] explored an approach for hydrogen production by utilizing sprayed oil-water microdroplets as a hydrogen source. This study demonstrated that CE between oil and water microdroplets could lead to electron extraction from water molecules. Such electron transfer could result in a significant enhancement of charge separation, which enables the formation of hydrogen species at the oil-water interface. Moreover, these produced hydrogen species are able to drive CO_2 hydrogenation to CO under ambient conditions. Charge measurements and electron paramagnetic resonance spectroscopy were conducted to confirm the charge transfer mechanism and observed a ~ 13-fold increase in charge carriers compared to ultrapure water sprays. This research can also be interpreted within the framework of CE. Frequent contact and separation cycles between oil and water are caused by the swift volumetric changes of the mixed liquids during the atomization process. A strong electric field is created at the interface, promoting electron transfer. This CE-driven electron transfer then facilitates subsequent reactions. This is consistent with the principle of CEC, inspiring deeper research into applying microdroplet chemistry to CEC.

Despite the promising future, several key questions remain open for further investigation. For example, the exact mechanisms by which CE induces electron transfer and the subsequent formation of free radicals in microdroplets are still not fully understood. Additionally, the efficiency and selectivity of the chemical reactions that occur in microdroplets need to be optimized for practical applications. Moreover, there is a growing interest in exploring how the properties of microdroplets, such as size, composition, and environmental conditions, affect the outcome of chemical reactions. Continued research is needed to address these remaining challenges and to fully harness the potential of combining microdroplet chemistry with CEC.

8.2.5 Influencing Factors of CEC

A series of investigations have been conducted to obtain the relationship between experimental parameters and overall CEC efficiency. These studies have in turn deepened our understanding of the mechanism of CEC and inspired us to devise the CEC in a more effective way.

8.2.5.1 Influencing Factors of Ultrasonication on CEC

Ultrasonication is the most conventional method to initiate CEC. The effectiveness of ultrasonic waves in promoting CEC is influenced by several key parameters, including frequency, power, and temperature. Understanding these factors is crucial for optimizing the performance of ultrasonication-based CEC processes.

The ultrasonic frequency and power would influence the generation of cavitation bubbles, which could in turn significantly impact the efficiency of CEC. Taking the degradation of methyl orange (MO) aqueous solution as a model reaction, Dong et al. [95] have investigated the relationship between ultrasonic parameters and CEC efficiency by utilizing a ultrasonic bath with varying frequencies and powers. The results shown in Fig. 8.9a and b indicates that when the power of ultrasonication is fixed, at lower frequencies (below 40 kHz), the degradation rate of MO positively correlates with the frequency. This correlation can be attributed to an increase in the number of cavitation bubbles due to the increase in frequency. However, when the frequency exceeds 40 kHz, the continued increase in frequency not only fails to further enhance the degradation rate of MO but also leads to a decrease. This decrease can be ascribed to the reduction in the size of cavitation bubbles due to the increase in frequency. The implosion of smaller cavitation bubbles releases less energy, which is detrimental to the electron transfer driven by CE. Therefore, there is a trade-off between the number of and energy released from cavitation bubbles, and 40 kHz represents an optimized frequency for achieving the CEC degradation of MO.

Besides ultrasonic frequency, changes in power can also affect the CEC efficiency. When the power is below 200 W, the degradation rate of MO increases significantly with the increase in power. However, when the power exceeds 200 W, although the increase in power still leads to an increase in the degradation rate, this increase is not as significant as when the power is below 200 W. This could be attributed to the fact that a significant amount of energy is used for heating at high power levels, leading to severe energy waste. Heat generation is a problem not only for

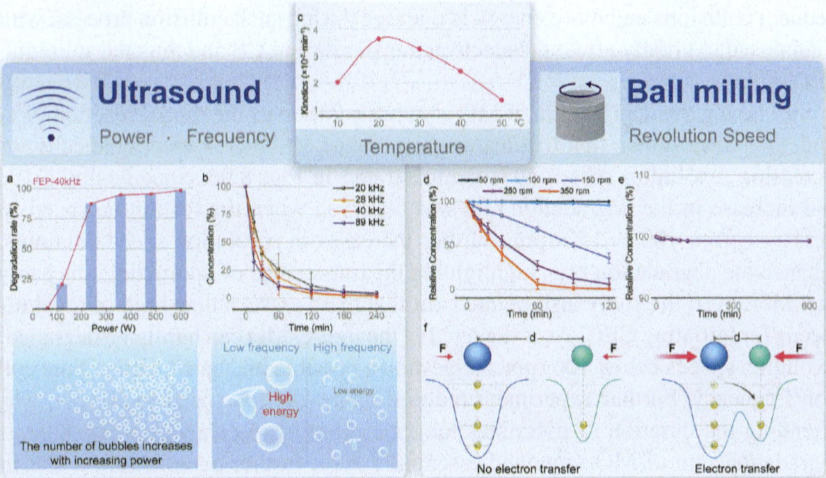

Fig. 8.9 Investigations on the experimental influence factors on the efficiency of contact-electro-catalysis. **a** Degradation rate of MO under various ultrasonic powers with a fixed frequency at 40 kHz. **b** Evolution of MO in condition of various ultrasonic frequencies and fixed power of 600 W. **c** Correlation between temperature and kinetics of degrading MO during ultrasonication. Reproduced with permission from Ref. [95]. Copyright 2022, Elsevier. **d** Evolution of MO absorbance under different revolution speeds of ball milling. **e** Degradation rate of MO under 50 RPM for 600 min. **f** Proposed mechanism for illustrating the existence of speed threshold for initiating CEC in ball milling. Reproduced with permission from Ref. [71]. Copyright 2022, Springer Nature

energy waste, but also because temperature itself is an influencing factor. On the one hand, high temperatures could facilitate target reactions by reducing the activation energy. On the other hand, the temperature could also significantly change the CE ability of employed CEC catalysts that are usually polymers. Figure 8.9c illustrates the variation of CEC efficiency with temperature, showing the highest efficiency when the temperature is between 20 and 30 °C. This can be attributed to the glass transition of polymer materials that can occur when the temperature is as low as 15 °C. Such glass transition is believed to be one of the main reasons for decreased CE ability of polymers at elevated temperatures. This inspires researchers to develop a more effective strategy to initiate CEC with limited heat generation. Moreover, further study on developing novel CEC catalysts that can maintain satisfactory CE characteristics at high temperatures is also very promising.

8.2.5.2 Influencing Factors of Ball Milling on CEC

Ball milling is another widely recognized technique for CEC due to its ability to induce frequent contact-separation cycles even at mild conditions. The revolution speed of the ball mill is a critical parameter that significantly affects the impact mode and energy of the milling process. Higher revolution speeds result in more frequent collisions and more energy is released during such collision process, which is supposed to greatly affect the electron transfer during CE and thus the subsequent CEC process.

Still taking the degradation of MO aqueous solution as the model reaction, Wang et al. [71] have shown that the degradation rate of MO increases dramatically with increasing revolution speed, as is demonstrated in Fig. 8.9d. Specifically, a 19.5-fold increase in the degradation rate was observed when the revolution speed was increased from 100 to 150 rpm. Further increases in revolution speed continue to enhance the degradation rate, highlighting the importance of optimizing this parameter. Moreover, the study also pointed out that there exists a threshold of revolution speeds for initiating CEC. As is depicted in the figure, MO can hardly be degraded at revolution speeds below 100 rpm, suggesting a possible mechanism other than collision frequency. Further experiment reduced the revolution speed to 50 rpm, while extending the duration to five times longer (600 min). As shown in Fig. 8.9e, the degradation rate of MO remained essentially zero, further demonstrating that this low degradation rate under low revolution speed is less dependent of the frequency. The electron-cloud-potential-well model presented in Fig. 8.9f illustrates the role of impact energy in the context of CE. Upon the contact of two interfaces through an applied external force, an overlap of electron clouds at the atomic level is expected to occur. This overlap is anticipated to be more significant when the revolution speed (applied force) increases, lowering the energy barrier and thereby facilitating the transfer of electrons between the atoms. However, below a certain revolution speed, the impact energy is inadequate to overcome the energy barrier necessary for electron transfer, since the energy barrier is high and the input energy is insufficient. In contrast, above this certain speed, the energy barrier is lowered and the input energy is sufficient to drive the electron transfer. This model introduces the concept of the energy barrier in CEC for the first time, highlighting the difference between the prerequisites of CE and CEC. While CE can spontaneously occur under various conditions, the initiation of CEC requires electrons to be excited to a sufficiently high energy level. As a result, the occurrence of CE does not automatically lead to CEC. Instead, certain energy barrier specifications must be satisfied.

8.2.5.3 Other Factors

CEC is also influenced by various factors inherent to the reaction system itself. These include the nature of the surface functional groups on the catalyst, the pH value of the solution, and the addition of supporting agents. Each of these components plays a distinct role in modulating the electron transfer and the generation of reactive ROS, which are central to the CEC process. These factors are crucial for optimizing the CEC system's performance in degrading emerging pollutants.

The efficiency of CEC is significantly modulated by the nature of surface functional groups present on the catalyst. Wang et al. have [48] employed PTFE, PVDF, Nylon-6,6 and NBR Rubber micro-powders as model catalysts with varying surface functional groups. They conducted a comparative experiment to evaluate their CEC performance. The result of UV-Vis and Mass spectra revealed that the catalytic efficiency for Mo degradation followed the order: PTFE > PVDF > Nylon-6,6 > NBR Rubber, which aligns with the electron-withdrawing or electron-donating abilities of their surface functional groups. Specifically, PTFE and PVDF, which are electron-withdrawing materials, achieved higher degradation rate than the electron-donating Nylon-6,6 and NBR Rubber. Due to higher density of electronegative fluorinated function groups (-F), PTFE exhibited the highest MO degradation rate. In addition to pristine polymers, Chen et al. have [66] also reported the influence of surface functionalities on CEC by modifying magnetic nanoparticles with different terminal groups (-F, $-CH_3$, and $-NH_2$). A clear correlation between the electronegativity of the surface groups and the CEC performance was observed. The R-F group, with its high electronegativity, facilitated the strongest CE. This thus enable the most effective degradation of MO, achieving a removal efficiency of over 95% within 3 h. In contrast, the $R-NH_2$ group, which acted as an electron donor rather than an acceptor, exhibited the weakest CEC performance due to its limited ability to initiate the electron transfer process essential for CEC. Moreover, surface functional groups also influence CEC by passivating electronic defects. In a recent study, Li et al. have [96] demonstrated the effectiveness of this approach by modifying the surface of FEP films with functional groups like $-CF_3$ and -O for passivation. Through plasma etching with oxygen and carbon tetrafluoride, they achieved a substantial reduction in deep-level defect density. As a consequence, the kinetic rate of MO degradation in CEC improved by 398%. It is evident that the surface functional groups, with diverse electron-withdrawing or donating nature, are critical to the efficiency of the CEC catalytic processes.

The pH value is another critical parameter in CEC reactions, as it significantly influences the ion concentration in the solution, which in turn impacts the electron transfer process during CEC. Berbille et al. have [97] explored the influence of pH on CEC by conducting parallel experiments with FEP to synthesize H_2O_2 under different pH conditions. The results revealed a significant decrease in the CEC rate when the solution pH deviated from the neutral pH of deionized water. This phenomenon could be attributed to the alteration of ion species and concentrations of ions at different pH levels, which subsequently influenced the CE process between FEP and water. In acidic/alkaline conditions, H^+/OH^- ions are in competition with the FEP surface for electrons, potentially limiting the availability of electrons for CEC reactions and thus reducing overall efficiency. Additionally, the altered charge distribution and structure of the EDL can lead to a less stable interface, further hindering electron transfer. Furthermore, competing redox reactions, such as the reduction of protons to hydrogen gas in acidic conditions and the oxidation of OH- ions to oxygen gas in alkaline conditions, can consume electrons and reduce the overall efficiency. Therefore, controlling the pH of the reaction environment is essential for optimizing the function of CEC catalysts.

Supporting agents also play a crucial role in CEC processes, significantly influencing electron transfer efficiency and reactive oxygen species (ROS) generation. Ma et al. [98] introduced fluorocarbon polymers (FCPs) as supporting agents to activate persulfate (PMS) to degrade emerging pollutant (EP) through CEC. The FCPs/

PMS/US system has exhibited a superior degradation rate, which is markedly higher than that in the PMS/US system without FCPs. This indicates that the incorporation of FCPs leads to a substantial increase in the degradation rate, improving the efficiency of the CEC process. They further conducted comparative experiments to investigate the performance of three different FCPs, including FEP, PTFE and PVDF, in activating PMS for EP degradation within a CEC system. The degradation rate followed the order: FEP > PTFE > PVDF, revealing a positive correlation between the degradation rate and the -F density of FCPs.

8.2.6 Comparison with Tribocatalysis

Under the stimulation of external mechanical forces, the interaction between two surfaces induces a series of physical and chemical effects. This phenomenon lays the foundation for the development of catalytic strategies. Tribocatalysis and CEC represent two distinct approaches within the realm of mechanocatalysis (Fig. 8.10), which harnesses the power of mechanical forces to drive chemical reactions, offering environmentally friendly alternatives to conventional catalytic processes. Tribocatalysis is an approach that primarily capitalizes on the temperature increase or bond breaking caused by intense friction to accelerate chemical reactions. While the CE effect is expected to occur during a standard tribocatalytic process, its influence may be reduced due to the thermionic emission of electrons. Under friction-induced elevated temperature, the majority of the electrons are more likely to be emitted from the interface instead of transferring to the adjacent surface. In contrast, CEC is promoted by CE-driven electron transfer. Given that the CE effect can occur without intense friction during contact, the reaction conditions for CEC are milder. CEC can typically be achieved through simple ultrasonication, whereas tribocatalysis often involves stronger frictional forces [99–101] that may lead to higher local temperatures and more intensive charge separation.

Owing to the differences in their reaction conditions, the choices of catalytic materials are different. CEC can utilize a wide spectrum of materials, including dielectric polymers, oxides, composites, and even liquids. Theoretically, all inorganic and organic materials can potentially be used as CEC catalysts. In contrast, tribocatalysis is more limited to solid materials such as metals or metal oxides, with their ability to withstand the mechanical stress of friction without significant degradation [102–104]. For example, materials like FEP [48] is commonly used in CEC, while metals like palladium [100] and copper [104] are often employed in tribocatalysis. Despite the ability to withstand fierce mechanical stress, the intense friction in tribocatalysis inevitably introduces defects and morphological changes to the catalyst, shortening its lifespan. In contrast, the CEC's mild reaction conditions contribute to the improved reusability of the catalyst, as there is less wear and tear on the material. In addition, CEC's mild operating conditions render it more environmentally friendly, as it does not produce wear particles that can contaminate the reaction medium. For

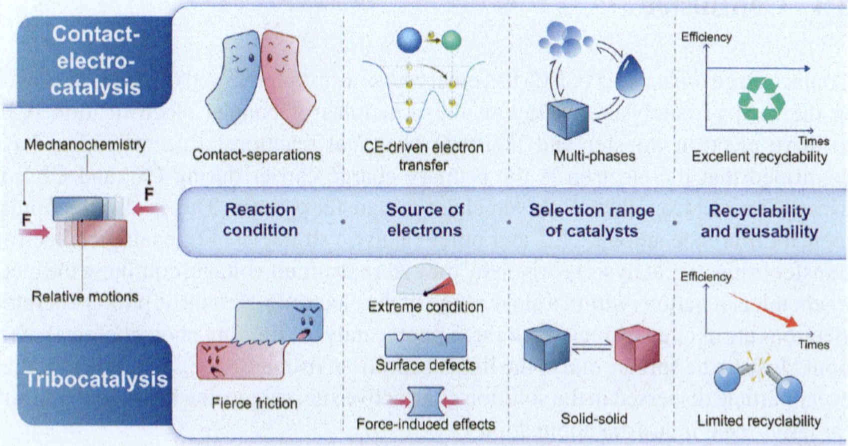

Fig. 8.10 Connection and difference between CEC and tribocatalysis. Reproduced with permission from Ref. [59]. Copyright 2024, Royal Society of Chemistry

instance, the wear particles generated in tribocatalysis could contaminate wastewater and require additional treatment steps, whereas CEC processes minimize such environmental impacts especially when utilizing films [74] or bulk materials [73] as catalysts.

Despite their differences, tribocatalysis and CEC share several similarities. For example, both tribocatalysis [101] and CEC [59] leverage mechanical energy to catalyze chemical reactions. They both involve relative motion at interfaces that leads to charge separation. Moreover, both tribocatalysis and CEC generate reactive species that enhance the rate of chemical reactions. In tribocatalysis, low-energy electrons emitted during friction can interact with molecules adsorbed on the catalyst's surface, resulting in the formation of anion radicals that participate in further chemical reactions [99]. In CEC, the CE-driven electrons directly participate in chemical reactions, leading to the production of reactive oxygen species (ROS) such as hydroxyl and superoxide radicals. Based on those similarities, it is possible to achieve synergistic effects by leveraging the strengths of both approaches. For example, ball milling involves the coexisting contact-separation and fierce friction, allowing the combination of CEC and tribocatalysis. While excessively low frequency of contact-separation may hinder the efficiency of both processes, excessively intense friction may impede CEC through thermionic emission. Therefore, achieving the highest combined efficiency of both catalytic methods is possible through optimizing the revolution speed.

8.3 Conclusion

Contact-electro-catalysis (CEC) has emerged as a sustainable and effective approach in the field of catalysis, leveraging the principles of contact electrification (CE) to drive electron transfer and facilitate chemical reactions. Recent studies have confirmed that the electron is the primary charge carrier during CE, and CEC is exactly derived from the CE-driven electron transfer process. This is also the fundamental difference among CEC and other catalytic strategies. For example, electron transfer in electrocatalysis is driven by the external circuit voltage, confining the electrochemical reactions within a close range of the electrode. Similarly, photogenerated electrons are excited by incident light in photocatalysis, limiting photochemical reactions close to the surface due to the light absorption. In contrast, CEC is able to utilize every particle dispersed in the solution as an active site, suggesting a promising future for large-scale industrial applications.

Owing to the CE effect widely existing among various interfaces (including but not limited to liquid-solid, liquid-liquid, solid-gas, liquid-gas), a broad selection range of catalysts are available for CEC. CEC catalysts are not confined to a certain scale, with powders offering the largest surface area to enhance the efficiency, while films are easier to separate from the system and reusable after cleaning. Therefore, CEC catalysts of different forms can be selected according to the requirements of the reaction. In addition, CEC catalysts are neither confined to a certain category of materials, with polymers exhibiting the best CE properties and oxides can withstand elevated temperatures. More significantly, the recent introduction of liquid catalysts further broadened the selection range of CEC, inspiring attention on the potential CEC phenomena that may occur at various interfaces, laying foundation for liquid-gas and solid-gas CEC.

CEC can be initiated through various mechanical stimuli, with ultrasonication and ball milling being the most prevalent. During ultrasonication, the implosion of numerous cavitation bubbles results in the frequent contact-separation. In these contact-separation cycles, CE-driven electron transfer induces the production of reactive species, accelerating redox reactions. While ultrasonication offering an easy-operation platform for CEC, the heat generation presents a problem as CEC could be hindered under elevated temperature. Therefore, ball-milling is introduced as an energy-efficient approach for CEC. However, ball milling imposes higher requirements on the catalytic material and introduces parameters such as catalyst size that may affect the CEC process. Owing to this, further research into optimizing ball milling CEC is necessary. In addition, the combination and coupling of multiple mechanical stimuli holds promise for the utilization of ubiquitous mechanical forces in nature. Furthermore, combining CEC with electrocatalysis or photocatalysis yields a promising future, as the CEC may activate the electrocatalysts or photocatalysts, allowing an enhanced performance.

In summary, CEC stands as a significant advancement in mechanocatalysis, offering a versatile and eco-friendly avenue for catalyzing chemical reactions. The most significant advantage of CEC lies in its extensive range of catalyst selections.

The universality of CE allows CEC to utilize catalysts with various materials, shapes, structures, and even states. This represents its unique capability, unreported in any other catalytic method. Another significant advantage lies in its mild reaction conditions, which not only conserve energy but also minimize the wear and deactivation of catalysts, reducing secondary pollution. Several key problems about CEC are not yet solved. These include the improvement of CE performance, lack of structural optimization, limited strategies for CEC initiation, and insufficient understanding of CEC mechanisms. To lay the foundation for the further advancements, a roadmap was proposed in Fig. 8.11 [59]. Recently, numerous studies have been proposed in line with this roadmap. For enhancing material performance, new materials like PDA films, modified EPDM rubber have been developed, and the first liquid CEC catalyst has been introduced. In terms of structural optimization, new structures such as polymer/metal Janus composite catalysts and RGO/ZnO nanocomposites have been utilized. For improving CEC initiating methods, auxiliary reagent is used in ultrasonication, and the rapid-developing microdroplet chemistry suggesting spraying as a promising initiating method. In enriching the theoretical framework, computational studies, experimental observations have been conducted. Broadening CEC applications in various chemical reactions also deepens the understanding of its unique mechanism.

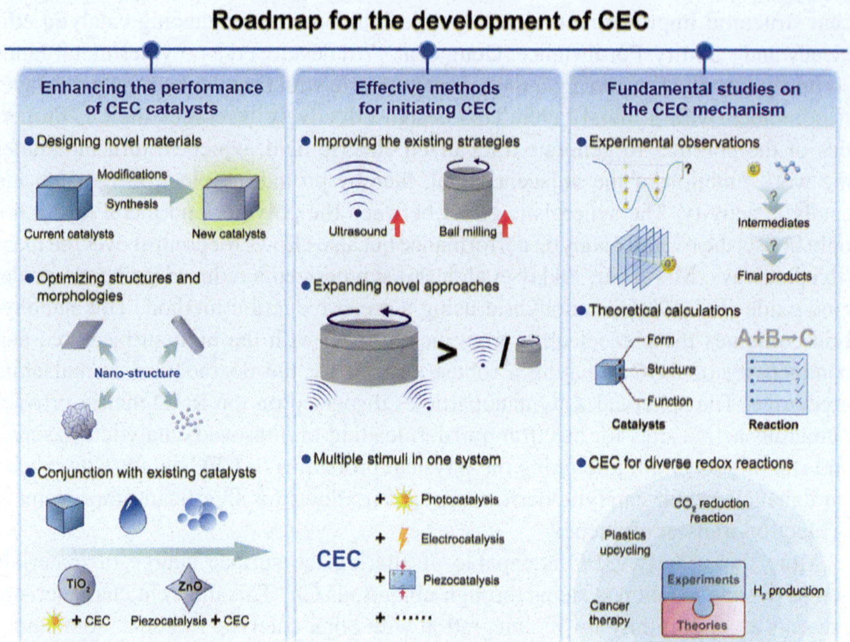

Fig. 8.11 Roadmap of contact-electro-catalysis. Reproduced with permission from Ref. [59]. Copyright 2024, Royal Society of Chemistry

8.3.1 Enhancement of Catalytic Performance of CEC Catalysts

There are multiple approaches to enhance the catalytic performance of CEC catalysts. This involves the development of innovative materials, optimization of structural/morphological characteristics, and integration with established catalysts.

In recent years, the development of novel materials for CEC has garnered worldwide attention, aiming to enhance catalytic efficiency and broaden the applicability of this technology. For example, Jia et al. [105] used polydopamine (PDA) films in CEC, leveraging their photosensitive and reversible redox properties to enhance the generation of reactive oxygen species (ROS) and improve substrate conversion rates. Moreover, Liu et al. [75] developed modified ethylene-propylene-diene monomer (EPDM) rubber, which effectively degrades pollutants like tetracycline due to its excellent dielectric properties and stability. More importantly, Li et al. [78] utilized liquid-liquid CE-driven electron transfer between perfluorocarbon-water interfaces to generate ROS for cancer therapy, broadening CEC catalysts to liquids. Each of these materials offers unique advantages for the adaptation to various reaction conditions.

Finetuning the physical properties of catalysts is essential for optimizing their catalytic performance. Recent advancements in CEC have also witnessed significant structural improvements in catalyst design, aimed at enhancing catalytic efficiency and stability. For instance, Dong et al. [76] developed a polymer/metal Janus composite catalyst, featuring a unique asymmetric structure where a polymer layer is combined with a metal layer. This design effectively leverages the CE properties of the polymer to generate a localized electric field, which in turn modulates the work function of the adjacent metal, thereby promoting electron transfer and catalytic activity. The synergistic effect between the polymer and metal layers not only boosts the overall catalytic performance but also allows for control over the reaction pathways. Moreover, Yadav et al. [106] synthesized a reduced graphene oxide/zinc oxide (RGO/ZnO) nanohybrid using a green reduction method. This nanohybrid combines the piezoelectric properties of ZnO with the high surface area and conductivity of RGO, creating a robust catalyst for the degradation of malachite green dye. The spherical ZnO nanoparticles dispersed on the RGO matrix provide numerous active sites for electron transfer, leading to improved catalytic efficiency and stability. Overall, finetuning the physical properties of CEC catalysts is crucial for enhancing their catalytic performance, as it allows for significant improvement in electron transfer efficiency.

More importantly, CEC is capable of altering the surface charge of materials across diverse reaction systems through ubiquitous CE. This inherent characteristic presents an opportunity for its integration with other catalytic systems, broadening the application of CEC in various scenarios. For instance, Li et al. [77] reported a significant advancement in the degradation of methyl orange dye using a novel Fe^{3+}-initiated self-cycling Fenton system integrated with CEC by loading PTFE onto ZSM-5 as a catalyst. While the traditional Fenton reaction is a well-established advanced oxidation process (AOP) that utilizes ferric (Fe^{3+}) or ferrous (Fe^{2+}) ions to catalyze the decomposition of H_2O_2 into highly reactive ·OH, this CEC-integrated system demonstrated a MO degradation rate six times faster than traditional CEC systems, achieving a fastest known ultrasound-induced degradation rate of nearly 99% within 10 min. The integration of CEC with various catalytic systems showcases its potential in improving reaction efficiency, and expanding the range of applications.

8.3.2 *Effective Strategies for Initiating CEC*

Developing effective strategies for initiating CEC is also important. This encompasses the refinement of current methodologies, the exploration of novel approaches, and the implementation of multiple stimuli within a single system.

The continuous improvement of existing methodologies is crucial for enhancing the efficiency and sustainability of CEC processes. For example, a recent research by Ma et al. [98] explored the enhancement of ultrasonic CEC systems through the addition of auxiliary agents in PFAS degradation and defluorination. Compared to systems without auxiliary agents, the inclusion of ROS generators not only increased the yield of reactive species but also facilitated the formation of a stronger interfacial electrostatic field, thereby accelerating the cleavage of C-F bonds in PFAS molecules. This research further involved a series of experiments comparing impacts of different auxiliary agents on the system. This work innovates the use of auxiliary agents in more CEC reactions.

Apart from ultrasonication or ball milling, entirely new initiation approaches may greatly expand the application scenarios of CEC. The emerging field of microdroplet chemistry offers a compelling inspiration, suggesting that utilizing sprays might serve as an effective method to initiate CEC. This proposition is supported by the unique properties of microdroplets, such as their high surface-to-volume ratio and the intense electric fields generated at the droplet interfaces. These characteristics facilitate efficient charge separation and electron transfer, which are essential processes for CEC. Furthermore, droplet formation and collision in sprays can induce frequent contact and separation events, enhancing the likelihood of charge generation and subsequent chemical reactions. Thus, integrating microdroplet chemistry with CEC could open new avenues for realizing mild and efficient chemical transformations under ambient conditions.

Another notable point is that incorporating multiple stimuli can significantly enhance the applicability of the catalytic processes. For instance, Song et al. [107] have successfully combined the effects of mechanical forces and visible light in a CEC system based on droplet triboelectric nanogenerators (TENGs), resulting in the efficient degradation of organic dyes. Similarly, the integration of ultrasonic agitation with continuous stirring, as demonstrated by Wang et al. [73], has led to a boost in hydrogen peroxide (H_2O_2) production. These approaches highlight the potential of integrating diverse stimuli to achieve synergistic effects in catalytic activities, thereby opening new avenues for the development of efficient and eco-friendly catalytic technologies.

The pursuit of effective strategies for initiating CEC unlocks the full potential of this technology. Refining existing methodologies has demonstrated significant improvements in efficiency, as evidenced by the enhanced PFAS degradation. Moreover, the exploration of novel initiation approaches promises to expand the application scope of CEC. Furthermore, the integration of multiple stimuli within a single CEC system has proven to be a powerful strategy for achieving synergistic enhancements in catalytic performance. These advances not only broaden the applicability of CEC but also pave the way for the development of more sustainable and efficient chemical processes.

8.3.3 Fundamental Investigations of the CEC Mechanism

To deepen the understanding of the CEC mechanism, a series of fundamental investigations have been conducted. These investigations focus on theoretical calculations, experimental observations, and the exploration of CEC applications in various redox reactions.

Theoretical calculations have provided valuable insights into how materials perform and interact during CEC. For example, Wang et al. [108] conducted a series of computational studies focusing on the impact of frontier molecular orbitals (FMOs) on the electron transfer process during CE. Through the calculation of vertical attachment energies (VAEs), this work highlighted that polymers with a lower LUMO require less energy for electron transfer, thereby enhancing the generation of ROS crucial for catalytic reactions. Moreover, Li et al. [96] took a different approach by investigating the impact of electronic defects on the performance of fluorinated polymers in liquid-solid CE and CEC. Using first-principle calculations, they showed that adding specific surface functional groups to FEP can adjust the positions of molecular orbitals and remove deep-level defect states from the HOMO-LUMO gap. This modification improves the alignment of energy levels, making electron transfer more efficient. By simulating the electronic structures and energy levels of materials, key factors that influence electron transfer and catalytic activity can be identified. These theoretical calculations not only help in explaining observed phenomena but also provide a basis for designing materials with tailored properties for specific applications.

Moreover, numerous experiments have been done in order to elucidate the fundamental principles and mechanisms of CEC. For instance, Berbille et al. [97] performed a series of experiments to systematically investigate the process of CEC in generating H_2O_2 through quantitative detection methods, such as nuclear magnetic resonance (NMR) spectroscopy. They tested several dielectric polymers (FEP, PTFE, PVDF, and HDPE) on their performances in H_2O_2 production. They also examined the effect of diverse amount of catalyst, demonstrating that reducing the catalyst weight from 10 to 5 mg could increase the yield per gram of catalyst by 56%, suggesting an optimal economic use of the catalyst. Furthermore, the team conducted control experiments without any catalyst to establish baseline levels of H_2O_2 production, which helped in assessing the effectiveness of the catalysts. These experiments collectively provided valuable insights into optimizing the conditions for H_2O_2 generation via CEC.

Last but not least, recent studies have demonstrated the potential of CEC in facilitating a wide array of chemical reactions. For instance, Cao et al. [109] utilized CEC to degrade trace antibiotics in wastewater, demonstrating the generation of ROS such as ·OH and $1O_2$ for efficient antibiotic removal. Moreover, Li et al. [110] applied CEC to degrade pentachlorophenol (PCP) using FEP as a catalyst, achieving dechlorination and oxidative degradation through the formation of ROS. Additionally, Li et al. [78] employed CEC at a liquid-liquid interface to generate ROS for cancer therapy, effectively triggering immunogenic cell death and activating antitumor immune responses. Furthermore, CEC is also utilized in inactivating Microcystis aeruginosa [72], reducing CO_2 [111], and controlling membrane fouling [74]. These rapid growing applications of CEC highlight its superior practicability.

By focusing on these fundamental investigations—conducting experimental observations, performing theoretical calculations, and exploring CEC applications in various redox reactions—researchers can gain a comprehensive understanding of the CEC mechanism. This knowledge will guide the development of more efficient and sustainable catalytic systems for energy conversion and environmental remediation.

8.4 Summary and Perspectives

Recent breakthroughs are in accord with the comprehensive roadmap. Strategies to enhance the performance of CEC catalysts encompass the development of innovative materials, optimization of catalyst structures, and integration with other catalytic systems. By leveraging new materials such as PDA films and EPDM rubber, along with the unique structural design of polymer/metal Janus composites and RGO/ZnO nanohybrids, the ROS generation efficiency have been significantly enhanced. Additionally, exploring effective CEC initiation strategies is crucial, which includes the improvement of existing methods, exploration of new approaches like spraying, and the integration of multiple stimuli, such as combing mechanical forces with visible light within a system. These strategies have greatly improved the efficiency and application range of CEC. In terms of basic research, a deeper understanding of CEC mechanism has been achieved through theoretical calculations and experimental observations. These studies not only explain the observed phenomena but also provide a foundation for designing materials with specific properties. The application of CEC in various reactions such as wastewater treatment, cancer therapy, and CO_2 reduction further demonstrates its practicality.

Looking forward, various new breakthroughs will further enhance the efficiency of CEC and expand its applications. Moreover, fundamental research on CEC also provides a new perspective for researchers to understand many phenomena in nature.

For instance, the ubiquity of CE in water bodies may lead to the occurrence of CEC, which could offer an explanation why running water does not decay. Additionally, CEC could be closely related to the origin of life on earth as CE could be a driving forth for the formation of primitive protein and nucleic acid molecules.

References

1. D. Linke, Johann Wolfgang Döbereiner (1780–1849) —Chemiker, Pharmazeut und Technologe. Nachr. Chem. Tech. Lab. **47**(3), 326–327 (2010)
2. G. Ertl, Wilhelm Ostwald: founder of physical chemistry and Nobel Laureate 1909. Angew. Chem. Int. Ed. **48**(36), 6600–6606 (2009)
3. L.G. Otten, C.F. Sio, A.M. van der Sloot, R.H. Cool, W.J. Quax, Mutational analysis of a key residue in the substrate specificity of a cephalosporin acylase. ChemBioChem **5**(6), 820–825 (2004)
4. S.S.Y. Ho, P.L. Chan, P.K. Wong, G.E. Antonio, K.T. Wong, D.J. Lyon, K.S.C. Fung, C.K. Li, A.F.B. Cheng, A.T. Ahuja, Perspective—eye of the storm: the roles of a radiology department in the outbreak of severe acute respiratory syndrome. AJR **181**(1), 19–24 (2003)
5. E.A. Karakhanov, A.L. Maksimov, Biphasic catalysis in petrochemical processes. Russ. J. Gen. Chem. **79**(6), 1370–1383 (2009)
6. P.A. Sermon, Catalysis of the petrochemical economy. Chem. Ind. **20**, 681–690 (1979)
7. M. Che, Nobel Prize in chemistry 1912 to Sabatier: organic chemistry or catalysis? Catal. Today **218–219**, 162–171 (2013)
8. A.A. Balandin, Modern state of the multiplet theor of heterogeneous catalysis. Adv. Catal. **19**, 1–210 (1969)
9. T. von Muenchow, S. Dana, Y. Xu, B. Yuan, L. Ackermann, Enantioselective electrochemical cobalt-catalyzed aryl C-H activation reactions. Science **379**(6636), 1036–1042 (2023)
10. V.R. Stamenkovic, D. Strmcnik, P.P. Lopes, N.M. Markovic, Energy and fuels from electrochemical interfaces. Nat. Mater. **16**(1), 57–69 (2017)
11. J.H. Montoya, C. Tsai, A. Vojvodic, J.K. Norskov, The Challenge of electrochemical ammonia synthesis: a new perspective on the role of nitrogen scaling relations. Chemsuschem **8**(13), 2180–2186 (2015)
12. V. Percec, T. Guliashvili, J.S. Ladislaw, A. Wistrand, A. Stjerndahl, M.J. Sienkowska, M.J. Monteiro, S. Sahoo, Ultrafast synthesis of ultrahigh molar mass polymers by metal-catalyzed living radical polymerization of acrylates, methacrylates, and vinyl chloride mediated by SET at 25 °C. JACS **128**(43), 14156–14165 (2006)
13. T.F. Degnan, Recent progress in the development of zeolitic catalysts for the petroleum refining and petrochemical manufacturing industries, in *15th International Zeolite Conference*, Beijing, People's Republic of China, Vol. 170 (2007), pp. 54–65
14. J.Y. Cha, T.W. Kim, J.H. Choi, K.S. Jang, L. Khaleda, W.Y. Kim, J.R. Jeon, Fungal laccase-catalyzed oxidation of naturally occurring phenols for enhanced germination and salt tolerance of *Arabidopsis thaliana*: a green route for synthesizing humic-like fertilizers. J. Agric. Food Chem. **65**(6), 1167–1177 (2017)
15. Y.N. Shang, X. Xu, B.Y. Gao, S.B. Wang, X.G. Duan, Single-atom catalysis in advanced oxidation processes for environmental remediation. Chem. Soc. Rev. **50**(8), 5281–5322 (2021)
16. J.K. Lin, W.J. Tian, Z.Y. Guan, H.Y. Zhang, X.G. Duan, H. Wang, H.Q. Sun, Y.F. Fang, Y.P. Huang, S.B. Wang, Functional carbon nitride materials in photo-fenton-like catalysis for environmental remediation. Adv. Funct. Mater. **32**(24), 2201743 (2022)
17. S.Y. Fang, Y.H. Hu, Thermo-photo catalysis: a whole greater than the sum of its parts. Chem. Soc. Rev. **51**(9), 3609–3647 (2022)

18. D.Y. Tang, G.L. Lu, Z.W. Shen, Y.Z. Hu, L. Yao, B.F. Li, G.X. Zhao, B.X. Peng, X.B. Huang, A review on photo-, electro- and photoelectro-catalytic strategies for selective oxidation of alcohols. J. Energy Chem. **77**, 80–118 (2023)
19. M. Hassan, Y. Slimani, M.A. Gondal, M.J.S. Mohamed, S. Guener, M.A. Almessiere, A.M. Surrati, A. Baykal, S. Trukhanov, A. Trukhanov, Structural parameters, energy states and magnetic properties of the novel Se-doped $NiFe_2O_4$ ferrites as highly efficient electrocatalysts for HER. Ceram. Int. **48**(17), 24866–24876 (2022)
20. F. Zeng, C. Mebrahtu, L.F. Liao, R. Palkovits, A.K. Beine, Stability and deactivation of OER electrocatalysts: a review. J. Energy Chem. **69**, 301–329 (2022)
21. L.J. Yang, J.L. Shui, L. Du, Y.Y. Shao, J. Liu, L.M. Dai, Z. Hu, Carbon-based metal-free ORR electrocatalysts for fuel cells: past, present, and future. Adv. Mater. **31**(13), 1804799 (2019)
22. Z.F. Xin, J.J. Liu, X.J. Wang, K.J. Shen, Z.B. Yuan, Y.F. Chen, Y.Q. Lan, Implanting polypyrrole in metal-porphyrin MOFs: enhanced electrocatalytic performance for CO_2RR. ACS Appl. Mater. Interfaces **13**(46), 54949–54956 (2021)
23. Q.Q. Li, Y.L. Guo, Y. Tian, W.M. Liu, K. Chu, Activating VS_2basal planes for enhanced NRR electrocatalysis: the synergistic role of S-vacancies and B dopants. J. Mater. Chem. A **8**(32), 16195–16202 (2020)
24. L. Li, P. Wang, Q. Shao, X. Huang, Metallic nanostructures with low dimensionality for electrochemical water splitting. Chem. Soc. Rev. **49**(10), 3072–3106 (2020)
25. N. Mahmood, Y. Yao, J.W. Zhang, L. Pan, X. Zhang, J.J. Zou, Electrocatalysts for hydrogen evolution in alkaline electrolytes: mechanisms, challenges, and prospective solutions. Adv. Sci. **5**(2), 1700464 (2018)
26. S. Cherevko, A.R. Zeradjanin, A.A. Topalov, N. Kulyk, I. Katsounaros, K.J.J. Mayrhofer, Dissolution of noble metals during oxygen evolution in acidic media. ChemCatChem **6**(8), 2219–2223 (2014)
27. J.Y. Han, J.R. Sun, S.Y. Chen, S.Y. Zhang, L.L. Qi, A. Husile, J.Q. Guan, Structure-activity relationships in oxygen electrocatalysis. Adv. Mater. **36**(47), 2408139 (2024)
28. A. Fujishima, K. Honda, Electrochemical photolysis of water at a semiconductor electrode. Nature **238**(5358), 37–38 (1972)
29. Q. Wang, K. Domen, Particulate photocatalysts for light-driven water splitting: mechanisms, challenges, and design strategies. Chem. Rev. **120**(2), 919–985 (2020)
30. E. Gong, S. Ali, C.B. Hiragond, H.S. Kim, N.S. Powar, D. Kim, H. Kim, S.I. In, Solar fuels: research and development strategies to accelerate photocatalytic CO_2 conversion into hydrocarbon fuels. Energy Environ. Sci. **15**(3), 880–937 (2022)
31. L.Q. Xiong, J.W. Tang, Strategies and challenges on selectivity of photocatalytic oxidation of organic substances. Adv. Energy Mater. **11**(8), 2003216 (2021)
32. V. Hasija, V.H. Nguyen, A. Kumar, P. Raizada, V. Krishnan, A.A.P. Khan, P. Singh, E. Lichtfouse, C.Y. Wang, P.T. Huong, Advanced activation of persulfate by polymeric g-C_3N_4 based photocatalysts for environmental remediation: a review. J. Hazard. Mater. **413**, 125324 (2021)
33. Q.Q. Lin, G.Y. Zeng, G.L. Yan, J.Q. Luo, X.J. Cheng, Z.Y. Zhao, H. Li, Self-cleaning photocatalytic MXene composite membrane for synergistically enhanced water treatment: oil/water separation and dyes removal. Chem. Eng. J. **427**, 131668 (2022)
34. Z. Liang, C.F. Yan, S. Rtimi, J. Bandara, Piezoelectric materials for catalytic/photocatalytic removal of pollutants: recent advances and outlook. Appl. Catal. B-Environ. **241**, 256–269 (2019)
35. K.S. Hong, H.F. Xu, H. Konishi, X.C. Li, Direct water splitting through vibrating piezoelectric microfibers in water. J. Phys. Chem. Lett. **1**(6), 997–1002 (2010)
36. M.B. Starr, J. Shi, X. Wang, Piezopotential-driven redox reactions at the surface of piezoelectric materials. Angew. Chem. Int. Ed. **51**(24), 5962–5966 (2012)
37. J. Wu, N. Qin, D.H. Bao, Effective enhancement of piezocatalytic activity of $BaTiO_3$ nanowires under ultrasonic vibration. Nano Energy **45**, 44–51 (2018)
38. X.X. Chen, J.X. Wang, Z.C. Wang, H.W. Xu, C. Liu, B.J. Huo, F.Q. Meng, Y.L. Wang, C.Y. Sun, Low-frequency mechanical energy in the environment for energy production and piezocatalytic degradation of organic pollutants in water: a review. J. Water Process Eng. **56**, 104312 (2023)

39. T. Hu, J.X. Feng, W. Yan, S.H. Tian, J.X. Sun, X.S. Liu, D. Wei, Z.M. Wang, Y. Yu, J.C.H. Lam, S.R. Liu, Z.L. Wang, Y. Xiong, Piezocatalysis for chemical-mechanical polishing of SiC: dual roles of t-$BaTiO_3$ as a piezocatalyst and an abrasive. Small **20**(21), 2310117 (2024)
40. Y. Zhao, T. Huang, X. Zhang, Y. Cui, L. Zhang, L. Li, Z.L. Wang, Piezotronic and piezo-phototronic effects on sonodynamic disease therapy. Bmemat **1**(1), e12006 (2023)
41. R. Su, A.L. Hsain, M. Wu, D.W. Zhang, X.H. Hu, Z.P. Wang, X.J. Wang, F.T. Li, X.M. Chen, L.N. Zhu, Y. Yang, Y.D. Yang, X.J. Lou, S.J. Pennycook, Nano-ferroelectric for high efficiency overall water splitting under ultrasonic vibration. Angew. Chem. Int. Ed. **58**(42), 15076–15081 (2019)
42. P.T.T. Phuong, D.V.N. Vo, N.P.H. Duy, H. Pearce, Z.M. Tsikriteas, E. Roake, C. Bowen, H. Khanbareh, Piezoelectric catalysis for efficient reduction of CO_2 using lead-free ferroelectric particulates. Nano Energy **95**, 107032 (2022)
43. K. Kubota, Y.D. Pang, A. Miura, H. Ito, Redox reactions of small organic molecules using ball milling and piezoelectric materials. Science **366**(6472), 1500–1504 (2019)
44. Y. He, G.F. Wang, W.B. Hu, D.D. Wei, J.J. Jia, H. Li, B.X. Yuan, Piezocatalyzed decarboxylative acylation of quinoxalin-2(1*H*)-ones using ball milling. ACS Sustain. Chem. Eng. **11**(3), 910–920 (2023)
45. F.R. Fan, Z.Q. Tian, Z.L. Wang, Flexible triboelectric generator! Nano Energy **1**(2), 328–334 (2012)
46. J. Zhang, S. Lin, M. Zheng, Z.L. Wang, Triboelectric nanogenerator as a probe for measuring the charge transfer between liquid and solid surfaces. ACS Nano **15**(9), 14830–14837 (2021)
47. S. Lin, L. Xu, A. Chi Wang, Z.L. Wang, Quantifying electron-transfer in liquid-solid contact electrification and the formation of electric double-layer. Nat. Commun. **11**(1), 399 (2020)
48. Z. Wang, A. Berbille, Y. Feng, S. Li, L. Zhu, W. Tang, Z.L. Wang, Contact-electro-catalysis for the degradation of organic pollutants using pristine dielectric powders. Nat. Commun. **13**(1), 130 (2022)
49. J. Zhao, X. Zhang, J. Xu, W. Tang, Z.L. Wang, F.R. Fan, Contact-electro-catalysis for direct synthesis of H_2O_2 under ambient conditions. Angew. Chem. Int. Ed. **62**(21), 202300604 (2023)
50. H. Li, A. Berbille, X. Zhao, Z. Wang, W. Tang, Z.L. Wang, A contact-electro-catalytic cathode recycling method for spent lithium-ion batteries. Nat. Energy **8**(10), 1137–1144 (2023)
51. J. Li, Y. Xia, X.W. Song, B.L. Chen, R.N. Zare, Continuous ammonia synthesis from water and nitrogen via contact electrification. Proc. Natl. Acad. Sci. U.S.A. **121**(4), e2318408121 (2024)
52. E. Roduner, Understanding catalysis. Chem. Soc. Rev. **43**(24), 8226–8239 (2014)
53. C.L. Wang, J. Wang, J.K. Jin, B. Li, Y.L. Phang, F.L. Zhang, T. Ye, H.M. Xia, L.W. Hui, J.H. Su, Y. Fu, Y.F. Wang, Boryl radical catalysis enables asymmetric radical cycloisomerization reactions. Science **382**(6674), 1056–1065 (2023)
54. M. Zhang, K. Zhang, X. Ai, X. Liang, Q. Zhang, H. Chen, X. Zou, Theory-guided electrocatalyst engineering: from mechanism analysis to structural design. Chin. J. Catal. **43**(12), 2987–3018 (2022)
55. Y. Ding, S. Maitra, S. Halder, C. Wang, R. Zheng, T. Barakat, S. Roy, L.H. Chen, B.L. Su, Emerging semiconductors and metal-organic-compounds-related photocatalysts for sustainable hydrogen peroxide production. Matter **5**(7), 2119–2167 (2022)
56. Z. Kang, N. Qin, E. Lin, J. Wu, B. Yuan, D. Bao, Effect of Bi_2WO_6 nanosheets on the ultrasonic degradation of organic dyes: Roles of adsorption and piezocatalysis. J. Cleaner Prod. **261**, 121125 (2020)
57. G. Singh, M. Sharma, R. Vaish, Transparent ferroelectric glass-ceramics for wastewater treatment by piezocatalysis. Commun. Mater. **1**(1), 100 (2020)
58. X. Liao, H. Xie, B. Liao, S. Hou, Y. Yu, X. Fan, Ball milling induced strong polarization electric fields in $Cu_3B_2O_6$ crystals for high efficiency piezocatalysis. Nano Energy **94**, 106890 (2022)
59. Z. Wang, X. Dong, W. Tang, Z.L. Wang, Contact-electro-catalysis (CEC). Chem. Soc. Rev. **53**(9), 4349–4373 (2024)

60. S. Lin, L. Xu, L. Zhu, X. Chen, Z.L. Wang, Electron transfer in nanoscale contact electrification: photon excitation effect. Adv. Mater. **31**(27), 1901418 (2019)
61. S. Lin, L. Zhu, Z. Tang, Z.L. Wang, Spin-selected electron transfer in liquid-solid contact electrification. Nat. Commun. **13**(1), 5230 (2022)
62. Z.L. Wang, A.C. Wang, On the origin of contact-electrification. Mater. Today **30**, 34–51 (2019)
63. Z.H. Loh, G. Doumy, C. Arnold, L. Kjellsson, S.H. Southworth, A. Al Haddad, Y. Kumagai, M.F. Tu, P.J. Ho, A.M. March, R.D. Schaller, M.S. B.M. Yusof, T. Debnath, M. Simon, R. Welsch, L. Inhester, K. Khalili, K. Nanda, A.I. Krylov, S. Moeller, G. Coslovich, J. Koralek, M.P. Minitti, W.F. Schlotter, J.E. Rubensson, R. Santra, L. Young, Observation of the fastest chemical processes in the radiolysis of water. Science **367** (6474), 179–182 (2020)
64. J. Ma, F. Wang, M. Mostafavi, Ultrafast chemistry of water radical cation, $H_2O^{\cdot+}$, in aqueous solutions. Molecules **23**(2), 244 (2018)
65. Y. Zhao, Y. Liu, Y. Wang, S. Li, Y. Liu, Z.L. Wang, P. Jiang, The process of free radical generation in contact electrification at solid-liquid interface. Nano Energy **112**, 108464 (2023)
66. Z. Chen, Y. Lu, X. Liu, J. Li, Q. Liu, Novel magnetic catalysts for organic pollutant degradation via contact electro-catalysis. Nano Energy **108**, 108198 (2023)
67. L. Lin, V.G.N. Thyagaraja, R. Ranjith, R. Yang, S. Ciampi, J. Chen, J. Liu, Degradation of organic molecules by tribovoltaic mechano-chemistry. Nano Energy **107**, 108163 (2023)
68. Y. Zhang, T. Kang, X. Han, W. Yang, W. Gong, K. Li, Y. Guo, Molecular-functionalized metal-organic frameworks enabling contact-electro-catalytic organic decomposition. Nano Energy **111**, 108433 (2023)
69. B. Jiang, X. Xue, Z. Mu, H. Zhang, F. Li, K. Liu, W. Wang, Y. Zhang, W. Li, C. Yang, K. Zhang, Contact-piezoelectric bi-catalysis of an electrospun ZnO@PVDF composite membrane for dye decomposition. Molecules **27**(23), 8579 (2022)
70. Y. Su, A. Berbille, X.F. Li, J. Zhang, M. PourhosseiniAsl, H. Li, Z. Liu, S. Li, J. Liu, L. Zhu, Z.L. Wang, Reduction of precious metal ions in aqueous solutions by contact-electro-catalysis. Nat. Commun. **15**(1), 4196 (2024). From NLM PubMed-not-MEDLINE
71. Z. Wang, X. Dong, X.F. Li, Y. Feng, S. Li, W. Tang, Z.L. Wang, A contact-electro-catalysis process for producing reactive oxygen species by ball milling of triboelectric materials. Nat. Commun. **15**(1), 757 (2024)
72. P. Wei, M. Tang, Y. Wang, B. Hu, X. Qu, Y. Wang, G. Gao, Low-frequency ultrasound assisted contact-electro-catalysis for efficient inactivation of Microcystis aeruginosa. J. Hazard. Mater. **478**, 135537 (2024). From NLM Medline
73. Y. Wang, Y.F. Wang, B.W. Hu, M.Q. Qiu, G.D. Gao, P.Y. Wei, Catalyst-free contact-electro-catalytic H_2O_2 synthesis via simple combination of a poly(tetrafluoroethylene) stir bar and ultrasound. Chem. Commun. **60**(57), 7331–7334 (2024)
74. H. Liu, Y. Wang, Contact-electro-catalysis-assisted separation via a dancing PTFE membrane for fouling control. ACS Appl. Mater. Interfaces **16**(1), 1826–1836 (2024). From NLM PubMed-not-MEDLINE
75. C. Liu, L. Zhao, J. Li, J. Wang, H. Xu, X. Chen, J. Qi, C. Sun, Z. Zhu, Y. Wang, F. Meng, Peroxymonosulfate activation by low-cost modified rubber during contact electrification for antibiotics efficient degradation at circumneutral pH: mechanism and toxicity assessment. Chem. Eng. Sci. **285**, 119642 (2024)
76. X. Dong, Z. Wang, Y. Hou, Y. Feng, A. Berbille, H. Li, Z.L. Wang, W. Tang, Regulating contact-electro-catalysis using polymer/metal Janus composite catalysts. JACS **146**(41), 28110–28118 (2024)
77. W. Li, J. Tu, J. Sun, Y. Zhang, J. Fang, M. Wang, X. Liu, Z.-Q. Tian, F. Ru Fan, Boosting reactive oxygen species generation via contact-electro-catalysis with Fe^{III}-initiated self-cycled fenton system. Angew. Chem. Int. Ed. e202413246 (2024)
78. H.M. Li, Z.C. Wang, X. Chu, Y. Zhao, G.Q. He, Y.L. Hu, Y. Liu, Z.L. Wang, P. Jiang, Free radicals generated in perfluorocarbon-water (liquid-liquid) interfacial contact electrification and their application in cancer therapy. JACS **146**(17), 12087–12099 (2024)
79. F. Wang, P. Yang, X. Tao, Y. Shi, S. Li, Z. Liu, X. Chen, Z.L. Wang, Study of contact electrification at liquid-gas interface. ACS Nano **15**(11), 18206–18213 (2021)

80. X. Li, X. Nong, C. Zhu, X. Gao, H. Chen, X. Yuan, D. Xing, L. Liu, C. Liang, D. Zang, X. Zhang, Atomization by acoustic levitation facilitates contactless microdroplet reactions. JACS **146**(43), 29267–29271 (2024)
81. X. Song, C. Basheer, R.N. Zare, Making ammonia from nitrogen and water microdroplets. Proc. Natl. Acad. Sci. **120**(16), e2301206120 (2023)
82. M. Gągol, A. Przyjazny, G. Boczkaj, Wastewater treatment by means of advanced oxidation processes based on cavitation—a review. Chem. Eng. J. **338**, 599–627 (2018)
83. K.S. Suslick, Sonochemistry. Science **247**(4949), 1439–1445 (1990)
84. X. Chen, Y. Xia, Z. Zhang, L. Hua, X. Jia, F. Wang, R.N. Zare, Hydrocarbon degradation by contact with anoxic water microdroplets. JACS **145**(39), 21538–21545 (2023)
85. X. Song, C. Basheer, Y. Xia, J. Li, I. Abdulazeez, A.A. Al-Saadi, M. Mofidfar, M.A. Suliman, R.N. Zare, One-step formation of urea from carbon dioxide and nitrogen using water microdroplets. JACS **145**(47), 25910–25916 (2023)
86. P. Lenard, Ueber die Electricität der Wasserfälle. Ann. Phys. **282**(8), 584–636 (1892)
87. S.Q. Lin, L.N.Y. Cao, Z. Tang, Z.L. Wang, Size-dependent charge transfer between water microdroplets. Proc. Natl. Acad. Sci. U.S.A. **120**(31), e2307977120 (2023)
88. Y. Xia, J.H. Xu, J. Li, B.L. Chen, Y.F. Dai, R.N. Zare, Visualization of the charging of water droplets sprayed into air. J. Phys. Chem. A **128**(28), 5684–5690 (2024)
89. X. Song, Y. Meng, R.N. Zare, Spraying water microdroplets containing 1,2,3-triazole converts carbon dioxide into formic acid. JACS **144**(37), 16744–16748 (2022)
90. Y. Meng, R.N. Zare, E. Gnanamani, One-step, catalyst-free formation of phenol from benzoic acid using water microdroplets. JACS **145**(35), 19202–19206 (2023)
91. M.A. Mehrgardi, M. Mofidfar, R.N. Zare, Sprayed water microdroplets are able to generate hydrogen peroxide spontaneously. JACS **144**(17), 7606–7609 (2022)
92. L.L. Zhao, X.W. Song, C. Gong, D.M. Zhang, R.J. Wang, R.N. Zare, X.X. Zhang, Sprayed water microdroplets containing dissolved pyridine spontaneously generate pyridyl anions. Proc. Natl. Acad. Sci. U.S.A. **119**(12), e2200991119 (2022)
93. L.Q. Qiu, R.G. Cooks, Simultaneous and spontaneous oxidation and reduction in microdroplets by the water radical cation/anion pair. Angew. Chem. Int. Ed. **61**(41), e202210765 (2022)
94. X.K. Chen, Y. Xia, Y.F. Wu, Y.P. Xu, X.Q. Jia, R.N. Zare, F. Wang, Sprayed oil-water microdroplets as a hydrogen source. JACS **146**(15), 10868–10874 (2024)
95. X. Dong, Z. Wang, A. Berbille, X. Zhao, W. Tang, Z.L. Wang, Investigations on the contact-electro-catalysis under various ultrasonic conditions and using different electrification particles. Nano Energy **99**, 107346 (2022)
96. X.F. Li, A. Berbille, T.Y. Wang, X. Zhao, S. Li, Y. Su, H. Li, G. Zhang, Z. Wang, L. Zhu, J. Liu, Z.L. Wang, Defect Passivation Toward Designing high-performance fluorinated polymers for liquid-solid contact-electrification and contact-electro-catalysis. Adv. Funct. Mater. **34**(25), 2315817 (2024)
97. A. Berbille, X.F. Li, Y. Su, S. Li, X. Zhao, L. Zhu, Z.L. Wang, Mechanism for generating H_2O_2 at water-solid interface by contact-electrification. Adv. Mater. **35**(46), 2304387 (2023)
98. D. Ma, W. Li, J. Zhang, K. He, C. Zhang, G. Wang, X. Xin, Q. Liu, F. Cheng, S. Lv, D. Xing, Fluorocarbon polymers mediated contact-electro-catalysis activating peroxymonosulfate for emerging pollutants degradation: the key role of fluorine density in electron transfer. Chem. Eng. J. **497**, 154996 (2024)
99. C.K. Kajdas, Importance of the triboemission process for tribochemical reaction. Tribol. Int. **38**(3), 337–353 (2005)
100. K. Hiratsuka, T. Abe, C. Kajdas, Tribocatalytic oxidation of ethylene in the rubbing of palladium against aluminum oxide. Tribol. Int. **43**(9), 1659–1664 (2010)
101. C. Kajdas, A. Kulczycki, D. Ozimina, A new concept of the mechanism of tribocatalytic reactions induced by mechanical forces. Tribol. Int. **107**, 144–151 (2017)
102. K. Hiratsuka, C. Kajdas, M. Yoshida, Tribo-catalysis in the synthesis reaction of carbon dioxide. Tribol. Trans. **47**(1), 86–93 (2004)

103. D. Yang, X. Liu, Y. Tian, Suppressed $MoO_3/FeMoO_4$ formation in molybdenum dialkyl-dithiocarbamate-derived tribofilms on nitrogen-incorporated Al-Mg-Ti-B coatings. Tribol. Int. **167**, 107374 (2022)
104. W. Song, J. Li, C. Zeng, C. Ouyang, S. Sun, K. Wang, J. Li, J. Luo, Tribo-catalysis triggered the in-situ formation of amphiphilic molecules to reduce friction and wear. Tribol. Int. **185**, 108541 (2023)
105. T. Jia, W. Wang, C. Zhang, L. Zhang, W. Wang, Polydopamine-mediated contact-electro-catalysis for efficient partial oxidation of methane. Angew. Chem. Int. Ed Engl. e202413343 (2024). From NLM Publisher
106. P. Yadav, S. Manori, R.K. Shukla, Contact electro catalysis driven degradation of malachite green dye by RGO/ZnO nanohybrid. Solid State Commun. **389**, 115578 (2024)
107. W.Z. Song, M. Zhang, H.J. Qiu, C.L. Li, T. Chen, L.L. Jiang, M. Yu, S. Ramakrishna, Z.L. Wang, Y.Z. Long, Insulator polymers achieve efficient catalysis under visible light due to contact electrification. Water Res. **226**, 119242 (2022)
108. Z. Wang, X. Dong, F.J. Lv, W. Tang, A perspective on contact-electro-catalysis based on frontier molecular orbitals. Mater. Adv. **5**(16), 6373–6377 (2024)
109. D.Q. Cao, R.K. Fang, Y.X. Song, M.G. Ma, H. Li, X.D. Hao, R. Wu, X. Chen, Contact-electro-catalysis for degradation of trace antibiotics in wastewater. Chem. Eng. J. **487**, 150531 (2024)
110. K. Li, Y. Lai, S. Lin, L. Zhou, M. He, H. Lin, Y. Yuan, Contact-electro-catalysis for the degradation of pentachlorophenol using inert fluorinated ethylene propylene powders. ACS ES&T Eng. **4**(10), 2485–2494 (2024)
111. N.N. Wang, W.B. Jiang, J. Yang, H.S. Feng, Y.B. Zheng, S. Wang, B.F. Li, J.Z.X. Heng, W.C. Ong, H.R. Tan, Y.W. Zhang, D.A. Wang, E.Y. Ye, Z.B. Li, Contact-electro-catalytic CO_2 reduction from ambient air. Nat. Commun. **15**(1), 5913 (2024)

Part II
Applications of Contact-Electrification

Chapter 9
Triboelectric Nanogenerator Using Solid–Solid Contact Electrification

Abstract This chapter delves into the technology of triboelectric nanogenerators (TENGs) based on solid-to-solid contact electrification (CE). We begin with an introduction to the two primary types of TENGs: AC-type and DC-type. The chapter then examines the five working modes of AC-type TENGs and the two material types used in DC-TENGs. Key aspects of performance optimization, material selection, environmental control, and charge excitation in solid–solid TENGs are also discussed. We then provide examples of TENG applications, emphasizing their potential as micro/nano power sources (e.g., from body motion energy, mechanical energy under in vivo conditions, vibration energy, and wind energy), self-powered sensors (including tactile, biomechanical, acoustic, human physiological, and chemical sensing), blue energy harvesting, and high-voltage sources. Finally, the chapter concludes with a forward-looking summary of future perspectives in TENG technology.

9.1 Introduction

The extensive use of fossil fuels has led to global warming, environmental degradation, energy shortages, and a proliferation of disorganized, low-quality, high-entropy energy in the environment, prompting researchers to explore green energy sources and develop advanced energy harvesting techniques [1]. Especially in the new era of artificial intelligence and the Internet of Things (IoT), characterized by billions of distributed sensors, it is highly challenging to sustain these vast, widely dispersed, low-power electronics using traditional methods like batteries, supercapacitors, and solar cells. Triboelectric nanogenerators (TENGs), known for their diverse material options, simplicity, ease of fabrication, affordability, lightweight nature, flexibility, and high efficiency even at low frequencies, have emerged as a groundbreaking technology for capturing low-amplitude, low-quality, high-entropy energy [2, 3]. TENGs have demonstrated versatility across multiple applications, including serving as micro/nano power sources [4, 5], self-powered sensors [6–8], blue energy harvesters [9–11], and novel high-voltage generators [12, 13]. Since their invention in 2012,

Z. L. Wang, *Contact-Electrification of Matter*, Springer Series in Surface Sciences 75,
https://doi.org/10.1007/978-3-031-88504-4_9

TENGs have garnered international interest from over 80 countries and engaged more than 10,000 researchers, fostering rapid advancements in this field [14].

As energy harvesters, TENGs convert mechanical energy into electricity through the combination of contact electrification (CE) and electrostatic induction (EI) or electrostatic breakdown (EB) [3, 15]. Over the past decade, significant efforts have been directed towards understanding TENGs' fundamental mechanisms and enhancing their output capabilities, broadening their impact on energy and sensing technologies. To date, TENGs can achieve peak power densities of 10 MW m^{-2} [16] and average power densities of 87.26 W m^{-2} Hz^{-1} [17], with energy conversion efficiencies exceeding 50% in specific conditions [18].

Solid–solid contact electrification-based TENGs, being among the earliest discovered and most widely adopted types, have driven extensive research due to their broad applicability [19–21]. This section provides an overview of solid–solid TENGs, emphasizing their operational modes, performance optimization strategies, and applications. The goal is to offer a comprehensive, systematic guide to TENGs, from theoretical foundations to practical experiments, and from basic operational modes to technological implementations, highlighting their dual roles as innovative energy technology and self-powered active sensors.

9.2 Working Modes of Solid–Solid TENG

Based on the type of output signals, TENGs using solid-to-solid materials can be classified into two categories: AC-TENGs and DC-TENGs. Below, we introduce the configurations of both types [14].

9.2.1 AC-TENG Based on Contact Electrification and Electrostatic Induction

AC-TENGs generate alternating current (AC) signals through the combined mechanisms of contact electrification (CE) and electrostatic induction (EI). Depending on their operational modes, they can be further classified into five categories: contact-separation, lateral-sliding, single-electrode, free-standing, and rolling modes, as shown in Fig. 9.1 [2, 14, 22, 23].

9.2.1.1 Contact-Separation Mode

Figure 9.1a illustrates the configuration of the contact-separation mode TENG which operates through vertical contact and separation of two distinct dielectric films [3, 24, 25]. The two dielectric films are positioned facing each other, with electrodes

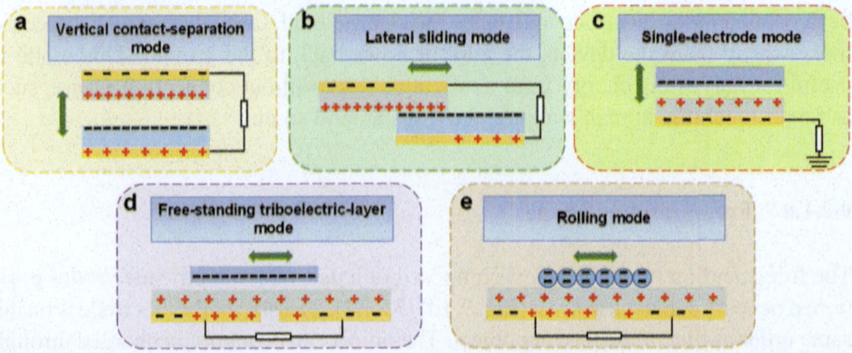

Fig. 9.1 The operation modes of TENGs: **a** vertical contact-separation mode, **b** lateral-sliding mode, **c** single-electrode mode, **d** free-standing triboelectric-layer mode, and **e** rolling mode TENGs, respectively

connected to a load deposited on their opposing surfaces. The differing electron affinities of the materials lead to the generation of opposite triboelectric charges at their surfaces upon contact. When the films are separated, a potential difference arises between the electrodes. This causes free electrons in one electrode to flow to the other, balancing the electrostatic field and converting mechanical energy into electrical energy. Once the films are brought back together, the created potential drop disappears, and the induced electrons would flow back. Continuous contact and separation of the two dielectric films drive the free electrons to flow forth and back between the two electrodes, resulting in an AC output in the external circuit.

9.2.1.2 Lateral-Sliding Mode

The configuration of lateral-sliding TENG is the same as that of the vertical contact-separation mode (Fig. 9.1b) [26, 27]. Surface charge transfer occurs when the two dielectric films make contact. As an externally applied force causes them to slide relative to each other, opposite triboelectric charges on the films generate lateral polarization along the sliding direction. This creates a potential difference across the two electrodes, driving electron flow between the top and bottom electrodes to balance the electric field generated by the triboelectric charges. Continuous sliding apart and coming together of the films produce an AC output. Sliding can be planar movement, cylindrical rotation, or disc rotation.

9.2.1.3 Single-Electrode Mode

The configuration of single-electrode TENG consists of a dielectric film with an electrode, as shown in Fig. 9.1c [28–30]. The electrode is connected to the ground. When there is a charged dielectric film approaches the device, induction current can be generated in the electrode to balance the electric field. Once the charged dielectric moves away from the device, the current flows back to the ground. This mode is useful for harvesting energy from a moving object without connecting a wire, such as finger touching, human walking, moving car, and so on.

9.2.1.4 Free-Standing Mode

The free-standing mode TENG is composed of a pair of symmetric electrodes positioned beneath a dielectric layer (Fig. 9.1d) [31, 32]. The size of the electrode is on the same order as that of the moving object. The object can be naturally charged through a contact electrification process. There is a small gap between the object and the device. As the object moves closer to and then away from the electrodes, it induces an asymmetric charge distribution in the system, leading to current flow between the electrodes to balance the potential distribution. When the object continuously moves back and forth, the current flow between the pair of symmetric electrodes, generating an AC current output. Because the moving object does not need to physically contact the dielectric layer, surface wear is minimized, effectively extending the durability of TENGs. This mode also highlights the advantages of harvesting energy from a moving object without the need for grounding.

9.2.1.5 Rolling Mode

The rolling-mode TENG (Fig. 9.1e) is a composite model that combines the above four modes [33]. It primarily relies on rolling contact electrification between the spheres and the dielectric surface. The strategically arranged electrodes beneath facilitate this process. As the charged spheres roll and disrupt the charge equilibrium, an electric field is generated, prompting electrons in the electrodes to flow. This approach effectively minimizes material wear during contact electrification while maintaining a high output.

9.2.2 DC-TENG Based on Contact Electrification and Electrostatic Breakdown

TENGs resulting from contact electrification (CE) and electrostatic breakdown (EB) are an effective method for converting mechanical energy into electricity, exhibiting high surface charge density, average power density, and output voltage. Typically, an air gap ranging from microns to millimeters is required to induce a high electric field, which disrupts the insulation of the gap and creates a conductive pathway for charge transfer [34]. Notably, TENGs generated through electrostatic breakdown can produce direct current (DC) output and even constant current output, which is named as DC-TENGs. Given their DC and constant current capabilities, DC-TENGs can directly power electronics, thereby eliminating the need for rectifiers and energy storage units. Based on the material properties of the two tribo-layers, TENGs resulting from electrostatic breakdown can be categorized into metal-dielectric and dielectric–dielectric modes.

9.2.2.1 Metal-Dielectric DC-TENG

In 2019, Liu et al. [15] first designed the DC-TENG arising from electrostatic breakdown that can produce continuous DC output (Fig. 9.2a). This DC-TENG mainly contains one triboelectric layer (TL) and two electrodes: a friction electrode (FE) as another friction layer, and an electrode as a charge collecting electrode (CCE) attached to the side of the slider. Note that a certain air gap between the CCE and the triboelectric layer is necessary to induce electrostatic breakdown, and subsequently electrons can transfer from the surface of the triboelectric layer to the CCE and then to the FE through an external circuit, forming an electric loop. Compared to the TENG arising from electrostatic induction, this DC-TENG shows a different physical model, with an electric charge source and a broken-down capacitor in series [15]. Subsequently, they proposed many methods to optimize the performance of DC-TENG, such as materials design and selection [35], configuration optimization [36, 37], interface lubrication [38] and environment control [39–41].

Utilizing the DC-TENG and the high efficiency of sliding triboelectrification, a microstructure electrode DC-TENG achieved a high effective charge density of 8.6 mC m^{-2} (Fig. 9.2b) [47]. However, despite this substantial increase in output charge density, there was no notable enhancement in power density. The DC-TENG's unique structural features result in multiple breakdown domains: an artificially created first domain between the charge collecting electrode (CCE) and the triboelectric layer (TL), a second domain between the friction electrode (FE) and the TL that hinders the increase of output charge, and a third domain between the CCE and the FE that limits the maximum output voltage of the DC-TENG (Fig. 9.2c) [42, 48]. To investigate the output characteristics of the DC-TENG, the researchers developed a dynamic model for the ideal condition, incorporating a cascaded capacitor-breakdown approach [42]. Subsequently, they introduced the "cask model" to elucidate the regulatory relationships among the three breakdown domains, highlighting the suppression of the second breakdown domain as a potential means to enhance DC-TENG performance [42]. Recently, their team proposed an effective strategy that employs a spontaneously established reverse electric field to modulate the electric field strength within the breakdown domains, thereby suppressing electrostatic breakdown at the electrode edges (Fig. 9.2d) [43]. By attaching an insulating layer to the edge of the FE, they successfully demonstrated a significant suppression of electrostatic breakdown in the second domain and an elevation of the breakdown threshold in the third domain. This approach optimizes the short-circuit charge, open-circuit voltage, and Coulomb efficiency, resulting in a substantial improvement in power

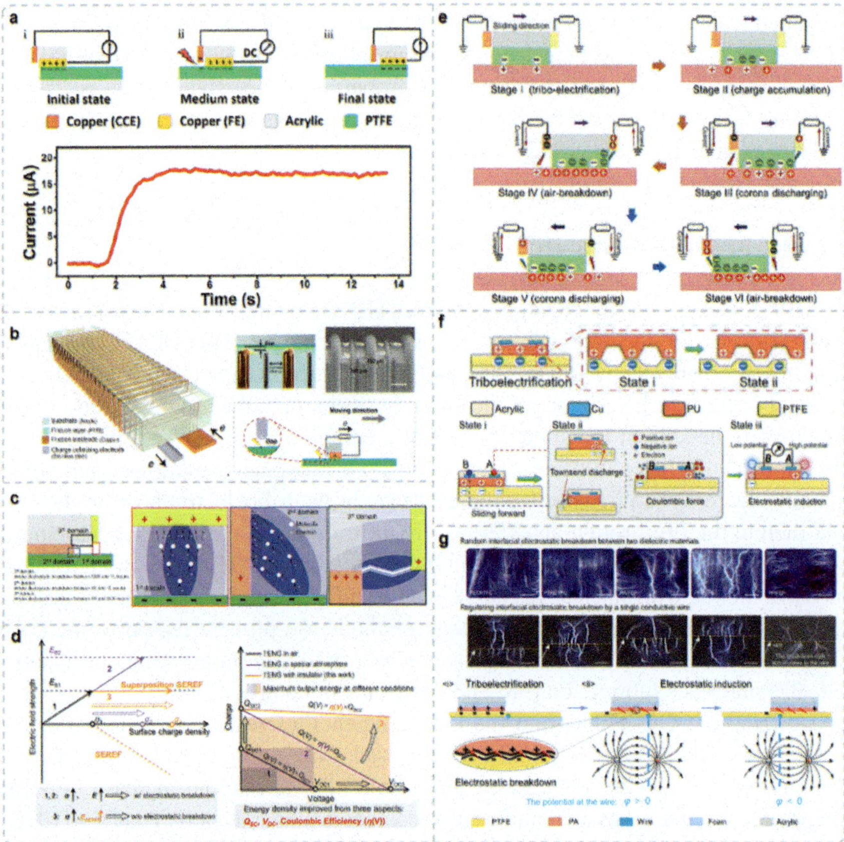

Fig. 9.2 TENG based on contact electrification and electrostatic breakdown. **a** Schematic diagram shows the working mechanism of DC-TENG arising from electrostatic breakdown and the corresponding constant current output [15]. **b** Microstructure electrode DC-TENG [36]. **c** Three discharge domains in DC-TENG [42]. **d** The spontaneously established reverse electric field to enhance the performance of TENG [43]. **e** A bidirectional and double-channel corona discharge TENG based on dielectric-dielectric material pairs [44]. **f** A sliding mode TENG based on the coupling of triboelectrification, electrostatic breakdown, and electrostatic induction [45]. **g** A single wire used for harvesting dynamic interfacial electrostatic energy [46]

density. Furthermore, this strategy is applicable to controlling edge breakdown in triboelectric nanogenerators based on electrostatic induction.

9.2.2.2 Dielectric–Dielectric DC-TENG

Probably drawing inspiration from the metal-dielectric DC-TENG, Shan et al. constructed a dual CCE structure on both sides of the slider to harvest positive and negative charges on the dual dielectric layers, respectively (Fig. 9.2e) [44,

49]. Although this structure loses DC output feature during the reciprocating linear motion, it has advantages in both power density and charge density, which could be attributed to the high triboelectric performance of dual dielectric triboelectric layers.

Beyond the artificially constructed fixed breakdown domains for collecting electrostatic breakdown charges, utilizing the interfacial electrostatic breakdown between dielectric materials emerges as a new approach in TENG design. Zeng et al. [45] proposed a novel energy conversion method that integrates triboelectrification, electrostatic breakdown, and electrostatic induction effects (Fig. 9.2f). Here, dual electrodes are fixed to the back of the slider. As the slider moves, the unevenly distributed interfacial charge resulting from electrostatic breakdown at the interface leads to varying charge accumulation near the two electrodes, thus creating an induced voltage between the electrodes. Very recently, He et al. [46] introduced a remarkably simple design, embedding a conductive wire beneath the dielectric materials. This ingenious method regulates the originally chaotic and distributed interfacial electrostatic energy, enabling the capture of energy released by interface electrostatic breakdown (Fig. 9.2g). This simple structure enables it an excellent candidate for seamless integration into wearable fabrics, which can be used for bioenergy harvesting and self-powered sensors. More importantly, they developed a point-charge physical model, different from the traditional TENG capacitor model, to explain the characteristics of electrical signals, providing crucial guidance for future works.

9.3 Performance Optimization of Solid–Solid TENG

To evaluate the performance of TENGs with various structures and materials, Figure-of-Merit (FOM) metrics were introduced in 2015 [50]. These FOMs were derived from the operational energy output cycle, considering the relationship between the built-up voltage (V) and the transferred charges (Q). Subsequently, a comprehensive performance FOM (FOMp) was established, incorporating both structural (FOMS) and material (FOMm) aspects. Notably, the FOMm is indicative of the square of the surface charge density, a crucial characteristic of the materials used. As a pivotal indicator for assessing TENG performance, power density exhibits a quadratic relationship with surface charge density. Therefore, significant research efforts have been directed towards enhancing surface charge density, primarily through three methods: material optimization, environmental control, and charge excitation.

9.3.1 Materials Choice

Materials optimization is a prevalent method for selecting suitable triboelectric materials to enhance the performance of TENGs. This process primarily relies on the triboelectric series, which is established through the mutual rubbing of two materials. The triboelectric series ranks materials based on their propensity to lose or gain electrons, placing those materials prone to losing electrons at the positive end and those prone to gaining electrons at the negative end. Notably, the traditional triboelectric series is empirical and solely focuses on charge polarity. In 2019, a quantified triboelectric series was introduced [51] using a controlled platform and environment. This series considers both charge polarity and quantity, offering a valuable reference for material selection in TENGs. However, the triboelectric performance remains environmentally dependent due to the complex interactions involving gas breakdown between triboelectric layers and external factors, such as humidity and gas molecules. In 2022, drawing inspiration from the integration of TENG technology with vacuum environments, researchers introduced a standardized methodology to assess the maximum triboelectric charge of dielectric materials [52]. This method circumvents the constraints imposed by environmental variables and TENG structures, thereby enhancing the reliability and precision of triboelectric performance evaluations. By analyzing the triboelectric charge density of a diverse selection of over forty materials, an updated triboelectric series was established (Fig. 9.3a), furnishing a thorough understanding of the optimal triboelectric characteristics of various materials. By minimizing the influence of air breakdown and environmental perturbations, vacuum-based triboelectric charge density measurements offer a more accurate portrayal of the intrinsic material properties. This study underscores the promising potential of optimizing triboelectric behavior through the consideration of triboelectric material pairs, rather than focusing exclusively on individual dielectric materials, thereby facilitating novel approaches to enhance triboelectric performance.

9.3.2 *Environmental Control*

Environmental factors, notably atmospheric pressure, humidity, and temperature, exert significant influence on surface charge accumulation. Manipulation of these parameters offers a pathway to augment surface charge density. In a contact-separation triboelectric nanogenerator (CS TENG), the triboelectric layers, upon separation, produce a strong electrostatic field spanning the intervening air gap. However, this mechanism is concurrently plagued by the depletion of surface charges due to the phenomenon of air breakdown. By establishing a high-vacuum atmosphere, the detrimental effect of air breakdown on charge retention can be substantially mitigated, enabling the attainment of a substantial charge density of 1003 $\mu C\ m^{-2}$ when employing materials such as Polytetrafluoroethylene (PTFE) and ferroelectric $BaTiO_3$ [53] (Fig. 9.3b). Furthermore, Fu et al. [54] conducted a comprehensive investigation into the performance of TENGs across a spectrum of atmospheric pressures. Their results underscore that in environments characterized by elevated atmospheric pressure, the suppression of air breakdown contributes to a marked enhancement in charge density (Fig. 9.3c).

Humidity plays a crucial role in altering surface properties, which in turn affects the dynamics of charge transfer processes. Such alterations can lead to enhancements in surface conductivity, thereby accelerating the rate of charge depletion. Importantly,

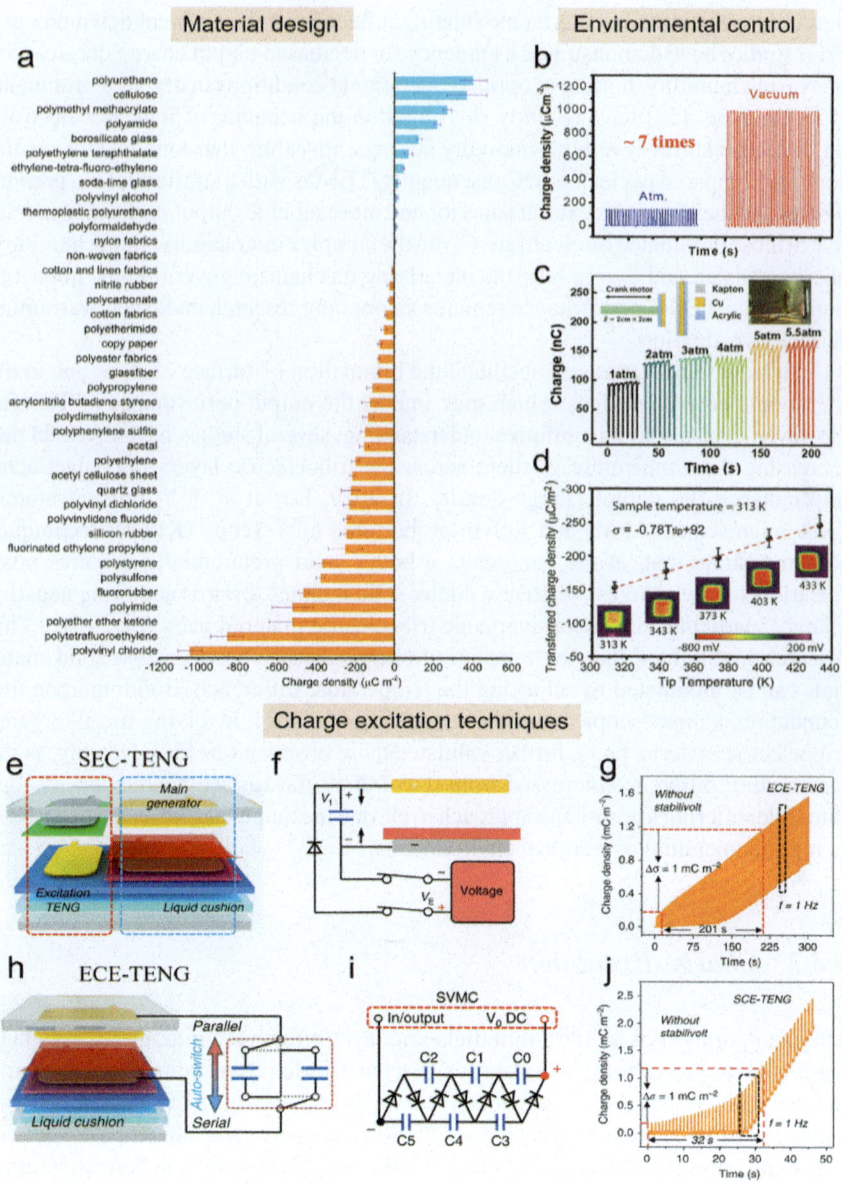

Fig. 9.3 Strategies for enhancing the output of triboelectric nanogenerators. **a** The output performance of TENG optimized by material design [52]. **b** Decreasing the pressure to vacuum conditions to optimize the surface charge density [53]. **c** Increasing the air pressure to enhance the surface charge density [54]. **d** Controlling the temperature difference between two materials to optimize the transferred charge density [55]. **e** Structural illustration of ECE TENG. **f** Simplified working components of ECE TENG. **g** Dynamic output charge accumulation process of ECE TENG. **h** Fundamental scheme of SCE TENG, the auto-switch can change capacitors from parallel to serial connection during the operation cycle. **i** Input/output node and scheme of voltage-multiplying circuit. **j** Dynamic output charge process of SCE TENG [56]

humidity variations also exert a modulating influence on the air breakdown process. Prior studies have demonstrated a tendency for decreased output charge density with increasing humidity, hinting at optimal operational conditions in dryer environments [57]. Liu et al. [58] have recently shed light on the behavior of ions and electrons on dielectric surfaces in high-humidity settings, revealing that ions exhibit greater stability compared to electrons. Consequently, TENGs with a substantial ion preponderance on their surfaces exhibit superior and more reliable output performance relative to those dominated by electrons. Given the complex interactions among humidity, charge transfer, and charge loss, the underlying mechanism governing the impact of humidity on TENG performance remains an ongoing research endeavor, warranting further investigation.

Elevating temperature can facilitate the dissipation of surface charges due to the thermionic emission effect, which may impact the output performance of TENGs. However, contrary to this intuitive understanding, several studies have revealed that increasing the temperature gradient across the triboelectric layers in contact actually enhance the output charge density. In 2019, Lin et al. [55] utilized atomic force microscopy (AFM) and Kelvin probe force microscopy (KPFM) techniques to demonstrate that, at the nanoscale, a hotter solid preferentially acquires positive triboelectric charges, whereas a cooler solid inclines towards acquiring negative charges, particularly in metal-inorganic triboelectric material pairs (Fig. 9.3d). This research highlighted that the transferred charge density between two solid materials can be modulated by adjusting the temperature difference. Building upon this foundation, a macroscopic experiment conducted in 2021, involving metal–organic triboelectric material pairs, further validated these observations [59]. Notably, as the temperature difference increased from 0 to 145 K, the surface charge density rose threefold, offering a promising approach to elevate the output performance of TENGs in high-temperature operational environments.

9.3.3 Charge-Excitation

The incorporation of an ultra-thin dielectric layer effectively mitigates the voltage across the air gap during the separation of triboelectric layers, consequently raising the breakdown threshold and notably augmenting the ultimate surface charge density of triboelectric nanogenerators (TENGs). However, the inherent triboelectric constraints of tribo-materials, coupled with the delicate equilibrium between charge accumulation and dissipation, pose substantial hurdles in achieving an optimal output solely through triboelectrification. To address these challenges, charge excitation technology has emerged as a promising approach. By leveraging the high output voltage produced by a pump TENG to efficiently charge a large capacitor serving as the primary TENG, this method decouples the maximum output performance of TENGs from the triboelectric properties of the tribo-materials involved. This decoupling offers novel pathways for improving the efficiency and overall performance of TENGs.

In 2018, two independent research groups led by Xu et al. [60] and Cheng et al. [61] made a pioneering discovery by introducing the first charge excitation method into TENGs. By utilizing thinner dielectric layers or dielectric materials with higher relative permittivity to enhance the capacitance of the main TENG, they achieved an exceptional effective surface charge density of up to 1020 $\mu C\ m^{-2}$ under ambient conditions. In 2019, Liu et al. [56] further advanced this field by introducing two novel TENG variants: the external-charge-excitation TENG (ECE TENG) and the self-charge-excitation TENG (SCE TENG), both incorporating a voltage multiplier circuit (VMC) (Fig. 9.3e–j). This integration allowed for the replacement of the traditional parallel plate capacitor structure of TENGs with commercial capacitors, significantly boosting charge storage capacity and enabling the generation of higher excitation voltages. Furthermore, the dynamic switching of capacitor modes within the SCE TENG introduced a new approach for multiplying output charge, eliminating the need for additional TENGs. Consequently, these innovative devices exhibited a remarkable increase in effective surface charge density, reaching 1.25 $mC\ m^{-2}$ with a 5 μm Kapton film. The use of a carbon/silicone gel electrode to enhance contact efficiency further improved the effective surface charge density to 2.38 $mC\ m^{-2}$ with a 4 μm polyimide (PI) film [62]. Following this, a composite film combining lead zirconate titanate and poly(vinylidene fluoride), noted for its exceptional relative permittivity, elevated the charge density to 3.53 $mC\ m^{-2}$ [62]. Subsequently, advancements led to a carbon powder–polyvinylidene fluoride composite film achieving a higher charge density of 4.3 $mC\ m^{-2}$, attributed to modulation of the carrier trap state within the dielectric film, which induced charge trapping failure [63]. Recently, Liu et al. [64] introduced poly(vinylidene fluoride–trifluoroe thylene–chlorofluoroethylene) [P(VDF-TrFE-CFE)], a polymer with high dielectric permittivity, to prevent air breakdown and thereby elevate the charge density to 8.6 $mC\ m^{-2}$. Furthermore, by engineering the trap density of P(VDF-TrFE-CFE) through the use of high-polarity interfaces from barium titanate (BTO) nanoparticles and dense chain segment stacking induced by electrostatic interaction with polyetherimide (PEI), both the air breakdown and charge loss effect were largely inhibited, therefore a record-high charge density of 9.23 $mC\ m^{-2}$ was achieved in ECE TENG using a 0.2 vol% PEI/P(VDF-TrFE-CFE) film [65].

9.4 Applications of Solid–Solid TENG

The applications of TENGs are diverse and extensive. In this section, we will provide an overview and highlight examples of solid–solid TENGs in various applications, including as micro/nano power sources, self-powered sensors, blue energy harvesters, and high-voltage sources.

9.4.1 As Micro- and Nano-power Sources

TENGs offer a significant advantage in converting low- and random-frequency mechanical energy into usable electricity, making them ideal for micro- and nano-power sources. In this section, we will highlight examples of TENGs designed to harvest energy from body motion, mechanical energy in vivo, vibration energy, and wind energy. The discussion will conclude with an outlook on the future developments and potential advancements in energy harvesting technology.

9.4.1.1 Harvesting Body Motion Energy

The movements of human bodies are readily accessible and omnipresent sources of energy that are unaffected by weather conditions and environmental factors. Harvesting human motion energy through TENGs represents a promising approach for wearable power sources and flexible electronics. The advantages of TENGs for harvesting body motion energy include materials diversity, stability, scalability, along with a good cost-effectiveness and manufacturability. Through embedding the devices in clothing, footwear or even attaching to the skin, this TENG-based power source can leverage the biomechanical motion of the human body, such as walking, stepping, hand swing, breathing, and even pulse, to generate electricity. One of the most straightforward and feasible methods used to harvest energy generated by human motion is the direct fabrication of wearable TENGs using fibers, textiles, and garments [66–70].

The first fiber-based TENG was proposed by Zhong et al. [66] (Fig. 9.4a–c). This fiber-based TENG is composed by two modified triboelectric fibers entangled to each other: carbon-nanotube (CNT)-coated cotton thread, and polytetrafluoroethylene (PTFE) and CNT coated cotton thread. The two triboelectric fibers operated in contact-separation mode. When the two fibers come into contact during stretching, they disrupt the original electrostatic balance, allowing charges in the CNT electrode to flow through an external circuit to restore this balance. As a result, an alternating current is generated between the threads. As shown in Fig. 9.4d and e when the two triboelectric fibers were stretched to 2.15% strain at 5 Hz, the CS TENG can generate AC-type current to 11.22 nA. An average output density of 0.1 $\mu W\ cm^{-2}$ can be obtained by the device.

In 2016, Lai et al. [67] reported the first single-electrode mode TENG thread using silicone rubber and a stainless-steel conductive thread (Fig. 9.5a and b). The TENG thread can generate electricity by contact and separation with human skin (Fig. 9.5c). With 5 cm in length, the TENG thread can produce AC-type output with a V_{oc}, Q_{tr} and I_{sc} of ~ 15 V, ~ 12 nC, and ~ 7 μA (Fig. 9.5d), respectively. The TENG thread can be stitched into a serpentine pattern on an elastic textile, enabling the efficient harvesting of body motion energy across a large area in a stretchable form. For a textile with a size of 6.5 × 3.5 in.^{2} sewing with the TENG-thread, it can generate a V_{oc}, Q_{tr} and I_{sc} of up to ~ 200 V, ~ 300 nC, and ~ 200 μA (Fig. 9.5e), respectively.

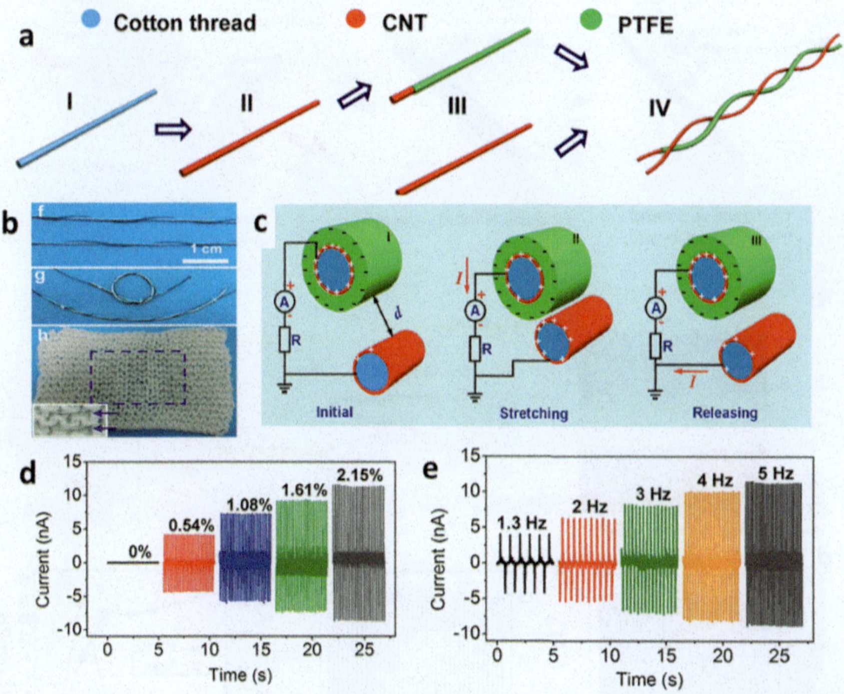

Fig. 9.4 **a** Fabrication of fiber-based TENG. **b** Photos of fiber-based TENG with linear shape, curved shape, and woven into fabric. **c** Power generation mechanism of fiber-based TENG. **d** Output current–time curve varied with stimulation strains. **e** Output current–time curve varied with stimulation frequencies [66]

The maximum instantaneous output power can achieve 14 mW for an external load resistance of 1 MΩ (Fig. 9.5f). The elastic TENG-thread textile can be worn on various parts of human bodies for collecting biomechanical energy from different types of body motions. For example, the TENG-thread textile worn on a human's elbow can scavenge energy from the flexion and extension of an elbow with output I_{sc} of 1.2 μA (Fig. 9.5g). The TENG-thread textile worn on a knee can harvest energy during knee flexion and extension and produce I_{sc} to ~ 2 μA (Fig. 9.5h). It can exploit walking energy and generate ~ 25 μA as it was worn under the foot. Through a power management circuit, the TENG-thread textile can sustainably power an electronic watch (Fig. 9.5i). After that, serious efforts have been reported to weave TENG fibers into various structural designs to enhance the efficiency of energy harvesting from human motion, such as 3D orthogonal woven TENG, 3D braided TENG, and so on [71–74].

Moreover, a multilayered waterproof multifunctional TENG textile (WPF-MTENG) was developed for harvesting energy from both body movements and natural tiny impacts (rain and wind) (Fig. 9.6a–c) [75]. The multilayered WPF-MTENG was mainly constructed of two laminated fabrics. The bottom combined

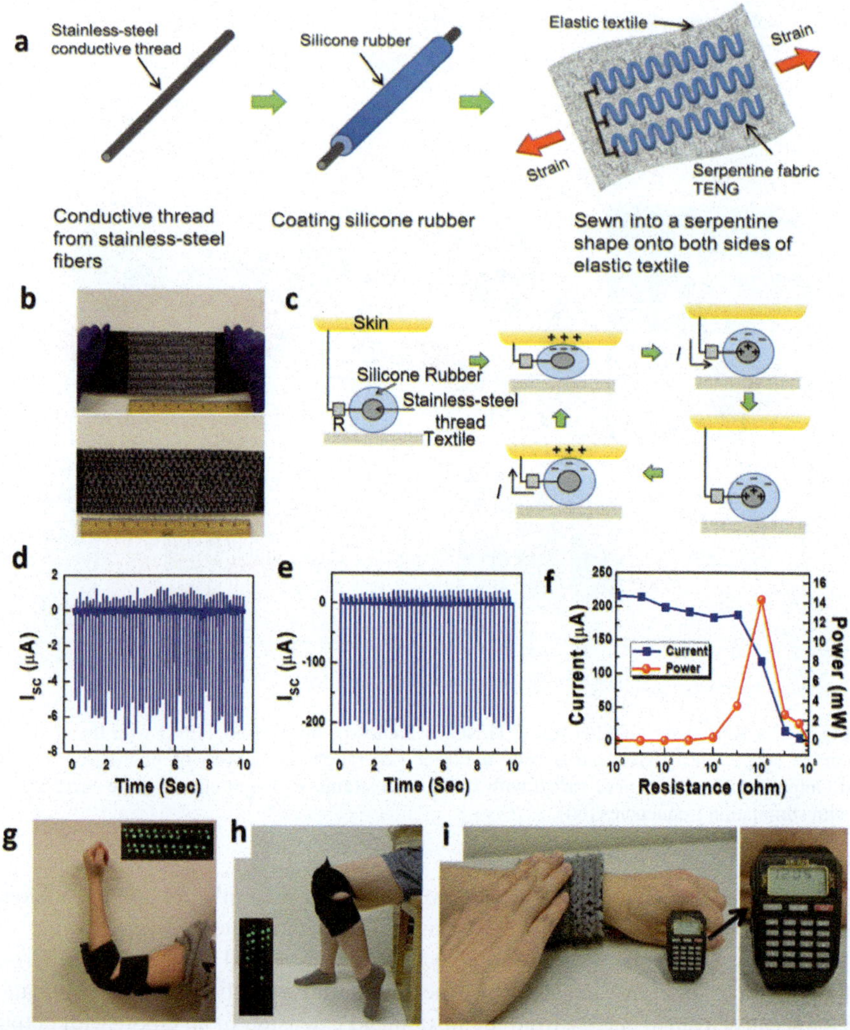

Fig. 9.5 **a** Fabrication of single thread-based TENG. **b** Photos of single thread-based TENG serpentine sewn onto an elastic fabric. **c** Operation mechanism of single thread-based TENG. **d** Output current from a 5-cm TENG thread. **e** Output current from a serpentine-sewn TENG textile with 6.5 × 3.5 in^2. **f** Relationship between the instantaneous power and resistance of external load. Photo demonstrating that TENG textile powers up LEDs by harvesting energy from **g** elbow motions and **h** knee motions. **i** Demonstration of continuously driving a smart watch by tapping on wrist-worn TENG textile [67]

fabric consisted of a conducting fabric sandwiched between an ethylene–vinyl acetate (EVA) film as underlay and a roughened rubber membrane. The top laminated fabric was composed of a conducting fabric sandwiched between a mesh fabric and an EVA film. It operates through contact separation of two internal active fabrics (Fig. 9.6d). Competing previous other TENG fibers can only extract energy from one designated source and suffer from malfunction caused by water from sweat, humid, or rainy weather [71, 76]. The WPF-MTENG can harvest various energy resources together, enabling to collect ambient energy more freely. A WPF-MTENG with a size of 8×11.6 in.^{2} can scavenging various body-movement energy at different parts of human body including the wrist, foot, elbow, and knee. For example, tapping and stepping the worn TENG textile can produce V_{oc} to over 2000 V m^{-2} and I_{sc} to over 500 μA m^{-2}. These results are comparable to with yarn-based TENG (~345 μA m^{-2}) and fabric-TENG (~500 μA m^{-2}) [71, 76]. When it was worn on the elbow and knee, it generated V_{oc} to ~ 1100 V m^{-2} and I_{sc} to ~ 25 μA m^{-2} during extending and flexing the arm and leg. Furthermore, the V_{oc} and I_{sc} of the optimized WPF-MTENG can reach ~ 850 V m^{-2} and 150 μA m^{-2} at a rainfall of 125 mL s^{-1}. The V_{oc} and I_{sc} reached ~ 900 V m^{-2} and ~ 150 μA m^{-2} at wind speed of 15.4 m s^{-1}, respectively. The converted energy can directly power up tens of LEDs (Fig. 9.6e–g) and be stored for driving small gadgets.

9.4.1.2 Harvesting Mechanical Energy Under In Vivo Conditions

Implantable biomedical devices are functional electronic systems surgically placed within the body to monitor physiological parameters, deliver therapy, or support bodily functions. Examples include pacemakers, implantable cardioverter defibrillators, and neurostimulators for the brain, peripheral nerves, and muscles, among others. Conventional devices are powered by batteries, which can be bulky, heavy, and have a short lifespan with limited capacity. In addition, issues such as potential toxicity, leakage, and biocompatibility are significant concerns when these batteries are used in the body. In particular, patients often face painful surgeries to replace batteries periodically, increasing both the risks and costs associated with long-term use of these devices.

Converting energy from body movements and physiological environments to power implantable electronics provides a sustainable solution for the long-term use of these biomedical systems. TENGs offer advantages such as diverse material options, exceptional low-frequency energy harvesting capabilities, and high output. Consequently, they reveal significant potential for capturing biomechanical energy from sources like muscle movements, blood circulation, and the contraction and relaxation of the heart and lungs, to power implantable biomedical devices.

The first demonstration of in vivo TENGs (iTENG) using for implantable applications was demonstrated by Zheng et al. [77] (Fig. 9.7a and b). The device was based on CS mode TENG structure using pyramid-patterned PDMS as the dielectric material, Au as the electrode, and a nano-surface Al foil as the contact material and electrode (Fig. 9.7c). The entire device was encapsulated in a thin layer of PDMS

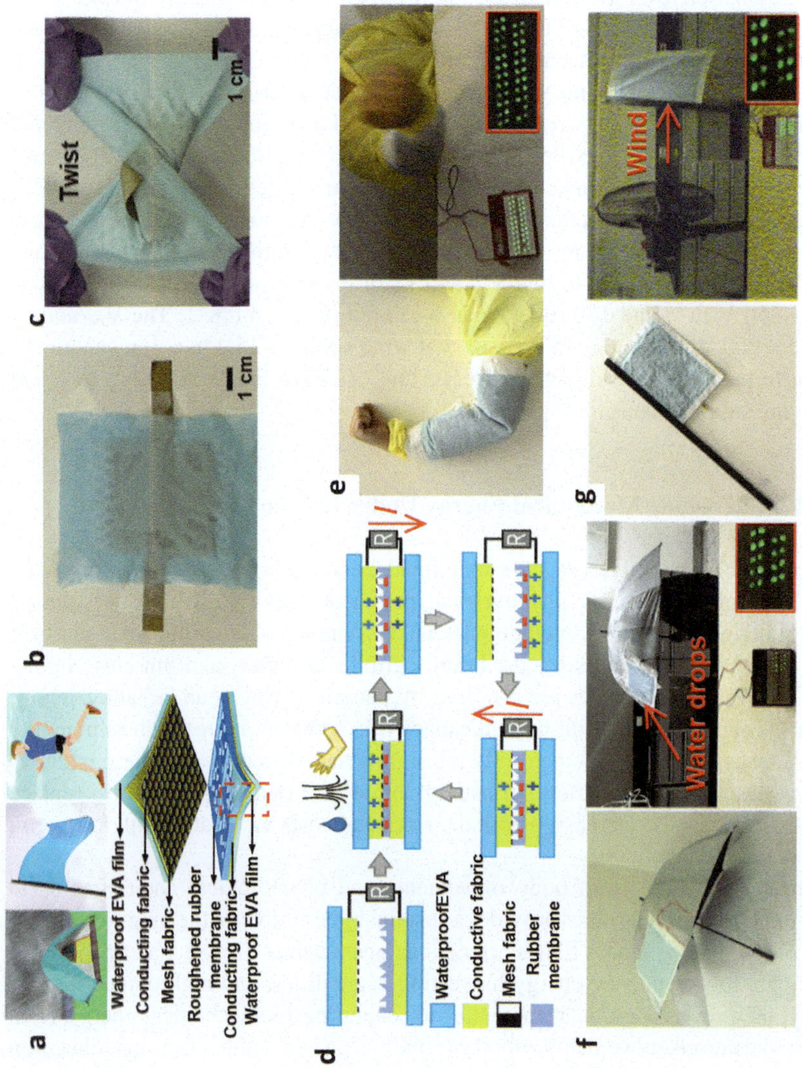

Fig. 9.6 **a** Schematic illustration of applications and exploded view of WPF-MTENG. **b** Photos of WPF-MTENG with demonstrations its **c** flexibility. **d** Working mechanism of WPF-MTENG. Photos demonstrating that WPF-MTENG powers up LEDs by harvesting energy from **e** elbow motions, **f** raindrops, and **g** wind [75]

to prevent biofluid infiltration and ensure safe insertion into the thorax of small rats (Fig. 9.7d and e). The iTENG can generate voltage and current signals of approximately 3.73 V and 0.14 μA, respectively, from the breath of a living rat (Fig. 9.7f). Its power density reached up to 8.44 mW m^{-2} (Fig. 9.7f). The converted electrical energy has been demonstrated to store in a capacitor to drive a pacemaker prototype to regulate the heart rate of a rat (Fig. 9.7h and i). Theoretical calculations suggest that the energy harvested from 5 breaths can stimulate a single pulse to regulate the rat's heart rate.

Zheng et al. [78] incorporated a memory alloy as a "keel structure" to improve the output of the iTENG, which optimizes the contact and separation of frictional layers under in vivo conditions (Fig. 9.8a). The device was firstly tested in a large animal modal (adult Yorkshire porcine, 30 kg), setting between the heart and pericardium (Fig. 9.8b). A periodical contraction and relaxation of the heart actuated the friction layers of the TENG and resulted in the contact and separation action, giving an in vivo V_{oc} and I_{sc} of 14 V and 5.5 μA, which are 3.5-fold and 25-fold higher than the previous one, respectively (Fig. 9.8c). In addition, the iTENG exhibited

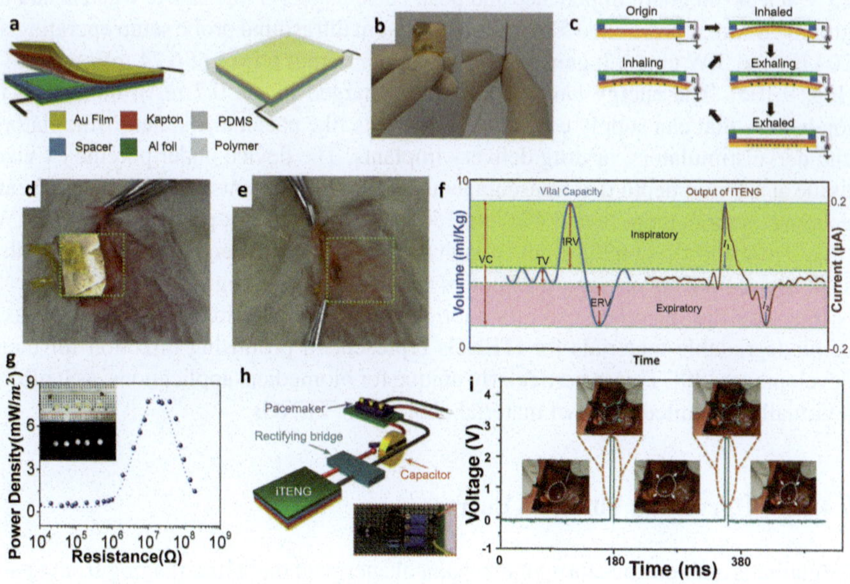

Fig. 9.7 **a** Structure and **b** photos of iTENG. **c** Working mechanism of iTENG. **d** Photos of WPF-MTENG with demonstrations its c flexibility. **e** Photos demonstrating that iTENG under in vivo conditions. **f** Correlation between the output of iTENG and respiratory movements (left: forced vital capacity curve; right: short-circuit current of the iTENG. *VC* vital capacity; *TV* tidal volume; *IRV* inspiratory reserve volume; *ERI* expiatory reserve volume). **g** Dependence of output power density (the insert shows the lighting up of 5 white LEDs) as a function of the applied external load. **h** Scheme diagram of a self-powered prototype pacemaker. **i** Stimulating the heart of the rat by self-powered pacemaker (insets: photographs of the heartbeat of the rat regulated by self-powered pacemaker) [77]

excellent stability after 72 h of implantation in the active animal. The output voltage from devices placed on different implantation sites including the outflow tract of the right ventricle, the auricle of the left atrium, the cardiac base, the lateral wall of the left ventricle, the inferior wall of the left ventricle, and the apex is shown in Fig. 9.8d. By integrating the iTENG with an implantable wireless transmitter, a self-powered data transmission system was developed for real-time remote cardiac monitoring (Fig. 9.8e).

Subsequently, Ouyang et al. [79] developed an iTENG utilized polarized PTFE and Al as contact electrification layers, successfully powering a commercial pacemaker with a maximum in vivo open-circuit voltage of 65.2 V (Fig. 9.9). Moreover, they demonstrated an implantable symbiotic pacemaker that can achieve cardiac pacing and sinus arrhythmia correction on a large animal mode.

Besides extracting energy from in vivo body motions, Hinchet et al. [5] presented an iTENG for harvesting ultrasound energy under in vivo conditions (Fig. 9.10a). The device working in single-electrode mode was structured using perfluoroalkoxy (PFA) as a triboelectric vibration membrane, Cu/Au as primary electrode and a small backside Cu as reference electrode (Fig. 9.10b). The implantable TENG can generate 25 V at a 40-megohm impedance and peak currents of 1.3 mA at 1 Ω when tested in grounded water, positioned 5 mm beneath a 3-cm ultrasound probe setup operating at 20 kHz and 3 W cm^{-2}. It generated a maximum output power of 6.74 mW at 33 kΩ (Fig. 9.10c). The energy harvesting can be charged into a 0.7 mAh thin-film Li-ion battery that can supply commercial implants like pacemakers, neurostimulators and nerve stimulators, or drug delivery implants. The device under porcine ex vivo tissue at a 5-mm depth that transcutaneous electronic implants typically located can generate outputs more than 2.4 V and 156 μA with an output power about 98.6 μW (Fig. 9.10e and f), which is high enough to recharge the batteries of small implants, such as pacemakers or neurostimulators (typically consuming ~ 1–100 μW).

Because the materials used for fabricating TENGs are highly diverse, the use of biodegradable materials for iTENGs represents a promising direction for their development [80]. This is particularly unique for biomedical applications, as it offers a virtually unlimited range of material choices for TENGs.

9.4.1.3 Harvesting Vibration Energy

Vibration is a ubiquitous form of mechanical energy in daily life, ranging from operating household appliances like washing machines and air conditioners to engines, bridges, railways, and roads. Vibration-based TENGs (V-TENGs) an effective tool for harvesting ambient vibration energy, with performance superior to other approaches of its kind. The operation of V-TENGs typically relies on the contact and separation of two different materials during vibrations, which results in charge transfer through the triboelectric effect. Another key factor that can significantly impact the performance of V-TENGs is the frequency of the vibrations.

Chen et al. [81] report the first harmonic-resonator-based V-TENG capable of harnessing random and small ambient vibrations as a sustainable power source. A

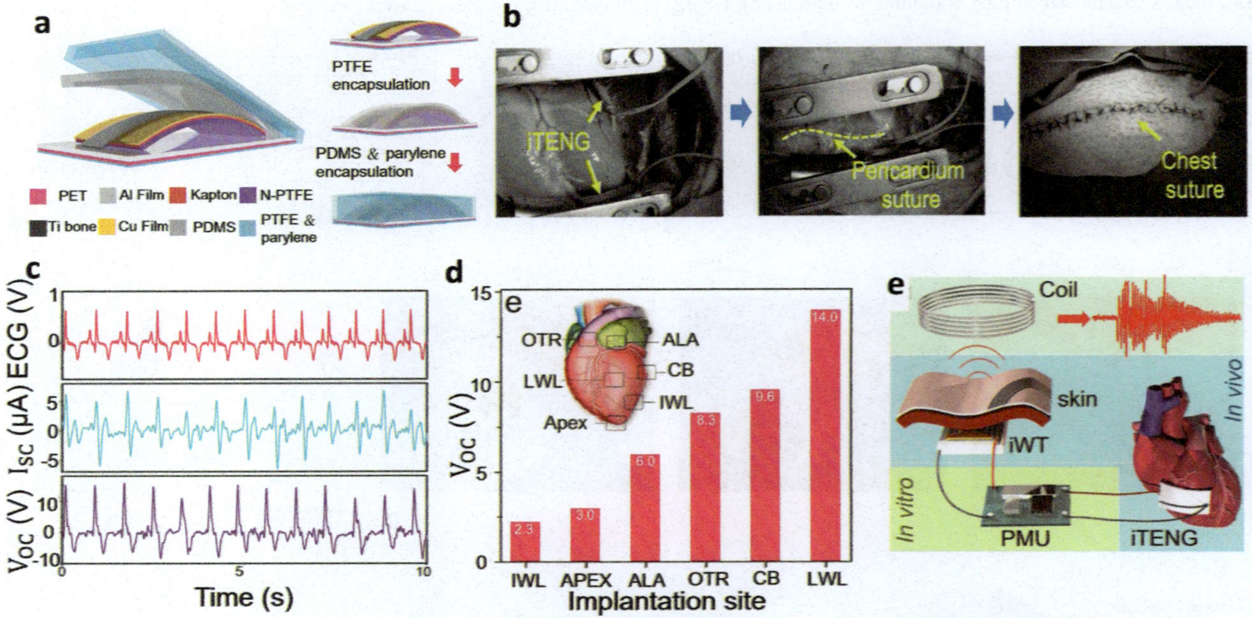

Fig. 9.8 **a** Structure of iTENG. **b** Photos demonstrating that iTENG under in vivo conditions. **c** Outputs of iTENG under in vivo conditions. **d** Output of the iTENG at a different implant site (*OTR* the right ventricular; *ALA* the auricle of the left atrium; *CB* the cardiac base; *LWL* the lateral wall of the left ventricular; *IWL* the inferior wall of the left ventricular). **e** Schematic diagram of the self-powered wireless transmission system based on the iTENG [78]

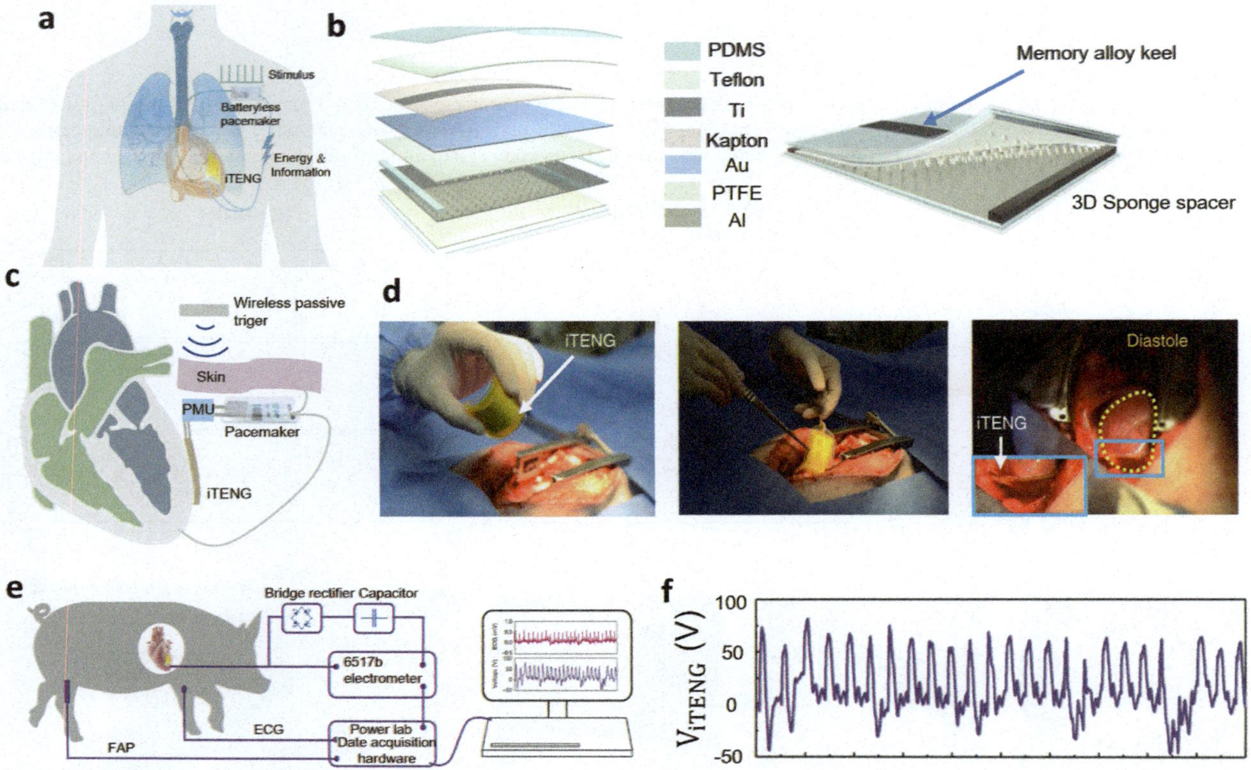

Fig. 9.9 **a** Illustration of symbiotic cardiac pacemaker system. **b** Structure of iTENG. **c** Illustration of symbiotic cardiac pacemaker system turned on by wireless passive trigger. **d** Photos demonstrating that iTENG under in vivo conditions. **e** Schematic of in vivo experimental electrical characterizations. **f** In vivo open-circuit voltage of the iTENG [79]

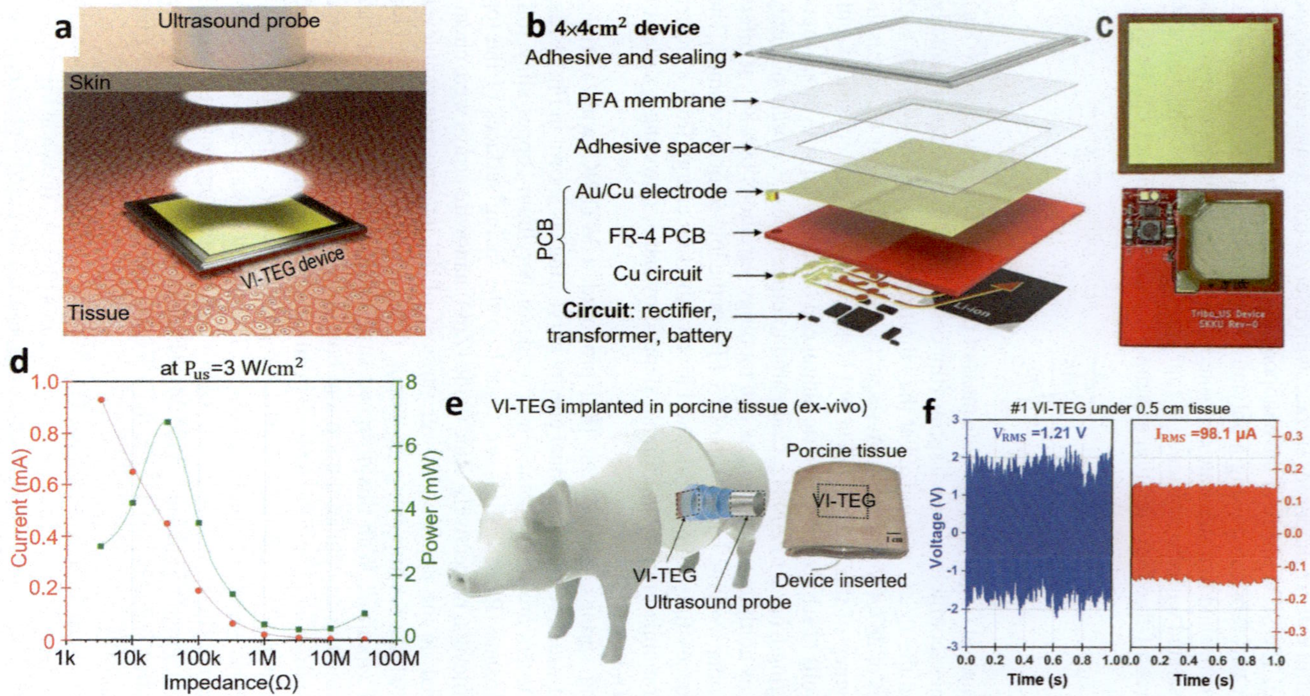

Fig. 9.10 **a** Illustration of ultrasound energy harvesting under skin using iTENG. **b** Exploded view of iTENG. **c** Front and back of the iTENG. **d** Power depending on electrical impedance at 5 mm from the ultrasound probe set at a power density of 3 W cm^{-2}. **e** Schematic of the porcine ex vivo showing the location of the implanted VI-TEG. **f** Voltage at 40-MΩ impedance, and current at 1-MΩ impedance generated by the VI-TEGs implanted at 1 cm under the porcine tissue. The US probe was set at 20 kHz and 1 W cm^{-2} [5]

schematic diagram of the harmonic-resonator-based V-TENG is shown in Fig. 9.11a and b. The device works on vertical contact-separation mode. The upper part consists of an aluminum film with a nanoporous surface, which serves both as an electrode and a contact surface. The SEM image of the nanopores on the Al is shown in Fig. 9.11c. The bottom part was fabricated using a polytetrafluoroethylene (PTFE)-nanowires film (Fig. 9.11c) attached to a substrate deposited with Cu as another electrode. The two parts are separated with the assistance of four springs (Fig. 9.11d). Theoretically, for a single degree-of-freedom vibration system, the natural frequency can be by

$$f_0 = \frac{1}{2\pi}\sqrt{\frac{4k}{m_0}}$$

where f_0 is the natural frequency, m_0 is the mass of the upper load (this case, 56.8 g), and k is the stiffness coefficient of each spring (this case, 112 N m^{-1}). Experimentally, both the V_{oc} and the I_{sc} are maximized at the vibration frequency of 14.5 Hz with maximum values of 287.4 V and 76.8 μA, respectively, indicating that 14.5 Hz is the resonance frequency of the V-TENG (Fig. 9.11f–h), which is well consistent with the theorical results (14.1 Hz). The electric output can be measured with broad input frequencies ranging from 2 to 200 Hz. Compared to other types of vibration energy harvesters such as nonlinear mechanisms and topology variations [82–84], it offers a significantly wider working bandwidth of 13.4 Hz in the low-frequency range. The V-TENG was successfully demonstrated to harvest vibration energy from a shaker (60 Hz), an automobile engine, and a table slapped by hand, powering a series of LEDs (Fig. 9.11i–k).

Besides using springs, Yang et al. [85] proposed a V-TENG with a triple-cantilever-based structure for vibrational energy harvesting, as shown in Fig. 9.12a. The three cantilevers are three metal plates made of beryllium-copper alloy foil. Polydimethylsiloxane (PDMS) films are coated on the facing surfaces of the top and bottom cantilevers, respectively. Zinc oxide (ZnO) nanowire arrays were chemically synthesized on the upper and lower surfaces of the intermediate cantilever and deposited with Cu for improving the output. A mass was attached at its end for effectiveness of vibration. With this designed structure, the middle cantilever can have two opportunities to contact the top and bottom cantilevers when the device is operating, resulting in a two-fold increase in energy conversion efficiency. At the resonance frequency of 3.7 Hz, the V-TENG can produces an V_{OC} up to 101 V, and a I_{sc} of 55.7 μA, as shown in Fig. 9.12b and c. The output voltage and current to the external load are shown in Fig. 9.12d, resulting in a maximum output power of 252.3 mW m^{-2} at a load resistance of 0.25 MΩ (Fig. 9.12e). When triggered by the shaker at a vibration frequency of 3.7 Hz, the V-TENG can directly power 40 LEDs, as shown in Fig. 9.12f.

Moreover, an internal wavy-structured Cu–Kapton–Cu film was introduced to develop self-restorable V-TENG without using springs [86]. As shown in Fig. 9.13a, the V-TENG consists of a wavy-structured Cu-Kapton-Cu film sandwiched between two flat PTFE nanowire films, with Cu electrodes deposited on the backside of the

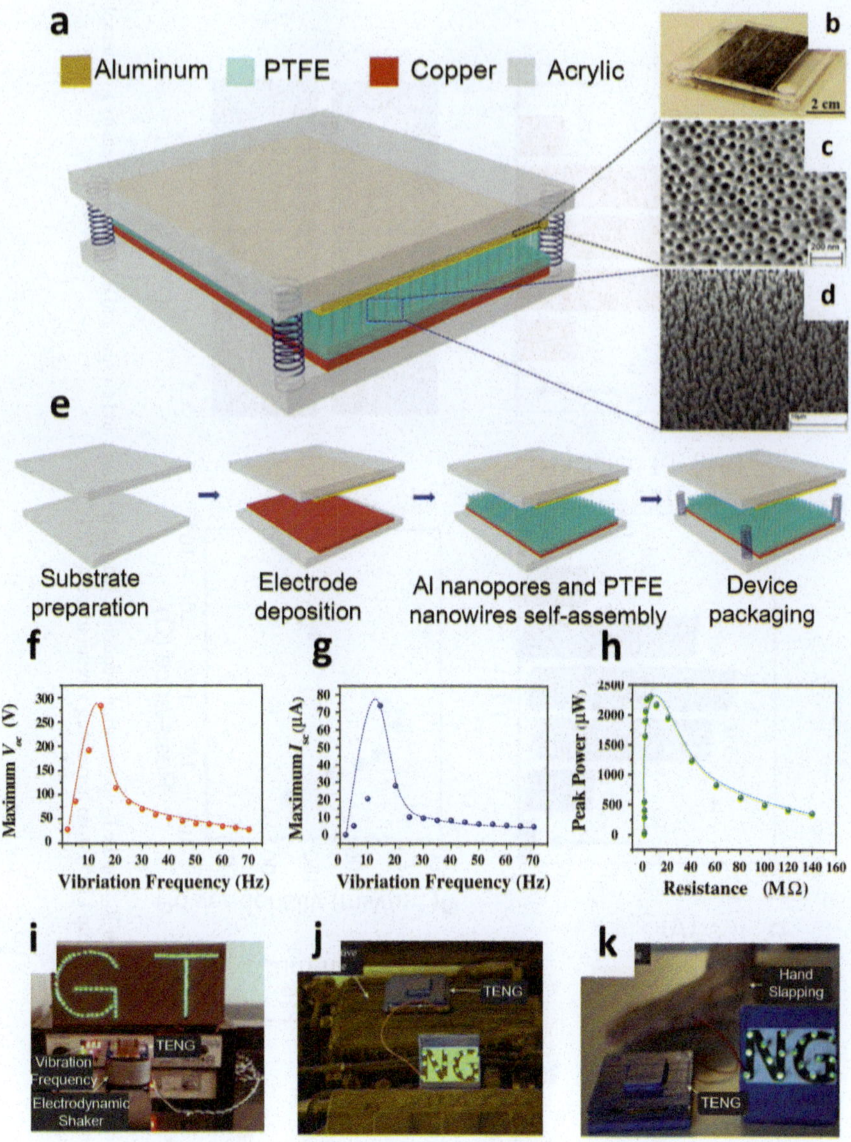

Fig. 9.11 **a** Schematic and **b** photo of V-TENG. SEM image of **c** porous Al film and **d** PTFE nanowire film. **e** Schematic of the fabrication process. Output **f** V_{OC}, and **g** I_{sc} to vibration frequency. **h** Peak power to external resistance. Photos demonstrating V-TENG to harvest vibration energy from a **i** shaker (60 Hz), **j** an automobile engine, and **k** a table slapped by hand [81]

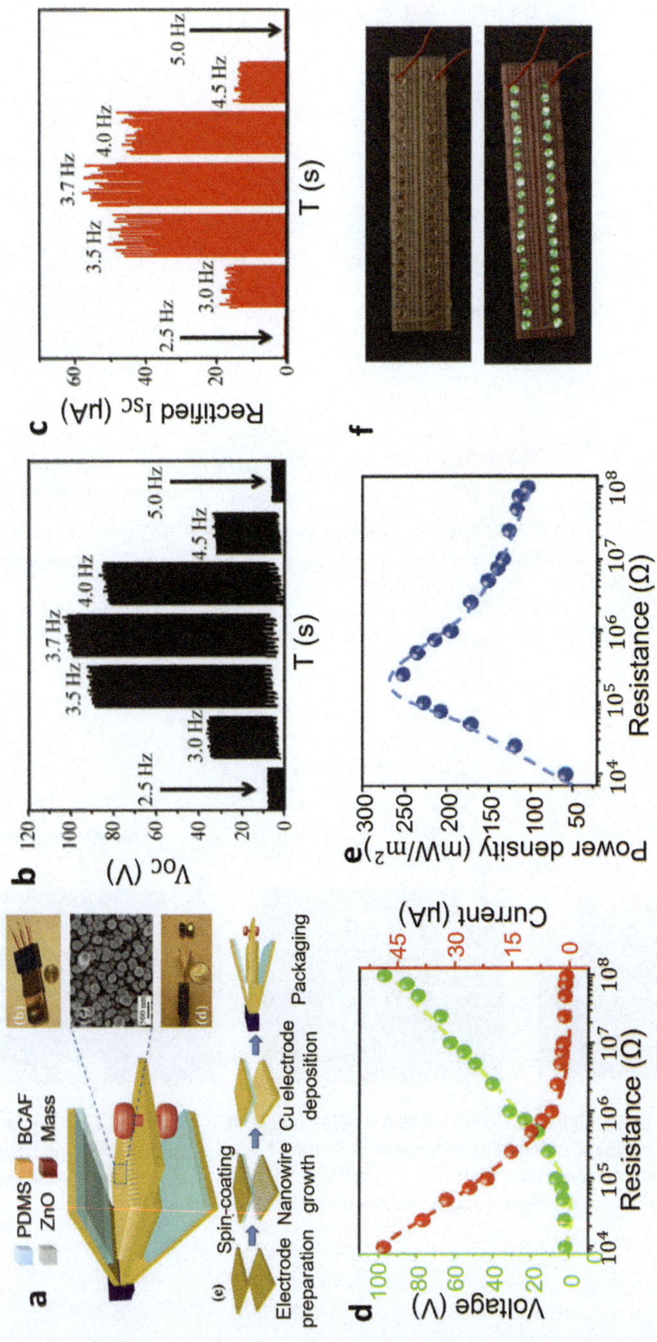

Fig. 9.12 **a** Schematic and photo of V-TENG. Output **b** V_{oc}, and **c** I_{sc} to vibration frequency. Output **d** voltage and current and **e** peak power density to external resistance. **f** Photos demonstrating V-TENG powering up LEDs from harvesting vibration energy [85]

PTFE. This configuration integrates two CS mode TENGs, where the Cu on the Kapton is electrically connected to the Cu electrodes on the back of the PTFE. A photo of the V-TENG is exhibited in Fig. 9.13b. This wavy structure of the internal Cu–Kapton–Cu film can efficiently convert vibration into lateral sliding, providing a much more efficient friction mode for energy harvesting, as shown in Fig. 9.13c. Such V-TENG can harvest vibrational energy from 5 to 500 Hz. And, it can generate V_{oc} of up to 72 V and I_{sc} of up to 32 μA at a resonance frequency of approximately 100 Hz (Fig. 9.13d and e). Moreover, its resonance exhibited a broad full width at half-maximum (FWHM) of over 100 Hz. At the resonance frequency, the device can produce a peak power density of 0.4 W/m^2 when the external load was 5 MΩ (Fig. 9.13f). The V-TENG can not only harvest vibrational energy but also be easily packaged for harvesting impact energy from water waves, as demonstrated in Fig. 9.13g and h.

9.4.1.4 Harvesting Wind Energy

Wind energy is a plentiful, renewable, and clean resource derived from nature. Conventional wind turbines are constrained by their large size, high costs, and the variability of wind direction. In contrast, wind-driven TENGs (WD-TENGs) can offer a lightweight, cost-effective, and simple structure with a small size, making them capable of efficiently harvesting wind energy for decentralized and distributed wind energy generation. The output voltage and current of various WD-TENGs can reach up to 400 V (with dimensions of $22 \times 10 \times 67$ mm^3) and 3.4 mA (with a radius of 70 mm), respectively [87, 88]. These outputs can be regulated using a transformer and power management circuit to meet the demands of different electrical equipment. Combining WD-TENG with other types of generators enhances adaptability for diverse environmental needs [89–91]. The structures of WD-TENGs are typically categorized into two types: flutter-driven and rotational structures.

The flutter-driven WD-TENG structure was first introduced by Yang et al. [92], as shown in Fig. 9.14a and b. The WD-TENG configuration, comprising two Al foil layers with a fluorinated ethylene propylene (FEP) film positioned between them. The FEP film is fixed on one side, leaving the other side free to flutter in response to wind. The Al foils serve as both triboelectric surfaces and electrodes, each connected to the ground (Fig. 9.14c). As a result, the WD-TENG consisted of two single-electrode TENGs. When wind flows across the structure, the freestanding side of the FEP film can vibrate up and down to contact the surfaces of both Al electrodes, leading to an output voltage and current to the external loads. The rectified output voltage and current from the WD-TENG (2.5 cm × 2.5 cm × 5.5 cm) at a wind speed of about 10 m s^{-1} can reach 20 V and 0.5 μA under a loading resistance of 100 MΩ (Fig. 9.14d and e). An integration of four WD-TENGs can reach output voltage and current of up to 100 V and 1.6 μA, respectively, corresponding to an output power of 0.16 mW. The flutter-driven WD-TENG can be developed using a CS mode, as shown in Figs. 9.14f, 9.15, 9.16 and 9.17a–g. The produced electricity energy can be used to directly drive 10 LEDs by scavenging the human mouth blowing induced wind energy, as shown

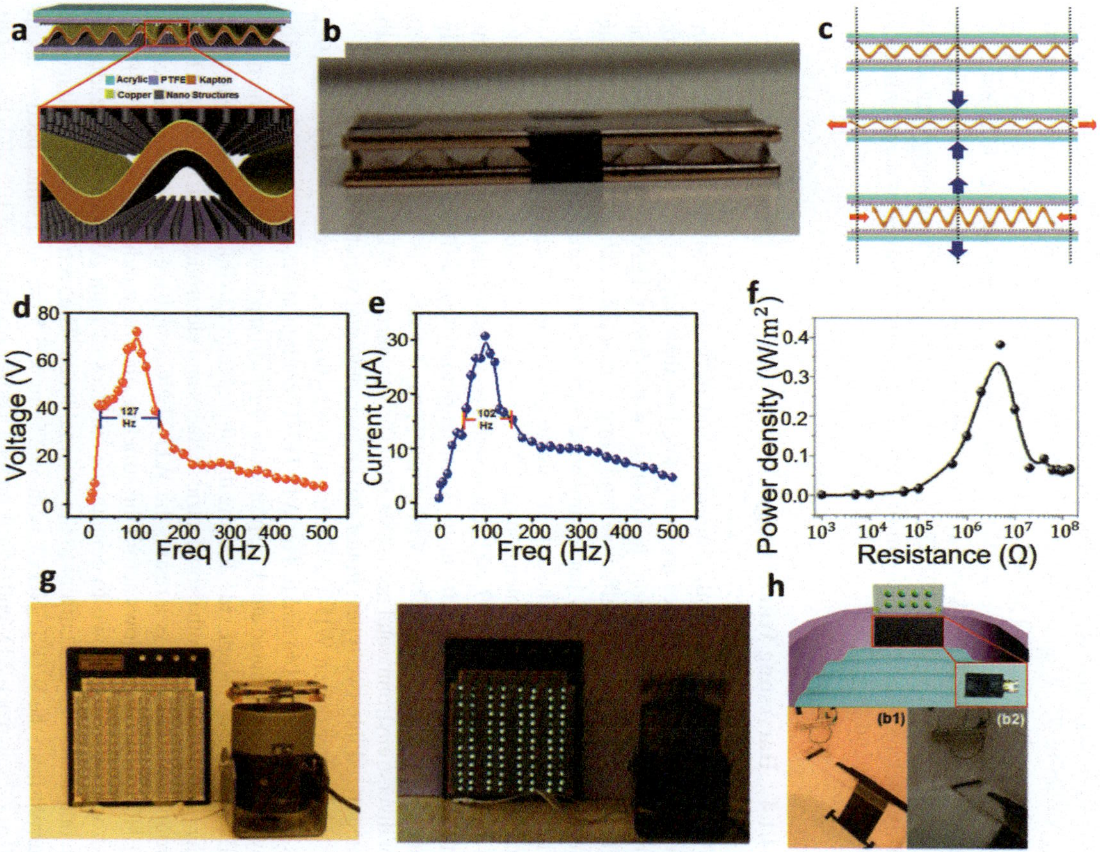

Fig. 9.13 **a** Schematic and **b** photo of V-TENG. **c** Working principle of V-TENG. Output **d** V_{oc}, and **e** I_{sc} to vibration frequency. Output **f** peak power density to external resistance. Photos demonstrating V-TENG powering up LEDs from **g** harvesting vibration energy and **h** wave energy [86]

in Fig. 9.14h. The CS-mode flutter-driven WD-TENG consists of a Cu electrode, a PTFE film, and an Al electrode on a Kapton film as the wind-responding unit. The electrodes were electrically connected to the load. For a CS-mode flutter-driven WD-TENG with a dimensions 2.5 cm × 0.6 cm × 5 cm, it achieved output voltage of up to 110 V and a current of 80 μA under a wind speed of about 10 m s^{-1} (Fig. 9.14i and j).

The rotational structures pair with wind turbines or wind cups are also widely used for the development of WD-TENG. For example, a rotational sweeping based WD-TENG was developed for harvesting weak wind energy in the environment, as shown in Fig. 9.15a and b [93]. The main structure of the rotational WD-TENG consists of a framework, shaft, flexible rotor blade, and two stators. The WD-TENG was operated using conventional wind cups. In this wind energy harvesting process, the rotor and stators undergo continuous face-to-face contact (over an area of 2.5 cm × 6.4 cm) and separation, which can serve as the basic TENG-based process for generating electricity (Fig. 9.15c). When wind blows on the wind cups, the flexible polyester (PET) rotor blade, with a polytetrafluoroethylene (PTFE) film adhered to its tip, would periodically sweep across the Al electrode. In this process, the PTFE film consecutively contacts and separates from Al electrodes, serving as the basic process for generating electricity. The working mechanism is shown in Fig. 9.15c. With the wind speeds being raised from 6.3 to 20.1 m/s, the V_{oc} generated from the WD-TENG reached a saturated value of 250 V (Fig. 9.15d), which can be attributed to the saturated surface charge density on the PTFE. While the I_{sc} show a very obvious increasing tendency with the increased wind speeds and reached 0.25 mA at the wind speed of 20.1 m/s (Fig. 9.15e). The increasing tendency of I_{sc} can be attributed to the fact that higher wind speeds can enhance higher charge transfer rates. The Wd-TENG can provide a maximum power of 62.5 mW at the wind speed of 15 m s^{-1}.

In addition to flutter-driven and rotational structures, other novel WD-TENG designs have been developed to meet the requirements of various conditions. For example, a thin membrane based WD-TENG was developed for miniaturization of the device with high output (Fig. 9.16a and b) [94]. It was structured by two rigid Al electrodes and a vibrating membrane that consisted of an Al film sandwiched by two micro-pyramid-exposed PDMS on both the top and bottom sides (Fig. 9.16c). Kapton films as spacers are used for creating air gaps between the Al electrodes and inner PDMS membrane. The create of the narrow air gap is critical for reducing device size, enhancing effective contact area, and improving mechanical robustness.

Moreover, a lawn structured, flexible and transparent WD-TENG was reported to harvest natural wind energy at arbitrary wind blowing direction on rooftops (Fig. 9.16d) [95]. A unit of vertically freestanding polymer strips was constructed using a PET thin film with an indium tin oxide (ITO) coating on one side. The laminar WD-TENGs are arranged in a kelp forest-like morphology, where each strip can sway independently, enabling a contact-separation process between adjacent strips as natural wind flows through. This movement generates output signals with each contact, and the numerous standing strips together form an environmental wind energy harvesting system (Fig. 9.16e).

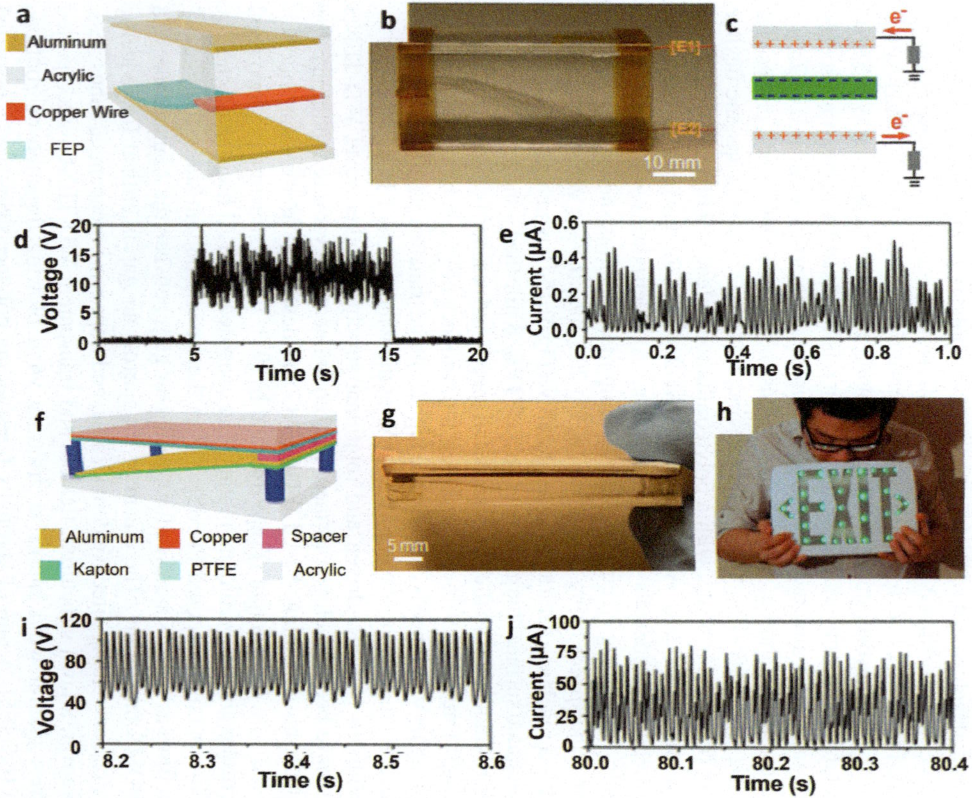

Fig. 9.14 **a** Schematic and **b** photo of flutter-driven WD-TENG. **c** Working principle of V-TENG. Output **d** voltage, and **e** current at a wind speed of about 10 m s^{-1}. **f** Schematic and **g** photo of CS-mode flutter-driven WD-TENG. **h** Photos demonstrating WD-TENG powering up LEDs by scavenging the human mouth blowing induced wind energy. Output **i** voltage, and **j** current of CS-mode flutter-driven WD-TENG at a wind speed of about 10 m s^{-1} [92]

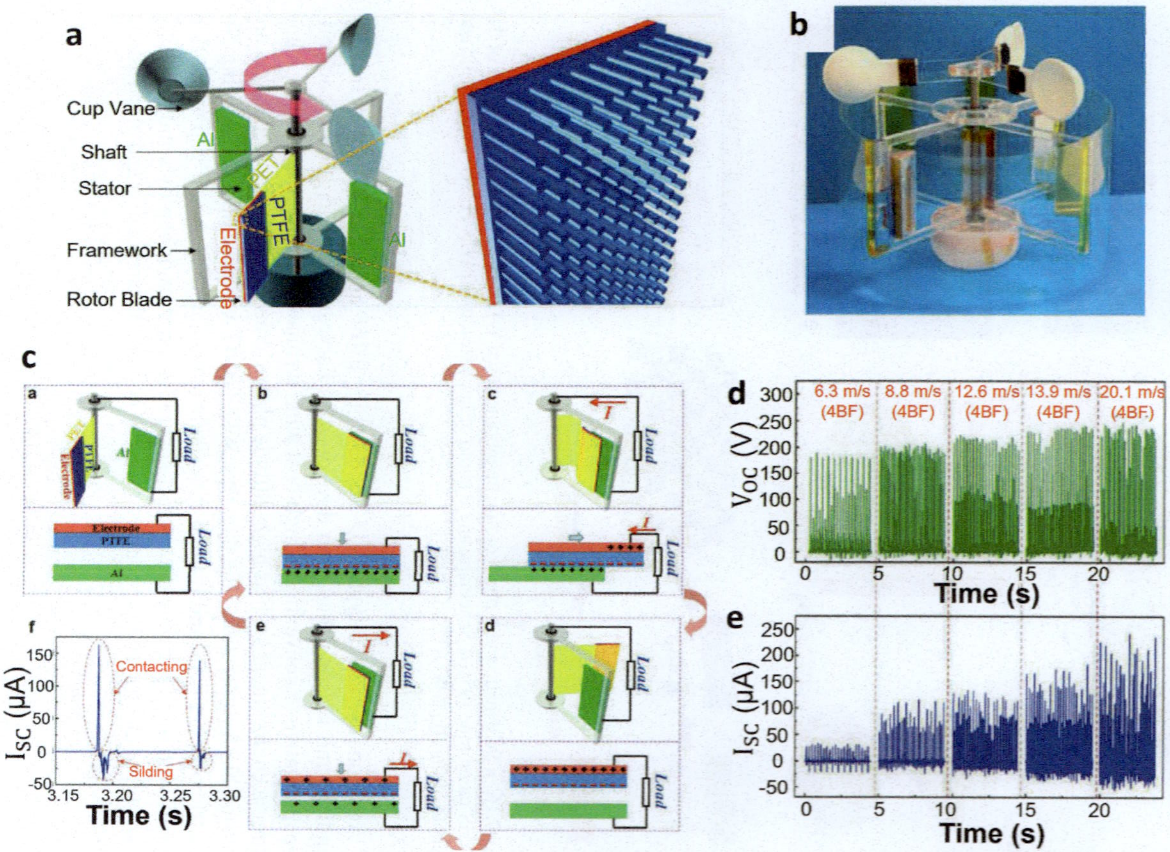

Fig. 9.15 **a** Schematic and **b** photo of rotational WD-TENG. **c** Working principle of rotational WD-TENG. Output **d** V_{oc}, and **e** I_{sc} at different wind speed [93]

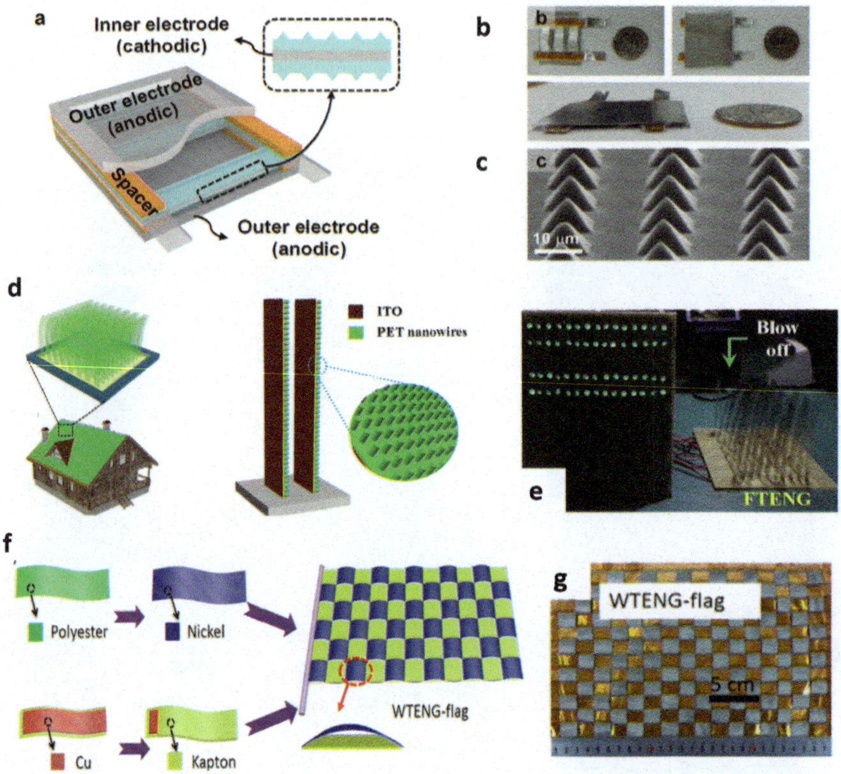

Fig. 9.16 **a** Schematic and **b** photo of thin membrane based WD-TENG [94]. **c** SEM image of the PDMS membrane [95]. **d** Schematic and **e** photo of lawn structured WD-TENG. **f** Schematic and **g** photo of flag-type WD-TENG [96]

Additionally, a flag-type WD-TENG (Fig. 9.16f and g), capable of folding, bending, and twisting, has been developed to harvest wind energy as a competitive alternative to conventional high-cost wind turbines [96]. In particular, for some extreme environment such as high-altitude places, conventional wind generators are not suitable for operation. The flag-type WD-TENG was developed by weaving Ni-coated polyester belts and Kapton film sandwiched Cu belts. An air gap was left in each woven unit to realize the contact-separation process driven by wind, resulting in an output electricity by a coupled effect of contact electrification and electrostatic induction.

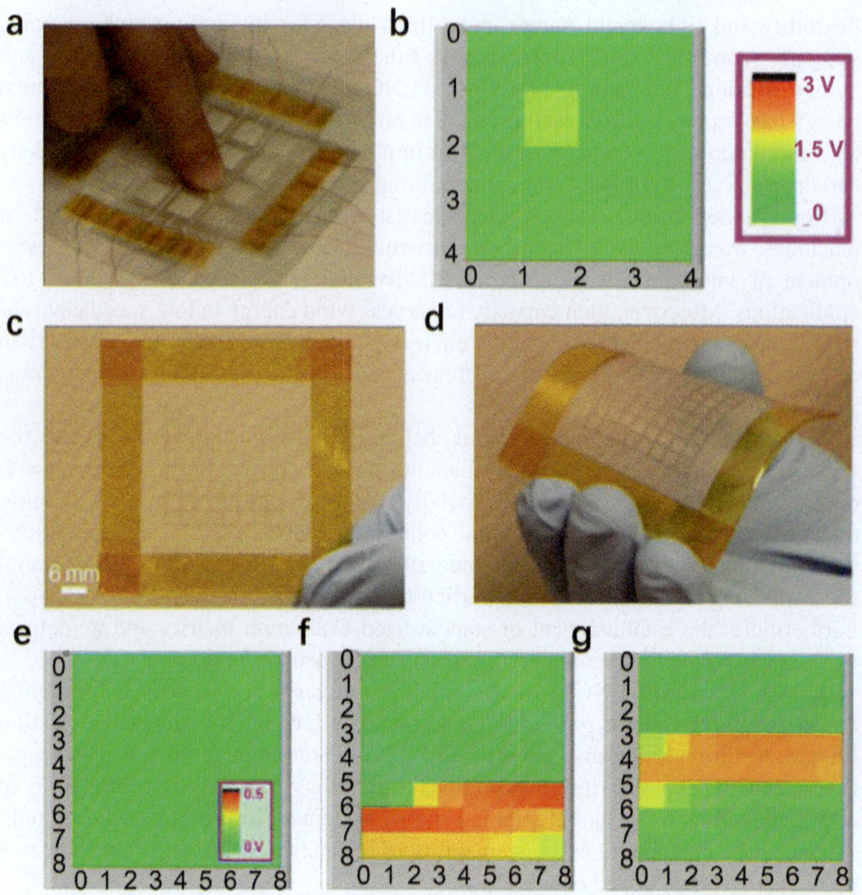

Fig. 9.17 **a** Photograph of the matrix touched by a human finger. **b** Measured positive output voltage mapping figure for the touching in (**a**). **c** Photograph of the fabricated 8 × 8 matrix with each pixel being 3 mm × 3 mm. **d** Photograph of the bent sensor system. **e** Mapping the figure of the sensor system without touching it. Measured mapping figures when the **f** bottom and **g** middle areas were touched by the hand [30]

9.4.1.5 Conclusions and Perspective

TENGs have established themselves as a promising technology for micro- and nano-power sources, offering a sustainable and efficient solution for harvesting mechanical energy from diverse sources. By converting mechanical stimuli into electrical energy through the triboelectric effect, TENGs have demonstrated remarkable versatility in applications such as harvesting body motion energy, capturing mechanical energy under in vivo conditions, utilizing vibration energy, and harnessing wind energy. These capabilities address the growing demand for self-powered systems in wearable devices, biomedical applications, and distributed energy solutions. Their flexibility and lightweight nature make them ideal for integrating with wearable systems, supporting applications such as fitness tracking, health monitoring, and human–machine interaction. Moreover, TENGs designed for in vivo mechanical energy harvesting highlight their promise in biomedical applications, offering a reliable and biocompatible energy source for implantable devices, such as pacemakers and drug delivery systems, that require continuous operation.

Beyond healthcare, TENGs can harvest vibration energy from industrial machines, transportation systems, and structural vibrations, facilitating the development of smart infrastructure and predictive maintenance within industrial IoT applications. Moreover, their capacity to harness wind energy in low-speed environments makes them ideal for powering environmental IoT sensors in remote or urban areas, contributing to applications such as air quality monitoring, weather stations, and smart agriculture.

Despite these achievements, several challenges must be addressed to enable the widespread adoption of TENGs. Enhancing energy conversion efficiency, developing durable and environmentally friendly materials, and improving the stability of TENGs under varying operational conditions remain critical research priorities. Miniaturization and seamless integration with existing systems are essential for expanding their applications, particularly in compact and portable devices. Furthermore, the establishment of standardized evaluation metrics and guidelines is necessary to facilitate commercialization and industry adoption.

Looking forward, TENGs are set to redefine the energy landscape of IoT, enabling a new generation of self-powered and sustainable devices. By addressing existing limitations through interdisciplinary collaboration among material scientists, engineers, and IoT system designers, TENG technology can play a central role in advancing autonomous and distributed energy systems. This progress will not only enhance the functionality and sustainability of IoT networks but also contribute to a more energy-efficient and connected world.

9.4.2 As Self-powered Sensors

Competing passive sensors rely on external electrical or optical signals to drive their operation. In contrast, TENGs generate electrical signals from mechanical stimuli, allowing them to function as self-powered sensors without the need for an external power source. In this section, we will highlight examples of self-powered TENG sensors, including tactile, biomechanical, acoustic, human physiological, and chemical sensing. Finally, we will conclude with an outlook on future developments and potential advancements in self-powered sensing technology.

9.4.2.1 Tactile Sensing

TENG-based self-powered tactile sensors are one the most important applications of TENG active sensors. A tactile sensor refers to a transducer that detects touch, force, or pressure, with wide-ranging applications in areas such as physiological monitoring, robotic skin, security systems, and human–machine interactions. Traditional passive sensors—such as resistive, capacitive, and optical types—require external power for their operation. These sensors are often constrained by their power consumption, design complexity, and limited material options, especially in large-area applications [97–99]. In contrast, TENGs offer extraordinary advantages including active sensing, streamlined design, material versatility, and cost-effectiveness, unlocking new potential for tactile sensors that could not be realized before [100–105].

Many works have been done on self-powered tactile sensors based on vertical contact-separation mode and single-electrode mode. In a CS-mode TENG, contact is made between two elements by touch, and active electrical signals are generated directly in response to the touch, enabling TENG-based tactile sensors to operate as self-powered devices [75, 106–108]. For a single-electrode-mode TENG, skin contact alters the electric field generated by the surface triboelectric charges and leads to free electrons flowing through the electrode, thus creating active electrical signals [30, 109–112].

Yang et al. [30] reported the first single-electrode TENG using micropyramid-surface PDMS as the triboelectric layer and indium tin oxide (ITO) as the electrode for sensing human skin touching. And, a sensing matrix can be realized by integrated 16 devices for mapping and tracking the location and pressure of touch. The tactile information including position and pressure information was obtained by monitoring the active voltage signals from the device. Figure 9.17a shows when one of the 16 devices was touched, a positive voltage signal of 1.4 V was observed in the measured mapping (Fig. 9.17b). To enhance the detection resolution, an 8×8 matrix with a pixel size of 3 mm $\times$ 3 mm was fabricated. Figure 9.17c and d shows a photograph of the matrix, where 64 pixels were created on a total area of 3 cm $\times$ 3 cm. By monitoring the positive voltage of each working pixel, the spatial profile of all pixels was easily displayed. Figure 9.17e shows no output signal was observed without applying pressure, while Fig. 9.17f and g demonstrates the bottom, middle, and top regions of the matrix were touched.

In order to achieve a deformable and stretchable tactile sensing matrix, inspired by an electric eel's skin, Lai et al. [113] reported a mechanical-durable and stretchable single-electrode TENG for deformable energy harvesting and self-powered tactile sensing (Fig. 9.18a–j). The device utilized a network of percolating silver nanowires as the electrode encapsulated in silicone rubber as the triboelectric layer. With mechanical durability and toughness, the device exhibited excellent biaxial stretch, uniaxial stretch at strains over 300%, and the ability to be twisted and folded multiple times (Fig. 9.18b–d). Figure 9.18e and f shows the V_{oc} and σ_{tr} under various strain levels, with Fig. 9.18g displaying the statistic results. Interestingly, the output of the stretchable TENG increased as the applied strain increased, which can be attributed to the Poisson's ratio of the silicone rubber. As the triboelectric silicone rubber was stretched, it became thinner, thereby inducing a greater flow of electrons in the electrode. These results showed that, regardless of various extreme mechanical conditions and deformations, the device can convert energy from skin touching into electricity. Due to its excellent mechanical tolerance, the skin-like device was applied to various non-planar surfaces to detect tactile stimuli through self-generating electricity. Moreover, a fully autonomous, body-attachable 3 × 3 sensing array system with intuitive visual signals was realized using the skin-like device (Fig. 9.18h and i).

The development of sensing skins for soft robots is extremely challenging because of continuously deformable bodies. With the unique merits of TENG-based tactile sensors, Lai et al. [105] reported the first self-powered, highly stretchable, and highly sensitive triboelectric robotic skin (tribo-skin) that can be integrated in various kinds of soft robots (Fig. 9.19a). Figure 9.19a showcases the exceptional deformability and mechanical compliance of the tribo-skin, demonstrating its ability to be folded, stretched, and twisted. The tribo-skins are made with silver nanoflakes (Ag NFs) as

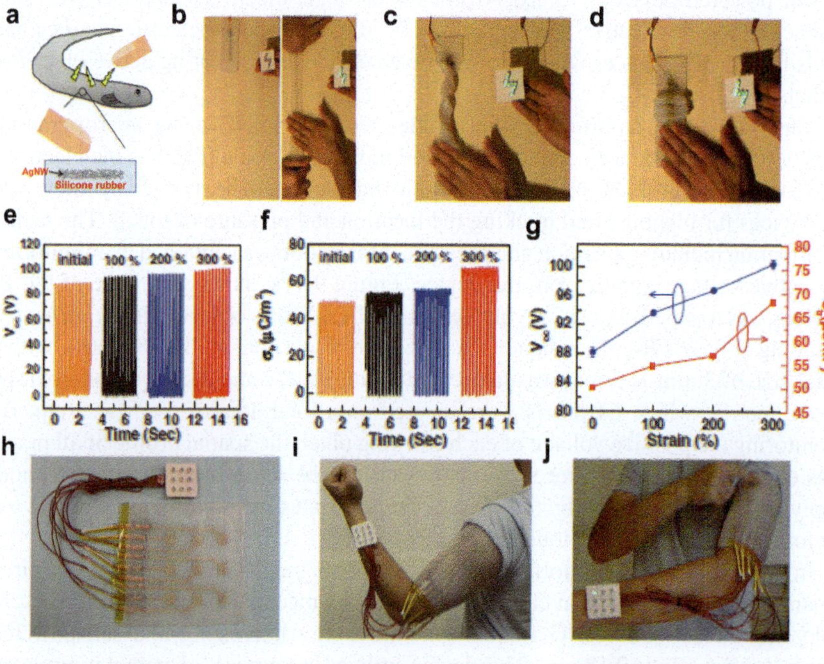

Fig. 9.18 **a** Illustration of concept of an electric eel skin-inspired TENG. **b** Photo demonstrating the device can well function and drive a load when it is extremely stretched in a long-axial direction. Photos demonstrating the device drove a load when it was **c** extremely twisted and **d** multiply folded. The measured **e** V_{oc} and **f** σ_{tr} after different strain levels. **g** Statistic results of V_{oc} and σ_{tr} under various strains. **h** Photograph of the fully autonomous and self-powered e-skin system. **i**, **j** Photo demonstrating the e-skin system conformally contacted on a human arm and the LED lit up corresponded to sensing unit that is being touched [113]

the electrode and silicone rubber with a triangular microprism surface as the triboelectric material. The device can actively sense proximity and contact pressure through self-generated electricity (Fig. 9.19b and c). It exhibits excellent stretchability up to 100% strain, a high sensitivity of 0.29 kPa^{-1} (9.54 V/kPa) in low-pressure ranges (< 5 kPa), and a minimum detection limit of 63 Pa (Fig. 9.19d). With these skins, soft robots can perform a range of sensing tasks, such as detecting their muscle motions and operational states (Fig. 9.19e and f), and monitoring subtle human physiological signals. The generated signals can also be processed by a computation module, enabling human–robot interaction (Fig. 9.19g).

A crucial challenge for single-electrode TENG sensing arrays is misrecognition of the signals between the sensing nodes and the conductive wires. To address the issue, a large-scale, multiplexing triboelectric tactile sensing array with an ultralow misrecognition rate of 0.20% was reported (Fig. 9.20) [114]. Figure 9.20a presents the exploded view of the multiplexing TENG tactile sensing array with a 16-channel sensing unit. Each TENG sensing unit was operated in single electrode mode, and a stretchable shielding layer consisting of a percolated carbon black-Ecoflex composite was designed to avoid the misrecognition between the sensing nodes and the conductive wires. The untethered TENG sensing array can be applied for diverse deformable tactile sensing devices, including electronic and robotic skins, stretchable keyboards, intelligent gloves, and smart insoles (Fig. 9.20b). Figure 9.20c shows the large-area stretchable tribo-skin (25 cm × 25 cm) with demonstrations of its exceptional deformability. Figure 9.20d shows the triboelectric sensing array without a shielding layer, which fails to localize the contact letters due to misidentification signals generated from the contact electrification on the conducting wires. However, the use of a top shielding layer can effectively prevent misrecognition interference (Fig. 9.20e). The results indicate that the use of a shielding layer can be beneficial for the development of multiplexed triboelectric sensing arrays.

9.4.2.2 Biomechanical Sensing and Monitoring

Leveraging the triboelectric effect to harvest biomechanical energy, TENGs have emerged as innovative self-powered devices for detecting biomechanical information, including gait analysis, motion detection, and activity monitoring [74, 115–119]. In motion detection, flexible TENG sensors excel due to their high sensitivity and adaptability to various types of motion.

Significant efforts have focused on developing wearable TENGs using fibers, textiles, and garments to harvest biomechanical energy and capture related data [68, 72, 74, 118, 120]. For instance, Dong et al. [74] reported stretchable yarn-based TENG with coaxial core-sheath and built-in spring-like spiral winding structures for biomechanical energy harvesting and self-powered body-motion sensing. The structure and the fabrication process of the proposed core-sheath TENG yarn are shown in Fig. 9.21a. By wearing a garment sewn with two TENG yarns onto the medial and lateral sides of a sleeve, the TENG-based clothing can detect the user's arm motion while playing golf (Fig. 9.21b). As shown in Fig. 9.21c–e, the TENG

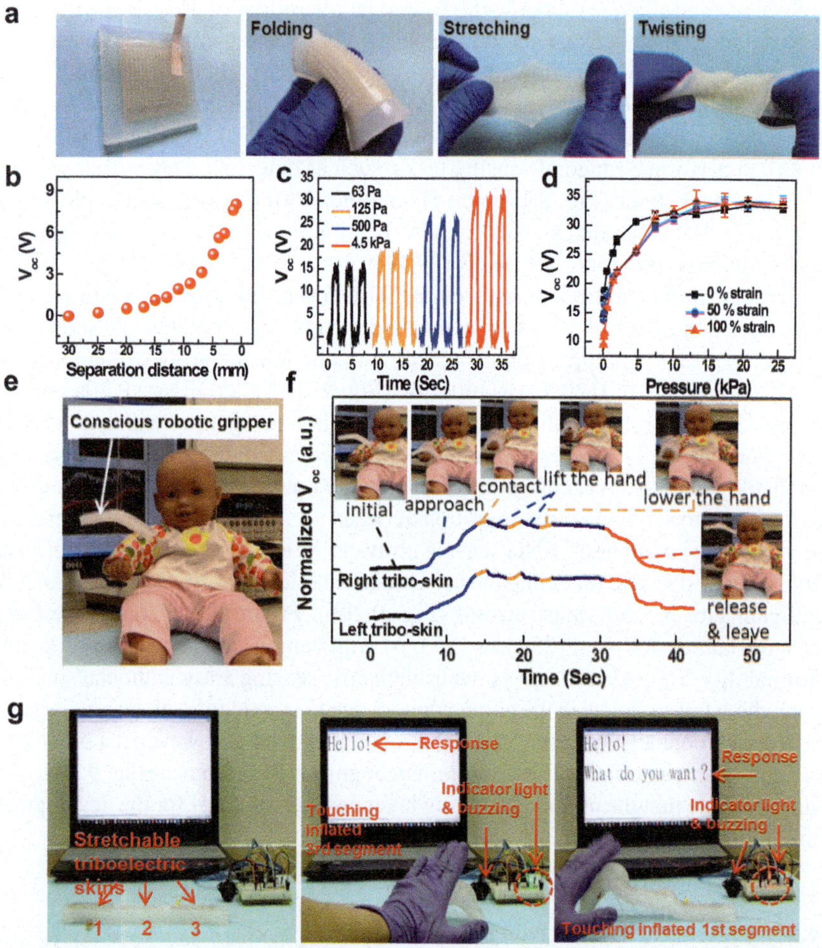

Fig. 9.19 **a** Photos of the as-prepared device with demonstrations under different mechanical deformations, including folding, stretching, and twisting. **b** Generated V_{OC} to the separation distance. **c** Real-time generated V_{OC} to different contact pressures. **d** Generated V_{OC} to difference pressures when the device was stretched. **e** Photos of the conscious gripper to hold the doll's hand. **f** Real-time outputs of right and left tribo-skins when conscious gripper held doll's hand. **g** Photos for demonstrating the actively sensed signals processed by a computing module to reactions with humans [105]

yarns can convert a series of standard actions into continuous electrical signals, which can provide a useful and convenient way for trainers to more efficiently master the skills and operational techniques of a sport.

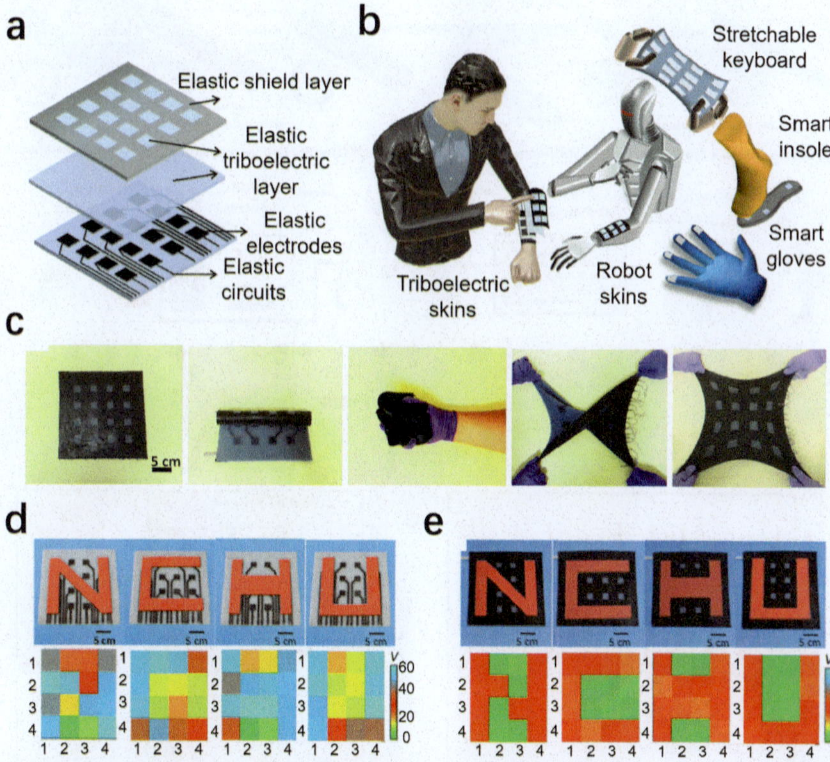

Fig. 9.20 **a** The exploded view of the triboelectric sensing array architecture involving elastic circuits, elastic electrodes, a triboelectric layer, and a shielding layer. **b** The broad applications of the tribo-skin in robotic skins, stretchable keyboards, smart insole, and smart gloves. **c** Photographs of triboelectric sensing array (left image) under folding, crumpling, twisting, and stretching. **d, e** Photographs (upper) and top views of 2D mapping for output voltage intensity (bottom) in response to different applied pressures, exerted using the contact letters "N", "C", "H", and "U" on the surfaces of 4 × 4 pixelated triboelectric sensing arrays d without and e with a top shielding layer [114]

Beside textile-based TENGs, on-skin TENG nanofibers can also extract biomechanical signals. For example, a permeable and stretchable monolithic hybrid tribo-piezo-electromagnetic nanogenerator-based electronic skin (TPEG-skin) was developed using a design featuring two-layer porous topological insulator-piezoelectric polymer composite nanofibers that sandwich liquid metal networks (Fig. 9.22a) [115]. This hybrid TPEG-skin can collect both biomechanical energy (from body motions) and electromagnetic energy (from ambient devices), serving as both a highly efficient epidermal power source and a multifunctional self-powered sensor. Figure 9.22b illustrates the exceptional stretchability of the TPEG-skin, reaching up to 200%, and demonstrates its ability to adhere to various body locations for monitoring user movements and energy harvesting. It enables precise detection of whole-body physiological signals, cheek vibrations during inhalation/exhalation, eye

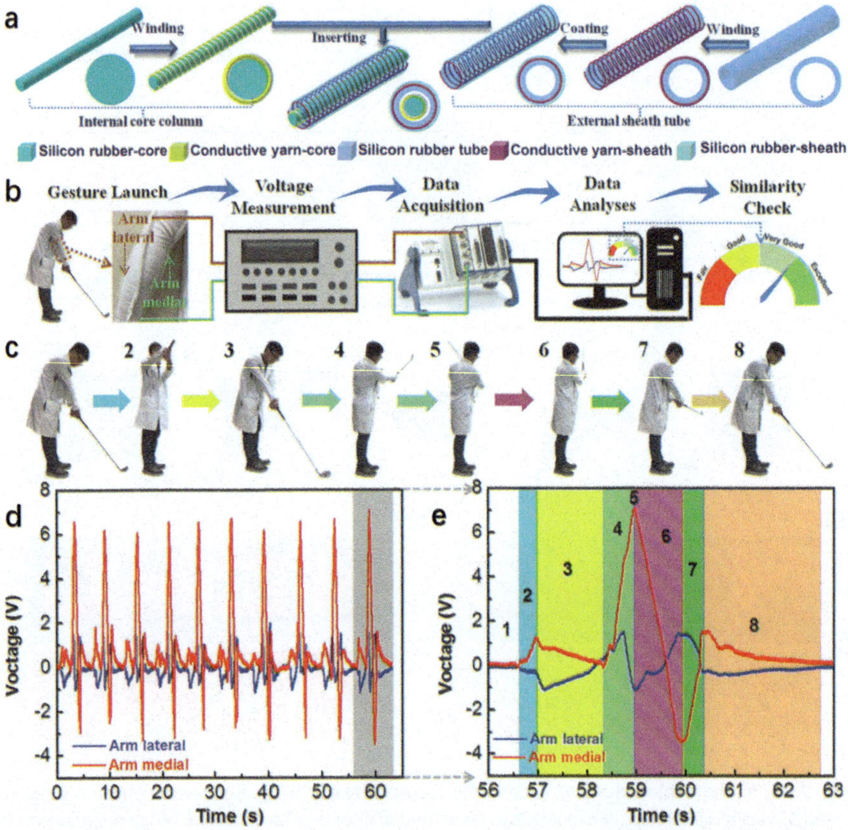

Fig. 9.21 **a** Fabrication process of the coaxial core–sheath yarn-based TENG. **b** Schematic of the real-time golf scoring system, which includes two yarn-based TENGs, a voltage measurement device, a data acquisition system, and specially developed data analysis and scoring software. The two yarn-based TENGs are sewn onto the medial and lateral sides of a sleeve. **c** Eight consecutive snapshots selected from a complete golf swing action. **d** Real-time voltage output signals from the two yarn-based TENGs during the golf swing process. **e** Amplified voltage signals of the two yarn-based TENGs throughout a complete golf swing, with the voltage curve divided into eight sections corresponding to the eight steps of the golf swing shown in (**c**) [74]

blinking, facial expressions, joint movements, and self-powered human–machine interactions, as demonstrated in Fig. 9.22c.

Non-contact TENGs are promising for biomechanical energy harvesting and sensing because they can operate without physical contact, ensuring the durability of the TENGs. A stretchable contactless TENG textile, using a nanocomposite coating (triboelectric layer) and a graphite-like fabric (tribo-charges reservoir), was reported for noncontact biomechanical energy harvesting and sensing (Fig. 9.23a) [121]. This design facilitates the capture, transport, and storage of tribo-charges, resulting in improved device performance (452 V, 1.96 mA m^2, and 179 mW m^2) and extended

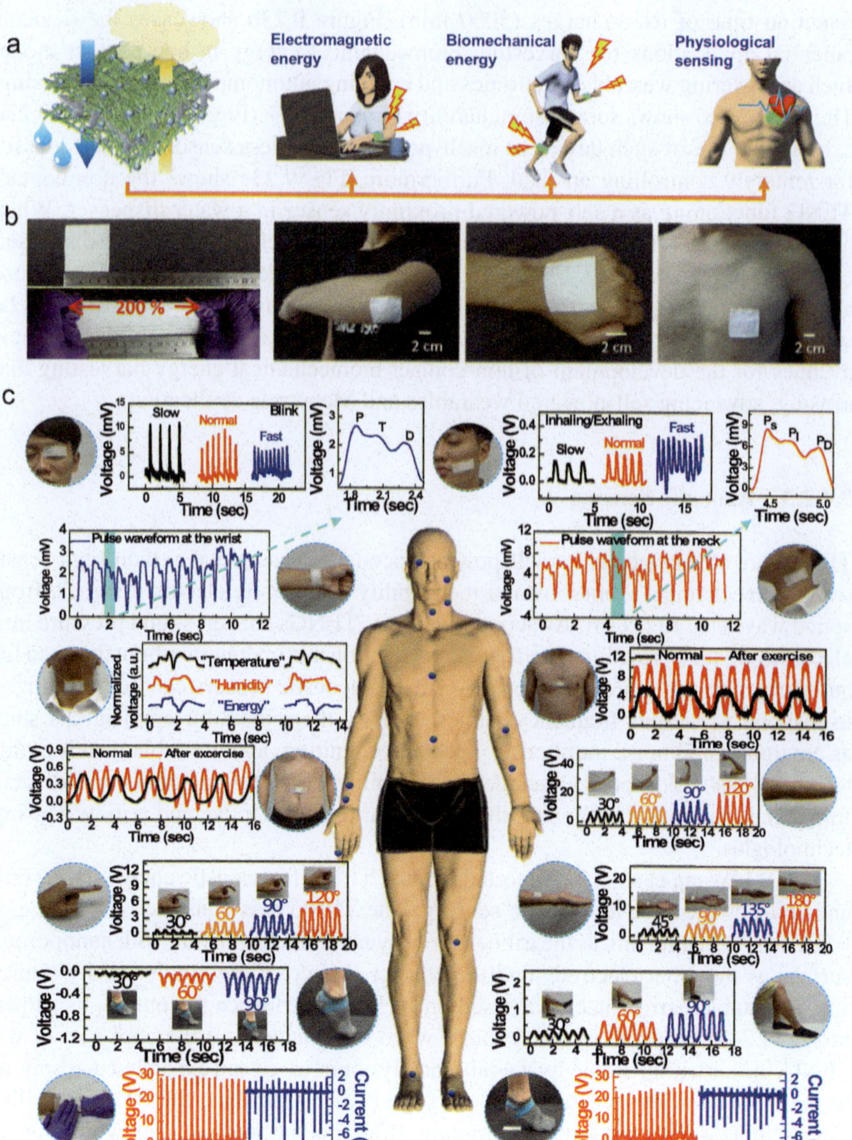

Fig. 9.22 **a** Schematic illustration of a permeable and stretchable TPEG-skin and its application scenarios for hybrid electromagnetic and biomechanical energy harvesting and self-powered physiological sensing. **b** Photographs of TPEG-skin at 0 and 200% strain and demonstrations of excellent adhesion on the body. **c** Demonstrations of TPEG-skins as whole-body self-powered biomechanical sensors [115]

retention time of tribo-charges (5000 min). Figure 9.23b showcases the device's potential applications for harvesting biomechanical energy in non-contact mode, such as powering wearable electronics and enabling autonomous proximity sensing. The device also shows superb stretchability of over 200% (Fig. 9.23c). Figure 9.23d exhibits the use of such device as a self-powered touchless sensor in smart glasses for remotely controlling an iPad. Furthermore, Fig. 9.23e shows the non-contact TENG functioning as a self-powered proximity sensor in a water dispenser. When a cup approaches the device, it generates a signal to activate the water dispenser, causing it to release water. Figure 9.23f demonstrates the non-contact TENG's use as a self-powered proximity sensor in a vending machine, where an item can be selected by simply approaching the sensor with the hand. These results open new avenues for the development of non-contact biomechanical energy harvesting and sensing, advancing self-powered wearables and Metaverse applications.

9.4.2.3 Acoustic Sensing

TENGs are advantageous for self-powered acoustic sensing due to their high sensitivity to mechanical vibrations and their ability to generate electrical signals from sound waves [6, 122, 123]. As acoustic sensors, TENGs convert sound pressure into electrical signals, providing a self-powered solution that operates without the need for an external power source. These TENG-based acoustic sensors can be customized to respond to various frequency ranges, making them ideal for applications such as environmental noise monitoring, speech recognition, and wearable health monitoring devices. Additionally, their design flexibility allows for integration into diverse forms and materials, broadening their potential use in adaptive and remote sensing technologies.

In 2014, Wang et al. [122] reported the first TENG for acoustic energy harvesting and self-powered active acoustic sensing. The TENG was constructed with a Cu/PTFE nanowire thin film as the triboelectric layer and an Al thin film with nanoporous surface as the other electrode and the contact surface. A Helmholtz cavity with a size-adjustable narrow neck was used on its back to enhance the output, as shown in Fig. 9.24a. When an external sound wave was incident on the core part of the TENG, the air within the cavity was alternately compressed and expanded in response to the magnitude and frequency of the sound wave, and thus, the PTFE thin film can oscillate due to the initiated pressure difference on its two sides while the Al film stays still. This configuration enables efficient conversion of acoustic energy into electrical energy through triboelectric transduction. The frequency components of the reconstructed signal span a broad range, covering frequencies from 10 to 1700 Hz, as illustrated in Fig. 9.24b. Additionally, Fig. 9.24c demonstrates that the fabricated TENG acoustic sensor can function as a self-powered microphone for sound recording.

Moreover, a paper-based TENG was reported to act as an affordable, high-quality, and versatile voice recorder [123]. Figure 9.25 illustrates the design in which a perforated paper layer faces another paper layer, separated by a small gap. The ultrathin

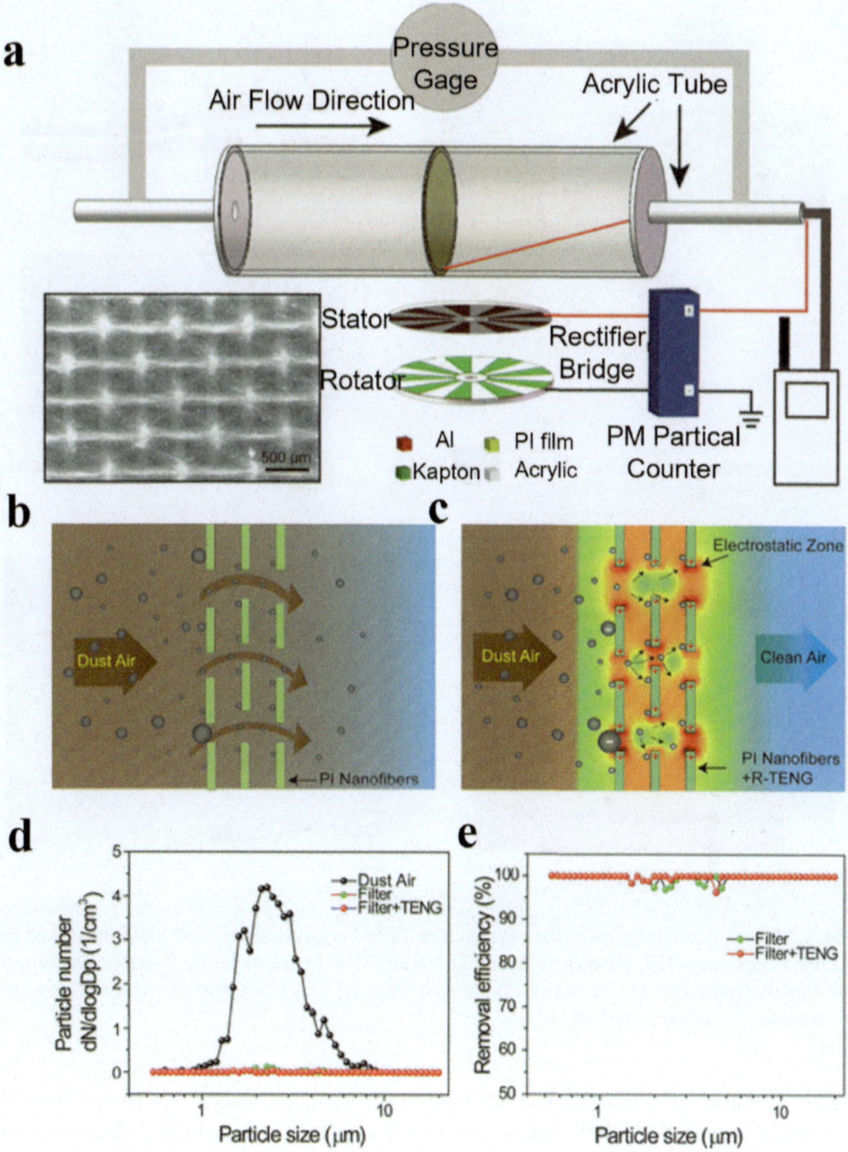

Fig. 9.23 **a** Schematic diagram of the design concept of the stretchable non-contact TENG textile. **b** Illustration demonstrating non-contact energy harvesting and self-powered sensing. **c** Photo demonstrating the stretchability of the non-contact TENG textile. Photos demonstrating the use of the non-contact TENGs as self-powered proximity sensors in **d** a pair of smart glasses, **e** a water dispenser, and **f** a vending machine [121]

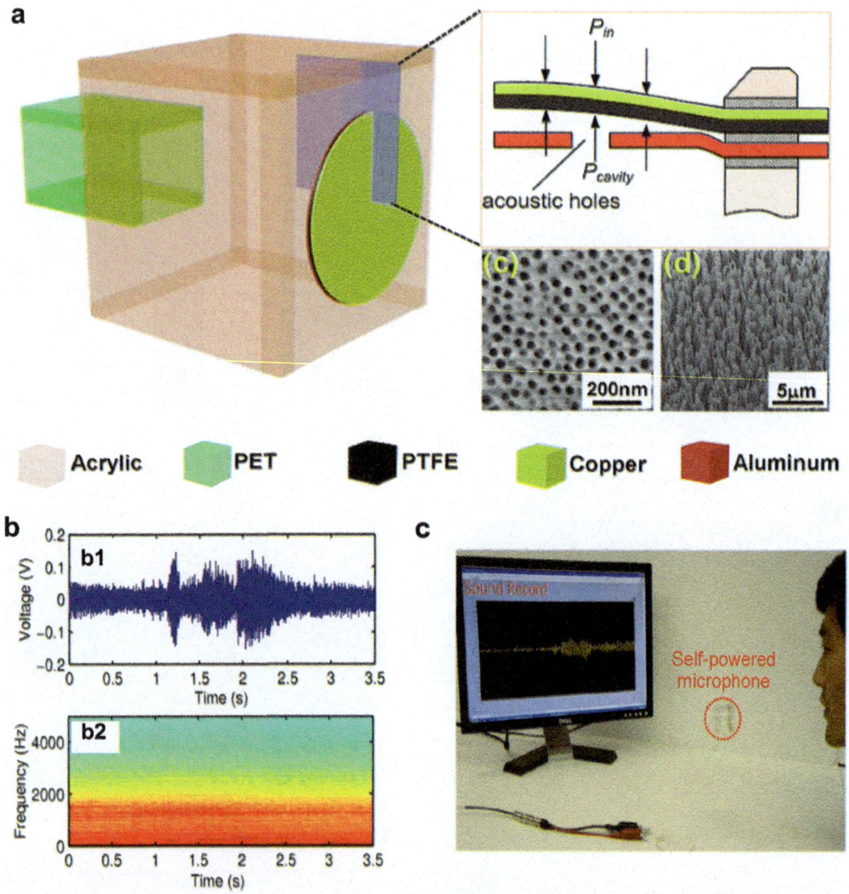

Fig. 9.24 **a** Structural design of the organic film TENG acoustic sensor and SEM images of the Al nanopores and PTFE nanowires. **b** Sound waveform and corresponding Fourier transform of the signal acquired by the TENG. **c** Photograph showing an TENG working as a self-powered microphone for sound recording [122]

TENG features a multilayered structure with vertically laminated thin film materials, consisting of a multi-holed paper layer, which is coated with a copper electrification layer (Fig. 9.25a and b). This layer generates triboelectric charges when it comes into contact with a thin PTFE membrane. Figure 9.25c shows that the deformation regions and magnitudes of the PTFE membrane are closely related to the external sound excitation frequencies, which are attributed to varying resonance frequencies under different vibration modes. The output signals corresponding to the sound frequency using paper and plastic films are shown in Fig. 9.25d. Finally, the paper-based TENG was developed as a low-cost and ultrathin self-powered microphone, capable of efficiently converting human voice into electrical signals for recording purposes, as demonstrated in Fig. 9.25e.

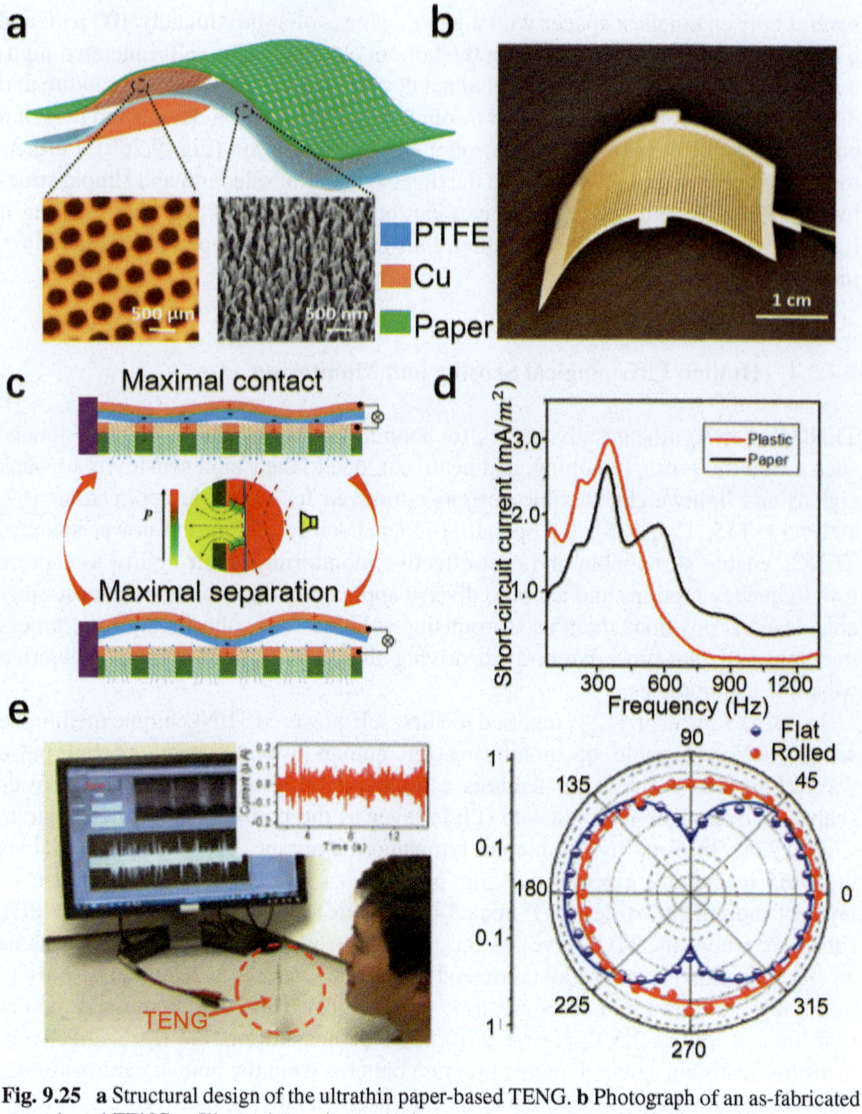

Fig. 9.25 **a** Structural design of the ultrathin paper-based TENG. **b** Photograph of an as-fabricated paper-based TENG. **c** Illustration to interpret the sound wave-induced PTFE membrane vibration and electricity generation. **d** Output performance of paper- or plastic-based TENG acoustic sensors as a function of sound frequency. **e** Photograph that shows a paper-thin triboelectric nanogenerator as a self-powered microphone for sound recording. Inset is the acquired electrical signals [123]

In 2018, Guo et al. [6] developed a highly sensitive, self-powered triboelectric auditory sensor for social robotics and hearing aids. Figure 9.26a presents the triboelectric cochlea device for an intelligent auditory system, which is integrated inside the artificial ear. The circular-type device primarily consists of an upper electrode covered with fluorinated ethylene propylene (FEP) and featuring several hole channels, a spacer with a gap thickness of approximately 100 μm, and a Kapton membrane connected with the bottom electrode. The self-generated high-output electrical signal and single-channel design enable easy real-time capture and display of the recorded information through a data acquisition device and personal computer, which helps prolong the robot's operational time (Fig. 9.26b). Furthermore, as shown in Fig. 9.26c, due to the diverse material selection and simple structure of TENG techniques, the receiver can be designed as an ear stud, making it lightweight and aesthetically appealing, thus creating a new type of hearing aid for medical applications.

9.4.2.4 Human Physiological Sensing and Monitoring

TENGs offer significant advantages for monitoring human physiological signals, such as arterial pulse, breathing, and heartbeat, due to their high sensitivity to weak signals and lightweight, flexible designs optimized for wearable applications [67, 105, 111, 115, 124, 125]. By operating independently of external power sources, TENGs enable sustainable and cost-effective monitoring. Their ability to capture low-frequency motions and adapt to diverse applications, combined with customizable designs, positions them as a promising solution for health monitoring, fitness tracking, and real-time diagnostics, driving the advancement of next-generation wearable technologies.

In 2015, Yang et al. [124] reported the first self-powered TENG bionic membrane sensor (BMS) capable of monitoring the human low-frequency arterial pulse wave (Fig. 9.27). The BMS features a multilayered structure with a thin, oval-shaped polyethylene terephthalate (PET) layer as the bottom supporting substrate (Fig. 9.27a). Inspired by the human tympanic membrane, the oval shape enables the BMS to monitor external dynamic pressure across a wide frequency range. A layer of indium tin oxide (ITO)-coated nylon thin film is laminated onto the PET substrate, where the ITO serves as the back electrode and the nylon functions as an electrification layer. The as-fabricated BMS demonstrates exceptional sensitivity of 51 mV Pa^{-1}, a rapid response time of less than 6 ms, and a pressure detection limit as low as 2.5 Pa (Fig. 9.27b). As a demonstration, the BMS effectively measures distinguishable dynamic pressure patterns from the human cardiovascular system, enabling its use as a wearable, mobile, user-friendly, and self-powered arterial pulse wave monitor for continuous, noninvasive health assessment and monitoring (Fig. 9.27c–e). Three BMSs were directly attached to simultaneously monitor the pulse waves of the participant from the carotid artery, chest, and wrist. Key arterial health information, including pulse rate, pulse wave velocity (PWV), augmentation index (AIx), and reflection index (RI), can be systematically obtained for further medical diagnosis (Fig. 9.27f–h).

Respiration is a critical vital sign, and monitoring respiratory function is essential for effective physical health management. Wang et al. [126] reported an air flow-driven TENG for self-powered real-time respiratory monitoring by converting the mechanical energy of human respiration into electric output signals (Fig. 9.28). The

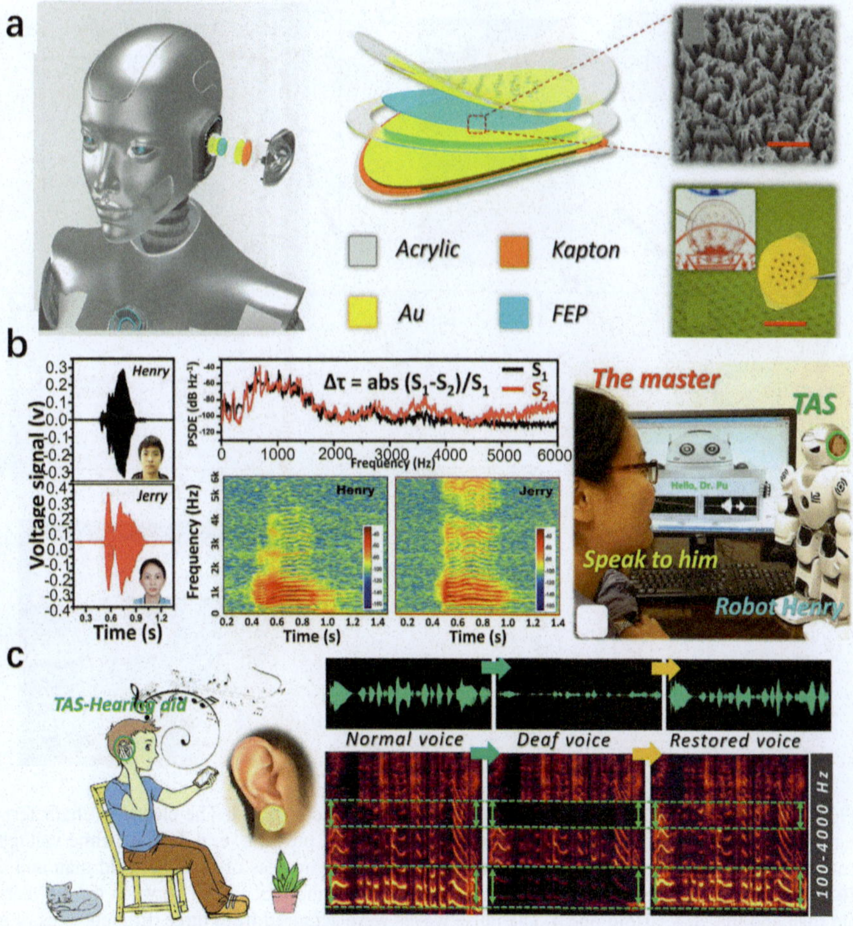

Fig. 9.26 **a** Proposed image of triboelectric auditory system for a robot (left), basic structure scheme of the device (middle), and SEM images (right) of the FEP surface (scale bar, 1 mm). **b** Voices recorded from two people saying "Hello." (left), power spectral and voice spectrogram of the recorded voices for identification (middle), and demonstration of the TENG-based voice recognition system (right). **c** Proposed medical application of TENG acoustic sensor in hearing aids to allow a hearing-impaired person to fully hear music (left) and sound waves and corresponding acoustic spectrograms of normal, weakened, and restored voices (right) [6]

TENG sensor was fabricated using a nanostructured PTFE (n-PTFE) film as the triboelectric layer, as shown in Fig. 9.28a. The n-PTFE film consists of numerous intertwined nanofibers, resulting in high porosity and an exceptionally large surface area both inside and on the surface of the film. Figure 9.28b shows the corresponding output voltage of the n-PTFE film, with an average peak value of 2.4 V. When a n-PTFE TENG was installed in a regular mask, it was able to provide a real-time display of different breathing patterns, as demonstrated in Fig. 9.28c and d.

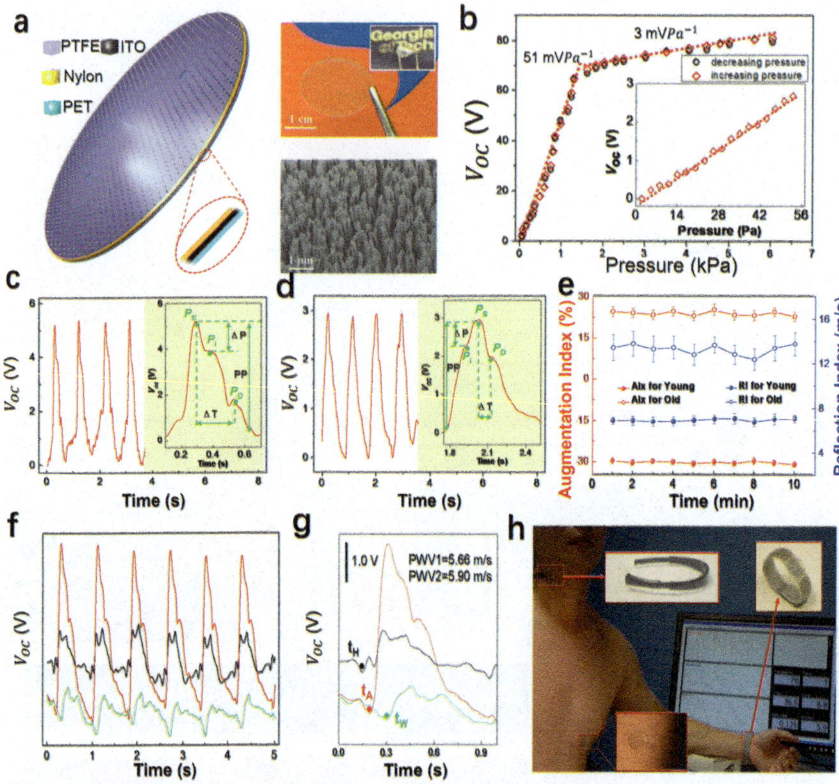

Fig. 9.27 **a** Structural design of the TENG bionic membrane sensor. **b** The electrical characterization of the bionic membrane sensor in response to applied pressure. **c, d** The real-time voltage outputs when the TENG sensors are placed over the carotid arteries of a 30-year-old man and a 70-year-old man, respectively. **e** Quantitative comparison of the AIx and RI between the 30-year-old man and the 70-year-old one. **f** The pulse waves were acquired from three different sites of a participant: the carotid artery, the left wrist, and the chest. **g** An enlarged view of a cycle of output signals corresponding to three positions. **h** A photograph showing the bionic membrane sensors was directly attached to simultaneously monitor the pulse waves of the participant from his carotid artery, chest, and wrist [124]

Heart rate monitoring is essential for personal healthcare management, and there is a growing demand for a low-cost, non-invasive, and user-friendly system. Lin et al. [127] developed a self-powered wireless Body Sensor Network (BSN) system for heart-rate monitoring by integrating a downy-structure-based TENG (D-TENG), a power management circuit, a TENG-based heart-rate sensor, a signal processing unit, and a Bluetooth module for wireless data transmission (Fig. 9.29). The D-TENG features a multilayer structure with an acrylic supporting backbone (Fig. 9.29a, b). It consists of two functional units: the first unit includes copper-coated PTFE and copper thin films, which are stationary layers anchored at one end to the acrylic frame, with the other end left free-standing. The second unit consists of acrylic sheets,

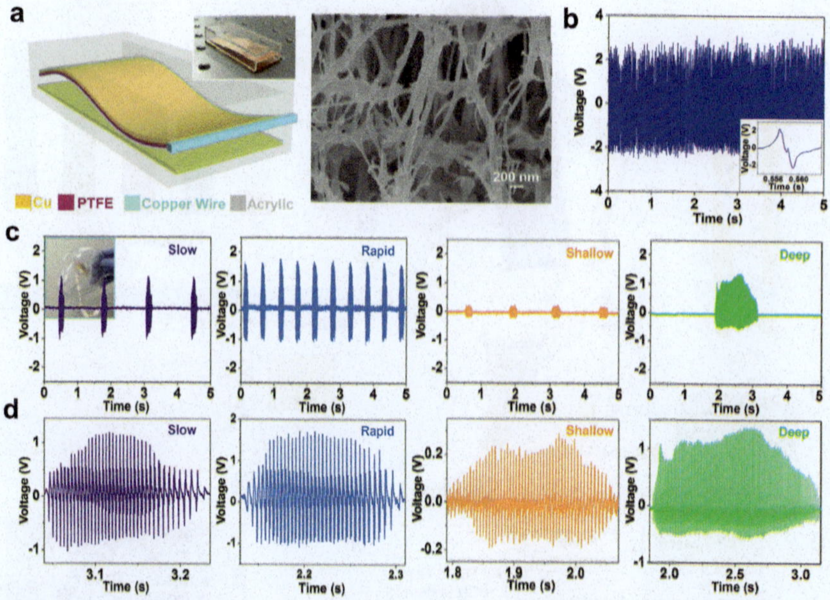

Fig. 9.28 **a** Schematic configuration of the air flow-driven TENG (left) with SEM image of the n-PTFE thin film (right). **b** The output voltage was recorded from the TENG at an airflow rate of 120 L/min. **c** The real-time respiratory signals of the TENG were recorded from four different human breathing behaviors. **d** The corresponding enlarged 1 V output envelope [126]

segmentally bonded to the front and back surfaces with copper-coated PTFE and copper thin films as freestanding triboelectric layers, all connected to the outer frame with stretchable rubber. Figure 9.29c demonstrates the D-TENG capability to harvest mechanical energy with high efficiency. The acquired heart-rate signal is processed by the signal processing circuit, transmitted to an external device via the Bluetooth module, and displayed in real-time on a personal cellphone (Fig. 9.29d). The self-powered sensing system is illustrated in Fig. 9.29e. Figure 9.29f demonstrates the practical application of the system, which integrates a TENG-based generator and a TENG-based sensor. This all-TENG-based wireless BSN system enables continuous, self-powered heart rate monitoring. Figure 9.29g compares the heart-rate signals received from the (I) TENG-based sensor and (II) commercial sensor.

9.4.2.5 Chemical Sensing and Monitoring

Triboelectric charge density is a critical factor influencing the output performance of TENGs [128]. The surface charge density of triboelectric materials can be adjusted by altering the chemical or physical properties of the surfaces. As a result, TENGs can serve as self-powered sensors for detecting specific chemical species when

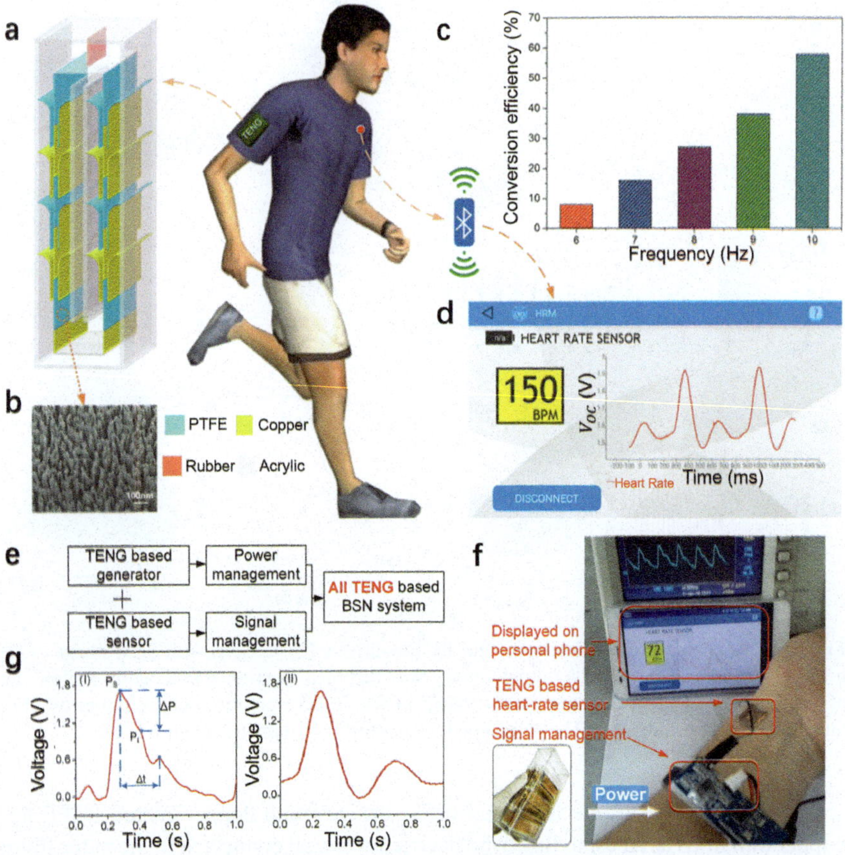

Fig. 9.29 **a** Schematic illustration of the device structure of the D-TENG. **b** A SEM image of the PTFE polymer nanowire array. **c** The electromechanical conversion efficiencies of D-TENG. **d** An enlarged view to show the whole software interface of the real-time acquired heart-rate signals. **e** Diagram of all TENG-based wireless BSN system. **f** Demonstration of an all-TENG-based wireless BSN system and the acquired real-time heart-rate information displayed on a personal phone. **g** Heart-rate signal received by a i TENG-based sensor and (II) commercial sensor [127]

the triboelectric surfaces are customized to respond to particular environmental conditions.

Lin et al. [129] demonstrated a TENG-based chemical sensor for the detection of Hg^{2+} ions by using 3-mercaptopropionic acid (3-MPA)-modified gold nanoparticles (AuNPs) as electrical performance enhancer and recognition element (Fig. 9.30a). The TENG performance is enhanced by assembling AuNPs onto the metal plate. These assembled AuNPs not only serve as stable spacers between the two plates in a strain-free condition but also enlarge the contact area between them, thereby increasing the electrical output of the TENG. Further modification of 3-MPA molecules on the assembled AuNPs enables the high-output TENG to function as a highly sensitive and selective sensor for Hg^{2+} ion detection, due to the differing triboelectric polarities of AuNPs and Hg^{2+} ions. Figure 9.30b shows the device selectively detects Hg^{2+} ions with a detection limit of 30 nM and a linear range from 100 nM to 5 μM. Figure 9.30c reveals the device has a high specific sensitivity for Hg^{2+} ion compared to other ions. This study presents a novel approach for the self-powered detection of Hg^{2+} ions.

Li et al. [130] reported contact-separation TENGs for selectively capturing and detecting Pb^{2+}, Cr^{3+}, and Cu^{2+}, which are commonly existing toxic heavy metal ions in industrial waste water, in a sensing range of 0–200 μM with a sensitivity of 0.005,

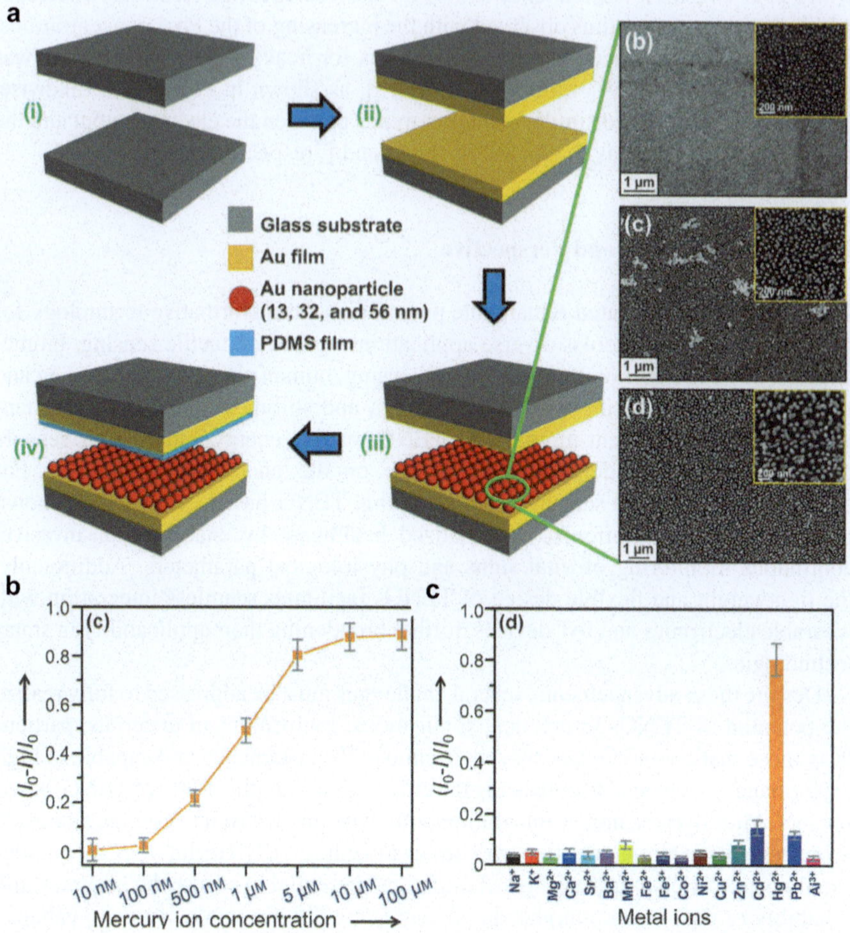

Fig. 9.30 **a** Fabrication process of the TENG-based chemical sensor (left), and SEM images of the Au film modified with (top) 13 nm, (middle) 32 nm, and (bottom) 56 nm AuNPs (right). **b** Sensitivity and **c** selectivity of the as-developed TENG for the detection of Hg^{2+} ions. The concentration of all metal ions tested in the selectivity experiment was 5 μM [129]

0.003, and 0.004 μM^{-1}, respectively. The contact-separation TENG, as shown in Fig. 9.31a and b, was developed using modified anodic aluminum oxide (AAO) a nanoporous surface with a layer of appropriate ligand molecules that are dithizone for Pb^{2+}, diphenylcarbazide for Cr^{3+}, and sodium diethyldithiocarbamate for Cu^{2+} for recognition marks. The other triboelectric layer was the PTFE nanowires. Figure 9.31c demonstrates the dependence of the output V_{oc} on the Pb^{2+} concentration. In a certain Pb^{2+} concentration region of 0 to 200 μM, the V_{oc} is a monotonically decreasing function of the Pb^{2+} concentrations throughout the experimental time windows. A decrease of the electric output is mainly attributed to the surface-adsorbed Pb^{2+} molecules, which can partially replace the position of AAO to contact with PTFE. Given a lower tendency of the Pb^{2+} molecules to lose electrons, a decrease of the electric output is thus observed with the increasing of the Pb^{2+} concentrations. The sensing performance of the TENG sensors for heavy metal ion detection was evaluated in terms of V_{oc} ratio $[(V_0 - V)/V_0]$, as shown in Fig. 9.31d. Likewise, Cr^{3+}, and Cu^{2+} exhibited similar changing trends between the electric output and the applied concentration, as shown in Fig. 9.31e and f, respectively.

9.4.2.6 Conclusions and Perspective

TENGs have demonstrated remarkable potential as a transformative technology for self-powered sensing across diverse applications, including tactile sensing, biomechanical sensing and monitoring, acoustic sensing, human physiological sensing and monitoring, chemical sensing and monitoring, and so on. By harnessing mechanical energy from ambient movements, TENGs enable energy-autonomous sensors that meet the growing demands for real-time, on-site, and remote monitoring. For example, in biomedical sensing and monitoring, TENG-based systems have shown the potential to revolutionize personalized healthcare by enabling non-invasive, continuous monitoring of vital signs and physiological parameters. Additionally, the lightweight and flexible design of TENGs facilitates seamless integration with wearable electronics and IoT devices, further broadening their applicability in smart technologies.

Despite these advancements, several challenges must be addressed to fully realize the potential of TENGs in practical applications. Innovations in materials, particularly those that are biocompatible, environmentally sustainable, and capable of high performance, are crucial to enhancing the efficiency and applicability of TENG-based sensors. Moreover, seamless integration with existing electronic systems, alongside advancements in miniaturization, will be key to scaling TENG technology for broader use. Improved signal processing techniques are essential to ensure the accuracy and reliability of data under complex and dynamic conditions, while efforts to enhance the durability and operational stability of TENGs are critical for their long-term deployment. Furthermore, the establishment of standardized evaluation metrics will facilitate the commercialization and adoption of TENG-based sensors.

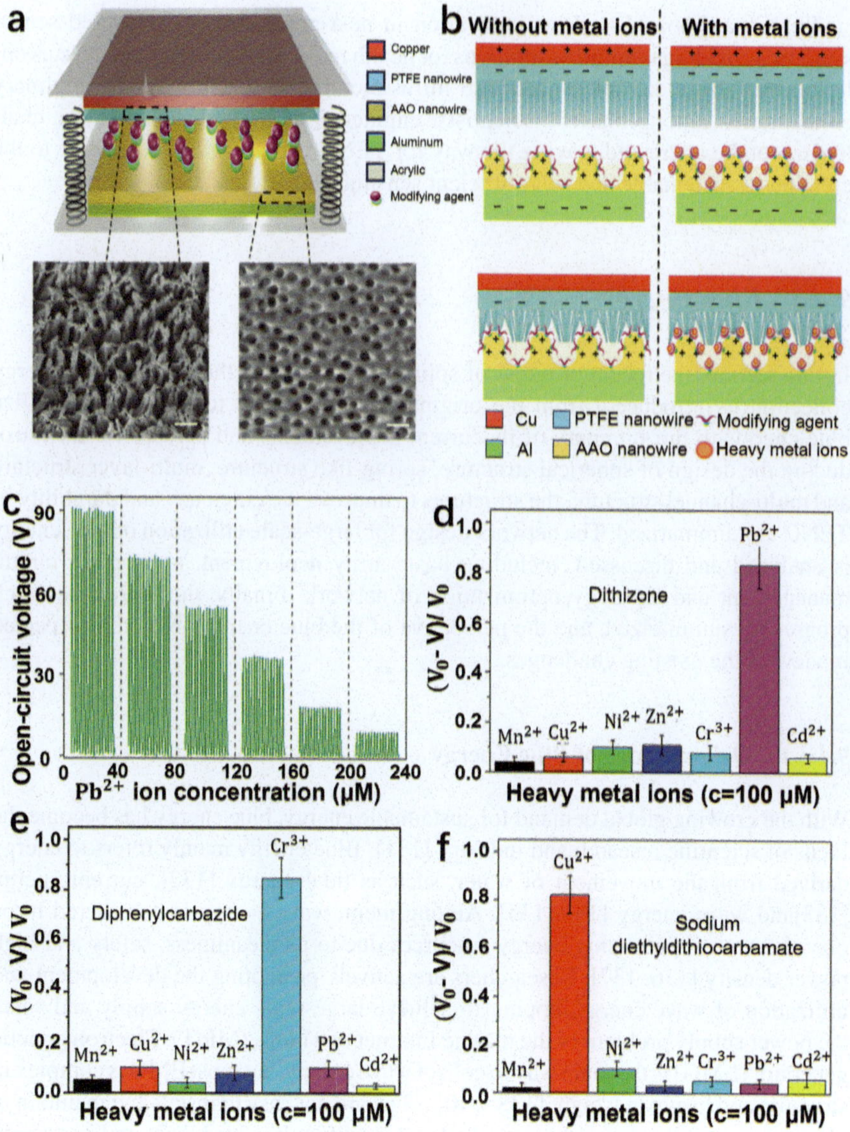

Fig. 9.31 **a** Device structure of the TENG-based chemical sensor (upper), and SEM images of the AAO nanopores (bottom left) and PTFE nanowires (bottom right). **b** Schematic illustration of working mechanism. **c** Output V_{oc} depending on the Pb^{2+} concentrations. **d–f** Selectivity of the TENGs for detection of Pb^{2+}, Cr^{3+}, and Cu^{2+} ions, respectively [130]

TENGs are poised to drive innovation in next-generation self-powered sensor systems, providing innovative solutions for health monitoring, environmental surveillance, wearable electronics, and smart infrastructure. By fostering interdisciplinary collaboration among material scientists, engineers, and data analysts, these challenges can be addressed, paving the way for TENG technology to contribute to the creation of sustainable and self-sufficient sensing ecosystems.

9.4.3 Blue Energy

In this section, the recent progress of solid–solid TENG in the field of blue energy collection is introduced, from the original idea of TENG to develop and utilize blue energy, to the summary of its current achievements and perceptive. By introducing the design of spherical structure, spring-like structure, multi-layer structure and multi-channel structure, the structures to improve the efficiency and durability of TENG are summarized. The network design for large-scale utilization of blue energy is analyzed and discussed, including basic array deployment, networking circuit management and topology optimization of network. Finally, the current research progress is summarized, and the perceptive of the blue energy field are prospected in view of the existing challenges.

9.4.3.1 Original Idea of Blue Energy

With the growing global demand for sustainable energy, blue energy has become the focus of scientific research and industry [131]. Blue energy mainly refers to energy derived from the movement of water, such as tidal energy [132], current energy [133]and wave energy [134, 135]. Among them, wave energy is considered to be one of the most promising energy resources due to its cleanliness, safety and high power density [136, 137]. Researchers are actively promoting the development and utilization of wave energy, hoping to achieve large-scale energy supply and solve the power supply problem of the marine Internet of Things (MIoT). Electromagnetic generator (EMG) is the main way to collect wave energy, but most EMG is complex in structure and heavy in weight [138–140]. They are not easy to install and maintain in the marine environment and are costly. It is also difficult to load them on lightweight MIoT sensors. Moreover, the output of EMG is not advantageous at low frequencies and low amplitudes, which is limited by the law of electromagnetic induction and cannot be changed [141].

On the contrary, the TENGs can effectively harvest mechanical energy from the environment [3, 142]. TENG has shown great potential in the field of blue energy with its unique energy conversion method. TENG converts tiny mechanical energy into electrical energy through the electrostatic effect caused by the continuous contact and separation of two different materials. This process has a wide range of material selections and is easy to manufacture and fabricate [143]. Therefore, it provides a

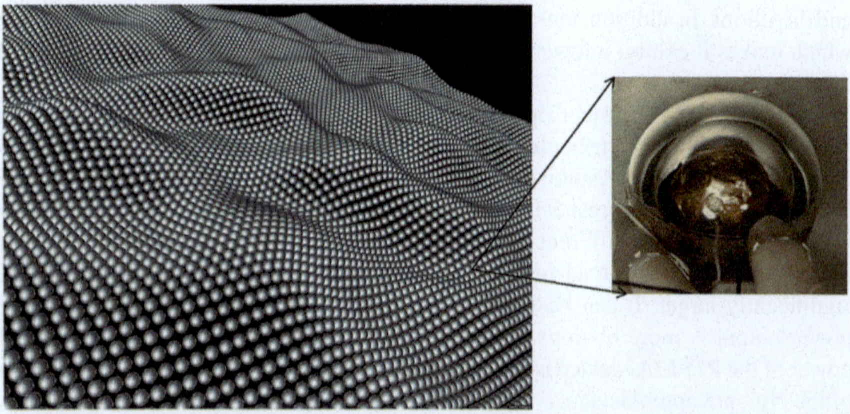

Fig. 9.32 The original idea for harvesting blue energy from ocean using an integrated assembly of small unit TENGs in a 3D network [2]

new way to collect energy brought by the impact of waves, the flow of seawater, etc. In addition, research has shown that TENG can work effectively at low frequencies, making it very suitable for capturing large-scale but low-frequency blue energy [141, 143].

TENG has the ability to efficiently collect energy in the low-frequency range, so it is quite suitable for wave energy, which has a low frequency and strong irregularity. In complex marine environments, wave energy can be effectively captured by rationally designing the structure of TENG. TENG materials are generally lightweight and low-cost, so they are easy to process and manufacture on a large scale. In 2014, Wang [2] proposed the original idea of realizing large-scale blue energy collection by deploying a large number of TENG units in a wide ocean area. As shown in Fig. 9.32, a single TENG power generation unit adopts a spherical structure, which is covered on the surface of the ocean in a network shape, and is connected by cables to make it move with the waves to achieve wave energy collection. The ocean covers about 71% of the earth's area. If it is deployed on a large scale in some sea areas without affecting navigation and the life of marine life, and energy is supplied to the land through the power grid, it will be expected to directly solve the current energy shortage problem.

9.4.3.2 Structures of TENGs for Blue Energy

The first TENG for blue energy collection proposed in 2013 adopted a spherical structure, and subsequent research has continuously discovered the advantages of its structure [142, 144, 145]. The spherical structure can face multi-directional wave impacts with the same posture and efficiently capture energy. Its unique geometric shape provides a larger surface area for contact with waves and can effectively cope with mechanical forces applied from different angles. This design is particularly suitable for ocean energy collection because waves have irregular motion trajectories and directions. In addition, the spherical structure often adopts a rolling design inside, which makes it exhibit a lower natural frequency and can effectively resonate with the waves.

Wang et al. [145] further optimized the initial design of the spherical structure by changing the paired triboelectric materials. As shown in Fig. 9.33a–f, dielectric (nylon ball)-dielectric (Kapton film) pairing and conductor (Al foil)-dielectric (PTFE ball) pairing were used respectively. Their working principles are the same, but the output performance is different due to the change of materials. As shown in Fig. 9.33b and e, under the same load resistance, the current of the Nylon/Kapton device is significantly larger. It can be seen in Fig. 9.33c and f that the difference in peak power output is more obvious. Under the same experimental conditions, the peak power of the PTFE/Al device is more than 10 times lower than the peak power output of the Nylon/Kapton device.

Zhu et al. [146] proposed a 3D fractal structure dual-mode spherical TENG (FDS-TENG) to harvest wave energy. The schematic diagram of the overall structure of FDS-TENG deployed in the ocean is shown in Fig. 9.34a. FDS-TENG is mainly composed of a sphere made of resin material and a spherical shell encapsulated by acrylic. By introducing a three-module multiphase coupling strategy, a 3D spatial geometrical isomorphic design is realized, thereby improving the space utilization inside the spherical structure. The internal units can couple motion and realize the simultaneous output of contact-separation mode and lateral sliding mode. The electrical output performance of FDS-TENG was tested at different overturning angles and swing frequencies, and the results are shown in Fig. 9.34b and c. FDS-TENG achieved significant electrical output at different swing frequencies (0.5–2.5 Hz) and different overturning angles (10°–25°), proving the ability of TENG to effectively capture energy under both weak and strong excitation wave conditions. In view of the random characteristics of wave direction, the output performance of FDS-TENG under multi-directional wave excitations in the horizontal circumferential range was tested. The TENG sphere was mounted on an acrylic frame with the overturning angle of 5°–20°, the frequency of the swing motor was fixed at 2 Hz, and the deflection angle was set at intervals of 30° within a 360° circumference. The change of the transferred charge with the flip angle and the deflection angle is shown in Fig. 9.34d. When the overturning angle is increased to 20°, TENG achieves a coupled motion with the maximum sliding distance in any direction of the circumference and generates the optimal electrical output. Furthermore, the electrical output performance and stability performance of different triboelectric materials were tested, and the results are shown in Fig. 9.34e and f. In summary, PTFE has both strong electrical output effect and durability, and is an effective and stable material to be used as the triboelectric layer material.

The spring-like structure is introduced into the design of TENG in consideration of its unique advantages in energy storage. The presence of the spring-like structure can amplify the low-frequency wave motion into high-frequency motion of the energy collection component. The characteristic of converting externally applied impact force into stored energy through elastic deformation can significantly improve the stability of TENG. When the wave impacts the device, the spring-like structure

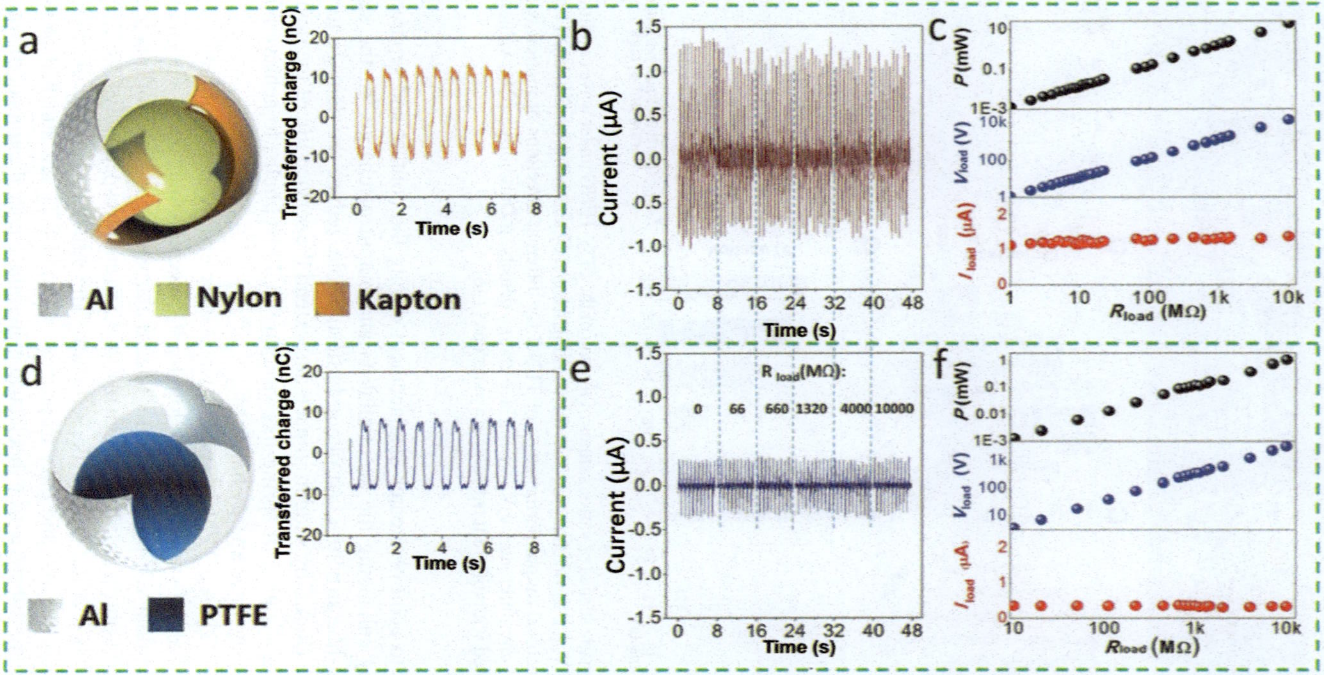

Fig. 9.33 **a** The schematic diagram and measured transferred charges (Q_{sc}) for the Nylon/Kapton device. **b, c** The dependence of the electrical outputs (the current, the voltage, and the power) on the load resistance for the Nylon/Kapton device. **d** The schematic diagram and measured transferred charge (Q_{sc}) for PTFE/Al device. **e, f** The dependence of the electrical outputs (the current, the voltage, and the power) on the load resistance for PTFE/Al device [145]

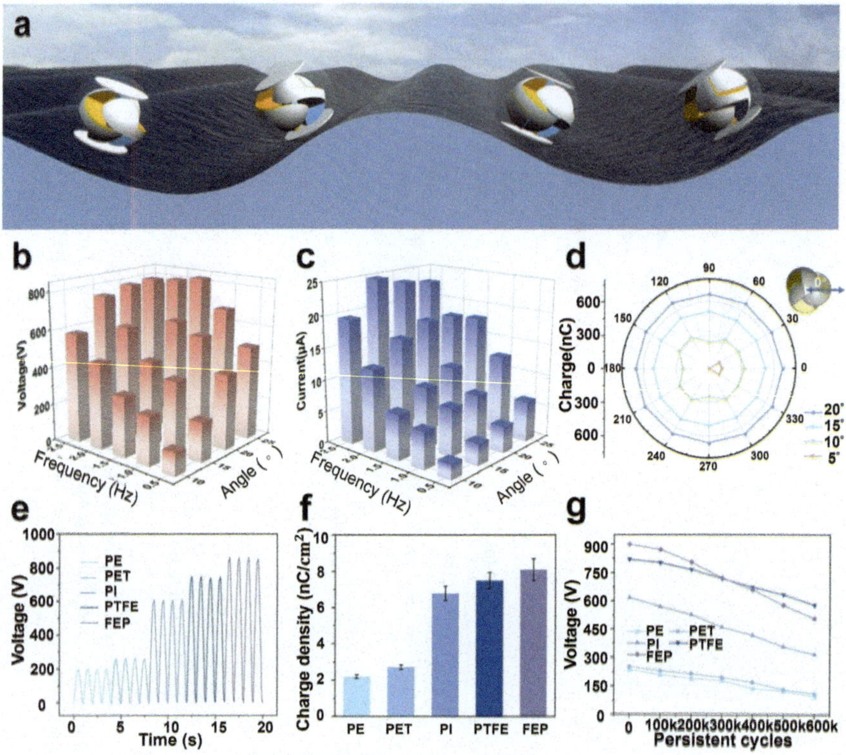

Fig. 9.34 **a** Schematic scenes of various states of motion of the FDS-TENG under wave conditions. **b** 3D histogram of V_{OC}. **c** I_{SC} on changing both the working frequency and swing amplitude. **d** Radar diagram of the electrical output performance of TENG for different directions of motion within 360° of horizontal circumference. **e** V_{OC}. **f** Surface charge density. **g** Durability of V_{OC} values for the TENG fabricated with different triboelectric materials [146]

enables TENG to achieve greater displacement during the contact and separation process, thereby enhancing the accumulation and output of charge. At the same time, the spring-like structure can play a buffering role, reduce the damage to the overall structure caused by sudden impact, and extend the service life of the device. Therefore, the application of spring-like structure in TENG helps to achieve more efficient and lasting wave energy harvesting.

Liang et al. [147] used TENG to collect water wave energy and combined it with the water hazard alarm systems to develop a self-powered smart buoy. This device can be used to build a self-powered water hazard alarm system. The working scene diagram and workflow diagram are shown in Fig. 9.35a and b respectively. The TENG unit is designed as a spiral spring-like structure. The structure is made of silicone material, which is flexible and easy to process. Cu and PET are attached to both sides of the silicone substrate as triboelectric materials. Under the impact of waves, the spiral TENG unit is compressed and elongated, and a contact-separation mode is formed between the electrode and the dielectric layer to generate electrical output. This solution directly solves the difficulty of integrating TENG units with ordinary spring elements and effectively improves the spatial structure and utilization rate. As shown in Fig. 9.35c, four spiral TENG units are arranged in a regular tetrahedron inside a single sphere. A Cu ball is placed at the center of the spiral structure to drive the vibration of the spiral TENG unit when triggered by water waves in any direction. To improve the output performance of the spiral TENG unit, they designed a charge excitation module (CEM) to accumulate the charge generated by TENG and output electrical energy through series–parallel switches. The principle is shown in Fig. 9.35d.

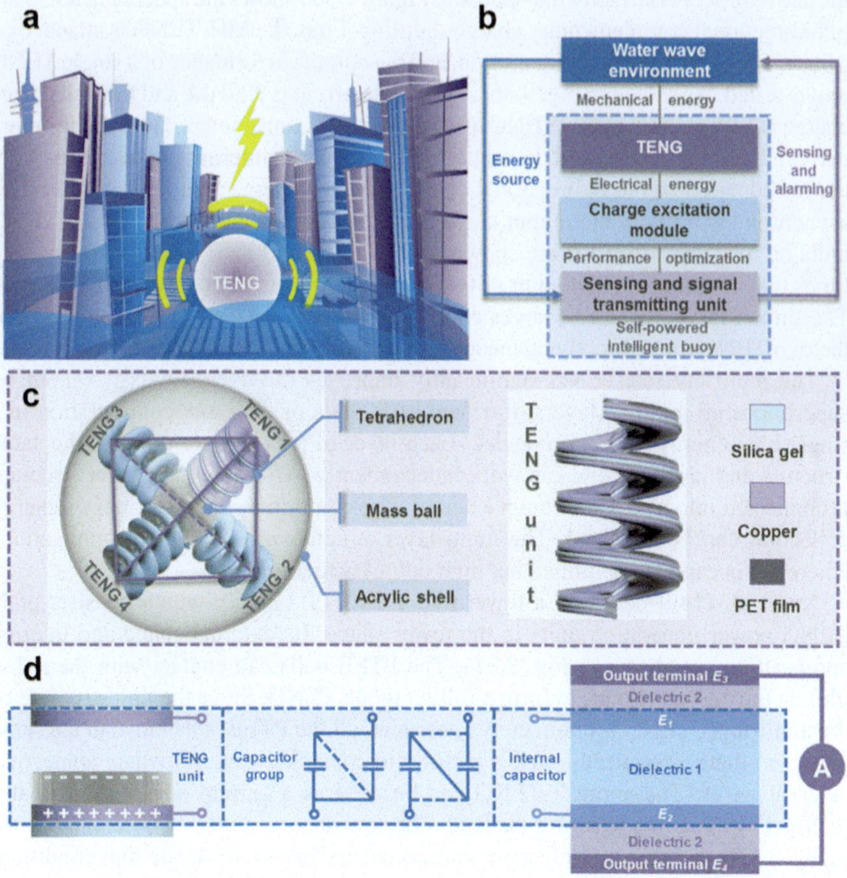

Fig. 9.35 **a** Schematic diagram. **b** Workflow framework of TENG harvesting water wave energy for self-powered water hazard alarming. **c** Structure of the spherical TENG and enlarged view of one spiral TENG unit inside. **d** Schematic diagram of a contact-separation TENG integrated with the CEM [147]

Liu et al. [148] further enhanced the space utilization and designed a multilayered helical spherical TENG (MH-TENG) for water wave energy harvesting. The two internal semispherical TENG units are mirror-symmetrical, and the internal space utilization of MH-TENG reaches 92.5%. The asynchronous motion of the two units enables the device to successfully realize the bidirectional complementary charge shuttling mechanism, as shown in Fig. 9.36a. The two TENG units will alternately contact and separate under the impact of waves, at this time their capacitance changes, resulting in a voltage difference and directional charge flow in the external circuit. Specifically, the right TENG unit is in a separated state, and the charge will flow from the right TENG unit to the left TENG unit. This process passes through the load and outputs power until the voltage difference disappears. When the left unit is separated, the above process is exactly the opposite. Figure 9.36b shows the specific mechanism of bidirectional complementary charge shuttling. First, the MH-TENG is attached on a linear motor for reciprocating motion, and the output performance of a single TENG unit is tested. As shown in Fig. 9.36c, its output current is 52.0 μA and the transferred charge is 1.26 μC. When two TENG units are directly connected in parallel, the output current does not increase and the output charge shows an intermittent decrease. After the introduction of the bidirectional complementary charge shuttle mode based on asynchronous motion, the output current and transferred charge of the two TENG units on load A and load B are shown in Fig. 9.36d and e. The output current is 3 times that of a single TENG unit or two TENG units directly connected in parallel. The sum of the transferred charges on the two loads is 1.3 times higher than that of the two TENG units directly connected in parallel, and there is no intermittent loss.

The multi-layered TENG significantly improves the overall energy output by superimposing multiple layers of triboelectric units of the same configuration in a single blue energy collection device. Each layer of triboelectric unit has the same structure and principle and can work independently. After the multi-layer structure is connected into a whole through external wires, multiple superpositions of energy collection can be achieved. The multi-layer structure is suitable for applications where the ocean area is limited but high output is required.

Xu et al. [149] designed a tower-like TENG (T-TENG), which consists of N TENG power generation units in the tower shape. Its external shape and internal composition are shown in Fig. 9.37a. The PTFE balls can contact with the nylon film on the curved surface to form a rolling mode TENG. Since the unit structure of the multi-layer TENG is completely consistent, all the PTFE balls move in the same phase, so there is no need to use a rectifier to connect every unit when connecting external wires. The entire T-TENG can be used as a current source, and finally rectified by a rectifier to power the load. The equivalent circuit of T-TENG is shown in Fig. 9.37b, and the multi-layer units are connected in parallel. Under this condition, the voltage across the resistance (V_R), short-circuit current i and transferred charge (Q_c) at different numbers of layers n were measured, and the results are shown in Fig. 9.37c–e. All three values increase with the increase in the number of units. From the linear fitting results of the voltage in Fig. 9.37f, the V_R increases proportionally with the number of power generation units. Figure 9.37g measures the current and voltage of 10 TENG units under different loads, and then obtains their power density.

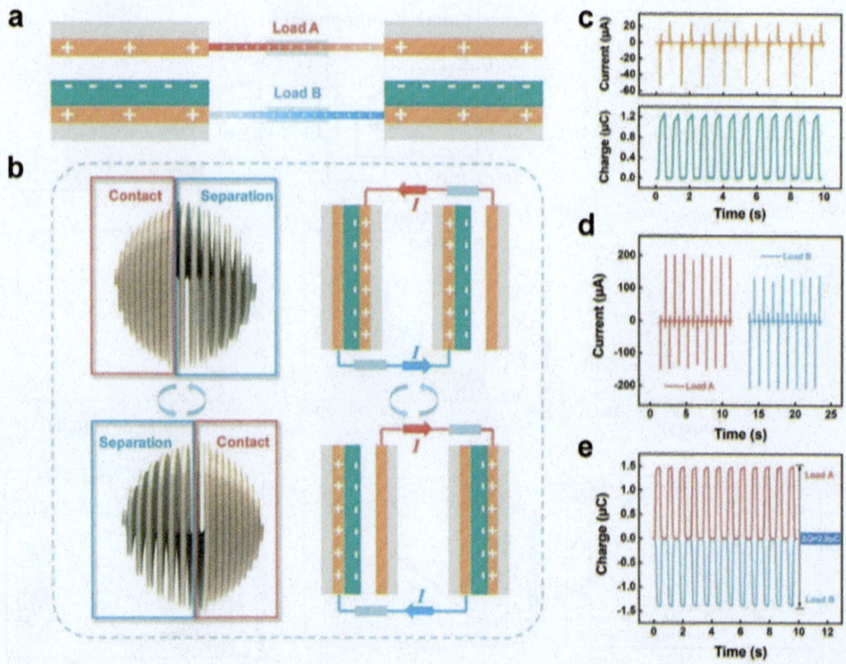

Fig. 9.36 **a** Working mode of bidirectional charge shuttling scheme illustrated by two simple contact-separation TENGs. **b** Working principle and process of bidirectional charge shuttling scheme between the two semispherical TENG units inside the MH-TENG. **c** Output current and transferred charge of a single semispherical TENG unit. **d** Output current and **e** transferred charge of the MH-TENG across Load A and Load B [148]

As the number of units increases, the I and V_R increase proportionally (Fig. 9.37h), while the internal impedance decreases. As can be seen from Fig. 9.37i, the power density increases from 1.03 W/m^3 for a single power generation unit to 10.6 W/m^3 for 10 units. Compared with previous works, a new stage has been reached [86, 150, 151]. This work provides a detailed exploration of the impact of multilayer structures on output performance and provides theoretical support for the design of subsequent multi-layer structures.

In addition to setting up multi-layer structures in the vertical direction, similar to the spring-like structure introduced in the previous section, multi-layer structures can also be expanded in the horizontal direction. Chen et al. [152] designed a one-meter TENG (OM-TENG) that is 1 m long in the horizontal direction, which is the largest single-body TENG in the industry (Fig. 9.38a). 12 TENG units are arranged horizontally inside, and the inside of the unit is a dense multi-layer arch. There are also 11 sliders inside, which move along the axis with low resistance in the cavity. The connection between the TENG unit and the slider is shown in Fig. 9.38b. The sliding of the slider drives the TENG units connected on both sides to achieve contact and separation. The two adjacent sliders pull or squeeze the multi-layer arch in the middle

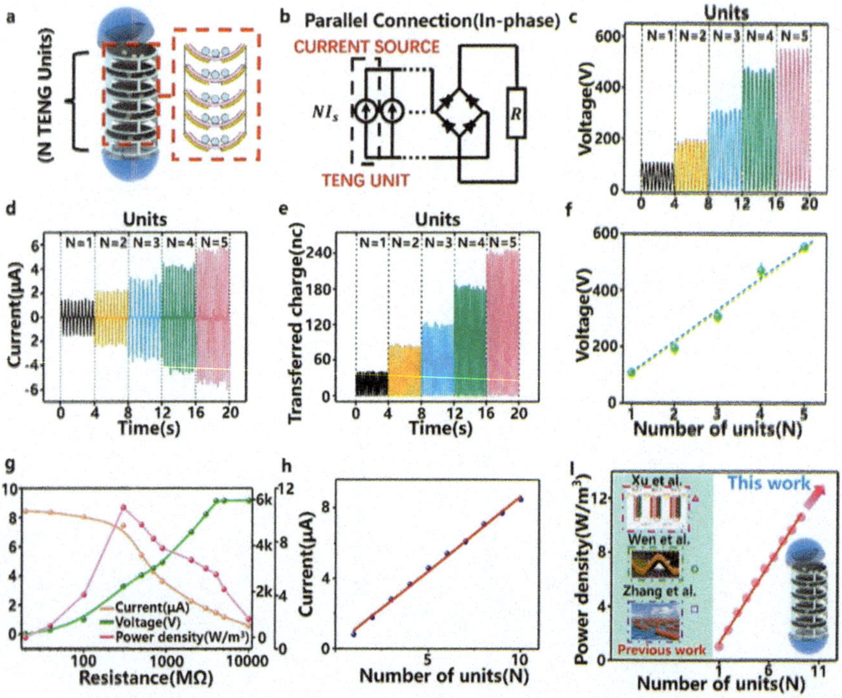

Fig. 9.37 **a** The diagram of the T-TENG structure. **b** The corresponding equivalent circuit diagram. **c** The voltage across the resistance V_R, **d** transferred charge Q_c. **e** induced current I of the T-TENG with the number of units increasing. **f** Variation of the voltage across the resistance output of the T-TENG with the number of units. **g** Dependence of the output voltage, current and power density for T-TENG with 10 units on the resistance of the load. **h** Variation of the current, voltage, and internal resistance with the number of units. **i** Comparison of the power density of the present T-TENG and previous TENG [149]

laterally to ensure that the friction layer is synchronously contacted and separated and reduce the risk of collapse. The fixed limiter is used to control the movement amplitude of each slider to prevent sudden strong waves from causing damage to the structure. An enlarged view of a single TENG unit is shown in Fig. 9.38c. Each TENG unit consists of 20 layers of arch structures. The slider structure is shown in Fig. 9.38d. Its main body is a cylinder with a central depression for placing counterweights. The single-layer arch structure is shown in Fig. 9.38e. The area density of the friction layer of the structure is increased to 1.76 cm^{-1}, which is the highest value reported so far for blue energy TENG. A photo of the OM-TENG is shown in Fig. 9.38f. As shown in Fig. 9.38g, the size of the structure has been expanded to the meter level, which is 2–10 times higher than some representative floating wave energy TENGs in previous studies, which means that more TENG units can be accommodated, which is conducive to improving output and space utilization. To study the changes in fluid pressure on shells of different sizes in water, the stress on the shell was simulated by fluid mechanics using COMSOL software (Fig. 9.38h).

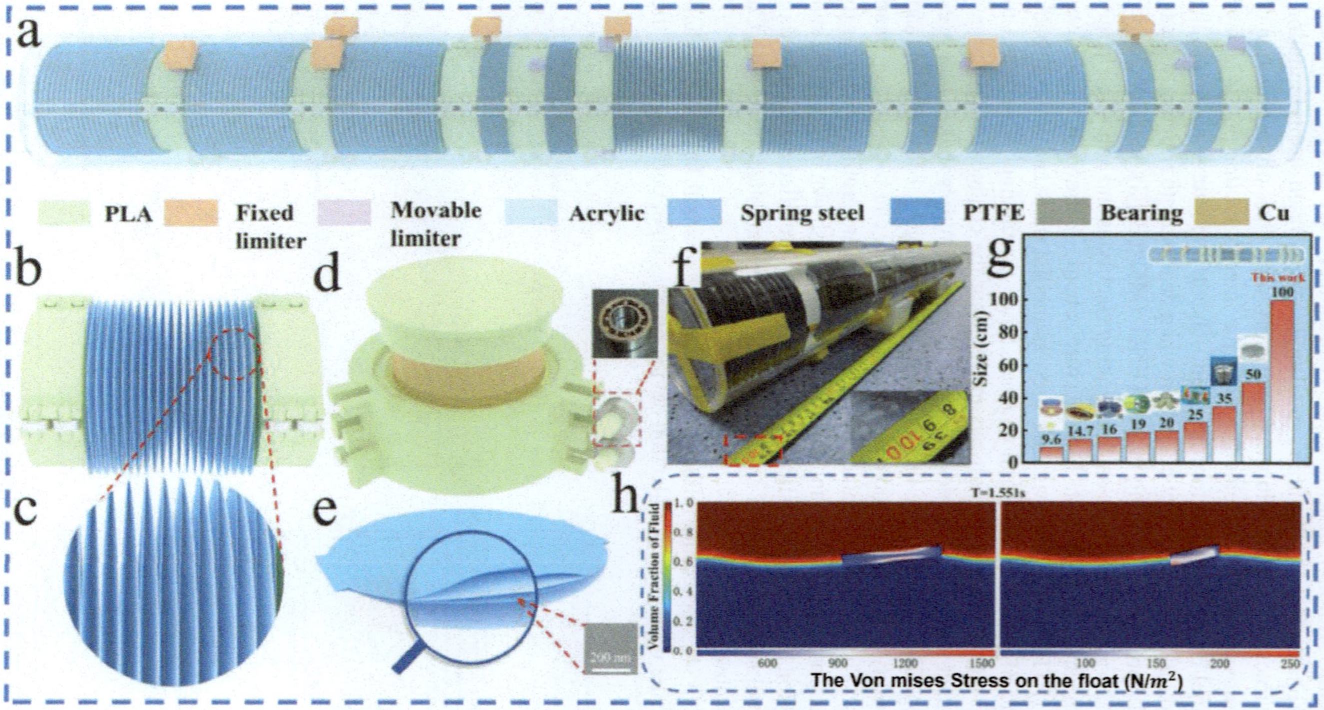

Fig. 9.38 Structure of OM-TENG. **a** Schematic diagram of the overall structure of the OM-TENG. **b** Schematic diagram of the TENG unit as well as the slider connection. **c** Schematic diagram of the enlarged TENG unit. **d** Schematic diagram of the slider. **e** Schematic diagram of the basic TENG unit. **f** Physical drawing of the overall structure of OM-TENG. **g** Comparison diagram of the dimensions of OM-TENG with the previous work. **h** Simulated diagrams of fluid pressure on two-sized objects in a wave [152]

In the structural design of various wave energy TENGs, the influence of wave angle is significant. Although many structures can collect wave energy in multiple directions, they only perform well within a small angle range. Based on this limitation, multi-channel structures were developed. Specifically, regardless of the direction of external wave movement, the movement path of the internal motion part is always along the channel direction. Therefore, the multi-channel structure can ensure the consistency and efficiency of energy collection when dealing with irregular waves.

Wang et al. proposed a rolling-mode TENG (MO-TENG) that effectively harvests wave energy through multi-channel grating electrodes and opposite-charge-enhancement mechanism [30, 153]. The structure scheme is shown in Fig. 9.39a. The dielectric (PTFE and nylon) balls are used as inertial parts and can roll back and forth in the 3D printed PLA shell channel. Since the channel is designed to be straight, the internal small balls always move along the channel to form effective rolling contact with the electrodes when the entire structure is shaken by waves in any direction. The rolling mode design has the advantages of low friction and improved durability (Fig. 9.39b), and ensures that the TENG can be easily activated by wave impact. This method combines multiple channels and grating electrodes to enhance the electrostatic induction and the current output, as shown in Fig. 9.39c and d. PTFE and nylon balls with opposite-charge polarization are filled and arranged in the multi-channel structure to form a ternary electrification architecture (PTFE, Cu and nylon), as shown in Fig. 9.39e.

Duan et al. [154] studied the influence of the filling of small balls in the shell and channel of the wave energy collection device on the output performance of the

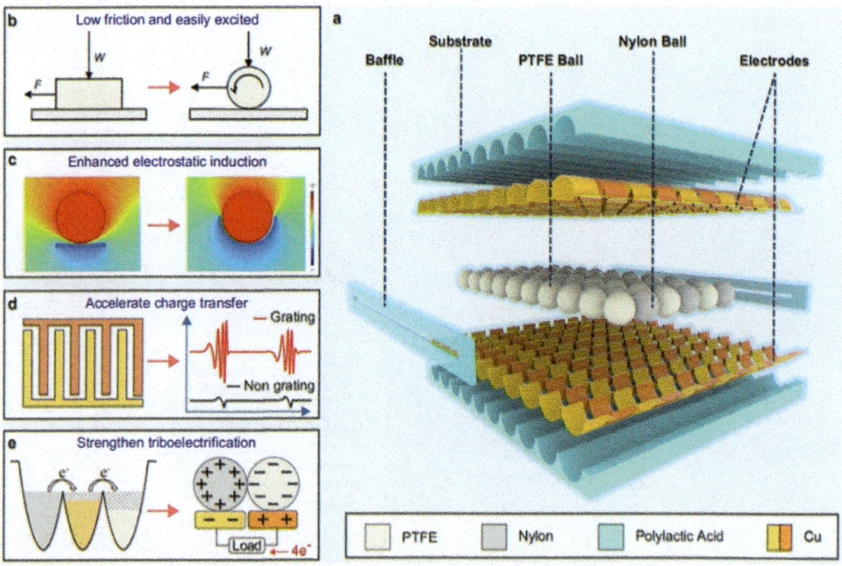

Fig. 9.39 **a** Structure scheme of the MO-TENG. **b** The rolling mode. **c** Tunnel-shaped electrode. **d** Grating electrodes. **e** Opposite-charge enhancement strategies [153]

channel structure TENG. They designed four common shapes for study: sphere, cylinder, regular hexahedron prism (RHP), and cube, as shown in Fig. 9.40a. Not only the characteristic scale of the shape is consistent, but also the internal channel size, small ball diameter and number of grids are the same. The output performance was tested on the experimental bench (Fig. 9.40b), and the internal structure parameters were adjusted and tested. As shown in Fig. 9.40d, the transfer charge increases with the increase of the mass of the filling balls. When the mass is 75 g, the corresponding volume filling rate is 9.52%, and the transfer charge reaches the maximum value. It can be seen from Fig. 9.40e that when the displacement amplitude in the channel changes in the range of 30–90 mm, the charge output first increases and then decreases with the increase of the displacement amplitude. The transferred charge is positively correlated with the increases in the acceleration of translational impact agitations and frequency of translational harmonic agitations (Fig. 9.40f and g). As the PTFE balls continue to contact the electrode, the charge generated on the PTFE balls and the electrode eventually reaches saturation. The output performance of the four TENG shapes was optimized based on the effects of the above factors. After optimization, the peak power density and average power density of the four TENGs are compared in Fig. 9.40c. The results show that the cube structure has the highest output performance. The peak power density is 52.87 W m^{-3} and the average power density is 10.08 W m^{-3}.

9.4.3.3 Network for Blue Energy Harvesting

As mentioned above, to realize the original idea of large-scale blue energy collection, tens of thousands of TENG power generation units need to be covered can connected on the ocean surface. Its layout is shown in Fig. 9.41a [155]. The TENG units are connected in network and then rectified and aggregated to supply energy to island or marine power equipment or store electrical energy. However, how to effectively connect these TENG units has always been a challenge. Previous studies often used flexible wires for connection, but in complex sea conditions, it is inevitable that the units will be entangled and unable to operate [156, 157]. And, because vertical and deflection movements are required, fully rigid connections cannot be used.

Liu et al. [155] designed a symmetrical structure power cable consisting of two spring steel tapes and three polymer films which are bounded by mucilage glue. The external PTFE films serves as a triboelectric layer and has excellent mechanical properties and anti-corrosion properties. The internal spring steel tape has good elasticity, flatness and fatigue-resistance. It acts as a skeleton and as an electrode of TENG. Inside the power cable, space is reserved between the two steel tapes for the wires that transmit power. And the TENG unit adopts a honeycomb topology connection as shown in Fig. 9.41b, which brings the following advantages: minimized cable consumption, less constraint on unit movement, and higher stability. In short, under this structural design, the power cable meets the basic mechanical requirements and in a certain extent improves the output performance of the TENG network.

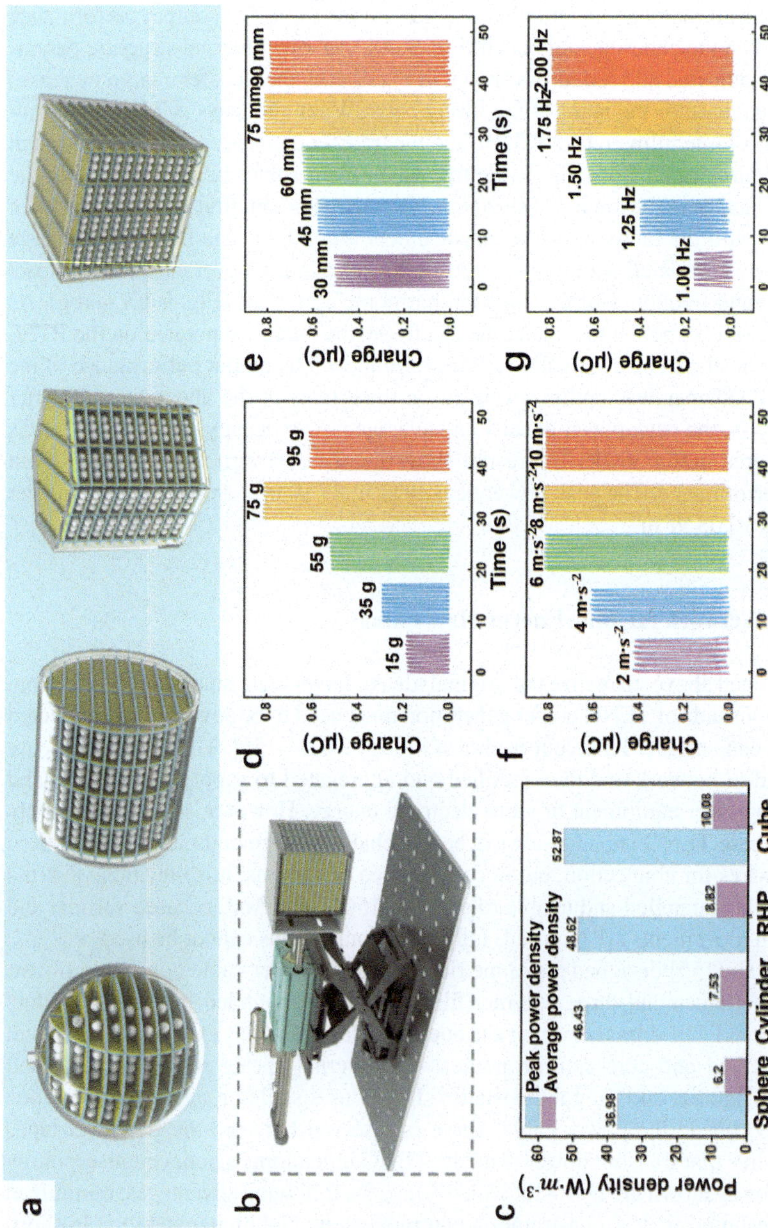

Fig. 9.40 **a** Schematic diagrams of the four structures: sphere, cylinder, RHP, and cube. **b** Schematic diagram of the test setup. **c** Comparison of the power densities for the four shapes of TENGs. **d** Dependences of transferred charges on a filling amount of PTFE balls. **e** Displacement amplitude, **f** acceleration of translational impact agitations, and **g** frequency of translational harmonic agitations [154]

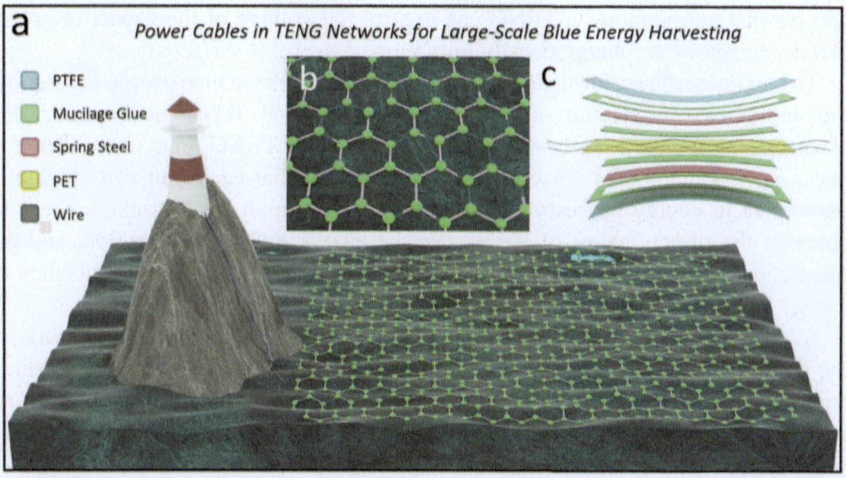

Fig. 9.41 **a** Thousands of TENGs connected by power cables for large-scale blue energy harvesting to power electrical devices. **b** Enlarged structure of a TENG array with honeycomb-like topology. **c** Breakdown drawing of the power cable [155]

In the circuit management of TENG array networking, the circuit connection mode and wire optimization between each unit play a key role in the overall energy output. According to different application requirements, series, parallel or mixed circuit connection modes can be used to optimize the output voltage and current characteristics of the system. Adjusting the circuit connection mode can not only adjust the power distribution between units, but also improve the overall energy collection and transmission efficiency. Through networking circuit management, the adaptability and durability of the TENG array can be further improved.

Liang et al. [158] designed a new charge excitation system based on voltage-multiplying circuit, which not only improved the output performance of a single TENG by several times, but also proposed a scheme to integrate with charge excitation circuits (CEC) to realize high-output TENG network. A single TENG unit floating on the sea surface is shown in Fig. 9.42a, which adopts a multi-layer structure inside. The CEC they designed can increase the output current of the TENG unit by several times and convert it into DC output. Figure 9.42b shows a photo of the hexagonal TENG network connected to the CEC. They used the hexagonal TENG network as the basic module to establish a multi-module TENG network shown in Fig. 9.42c, which is equivalent to two-level circuit management. More importantly, they proposed a self-powered system framework for TENG networks with circuit management, which provides a standard process guide for subsequent TENG network energy supply research (Fig. 9.42d). The DC output of the TENG network with enhanced performance through CEC can power various functional circuits, greatly expanding the application scope of TENG in blue energy collection. The working principle of the charge excitation is illustrated in Fig. 9.42e. CEC has a capacitor group consisting of two identical capacitors that can switch between series and parallel autonomously. TENG can control the voltage of the capacitor group, thereby breaking the charge density limitation of the tribo-surfaces.

To maximize the performance of TENG array networks in blue energy harvesting, the topological optimization of the layout is crucial [159]. Topological optimization involves how to select the best networking method of TENG units under different ocean environments and wave conditions to ensure that each unit can efficiently participate in energy harvesting. The design of the topological structure needs to consider the directionality of the waves, the energy density distribution, and the interaction between the units, with the aim of increasing the overall efficiency of energy harvesting through the optimal layout scheme.

Li et al. [159] designed a novel three-dimensional (3D) chiral TENG network for the first time to harvest blue energy. The network adopts a distributed architecture, and the chiral connections between TENG units are similar to mechanical chiral metamaterials. This gives the network flexibility, hyper-elasticity in water, and good wave absorption behavior. This design makes it feasible for TENG to establish a deep network. Figure 9.43a shows the prospect of large-scale deployment of 3D chiral TENG networks in the ocean as marine power stations. The TENG network consists of multiple sub-networks, which are assembled into a tubular module array. Figure 9.43b clearly shows the response of the large-scale chiral TENG network under the action of waves. When waves propagate in the network, mechanical energy is absorbed by the network and the TENG units are rotated to output electrical energy through chiral coupling. When the network is large enough, local agitations will also propagate like waves inside. The network flexibly responds to various disturbances, making it more adaptable in harsh marine environments. Figure 9.43c intuitively shows the output of the 3D chiral network, which can easily light up 348 LEDs. The self-powered system shown in Fig. 9.43d is usually composed of a 3D chiral network, a power management circuit, and a sensor. Due to the large scale of the network and excellent output performance, the acquired data can be directly transmitted to a remote terminal wirelessly. The demonstration of the self-powered temperature and humidity monitoring system is shown in Fig. 9.43e and f.

The application of TENG in the field of blue energy harvesting has shown great potential, especially in the efficient collection and conversion of wave energy, which has significant advantages. This section mainly introduces the classic configurations of spherical structure, spring-like structure, multi-layer structure and multi-channel structure. The shape design and internal configuration of these structures are continuously optimized, allowing TENG to achieve more efficient and stable energy collection under different wave intensities and frequencies. To realize the large-scale use of TENG to collect wave energy, it is possible to integrate multi-unit power generation through networking to increase the overall energy output. This section briefly introduces the upgrade of basic array deployment, the regulation of circuit management, and the topology optimization of network layout during the networking process. The upgrade of cable connection form can achieve stable power output and efficient energy transmission. Array networking significantly improves the energy collection efficiency and availability of the system through complete circuit management.

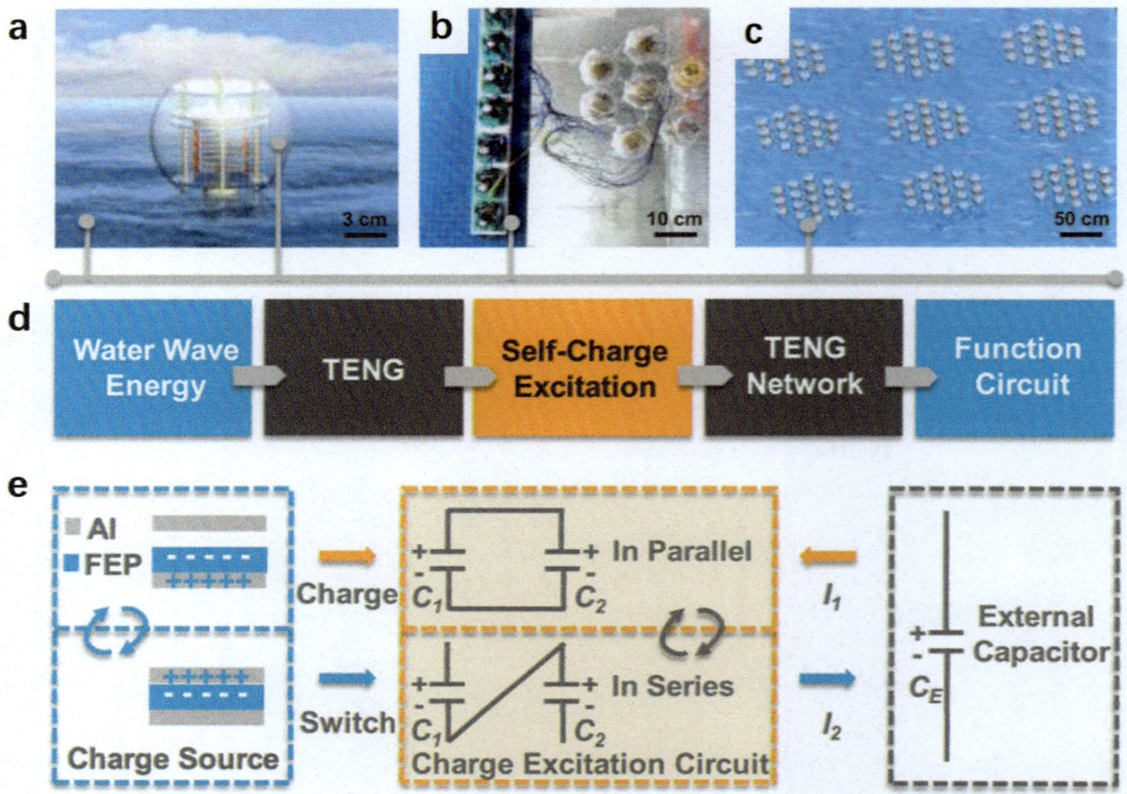

Fig. 9.42 **a** Schematic diagram of a single spherical TENG floating on the ocean surface. **b** The photo of the hexagonal TENG network connected to the CEC. **c** Schematic illustration of a multi-module TENG network for large-scale blue energy harvesting. **d** Framework for the self-powered system. **e** Working principle of a contact-separation TENG integrated with the CEC [158]

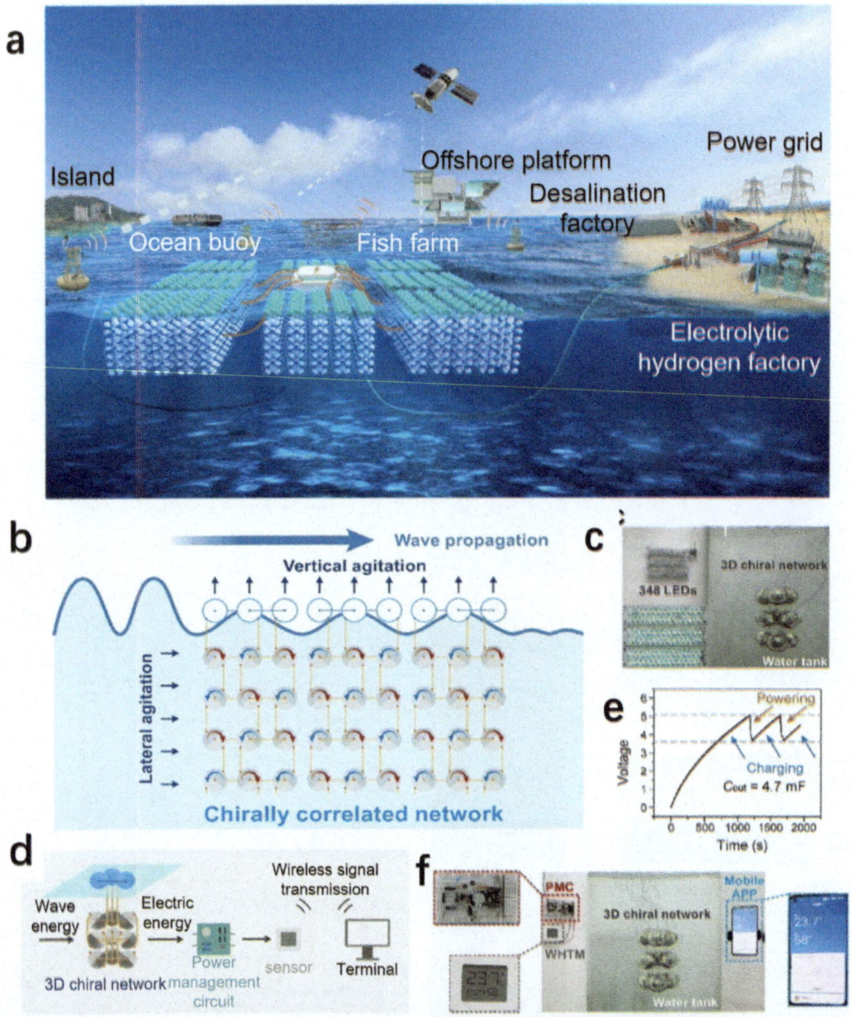

Fig. 9.43 **a** The schematic prospect of application systems based on a large-scale 3D chiral TENG network as a marine power station. **b** Schematic diagram of the wave energy harvesting behavior of the large-scale chiral TENG network. **c** Photograph of 348 LEDs powered by the 3D chiral network. **d** Schematic diagram of self-powered systems based on the network. **e** Voltage curve of the output capacitor. **f** Photograph of the 3D chiral network to power a wireless hygro-thermometer [159]

The rational design of the topology enables the system to adapt to changing marine environments and ensure long-term stable operation.

The TENG has developed rapidly in the field of blue energy harvesting due to its light weight, flexibility and efficient low-frequency energy harvesting capabilities. To truly achieve the goal of sustainability and efficiency in blue energy applications in the future, research must delve into more challenging technical and environmental issues. Not only is it necessary to further optimize the electrical output performance of the TENG inside the structure, but it is also necessary to consider the hydrodynamic performance of the overall structure in the actual marine environment. And, according to the structure of the hydrodynamic analysis, the external structure is modified to further iterate the internal power generation unit. The well-adaptable external design helps to improve the overall energy capture capability and ensure long-term stable operation in a complex and changing marine environment. In this process, some standard parameters of blue energy TENG should be formulated, such as the calculation of power density and energy conversion efficiency. So that TENG units of different structures and different modes can be fully compared. Regarding the large-scale TENG array networking for large-scale blue energy collection, current research has proved its feasibility at the concept verification level. However, to achieve real large-scale application, it is necessary to deeply analyze the large-scale processing and assembly technology of TENG to ensure efficient and economically feasible manufacturing and installation methods. In addition, the research should also focus on how to improve the reliability and survivability of the system in long-term marine exposure environments and solve possible equipment corrosion, mechanical wear and other problems. In conjunction with this, it is also necessary to ensure its environmental friendliness and avoid large-scale deployment from having adverse effects on the marine ecosystem.

TENG for the collection of blue energy is a brand-new technology, with only more than ten years of research history. It is believed that by combining advanced materials science, structural mechanics, fluid dynamics and ecological engineering technology, the research and application of blue energy TENG is expected to achieve a leap from concept verification to large-scale commercialization, making an important contribution to the diversification of global clean energy.

9.4.4 *High-Voltage Source*

TENGs have emerged as a pivotal technology in the development of high-voltage (HV) power sources, achieving significant advancements in both applications and industrialization [100]. Characterized by their high V_{oc} and low short-circuit current outputs, TENGs possess an exceptionally high best-matching impedance (~MΩ), which aligns well with electronic devices exhibiting high internal resistance [160–162]. This distinctive attribute offers numerous advantages compared to traditional high-voltage (HV) power sources, including portability, self-powered ability, safety, adjustable output and high integration capabilities [163, 164].

9.4.4.1 The Methods to Enhance Voltage of TENG

To broaden the application of TENGs as HV power sources, considerable efforts have been dedicated to further elevating their voltage output. For sliding-mode alternating current TENG (AC-TENGs), the V_{oc} can be expressed as [165]:

$$V_{oc} = \frac{Q}{C_{TENG}} \tag{9.1}$$

where Q is the short-circuit charge (Q_{sc}) of the slider, and the C_{TENG} is the intrinsic capacitor of AC-TENG. To enhance the output voltage, primary strategies involve increasing the charge density of the dielectric layers and reducing the intrinsic capacitance of the TENG. Liu et al. [165] achieved a significant reduction in C_{TENG} to 46 pF by augmenting the thickness of the packing layers (Fig. 9.44a–e), while simultaneously increasing the charge density to 163 μC m^{-2} using polyamide (PA)-Kapton as tribo-pairs (Fig. 9.44f and g). This resulted in a record output voltage of 12.7 kV. To further broaden the application scope, Gao et al. [166] reported a TENG that simultaneously obtained high-voltage and high-charge outputs. In theory, once materials are selected, material parameters such as relative permittivity and Q_{sc} become less adjustable (Fig. 9.44h and i). Additionally, increasing the air gap, the thickness of the bottom triboelectric material, or reducing the electrode overlap area can also boost V_{oc}. However, these adjustments may lead to a substantial decrease in Q_{sc} or a sharp increase in electric field intensity, potentially compromising stable charge output and long-term device functionality. Therefore, increasing the thickness of the top triboelectric layer is a viable strategy. When the thickness of the top triboelectric layer, such as PTFE, is increased from 30 μm to 150 μm, the V_{oc} of the sliding-mode AC-TENG reaches 7.2 kV, while Q_{sc} remains relatively a high value of 1.24 mC m^{-2} (Fig. 9.44j–l). This effect occurs because increasing the thickness of the top triboelectric layer attenuates the electric field intensity within the air gap, thereby preventing charge loss due to excessively high electric field intensities.

Different from AC-TENG, the V_{oc} of DC-TENG is governed by the following equation [167]:

$$V_{oc} = min\{V_{source}, V_{3,breakdown}, V_{leak}\},$$

where the V_{source} is depend on the triboelectric charging ability between the TL and FE, the $V_{3,breakdown}$ is the maximum breakdown voltage between the CCE and FE regions, and V_{leak} is the voltage leakage current occurs. According to reports [167], the theoretical maximum V_{oc} of DC-TENGs can be attained when the first domain is in a breakdown state while the second domain remains unbroken. To maintain this condition, isolation layers are employed to inhibit secondary breakdown, coupled with minimizing the vertical distance between the FE and CCE to ensure the initiation of the first breakdown. Addressing charge leakage is another challenge. To mitigate this, an optimized "fully isolated" DC-TENG configuration is introduced, utilizing two layers of polyimide (PI) films to fully fill the space between the FE and CCE. This configuration suppresses the second breakdown more effectively, resulting in a significant linear increase in V_{oc} as the horizontal distances of "fully isolated" components increase (Fig. 9.44m). Consequently, a record output voltage exceeding 10 kV was successfully achieved at a distance of 10 mm.

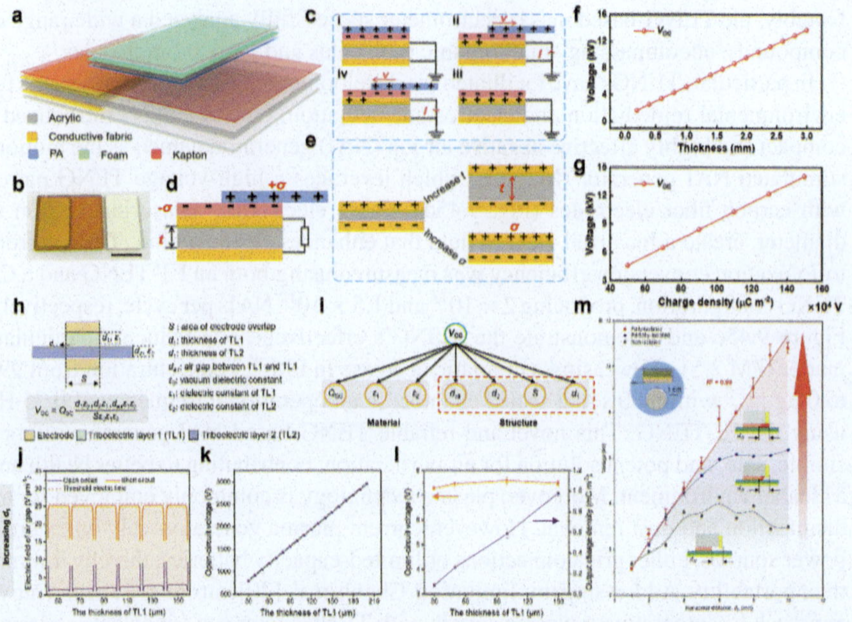

Fig. 9.44 Strategies for enhancing open-circuit voltage of triboelectric nanogenerators. Enhancing open-circuit voltage of single electrode mode (SE-TENG) [165]: **a** The cross section view of SE-TENG. **b** The photograph of SE-TENG. The scale bar is 5 cm. **c** The charge transfer schematic diagram of SE-TENG. **d** The theoretical mode of SE-TENG. **e** The method to increase V_{oc}. **f** The simulated V_{oc} with different thicknesses and **g** charge densities. Enhancing open-circuit voltage of S-TENG: **h** The structure parameters of S-TENG [166]: **i** The output charge of S-TENG when the top layer with or without grounded electrode. **j** The electric field at air gap of S-TENG. **k** The simulation results of output voltage of S-TENG with different thicknesses PTFE. **l** The experiment results of the output voltage and output charge of STENG with different thicknesses PTFE. Enhancing open-circuit voltage of DC-TENG [167]: **m** Open-circuit voltage of "non-isolated", "partially-isolated", and "fully-isolated" DC-TENG configurations under varying horizontal distances (d_H). The insets are the photo of "fully-isolated" configuration and diagrams of the three configurations

9.4.4.2 The Application of TENG as High-Voltage Source

The high output voltage of the TENG qualifies it as a novel high-voltage power source suitable for nano-electrospray ionization in mass spectrometric analysis. Li et al. [19] demonstrated a pioneering approach for self-powered mass spectrometric analysis by integrating TENGs with a nano-electrospray ionization emitter. This integration enables the production of single-polarity and alternating-polarity charge pulses in both contact-separation and freestanding modes (Fig. 9.45a). Despite the TENG's limited charge output, its efficient utilization and high voltage enable high sensitivity in nano-electrospray ionization. Figure 9.45b illustrates a TENG-powered nano-electrospray ionization emitter capable of facilitating plasma discharge ionization via a needle electrode. The consistent charge supplied by the TENG guarantees highly reproducible ionization pulses, as evidenced by the charge-time graph in Fig. 9.45c.

Notably, the TENG-based mass spectrometer successfully analyzed a wide range of compounds, encompassing small organic molecules and large biomolecules.

In particular, TENGs have facilitated the development of self-powered systems for environmental remediation and electrostatic actuation. Guo et al. [21] introduced a compact and highly effective negative air ion (NAI) generator, named as the motion-stimulated NAI generator (MSNG), which leverages a high-voltage TENG paired with carbon fiber electrodes (Fig. 9.45d). These electrodes, measuring 10 mm in diameter, create a localized electric field that enhances air ionization. The electron-to-ionization conversion efficiency was measured using both an FT-TENG and a CS TENG configuration, producing 2×10^{13} and 1.5×10^{12} NAIs per cycle, respectively. Figure 9.45e and f demonstrate the MSNG's effectiveness in reducing particulate matter (PM 2.5), showcasing a dramatic decrease in PM 2.5 concentration from 999 to 0 $\mu g\ m^{-3}$ within 80 s in a sealed chamber at an operational frequency of 0.25 Hz using the FT-TENG. This novel and reliable TENG-based NAI generator offers a simple, safe, and potent solution for air purification, contributing to better health and a cleaner environment. Moreover, plasma technology is commonly employed for air disinfection and dust removal. However, current plasma generators rely on external power sources, either grid connections or limited-capacity batteries, thereby limiting their portability. Addressing this limitation, Cheng et al. [20] introduced an innovative approach by integrating a plasma source with TENGs to create triboelectric plasma (Fig. 9.45e and f). By exploiting the high voltage output of TENGs, these devices exhibit remarkable efficiency in initiating gas breakdown. The demonstration of prepared TENG-powered plasma device in graphic luminescence (Fig. 9.45i) offers a novel solution that enhances the versatility of plasma in air purification applications, including bacterial growth inhibition, suspended particle adsorption, and ammonia gas decomposition [168–170].

Liquid manipulation is pivotal in numerous biological and chemical applications, including chemical reactions, drug delivery, and biochemical analysis [171]. However, these applications frequently encounter obstacles due to their dependence on HV sources, which can destroy safety, portability, and integration within biochemical analysis systems. To mitigate these challenges, researchers have integrated TENGs into the realm of electronically controlled droplets, resulting in the creation of diverse and adaptable droplet operating systems [172, 173]. Sun et al. [174] developed a motion-controlled, highly efficient micro-flow pump that harnesses triboelectricity-driven electroosmosis. This pump employs the high voltage produced by FS-TENG to establish a strong electric field between two electrodes within a microfluidic channel, enabling precise control over electroosmotic flow (Fig. 9.45j). Notably, the device achieved a microflow rate of approximately 600 $nL\ min^{-1}$ with minimal Joule heating of 1.76 $J\ cm^{-3}\ nL^{-1}$, significantly surpassing conventional sources that delivered a flow rate of approximately 50 $\mu l\ min^{-1}$ with a Joule heating of 8.12 $J\ cm^{-3}\ nL^{-1}$. Furthermore, the research revealed that adjusting the sliding distance of TENG allows for accurate modulation of the output charge, thereby controlling the flow rate of the electroosmotic pump (Fig. 9.45k). To address the limitation of droplet charge deficiency impacting its velocity, Sun et al. [175] presented a direct charge injection (DCI) method that utilizes FS-TENG as a source of charges

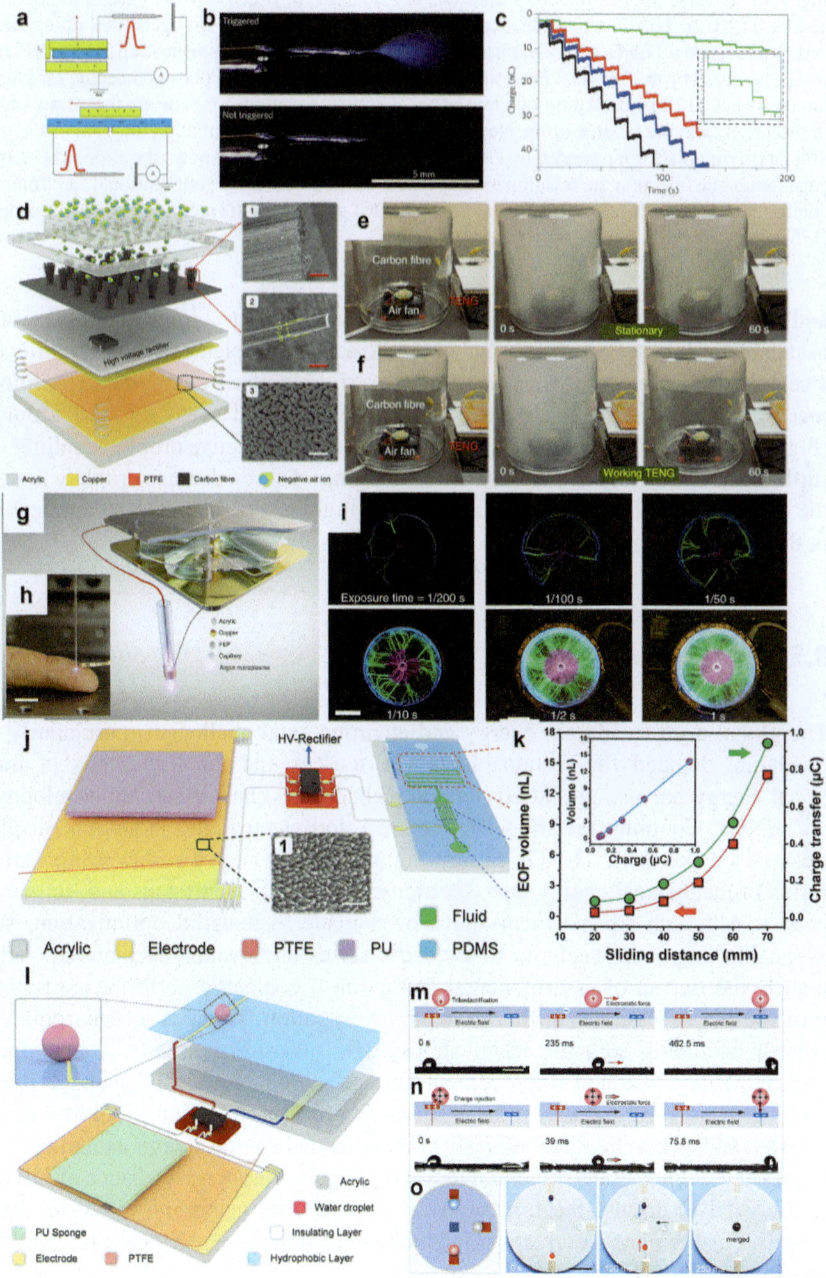
a
b
Triggered
Not triggered
5 mm
c
Charge (nC)
Time (s)
d
High voltage rectifier
Acrylic
Copper
PTFE
Carbon fibre
Negative air ion
e
Carbon fibre
Air fan
TENG
0 s
Stationary
60 s
f
Carbon fibre
Air fan
TENG
0 s
Working TENG
60 s
g
h
Acrylic
Copper
FEP
Capillary
Argon microplasma
i
Exposure time = 1/200 s
1/100 s
1/50 s
1/10 s
1/2 s
1 s
j
HV-Rectifier
1
Acrylic
Electrode
PTFE
PU
Fluid
PDMS
k
EOF volume (nL)
Charge transfer (μC)
Sliding distance (mm)
Volume (nL)
Charge (μC)
l
Acrylic
Water droplet
Insulating Layer
Hydrophobic Layer
PU Sponge
Electrode
PTFE
m
0 s
235 ms
462.5 ms
n
0 s
39 ms
75.8 ms
o
merged

◂**Fig. 9.45** Diverse applications of TENGs as HV sources. **a, b** ESI devices based on CS TENG and FT-TENG. **c** TENG-driven electrospray ionization effect [19]. **d** NAI generator composed of carbon fibers and a TENG. **e, f** Demonstration of MSNG's ability in air purification [21]. **g, h** Plasma source powered by the TENG. **i** Demonstration of the TENG-powered plasma machine for graphic luminescence [20]. **j** Installation diagram of the TEOP. **k** Relationship between the output charge of the TENG and the volume of the electroosmotic pump [174]. **l** Installation diagram of the basic droplet driving platform powered by DCI method. The enlarged area shows the exposed electrode edge for charge injection. **m, n** Comparison of the DCI method with the conventional electronically controlled droplet method. **o** Demonstration of the DCI method for the merging of multiple droplets [175]

and electric fields to achieve efficient droplet manipulation (Fig. 9.45l). This system is capable of injecting 0.25 nC of charge into a 10 μl droplet, enabling it to move at a velocity of 255 mm s^{-1}. This represents a sixfold increase compared to traditional methods, which achieved a charge injection of 0.03 nC and a velocity of 43.2 mm s^{-1} (Fig. 9.45m and n). The capability for meticulous and effective droplet handling was further exemplified by the simultaneous merging of three distinct droplets towards the center during TENG sliding (Fig. 9.45o), highlighting its potential applications in the liquid manipulation fields.

9.5 Conclusion

The global need to address energy and environmental challenges, particularly the escalating demand for sustainable power sources and the limitations of traditional energy harvesting and conversion technologies, has driven the development of TENGs. Coupling the triboelectric effect and electrostatic induction (or electrostatic breakdown), TENGs exhibit significant potential for converting various forms of mechanical energy into electricity. However, several gaps and limitations persist. Although advancements have been made in material optimization, such as establishing more accurate triboelectric series and evaluating material performance, the impact of environmental factors on triboelectric performance remains intricate and necessitates further in-depth investigation. Methods for environmental control, including adjustments to atmospheric pressure, humidity, and temperature, have demonstrated improvements in surface charge density, yet the underlying mechanisms, especially the intricate interactions between humidity, charge transfer, and charge loss, are not fully comprehended. While charge-excitation techniques have demonstrated potential for improving the output performance of TENGs, it is crucial to discuss the limitations associated with maximum effective charge density. Furthermore, a corresponding method to address these limitations should be proposed.

Despite these limitations, this chapter offers a significant contribution by providing a comprehensive and systematic overview of solid–solid TENGs. It encompasses their working modes, performance optimization strategies, and a broad array of applications. The detailed analysis of TENGs, spanning from theoretical foundations to practical experiments and from basic operational modes to technological implementations, offers valuable insights and guidance for future research and development in the field. This chapter aids researchers in understanding the current state of TENGs and paves the way for further exploration and innovation, aiming to fully realize their potential across various domains such as energy, sensing, and environmental protection.

References

1. Z.L. Wang, Entropy theory of distributed energy for internet of things. Nano Energy **58**, 669–672 (2019)
2. Z.L. Wang, Triboelectric nanogenerators as new energy technology and self-powered sensors–Principles, problems and perspectives. Faraday Discuss. **176**, 447–458 (2014)
3. F.-R. Fan, Z.-Q. Tian, Z.L. Wang, Flexible triboelectric generator. Nano energy **1**(2), 328–334 (2012)
4. G. Zhu, J. Chen, T. Zhang, Q. Jing, Z.L. Wang, Radial-arrayed rotary electrification for high performance triboelectric generator. Nat. Commun. **5**(1), 3426 (2014)
5. R. Hinchet, H.-J. Yoon, H. Ryu, M.-K. Kim, E.-K. Choi, D.-S. Kim, S.-W. Kim, Transcutaneous ultrasound energy harvesting using capacitive triboelectric technology. Science **365**(6452), 491–494 (2019)
6. H. Guo, X. Pu, J. Chen, Y. Meng, M.-H. Yeh, G. Liu, Q. Tang, B. Chen, D. Liu, S. Qi, A highly sensitive, self-powered triboelectric auditory sensor for social robotics and hearing aids. Sci. Robot. **3**(20), eaat2516 (2018)
7. S. Li, Z. Zhao, D. Liu, J. An, Y. Gao, L. Zhou, Y. Li, S. Cui, J. Wang, Z.L. Wang, A self-powered dual-type signal vector sensor for smart robotics and automatic vehicles. Adv. Mater. **34**(14), 2110363 (2022)
8. W. Qiao, L. Zhou, J. Zhang, D. Liu, Y. Gao, X. Liu, Z. Zhao, Z. Guo, X. Li, B. Zhang, A highly-sensitive omnidirectional acoustic sensor for enhanced human–machine interaction. Adv. Mater. **36**(48), 2413086 (2024)
9. Q. Zhang, Q. Liang, D. Nandakumar, H. Qu, Q. Shi, F. Alzakia, D. Tay, L. Yang, X. Zhang, L. Suresh, Shadow enhanced self-charging power system for wave and solar energy harvesting from the ocean. Nat. Commun. **12**(1), 616 (2021)
10. C. Zhang, L. He, L. Zhou, O. Yang, W. Yuan, X. Wei, Y. Liu, L. Lu, J. Wang, Z.L. Wang, Active resonance triboelectric nanogenerator for harvesting omnidirectional water-wave energy. Joule **5**(6), 1613–1623 (2021)
11. X. Liang, T. Jiang, G. Liu, Y. Feng, C. Zhang, Z.L. Wang, Spherical triboelectric nanogenerator integrated with power management module for harvesting multidirectional water wave energy. Energy Environ. Sci. **13**(1), 277–285 (2020)
12. Y. Li, M. Bouza, C. Wu, H. Guo, D. Huang, G. Doron, J. Temenoff, A. Stecenko, Z.L. Wang, F. Fernández, Sub-nanoliter metabolomics via mass spectrometry to characterize volume-limited samples. Nat. Commun. **11**(1), 5625 (2020)
13. H. Yang, Y. Pang, T. Bu, W. Liu, J. Luo, D. Jiang, C. Zhang, Z.L. Wang, Triboelectric micromotors actuated by ultralow frequency mechanical stimuli. Nat. Commun. **10**(1), 2309 (2019)
14. Z.L. Wang, From contact electrification to triboelectric nanogenerators. Rep. Prog. Phys. **84**(9), 096502 (2021)
15. D. Liu, X. Yin, H. Guo, L. Zhou, X. Li, C. Zhang, J. Wang, Z.L. Wang, A constant current triboelectric nanogenerator arising from electrostatic breakdown. Sci. Adv. **5**(4), eaav6437 (2019)

16. H. Wu, S. Wang, Z. Wang, Y. Zi, Achieving ultrahigh instantaneous power density of 10 MW/m^2 by leveraging the opposite-charge-enhanced transistor-like triboelectric nanogenerator (OCT-TENG). Nat. Commun. **12**(1), 5470 (2021)
17. W. He, W. Liu, S. Fu, H. Wu, C. Shan, Z. Wang, Y. Xi, X. Wang, H. Guo, H. Liu, Ultrahigh performance triboelectric nanogenerator enabled by charge transmission in interfacial lubrication and potential decentralization design. Research (2022). https://doi.org/10.34133/2022/9812865
18. W. He, C. Shan, S. Fu, H. Wu, J. Wang, Q. Mu, G. Li, C. Hu, Large harvested energy by self-excited liquid suspension triboelectric nanogenerator with optimized charge transportation behavior. Adv. Mater. **35**(7), 2209657 (2023)
19. A. Li, Y. Zi, H. Guo, Z.L. Wang, F. Fernández, Triboelectric nanogenerators for sensitive nano-coulomb molecular mass spectrometry. Nat. Nanotechnol. **12**(5), 481–487 (2017)
20. J. Cheng, W. Ding, Y. Zi, Y. Lu, L. Ji, F. Liu, C. Wu, Z.L. Wang, Triboelectric microplasma powered by mechanical stimuli. Nat. Commun. **9**(1), 3733 (2018)
21. H. Guo, J. Chen, L. Wang, A.C. Wang, Y. Li, C. An, J.-H. He, C. Hu, V.K. Hsiao, Z.L. Wang, A highly efficient triboelectric negative air ion generator. Nat. Sustain. **4**(2), 147–153 (2021)
22. Z.L. Wang, Fundamentals of triboelectric nanogenerators, in *Handbook of triboelectric nanogenerators*. ed. by Z.L. Wang, Y. Yang, J. Zhai, J. Wang (Springer, Cham, 2023), pp.1–30
23. Z.L. Wang, J. Chen, L. Lin, Progress in triboelectric nanogenerators as a new energy technology and self-powered sensors. Energy Environ. Sci. **8**(8), 2250–2282 (2015)
24. G. Zhu, C. Pan, W. Guo, C.-Y. Chen, Y. Zhou, R. Yu, Z.L. Wang, Triboelectric-generator-driven pulse electrodeposition for micropatterning. Nano Lett. **12**(9), 4960–4965 (2012)
25. S. Wang, L. Lin, Z.L. Wang, Nanoscale triboelectric-effect-enabled energy conversion for sustainably powering portable electronics. Nano Lett. **12**(12), 6339–6346 (2012)
26. S. Wang, L. Lin, Y. Xie, Q. Jing, S. Niu, Z.L. Wang, Sliding-triboelectric nanogenerators based on in-plane charge-separation mechanism. Nano Lett. **13**(5), 2226–2233 (2013)
27. G. Zhu, J. Chen, Y. Liu, P. Bai, Y.S. Zhou, Q. Jing, C. Pan, Z.L. Wang, Linear-grating triboelectric generator based on sliding electrification. Nano Lett. **13**(5), 2282–2289 (2013)
28. Y. Yang, Y.S. Zhou, H. Zhang, Y. Liu, S. Lee, Z.L. Wang, A single-electrode based triboelectric nanogenerator as self-powered tracking system. Adv. Mater. **25**(45), 6594–6601 (2013)
29. Y. Yang, H. Zhang, J. Chen, Q. Jing, Y.S. Zhou, X. Wen, Z.L. Wang, Single-electrode-based sliding triboelectric nanogenerator for self-powered displacement vector sensor system. ACS Nano **7**(8), 7342–7351 (2013)
30. Y. Yang, H. Zhang, Z.-H. Lin, Y.S. Zhou, Q. Jing, Y. Su, J. Yang, J. Chen, C. Hu, Z.L. Wang, Human skin based triboelectric nanogenerators for harvesting biomechanical energy and as self-powered active tactile sensor system. ACS Nano **7**(10), 9213–9222 (2013)
31. S. Wang, Y. Xie, S. Niu, L. Lin, Z.L. Wang, Freestanding triboelectric-layer-based nanogenerators for harvesting energy from a moving object or human motion in contact and non-contact modes. Adv. Mater. **26**(18), 2818–2824 (2014)
32. L. Lin, S. Wang, S. Niu, C. Liu, Y. Xie, Z.L. Wang, Noncontact free-rotating disk triboelectric nanogenerator as a sustainable energy harvester and self-powered mechanical sensor. ACS Appl. Mater. Interfaces **6**(4), 3031–3038 (2014)
33. L. Lin, Y. Xie, S. Niu, S. Wang, P.-K. Yang, Z.L. Wang, Robust triboelectric nanogenerator based on rolling electrification and electrostatic induction at an instantaneous energy conversion efficiency of $\sim$ 55%. ACS Nano **9**(1), 922–930 (2015)
34. D. Liu, L. Zhou, Z.L. Wang, J. Wang, Triboelectric nanogenerator: from alternating current to direct current. Iscience **24**(1), 102018 (2021)
35. S. Cui, L. Zhou, D. Liu, S. Li, L. Liu, S. Chen, Z. Zhao, W. Yuan, Z.L. Wang, J. Wang, Improving performance of triboelectric nanogenerators by dielectric enhancement effect. Matter **5**(1), 180–193 (2022)
36. Z. Zhao, Y. Dai, D. Liu, L. Zhou, S. Li, Z.L. Wang, J. Wang, Rationally patterned electrode of direct-current triboelectric nanogenerators for ultrahigh effective surface charge density. Nat. Commun. **11**(1), 6186 (2020)

37. Y. Gao, D. Liu, L. Zhou, S. Li, Z. Zhao, X. Yin, S. Chen, Z.L. Wang, J. Wang, A robust rolling-mode direct-current triboelectric nanogenerator arising from electrostatic breakdown effect. Nano Energy **85**, 106014 (2021)
38. L. Zhou, D. Liu, Z. Zhao, S. Li, Y. Liu, L. Liu, Y. Gao, Z.L. Wang, J. Wang, Simultaneously enhancing power density and durability of sliding-mode triboelectric nanogenerator via interface liquid lubrication. Adv. Energy Mater. **10**(45), 2002920 (2020)
39. D. Liu, L. Zhou, S. Li, Z. Zhao, X. Yin, Z. Yi, C. Zhang, X. Li, J. Wang, Z.L. Wang, Hugely enhanced output power of direct-current triboelectric nanogenerators by using electrostatic breakdown effect. Adv. Mater. Technol. **5**(7), 2000289 (2020)
40. Z. Yi, D. Liu, L. Zhou, S. Li, Z. Zhao, X. Li, Z.L. Wang, J. Wang, Enhancing output performance of direct-current triboelectric nanogenerator under controlled atmosphere. Nano Energy **84**, 105864 (2021)
41. L. Liu, Z. Zhao, Y. Li, X. Li, D. Liu, S. Li, Y. Gao, L. Zhou, J. Wang, Z.L. Wang, Achieving ultrahigh effective surface charge density of direct-current triboelectric nanogenerator in high humidity. Small **18**(24), 2201402 (2022)
42. J. Zhang, Y. Gao, D. Liu, J.S. Zhao, J. Wang, Discharge domains regulation and dynamic processes of direct-current triboelectric nanogenerator. Nat. Commun. **14**(1), 3218 (2023)
43. Y. Gao, L. He, D. Liu, J. Zhang, L. Zhou, Z.L. Wang, J. Wang, Spontaneously established reverse electric field to enhance the performance of triboelectric nanogenerators via improving Coulombic efficiency. Nat. Commun. **15**(1), 4167 (2024)
44. C. Shan, W. He, H. Wu, S. Fu, Q. Tang, Z. Wang, Y. Du, J. Wang, H. Guo, C. Hu, A high-performance bidirectional direct current TENG by triboelectrification of two dielectrics and local corona discharge. Adv. Energy Mater. **12**(25), 2200963 (2022)
45. Q. Zeng, A. Chen, X. Zhang, Y. Luo, L. Tan, X. Wang, A dual-functional triboelectric nanogenerator based on the comprehensive integration and synergetic utilization of triboelectrification, electrostatic induction, and electrostatic discharge to achieve alternating current/direct current convertible outputs. Adv. Mater. **35**(7), 2208139 (2023)
46. L. He, Y. Gao, D. Liu, Y. Hu, J. Shi, J. Zhang, X. Li, B. Jin, B. Zhang, Z.L. Wang, Dynamic interfacial electrostatic energy harvesting via a single wire. Sci. Adv. **10**(24), eado5362 (2024)
47. Z. Zhao, L. Zhou, S. Li, D. Liu, Y. Li, Y. Gao, Y. Liu, Y. Dai, J. Wang, Z.L. Wang, Selection rules of triboelectric materials for direct-current triboelectric nanogenerator. Nat. Commun. **12**(1), 4686 (2021)
48. K. Dai, D. Liu, Y. Yin, X. Wang, J. Wang, Z. You, H. Zhang, Z.L. Wang, Transient physical modeling and comprehensive optimal design of air-breakdown direct-current triboelectric nanogenerators. Nano Energy **92**, 106742 (2022)
49. W. He, C. Shan, H. Wu, S. Fu, Q. Li, G. Li, X. Zhang, Y. Du, J. Wang, X. Wang, Capturing dissipation charge in charge space accumulation area for enhancing output performance of sliding triboelectric nanogenerator. Adv. Energy Mater. **12**(31), 2201454 (2022)
50. Y. Zi, S. Niu, J. Wang, Z. Wen, W. Tang, Z.L. Wang, Standards and figure-of-merits for quantifying the performance of triboelectric nanogenerators. Nat. Commun. **6**(1), 8376 (2015)
51. H. Zou, Y. Zhang, L. Guo, P. Wang, X. He, G. Dai, H. Zheng, C. Chen, A.C. Wang, C. Xu, Quantifying the triboelectric series. Nat. Commun. **10**(1), 1427 (2019)
52. D. Liu, L. Zhou, S. Cui, Y. Gao, S. Li, Z. Zhao, Z. Yi, H. Zou, Y. Fan, J. Wang, Standardized measurement of dielectric materials' intrinsic triboelectric charge density through the suppression of air breakdown. Nat. Commun. **13**(1), 6019 (2022)
53. J. Wang, C. Wu, Y. Dai, Z. Zhao, A. Wang, T. Zhang, Z.L. Wang, Achieving ultrahigh triboelectric charge density for efficient energy harvesting. Nat. Commun. **8**(1), 88 (2017)
54. J. Fu, G. Xu, C. Li, X. Xia, D. Guan, J. Li, Z. Huang, Y. Zi, Achieving ultrahigh output energy density of triboelectric nanogenerators in high-pressure gas environment. Adv. Sci. **7**(24), 2001757 (2020)
55. S. Lin, L. Xu, C. Xu, X. Chen, A.C. Wang, B. Zhang, P. Lin, Y. Yang, H. Zhao, Z.L. Wang, Electron transfer in nanoscale contact electrification: effect of temperature in the metal–dielectric case. Adv. Mater. **31**(17), 1808197 (2019)

56. W. Liu, Z. Wang, G. Wang, G. Liu, J. Chen, X. Pu, Y. Xi, X. Wang, H. Guo, C. Hu, Integrated charge excitation triboelectric nanogenerator. Nat. Commun. **10**(1), 1426 (2019)
57. K. Wang, Z. Qiu, J. Wang, Y. Liu, R. Chen, H. An, J.H. Park, C.H. Suk, C. Wu, J. Lin, Effect of relative humidity on the enhancement of the triboelectrification efficiency utilizing water bridges between triboelectric materials. Nano Energy **93**, 106880 (2022)
58. L. Liu, L. Zhou, C. Zhang, Z. Zhao, S. Li, X. Li, X. Yin, J. Wang, Z.L. Wang, A high humidity-resistive triboelectric nanogenerator via coupling of dielectric material selection and surface-charge engineering. J. Mater. Chem. A **9**(37), 21357–21365 (2021)
59. B. Cheng, Q. Xu, Y. Ding, S. Bai, X. Jia, Y. Yu, J. Wen, Y. Qin, High performance temperature difference triboelectric nanogenerator. Nat. Commun. **12**(1), 4782 (2021)
60. L. Xu, T.Z. Bu, X.D. Yang, C. Zhang, Z.L. Wang, Ultrahigh charge density realized by charge pumping at ambient conditions for triboelectric nanogenerators. Nano Energy **49**, 625–633 (2018)
61. L. Cheng, Q. Xu, Y. Zheng, X. Jia, Y. Qin, A self-improving triboelectric nanogenerator with improved charge density and increased charge accumulation speed. Nat. Commun. **9**(1), 3773 (2018)
62. Y. Liu, W. Liu, Z. Wang, W. He, Q. Tang, Y. Xi, X. Wang, H. Guo, C. Hu, Quantifying contact status and the air-breakdown model of charge-excitation triboelectric nanogenerators to maximize charge density. Nat. Commun. **11**(1), 1599 (2020)
63. H. Wu, J. Wang, W. He, C. Shan, S. Fu, G. Li, Q. Zhao, W. Liu, C. Hu, Ultrahigh output charge density achieved by charge trapping failure of dielectric polymers. Energy Environ. Sci. **16**(5), 2274–2283 (2023)
64. X. Liu, Z. Zhao, Y. Gao, Y. Nan, Y. Hu, Z. Guo, W. Qiao, J. Wang, L. Zhou, Z.L. Wang, Triboelectric nanogenerators exhibiting ultrahigh charge density and energy density. Energy Environ. Sci. **17**(11), 3819–3831 (2024)
65. X. Liu, Z. Zhao, B. Zhang, Y. Hu, W. Qiao, Y. Gao, J. Wang, Z. Guo, L. Zhou, Z.L. Wang, Superior charge density of triboelectric nanogenerator via trap engineering. Adv. Funct. Mater. **35**(14), 2416944 (2024)
66. J. Zhong, Y. Zhang, Q. Zhong, Q. Hu, B. Hu, Z.L. Wang, J. Zhou, Fiber-based generator for wearable electronics and mobile medication. ACS Nano **8**(6), 6273–6280 (2014)
67. Y. Lai, J. Deng, S. Zhang, S. Niu, H. Guo, Z.L. Wang, Single-thread-based wearable and highly stretchable triboelectric nanogenerators and their applications in cloth-based self-powered human-interactive and biomedical sensing. Adv. Funct. Mater. **27**(1), 1604462 (2017)
68. K. Dong, X. Peng, Z.L. Wang, Fiber/fabric-based piezoelectric and triboelectric nanogenerators for flexible/stretchable and wearable electronics and artificial intelligence. Adv. Mater. **32**(5), 1902549 (2020)
69. J. Li, J. Cai, J. Yu, Z. Li, B. Ding, The rising of fiber constructed piezo/triboelectric nanogenerators: from material selections, fabrication techniques to emerging applications. Adv. Funct. Mater. **33**(44), 2303249 (2023)
70. W. Paosangthong, R. Torah, S. Beeby, Recent progress on textile-based triboelectric nanogenerators. Nano Energy **55**, 401–423 (2019)
71. K. Dong, Y.-C. Wang, J. Deng, Y. Dai, S. Zhang, H. Zou, B. Gu, B. Sun, Z.L. Wang, A highly stretchable and washable all-yarn-based self-charging knitting power textile composed of fiber triboelectric nanogenerators and supercapacitors. ACS Nano **11**(9), 9490–9499 (2017)
72. K. Dong, J. Deng, Y. Zi, Y. Wang, C. Xu, H. Zou, W. Ding, Y. Dai, B. Gu, B. Sun, 3D orthogonal woven triboelectric nanogenerator for effective biomechanical energy harvesting and as self-powered active motion sensors. Adv. Mater. **29**(38), 1702648 (2017)
73. K. Dong, X. Peng, J. An, A. Wang, J. Luo, B. Sun, J. Wang, Z.L. Wang, Shape adaptable and highly resilient 3D braided triboelectric nanogenerators as e-textiles for power and sensing. Nat. Commun. **11**(1), 2868 (2020)
74. K. Dong, J. Deng, W. Ding, A. Wang, P. Wang, C. Cheng, Y. Wang, L. Jin, B. Gu, B. Sun, Versatile core–sheath yarn for sustainable biomechanical energy harvesting and real-time human-interactive sensing. Adv. Energy Mater. **8**(23), 1801114 (2018)

75. Y. Lai, Y. Hsiao, H. Wu, Z.L. Wang, Waterproof fabric-based multifunctional triboelectric nanogenerator for universally harvesting energy from raindrops, wind, and human motions and as self-powered sensors. Adv. Sci. **6**(5), 1801883 (2019)
76. X. Pu, L. Li, M. Liu, C. Jiang, C. Du, Z. Zhao, W. Hu, Z.L. Wang, Wearable self-charging power textile based on flexible yarn supercapacitors and fabric nanogenerators. Adv. Mater. **28**(1), 98–105 (2016)
77. Q. Zheng, B. Shi, F. Fan, X. Wang, L. Yan, W. Yuan, S. Wang, H. Liu, Z. Li, Z.L. Wang, In vivo powering of pacemaker by breathing-driven implanted triboelectric nanogenerator. Adv. Mater. **26**(33), 5851–5856 (2014)
78. Q. Zheng, H. Zhang, B. Shi, X. Xue, Z. Liu, Y. Jin, Y. Ma, Y. Zou, X. Wang, Z. An, In vivo self-powered wireless cardiac monitoring via implantable triboelectric nanogenerator. ACS Nano **10**(7), 6510–6518 (2016)
79. H. Ouyang, Z. Liu, N. Li, B. Shi, Y. Zou, F. Xie, Y. Ma, Z. Li, H. Li, Q. Zheng, Symbiotic cardiac pacemaker. Nat. Commun. **10**(1), 1821 (2019)
80. Q. Zheng, Y. Zou, Y. Zhang, Z. Liu, B. Shi, X. Wang, Y. Jin, H. Ouyang, Z. Li, Z.L. Wang, Biodegradable triboelectric nanogenerator as a life-time designed implantable power source. Sci. Adv. **2**(3), e1501478 (2016)
81. J. Chen, G. Zhu, W. Yang, Q. Jing, P. Bai, Y. Yang, T. Hou, Z.L. Wang, Harmonic-resonator-based triboelectric nanogenerator as a sustainable power source and a self-powered active vibration sensor. Adv. Mater. **25**(42), 6094–6099 (2013)
82. L. Tang, Y. Yang, A nonlinear piezoelectric energy harvester with magnetic oscillator. Appl. Phys. Lett. **101**(9), 094102 (2012)
83. X. Wang, S. Palagummi, L. Liu, F. Yuan, A magnetically levitated vibration energy harvester. Smart Mater. Struct. **22**(5), 055016 (2013)
84. K. Ashraf, M. Khir, J. Dennis, Z. Baharudin, A wideband, frequency up-converting bounded vibration energy harvester for a low-frequency environment. Smart Mater. Struct. **22**(2), 025018 (2013)
85. W. Yang, J. Chen, G. Zhu, X. Wen, P. Bai, Y. Su, Y. Lin, Z. Wang, Harvesting vibration energy by a triple-cantilever based triboelectric nanogenerator. Nano Res. **6**, 880–886 (2013)
86. X. Wen, W. Yang, Q. Jing, Z.L. Wang, Harvesting broadband kinetic impact energy from mechanical triggering/vibration and water waves. ACS Nano **8**(7), 7405–7412 (2014)
87. S. Wang, X. Mu, Y. Yang, C. Sun, A. Gu, Z.L. Wang, Flow-driven triboelectric generator for directly powering a wireless sensor node. Adv. Mater. **27**(2), 240–248 (2014)
88. S. Chen, C. Gao, W. Tang, H. Zhu, Y. Han, Q. Jiang, T. Li, X. Cao, Z. Wang, Self-powered cleaning of air pollution by wind driven triboelectric nanogenerator. Nano Energy **14**, 217–225 (2015)
89. H. Guo, X. He, J. Zhong, Q. Zhong, Q. Leng, C. Hu, J. Chen, L. Tian, Y. Xi, J. Zhou, A nanogenerator for harvesting airflow energy and light energy. J. Mater. Chem. A **2**(7), 2079–2087 (2014)
90. X. Wang, S. Wang, Y. Yang, Z.L. Wang, Hybridized electromagnetic–triboelectric nanogenerator for scavenging air-flow energy to sustainably power temperature sensors. ACS Nano **9**(4), 4553–4562 (2015)
91. X. Wang, Y. Yang, Effective energy storage from a hybridized electromagnetic-triboelectric nanogenerator. Nano Energy **32**, 36–41 (2017)
92. Y. Yang, G. Zhu, H. Zhang, J. Chen, X. Zhong, Z.-H. Lin, Y. Su, P. Bai, X. Wen, Z.L. Wang, Triboelectric nanogenerator for harvesting wind energy and as self-powered wind vector sensor system. ACS Nano **7**(10), 9461–9468 (2013)
93. Y. Xie, S. Wang, L. Lin, Q. Jing, Z.-H. Lin, S. Niu, Z. Wu, Z.L. Wang, Rotary triboelectric nanogenerator based on a hybridized mechanism for harvesting wind energy. ACS Nano **7**(8), 7119–7125 (2013)
94. M.-L. Seol, J.-H. Woo, S.-B. Jeon, D. Kim, S.-J. Park, J. Hur, Y.-K. Choi, Vertically stacked thin triboelectric nanogenerator for wind energy harvesting. Nano Energy **14**, 201–208 (2015)
95. L. Zhang, B. Zhang, J. Chen, L. Jin, W. Deng, J. Tang, H. Zhang, H. Pan, M. Zhu, W. Yang, Lawn structured triboelectric nanogenerators for scavenging sweeping wind energy on rooftops. Adv. Mater. **28**(8), 1650–1656 (2016)

96. Z. Zhao, X. Pu, C. Du, L. Li, C. Jiang, W. Hu, Z.L. Wang, Freestanding flag-type triboelectric nanogenerator for harvesting high-altitude wind energy from arbitrary directions. ACS Nano **10**(2), 1780–1787 (2016)
97. A. Chortos, J. Liu, Z. Bao, Pursuing prosthetic electronic skin. Nat. Mater. **15**(9), 937–950 (2016)
98. Y. Lai, B. Ye, C. Lu, C. Chen, M. Jao, W. Su, W. Hung, T. Lin, Y. Chen, Extraordinarily sensitive and low-voltage operational cloth-based electronic skin for wearable sensing and multifunctional integration uses: a tactile-induced insulating-to-conducting transition. Adv. Funct. Mater. **26**(8), 1286–1295 (2016)
99. X. Wang, L. Dong, H. Zhang, R. Yu, C. Pan, Z.L. Wang, Recent progress in electronic skin. Adv. Sci. **2**(10), 1500169 (2015)
100. Z.L. Wang, Triboelectric nanogenerator (TENG)—sparking an energy and sensor revolution. Adv. Energy Mater. **10**(17), 2000137 (2020)
101. S. Wang, L. Lin, Z.L. Wang, Triboelectric nanogenerators as self-powered active sensors. Nano Energy **11**, 436–462 (2015)
102. G. Zhu, B. Peng, J. Chen, Q. Jing, Z.L. Wang, Triboelectric nanogenerators as a new energy technology: From fundamentals, devices, to applications. Nano Energy **14**, 126–138 (2015)
103. Y. Lai, H. Wu, H. Lin, C. Chang, H. Chou, Y. Hsiao, Y. Wu, Entirely, intrinsically, and autonomously self-healable, highly transparent, and superstretchable triboelectric nanogenerator for personal power sources and self-powered electronic skins. Adv. Funct. Mater. **29**(40), 1904626 (2019)
104. A. Khan, S. Ginnaram, C.-H. Wu, H.W. Lu, Y.F. Pu, J. Wu, D. Gupta, Y.C. Lai, H.C. Lin, Fully self-healable, highly stretchable, and anti-freezing supramolecular gels for energy-harvesting triboelectric nanogenerator and self-powered wearable electronics. Nano Energy **90**, 106525 (2021)
105. Y.C. Lai, J. Deng, R. Liu, Y. Hsiao, S. Zhang, W. Peng, H. Wu, X. Wang, Z.L. Wang, Actively perceiving and responsive soft robots enabled by self-powered, highly extensible, and highly sensitive triboelectric proximity-and pressure-sensing skins. Adv. Mater. **30**(28), 1801114 (2018)
106. L. Lin, Y. Xie, S. Wang, W. Wu, S. Niu, X. Wen, Z.L. Wang, Triboelectric active sensor array for self-powered static and dynamic pressure detection and tactile imaging. ACS Nano **7**(9), 8266–8274 (2013)
107. F.-R. Fan, L. Lin, G. Zhu, W. Wu, R. Zhang, Z.L. Wang, Transparent triboelectric nanogenerators and self-powered pressure sensors based on micropatterned plastic films. Nano lett. **12**(6), 3109–3114 (2012)
108. M. Li, H.-W. Lu, S.-W. Wang, R.-P. Li, J.-Y. Chen, W.S. Chuang, F.S. Yang, Y.F. Lin, C.Y. Chen, Y.C. Lai, Filling the gap between topological insulator nanomaterials and triboelectric nanogenerators. Nat. Commun. **13**(1), 938 (2022)
109. G. Zhu, W. Yang, T. Zhang, Q. Jing, J. Chen, Y. Zhou, P. Bai, Z.L. Wang, Self-powered, ultrasensitive, flexible tactile sensors based on contact electrification. Nano Lett. **14**(6), 3208–3213 (2014)
110. S. Chen, X. Cao, N. Wang, L. Ma, H. Zhu, M. Willander, Y. Jie, Z.L. Wang, An ultrathin flexible single-electrode triboelectric-nanogenerator for mechanical energy harvesting and instantaneous force sensing. Adv. Energy Mater. **7**(1), 1601255 (2017)
111. Y. Lai, S. Ginnaram, S. Lin, F. Hsu, T. Lu, M. Lu, Breathable and stretchable multifunctional triboelectric liquid-metal e-skin for recovering electromagnetic pollution, extracting biomechanical energy, and as whole-body epidermal self-powered sensors. Adv. Funct. Mater. **34**(10), 2312443 (2024)
112. Y. Lai, H. Lu, H. Wu, D. Zhang, J. Yang, J. Ma, M. Shamsi, V. Vallem, M. Dickey, Elastic multifunctional liquid–metal fibers for harvesting mechanical and electromagnetic energy and as self-powered sensors. Adv. Energy Mater. **11**(18), 2100411 (2021)
113. Y. Lai, J. Deng, S. Niu, W. Peng, C. Wu, R. Liu, Z. Wen, Z.L. Wang, Electric eel-skin-inspired mechanically durable and super-stretchable nanogenerator for deformable power source and fully autonomous conformable electronic-skin applications. Adv. Mater. **28**(45), 10024–10032 (2016)

114. B. Shao, M.H. Lu, T.C. Wu, W.C. Peng, T.Y. Ko, Y.C. Hsiao, J.Y. Chen, B. Sun, R. Liu, Y.C. Lai, Large-area, untethered, metamorphic, and omnidirectionally stretchable multiplexing self-powered triboelectric skins. Nat. Commun. **15**(1), 1238 (2024)
115. B. Shao, T. Lu, M. Lu, Y. Chen, T. Wu, W. Peng, T. Ko, J. Chen, B. Sun, C. Chen, Efficient permeable monolithic hybrid tribo-piezo-electromagnetic nanogenerator based on topological-insulator-composite. Adv. Mater. **36**(41), 2408936 (2024)
116. C. Zhao, Z. Wang, Y. Wang, Z. Qian, Z. Tan, Q. Chen, X. Pan, M. Xu, Y. Lai, MXene-composite-enabled ultra-long-distance detection and highly sensitive self-powered noncontact triboelectric sensors and their applications in intelligent vehicle perception. Adv. Funct. Mater. **33**(52), 2306381 (2023)
117. D.V. Anaya, K. Zhan, L. Tao, C. Lee, M.R. Yuce, T. Alan, Contactless tracking of humans using non-contact triboelectric sensing technology: enabling new assistive applications for the elderly and the visually impaired. Nano Energy **90**, 106486 (2021)
118. S. Ginnaram, Y.T. Chen, Y.C. Lai, Solid-state intrinsically-superstretchable multifunctional nanogenerator fiber for biomechanical and ambient electromagnetic energy harvesting and self-powered sensing. Nano Energy **95**, 107035 (2022)
119. Y. Xi, J. Hua, Y. Shi, Noncontact triboelectric nanogenerator for human motion monitoring and energy harvesting. Nano Energy **69**, 104390 (2020)
120. L. Li, Y.T. Chen, Y.C. Hsiao, Y.C. Lai, *Mycena chlorophos*-inspired autoluminescent triboelectric fiber for wearable energy harvesting, self-powered sensing, and as human–device interfaces. Nano Energy **94**, 106944 (2022)
121. S.H. Chou, Y.T. Chen, Z.X. Yan, T.C. Lu, T.C. Wu, M.H. Lu, T. Ko, W. Peng, J. Chen, F. Hsu, Topological-insulator nanocomposite and graphite-like tribo-charge-accumulating fabric enabling high-performance non-contact stretchable and textile-based triboelectric nanogenerators with robust charge retention. Adv. Energy Mater. **14**(37), 2402169 (2024)
122. J. Yang, J. Chen, Y. Liu, W. Yang, Y. Su, Z.L. Wang, Triboelectrification-based organic film nanogenerator for acoustic energy harvesting and self-powered active acoustic sensing. ACS Nano **8**(3), 2649–2657 (2014)
123. X. Fan, J. Chen, J. Yang, P. Bai, Z. Li, Z.L. Wang, Ultrathin, rollable, paper-based triboelectric nanogenerator for acoustic energy harvesting and self-powered sound recording. ACS Nano **9**(4), 4236–4243 (2015)
124. J. Yang, J. Chen, Y. Su, Q. Jing, Z. Li, F. Yi, X. Wen, Z. Wang, Z.L. Wang, Eardrum-inspired active sensors for self-powered cardiovascular system characterization and throat-attached anti-interference voice recognition. Adv. Mater. **27**(8), 1316–1326 (2015)
125. M. Li, W.Y. Cheng, Y.C. Li, H.-M. Wu, Y.C. Wu, H.W. Lu, S.L. Cheng, L. Li, K.C. Chang, H.J. Liu, Deformable, resilient, and mechanically-durable triboelectric nanogenerator based on recycled coffee waste for wearable power and self-powered smart sensors. Nano Energy **79**, 105405 (2021)
126. M. Wang, J. Zhang, Y. Tang, J. Li, B. Zhang, E. Liang, Y. Mao, X. Wang, Air-flow-driven triboelectric nanogenerators for self-powered real-time respiratory monitoring. ACS Nano **12**(6), 6156–6162 (2018)
127. Z. Lin, J. Chen, X. Li, Z. Zhou, K. Meng, W. Wei, J. Yang, Z.L. Wang, Triboelectric nanogenerator enabled body sensor network for self-powered human heart-rate monitoring. ACS Nano **11**(9), 8830–8837 (2017)
128. C. Huang, G. Chen, A. Nashalian, J. Chen, Advances in self-powered chemical sensing via a triboelectric nanogenerator. Nanoscale **13**(4), 2065–2081 (2021)
129. Z. Lin, G. Zhu, Y. Zhou, Y. Yang, P. Bai, J. Chen, Z.L. Wang, A self-powered triboelectric nanosensor for mercury ion detection. Angew. Chem. **125**(19), 5169–5173 (2013)
130. Z. Li, J. Chen, H. Guo, X. Fan, Z. Wen, M.H. Yeh, C. Yu, X. Cao, Z.L. Wang, Triboelectrification-enabled self-powered detection and removal of heavy metal ions in wastewater. Adv. Mater. **28**(15), 2983–2991 (2016)
131. R. Pelc, R. Fujita, Renewable energy from the ocean. Mar. Policy **26**(6), 471–479 (2002)
132. C. Shetty, A. Priyam, A review on tidal energy technologies. Mater. Today Proc. **56**, 2774–2779 (2022)

133. Y. Wen, P. Lin, Exploitation potential of tidal current energy in Southern China seas. Energy Convers. Manag. **267**, 115901 (2022)
134. Z.L. Wang, Catch wave power in floating nets. Nature **542**(7640), 159–160 (2017)
135. B. Kamranzad, P. Lin, G. Iglesias, Combining methodologies on the impact of inter and intra-annual variation of wave energy on selection of suitable location and technology. Renew. Energy **172**, 697–713 (2021)
136. S. Salter, Wave power. Nature **249**(5459), 720–724 (1974)
137. U. Henfridsson, V. Neimane, K. Strand, R. Kapper, H. Bernhoff, O. Danielsson, M. Leijon, J. Sundberg, K. Thorburn, E. Ericsson, Wave energy potential in the Baltic Sea and the Danish part of the North Sea, with reflections on the Skagerrak. Renew. Energy **32**(12), 2069–2084 (2007)
138. Y. Zhang, Y. Zhao, W. Sun, J. Li, Ocean wave energy converters: Technical principle, device realization, and performance evaluation. Renew. Sust. Energ. Rev. **141**, 110764 (2021)
139. J. Henriques, J. Portillo, L. Gato, R. Gomes, D. Ferreira, A. Falcão, Design of oscillating-water-column wave energy converters with an application to self-powered sensor buoys. Energy **112**, 852–867 (2016)
140. Y. Li, Q. Guo, M. Huang, X. Ma, Z. Chen, H. Liu, L. Sun, Study of an electromagnetic ocean wave energy harvester driven by an efficient swing body toward the self-powered ocean buoy application. IEEE Access **7**, 129758–129769 (2019)
141. Y. Zi, H. Guo, Z. Wen, M.-H. Yeh, C. Hu, Z.L. Wang, Harvesting low-frequency (< 5 Hz) irregular mechanical energy: a possible killer application of triboelectric nanogenerator. ACS Nano **10**(4), 4797–4805 (2016)
142. C. Zhu, M. Wu, C. Liu, C. Xiang, R. Xu, H. Yang, Z. Wang, Z. Wang, P. Xu, F. Xing, Highly integrated triboelectric-electromagnetic wave energy harvester toward self-powered marine buoy. Adv. Energy Mater. **13**(37), 2301665 (2023)
143. C. Zhu, C. Xiang, M. Wu, C. Yu, S. Dai, Q. Sun, T. Zhou, H. Wang, M. Xu, Recent advances in wave-driven triboelectric nanogenerators: from manufacturing to applications. Int. J. Extrem. Manuf. (2024)
144. Y. Yang, H. Zhang, R. Liu, X. Wen, T.C. Hou, Z.L. Wang, Fully enclosed triboelectric nanogenerators for applications in water and harsh environments. Adv. Energy Mater. **3**(12), 1563–1568 (2013)
145. X. Wang, S. Niu, Y. Yin, F. Yi, Z. You, Z.L. Wang, Triboelectric nanogenerator based on fully enclosed rolling spherical structure for harvesting low-frequency water wave energy. Adv. Energy Mater. **5**(24), 1501467 (2015)
146. B. Zhu, H. Wu, H. Wang, Z. Quan, H. Luo, L. Yang, R. Liao, J. Wang, Spherical 3D fractal structured dual-mode triboelectric nanogenerator for multidirectional low-frequency wave energy harvesting. Nano Energy **124**, 109446 (2024)
147. X. Liang, S. Liu, Z. Ren, T. Jiang, Z.L. Wang, Self-powered intelligent buoy based on triboelectric nanogenerator for water level alarming. Adv. Funct. Mater. **32**(35), 2205313 (2022)
148. S. Liu, X. Liang, P. Chen, H. Long, T. Jiang, Z.L. Wang, Multilayered helical spherical triboelectric nanogenerator with charge shuttling for water wave energy harvesting. Small Methods **7**(3), 2201392 (2023)
149. M. Xu, T. Zhao, C. Wang, S.L. Zhang, Z. Li, X. Pan, Z.L. Wang, High power density tower-like triboelectric nanogenerator for harvesting arbitrary directional water wave energy. ACS Nano **13**(2), 1932–1939 (2019)
150. D.Y. Kim, H.S. Kim, D.S. Kong, M. Choi, H.B. Kim, J.H. Lee, G. Murillo, M. Lee, S. Kim, J. Jung, Floating buoy-based triboelectric nanogenerator for an effective vibrational energy harvesting from irregular and random water waves in wild sea. Nano Energy **45**, 247–254 (2018)
151. L. Xu, Y. Pang, C. Zhang, T. Jiang, X. Chen, J. Luo, W. Tang, X. Cao, Z.L. Wang, Integrated triboelectric nanogenerator array based on air-driven membrane structures for water wave energy harvesting. Nano Energy **31**, 351–358 (2017)

152. C. Chen, D. Guo, L. Tuo, Y. Wen, J. Li, H. Qu, H. Wen, L. Wan, G. Liu, H. Guo, One meter triboelectric nanogenerator for efficient harvesting of meter-scale wave energy. Adv. Funct. Mater. **34**(42), 2406775 (2024)
153. X.J. Zhai, M.Y. Luo, X.-M. Luo, X.Y. Dong, Y. Si, C. Zhang, Z. Han, R. Han, S.Q. Zang, T. Mak, Hierarchical assembly of Ag40 nanowheel ranging from building blocks to diverse superstructure regulation. Nat. Commun. **15**(1), 9155 (2024)
154. Y. Duan, H. Xu, S. Liu, P. Chen, X. Wang, L. Xu, T. Jiang, Z.L. Wang, Scalable rolling-structured triboelectric nanogenerator with high power density for water wave energy harvesting toward marine environmental monitoring. Nano Res. **16**(9), 11646–11652 (2023)
155. G. Liu, L. Xiao, C. Chen, W. Liu, X. Pu, Z. Wu, C. Hu, Z.L. Wang, Power cables for triboelectric nanogenerator networks for large-scale blue energy harvesting. Nano Energy **75**, 104975 (2020)
156. L. Xu, T. Jiang, P. Lin, J. Shao, C. He, W. Zhong, X. Chen, Z.L. Wang, Coupled triboelectric nanogenerator networks for efficient water wave energy harvesting. ACS Nano **12**(2), 1849–1858 (2018)
157. X. Liang, T. Jiang, G. Liu, T. Xiao, L. Xu, W. Li, F. Xi, C. Zhang, Z.L. Wang, Triboelectric nanogenerator networks integrated with power management module for water wave energy harvesting. Adv. Funct. Mater. **29**(41), 1807241 (2019)
158. X. Liang, T. Jiang, Y. Feng, P. Lu, J. An, Z.L. Wang, Triboelectric nanogenerator network integrated with charge excitation circuit for effective water wave energy harvesting. Adv. Energy Mater. **10**(40), 2002123 (2020)
159. X. Li, L. Xu, P. Lin, X. Yang, H. Wang, H. Qin, Z.L. Wang, Three-dimensional chiral networks of triboelectric nanogenerators inspired by metamaterial's structure. Energy Environ. Sci. **16**(7), 3040–3052 (2023)
160. Z.L. Wang, On Maxwell's displacement current for energy and sensors: the origin of nanogenerators. Mater. today **20**(2), 74–82 (2017)
161. S. Shian, K. Bertoldi, D.R. Clarke, Dielectric elastomer based "grippers" for soft robotics. Adv. Mater. (2015)
162. C. Zhang, W. Tang, Y. Pang, C. Han, Z.L. Wang, Active micro-actuators for optical modulation based on a planar sliding triboelectric nanogenerator. Adv. Mater. **27**(4), 719–726 (2015)
163. C.B. Han, T. Jiang, C. Zhang, X. Li, C. Zhang, X. Cao, Z.L. Wang, Removal of particulate matter emissions from a vehicle using a self-powered triboelectric filter. ACS Nano **9**(12), 12552–12561 (2015)
164. Y. Zi, C. Wu, W. Ding, X. Wang, Y. Dai, J. Cheng, J. Wang, Z. Wang, Z.L. Wang, Field emission of electrons powered by a triboelectric nanogenerator. Adv. Funct. Mater. **28**(21), 1800610 (2018)
165. J. Liu, L. Zhou, Y. Gao, P. Yang, D. Liu, W. Qiao, B. Zhang, Z. Zhao, Z.L. Wang, J. Wang, Achieving ultra-high voltage ($\approx$ 10 kV) triboelectric nanogenerators. Adv. Energy Mater. **13**(21), 2300410 (2023)
166. Y. Gao, J. Liu, L. Zhou, L. He, D. Liu, P. Yang, B. Jin, Z.L. Wang, J. Wang, Achieving high performance triboelectric nanogenerators simultaneously with high-voltage and high-charge energy cycle. Energy Environ. Sci. **17**(22), 8734–8744 (2024)
167. J. Zhang, D. Liu, J. Shi, P. Yang, S. Li, Z. Zhao, Z. Guo, Y. Gao, L. He, J.-S. Zhao, An ultra-high voltage ($>$ 10 kV) direct-current triboelectric nanogenerator realized by structural and material optimizations. Nano Energy **124**, 109517 (2024)
168. M. Rong, J. Liu, X. Wang, X. Yuan, Research on air purification efficiency by nonthermal plasma along with the application of magnetic field. IEEE Trans. Plasma Sci. **34**(4), 1345–1350 (2006)
169. T. Ochiai, K. Nakata, T. Murakami, Y. Morito, S. Hosokawa, A. Fujishima, Development of an air-purification unit using a photocatalysis-plasma hybrid reactor. Electrochemistry **79**(10), 838–841 (2011)
170. J.S. Jung, J.G. Kim, An indoor air purification technology using a non-thermal plasma reactor with multiple-wire-to-wire type electrodes and a fiber air filter. J. Electrostat. **86**, 12–17 (2017)

171. M. Liu, S. Wang, L. Jiang, Nature-inspired superwettability systems. Nat. Rev. Mater. **2**(7), 1–17 (2017)
172. J. Nie, Z. Ren, J. Shao, C. Deng, L. Xu, X. Chen, M. Li, Z.L. Wang, Self-powered microfluidic transport system based on triboelectric nanogenerator and electrowetting technique. ACS Nano **12**(2), 1491–1499 (2018)
173. X. Sun, Y. Feng, B. Wang, Y. Liu, Z. Wu, D. Yang, Y. Zheng, J. Peng, M. Feng, D. Wang, A new method for the electrostatic manipulation of droplet movement by triboelectric nanogenerator. Nano Energy **86**, 106115 (2021)
174. J. Sun, L. Zhang, Z. Li, Q. Tang, J. Chen, Y. Huang, C. Hu, H. Guo, Y. Peng, Z.L. Wang, A Mobile and Self-Powered Micro-Flow Pump Based on Triboelectricity Driven Electroosmosis. Adv. Mater. **33**(34), 2102765 (2021)
175. J. Sun, L. Zhang, Y. Zhou, Z. Li, A. Libanori, Q. Tang, Y. Huang, C. Hu, H. Guo, Y. Peng, Highly efficient liquid droplet manipulation via human-motion-induced direct charge injection. Mater. Today **58**, 41–47 (2022)

Chapter 10
Triboelectric Nanogenerator Using Liquid–Solid CE

Abstract Contact-electrification (CE) is not limited to solid–solid cases, but also occurs at the liquid–solid interface. The charge transfer phenomenon between liquid and solid was widely studied by advanced technologies, such as KPFM, and has shown that electron transfer and ion transfer are both involved in liquid–solid situations. Building on recent years' understanding of the fundamentals of liquid–solid CE, the applications of triboelectric nanogenerator (TENG) based on CE between liquid and solid range from water energy harvesting to sensing systems, and TENG probes are developed. In this chapter, recent advancements in the applications of TENGs utilizing liquid–solid CE are introduced in detail, the application mainly focuses on their roles as energy harvesters, sensors, and TENG probes, which can realize liquid kinetic energy converted into electricity, self-powered microfluidic or chemical sensing, and the liquid–solid interfaces intensive study. For energy harvesting, the liquid droplet-based TENGs are highlighted in detail, and a maximum charge density of 5.4 mC m^{-2} is achieved with an individual droplet. During the application of sensing using liquid–solid CE, the demonstration of detecting or monitoring liquid motion, mechanical vibration, ion concentration, sediment, and wave state is presented. As for a probe, the studies of chemical analysis in liquids, fluid status monitoring, and taste-sensing are discussed in detail. At last, some perspectives on the future development of this research direction are presented.

10.1 Introduction

When two materials come into contact and then subsequently separated, they become statically charged. One surface gains a net positive charge, while the other acquires a net negative charge. This process, known as contact electrification or triboelectrification, has been studied for centuries and finds extensive use in practical applications. Interestingly, recent studies show this phenomenon occurs not only at solid–solid interfaces but also at liquid–solid interfaces and such type of triboelectrification can be measured by an electrometer [1]. Moreover, advanced technologies, including

Z. L. Wang, *Contact-Electrification of Matter*, Springer Series in Surface Sciences 75,
https://doi.org/10.1007/978-3-031-88504-4_10

nanoscale studies with KPFM or charge transfer studies using triboelectric nanogenerators probe (TENG probe), have shown that electron transfer and ion transfer are both in liquid–solid situations [2]. When a liquid droplet contacts and separates with a dielectric solid surface, electron transfer will first occur between solid and liquid. After that, the ions in the liquid will adsorb to the surface of the charged solid due to electrostatic interactions [3]. In this way, the measurable static charges generated between liquid and solid is a net charge, which is the sum of electron and ion transfer.

Building on the recent years understanding of the fundamentals at liquid–solid CE, the applications of TENG based on CE between Liquid and solid various from water energy harvesting, sensing systems to TENG probe are developed. When droplets collide with or moves on a solid surface, for example, raindrops falling, TENG can convert such liquid kinetic energy or liquid potential energy into electricity [4]. Besides harvesting energy from water, TENGs can also enable self-powered sensing based on triboelectrification and electrostatic induction at liquid–solid interfaces [5]. This functionality is influenced by the chemical and physical properties of both the liquid and the solid. Moreover, when an electrode array is applied, the charge transfer dynamics generated by the movement of droplets along their trajectory-referred to as the "TENG probe"-can also be utilized to study liquid–solid interfaces, and to sense various liquids and chemicals [6].

10.2 Liquid–Solid TENGs for Energy Harvesting

The liquid–solid TENGs can realize energy harvesting by converting various liquid kinetic energy or liquid potential energy into electricity. Due to the diverse liquid energy source with irregular motion, rich liquid–solid TENG prototypes, and device designs for harvesting liquid mechanical energy have been demonstrated [7–9]. Here, typical liquid–solid TENG forms are introduced in detail, including liquid droplet-based, liquid flow-based, and liquid wave-based. Under the action of these types of liquid–solid TENGs, the harvesting of different forms of liquid kinetic energy or liquid potential energy has been achieved.

10.2.1 Liquid Droplet-Based TENG

The liquid droplet-based TENG is the most typical type for liquid kinetic energy or liquid potential energy harvesting. Generally, the liquid droplet-based TENG can be suited to the traditional TENG mode of single-electrode, contract-separation, and free-stranding. Figure 10.1a shows the basic structure and working mechanism of the single-electrode liquid droplet-based TENG, the prototype for droplet energy harvesting proposed in 2014 [10]. As shown in Fig. 10.1a, the structure of the device consists of a poly(methyl methacrylate) (PMMA) substrate coated with a copper

(Cu) electrode and a superhydrophobic micro-/nanostructured polytetrafluoroethylene (PTFE) superhydrophobic film, the PTFE film is attached on the top of the Cu electrode. With the falling water droplet contact and separation with the PTFE film, the current will be generated and flow during the Cu electrode and the ground. The detailed working mechanism is shown in Fig. 10.1a i–viii. In the initial stage (Fig. 10.1a i–iv), there are few charges on the PTFE film, and the uncharged water droplet falls to collide with the PTFE film. Due to the CE effect, the impact zone of the PTFE film and water droplet will be electrified with opposite charges, and the surface of the PTFE film will normally be negatively charged. Subsequently, the water droplet falls from the device with positive charges, leaving negative charges on the PTFE film surface, which will cause the transfer of free electrons between the Cu electrode and ground due to electrostatic induction. With a series of water droplets, the charges on the PTFE film surface will reach a saturated state. In the following stage (Fig. 10.1a v–viii), the falling water droplet will be polarized when close to contact with the charged PTFE film surface and cause opposite charge transfer in the back electrode. Then, as the water droplet leaves the device surface, inducing corresponding charge transfer in the back electrode. Additionally, the water droplet may already contain triboelectric charges because of the friction with air/pipes. With a 30 μL tap water droplet falling from a height of 90 cm, the output current of the device reaches a value of 17 μA and exhibits alternating current (AC) behavior (Fig. 10.1b), the result validates the working mechanism proposed in Fig. 10.1a. By attaching a full-wave rectifying bridge, The AC output of the single-electrode liquid droplet-based TENG could be converted to unidirectional pulse output and drive 20 commercial LEDs instantaneously (Fig. 10.1c). Moreover, the rectifying device can charge capacitors to power electrics. Besides the single-electrode mode, the free-standing liquid droplet-based TENG can also be used to harvest droplet energy, which possesses the advantage of desirable scalability and wide applicability. The free-standing liquid droplet-based TENG is typically a three-layer structure. Here, as shown in Fig. 10.1d, the pre-patterned Indium Tin Oxide (ITO) serves as an electrode with the glass substrate, and the poly-4-vinyl phenol (PVP) dielectric material is spin-coated onto the patterned ITO glass substrate as a dielectric layer. Additionally, to realize smooth sliding of the water droplets, EKG6015N is coated on the PVP dielectric layer as a hydrophobic layer [11]. Different from the single-electrode mode, the electrical output signal of the free-standing liquid droplet-based TENG is induced by the charge interaction between the water droplets and the dielectric layer, which generates a potential difference between the two electrodes and drives the flow of electrons in the outer circuit. Figure 10.1e and f show the free-standing liquid droplet-based TENG's output current and lighted LED image. With water droplets, a volume of 30 μL, falling and passing through the electrodes at a speed of 0.4 m/s, a peak current of 5.3 μA is generated, which can be larger enough to turn on a green LED. For the single-electrode and free-standing liquid droplet-based TENG, the water droplets, as free objects, can directly convert their kinetic energy into electrical energy. However, for the contact-separation liquid droplet-based TENG, water is generally used as the triboelectric material, relying on external mechanical motion to achieve contact electrification between the dielectric layer and water droplets.

Figure 10.1g shows the photograph and structure of the contact-separation liquid droplet-based TENG [12]. A layer of aluminum (Al, thickness: 70 nm) is covered on a piece of fluorinated ethylene propylene (FEP: thickness 25 μm) film using an electron beam evaporator, and subsequently cleaned with deionized water before use. A drop of ultrapure water was dispensed onto another aluminum (Al) electrode with a polyimide film as the substrate. With the metal electrode moving up and down using a screw mechanism, the FEP and water achieve contact and separation motion, and generate a potential difference between the two electrodes based on CE and electrostatic induction. Here, as the lower electrode is displaced 3 mm into the droplet with a volume of 200 μL, the short-circuit current (I_{SC}) reaches a peak of about 0.5 μA and the open-circuit voltage (V_{OC}) reaches a peak of approximately 16 V (Fig. 10.1h). In addition, as motion amplitude and frequency reach 8 mm and 5 Hz, respectively, a maximum peak and average power of about 5 and 0.7 μW is delivered, this power is enough to light up an LED instantaneously (Fig. 10.1i).

For highly effective droplet energy harvesting, the liquid droplet-based TENG should possess high-performance characteristics, which generally mean higher electric output, better stability, and superior durability. Typically, the methods for developing high-performance liquid droplet-based TENGs are composed of three aspects: changing or modifying the surface of the triboelectric layer, implantation charges into the triboelectric layer, and designing new device structures. Firstly, by changing and modifying the surface of the triboelectric layer to improve wettability, the water droplet can be left from the triboelectric layer thoroughly, thus the electrical output of the TENG will be enhanced with the highly effective separation of the electric double layer (EDL). A detailed realization of the surface modifying strategy is shown in Fig. 10.2a [13]. The device is expressed as a hierarchical structure, an Al plate is chemically etched with hydrochloric acid, and $Al(OH)_3$ nanostructures were entirely formed on the microroughened Al surface, then, a high fluorine-containing material (heptadecafluoro-1,1,2,2-tetrahydrodecyl-trichlorosilane, HDFS) is coated on the $Al(OH)_3$ layer to make the surfaces superhydrophobic (contact angle, CA > 160°). Here, the HDFS is a self-assembled monolayer, which can maintain the surface roughness, and the root-mean-square roughness (R_q) of hierarchically structured surfaces reaches 6.24 ± 0.26 μm, while another nanostructure surface (smooth Al plate with chemical reactions of alkali treatment and a boiling process, then covered with HDFS) is 639.23 ± 19.17 nm. Figure 10.2b shows the sequential images of impacting droplets on the superhydrophobic surfaces with hierarchical structures and nanostructures. The droplet impact behavior can be characterized by a dimensionless parameter, Weber number (*We*). With different *We*, the dynamic variation of water droplets on the hierarchical structures and nanostructures superhydrophobic surfaces express difference, thus influencing the electric output of the liquid droplet-based TENG. For example, on the hierarchically structured surface with a relatively high value of *We* (*We* > ~ 50), the droplets will partially be rebounded with small droplets remaining. However, for the nanostructured surface, the droplets will be deposited without any rebound as *We* slightly increased (~10 < *We* < ~ 30). These phenomena can be attributed to the layer of air trapped between impacting droplets and surfaces. On hierarchical structures, a relatively thick air layer can act as an air

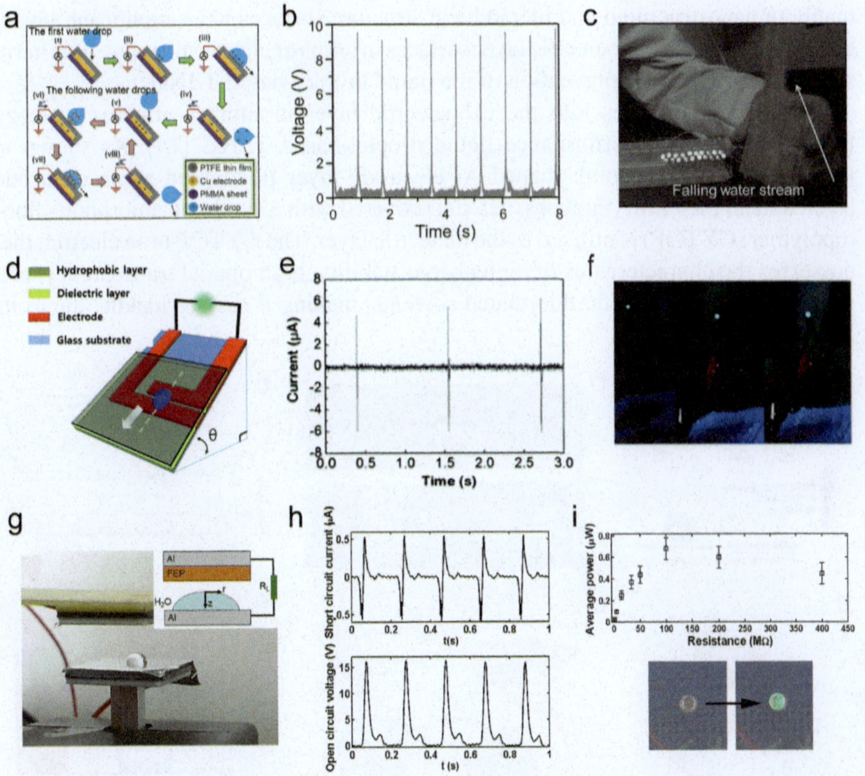

Fig. 10.1 Basic modes of the liquid droplet-based TENG. **a** Schematic and working mechanism of the single-electrode liquid droplet-based TENG. **b** Open-circuit voltage of the single-electrode liquid droplet-based TENG. **c** Directly powering LEDs by single-electrode liquid droplet-based TENG [10]. **d** Schematic of the free-stranding liquid droplet-based TENG. **e** Short-circuit current of the free-stranding liquid droplet-based TENG. **f** LEDs lighted up by free-stranding liquid droplet-based TENG [11]. **g** Schematic and photograph of the contact-separation liquid droplet-based TENG. **h** Short-circuit current and open-circuit voltage of the contact-separation liquid droplet-based TENG. **i** Output power and lighted LEDs with the contact-separation liquid droplet-based TENG [12]

cushion, preventing a droplet from penetrating the structure, known as the Cassie–Baxter state. In contrast, the thin air layer on nanostructures is more vulnerable to breakdown under high Weber numbers, leading to droplet penetration into the structures, a condition referred to as the Wenzel state. Figure 10.2c shows the normalized V_{OC} with water droplets falling on the surfaces of the hierarchical structure. Compared with the nanostructured surface, the output voltage with the surfaces of the hierarchical structure shows superior stability, which means the surface of the hierarchical structure is beneficial for ensuring durable energy harvesting performance under harsh droplet impact conditions. Additionally, it should be emphasized that both nanostructured and hierarchical structure surfaces have significant advantages over common triboelectric layer surfaces in terms of electrical output, long-term stability, and pollution prevention of the liquid droplet-based TENG.

Implantation charges into the triboelectric layer is another effective strategy for developing high-performance liquid droplet-based TENG [14]. As shown in Fig. 10.2d and e, A comb-shaped Al electrode layer is adopted as an electrode layer, and an FEP film (thickness: 25 μm) covered with a synthetic amorphous fluoropolymer (CYTOP) is utilized as the dielectric layer. The CYTOP is an electret, that possesses the characteristics of high charge stability, high optical transparency, and good solubility in specific fluorinated solvents, making it easy to handle due to its

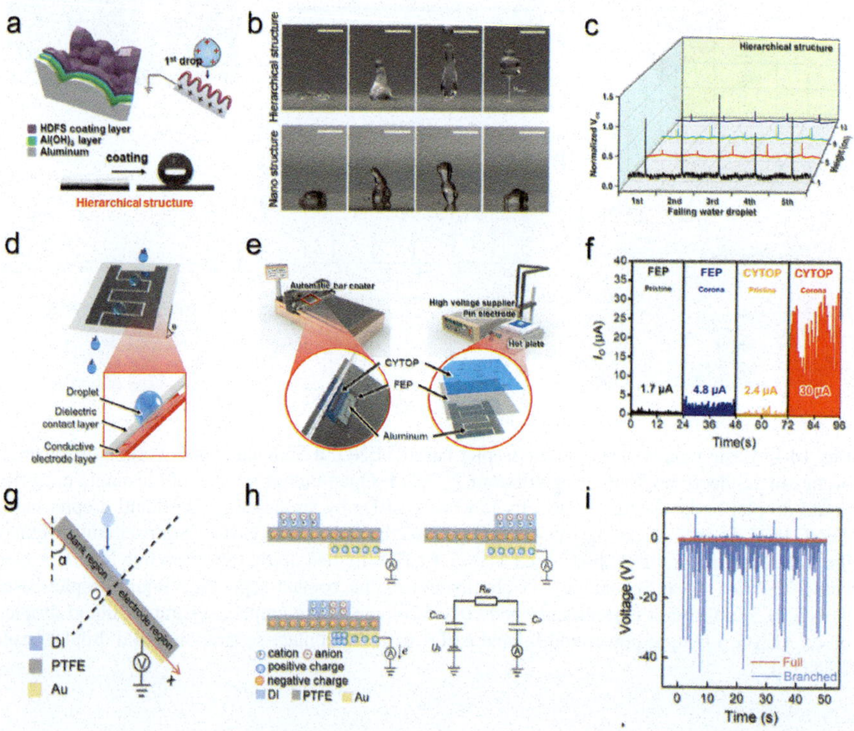

Fig. 10.2 Strategies for developing high-performance liquid droplet-based TENGs. **a** Schematic of the water droplet-based TENG with a hierarchical structure. **b** The moving state of the water droplet impacts the device with a hierarchical and nano structure. **c** Normalized open-circuit voltage of the device with hierarchical structure [13]. **d** Schematic of the water droplet-based TENG with a stronger electret layer. **e** Fabrication of the electret layer using bar coater and corona charging. **f** Comparison of output current with different dielectric layers [14]. **g** Schematic of the water droplet-based TENG with a branched electrode. **h** Schematics illustrate the charge distribution of the device for water droplets in different regions and the corresponding equivalent electrical circuit. **i** Output voltage of the device with a branched electrode (blue curves) and a controlled device with a full electrode (red curves) [15]

high processability. By applying high-voltage corona discharges, the CYTOP dielectric layer can be implanted with excess electric charges. The detailed corona charging is set by using a high-voltage power supply and pin electrode, the pin electrode is applied a voltage of − 10 kV with the Al electrode grounded and the pin electrode is placed 3 cm above the sample. The corona charging time is 25 min with 120 °C thermal annealing for the first 5 min. After the corona charging process, the device with CYTOP generated a current of 30 μA, which is more than 12 times compared with the pristine case, implying that the charges implantation to the dielectric layer can greatly enhance the electrical output performance of the liquid droplet-based TENG (Fig. 10.2f).

Apart from regulating the triboelectric layer, novel structure design can also improve the output ability of the liquid droplet-based TENG [15]. As shown in Fig. 10.2g, the novel working device comprises a dielectric layer (PTFE, thickness: 127 μm), a back electrode (Au deposited on the PTFE film), and falling water droplets based on the single-electrode TENG. Here, unlike the traditional single-electrode TENG, the back electrode is partially deposited on the dielectric layer, and the PTFE region with a bottom electrode is referred to as the electrode region, while the region without it is called the blank region. The area near the top edge of the electrode is referred to as the boundary region. When the droplets impact the dielectric layer, the electric output exhibits a noticeable dependence on position. In comparison to water striking the blank and electrode regions, both the output current and voltage are significantly enhanced when the droplets land and spread on the boundary region. Figure 10.2h shows the charge distribution of the device for water droplets in different regions. Water contains positive and negative charges, specifically cations and anions such as $H^+(H_3O^+)$ and OH^-. When a charged PTFE surface is present, it will influence the redistribution of these charged ions in neutral water. For droplets that are fully located either in the blank region or over the electrode region, the ions in the water will separate vertically. This creates a positively charged layer near the PTFE surface and a layer of negatively charged ions at the water film's surface. However, this vertical separation of ions is inefficient for driving electron movement toward the electrode because it does not produce a net charge close to the electrode. In contrast, when droplets impact the boundary region, the asymmetric distribution of the bottom electrode leads to the lateral movement of ions in water. The net negative charges in the blank region push anions toward the electrode side, which reduces the local potential at the electrode, this reduction then allows electrons to flow into the ground. In summary, the increased output of the novel partly electrode device is due to the formation of a water bridge between the blank region and the electrode region, facilitating charge transfer between the two areas. Figure 10.2i shows the output voltage of the novel partly electrode device with a branched electrode and a controlled device with a full electrode, respectively, which verify the efficiency of the proposed partly electrode structure.

For the above efforts to develop high-performance liquid droplet-based TENG, the number of charges generated in interfacial contact electrification limits the output performance to a certain degree. To overcome the limitations, a transistor-inspired droplet-based electricity generator (DEG) is proposed, which converts the interfacial effect to bulk effect, thereby increasing the instantaneous power density of the droplet-based TENG by several orders of magnitude [4]. As shown in Fig. 10.3a and b, the device comprises a PTFE film coated with an ITO electrode and an additional Al electrode, and the Al electrode is set on the upper of the PTFE film. When a water droplet falls and spreads on the surface of the PTFE film, it bridges the previously disconnected components, creating a closed-loop electrical system. As the interface effect transitions into a bulk effect, the output voltage and current of the DEG can reach 143.5 V and 270 μA, respectively. These values are approximately 295 times and 2600 times greater than the performance observed in the interface effect. Figure 10.3c illustrates the variation of surface charges measured on the PTFE film of an as-fabricated device. As the number of tap water droplets impacting the surface increases, the amount of surface charges gradually rises. After approximately 16,000 droplets, the surface charge stabilizes at about 49.8 nC. In addition, the surface charge largely influences the output of the DEG device.

Based on the outstanding characteristics of DEG, a series of studies have been conducted, which further improve the output performance of the device. Firstly, similar to the traditional liquid droplet-based TENG, the functionalization of the dielectric layer surface can also enhance the electric output of the DEG device. As shown in Fig. 10.3d, by inductively coupled plasma (ICP) etching, numerous biomimetic structures consisting of micro- and nanocolumnar protrusions are created on the PTFE dielectric layer. These bionic structures, resembling the surface texture of lotus leaves, serve as a triboelectric layer on the PTFE and contribute to the superhydrophobic performance of the device [16]. With the micro-sized conical structures on the PTFE hydrophobic surface, the device can achieve functions of increasing the surface area of friction materials, changing the local energy band junction of friction materials at the interface, and reducing the amount of residual liquid at the interface and residual dust and minerals on the friction surface. Combined with the dielectric layer structure and the bulk effect of DEG, the transferred charge of the device can reach 150 nC (Fig. 10.3e). Secondly, maintaining the stability of the charges in the dielectric layer or inducing more charges on the dielectric surface is also applicable to DEG devices. Charges at the solid–liquid interface are more likely to dissipate when water flows away. One potential method is to use an intermediate-layer approach to reduce electron scattering and enhance surface charge retention. This method could optimize electron quantities by increasing the polarization of dielectric materials and creating a charge-storage layer, which would improve surface charge management effectively. As illustrated in Fig. 10.3f, molybdenum selenide ($MoSe_2$) nanosheets are chosen to serve as intermediate layers [17]. The layered structure of $MoSe_2$ is recognized for its presence of electron-trapping sites, which help retain surface charges and enhance the output of the DEG, and the I_{SC} of the DEG can reach up to 1.2 mA (Fig. 10.3g). Figure 10.3h and i show another charge injection strategy to enhance the output of the DEG device. Here, the hydrophobic fluoropolymer films are pre-charged using a homogeneous electrowetting-assisted charge injection (h-EWCI) method, which offers the advantage of generating a highly stable charge density, eliminating the need for additional charge excitation processes

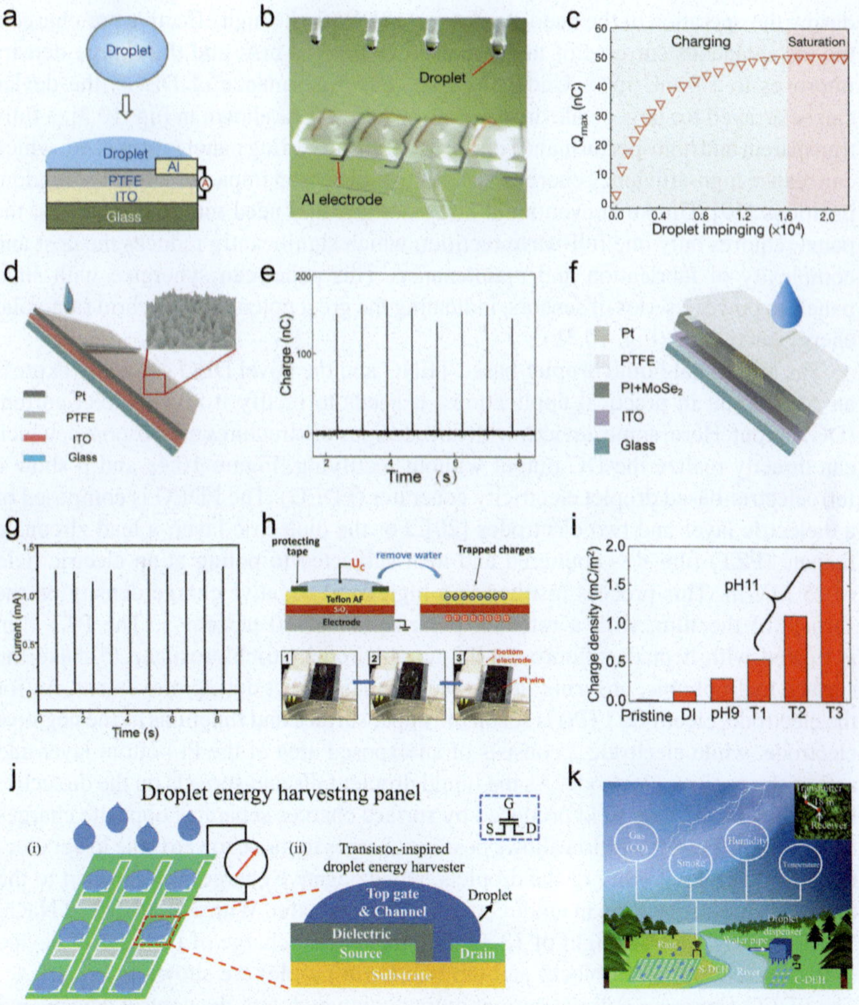

Fig. 10.3 Novel DEG device. **a** Schematic of the DEG. **b** Photograph of the DEG devices fabricated on a glass substrate. **c** Dependence of surface charges on water droplet number [4]. **d** Schematic of a DEG device with micro- and nanocolumnar dielectric layer. **e** Transferred charges of a DEG device with micro- and nanocolumnar dielectric layer [16]. **f** Schematic of a DEG device with a $MoSe_2$ intermediate layer. **g** Short-circuit current of a DEG device with a $MoSe_2$ intermediate layer [17]. **h** Schematics of electrowetting-assisted charge injection and photograph of the DEG. **i** Charge density of the device with electrowetting-assisted charge injection [18]. **j** Schematics of DEG panel and the DEG cell. **k** Schematic of the self-powered wireless forest monitoring system based on DEG panel [19]

during the operation of the nanogenerators [18]. With the high effective pre-charged, the instantaneous currents of the device reach up to 2 mA, and the charge density improves to 1.8 mC m^{-2}. Under the excellent performance of DEGs, the device can be arrayed for large-scale droplet energy harvest. As shown in Fig. 10.3j, a fully transparent and high-performance droplet energy harvesting panel is proposed, which can realize high-efficiency energy harvesting with droplet impacts on fixed or random positions [19]. Unlike conventional array designs that need multiple rectifiers, the panel requires only one full-wave rectifier, which significantly reduces the cost and complexity of fabrication and maintenance. This panel can synergize with solar panels to power a series of sensors, indicating the great potential for hybrid rain-solar energy harvesting (Fig. 10.3k).

The traditional liquid droplet-based TENG and the novel DEG generally express an AC output, in practical applications, it needs to rectify it into a direct-current (DC) output. Here, some devices with new designs or structures are proposed, which can directly realize the DC output without rectifying. Figure 10.4a and b show a ferroelectric-based droplet electricity generator (FDEG). The FDEG is composed of a dielectric layer and two electrodes [20]. For the dielectric layer, a lead zirconate titanate (PZT) film was sputtered and then subjected to poling at an electric field of 25 MV/m. This process resulted in a high fixed negative charge density on the surface of the film, with a remnant polarization of 50 μC cm^{-2}. The PZT film is treated with heptadecafluoro-1,1,2,2-tetrahydrodecyltrichlorosilane to make the surface hydrophobic, ensuring the droplets remain intact during movement. As for the electrode, electrode 1 (Pt) is set on the upper surface and functions as the negative electrode, while electrode 2 consists of an exposed area of the Pt bottom layer and acts as the positive electrode. As the liquid droplet falls and spreads on the dielectric layer, the electrostatic field produced by surface charges separates opposite charges in the liquid. This separation allows positive charges to move towards the lower electrode through the fluidity of the droplets, while negative charges are directed to the upper electrode, resulting in unidirectional charge transfer. With a droplet (1 M NaCl, 55 μL) falling from a height of 12 cm, the transferred charge of the control device (without electrode 2, in black) and the FDEG (in purple) are shown in Fig. 10.4c. The FDEG expresses a direct current output compared with the control device, and the transferred charge is up to 0.65 μC, corresponding to a maximum charge density of 5.4 mC m^{-2}. Figure 10.4d and e demonstrate another DC droplet-based electricity generator (DC-DEG), the DC-DEG is created by attaching two copper electrodes to the top and bottom surfaces of a PTFE film, with a charge collector needle connected to the upper electrode [21]. As for harvesting the mechanical energy of droplets, the DC-DEG is tilted at 50° for rapid liquid detachment. The working mechanism of the device is similar to the high voltage discharge phenomenon between cloud and cloud. Due to the electrostatic induction effect, when a water droplet falls on the device, the positive and negative charges in the water temporarily move apart from each other, perpendicular to the PTFE surface, subsequently, gravity causes these charges to separate, moving parallel to the PTFE surface, as illustrated in Fig. 10.4e. The top electrode collects the negative charges, while the bottom electrode returns electrons to the droplet, restoring the droplet's neutrality. As a result, DC electricity, featuring two unidirectional pulse peaks generated from a single droplet, is delivered through these two electrodes to an external load. Figure 10.4f illustrates the current curve featuring two unidirectional peaks that result from a single droplet striking the PTFE surface. Additionally, the output voltage achieves over 1600 V. In subsequent research, by successfully realizing the voltage-controlled electric shock process, the output voltage of the DC-DEG is improved to as high as 3000 V (Fig. 10.4g), demonstrating its unparalleled capability for harvesting high-entropy water energy [22].

Under harsh environments, the·output of the liquid droplet-based TENG will be decreased accompanied by the fast degradation of the physiochemical properties of interfacial materials. To overcome this problem, a slippery lubricant-impregnated porous surface (SLIPS) based TENG (SLIPS-TENG) is proposed, which exhibits distinctive advantages of self-cleaning, optical transparency, power

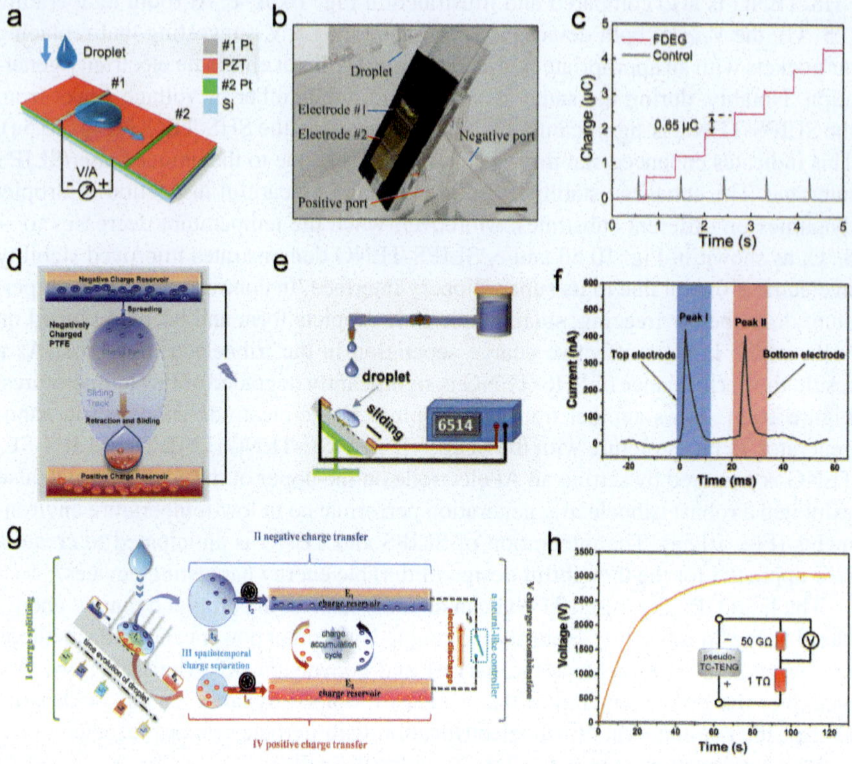

Fig. 10.4 Liquid droplet-based TENGs with a DC output. **a** Schematic of the FDEG with a DC output. **b** Photograph of the FDEG. **c** Transferred charge of the FDEG (in purple) and the control device (without electrode 2, in black) [20]. **d** The design of DC-DEG to mimic the process of thunderbolt. **e** Experimental platform and test system of the DC-DEG. **f** Output current of DC-DEG [21]. **g** Stages of DC-DEG along with the time evolution of droplet. **h** Open-circuit voltage of the DC-DEG [22]

generation stability, and a wide range of working environments. As shown in Fig. 10.5a, to construct the SLIPS-TENG, the patterned ITO electrodes are prepared on a transparent glass slide, then the SLIPS layer is deposed upon the electrode gently. For the SLIPS fabrication, a porous PTFE membrane (thickness: 20 μm; average pore size: 200 nm) is selected as the matrix, and the perfluorinated liquid (DuPont KrytoxGPL103) is chosen as the lubricant [23]. The lubricant possesses the characteristics of high wetting affinity with PTFE membrane, high-temperature stability, immiscibility with most liquids, chemical inertness, low vapor pressure, and compatibility with various materials, and it can wet into the membrane spontaneously via capillary wicking with a thickness of ~ 2 μm. Figure 10.5b shows the enhanced anti-icing capability of the SLIPS-TENG with droplets moving at freezing temperature. A droplet on superhydrophobic surfaces (SHS) is easily pinned due to the formation of an ice layer, which then freezes at the pinning position, while the impinging droplet slides off efficiently on SLIPS. The electric output of the SLIPS-TENG and SHS-TENG is also compared and illustrated in Fig. 10.5c–e. At room temperature (25 °C), the V_{OC} of both devices is approximately 1.2 V, suggesting that replacing air pockets with an appropriate lubricating liquid will not eliminate electricity generation. Notably, during the same testing period, the number of voltage peaks from the SLIPS-TENG is significantly higher than that from the SHS-TENG (Fig. 10.5c). This indicates enhanced stability in electrical output due to the unique water/SLIPS interface. The enhanced stability can be explained by careful inspection of droplet dynamics on different substrates. Moreover, when the temperature decreases to − 3 °C, as shown in Fig. 10.5d and e, SLIPS-TENG demonstrates improved stability in electrical output due to its super-slippery interface. In contrast, when the temperature drops below freezing, small condensate droplets form and become pinned on SHS, which hinders effective charge separation in the triboelectric process. As a result, the performance of SHS-TENG is significantly degraded at low temperatures. Figure 10.5f shows another transistor-inspired single-electrode triboelectric nanogenerator TENG structure with the SLIPS (SLIPS-SE-TENG) [24]. The SLIPS-SE-TENG is prepared by setting an Al electrode on the upper of the SLIPS, which also exhibited a robust triboelectric generation performance in low-temperature environments (Fig. 10.5g). The integration of SLIPS and TENG is anticipated to create a new approach for the thoughtful design of durable energy harvesting devices.

The liquid droplet-based TENG can not only harvest the droplet dynamic energy based on the liquid–solid CE but also be coupled with other power generation methods or energy sources to achieve the harvest and conversion of various energies. For example, the power generation that couples droplet's dynamic energy with wind energy, the combination of triboelectrification with pyroelectric or piezoelectricity, and the combination of liquid–solid CE with solar cells.

Figure 10.6a and b shows a humidity-responsive TENG based on the vapor-driven material of the perfluorosulfonic acid (PFSA) ionomer, which can reflect the sunlight in dry conditions, harvest wind energy in medium humidity, and collect energy from raindrops in rainy weather [25]. To fabricate the device, poly(ethylene terephthalate) (PET) is selected as the backbone of the TENG blade due to its excellent mechanical strength and flexibility. The Al-coated surface serves a dual purpose as both

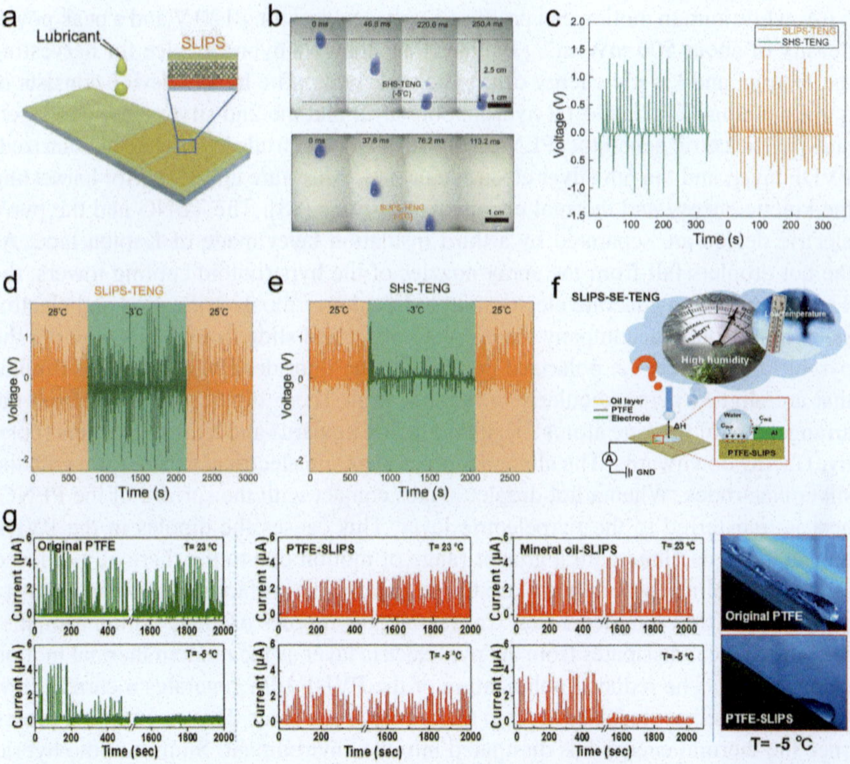

Fig. 10.5 Liquid droplet-based TENGs with SLIPS design. **a** Schematic of the SLIPS-TENG. **b** Comparison of anti-icing capability with water droplets moving on the SHS and SLIPS. **c** Comparison of open-circuit voltage between SLIPS-TENG with SHS-TENG at room temperature of 25 °C. **d** Dependence of open-circuit voltage of SLIPS-TENG on time at 25 and − 3 °C, respectively. **e** Dependence of open-circuit voltage of SHS-TENG on time at 25 and − 3 °C, respectively [23]. **f** Conceptual of SLIPS-TENG in harsh environments. **g** Influence of environmental temperature on the performance of the SLIPS-TENG and original device [24]

the output electrode and a positive tribo-layer of the TENG. FEP is chosen as the dielectric material and the negative tribo-layer. Additionally, the PFSA membrane is attached to the PET substrate, functioning as an "active joint" for the TENG blade. This membrane can provide a bending force for the blade under various humidity conditions [26]. The device can effectively harness energy from wind and raindrops (Fig. 10.6b), reaching a power density of 1.6 W m^{-2} at a wind speed of 25 m/s and 230 mW m^{-2} during rainy conditions, and can light serially connected LEDs to high brightness (Fig. 10.6c). Additionally, a flexible, waterproof, dual-mode textile TENG is also proposed to simultaneously scavenge multiple types of high-entropy kinetic energies from human motion, raindrops, and wind [27]. As illustrated in Fig. 10.6d, the device incorporates both a freestanding TENG and a contact-separation TENG. Raindrops can generate a voltage of up to approximately 4.3 V and a current of around

6 μA, while human motion can produce a voltage exceeding 120 V and a peak power density of about 500 mW m^{-2}. Figure 10.6e shows a hybrid device for harvesting the thermal and kinetic energy of the thermal fluids. The hybrid device consists of a freestanding TENG (the 1st hydrophobic layer and the 2nd silver electrode layer) and a pyroelectric generator [PENG, (the 4th silver electrode layer, the 5th polarized-PVDF layer, and the 6th silver electrode layer)], which are employed for harvesting the kinetic energy and thermal energy, respectively [28]. The TENG and the pyroelectric device are separated by a third insulation layer made of Kapton tape. As the hot droplets fall from the spray nozzles of the hyperboloid cooling towers, the TENG device can generate electric output based on triboelectrification and electrostatic induction to accompany the droplet impact and slide from the device. For the pyroelectric device, the polarized polyvinylidene fluoride (PVDF) features dipoles that are aligned perpendicularly to the thin film. These dipoles originate from the arrangement of fluorine atoms (negative) facing upwards and hydrogen atoms (positive) facing downwards. The aligned dipoles maintain electrical equilibrium with the silver electrodes. When a hot droplet makes contact with the surface of the PENG, heat is transferred to the pyroelectric layer. This causes the dipoles in the PVDF molecules to oscillate with a greater range of motion due to the thermally induced random wobbling around their axes. Consequently, the surface polarization intensity of the PVDF thin film decreases. Conversely, the dipoles oscillate within a smaller angle when heat dissipates from the pyroelectric layer as the random thermal motion is minimized. The reduced polarization of the PVDF film generates a current flow from the 4th silver electrode layer to the 6th electrode layer. This current flow ceases once the thermal energy has dissipated into the environment. Similar to the hybrid device with TENG and PENG, a leaf-mimic rain energy harvester, the synergy of the liquid–solid contact electrification and the piezoelectric effect, is proposed [29]. The device can simultaneously realize the harvest of electrostatic and kinetic energy induced by droplets. As illustrated in Fig. 10.6f and g, the Reconfigurable Energy Harvester (REH) consists of two main components: a top functional layer and a bottom functional layer. The top functional layer features a design inspired by transistors and is referred to as the Transistor-Inspired Liquid–Solid Generator (TLG). The bottom functional layer serves as a piezoelectric energy harvester (PEH). When a water droplet impacts the device, the TLG generates an output voltage of 261.2 V, while the PEH produces an output voltage of 19.6 V. This configuration results in a high-power density of 82.66 W m^{-2}. Lastly, Fig. 10.6h shows a mechanical buckling-based 4D printed elastic hybrid droplet-based electricity generator (HDEG), which integrates with a contact-separation solid–solid TENG and a DEG [30]. In the electricity generation process of the DEG, the HDEG structure is created using a simple 4D printing technique. This structure allows for the conversion of dissipated energy into elastic energy, leading to the output of a solid–solid TENG. The outputs of the HDEG are effectively integrated due to the simultaneous activation of a single droplet, resulting in an approximate 30% improvement over the output of a single DEG.

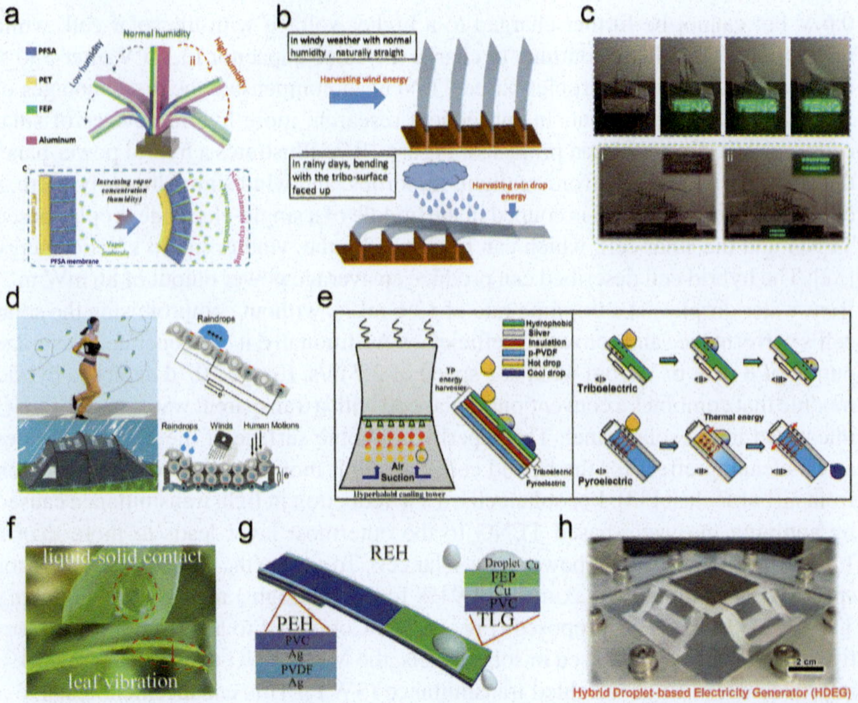

Fig. 10.6 Hybrid device with liquid–solid CE and other working modes or energy source. **a** Schematic and deformation mechanism of the humidity-responsive TENG. **b** Schematic of the TENG applied in windy and rainy weather with three different functions. **c** Lighted LEDs with wind and raindrops, respectively [25]. **d** Concept of dual-mode TENG to simultaneously scavenge energies from human motion, raindrops, and wind [27]. **e** Schematic of thermal fluids energy harvesting with a TENG and pyroelectric generator hybrid device [28]. **f** Liquid–solid contact and leaf vibration after droplet impinging. **g** Schematic of droplet impinging energy harvesting device utilizes both the liquid–solid contact electrification and the piezoelectric effect [29]. **h** Photograph of an HDEG configured with a DEG and a solid–solid TENG [30]

Combining liquid droplet-based TENG with solar cells is another hybrid power generation device. Solar cells, as promising devices for converting light into electricity, are highly dependent on weather conditions, which express a dramatically reduced performance on rainy days. To overcome this limitation, a hybrid energy cell is proposed, which is composed of a solar cell and a liquid droplet-based TENG, for harvesting solar and raindrop energy [31]. As shown in Fig. 10.7a, the bottom part of the hybrid energy cell features a silicon-based solar cell with micropyramids, the micropyramid surface morphology is designed to enhance light absorption and reduce reflectance. The top part is a single-electrode liquid droplet-based TENG, which consists of a PET substrate, an ITO electrode, and a superhydrophobic PTFE dielectric layer. Under a room light irradiation (1000 W m^{-2}) and water dripping at a rate of 0.116 mL/s, Fig. 10.7b shows the charging curve of the hybrid cell with TENG and solar cell connected in parallel. The capacitor can only be charged to

0.6 V but cannot be further charged to a higher voltage with the solar cell, while the rectified TENG can continue to charge the same capacitor to 3.5 V after 530 s, implying that the liquid droplet-based TENG can compensate the disadvantages of solar cell in voltage output. In subsequent research, more hybrid devices of solar cells and TENGs have been proposed. Figure 10.7c illustrates a hybrid power panel that can generate energy from sunlight, raindrops, and wind simultaneously. Here, a contact-separation TENG is couped in the middle of a single-electrode droplet-based TENG and the solar cell, which can be driven by the wind or liquid kinetic energy [32]. The hybrid cell described can produce an average power output of 86 mW m^{-2} from water droplets at a dripping rate of 13.6 mL/s, without compromising the solar cell's performance and conversion efficiency. Additionally, it can generate an average output of 8 mW m^{-2} from wind at a speed of 2.7 m/s. Figure 10.7d depicts a hybrid module that combines a conventional solar cell with a transparent water-based TENG integrated into a solar panel. The superhydrophobic surface of the TENG provides a self-cleaning effect for the hybrid cell, making it more resistant to contamination from airborne dust [33]. For solar cells, a 1% reduction in light transmittance caused by applying the water-based TENG to the outermost layer leads to more than a 1 mW cm^{-2} loss in output power for a solar cell. To reduce the power loss, a superior specular transmittance (maximum of 91% for visible light) moth's eye mimicking TENG (MM-TENG) is proposed (Fig. 10.7e). Compared to a conventional protective glass plate typically used in solar panels, the MM-TENG demonstrates a 0.01% improvement in solar-weighted transmittance (SWT). This enhancement leads to a 0.5% increase in the fill factor and a 0.17% improvement in the power conversion efficiency of the solar cell [34]. Moreover, Fig. 10.7f presents another hybrid cell that utilizes a transparent, solution-processable composite film made from carbon dots [35]. This film serves a dual purpose of enhancing the transmittance of the solar cell and functions as an ionic conductor for the TENG. With the optimal loading of carbon dots in the composite, there is a notable increase in transmittance within the visible spectrum. This improvement enhances the short-circuit current density of the solar cell, leading to an increase in its power conversion efficiency from 13.6 to 14.6%. Since the DEG can realize a higher electric output than the traditional TENG, combining DEG with solar cells will enhance the performance of the hybrid device. As shown in Fig. 10.7g, a silver/polydimethylsiloxane (Ag/PDMS) sub-cell is stacked on top of a monocrystalline Si solar cell [36]. By systematically optimizing the configuration of the device and the properties of the top nanogenerator, the bottom Si solar cell is used as a friction layer to control the charge distribution of the TENG. This setup achieved a short-circuit current of 7.59 μA and a peak open-circuit voltage of 37.19 V when stimulated by a single raindrop. Additionally, this configuration resulted in a 20% improvement in power conversion efficiency compared to a pristine Si solar cell. The liquid droplet-based TENG is an innovative energy compensation to the common solar cells, especially in rainy seasons or at night.

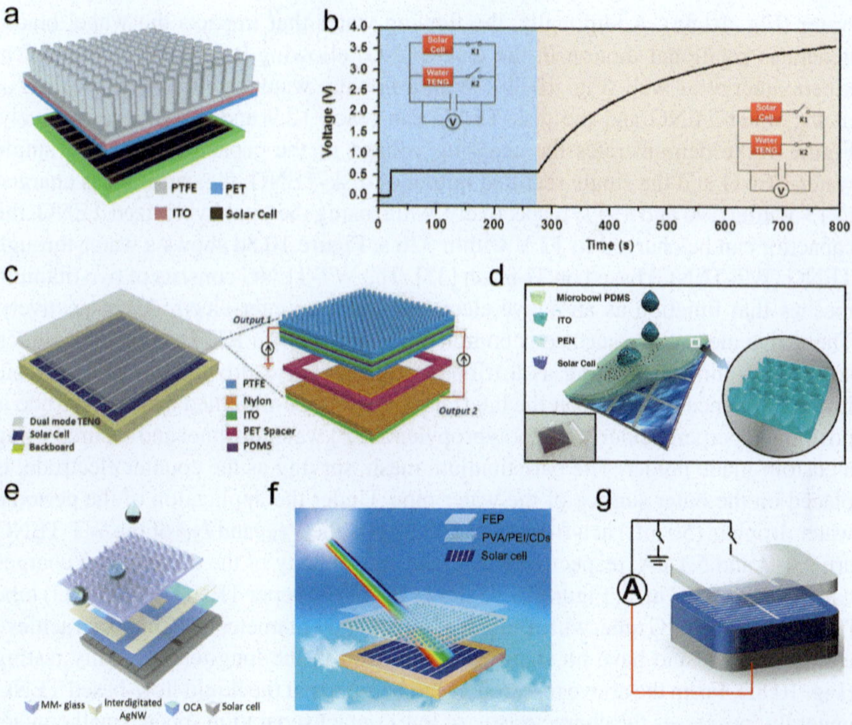

Fig. 10.7 Hybrid device with liquid droplet-based TENG and solar cells. **a** Schematic of the hybrid energy cell with a silicon-based solar cell and a single-electrode liquid droplet-based TENG. **b** Charging curve of the hybrid cell with TENG and solar cell connected in parallel [31]. **c** Schematic of a hybrid power panel that can generate energy from sunlight, raindrops, and wind simultaneously [32]. **d** Schematic of a hybrid module that combines a solar cell with a transparent water-based TENG [33]. **e** Schematic of a hybrid module with a solar cell and a superior specular transmittance moth's eye mimicking TENG [34]. **f** Schematic of a hybrid energy cell with a solar cell and a transparent, solution-processable composite film TENG [35]. **g** Schematic of a hybrid module that combines a solar cell with a DEG [36]

10.2.2 Liquid Flow-Based TENG

The liquid flow-based TENG is generally used for harvesting the energy of flowing water, which possesses mechanical kinetic energy and electrostatic energy owing to the triboelectric charges. The driving force of water flow is generally divided into two types: one relies on its potential energy or pressure, and the other relies on external mechanical forces. For the water flow induced by the potential energy or pressure, Fig. 10.8a shows a hybridized water wheel TENG, which combines a water-TENG component and a disk-TENG component to harvest both types of energy from flowing water simultaneously [9]. The wheel blades of the hybridized TENG are made from superhydrophobic PTFE films embedded with nanostructures. These materials are utilized as water TENGs to harvest electrostatic energy from flowing water (Fig. 10.8b). Additionally, the flowing water that impacts the wheel blades generates rotational motion in the disk-TENG, allowing it to harvest mechanical kinetic energy as well (Fig. 10.8b). With a flowing water rate of 54 mL/s, the I_{SC} of the water-TENG and the disk-TENG can reach 12.9 and 3.8 μA, respectively. Figure 10.8c demonstrates the charging voltage of the capacitor. With the single water-TENG and the single rectified output of disk-TENG, the capacitor is charged to 13 V after 596 and 435 s, respectively. while using the total hybridized TENG, the capacitor can be charged to 13 V within 326 s. Figure 10.8d shows a water-through TENG (WT-TENG) based on Ti-mesh [37]. The WT-TENG consists of two titanium meshes that function as an active electrode and a counter electrode, respectively. The active electrode's surface is completely covered with PTFE nanowires, which serve dual purposes: acting as a triboelectric layer and providing a hydrophobic layer to enhance liquid slip at the interface. The surface-modified active electrode is positioned perpendicularly to a polypropylene (PP) water channel and secured using a custom-made holder. The bare titanium mesh, serving as the counter electrode, is placed on the outer surface of the water pipe. Under the application of the periodic water droplets (50 mL) at a frequency of 0.25 Hz, the V_{OC} and I_{SC} of the WT-TENG are 9.4 V and 5.1 μA, respectively. And the area density of the triboelectric charges can reach 94.6 $\mu C\ m^{-2}$. Figure 10.8e shows a PTFE-copper TENG (PC-TENG) tube [38]. The PC-TENG tube, with a length of 40 cm and a diameter of 10 mm, can achieve a V_{OC} of 36 V and have no significant reduction in the long-term stability testing (Fig. 10.8f). From the above research, it can be seen that the liquid flow-based TENG generally possesses the characteristic of low contact separation speed, small contact area, and long contacting time during solid–liquid electrification, which severely limits their output properties. To overcome this problem, a gas–liquid two-phase flow-based TENG (GL-TENG) is proposed, which can achieve ultrahigh voltage and a volumetric charge density of 3789 V and 859 $mC\ m^{-3}$, respectively [39]. As shown in Fig. 10.8g, The GL-TENG comprises several components: the gas–liquid two-phase flow generating section, the water mist frictional electrification section, and the electricity discharge and energy collection section. These correspond to an airflow pipe, a vertical capillary tube submerged in a water tank, a horizontal PTFE tube, and a porous mesh conductive titanium electrode, respectively. In the gas–liquid two-phase flow generation process, a negative pressure is created at the top of the vertical capillary when high-speed external airflow is directed into the entrance of the horizontal PTFE tube. This pressure difference causes the liquid at the bottom to be drawn out. Simultaneously, the liquid being drawn in is dispersed and atomized due to the intense shear force of the high-speed airflow, resulting in a rapidly moving gas–liquid two-phase flow. Considering the high-speed airflow and the shearing force as well as the effect of wall collisions, the gas–liquid two-phase flow will undergo charge exchange when passing through the horizontal PTFE tube in the water mist frictional electrification section. During this process, the form of water and the contact state at the solid–liquid interface, including the contact area and relative separation speed, will change significantly. PTFE has an exceptional ability to capture charges, resulting in the liquid becoming positively charged after friction. Finally, the negative charges on the inner wall of the tube are transferred to the electrode through discharge in the electric energy collection section.

The liquid flow-based TENG can also harvest flowing water energy under the drive of external mechanical forces. Here, Liquid is usually placed in a sealed container, which serves as a TENG based on solid–liquid CE and designed structure to generate electricity. With external mechanical force driving the liquid to flow inside the device, a potential difference is generated between the electrodes. Figure 10.9a–f demonstrates various structures and designs of liquid flow-based TENG with the drive of external mechanical forces [40–45]. As shown in Fig. 10.9a, based on water electrification and rotating fluid inertia, a rotating water TENG is proposed [40]. The device primarily consists of an acrylic cylinder and multiple patterned Al electrodes that are covered with PTFE tape. The PTFE tape is stretched and attached to the electrodes to create wrinkled surfaces with micro-sized dimples near the flat PTFE surface. This

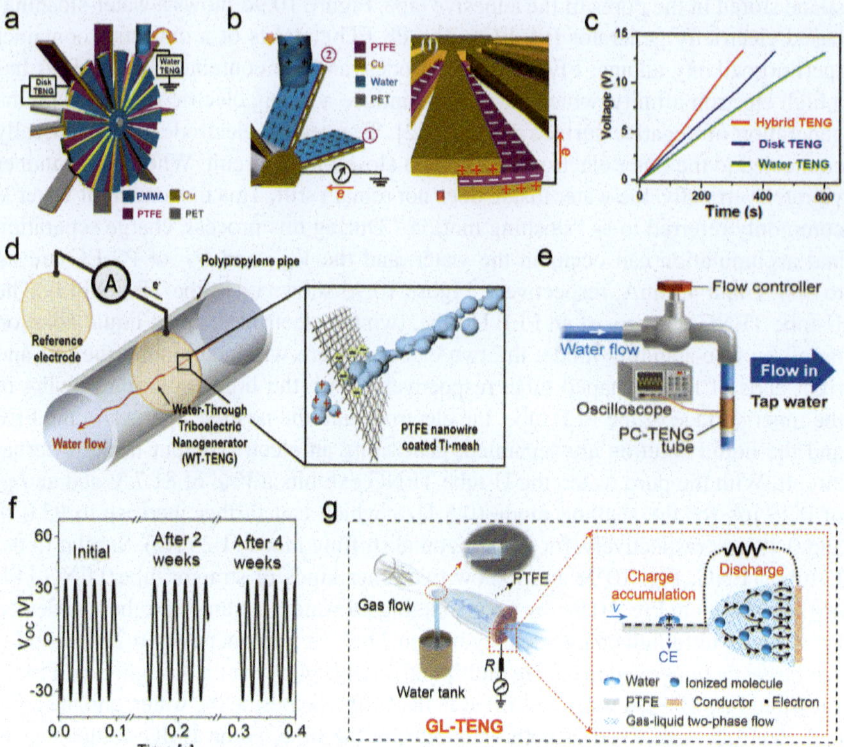

Fig. 10.8 Liquid flow-based TENG. **a** Structure schematic of hybridized water wheel TENG. **b** The working mechanism of the hybridized device with a water-TENG component and a disk-TENG component. **c** Charging curve of the capacitor with the device [9]. **d** Schematic of a water-through TENG (WT-TENG) based on Ti-mesh [37]. **e** Schematic of a PC-TENG tube. **f** Open-circuit voltage of the PC-TENG tube in long-term working [38]. **g** Schematic of a gas–liquid two-phase flow-based TENG [39]

increase in surface roughness reduces the sliding angle, which aids in the detachment of water from the PTFE surface. The PTFE-covered Al electrodes are positioned every 30° around the inner surface of the acrylic cylinder for electrical power generation. With tap water partially filled in the device, an electric output will be generated as the rolling of the device. Figure 10.9b shows a repairable water–solid cylinder TENG (W-TENG) [41]. The power generation cells of the W-TENG are placed on the section and both sides of the cylinder, with water flowing and covering one side of the power generation cell, an electrical output will be generated. The device can be integrated into an array, and the dielectric layer of the device possesses the characteristics of superhydrophobic and self-repaired, which is fabricated by adhering PVDF nanospheres to a PI film and followed by a fluorination process. Under the driving of mechanical vibration, the W-TENGs array could obtain a current output of 20 μA with a water flow rate of 0.3 m/s. Moreover, when a superhydrophobic surface is damaged during prolonged use, it can self-repair by releasing the fluorinated alkyl silane stored in the pores of the adhesive tape. Figure 10.9c shows a water-sloshing-based electricity generator (PS-EG). The PS-EG consists of a dielectric container (perfluoroalkoxy alkane, PFA), two electrodes, and the containing water. PFA has a high electron affinity, which allows it to induce a strong electric field through the generation of negative surface charges [42]. The central electrode was electrically connected to the outer electrode, forming a closed-loop circuit. When the container vibrates vertically, the water inside does not remain still. This movement of water is commonly referred to as "sloshing motion." During this process, charge separation and accumulation can occur in the water, and the V_{OC} and I_{SC} of PS-EG are up to 484 V and 4.1 mA, respectively. Figure 10.9d shows a U-tube TENG [43]. The U-tube TENG consists of an FEP U-tube, two Cu electrodes, and a liquid solution based on free-standing mode, the two electrodes are wrapped around the left and right ends of the U-shaped tube, respectively. With the liquid solution flowing in the inner surface of the FEP tube, the electrons and the positive charges in the FEP and the liquid solution are separated, generating an electric output in the external circuit. With the pure water, the U-tube TENG exhibits a V_{OC} of 81.7 V and an I_{SC} of 0.26 μA for the shaking mode (0.5 Hz), which can further increase to 93.0 V and 0.48 μA, respectively, for the horizontal shifting mode (1.25 Hz). Similar to the U-tube TENG, Fig. 10.9e and f show two other kinds of straight tube TENG [44, 45]. As shown in Fig. 10.9e, by encapsulating DI water into an FEP tube, the device achieves an ultrahigh charge density of 9 mC m^{-3} at a frequency of 0.25 Hz, and it can be easily integrated to realize multiplied outputs [44]. For the straight tube TENG shown in Fig. 10.9f, a magnet is filled in the FEP tube beside the water, and the coils and electrodes are fixed at both ends outside the tube, so an EMG component is formed beside the TENG component [45]. At a frequency of 1 Hz, the instantaneous power density and current density of the device are measured at 0.25 mW cm^{-3} and 5 mA cm^{-2}, respectively. Even at a very low operational frequency of 0.2 Hz, the device demonstrates impressive performance, achieving a peak current of nearly 15 mA.

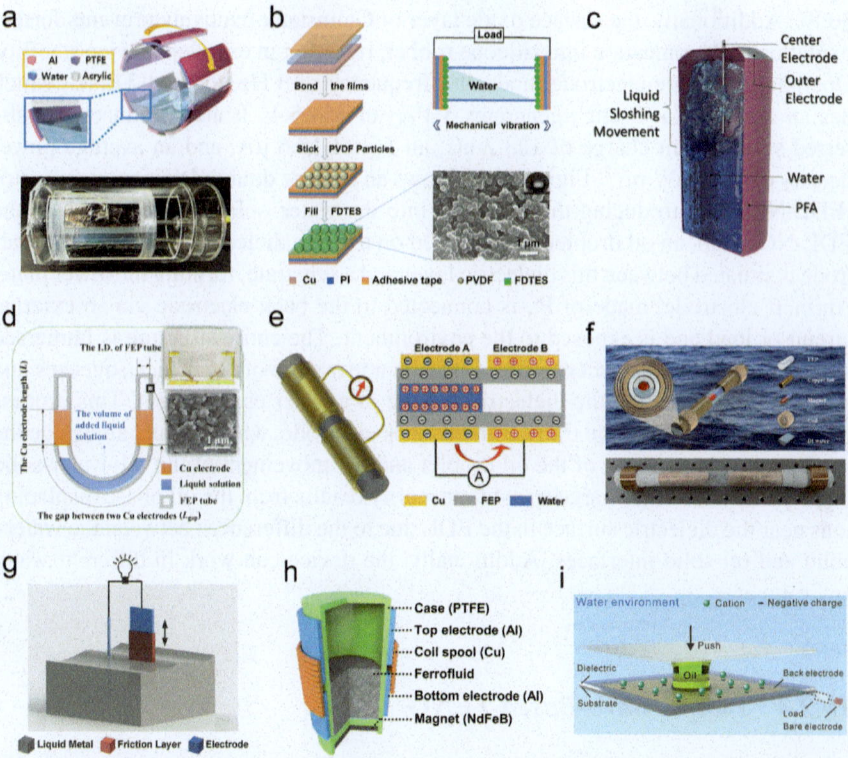

Fig. 10.9 Liquid flow-based TENG. **a** Schematic and photograph of the rotating water flow-based TENG [40]. **b** Schematic of a repairable cylinder water flow-based TENG [41]. **c** Schematic of a water-sloshing-based electricity generator [42]. **d** Schematic of U-tube water flow-based TENG [43]. **e** Schematic of a straight tube water flow-based TENG [44]. **f** Schematic of a straight tube water flow-based TENG combed with EMG [45]. **g** Schematic of a liquid–metal-based TENG [46]. **h** Schematic of a soft encapsulated liquid–metal-based TENG [47]. **i** Schematic of an electric double-layer nanogenerator [48]

Besides the typical water, other liquids are also suited for the liquid flow-based TENG. Figure 10.9g shows a liquid–metal-based TENG (LM-TENG) [46]. To fabricate the device, the Kapton film, copper, and mercury serve as the triboelectric material, induction electrode, and liquid metal, respectively. The LM-TENG, utilizing liquid metal, demonstrates an impressive output charge density of 430 μC m^{-2}, and the power density reaches 6.7 W m^{-2} and 133 kW m^{-3}. Most importantly, the instantaneous energy conversion efficiency is as high as 70.6%. Since liquid metal possesses the characteristics of deformable and conductive, using liquid metals as electrodes may provide a viable strategy for producing intelligent flexible electronics. As shown in Fig. 10.9h, by employing Galinstan as the electrode and silicone rubber as the triboelectric and encapsulation layer, another LM-TENG is proposed [47]. The low Young's modulus of liquid metal allows the electrode to maintain continuous conductivity even under deformation, with a strain capacity of up to approximately

300%. Additionally, the surface oxide layer of Galinstan effectively prevents further oxidation and permeation into silicone rubber, resulting in excellent device stability. Operating in single-electrode mode at a frequency of 3 Hz, the LM-TENG, which has an area of 6 × 3 cm^2, generates a V_{OC} of 354.5 V. It also produces a transferred short-circuit charge of 123.2 nC, an I_{SC} of 15.6 μA, and an average power density of 8.43 mW m^{-2}. Figure 10.9i shows an electric double-layer nanogenerator (EDL-NG) by introducing the oil phase into the water–solid interface [48]. In the EDL-NG setup, an oil droplet is positioned on top of a dielectric layer. A back electrode is situated between this dielectric layer and a substrate, forming the lower plate. An inert electrode, made of Pt, is connected to the back electrode via an external circuit or load and is exposed to the environment. The entire structure is immersed in water. Under the impact of water flow, the upper plate of the device squeezes and releases oil droplets on the dielectric surface in a water environment. This process leads to observable charge transfer in the back electrode, which occurs alongside the spreading and retracting of the oil droplet and the movement of the oil–water-solid three-phase line. This charge transfer primarily results from the oil phase displacing ions near the dielectric surface in the EDL due to the differences between the water–solid and oil-solid interfaces. Additionally, the device can work in different water environments.

10.2.3 Liquid Wave-Based TENG

The liquid wave-based TENG can be used for harvesting wave energy. Water wave, as a clean and renewable energy source, possesses the characteristics of large reserves, low frequency, and irregular motion. TENGs can effectively harvest wave energy based on their high low-frequency responsiveness. Figure 10.10 demonstrates a series of liquid wave-based TENGs with different structures and working modes. As shown in Fig. 10.10a, the first liquid wave-based TENG was proposed in 2013, which is also referred to as the first liquid–solid CE TENG. The device consists of two parts, the upper part is a PDMS film with a Cu electrode attached to the back, and the lower part is another Cu electrode covered with water [7]. With the fluctuation of water waves, the upper PDMS comes into contact and separates from the water, inducing the flow of charges between the two electrodes. However, the liquid wave-based TENG, based on the contact-separation mode, will have some limitations in wave energy harvesting. During the following research, other modes and structures of liquid wave-based TENG are developed. Figure 10.10b shows a free-standing liquid wave-based TENG. Here, aligned nanowires are formed on the FEP film of the device, which enhances its hydrophobic properties and increases the surface area [8]. During the processes of emerging and submerging with the water wave, an asymmetric screening effect is produced on the surface charges. This interaction causes the free electrons on the electrodes to flow through an external load, leading to power generation. With a relative velocity of 0.5 m/s of water wave, the device can generate an optimum average output power of 0.12 mW. The device can integrate into an array to obtain a higher current (Fig. 10.10c). However, the harvesting efficiency of water waves through simple device stacking integration is still relatively low. Figure 10.10d and e show two other kinds of free-standing liquid wave-based TENG with a superior integration approach. As shown in Fig. 10.10d, the device is designed using an array of surface-mounted bridge rectifiers to connect multiple parallel electrodes [49]. This configuration allows the induced charge flow between any pair of adjacent electrodes to be rectified and constructively combined, resulting in a pulsed direct current between an anode and a cathode. It also enables a significant increase in output power and allows for area-scalable integration of electrode arrays. For the device shown in Fig. 10.10e, a two-dimensional array of strip-shaped electrode units made from conductive textiles is laminated onto the substrate. Both ends of the electrode units are connected to rectifying chips based on p-n junctions. These two chips are connected in series at opposite ends of each electrode unit. The comb-shaped anode and cathode buses are positioned between the columns of the electrode units, connecting to the "P" and "N" sides of the rectifying chips, respectively. The chips function as "gates," controlling the flow of induced current from the anode to the cathode [50]. With the superior integrated design, the device can handle various water wave motions and produce stable electric output, regardless of the randomness of the waves. The device with an area of $100 \times 70\ mm^2$ can produce a stable I_{SC} of 13.5 μA and a corresponding electric power of 1.03 mW at a water wave height of 12 cm. Figure 10.10f and g show two novel liquid wave-based TENG based on the bulk effect, which minimizes the shielding effect from water and achieves more charge transfer. As shown in Fig. 10.10f, with the novel external U electrode, which includes a bar electrode (B electrode) and a U-shaped electrode (U electrode), the shielding effect from water is significantly minimized, leading to improved outputs [51]. For the device shown in Fig. 10.10g, when the wave impacts the PTFE surface and interacts with the 3D electrode, it creates a closed-loop system that bridges the 3D electrode, the bottom electrode, and the PTFE [52]. The design of the 3D electrode enables charge transfer in bulk water, rather than relying solely on electrostatic induction at the liquid–solid interface. After optimization, the device can output a voltage of up to 42 V and a current of 4 mA. Figure 10.10h shows another new working scheme liquid wave-based TENG [53]. Here, by designing opposite dynamic EDLs, the efficient conversion of water's kinetic energy into electrical energy is achieved. The device can be integrated into an array, under irregular water waves, the TENG array shows improved performance, featuring a stable output current of 60.0 μA, an output voltage of 60.0 V, and an average power density of $5.38\ W\ m^{-3}$.

The liquid wave-based TENG can also be fabricated to enclose the structure for wave energy harvesting. Figure 10.10i and j demonstrate two kinds of closed chamber devices. As shown in Fig. 10.10i, the device expresses a highly symmetric spherical-shaped structure. It features a spherical shell frame with a double-layer, liquid wave-based TENG on both its inner and outer surfaces [54]. This TENG, which contains water, is ideally suited for harvesting energy from water waves. The 3D symmetric design enhances its efficiency in capturing energy from random directions, while the double-layer device minimizes the risk of water leakage. which further improves the device robustness and efficiency of TENG for water wave energy harvesting. With

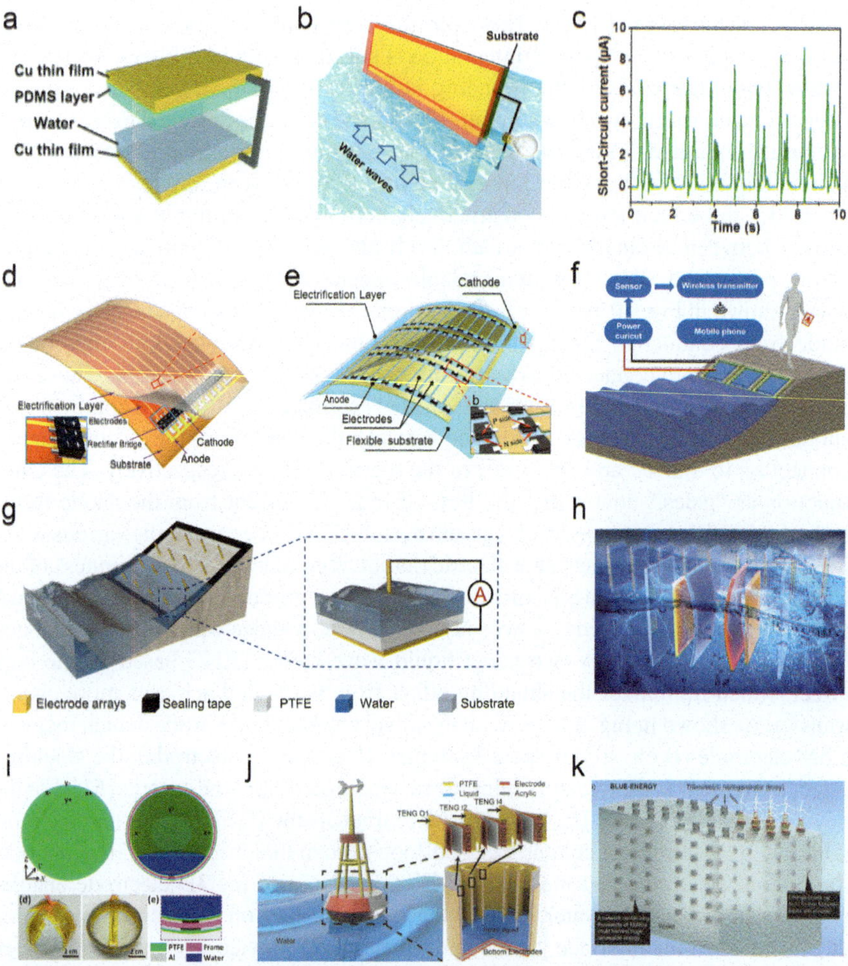

Fig. 10.10 Liquid wave-based TENG. **a** Schematic of the first proposed liquid–solid TENG [7]. **b** Schematic of a free-standing water wave-based TENG. **c** Output current of the integrated water wave-based TENG [8]. **d** Schematic of a multiple parallel electrodes water wave-based TENG [49]. **e** Schematic of a water wave-based TENG with a two-dimensional array of strip-shaped electrode units [50]. **f** Schematic of a water wave-based TENG with novel external U-shape electrode [51]. **g** Schematic of a 3D electrode water wave-based TENG based on bulk effect [52]. **h** Schematic of a new working scheme water wave-based TENG with designing opposite dynamic EDLs [53]. **i** Schematic of water wave-based TENG with highly symmetric spherical-shaped close structure [54]. **j** Schematic of buoy-like water wave-based TENG with a close structure. **k** Schematic of networking strategy of the device large-scale blue energy harvesting [55]

an enclosed structure, Fig. 10.10j shows the buoy-like liquid wave-based TENG [55]. The buoy liquid wave-based TENG can produce multiple sequential damping signals with just a single triggering pulse, demonstrating its ability to maximize the use of vibration energy, and it can harness energy from various movements, including up-and-down motions, shaking, and rotation. Moreover, the buoy liquid wave-based TENG with a network (number 18) can reach an electric output of 290 μA, 16,725 nC, and 300 V. Thus, the networking strategy of the device will be an innovative and effective approach toward large-scale blue energy harvesting (Fig. 10.10k).

10.3 Liquid–Solid TENGs for Sensing

The liquid–solid TENGs can realize self-powered sensing on the basis of the triboelectrification and electrostatic induction between liquid and solid interfaces. Due to the diverse modes and applications with liquid–solid contact, rich self-powered sensing based on liquid–solid CE has been demonstrated [5, 56, 57]. Here, typical self-powered sensing forms using liquid–solid CE are introduced in detail, including Self-powered microfluidic sensors, Self-powered chemical sensors, and other self-powered sensors. These sensors can detect or monitor liquid motion, mechanical vibration, ion concentration, sediment, wave state, and so on, showing greater potential value in healthcare, smart factory, and ocean exploration.

10.3.1 Self-powered Microfluidic Sensor

The self-powered microfluidic sensor has been developed and demonstrated to detect and characterize the liquid state based on liquid–solid CE. Compared with conventional technologies to measure the fluid with complex analysis and high-cost equipment, the self-powered microfluidic sensor using CE possesses the advantages of simple structure, low price, and facile fabrication craft. The self-powered microfluidic sensor is generally divided into two types: one is based on the motion of fluid in microchannels, and the other is based on microdroplets impacting the dielectrics. Figure 10.11a shows a self-powered microchannel triboelectric microfluidic nanosensor, which was first proposed in 2015 and based on the working principle of single-electrode TENG [58]. The device contains a PMMA substrate, a Cu electrode, and a Kapton triboelectric layer with etched nanostructure and coated 1*H*,1*H*,2*H*,2*H*perfluorodecyltriethoxysilane (PTFS) on the surface. Here, to meet the requirements of microfluidics applications, the surface of the microfluidic channel must be superhydrophobic. Therefore, the PMMA substrate is also coated with PFTS. Figures 10.11b and c show the working mechanism and output signal of the device. When the liquid begins to contact the triboelectric layer (Kapton), the positive charges in the fluid will neutralize the negative charges in the Kapton layer, causing electrons to flow from the ground to the electrode (Fig. 10.11b), and generating an electric output (Fig. 10.11c). With the variation of liquid movement, the output signal will express corresponding change, thus the sensor can detect the flowing water velocity, droplet position, etc. Additionally, with liquid flows in the microfluidic channel, a flexible microfluidic pressure sensor is proposed [59]. As shown in Fig. 10.11d, the microfluidic pressure sensor is composed of a flexible PET substrate, multiple Cu electrodes, and a thin PDMS triboelectric layer. The microfluidic channel and chamber are also fabricated with a PDMS layer, and a small PDMS disc is set on the upper of the chamber, which can effectively transfer the applied pressure to the chamber. One chamber of the microfluidic pressure sensor is filled with DI water, while the other chamber contains air. E1 to E6 denote the six copper electrodes in the water chamber extending to the air chamber. When the microfluidic pressure sensor is mounted on a human finger, the centre of the water chamber is positioned over the finger joint. As the finger bends at different angles, varying pressure is applied to the water chamber, which induces water flow through the channel. Consequently, different triboelectric outputs can be measured from the multiple copper electrodes located beneath the channel, and the sensitivity of force sensing can reach 0.0323 N^{-1}. Based on the above microfluidic pressure sensor, another flexible microfluidic triboelectric sensor (FMTS) was proposed with the fine distribution interdigitated ITO electrodes (Fig. 10.11e) [60]. With a microchannel width of 500 μm, the FMTS can reach a maximum sensitivity of 0.418 kPa^{-1} and a wide detection range from 2.38 to 58.12 kPa (Fig. 10.11f). The FMTS exhibits bendable, twistable, and conformable characteristics, allowing it to operate while attached to the skin and detect human motion combined with the machine learning approach. For example, an FMTS is attached to the finger, and the number of peaks is used to determine the angle of the finger's bending with a resolution of 10°. Figure 10.11g and h show two kinds of tube microfluidic sensors, which can realize the detection of liquid motion and pressure. As shown in Fig. 10.11g, the microfluidic sensor is assembled with a PTFE tube and the wrapped Cu electrodes around its exterior. The Cu electrodes (width: 10 mm), are attached to both ends of the PTFE tube, leaving a gap of 100 mm between them. The PTFE tube is filled with DI water, covering the copper electrodes partially [61]. Driving by a rotary actuator, the water moves back and forth inside the tube, and the generating current signal can be used for analysing movement speed. The results have been achieved with better linearity ~ $R^2 = 0.99$, which shows a distinct response with the change of the motion length and speed of the liquid. The tube microfluidic sensor, as shown in Fig. 10.11h, can serve as a portable self-powered smart fluid pressure sensor [62]. The sensor consists of a tube, a liquid column, electrodes, and an air reservoir. The liquid column acts as a piston within the tube and is connected to the air reservoir. By analysing the characteristic signals generated by the triboelectrification effect, electrostatic induction, and the ideal gas law, this sensor can quantitatively determine the fluid's pressure and flow status. The pressure accuracy of the device reaches 0.4 kPa with a pressure range from 0 to 30 kPa, and the electric performance achieves a resolution ratio of 10 mmHg, which is adequate to accurately identify the blood pressure level.

Besides realizing self-powered microfluidic sensing with fluid flowing in the microchannels, the microdroplets impacting the dielectrics with a liquid–solid CE

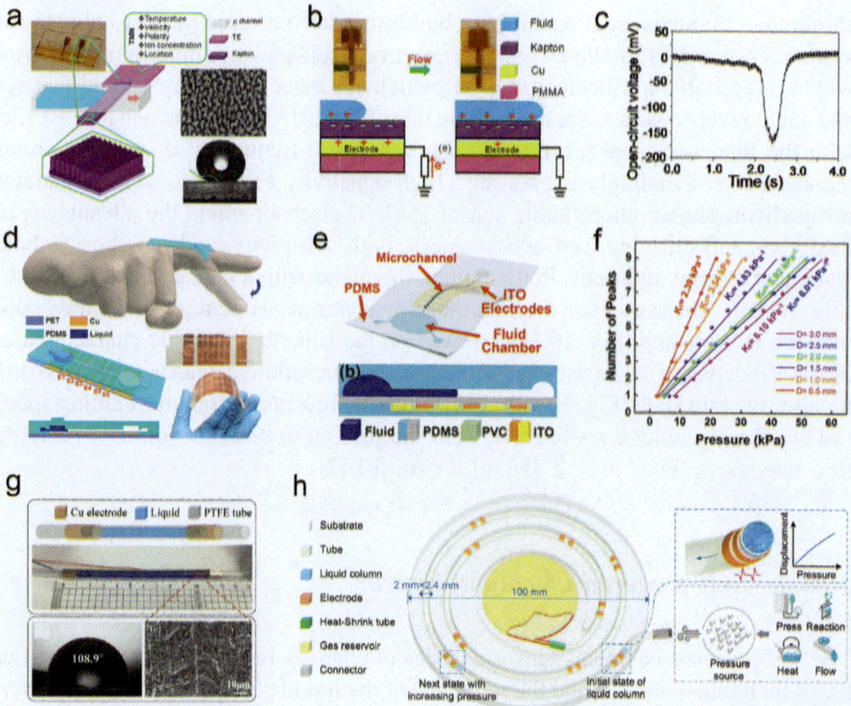

Fig. 10.11 Self-powered microfluidic sensor. **a** Schematic of the self-powered microchannel triboelectric microfluidic sensor. **b** Working mechanism of the self-powered microchannel triboelectric microfluidic sensor. **c** Output voltage of the self-powered microchannel triboelectric microfluidic sensor [58]. **d** Schematic of a self-powered microchannel microfluidic sensor for pressure monitoring [59]. **e** Schematic of the flexible microfluidic triboelectric pressure sensor with the fine electrodes. **f** Sensitivity and detection range of the device [60]. **g** Schematic of the tube microfluidic sensors for liquid motion monitoring [61]. **h** Schematic of the tube microfluidic sensors for liquid pressure monitoring [62]

can also applied in some fluidic sensing. As shown in Fig. 10.12a, by utilizing the signals produced from the droplet/bubble via the capillary and the triboelectrification effects on the liquid/solid interface, a self-powered triboelectric microfluidic sensor (TMS) is proposed for real-time liquid and gas flow detection [5]. The TMS is fabricated with a polyethylene (PE) tube substrate, a PTFE triboelectric layer, and a Cu electrode. The PTFE layer is placed on the internal surface of the PE tube, while the Cu layer is deposited on the outside surface of the PE tube. In the case of a liquid flow sensor, droplets are detached from the tip of a capillary as it passes through the TENG area, which generates an output signal. Conversely, for a gas flow sensor, bubbles form on the tip of the capillary in the TMS area, resulting in an induced voltage signal. By using capillaries of varying diameters, the sensor's detection range and sensitivity can be adjusted (Fig. 10.12b). Figure 10.12c demonstrates the monitoring of the transfusion process for a patient and the gas flow produced from an injector. It shows that TMS has great potential in building a self-powered micro total analysis system. Figure 10.12d shows a single-electrode self-powered microfluidic sensor with a p-type silicon triboelectric layer, which can detect very small liquid leakage and qualitatively characterize the leakage rate of liquid [63]. As shown in Fig. 10.12e, with the increasing leakage rates of tap water, the frequency of output currents increases correspondingly, expressing a high sensitivity. Figure 10.12f demonstrates a superhydrophobic microfluidic sensor [64]. The sensor offers the advantages of flexibility, self-cleaning, self-adhesiveness, high sensitivity, and repelling to both water and various solutions. For example, the blood with a contact angle of 158.6° (Fig. 10.12f). The sensor can monitor the clinical drainage operation and intravenous infusion in real-time (Fig. 10.12g). Based on the superhydrophobic characteristics of the triboelectric layer, another self-powered microfluidic sensor is proposed to monitor the rain state (Fig. 10.12h) [65]. With the increase of raindrop falling speed (mL/s), the V_{OC} increases linearly, And the minimum detection limit of raindrop flow rate is calculated to be 2.756 mL/s (Fig. 10.12i).

10.3.2 Self-powered Chemical Sensor

The self-powered chemical sensor can detect changes in a liquid's characteristics based on liquid–solid CE and the influence of the liquid phase on the output. Generally, the chemical characteristics of the liquid may affect the charges of the solid surface, changing the triboelectric charge density, thus reflecting different electrical outputs in the sensor. Here, a series of self-powered chemical sensors are demonstrated to detect ion liquid type and concentration, organic type and concentration, and biological response. Figure 10.13a shows a multifunctional active mechanical/chemical sensor, which can harvest vibration energy to detect rotation and in situ measurement of concentration in a specific range [56]. The device consists of a PE container with nanostructures on the inner surface, copper electrodes on the outside, and an aqueous solution encapsulated inside. When the aqueous solution is injected and comes into contact with the PE, an EDL forms. In this case, the PE surface that is covered by the solution becomes negatively charged, while the aqueous solution accumulates positive charges due to the properties of the materials involved. With the charged surfaces of the PE and the aqueous solution, any movement of the liquid can induce a flow of charges between the electrodes in the external circuit. Due to the screening effect of the ions remaining on the PE surface (Fig. 10.13b), the output of the device will be changed with the charges of the aqueous solution, thus the device can be used as an active chemical sensor for in situ detection of the concentration of the encapsulated liquid. As the concentration of NaCl solution increases from 0 to 4 mol L^{-1}, the vibration-caused RMS voltages monotonically decrease from 8.7 to 0.42 V (Fig. 10.13c). Utilizing this active chemical sensor, the concentration of a certain solution in a specific range (NaCl solution from 0 to 4 mol L^{-1}) can be in situ measured through the pre-determined data. With a similar detecting mechanism, another self-powered ion concentration sensor is also proposed based

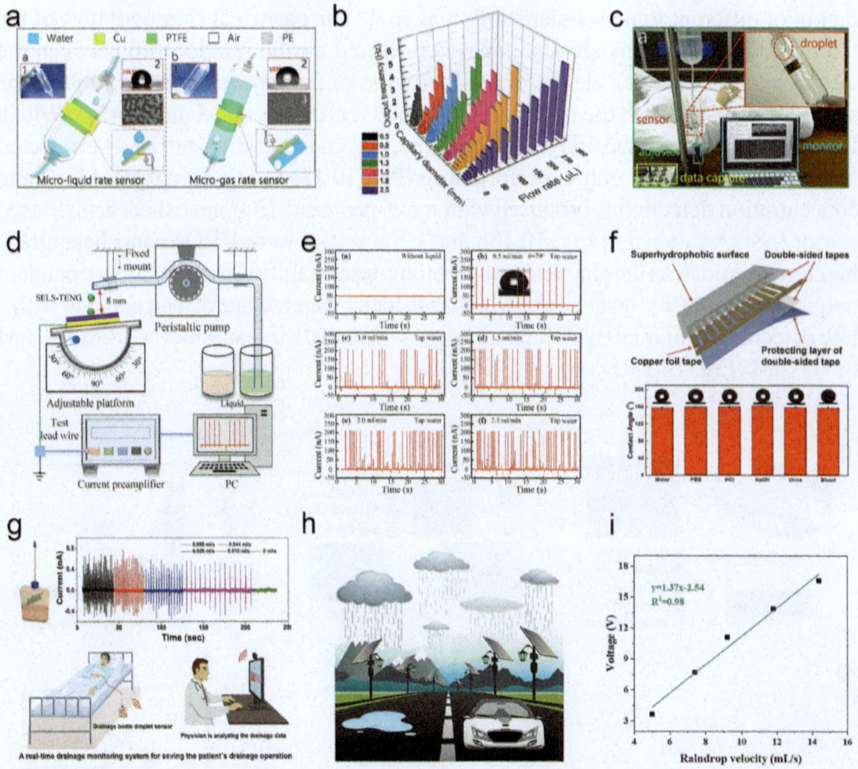

Fig. 10.12 Self-powered microfluidic sensor. **a** Schematic of the self-powered triboelectric microfluidic sensor for real-time liquid and gas flow detection. **b** Output frequency of device with various capillaries under different external flow rates. **c** Infusion monitor using the device [5]. **d** Schematic of self-powered microfluidic sensor for small liquid leakage detection. **e** Short-circuit currents under different leakage rates [63]. **f** Schematic of a superhydrophobic microfluidic sensor. **g** Demonstration of monitoring the clinical drainage operation and intravenous infusion in real-time [64]. **h** Schematic of a self-powered microfluidic sensor for rain state monitoring. **i** Detection limits of raindrop flow rates [65]

on the microfluidic system [66]. The sensor generates electrical signals through CE between the surface of a polymeric microfluidic channel and a flowing liquid. A simple setup that uses two remote electrodes, which do not directly contact the target liquid, enhances stability (Fig. 10.13d). This design also allows for scalability since the output signal is produced not from an external power source, but from the triboelectric charges induced by the target liquid within the microfluidic channel. Figure 10.13e shows the sensor's sensitivity versus the NaCl concentration (500×10^{-6} to 0.75 M). Figure 10.13f and g show two ion-type detection self-powered sensors. As shown in Fig. 10.13f, based on a self-powered microfluidic system, practical testing in which ten ionic standard solutions are measured, which include Na^+, Mg^{2+}, Zn^{2+}, Cu^{2+}, Sn^{2+}, Al^{3+}, NH^{4+}, Cl^-, NO_3^-, and SO_4^{2-}, thus obtain the electrical output of different ionic standard solutions [67]. The electrical conductivities of ten ionic standard solutions are also measured. Based on the relationship between the electrical output and the electrical conductivities of ten ionic standard solutions, the sensors can distinguish the special ion type and realize chemical monitoring. With a U-tube liquid flow-based TENG, the ion type, and concentration can also be detected based on the electrical output performance (Fig. 10.13g) [43]. The organic kind and concentration detection is proposed with a self-powered TiO_2 nanosheet array-based sensor [68]. As shown in Fig. 10.13h and i, the self-powered TiO_2 nanosheet array-based sensor possess the characteristics of long-term stability, frequency-independent output, and humidity-insensitive, which can detect water, ethanol, and acetone with a low detection limit of 30 nM. It also can provide a wide linear window from 100 nM to 100 μM (Fig. 10.13i).

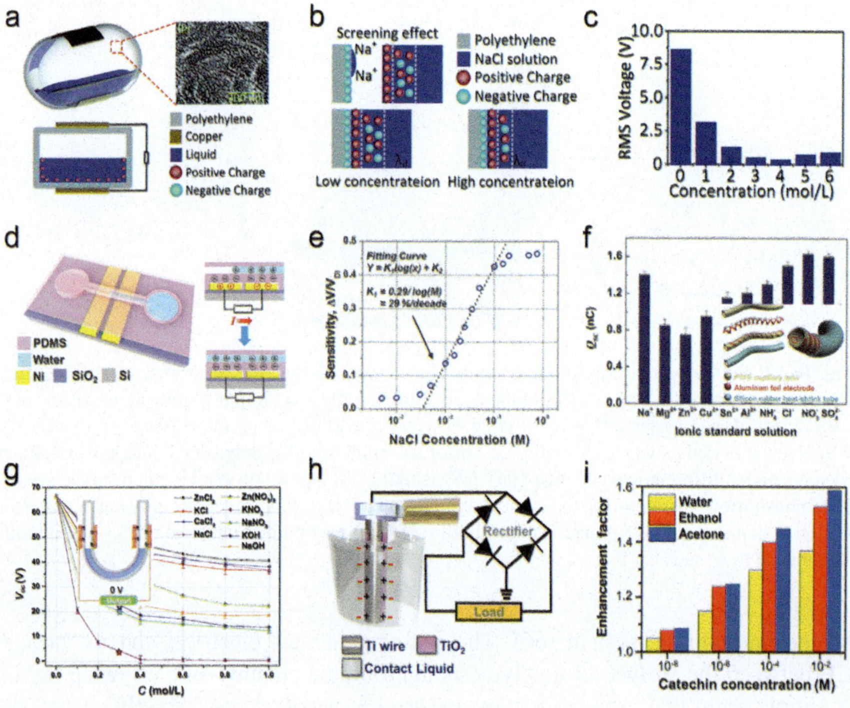

Fig. 10.13 Self-powered chemical sensor. **a** Schematic of a multifunctional active mechanical/chemical sensor for ion concentration detection. **b** Charge distribution at different conditions. **c** Detailed RMS voltage with NaCl concentrations [56]. **d** Schematic of a self-powered ion concentration sensor based on the microfluidic system. **e** Sensitivity of the device versus the NaCl concentration [66]. **f** Transferred charge of the microfluidic ion-type detection sensors with different ions. Insert: schematic of the device [67]. **g** Open-circuit voltage of the U-tube ion-type detection sensors with different ions. Insert: schematic of the device [43]. **h** Schematic of an organic type and concentration detection sensor. **i** Voltage enhancement factor for different organic liquids [68]

Based on the unique advantages of self-powered chemical sensors, it can be combined with robots to achieve chemical detection of some hazardous liquids, and combined with artificial intelligence, deep machine learning, and neural networks to achieve intelligent chemical sensing through human–computer interaction. Figure 10.14a shows an innovative self-powered triboelectric Hg^{2+} ions detection sensor based on robotics [69]. Using a single-electrode configuration, the sensor operates with a flexible Al substrate as the electrode, vertically aligned tellurium nanowires (Te NWs) were created on the Al substrate through a chemical reduction method. Due to the flexible nature of the substrate, the designed sensors can conform well to the shape of a robotic finger (Fig. 10.14b). The Te NWs, which act as both a solid triboelectric material and a sensing probe, experienced periodic contact and separation with the Hg^{2+} solution. This interaction resulted in the in-situ formation of mercury telluride nanowires (HgTe NWs) due to the selective binding affinity of Te NWs for Hg^{2+} ions (Fig. 10.14b). Utilizing a simple "touch and sense" mechanism, The Te NW arrays, which is mounted onto the robotic hands equipped with additional wireless transmission, can realize detection of Hg^{2+} ions in resource-limited settings and on-site rapidly (Fig. 10.14c). Figure 10.14d–f demonstrate an intelligent dual-mode solution detection sensor with a deep learning method, which can be capable of monitoring multiple solution parameters and may be applied in various industries [70]. The sensor consists of two sensing modes, in capacitive sensing mode, the sensor effectively identifies the type of solution by detecting changes in capacitance caused by the conductivity of the mixed solution. In triboelectric sensing mode, the sensor demonstrates high sensitivity to solution concentration by utilizing both the capacitive enhancement effect and the triboelectric effect. With an electric switch to control the signal acquisition of the dual-mode sensor, high recognition accuracies for both solution type and concentration are achieved by integrating deep learning methods with the dual-mode sensor (Fig. 10.14d) and the average accuracies exceed 95% (Fig. 10.14e). Figure 10.14f illustrates the critical role of the dual-mode sensor in monitoring liquids at point-of-care within a smart factory environment. Such research on integrating self-powered sensors with robotics and deep learning would lead to the development of a sensing system to realize intelligent monitoring.

10.3.3 Other Self-powered Sensors

Using solid–liquid CE, other self-powered sensors have also been proposed, which can monitor or detect human motion, mechanical vibration(acceleration), pressure variation, sediment, and wave state. Figure 10.15a–c show a highly deformable and stretchable self-powered body motion sensor, which can detect the bending angle of the forearm when doing bicep curls [71]. The sensor is fabricated with conductive liquid filled in a polymer cover, which can work in various modes and maintain its performance under a strain of as large as 300% (Fig. 10.15a and b). Due to its flexible characteristics, the sensor can conform to any three-dimensional and curvilinear surface. During the detection of the bending angle of the forearm, the volume of the

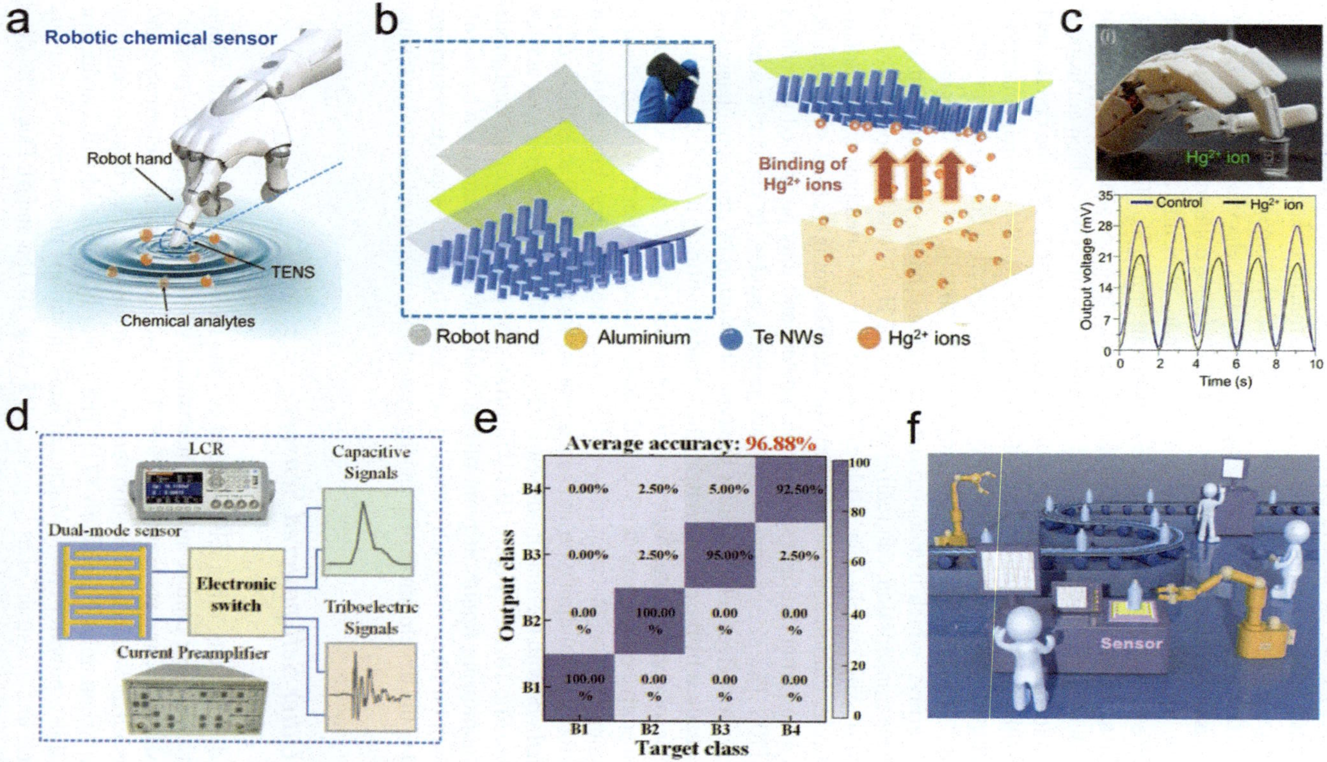

Fig. 10.14 Self-powered chemical sensor. **a** Schematic of robotic hand-based triboelectric sensor for Hg^{2+} ion sensing. **b** Structure schematic of the device and the sensing mechanism. **c** Robotic fingers with the sensing probe contact and separate with Hg^{2+} and its corresponding triboelectric output voltage signals [69]. **d** The intelligent dual-mode solution detection sensor with a deep learning method. **e** Confusion map results of different solutions recognition. **f** Schematic of the application of the intelligent dual-mode sensor [70]

biceps increases when the elbow is bent and decreases when the arm is extended. This change in volume leads to a variation in the circumference of the upper arm, which alters the contact area between the sensor and the skin. The periodic changes in this contact area cause a flow of electrons to move back and forth between the sensor's water electrode and the ground. As the bending angle increases, the contact area also increases, resulting in a greater number of transferred electrons and a higher output current for the same bending rate (Fig. 10.15c). Figure 10.15d, e and f show two self-powered acceleration sensors. As shown in Fig. 10.15d, the sensor consists of an inner mercury droplet and an outer acrylic shell, the structure of the shell is composed of a hemisphere-shaped acrylic pit, a Cu electrode deposited on the inner acrylic surface, and a layer of Cu-coated nanofiber-networked PVDF laminated on the top of the pit as the triboelectric layers [57]. The acceleration sensor has a broad detection range of 0 to 60 m/s^2, with a high sensitivity of 0.26 V s/m^2 (Fig. 10.15e). And the output voltage and current show a negligible decrease over 200,000 cycles, demonstrating excellent stability. As for the self-powered acceleration sensor shown in Fig. 10.15f, the device features a 3D structure with a detection range of 0–100 m/s^2 in the horizontal direction and 0–50 m/s^2 in the vertical direction, achieving a sensitivity of 800 mV/g [72]. Moreover, the 3D acceleration sensor is incorporated into the airbag system to detect the collision's position and force simultaneously, this technology has potential applications in equipment vibration monitoring and troubleshooting. Figure 10.15g and h show a self-powered sediment sensor [73]. Based on the single-electrode liquid droplet-based TENG and the deep learning method, the sensor can realize real-time sediment monitoring, which is very sensitive to the sand particle size and mass fraction, and high identifying accuracies of over 90% is achieved to identify the particle parameters based on the output signal. Additionally, another self-powered sediment sensor can detect and classify microplastics in liquids using a liquid–solid TENG [74]. As shown in Fig. 10.15i, with liquid–solid TENG combined with a deep learning model, the sensor can classify a series of microplastics into liquids, and high recognition accuracy is achieved. Figure 10.15j–l show a series of self-powered water and wave state sensors using liquid–solid CE. The sensor can be used to detect water level height in a container (Fig. 10.15j) [75] or the wave height in the ocean (Fig. 10.15k) [76]. And it also can be utilized as a robust and sensitive indicator for detecting the water level, the ship draft is successfully detected with an accuracy of 10 mm (Fig. 10.15l) [77]. Finally, with the ship is navigating in the ocean, a robust and self-powered tilt sensor is proposed [78], which can monitor the ship's attitude information (Fig. 10.15m). The tilt sensor shows consistent output performance and high sensitivity in low-frequency and low-amplitude inclination conditions. Therefore, the sensor with a TENG using liquid–solid CE is self-powered, robust, and accurate for extensive applications in the marine industry.

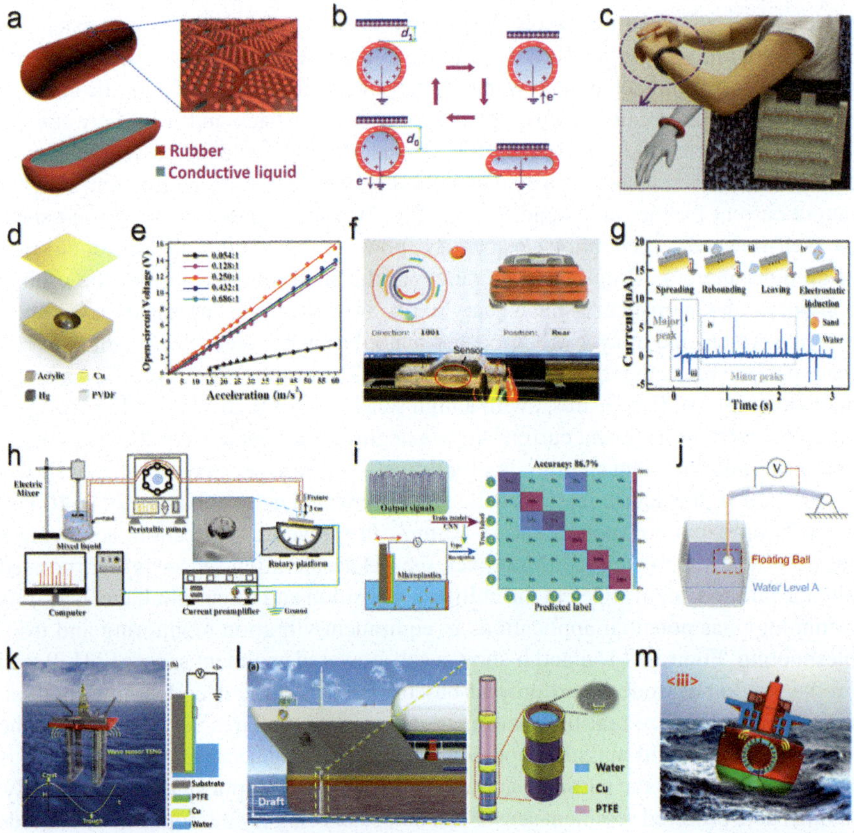

Fig. 10.15 Other self-powered sensors. **a** Schematic of the highly deformable and stretchable self-powered body motion sensor. **b** Working mechanism of the self-powered body motion sensor. **c** Demonstration of arm motion sensing [71]. **d** Schematic of a self-powered acceleration sensor. **e** Detection range and sensitivity of the acceleration sensor [57]. **f** Schematic of a self-powered 3D acceleration sensor [72]. **g** Output current and schematic for sediment detection. **h** The self-powered sediment sensing system uses a deep learning method [73]. **i** Recognition of microplastics into liquids with liquid–solid TENG combined with a deep learning model [74]. **j** Schematic of detecting water level height in a container [75]. **k** Schematic of detecting the wave height in the ocean [76]. **l** Schematic of detecting the wave level of ship draft in the ocean [77]. **m** Schematic of self-powered tilt sensor for ship attitude monitoring in the ocean [78]

10.4 Liquid–Solid TENGs Probe

When a liquid droplet moves on a dielectric solid surface, the measurable static charges will be generated at the liquid–solid interface. Such type of triboelectrification can be measured by Faraday pail, KPFM, and electrometer. The concept of "Triboelectric Nanogenerator probe (TENG probe)" was first proposed by Wang's group in 2020 to study liquid–solid interfaces [1]. When a drop of the liquid sample slides along an insulating surface, the local triboelectrification of the surface will be recorded, and the changes in liquid–solid interface information will directly affect the charge transfer dynamics, for example, flow rate, ion concentration, and chemical composition. After Wang proposed the concept of "TENG probe", a growing number of research groups have successfully exploited such a type of probe for studying charge transfer at various liquid–solid interfaces over the last five years, as well as for sensing systems [79–81]. Available proof-of-principle applications range now from charge transfer studies at various liquid–solid interfaces to chemical analysis in liquids, fluid status monitoring, curing progress analysis, and taste-sensing. Studying the liquid–solid interface by TENG probe will help improve technologies for sensors, energy storage, and environmental applications, as well as enable better control over reactions at the interfaces.

10.4.1 Structure Design of TENG Probe

In 2014, Wang's group developed a new type of TENG that uses triboelectrification between water droplets and insulating hydrophobic solids to collect water energy, which is called droplet TENG [10]. Based on such type of TENG, the concept of the "TENG probe" was introduced in 2020 [1]. The structure of TENG probe mainly consists of three parts: droplets, an insulating film used for charging upon contact with the droplets, and electrodes spread beneath the insulating film are connected to an external electrometer to obtain electrical signals generated by triboelectrification between droplets and solid a film. When the droplets slide over the insulating hydrophobic film, induced charges will generate on the electrodes due to the electrostatic induction (Fig. 10.16a), and such electric signals can be measured by electrometer. The first TENG probe was one electrode, which means it can only tell us the total charge transfer along the sliding path. To analyze charge transfer dynamics, i.e. whether the charge is evenly distributed throughout the path of the droplet, a dual-electrode system was developed in 2021 [82], where two separated copper electrodes were positioned beneath a FEP film and connected to two separate electrometers (Fig. 10.16b). By measuring the electric signals on two spatially arranged electrodes, this type of TENG probe was able to study the charge transfer at different sliding positions on the solid surface. However, the design of the two electrodes was limited in its ability to resolve charge transfer along the X-axis, as well as with low resolution. In 2023, a pixeled TENG probe with a high-density electrode array—comprising 432 separate copper electrodes—was developed to enhance both the resolution and sensitivity [83]. Such TENG probe allowed for the creation of a two-dimensional, high-resolution charge transfer mapping as the droplet traveled across the dielectric solid surface, achieving a spatial resolution of 400 μm (Fig. 10.16c). By analyzing such high-resolution charge transfer mapping, valuable physical and chemical information about both the solid and liquid can be obtained, including the dynamic behavior of the droplet, such as its velocity and acceleration. Moreover, this high-sensitivity charge transfer mapping is responsive to various factors, including droplet movement, light intensity, and pH levels, showing great potential for sensing applications. Based on the TENG probe with single electrode or electrode array, a series of structures were designed to meet the needs of different scenarios, for example, the cylindrical TENG probe is used for in-situ fluid monitoring in pipelines.

10.4.2 Detection Principle of TENG Probe

The TENG probe operates based on the triboelectric effect and electrostatic induction. These principles allow the TENG to convert mechanical energy into electrical energy, and thus the electric signals will be captured and used to sensing. When water droplet and dielectric solid come into contact and then separate, electron transfer will first occur [2, 84, 85]. In this way, the droplet becomes positively charged, and solid becomes negatively charged. This is because when water comes into contact with a solid material like FEP, it temporarily loses an electron, forming H_2O^+. Shortly after, the H_2O^+ interacts with a nearby water molecule, resulting in the formation of an OH radical and H_3O^+ [1, 86]. The magnitude of the electron transfer depends on the materials' triboelectric properties. The separation or movement of the charged surfaces creates a varying electric field: This induces an alternating current (AC) in the electrodes attached to the TENG. The generated voltage or current corresponds to the motion of the water droplet. Consider a TENG probe with two separated electrodes [82]: as a water droplet slides across the FEP film, electrons transfer from the droplet to the FEP due to FEP's strong electron affinity [87]. This process leaves the water droplet positively charged and the FEP negatively charged. The current peaks and induced charges generated during the droplet's contact and separation with the copper electrode are represented as j_c, Q_c and j_s, Q_s, respectively. When the first water droplet begins to touch the FEP film over the copper electrode, an induced current j_c is observed (Fig. 10.17). By integrating this current, the corresponding transferred charge Q_c is determined. Similarly, the charge Q_s is obtained when the droplet separates from the polymer surface at the copper electrode. The difference between Q_c and Q_s is defined as charge transfer q_1 for electrode 1 and q_2 for electrode 2, reflecting the transferred charges on each electrode. The principle of TENG probe is based on the changing of electric signals on different electrodes or charge transfer distribution. In addition to electron transfer, ion adsorption also affects the final measured charge transfer. The charged insulating hydrophobic film will absorb the oppositely charged ions in the solution and form an EDL at the liquid–solid interface, thus having a shielding effect on the interface electron transfer. Therefore, the charge transfer measured on the electrodes also depends on the chemical properties of liquid, such as chemical composition, ion concentration, pH, etc.

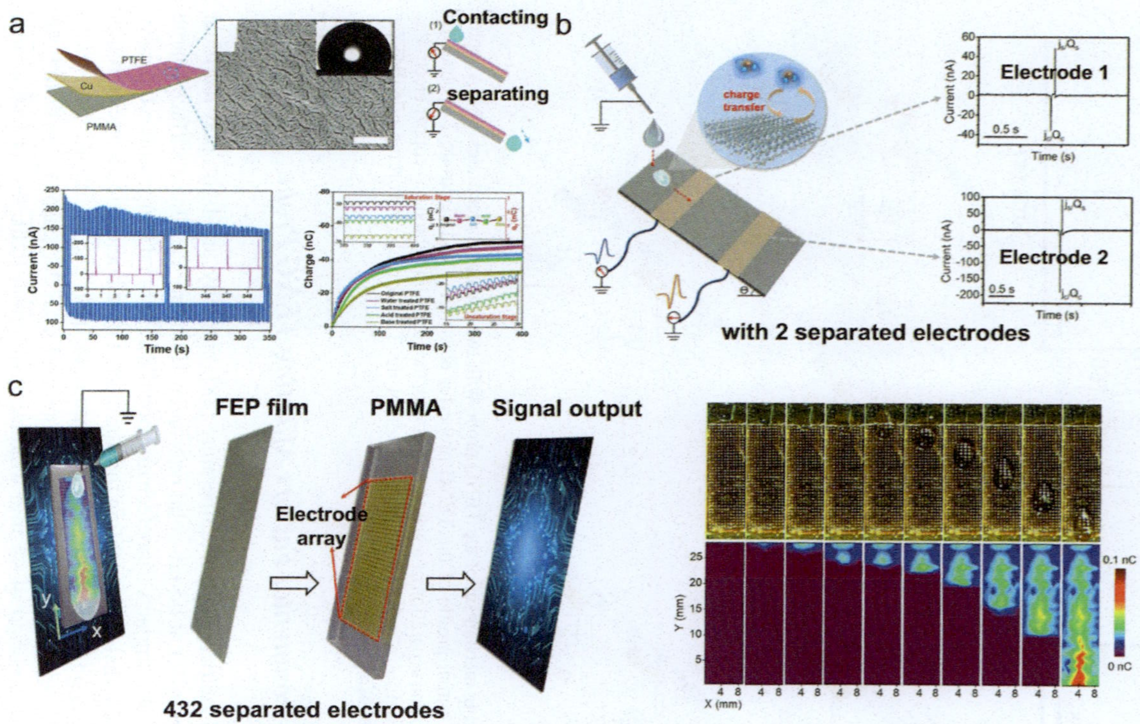

Fig. 10.16 TENG probe: from single electrode to electrode array. **a** TENG probe with one electrode. When the droplet slides over the insulating hydrophobic film, induced charges will be generated on the electrode, and the electric signals will be measured by an external electrometer [1]. **b** TENG probe with a dual-electrode system and by measuring the electric signals on two spatially arranged electrodes, such type of TENG probe was able to study the charge transfer at different sliding positions [82]. **c** TENG probe with a high-density electrode array and when a droplet flows through the FEP surface, the high-resolution charge transfer mapping can be obtained [83]

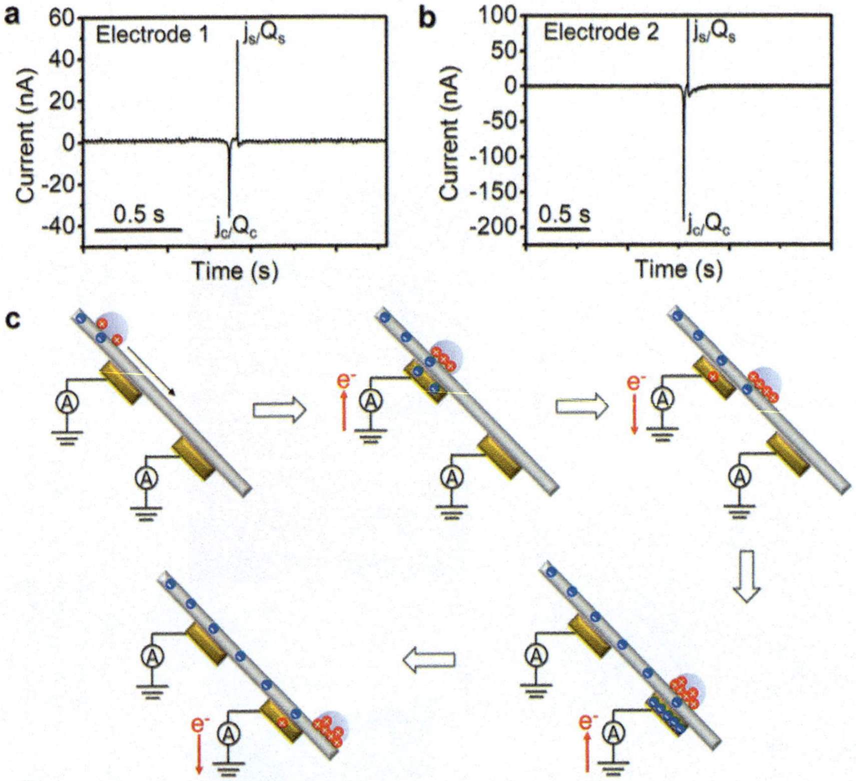

Fig. 10.17 Working mechanism of the TENG probe with two separated electrodes. **a, b** A representative current output profile of the TENG probe for electrode 1 and electrode 2 is observed during the contact and separation of a water droplet with the FEP film, which has copper electrodes positioned beneath it. **c** A schematic representation of the TENG probe's working principle for detecting a droplet using two electrodes [82]

10.4.3 Environmental Factors Affecting TENG Probe

Environmental factors, such as temperature [88], humidity, and light, play a critical role in charge transfer at liquid–solid interfaces. Humidity facilitates surface charge dissipation and also impacts triboelectrification [28]. Elevated temperatures typically decrease charge transfer by enabling the dissipation of triboelectric charges through thermionic emission [28]. As the temperature rises from 288 to 323 K, the current produced by triboelectrification between water and SiO_2 decreases from 17 to 9 nA [89]. Also, over a period of 45 min irradiation of xenon light, the surface charge density on SiO_2 generated by CE with PMMA dropped significantly from $-$ 160 to $-$ 10 $\mu C\ m^{-2}$, demonstrating the effectiveness of light irradiation in dissipating triboelectric charges [84]. In 2023, Wang's group utilized a pixeled TENG probe to

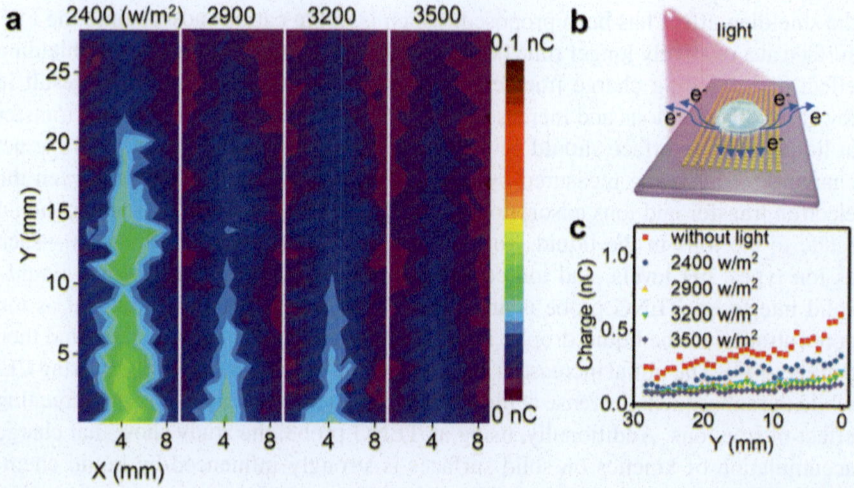

Fig. 10.18 A pixeled TENG probe to investigate the effect of light intensity on charge transfer distribution at the liquid–solid interface. **a** Charge transfer mappings of CE at water/FEP interface under different light intensities. **b** Schematic representation of the light exposure experiments. **c** Variation in charge transfer along the Y-axis under different light intensities [83]

investigate the effect of light intensity on charge transfer distribution at the liquid–solid interface [83]. Their experiments revealed that increasing the light intensity reduces the overall charge transfer on the droplet sliding path, but has no significant effect on the charge transfer distribution at liquid–solid interface (Fig. 10.18). This phenomenon also applies to changes in temperature and humidity. It was observed that as the temperature increased from 15 to 30 °C or the humidity rose from 20 to 40%, the charge transfer distribution generated by liquid droplets sliding over the FEP surface remained consistent. Only the total charge transfer was affected by variations in temperature or humidity. These results indicate that the TENG probe can be used under a normal testing environment.

10.4.4 Charge Transfer Mechanism Studied by TENG Probe at Liquid–Solid Interface

The sliding velocity of moving droplets is often overlooked when seeking to optimize the CE of liquid–solid contacts. In 2023, a TENG probe was used to measure charge transfer between water droplets and FEP at varying droplet velocities [3]. The study regulates the sliding speed of water droplets on the FEP surface by changing the drop height and revealed that the transferred charges at the liquid–solid interface are strongly influenced by the droplet's sliding speed: higher velocities resulted in greater charge transfer (Fig. 10.19a). A mechanism involving ion adsorption and the shielding effect has been proposed, which is, more cations adsorb onto the FEP surface at a relatively longer time (slower sliding velocities) enhancing the shielding effect and reducing charge transfer. In contrast, faster sliding velocities result in fewer adsorbed cations and increased charge transfer. Therefore, the charge transfer at liquid–solid interface should be related to the formation of an EDL, and the net charge on solid being measured by an electrometer, is the difference between the electron transfer and ions adsorption events. In this way, the charge transfer should relate to the ions in the liquid. To better understand how liquid chemistry—such as ion types, pH levels, and ion concentrations—affect charge transfer at liquid–solid interfaces, TENG probe is an efficient tool. Recent studies show that as ion concentration in the liquid droplet increases, charge transfer initially rises and then declines [90]. The initial increase is attributed to the enhanced ion transfer during CE, while the subsequent decrease at higher concentrations is likely due to the screening effect of free ions. Additionally, using as TENG probe, the study show that charge accumulation of kinetics on solid surfaces is strongly influenced by liquid chemistry, especially for ion diameter. That is, droplets containing cations with smaller ionic diameters and stronger ionic bonds between cations and anions exhibit faster charge accumulation rates (Fig. 10.19b) [91]. Apart from water solution, A TENG probe was also employed to study interfacial charge transfer during the CE process of colloid-solid system. Colloidal droplets are sliding on the PTFE film, and the corresponding currents signals were recorded. Based on the probe signal, it is noted that the charge transfer between colloid and solid is influenced by the ionic concentration of the colloidal dispersion medium, the wettability of the colloidal particles, and the structure of the EDL. Higher ionic concentration in the dispersion medium will dominate the interfacial charge transfer process. Additionally, the formation of hydrated protons from electron transfer between the colloid and solid promotes the deposition of nanoparticles on the solid surface, thus suppressing the charge transfer at colloid-solid interface (Fig. 10.19c) [80]. These findings offer valuable insights and pave the way for further exploration of colloid-solid interactions. These findings emphasize the importance role of TENG probes in studying charge transfer mechanisms.

10.4.5 Sensing Systems Enabled by TENG Probe

TENG probes are not only used for the basic charge transfer mechanisms at various liquid systems, but also attracted more and more attention on sensing systems in recent years. A droplet-tasting sensor was developed using a TENG probe with two separate electrodes (Fig. 10.20a) [92]. This sensor identifies different types of liquid food by analyzing the current curve patterns on the two electrodes when the liquid food slides across the FEP surface. Key parameters include the magnitudes of current signals, the time interval between a waveform's peak and valley, and the interval between the peaks of two successive waveforms, are used for food sensing (Fig. 10.20b), for example, liquor, coffee, and white vinegar. Combined with deep

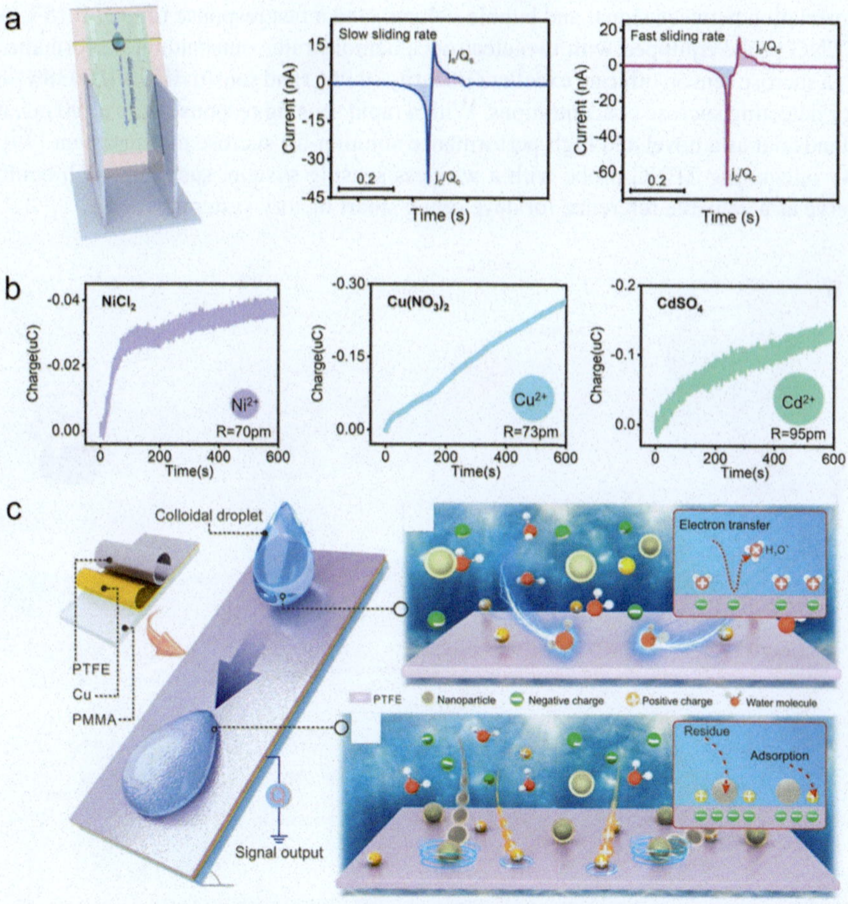

Fig. 10.19 Charge transfer mechanism studied by TENG probe at liquid–solid interface. **a** The velocity effect studied by TENG probe. The transferred charges at the liquid–solid interface are strongly influenced by the droplet's sliding speed: higher velocities resulted in greater charge transfer [3]. **b** The effect of ionic diameters studied by TENG probe: droplets containing cations with smaller ionic diameters exhibit faster charge accumulation rates on the solid surface [91]. **c** The interfacial charge transfer during the CE process of colloid-solid system was studied by TENG probe [80]

learning, an integrated image sensor enhances the liquid-sensing system by extracting visual features, such as liquid food color, boosting its recognition accuracy to 96.0%. The TENG probe is also capable of monitoring bubble states in liquid. For instance, a single-electrode TENG probe is used to monitor bubbles status in the liquid [93]. When liquid with bubbles slides through PTFE tube, the electrometer connected with Cu electrode outside the PTFE tube will record the current signals generated by CE between liquid and solid. The detailed investigations were conducted on how bubble fluid dynamics, including flow rates, volumes, and release intervals, influence liquid–solid CE. This sensor demonstrates a sensitivity of 13.2 V cm^{-3}, a strong correlation between signal and bubble volume, and a fast response time of 0.15 s. A TENG probe equipped with two electrodes demonstrates outstanding performance as a sucrose sensor, offering excellent linearity (0.966) and sensitivity ($-0.0038\%^{-1}$) for detecting sucrose concentrations. With a rapid sensing response time of 90 ms, it stands out as a novel and high-performance solution for sucrose measurement [79]. By integrating TENG probe with a wireless sensing system, such approach could serve as a valuable reference for developing smart factory systems.

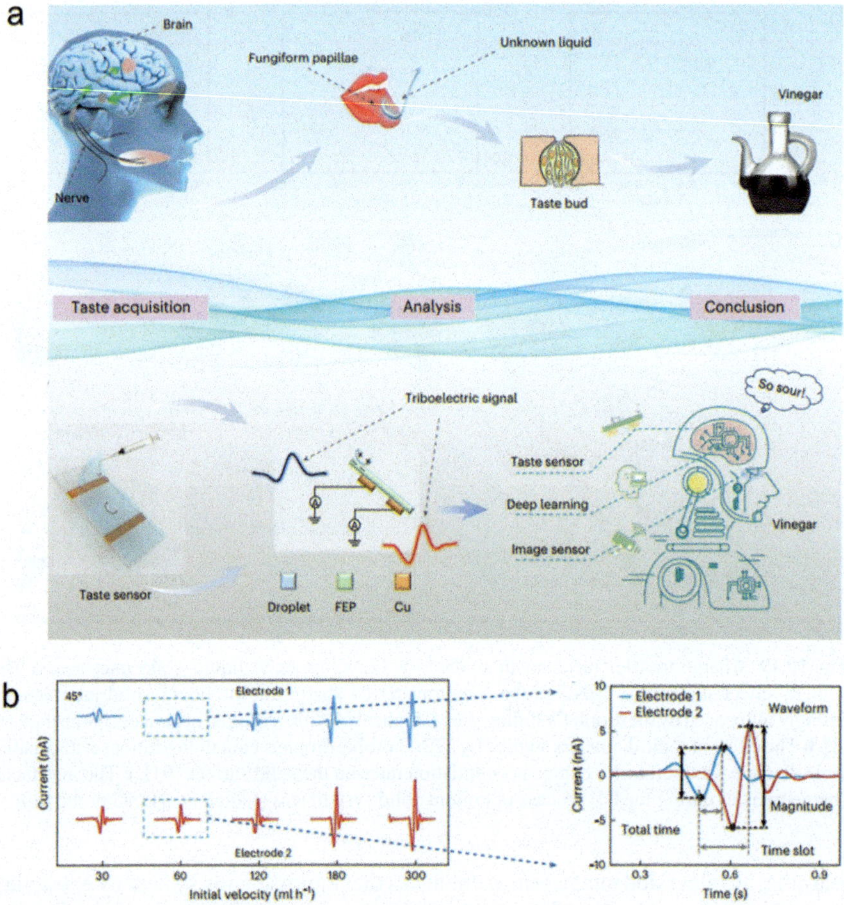

Fig. 10.20 A droplet-tasting sensor was developed using a TENG probe with two separate electrodes. This sensor identifies different types of liquid food by analyzing the current curve patterns on the two electrodes, including the magnitudes of current signals, the time interval between a waveform's peak and valley, and the interval between the peaks of two successive waveforms [92]

10.4.6 Chemical Analysis Enabled by TENG Probe

Chemical analysis of ions in water resources, is crucial for cultivated farmland, sustainability, economic development, and environmental systems. Building on recent breakthroughs in understanding the origins of triboelectricity between insulators and liquid, and based on the TENG probe, a novel technique called triboelectric spectroscopy (TES) was introduced in 2024 [6, 94]. This innovation represents a new form of "electrostatic" spectroscopy. 14 independent electrodes are laid under the hydrophobic FEP film along the direction of droplet movement in this work (Fig. 10.21a). When a liquid droplet containing dissolved analytes slides down an FEP surface, each electrode will generate a current signal (Fig. 10.21b–d), and the charge transfer on each electrode can be obtained by integrating the current curve at each electrode. Therefore, by plotting the relationship between the charge transfer and the movement time of the droplet, the corresponding TES is obtained (Fig. 10.21e–g). It is noted that the resulting charge distribution along the droplet's path is unique to each analyte, thus can be used to analysis chemicals in liquids. By constructing a database of more than 30 common salts, acids, bases, and organic compounds, and utilizing a simple automated identification system, such method achieves qualitative and quantitative accuracies of up to 93%, with detection limits as low as ppb level.

10.5 Summary

In this chapter, the recent advancements in triboelectric nanogenerator applications using liquid–solid CE are discussed, which involves energy harvesting, self-powered sensing, and TENG probes. For energy harvesting, typical liquid–solid TENG forms are introduced in detail, the form is mainly composed of liquid droplet-based, liquid flow-based, and liquid wave-based, which can realize diverse liquid kinetic energy or liquid potential energy harvesting and response to low-frequency motion highly. The liquid droplet-based TENGs are highlighted in detail, which include traditional TENG, novel DEG, DC-TENG, SLIPS-TENG, and hybrid devices combined with liquid–solid CE and other working mechanisms. A maximum charge density of $5.4\ \mathrm{mC\ m^{-2}}$ is achieved with an individual droplet. During the application of sensing using liquid–solid CE, self-powered microfluidic sensors, chemical sensors, and other self-powered sensors are presented, which can detect or monitor liquid motion, mechanical vibration, ion concentration, sediment, wave state, and so on, showing greater potential value healthcare, smart factory, and ocean exploration. The TENG probes, which were first proposed to study the liquid–solid interfaces, are also introduced. The applications of the TENG probe range from charge transfer studies at various liquid–solid interfaces to chemical analysis in liquids, fluid status monitoring, and taste-sensing, which are discussed in detail. Studying the liquid–solid interface by TENG probe will help improve technologies for sensors, energy storage, and environmental applications.

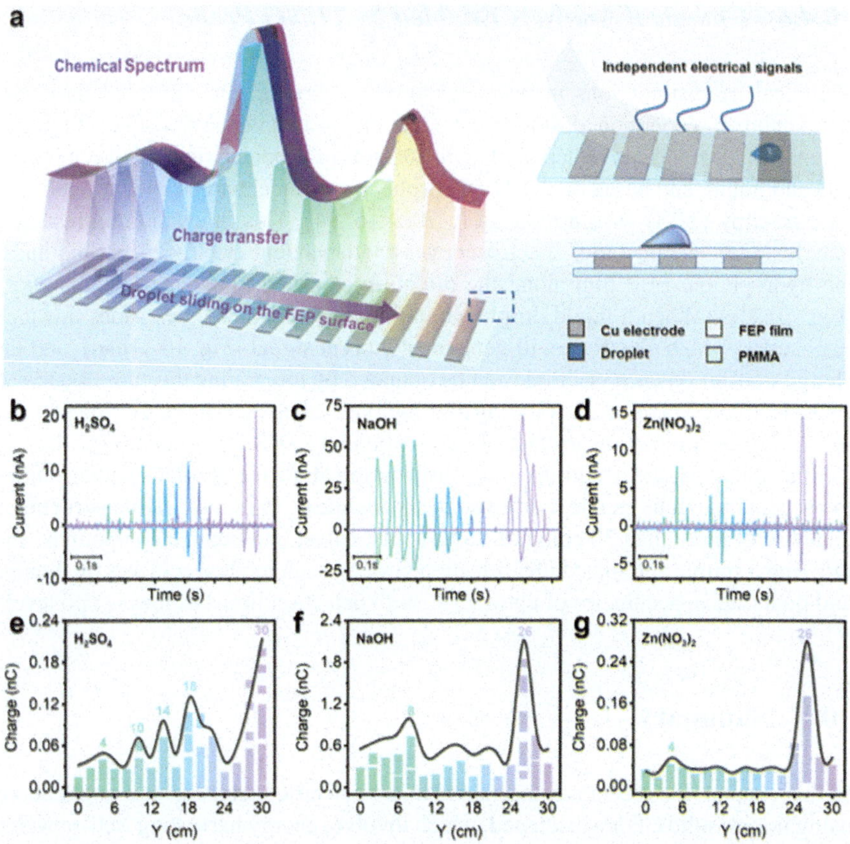

Fig. 10.21 Chemical analysis enabled by TENG probe. **a** Experimental design of TES. When a liquid droplet containing dissolved analytes slides down an FEP surface, the charge transfer measured by each electrode is recorded and used to build a TES profile. **b–d** Current output profile when the droplets with different chemicals [H_2SO_4 (pH = 3), NaOH (pH = 13), and $Zn(NO_3)_2$ (1 M)] slide across the FEP film. **e–g** Corresponding TES [6]

In future development of the research direction, the following aspects are suggested to pay special attention to:

(1) Surface charge density and output power density of the liquid–solid TENG. The power density is an essential factor for developing high-performance liquid–solid TENG, which is largely determined by the surface charge density of the device. Surface modification and charge injection methods can achieve certain effects in improving surface charge density, but they still face challenges in long-term device operation. Other approaches to enhance the power density or charge density should also be explored.
(2) Durability of the liquid–solid TENG. The liquid–solid TENG unit works in a natural environment with dust, oil, and possibly low temperatures. To realize a

device without maintenance, super high anti-fouling, durability, and anti-icing ability should be achieved.

(3) Integration and packing of the liquid–solid TENG. The future application of self-powered sensing with liquid–solid TENG should be stable and intelligent, thus it is necessary to effectively integrate the device with other terminals, and package the devices to maintain long-term working stability.
(4) Intensive research on the TENG probe. The in-depth study of TENG probes will elucidate the mechanism of solid–liquid contact electrification, providing support for the development of high-performance power generation devices or sensing units.

References

1. F. Zhan, A.C. Wang, L. Xu, S. Lin, J. Shao, X. Chen, Z.L. Wang, Electron transfer as a liquid droplet contacting a polymer surface. ACS Nano **14**(12), 17565–17573 (2020)
2. S. Lin, L. Xu, A.C. Wang, Z.L. Wang, Quantifying electron-transfer in liquid–solid contact electrification and the formation of electric double-layer. Nat. Commun. **11**, 1–8 (2020)
3. X. Wang, J. Zhang, X. Liu, S. Lin, Z.L. Wang, Studying the droplet sliding velocity and charge transfer at a liquid–solid interface. J. Mater. Chem. A **11**, 5696–5702 (2023)
4. W. Xu, H. Zheng, Y. Liu, X. Zhou, C. Zhang, Y. Song, X. Deng, M. Leung, Z. Yang, R.X. Xu, Z.L. Wang, X.C. Zeng, Z. Wang, A droplet-based electricity generator with high instantaneous power density. Nature **578**, 392–396 (2020)
5. J. Chen, H. Guo, J. Zheng, Y. Huang, G. Liu, C. Hu, Z.L. Wang, Self-powered triboelectric micro liquid/gas flow sensor for microfluidics. ACS Nano **10**(8), 8104–8112 (2016)
6. J. Zhang, X. Wang, L. Zhang, S. Lin, S. Ciampi, Z.L. Wang, Triboelectric spectroscopy for in situ chemical analysis of liquids. J. Am. Chem. Soc. **146**(9), 6125–6133 (2024)
7. Z.H. Lin, G. Cheng, L. Lin, S. Lee, Z.L. Wang, Water–solid surface contact electrification and its use for harvesting liquid-wave energy. Angew. Chem. Int. Ed. **52**(48), 12545–12549 (2013)
8. G. Zhu, Y. Su, P. Bai, J. Chen, Q. Jing, W. Yang, Z.L. Wang, Harvesting water wave energy by asymmetric screening of electrostatic charges on a nanostructured hydrophobic thin-film surface. ACS Nano **8**(6), 6031–6037 (2014)
9. G. Cheng, Z.-H. Lin, Z.-L. Du, Z.L. Wang, Simultaneously harvesting electrostatic and mechanical energies from flowing water by a hybridized triboelectric nanogenerator. ACS Nano **8**(2), 1932–1939 (2014)
10. Z.H. Lin, G. Cheng, S. Lee, K.C. Pradel, Z.L. Wang, Harvesting water drop energy by a sequential contact-electrification and electrostatic-induction process. Adv. Mater. **26**(27), 4690–4696 (2014)
11. S.-H. Kwon, J. Park, W.K. Kim, Y. Yang, E. Lee, C.J. Han, S.Y. Park, J. Lee, Y.S. Kim, An effective energy harvesting method from a natural water motion active transducer. Energy Environ. Sci. **7**, 3279–3283 (2014)
12. L.E. Helseth, X.D. Guo, Contact electrification and energy harvesting using periodically contacted and squeezed water droplets. Langmuir **31**(10), 3269–3276 (2015)
13. H. Cho, J. Chung, G. Shin, J.-Y. Sim, D.S. Kim, S. Lee, W. Hwang, Toward sustainable output generation of liquid–solid contact triboelectric nanogenerators: The role of hierarchical structures. Nano Energy **56**, 56–64 (2019)
14. S. Jang, M. La, S. Cho, Y. Yun, J.H. Choi, Y. Ra, S.J. Park, D. Choi, Monocharged electret based liquid–solid interacting triboelectric nanogenerator for its boosted electrical output performance. Nano Energy **70**, 104541 (2020)

15. L. Li, X. Li, X. Yu, C. Shen, X. Wang, B. Li, J. Li, L. Wang, J. Yin, W. Guo, Boosting the output of bottom-electrode droplets energy harvester by a branched electrode. Nano Energy **95**, 107024 (2022)
16. Y. Zheng, T. Liu, J. Wu, T. Xu, X. Wang, X. Han, H. Cui, X. Xu, C. Pan, X. Li, Energy conversion analysis of multilayered triboelectric nanogenerators for synergistic rain and solar energy harvesting. Adv. Mater. **34**(28), 2202238 (2022)
17. Y. Zheng, X. Li, M. Zheng, G. Cheng, Y. Zi, S. Cheng, H. Cui, X. Li, $MoSe_2$ enhanced raindrop triboelectric nanogenerators and its energy conversion analysis. Adv. Funct. Mater. **34**(1), 2307669 (2023)
18. H. Wu, N. Mendel, S. van der Ham, L. Shui, G. Zhou, F. Mugele, Charge trapping-based electricity generator (CTEG): an ultrarobust and high efficiency nanogenerator for energy harvesting from water droplets. Adv. Mater. **32**(33), 2001699 (2020)
19. X. Xu, P. Li, Y. Ding, W. Xu, S. Liu, Z. Zhang, Z. Wang, Z. Yang, Droplet energy harvesting panel. Energy Environ. Sci. **15**, 2916–2926 (2022)
20. H. Kuang, S. Huang, K. Zhang, L. Xu, X. Cai, C. Zhang, J. Chen, Y. Li, X. Wang, H. Jin, S. Dong, Z. Yang, T. Hasan, Z.L. Wang, J. Luo, Generating direct current electricity from ionic droplets by using ferroelectric material. ACS Energy Lett. **8**(9), 3832–3838 (2023)
21. J. Dong, C. Xu, L. Zhu, X. Zhao, H. Zhou, H. Liu, G. Xu, G. Wang, G. Zhou, Q. Zeng, Q. Song, A high voltage direct current droplet-based electricity generator inspired by thunderbolts. Nano Energy **90**, 106567 (2021)
22. J. Dong, L. Zhu, P. Guo, C. Xu, X. Zhao, S. Yang, X. He, G. Zhou, G. Ma, H. Guo, C. Hu, Q. Song, A bio-inspired total current nanogenerator. Energy Environ. Sci. **16**, 1071–1081 (2023)
23. W. Xu, X. Zhou, C. Hao, H. Zheng, Y. Liu, X. Yan, Z. Yang, M. Leung, X.C. Zeng, R.X. Xu, Z. Wang, SLIPS-TENG: robust triboelectric nanogenerator with optical and charge transparency using a slippery interface. Natl. Sci. Rev. **6**, 540–550 (2019)
24. Z. Chen, Y. Lu, R. Li, R.J. Orlando, R. Manica, Q. Liu, Liquid–solid triboelectric nanogenerators for a wide operation window based on slippery lubricant-infused surfaces (SLIPS). Chem. Eng. J. **439**, 135688 (2022)
25. Z. Ren, Y. Ding, J. Nie, F. Wang, L. Xu, S. Lin, X. Chen, Z.L. Wang, Environmental energy harvesting adapting to different weather conditions and self-powered vapor sensor based on humidity-responsive triboelectric nanogenerators. ACS Appl. Mater. Interfaces **11**(6), 6143–6153 (2019)
26. A. Kusoglu, M.H. Santare, A.M. Karlsson, Mechanics-based model for non-affine swelling in perfluorosulfonic acid (PFSA) membranes. Polymer **50**(11), 2481–2491 (2009)
27. X. Gang, Z.H. Guo, Z. Cong, J. Wang, C. Chang, C. Pan, X. Pu, Z.L. Wang, Textile triboelectric nanogenerators simultaneously harvesting multiple "high-entropy" kinetic energies. ACS Appl. Mater. Interfaces **13**(17), 20145–20152 (2021)
28. D. Jiang, Y. Su, K. Wang, Y. Wang, M. Xu, M. Dong, G. Chen, A triboelectric and pyroelectric hybrid energy harvester for recovering energy from low-grade waste fluids. Nano Energy **70**, 104459 (2020)
29. X. Xu, Y. Wang, P. Li, W. Xu, L. Wei, Z. Wang, Z. Yang, A leaf-mimic rain energy harvester by liquid–solid contact electrification and piezoelectricity. Nano Energy **90**, 106573 (2021)
30. D. Kam, G. Gwon, S. Jang, D. Yoo, S.J. Park, M. La, D. Choi, Advancing energy harvesting efficiency from a single droplet: a mechanically guided 4d printed elastic hybrid droplet-based electricity generator. Adv. Mater. **35**(48), 2303681 (2023)
31. L. Zheng, Z.-H. Lin, G. Cheng, W. Wu, X. Wen, S. Lee, Z.L. Wang, Silicon-based hybrid cell for harvesting solar energy and raindrop electrostatic energy. Nano Energy **9**, 291–300 (2014)
32. L. Zheng, G. Cheng, J. Chen, L. Lin, J. Wang, Y. Liu, H. Li, Z.L. Wang, A hybridized power panel to simultaneously generate electricity from sunlight, raindrops, and wind around the clock. Adv. Energy Mater. **5**(21), 1501152 (2015)
33. S.-B. Jeon, D. Kim, G.-W. Yoon, J.-B. Yoon, Y.-K. Choi, Self-cleaning hybrid energy harvester to generate power from raindrop and sunlight. Nano Energy **12**, 636–645 (2015)
34. D. Yoo, S.-C. Park, S. Lee, J.-Y. Sim, I. Song, D. Choi, H. Lim, D.S. Kim, Biomimetic anti-reflective triboelectric nanogenerator for concurrent harvesting of solar and raindrop energies. Nano Energy **57**, 424–431 (2019)

35. L. Wang, Y. Wang, H. Wang, G. Xu, A. Döring, W.A. Daoud, J. Xu, A.L. Rogach, Y. Xi, Y. Zi, Carbon Dot-Based composite films for simultaneously harvesting raindrop energy and boosting solar energy conversion efficiency in hybrid cells. ACS Nano **14**(8), 10359–10369 (2020)
36. L. Zhao, J. Duan, L. Liu, J. Wang, Y. Duan, L. Vaillant-Roca, X. Yang, Q. Tang, Boosting power conversion efficiency by hybrid triboelectric nanogenerator/silicon tandem solar cell toward rain energy harvesting. Nano Energy **82**, 105773 (2021)
37. H.-Y. Park, H.K. Kim, Y.-H. Hwang, D.-M. Shin, Water-through triboelectric nanogenerator based on Ti-mesh for harvesting liquid flow. J. Korean Phys. Soc. **72**, 499–503 (2018)
38. K. Munirathinam, D.-S. Kim, A. Shanmugasundaram, J. Park, Y.-J. Jeong, D.-W. Lee, Flowing water-based tubular triboelectric nanogenerators for sustainable green energy harvesting. Nano Energy **102**, 107675 (2022)
39. Y. Dong, S. Xu, C. Zhang, L. Zhang, D. Wang, Y. Xie, N. Luo, Y. Feng, N. Wang, M. Feng, X. Zhang, F. Zhou, Z.L. Wang, Gas-liquid two-phase flow-based triboelectric nanogenerator with ultrahigh output power. Sci. Adv. **8**, eadd0464 (2022)
40. T. Kim, J. Chung, D.Y. Kim, J.H. Moon, S. Lee, M. Cho, S.H. Lee, S. Lee, Design and optimization of rotating triboelectric nanogenerator by water electrification and inertia. Nano Energy **27**, 340–351 (2016)
41. Y. Liu, Y. Zheng, T. Li, D. Wang, F. Zhou, Water-solid triboelectrification with self-repairable surfaces for water-flow energy harvesting. Nano Energy **61**, 454–461 (2019)
42. J. Chung, D. Heo, K. Cha, Z.-H. Lin, J. Hong, S. Lee, A portable device for water-sloshing-based electricity generation based on charge separation and accumulation. iScience **24**, 102442 (2021).
43. L. Pan, J. Wang, P. Wang, R. Gao, Y.-C. Wang, X. Zhang, J.-J. Zou, Z.L. Wang, Liquid-FEP-based U-tube triboelectric nanogenerator for harvesting water-wave energy. Nano Res. **11**, 4062–4073 (2018)
44. H. Wu, Z. Wang, Y. Zi, Multi-mode water-tube-based triboelectric nanogenerator designed for low-frequency energy harvesting with ultrahigh volumetric charge density. Adv. Energy Mater. **11**, 2100038 (2021)
45. X. Sun, C. Shang, H. Ma, C. Li, L. Xue, Q. Xu, Z. Wei, W. Li, Y. Yalikun, Y.-C. Lai, Y. Yang, A tube-shaped solid–liquid-interfaced triboelectric–electromagnetic hybrid nanogenerator for efficient ocean wave energy harvesting. Nano Energy **100**, 107540 (2022)
46. W. Tang, T. Jiang, F.R. Fan, A.F. Yu, C. Zhang, X. Cao, Z.L. Wang, Liquid-metal electrode for high-performance triboelectric nanogenerator at an instantaneous energy conversion efficiency of 70.6%. Adv. Funct. Mater. **25**(24), 3718–3725 (2015)
47. M.-L. Seol, S.-B. Jeon, J.-W. Han, Y.-K. Choi, Ferrofluid-based triboelectric-electromagnetic hybrid generator for sensitive and sustainable vibration energy harvesting. Nano Energy **31**, 233–238 (2017)
48. H. Qin, L. Xu, S. Lin, F. Zhan, K. Dong, K. Han, H. Wang, Y. Feng, Z.L. Wang, Underwater energy harvesting and sensing by sweeping out the charges in an electric double layer using an oil droplet. Adv. Funct. Mater. **32**(18), 2111662 (2022)
49. X.J. Zhao, G. Zhu, Y.J. Fan, H.Y. Li, Z.L. Wang, Triboelectric charging at the nanostructured solid/liquid interface for area-scalable wave energy conversion and its use in corrosion protection. ACS Nano **9**(7), 7671–7677 (2015)
50. X.J. Zhao, S.Y. Kuang, Z.L. Wang, G. Zhu, Highly adaptive solid–liquid interfacing triboelectric nanogenerator for harvesting diverse water wave energy. ACS Nano **12**(5), 4280–4285 (2018)
51. L. Liu, Q. Shi, J.S. Ho, C. Lee, Study of thin film blue energy harvester based on triboelectric nanogenerator and seashore IoT applications. Nano Energy **66**, 104167 (2019)
52. H. Gu, N. Zhang, Z. Zhou, S. Ye, W. Wang, W. Xu, H. Zheng, Y. Song, J. Jiao, Z. Wang, X. Zhou, A bulk effect liquid–solid generator with 3D electrodes for wave energy harvesting. Nano Energy **87**, 106218 (2021)
53. X. Liang, S. Liu, S. Lin, H. Yang, T. Jiang, Z.L. Wang, Liquid–solid triboelectric nanogenerator arrays based on dynamic electric-double-layer for harvesting water wave energy. Adv. Energy Mater. **13**(24), 2300571 (2023)

54. Q. Shi, H. Wang, H. Wu, C. Lee, Self-powered triboelectric nanogenerator buoy ball for applications ranging from environment monitoring to water wave energy farm. Nano Energy **40**, 203–213 (2017)
55. X. Li, J. Tao, X. Wang, J. Zhu, C. Pan, Z.L. Wang, Networks of high performance triboelectric nanogenerators based on liquid–solid interface contact electrification for harvesting low-frequency blue energy. Adv. Energy Mater. **8**(21), 1800705 (2018)
56. M. Han, B. Yu, G. Qiu, H. Chen, Z. Su, M. Shi, B. Meng, X. Cheng, H. Zhang, Electrification based devices with encapsulated liquid for energy harvesting, multifunctional sensing, and self-powered visualized detection. J. Mater. Chem. A **3**, 7382–7388 (2015)
57. B. Zhang, L. Zhang, W. Deng, L. Jin, F. Chun, H. Pan, B. Gu, H. Zhang, Z. Lv, W. Yang, Z.L. Wang, Self-powered acceleration sensor based on liquid metal triboelectric nanogenerator for vibration monitoring. ACS Nano **11**(7), 7440–7446 (2017)
58. X. Li, M.-H. Yeh, Z.-H. Lin, H. Guo, P.-K. Yang, J. Wang, S. Wang, R. Yu, T. Zhang, Z.L. Wang, Self-powered triboelectric nanosensor for microfluidics and cavity-confined solution chemistry. ACS Nano **9**(11), 11056–11063 (2015)
59. Q. Shi, H. Wang, T. Wang, C. Lee, Self-powered liquid triboelectric microfluidic sensor for pressure sensing and finger motion monitoring applications. Nano Energy **30**, 450–459 (2016)
60. X. Ge, Z. Gao, L. Zhang, H. Ji, J. Yi, P. Jiang, Z. Li, L. Shen, X. Sun, Z. Wen, Flexible microfluidic triboelectric sensor for gesture recognition and information encoding. Nano Energy **113**, 108541 (2023)
61. C.P. Vo, M. Shahriar, C.D. Le, K.K. Ahn, Mechanically active transducing element based on solid–liquid triboelectric nanogenerator for self-powered sensing. INT J PR ENG MAN-GT **6**, 741–749 (2019)
62. T. Zhan, H. Zou, H. Zhang, P. He, Z. Liu, J. Chen, M. He, Y. Zhang, Z.L. Wang, Smart liquid-piston based triboelectric nanogenerator sensor for real-time monitoring of fluid status. Nano Energy **111**, 108419 (2023)
63. W. Zhang, P. Wang, K. Sun, C. Wang, D. Diao, Intelligently detecting and identifying liquids leakage combining triboelectric nanogenerator based self-powered sensor with machine learning. Nano Energy **56**, 277–285 (2019)
64. S. Hu, Z. Shi, R. Zheng, W. Ye, X. Gao, W. Zhao, G. Yang, Superhydrophobic liquid–solid contact triboelectric nanogenerator as a droplet sensor for biomedical applications. ACS Appl. Mater. Interfaces **12**(36), 40021–40030 (2020)
65. Y. Pan, M. Li, Y. Lu, M. Lu, B. Guo, X. Liu, K. Zhang, Y. Wang, L. Feng, A. Qin, Hydrophobic sisal cellulose paper-based TENG for collecting rain energy and raindrop-based sensor. Chem. Eng. J. **490**, 151590 (2024)
66. S.B. Jeon, M.L. Seol, D. Kim, S.J. Park, Y.K. Choi, Self-powered ion concentration sensor with triboelectricity from liquid–solid contact electrification. Adv. Electron. Mater. **2**(5), 1600006 (2016)
67. B.D. Chen, W. Tang, C. He, T. Jiang, L. Xu, L.P. Zhu, G.Q. Gu, J. Chen, J.J. Shao, J.J. Luo, Z.L. Wang, Ultrafine capillary-tube triboelectric nanogenerator as active sensor for microliquid biological and chemical sensing. Adv. Mater. Technol. **3**(1), 1700229 (2017)
68. S. Chatterjee, S. Saha, S.R. Barman, I. Khan, Y.-P. Pao, S. Lee, D. Choi, Z.-H. Lin, Enhanced sensing performance of triboelectric nanosensors by solid-liquid contact electrification. Nano Energy **77**, 105093 (2020)
69. S. RoyBarman, Y.-J. Lin, K.-M. Lee, A. Pal, N. Tiwari, S. Lee, Z.-H. Lin, Triboelectric nanosensor integrated with robotic platform for self-powered detection of chemical analytes. ACS Nano **17**(3), 2689–2701 (2023)
70. J. Yu, J.F. Tang, L. Wang, Y.J. Guo, W.Y. Ma, L. Yang, S.Y. Chen, Dual-mode sensor for intelligent solution monitoring: enhancing sensitivity and recognition accuracy through capacitive and triboelectric sensing. Nano Energy **118**, 109009 (2023)
71. F. Yi, X. Wang, S. Niu, S. Li, Y. Yin, K. Dai, G. Zhang, L. Lin, Z. Wen, H. Guo, J. Wang, M.-H. Yeh, Y. Zi, Q. Liao, Z. You, Y. Zhang, Z.L. Wang, A highly shape-adaptive, stretchable design based on conductive liquid for energy harvesting and self-powered biomechanical monitoring. Sci. Adv. **2**, e1501624 (2016)

72. B. Zhang, Z. Wu, Z. Lin, H. Guo, F. Chun, W. Yang, Z.L. Wang, All-in-one 3D acceleration sensor based on coded liquid–metal triboelectric nanogenerator for vehicle restraint system. Mater. Today **43**, 37–44 (2021)
73. L. Yang, Y. Wang, Z. Zhao, Y. Guo, S. Chen, W. Zhang, X. Guo, Particle-laden droplet-driven triboelectric nanogenerator for real-time sediment monitoring using a deep learning method. ACS Appl. Mater. Interfaces **12**(34), 38192–38201 (2020)
74. T. Huang, W. Sun, L. Liao, K. Zhang, M. Lu, L. Jiang, S. Chen, A. Qin, Detection of microplastics based on a liquid–solid triboelectric nanogenerator and a deep learning method. ACS Appl. Mater. Interfaces **15**(29), 35014–35023 (2023)
75. C. Li, X. Liu, D. Yang, Z. Liu, Triboelectric nanogenerator based on a moving bubble in liquid for mechanical energy harvesting and water level monitoring. Nano Energy **95**, 106998 (2022)
76. M. Xu, S. Wang, S.L. Zhang, W. Ding, P.T. Kien, C. Wang, Z. Li, X. Pan, Z.L. Wang, A highly-sensitive wave sensor based on liquid–solid interfacing triboelectric nanogenerator for smart marine equipment. Nano Energy **57**, 574–580 (2019)
77. X. Zhang, M. Yu, Z. Ma, H. Ouyang, Y. Zou, S.L. Zhang, H. Niu, X. Pan, M. Xu, Z. Li, Z.L. Wang, Self-powered distributed water level sensors based on liquid–solid triboelectric nanogenerators for ship draft detecting. Adv. Funct. Mater. **29**(41), 1900327 (2019)
78. S. Wang, Y. Wang, D. Liu, Z. Zhang, W. Li, C. Liu, T. Du, X. Xiao, L. Song, H. Pang, M. Xu, A robust and self-powered tilt sensor based on annular liquid–solid interfacing triboelectric nanogenerator for ship attitude sensing. Sensor Actuat A-Phys **317**, 112459 (2021)
79. T. Liu, W. Mo, X. Zou, B. Luo, S. Zhang, Y. Liu, C. Cai, M. Chi, J. Wang, S. Wang, Liquid–solid triboelectric probes for real-time monitoring of sucrose fluid status. Adv. Funct. Mater. **33**(45), 2304321 (2023)
80. B. Luo, C. Cai, T. Liu, S. Zhang, C. Gao, Y. Liu, M. Chi, J. Wang, S. Wang, S. Nie, Triboelectric probes for investigating charge transfer at the colloid-solid interface. Nano Energy **117**, 108874 (2023)
81. Y. Zhu, B. Luo, X. Zou, T. Liu, S. Zhang, M. Chi, Y. Sun, K. Jiang, L. Liu, F. Meng, Triboelectric probes integrated with deep learning for real-time online monitoring of suspensions in liquid transport. Nano Energy **122**, 109340 (2024)
82. J. Zhang, S. Lin, M. Zheng, Z.L. Wang, Triboelectric nanogenerator as a probe for measuring the charge transfer between liquid and solid surfaces. ACS Nano **15**(9), 14830–14837 (2021)
83. J. Zhang, S. Lin, Z.L. Wang, Triboelectric nanogenerator array as a probe for in situ dynamic mapping of interface charge transfer at a liquid–solid contacting. ACS Nano **17**(2), 1646–1652 (2023)
84. S. Lin, L. Xu, L. Zhu, X. Chen, Z.L. Wang, Electron transfer in nanoscale contact electrification: photon excitation effect. Adv. Mater. **31**(27), 1901418 (2019)
85. S. Lin, L. Zhu, Z. Tang, Z.L. Wang, Spin-selected electron transfer in liquid–solid contact electrification. Nat. Commun. **13**, 5230 (2022)
86. Y. Gauduel, S. Pommeret, A. Migus, A. Antonetti, Some evidence of ultrafast H_2O^+–water molecule reaction in femtosecond photoionization of pure liquid water: influence on geminate pair recombination dynamics. Chem. Phys. **149**, 1–10 (1990)
87. A. Chen, C. Zhang, G. Zhu, Z.L. Wang, Polymer materials for high-performance triboelectric nanogenerators. Adv. Sci. **7**, 2000186 (2020)
88. L. Liu, Z. Zhao, Y. Li, X. Li, D. Liu, S. Li, Y. Gao, L. Zhou, J. Wang, Z.L. Wang, Achieving ultrahigh effective surface charge density of direct-current triboelectric nanogenerator in high humidity. Small **18**(24), 2201402 (2022)
89. L. Zhang, X. Li, Y. Zhang, Y. Feng, F. Zhou, D. Wang, Regulation and influence factors of triboelectricity at the solid–liquid interface. Nano Energy **78**, 105370 (2020)
90. J. Nie, Z. Ren, L. Xu, S. Lin, F. Zhan, X. Chen, Z.L. Wang, Probing contact-electrification-induced electron and ion transfers at a liquid–solid interface. Adv. Mater. **32**(2), 1905696 (2020)
91. X. Liu, J. Zhang, X. Wang, S. Lin, Z.L. Wang, Charge accumulation kinetics at a liquid–solid interface depend on liquid chemistry. J. Mater. Chem. A **11**, 18964–18971 (2023)

92. X. Wei, B. Wang, X. Cao, H. Zhou, Z. Wu, Z.L. Wang, Dual-sensory fusion self-powered triboelectric taste-sensing system towards effective and low-cost liquid identification. Nat. Food **4**, 721–732 (2023)
93. B. Luo, X. Wang, T. Liu, C. Cai, Y. Liu, S. Zhang, M. Chi, C. Gao, J. Wang, Z. Liu, Liquid–solid triboelectric probes for bubbles status monitoring. Adv. Funct. Mater. **34**(32), 2315725 (2024)
94. C. Cai, S. Wang, S. Nie, Triboelectric charge at the liquid–solid interface for chemical analysis. Matter **7**(9), 2806–2809 (2024)

Chapter 11
Tribovoltaic Nanogenerator

Abstract The tribovoltaic nanogenerator (TVNG), as an innovative energy conversion device, has demonstrated great application potential in areas such as mechanical components, multi-source energy harvesting, intelligent flexible devices, and self-powered sensing systems in recent years. In mechanical components, the TVNG converts mechanical vibration and frictional energy into electrical energy, enhancing the energy efficiency and self-driving capabilities of traditional mechanical systems. In multi-source energy harvesting, the TVNG effectively integrates various forms of energy (such as mechanical, thermal, and light energy) and, through coupling with other energy conversion mechanisms (like the tribo-thermoelectric effect), significantly improves energy conversion efficiency, promoting the development of sustainable energy technologies. In intelligent flexible devices, the TVNG, with its flexible design and high current density, enables smart wearable devices to power themselves through human motion, achieving battery-free operation and enhancing the devices' flexibility and portability. In self-powered sensing systems, the direct current signal from the TVNG can reflect various operational and environmental parameters, making it suitable for real-time sensing. This provides efficient and reliable solutions for fields such as intelligent environmental monitoring, health monitoring, and industrial automation. These applications not only demonstrate the broad adaptability and high efficiency of the tribovoltaic effect across different scenarios but also guide the development of future intelligent systems and new energy technologies.

11.1 Introduction

In everyday human activities, approximately one-third of energy is lost through frictional processes, such as minor resistance during daily movements or energy dissipation in fundamental industrial components like bearings and gears. This energy loss is typically considered difficult to recover. However, harnessing this frictional energy could significantly enhance energy utilization efficiency and provide sustainable power for numerous sensor network nodes. With advancements in nanotechnology and tribology, researchers have increasingly recognized that this seemingly negligible

Z. L. Wang, *Contact-Electrification of Matter*, Springer Series in Surface Sciences 75,
https://doi.org/10.1007/978-3-031-88504-4_11

frictional energy can be converted into usable electricity, thus fostering the development of green energy and advancing energy recovery technologies. This realization led to the invention of the triboelectric nanogenerator (TENG), which operates based on the contact electrification (CE) effect. TENGs generate electric charges through friction and separation between materials, enabling them to capture weak mechanical energy from the environment. However, TENGs output alternating current (AC), requiring rectification for many applications. This increases system complexity and limits their direct application in micro- and nanoscale devices. To address this limitation, scientists developed a semiconductor-based triboelectric generator capable of directly outputting direct current (DC). In 2019, Wang et al. proposed the "tribovoltaic effect" theory, explaining the mechanism of electron–hole pair separation at the interface of p-type and n-type semiconductors during sliding. The tribovoltaic effect is a complex charge transfer phenomenon driven by atomic interactions at the semiconductor interface. During contact sliding, the formation and breaking of chemical bonds between interfacial atoms release energy, exciting electron–hole pairs. These pairs are separated by the built-in or interfacial electric field, generating a measurable DC current.

The DC tribovoltaic nanogenerator (TVNG), based on the tribovoltaic effect, represents a breakthrough in designing self-powered sensors and energy-harvesting devices. Unlike conventional TENGs, TVNGs output DC power directly, eliminating the need for rectifiers, simplifying the energy conversion process, and offering extensive application potential across various fields. In the field of energy recovery for mechanical components, TVNGs can generate electricity from friction during equipment operation, reducing energy loss while supporting self-powered sensing and monitoring functions. For example, TVNGs integrated into mechanical components like shafts and gears can convert friction into electrical energy, powering systems for status monitoring and fault detection, thereby improving equipment safety and operational efficiency. In multi-source energy harvesting, TVNGs can combine various forms of energy, such as solar, thermal, and mechanical energy, to fully utilize dispersed environmental energy. This capability extends the operational time of devices, making them particularly suitable for remote or resource-constrained environments. By complementing other power generation technologies, TVNGs further improve energy utilization efficiency, offering a novel pathway for long-term and high-frequency energy harvesting. TVNGs also exhibit high adaptability and flexibility, making them ideal for integration into intelligent flexible devices like wearable electronics or portable gadgets. For instance, a TVNG embedded in a smart wristband or fitness tracker can harness mechanical energy from everyday human movements, reducing battery dependency and extending device lifespan. This innovative approach provides new perspectives for the development of self-powered portable devices. Moreover, TVNGs have significant potential in self-powered sensing systems, which rely on small amounts of energy harvested from the environment for continuous monitoring and operation. A TVNG can convert mechanical vibrations, airflow, and other kinetic energy into electricity, enabling sensors to operate persistently without external power sources. This is especially advantageous for remote or hard-to-maintain locations. Applications include air quality monitoring, vibration detection, and other sensing tasks, where TVNGs harness subtle mechanical energy to power the system and deliver sustained, efficient monitoring capabilities. The ability of TVNGs to efficiently capture mechanical energy and directly convert it into a DC signal also enables real-time monitoring. For example, TVNGs installed on bearings or rotors in rotating machinery can generate electrical signals correlated to changes in friction frequency, allowing precise measurement of rotational speed.

This chapter systematically reviews the applications of TVNGs in various fields. Section 11.2 explores their use in energy recovery for mechanical components, highlighting their benefits in reducing energy loss and supporting self-powered devices. Section 11.3 focuses on TVNG performance in multi-source energy harvesting, emphasizing their potential to integrate different energy forms and enhance overall energy efficiency. Section 11.4 examines TVNG applications in intelligent flexible devices, demonstrating their adaptability and suitability for wearable and portable self-powered devices. Section 11.5 discusses the future prospects of TVNGs in self-powered sensing systems, showcasing their vast potential across diverse sensing applications. Together, these developments position TVNGs as a promising technology for efficient energy harvesting and intelligent device innovation in the future.

11.2 Application in Mechanical Components

11.2.1 Pneumatic Monitoring Sensor for Rotating Cylinders

With the rapid advancement of industrial automation and intelligence, pneumatic systems have become widely adopted across various applications. However, traditional monitoring and control methods for pneumatic devices often encounter challenges such as imprecise positioning, insufficient real-time responsiveness, and high-power consumption. Addressing these issues, Shi et al. [1] proposed a novel self-powered sensor system incorporating a tribovoltaic nanogenerator (TVNG) and a triboelectric nanogenerator (TENG). This innovative solution provides high-precision, high-sensitivity pneumatic monitoring and servo control. By integrating TENG and TVNG technologies, their dual-function pneumatic sensor system enables precise monitoring of the rotating disc's angle and speed within the cylinder, laying a foundation for the development of self-powered, efficient pneumatic systems.

In the dual-function pneumatic sensor system (SDS) described in this section, the TENG and TVNG are positioned on the inner and outer rings of the cylinder, respectively. The TENG is primarily responsible for monitoring the rotation angle, while the TVNG detects rotational speed. The system operates based on interface contact between PEDOT and PTFE materials. As the cylinder rotates, the TENG and

TVNG simultaneously generate electrical signals corresponding to the angle and speed, enabling accurate detection and control of the cylinder's motion.

The design features an output disc driven by a gear-and-rack mechanism, where the cylinder's stator and rotor are mounted on the disc and gear, respectively. When gas enters the cylinder, it pushes the rack, driving the gear and disc rotation. This motion activates the TENG and TVNG. The flexible hybrid thin-film structure of the system ensures high reliability in complex pneumatic environments, offering the advantages of self-powered and maintenance-free operation, even under harsh working conditions (Fig. 11.1a).

The TENG operates as an embedded triboelectric position sensor, utilizing voltage signals to represent the rotation angle. During rotation, triboelectric charges are generated between the PEDOT and PTFE films, and variations in the overlap area between these films produce a corresponding potential difference. As the disc rotates, changes in the potential difference accurately reflect the rotation angle. Experimental results demonstrate a strong linear relationship between the TENG's output voltage

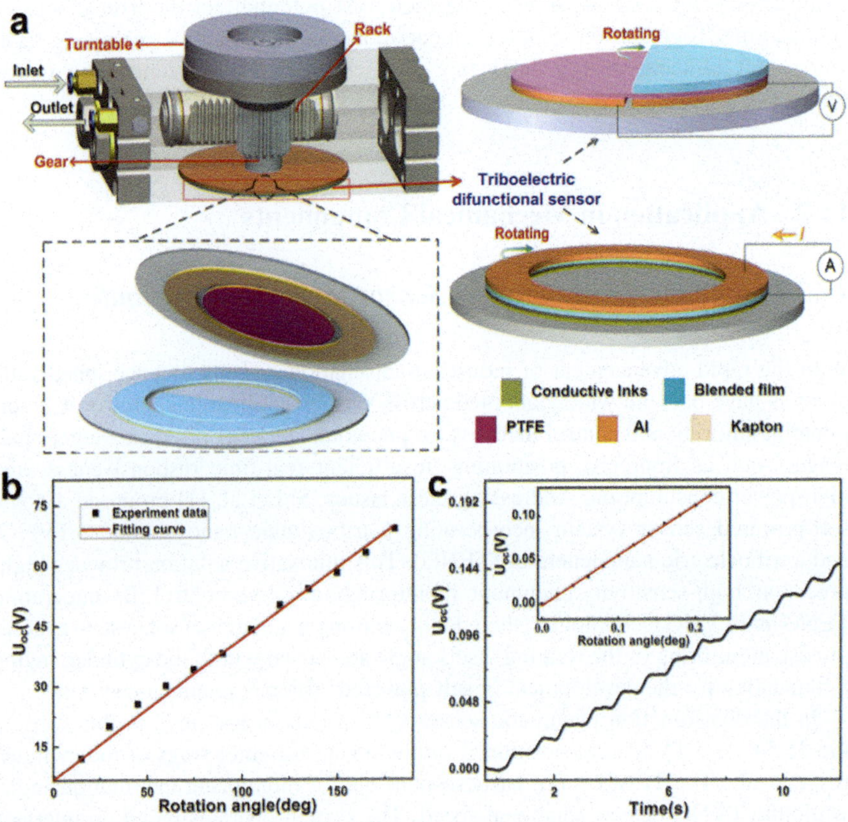

Fig. 11.1 **a** Structural design schematic of the SDS. **b** Relationship between open-circuit voltage and sliding distance. **c** Displacement resolution of the embedded triboelectric position sensor

and the rotation angle, with a sensitivity of 0.0389 V/deg and a displacement resolution exceeding 0.02°. These characteristics highlight its capability as a high-precision position sensor (Fig. 11.1b and c). Additionally, the TENG's low-impedance structure enhances sensing efficiency and facilitates maximum power output by optimizing load resistance, making it an excellent candidate for self-powered applications in pneumatic systems.

The TVNG's speed monitoring is primarily driven by the tribovoltaic effect. Through interface friction between the PEDOT and aluminum electrodes, frictional energy excites electrons to higher energy states within the hybrid film, producing a current under the influence of the built-in electric field. Experimental data reveal a linear relationship between the TVNG's output current and rotational speed, with a sensitivity of 0.008 μA/rpm and a linear correlation coefficient of 0.998. The speed resolution reaches 1 rpm, demonstrating superior detection performance in low-speed environments (Fig. 11.2a). Furthermore, the TVNG achieves an instantaneous power output of 0.09 μW under varying impedance loads, making it well-suited for self-powered sensing applications (Fig. 11.2b).

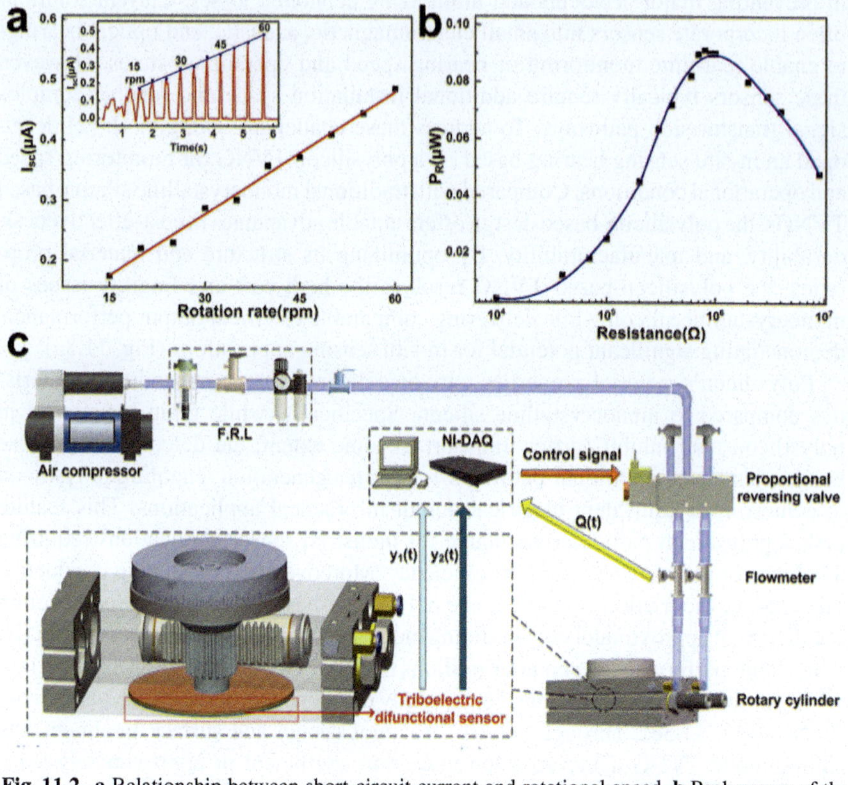

Fig. 11.2 **a** Relationship between short-circuit current and rotational speed. **b** Peak power of the embedded triboelectric active speed sensor with variable resistance. **c** Pneumatic circuit and control module of the SDS cylinder

The dual-function integration of TENG and TVNG provides a novel, flexible, and efficient solution for the monitoring and control of pneumatic systems. The cylinder control system based on TENG and TVNG not only achieves high-precision servo control with an overshoot range of 2% and control error of 1.5% but also lays a foundation for self-powered and multifunctional pneumatic intelligent systems. The SDS enables real-time monitoring of the disc's angle and speed, supporting precise control of pneumatic devices. This system's high sensitivity and high-precision response make it highly promising for applications in industrial automation, robotics, and medical rehabilitation equipment (Fig. 11.2c).

11.2.2 Tribovoltaic Bearing for Sensing

11.2.2.1 Polysilicon Bearings for Rotational Speed Sensing

As critical components in industrial equipment, smart bearings play a vital role in preventing major accidents and minimizing economic losses. Current solutions often incorporate sensors into small electromagnetic, acoustic, and optical bearings to enable real-time monitoring of bearing speed and operational status. However, these sensors typically require additional installation space and involve complex signal transmission pathways. To address these challenges, Gong et al. [2] developed an in-situ sensing bearing based on a polysilicon TVNG for monitoring speed and operational conditions. Compared with traditional monocrystalline silicon-based TVNGs, the polysilicon-based design offers notable advantages in cost-effectiveness, durability, and manufacturability. By optimizing its structure and material properties, the polysilicon-based TVNG resolves the high cost and fragility issues of monocrystalline silicon while delivering comparable electrical output performance, demonstrating significant potential for in-situ sensing applications (Fig. 11.3a).

Polysilicon's material properties introduce distinct charge transport characteristics compared to monocrystalline silicon. Specifically, while grain boundaries in polysilicon may inhibit carrier transport to some extent, the defects within these boundaries create additional pathways for carrier generation, enabling polysilicon to achieve near-equivalent electrical output in practical applications. This feature makes polysilicon particularly suitable for industrial sensor applications requiring a balance of performance and cost efficiency. Moreover, by introducing conductive oil to reduce the friction coefficient, the polysilicon-based TVNG achieves a friction coefficient of approximately 0.2 in sliding mode and maintains exceptional durability, with stable performance even after millions of operational cycles (Fig. 11.3b–d).

To further optimize performance, the study developed a rolling-mode polysilicon TVNG (Fig. 11.3e). Through precise structural design and surface treatment, the rolling-mode TVNG achieved a lower friction coefficient of approximately 0.14, significantly improving the linear relationship between electrical output and speed, providing more accurate data support for in-situ sensing (Fig. 11.3f and g). The advantage of the rolling mode lies in its design, which effectively reduces friction

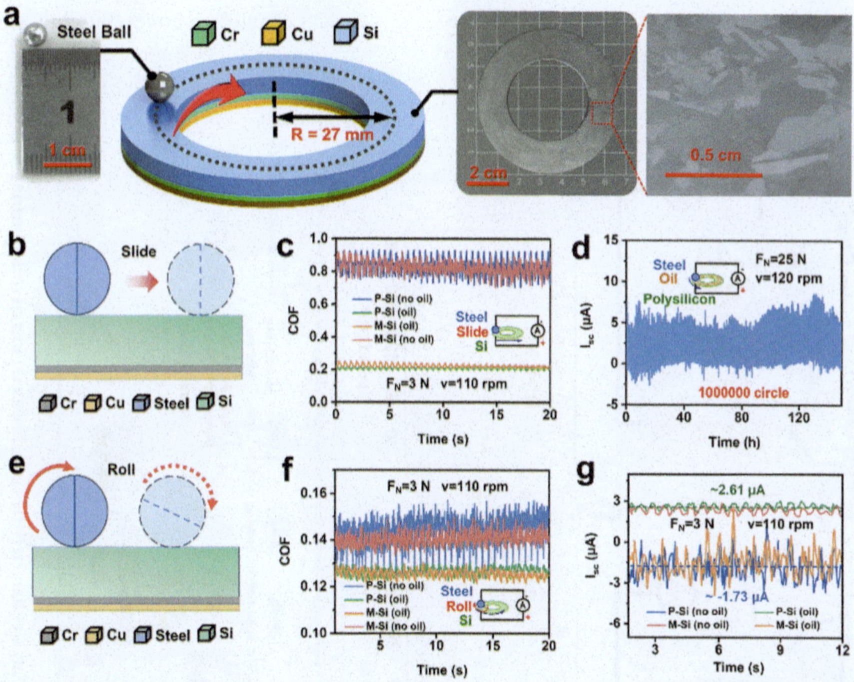

Fig. 11.3 **a** 3D structural schematic and geometric dimensions of the polysilicon-based TVNG. **b** Structure diagram of the sliding TVNG. **c** COF characteristics of the sliding TVNG. **d** I_{SC} durability of lubricated sliding TVNG. **e** Structure diagram of the rolling TVNG. **f** COF characteristics of the rolling TVNG. **g** I_{SC} characteristics of the rolling TVNG

loss, enhancing device durability and electrical output stability. Therefore, the rolling-mode polysilicon TVNG sensor better meets the industrial requirements for high durability and high-precision sensing.

Additionally, the study fabricated various types of polysilicon TVNG-based bearings, including tapered roller bearings, thrust ball bearings, and deep-groove ball bearings, using cutting and etching techniques. These bearings demonstrated excellent electrical output characteristics in application, with both open-circuit voltage and short-circuit current showing a linear relationship with bearing speed and a fitting coefficient close to 0.99. This linear relationship provides direct, precise feedback for speed sensing, making it widely applicable in industrial sensing (Fig. 11.4a–f).

The developed bearing speed monitoring system enables real-time monitoring, wireless transmission, and overload warning for deep-groove ball bearings. The system collects electrical signals generated by the bearing via a data acquisition module, transmitting them wirelessly to a data receiver module on a computer. The rotor's electrodes are connected through a conductive slip ring, with the data acquisition and wireless transmission modules powered by a small battery. Upon receiving the signals, the computer processes and analyzes the data in real time, displaying them on an interface. At a speed of 30 rpm, the monitoring system records average

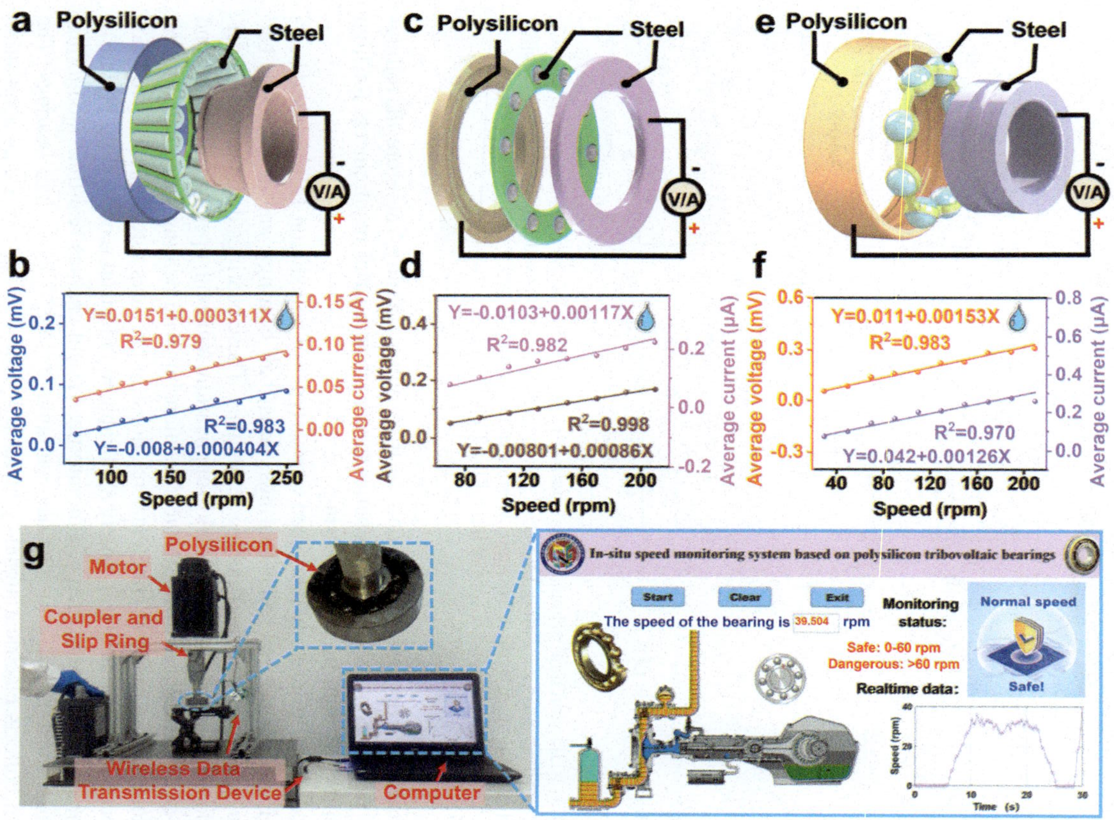

Fig. 11.4 **a** Structure of the tapered roller bearing based on rolling TVNG. **b** V_{oc} and I_{sc} characteristics of lubricated tapered roller bearing. **c** Structure of the thrust ball bearing based on rolling TVNG. **d** Voc and I_{sc} characteristics of lubricated thrust ball bearing. **e** Structure of the deep-groove ball bearing based on rolling TVNG. **f** V_{oc} and I_{sc} characteristics of lubricated deep-groove ball bearing. **g** Real-time speed monitoring system of tribovoltaic bearing

speed data over 5 s. when the motor speed exceeds 60 rpm, the system triggers an alert (Fig. 11.4g). Overall, the polysilicon TVNG proposed in this study provides a low-cost, structurally simple, and highly durable solution for designing smart bearings in the industrial IoT. This polysilicon-based TVNG, with substantially lower costs than monocrystalline silicon, along with excellent wear resistance and longevity, offers a convenient and efficient choice for real-time monitoring and in-situ speed sensing in operational equipment.

11.2.2.2 Freestanding Bearings for Rotational Speed Sensing

The aforementioned TVNG needs to connect electrodes to two relatively moving components, and this wiring dependence greatly limits the application of TVNG in mobile devices. To address this issue, Dong et al. [3] designed a freestanding TVNG based on metal, p-type silicon (p-Si), and n-type silicon (n-Si). This freestanding TVNG features a unique structure in which a freestanding metal component can frictionally contact both p-Si and n-Si simultaneously, producing opposite DC outputs that form a direct current from p-Si to n-Si in an external circuit. This innovative structure not only enhances the energy harvesting efficiency of the TVNG but also resolves the wiring connection issue for moving parts, broadening the application prospects of TVNG in modern mechanical components.

The TVNG design primarily includes two parts: a freestanding metal component and a semiconductor component for friction. The frictional part consists of two semicircular wafers, p-Si and n-Si, separated by acrylic plastic. The freestanding metal part comprises two stainless steel pins connected through leads. With the rotation or sliding of the frictional components, the steel pins contact both p-Si and n-Si simultaneously, facilitating charge transfer. This freestanding mode design enables the TVNG to generate DC output from p-Si to n-Si, forming a stable current path. This characteristic not only achieves continuous DC output but also meets the demand for energy harvesting during movement (Fig. 11.5a).

In experiments, the research team used steel pins to control the actual contact area, further exploring the electrical output characteristics of the TVNG under different frictional areas and speeds. For instance, in the sliding friction mode, the TVNG was connected to a linear motor and moved reciprocally at 40 cm/s. In this case, the TVNG's short-circuit current showed a pulsed waveform due to the periodic friction generated during reciprocal motion (Fig. 11.5b). In the rotating mode, when the TVNG rotated at 60 rpm, its short-circuit current exhibited a continuous constant current waveform, reflecting the differing effects of sliding and rotating modes on the TVNG's electrical output. Additionally, in the rotating mode, the TVNG demonstrated prolonged constant current output, providing data support for the TVNG's stability in practical applications (Fig. 11.5c). Further experiments analyzed the TVNG's output characteristics under different load pressures and contact areas. Results indicated that as load pressure increased, both short-circuit current and open-circuit voltage of the TVNG grew linearly. when load pressure reached a certain

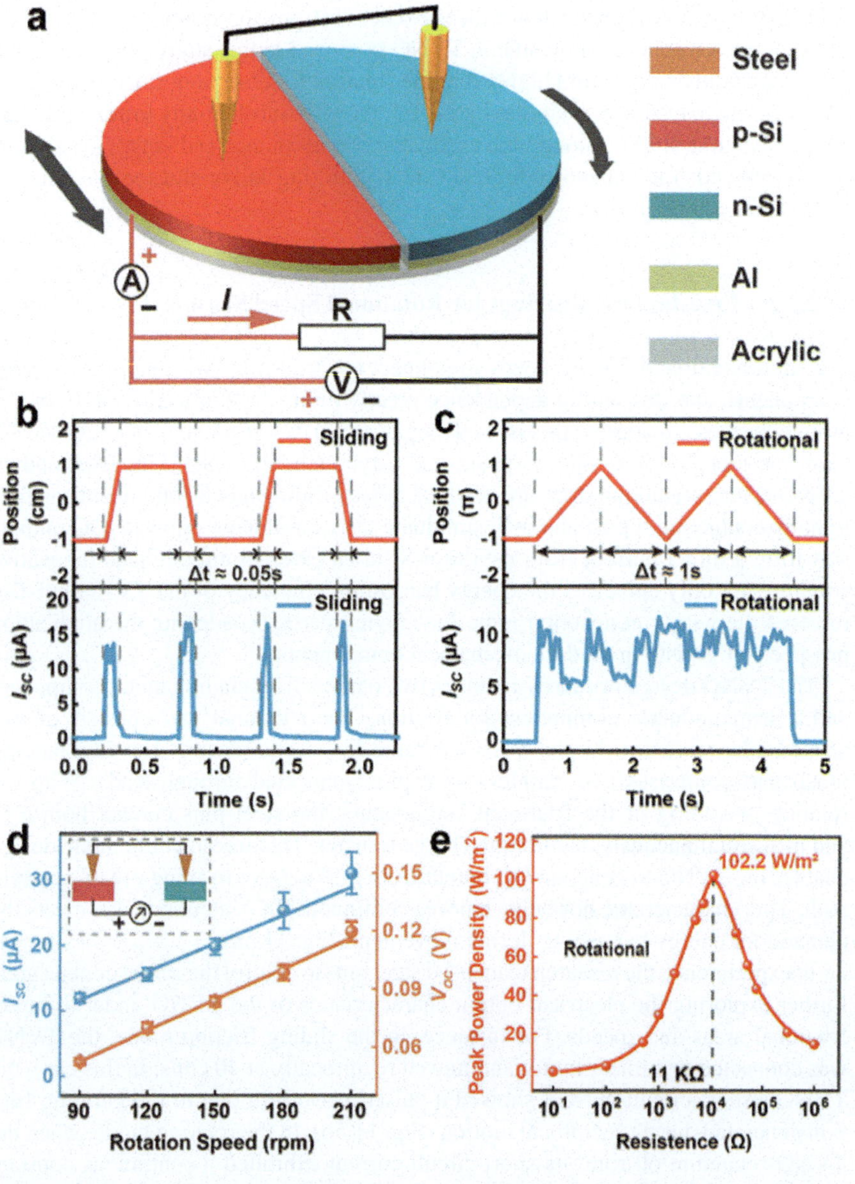

Fig. 11.5 **a** 3D schematic of the TVNG structure. **b** Positional curve and short-circuit current of sliding TVNG. **c** Positional curve and short-circuit current of rolling TVNG. **d** Output characteristics of I_{sc} and V_{oc} of rotating TVNG at different speeds. **e** Peak power density of rotating TVNG under various loads

value, open-circuit voltage plateaued, and short-circuit current stabilized. Additionally, when the steel pin contact area increased, short-circuit current rose linearly, while open-circuit voltage remained constant. This shows that the TVNG's electrical output is not only related to load pressure but also closely linked to the actual frictional contact area. This finding provides optimization direction for practical applications of TVNG by increasing contact area or load pressure to improve electrical output. Furthermore, in rotating mode, the TVNG demonstrated higher constant current characteristics, enhancing its stability in practical applications. Experimental data showed that at 90 rpm, the rotating TVNG's short-circuit current reached approximately 11 μA, and open-circuit voltage was 0.05 V. Testing TVNG's current and voltage characteristics at different speeds revealed a linear increase in both short-circuit current and open-circuit voltage as speed increased from 90 to 210 rpm (Fig. 11.5d). Under high load conditions, the TVNG showed a high peak power density of 102.2 W/m^2 (with 11 kΩ load), with an average power density of approximately 59.6 W/ m^2 (Fig. 11.5e).

To achieve frictional energy harvesting from basic mechanical components, the research team integrated the TVNG into a conventional rolling bearing, using it as a friction bearing. The specific structure is shown in Fig. 11.6a: a copper foam component is independently connected to the inner ring of the rolling bearing, while the machined p-type and n-type silicon semicircular rings are attached to the end cover as the friction pair. Electrical wires are drawn from the back sides of the p-Si and n-Si components, respectively. To validate the practical application of the friction bearing, the researchers further integrated it into a gearbox, as shown in Fig. 11.6b. A real-world photo of the TVNG-integrated friction bearing is presented in Fig. 11.6b, with an appearance nearly identical to that of a conventional bearing. As shown in Fig. 11.6c, the short-circuit current increases linearly as the bearing speed increases from 30 rpm to 120 rpm. However, when the speed exceeds 120 rpm, the current enhancement tends to slow down due to deteriorated contact conditions and surface wear. As illustrated in Fig. 11.6d, when charging a capacitor using the friction bearing at a speed of 180 rpm, the voltage can reach nearly 7 mV. Moreover, it takes less than 15 s to charge a 1 mF capacitor to 7 mV.

11.2.3 Self-powered Sensing Thrust Bearing System

The previously discussed TVNG devices face limitations in output power density and wear resistance, making them less suitable for high-intensity industrial environments. Zhang et al. [4] proposed a TVNG system based on a gallium nitride (GaN) and steel ball combination, successfully overcoming these barriers. This system not only excels in power output and wear resistance but also achieves the self-powered wireless transmission of temperature and pressure data in a thrust bearing application, showcasing the potential of TVNG in industrial IoT applications.

This study introduces a GaN-based TVNG with a ball-disc structure that uses tungsten carbide (WC) steel balls as the friction interface (Fig. 11.7a). Through

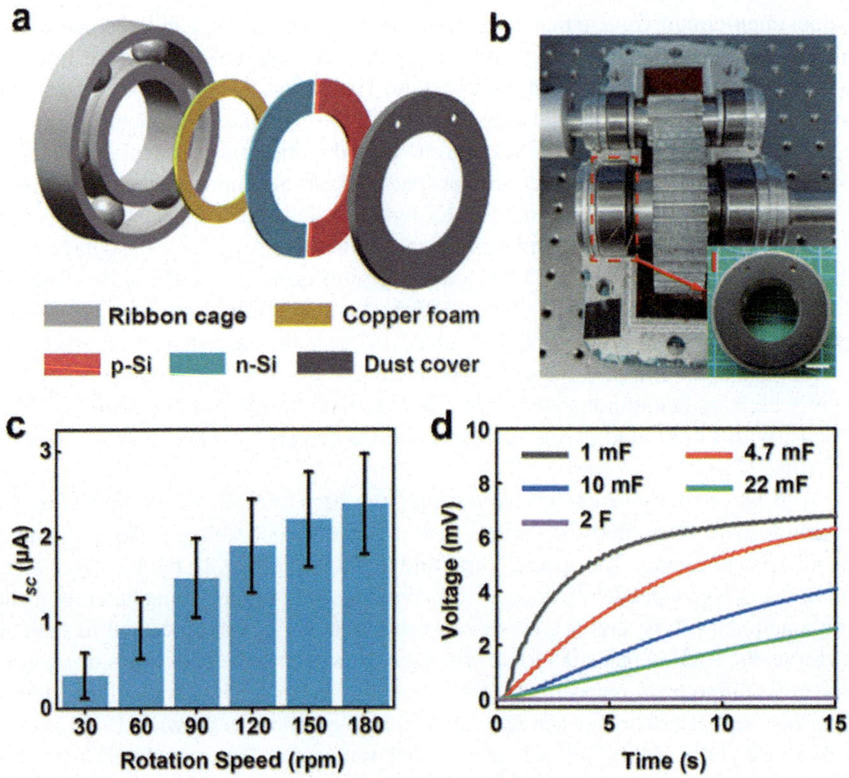

Fig. 11.6 **a** 3D schematic of TVNG integrated triboelectric bearing structure. **b** Detailed photo of speed reducer. **c** I_{sc} of TVNG integrated triboelectric bearing at different speeds. **d** Charging characteristics of TVNG integrated triboelectric bearing for different capacitors

innovative design, this system achieves an open-circuit voltage exceeding 130 V and an average power density of 24.6 kW m^{-2} Hz^{-1}, marking a 282-fold improvement over traditional devices (Fig. 11.7b and c). More importantly, this TVNG system maintains an ultra-low wear rate of 5×10^{-7} mm^3 N^{-1} m^{-1} under a maximum contact pressure of 906.6 MPa, which is three orders of magnitude higher in wear resistance than silicon-based TVNGs (Fig. 11.7d and e). This innovation opens new possibilities for efficient friction energy harvesting in industrial settings.

The ball-disc structure of the TVNG ensures high electrical output with a small contact area. Using Hertz contact theory for contact area estimation reveals that the GaN and steel ball interface forms a stable heterojunction that effectively transfers mechanical energy. Under 3N of pressure, this structure demonstrates typical Schottky contact characteristics and consistently generates an open-circuit voltage of 130 V and a short-circuit current of 11.8 μA at a rotation speed of 120 rpm. The short-circuit current density is measured at 4.5 kA m^{-2}, indicating a high energy conversion efficiency. After three rotation cycles, the TVNG produces a short-circuit

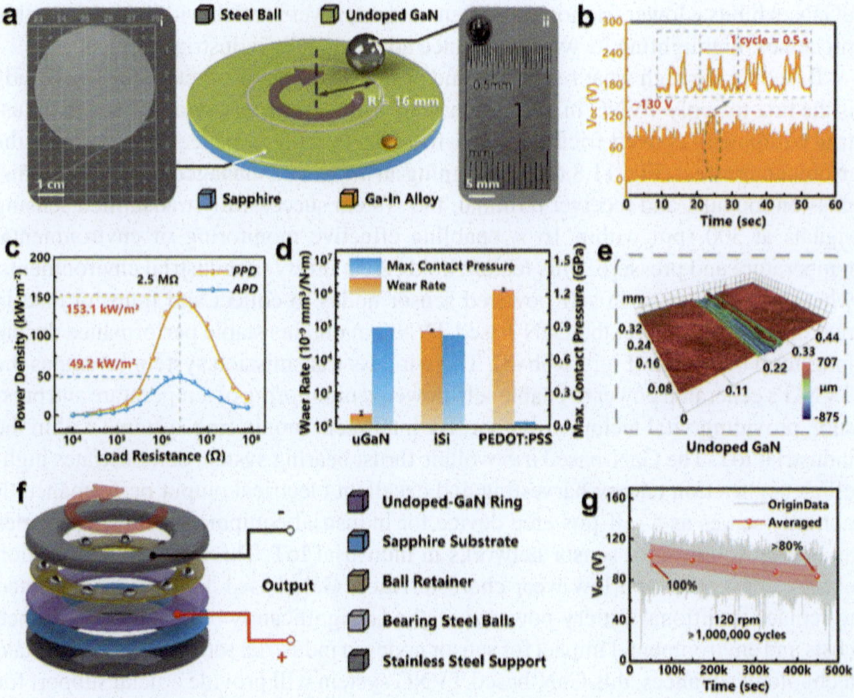

Fig. 11.7 **a** 3D structural schematic and dimensions of TVNG. **b** Open-circuit voltage of TVNG. **c** TVNG peak power density (PPD) and average power density (APD). **d** Wear rate and maximum contact pressure. **e** 3D surface morphology of uGaN after 40,000 working cycles. **f** 3D structure schematic of friction-sliding ball bearing. **g** Open-circuit voltage durability test over one million cycles

charge of 20.6 μC, with a transferred charge density reaching 2.6 kC m^{-2}. Such high electrical output characteristics make TVNG an ideal choice as a high-performance, self-powered energy harvesting device in industrial applications (Fig. 11.7f). Even after one million friction cycles, the TVNG's open-circuit voltage decreases by less than 20%, demonstrating its durability in high-intensity industrial environments (Fig. 11.7g).

In tribovoltaic nanogeneration, material wear resistance is crucial for device stability and longevity. This TVNG uses GaN material, which demonstrates superior wear resistance compared to traditional silicon-based or polymer materials. In experiments, the friction coefficient of uGaN gradually increased from 0.2 to 0.4 and remained stable, while the friction coefficient of Si- and PEDOT-based TVNGs performed poorly. After 20,000 friction cycles, the scratch width on the uGaN surface was only 0.092 mm with a depth of 0.231 μm, yielding a wear rate as low as 5×10^{-7} mm^3 N^{-1} m^{-1}, demonstrating excellent durability. In contrast, the wear rate of silicon-based TVNGs significantly increased, and polymer materials like PEDOT showed even greater wear and lower pressure resistance. These test results indicate that uGaN exhibits a lower friction coefficient and narrower scratch width during friction processes, highlighting its wear resistance advantage for industrial applications.

Based on the high wear resistance and electrical output characteristics, this study is the first to apply TVNG in a thrust bearing. Thrust bearings, widely used in industrial equipment, convert friction energy from the bearing into electricity through the tribovoltaic effect (Fig. 11.8a). By designing an integrated management circuit, transmission module, and receiver terminal, the system successfully transmitted sensing signals at 300 rpm within 16 s, enabling effective monitoring of environmental temperature and pressure. This feature holds great utility in industrial environments, where it can be used in self-powered sensor nodes to collect and transmit data in real time. Furthermore, the GaN-based TVNG maintains stable performance during long-term operation (Fig. 11.8b–d). The wireless transmission system leverages the TVNG's generated power to enable self-powered monitoring of temperature and pressure, providing vital technical support for intelligent monitoring systems within the industrial IoT. The GaN-based tribovoltaic thrust bearing system demonstrates high-efficiency friction energy harvesting and excellent electrical output performance. It not only serves as a self-powered device for industrial equipment but also provides an energy solution for sensor networks in industrial IoT. Through efficient friction energy conversion and low wear characteristics, GaN-based TVNGs are expected to replace traditional battery-powered methods, significantly reducing maintenance costs and environmental impact for sensor nodes in industrial settings. As tribovoltaic technology advances, this GaN-based TVNG system will provide crucial support for the development of smart manufacturing and self-powered sensor networks.

11.3 Application in Multisource Energy Harvesting

The tribovoltaic effect, a recently discovered phenomenon in semiconductors, enables the direct conversion of mechanical energy into electrical energy. However, semiconductor interfaces are highly sensitive to various environmental multiphysics factors, such as illumination, temperature, and humidity, which can significantly influence the energy conversion process. This research explores the application of tribovoltaic effects in energy harvesting under the influence of environmental multiphysics coupling. We provide a comprehensive overview of the electrical output characteristics of tribovoltaic nanogenerators (TVNGs) across diverse physical environments and analyze the mechanisms by which these environmental factors affect their electrical performance. The study highlights the ability of TVNGs to capture and convert various forms of mechanical energy, including wind energy, droplet energy, thermal energy, and solar energy. Furthermore, we delve into the fundamental principles underpinning these devices, their potential applications, and the prospects for future advancements. By adopting multiphysics energy harvesting strategies, we not only enhance the range and efficiency of energy capture but also drive the innovation of novel energy conversion devices. This approach offers promising insights into

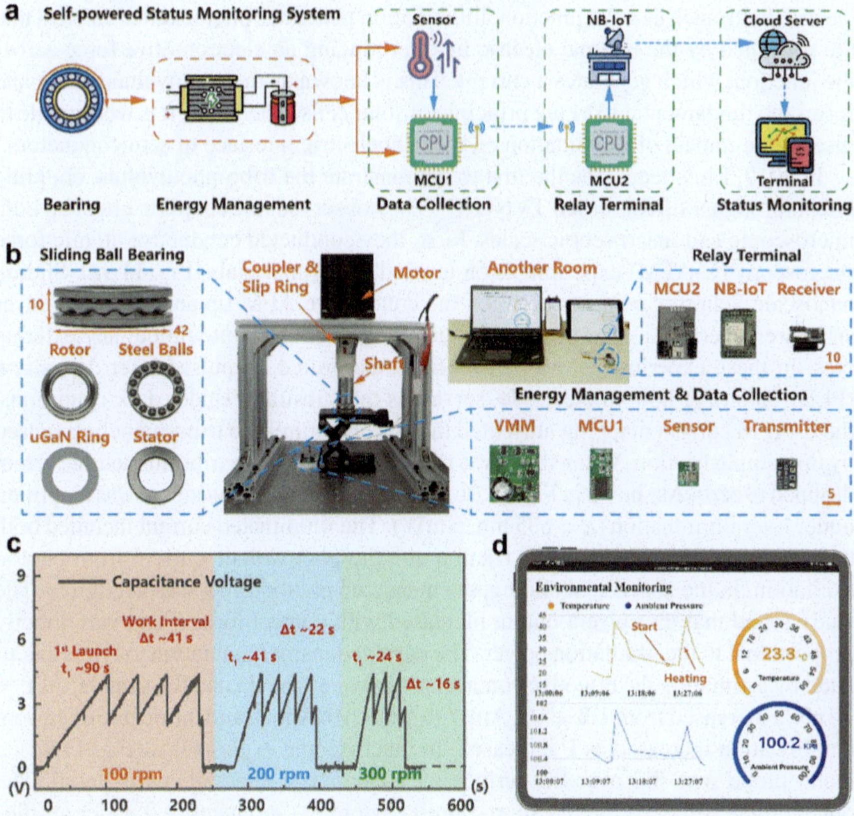

Fig. 11.8 **a** System framework. **b** Composition and experimental setup of a self-powered IoT sensing node for industrial condition monitoring. **c** Working voltage of the energy storage capacitor at different bearing speeds. **d** Display interface for multi-platform terminals

sustainable energy solutions, paving the way for advancements in energy technology and environmental adaptation.

11.3.1 Friction Energy and Light Energy Coupling Energy Device

When semiconductor materials are exposed to light, photons with energy greater than or equal to the semiconductor bandgap can be absorbed by the material, causing electrons to transition from the valence band to the conduction band and creating free electron–hole pairs. This process increases the concentration of charge carriers, thereby enhancing the conductivity of the semiconductor—a phenomenon known as the photoconductive effect. Additionally, in semiconductor structures with a built-in electric field, such as a PN junction, illumination generates photoexcited carriers that are separated by the internal electric field, producing an electromotive force across the junction, which generates a current. This is known as the photovoltaic effect and forms the fundamental working principle of solar cells. Therefore, it is worthwhile to discuss the impact of illumination on the triboelectric interface of semiconductors.

In 2019, Liu's team was the first to demonstrate the tribo-photovoltaic coupling effect through a silicon-based TVNG [5]. They observed this coupling effect on both microscopic and macroscopic scales. First, they conducted conductive atomic force microscopy (C-AFM) experiments on layered MoS_2 materials (Fig. 11.9a). Sliding nanoscale Schottky contacts between the conductive AFM tip and the MoS_2 layer structure generated a significant tribo-tunneling direct current without any external bias. In these experiments, MoS_2 thin films, deposited by pulsed laser deposition (PLD) on an Ag bottom electrode, served as the substrate. Under dark conditions, the C-AFM current mapping attributed the current entirely to tribo-tunneling caused by tip-sample friction. When there was no sliding motion, the tribo-tunneling current dropped to zero. As shown in Fig. 11.9b and c, the current output signal was amplified under laser illumination ($\lambda = 635$ nm, 5 mW). The illuminated current included both the tribo-tunneling current from friction and the photovoltaic current from photon excitation. In the absence of sliding, the measured photocurrent was negligible. The study found that the current output increased with contact force F and was directly proportional to the irradiation power. The current density J in relation to F was calculated by estimating the effective contact area between the tip and the sample. At $F = 30$ nN, J increased from 1.2×10^6 Am^{-2} to $5.6 \times 10^6 Am^{-2}$, and the degree of current amplification increased as F increased. In macroscopic experiments (Fig. 11.9d), a metal probe was slid over the surface of a monocrystalline silicon material. The open-circuit voltage V_{OC} and short-circuit current I_{SC} are measured under an illumination intensity of 900 Wm^{-2}. Due to tribo-tunneling transmission, a maximum V_{OC} of approximately 300 mV and I_{SC} of around 2 μA were generated as steady direct current. When the sliding contact was illuminated, the maximum V_{OC} and I_{SC} increased to approximately 600 mV and 8 μA, respectively. Notably, when light and friction were present simultaneously, the current enhancement ΔI significantly exceeded the photocurrent output without friction (Fig. 11.9e). Illumination significantly boosted both the voltage and current output. Compared with dark conditions, both tribo-voltage and tribo-current under illumination ($\lambda = 635$ nm) increased substantially. The decay time of the tribo-signal pulse peak, which was 0.22 s under illumination compared to 60 ms in the dark (Fig. 11.9f), indicated that the recombination of tribo-induced electron–hole pairs was significantly slowed by the photoelectric field. This coupling between the tribo-electric field and the photoelectric field enhanced the separation and transport of tribo- and photo-excited electron–hole pairs. To demonstrate the feasibility of broader applications, a planar prototype was developed using indium tin oxide (ITO) coated glass as the moving electrode and a silicon wafer as the semiconductor component (Fig. 11.9g). The ITO substrate's good conductivity (8–12 Ω/sq) and transparency allowed for large-area tribo-photovoltaic generation (demonstrated over a 1 × 1 cm area). The generator's V_{OC} and I_{SC} output were recorded as the Si substrate made sliding contact with the ITO glass at a speed of 3 cm/s. Under illumination, V_{OC} and I_{SC} increased from approximately 0.2 V to 0.35 V and from 0.07 μA to 0.25 μA, respectively.

Perovskite materials exhibit remarkable properties such as efficient light absorption, excellent charge carrier transport, low cost, and structural tunability, making them highly suitable for photovoltaic applications. Consequently, tribo-photovoltaic coupling based on perovskite-based tribovoltaic nanogenerators (TVNGs) holds significant research potential. Lin et al. [6] were the first to propose a perovskite-based TVNG and observed the tribo-photovoltaic coupling effect. Their generator was structured around a metal/perovskite Schottky junction (Fig. 11.10a). Perovskite is an excellent light-harvesting material, with a high absorption coefficient of up to 10^5 cm^{-1} and a broad absorption range from UV to visible light. Under AM1.5G sunlight, the static Al/perovskite Schottky junction did not show an output current. However, under AM1.5G sunlight, the output pulse current of a generator based on a dynamic Al/perovskite Schottky junction was − 1.80 μA, compared to − 0.55 μA in the dark, showing a significant three-fold nonlinear photo-enhancement effect (Fig. 11.10b). This "nonlinear" effect occurs because the output current of the dynamic metal/perovskite Schottky junction under illumination is more than the sum of the output current in dark conditions and the output current under static illumination. This nonlinear photo-enhancement effect is attributed to the strong interaction between mechanically induced bound background carriers and photo-generated carriers in the dynamic Schottky junction perovskite film, creating a platform for simultaneous harvesting of light and mechanical energy. This TVNG can collect both light and mechanical energy, demonstrating the nonlinear photo-enhancement effect, later termed the tribo-photovoltaic coupling effect. Under illumination, the photo-generated electrons and holes in the light absorption region are scattered and accelerated by bound background carriers and the holes in the depletion region, respectively, resulting in the nonlinear photo-enhancement effect (Fig. 11.10c). This effect presents a novel physical perspective on light-harvesting devices and offers a feasible way to improve output current, useful for sensors and generators with high power conversion efficiency. Under AM1.5G sunlight, the output voltage of the generator based on a dynamic Al/perovskite Schottky junction did not increase compared to the dark conditions. Compared to traditional semiconductors like silicon and gallium arsenide, generators based on dynamic metal/perovskite Schottky junctions demonstrated higher current enhancement under illumination. Due to the wide absorption range and high absorption coefficient (up to 10 5 cm^{-1} of the perovskite film, it effectively absorbs incident light and can generate more photo-generated carriers at the junction region on the perovskite film surface. These photo-generated electrons and holes in the junction region are immediately scattered and accelerated by bound background carriers in the dynamic Schottky junction, producing an enhanced output current. This dynamic metal/perovskite Schottky junction has potential applications as a photodetector under the nonlinear photo-enhancement effect.

To explore the potential of the generator based on a dynamic metal/perovskite Schottky junction as a portable flexible power source, a flexible generator composed of an Al film/perovskite on ITO/PET was demonstrated. The perovskite film, with

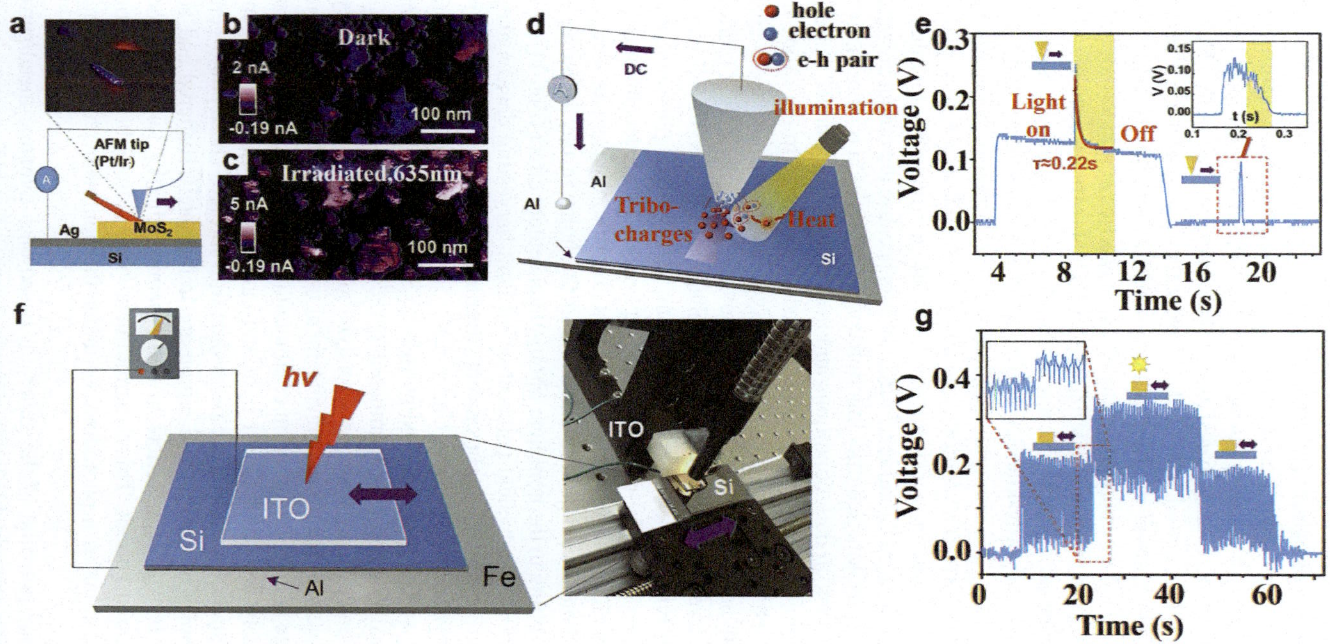

Fig. 11.9 Tribo-photovoltaic coupling effect of the Si-based TVNG. **a–c** Schematic of the C-AFM experiments. **d**, **e** Schematic illustration of the tribo-photovoltaic effect in the macroscopic metal/silicon TVNG. **f**, **g** Schematic of the prototype ITO-coated glass/silicon tribovoltaic generator

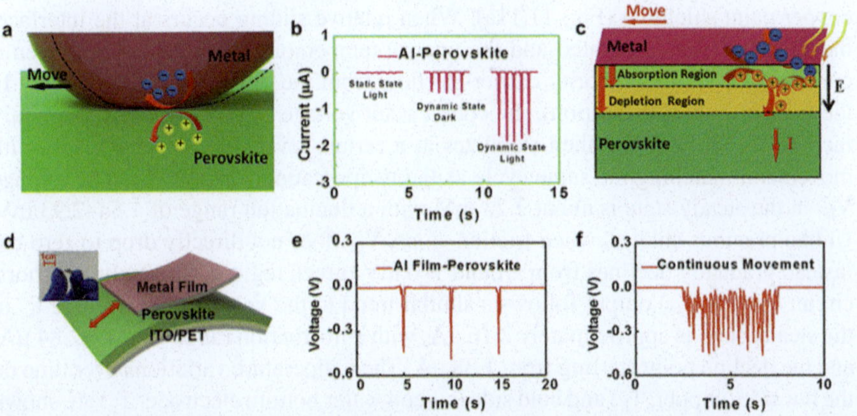

Fig. 11.10 **a** Schematic of the dynamic metal/perovskite TVNG. **b** Electrical output performance of the TVNG under different conditions. **c** Carrier processes in the dynamic metal/perovskite TVNG under illumination. **d** Structure diagram based on perovskite flexible substrate. **e, f** open circuit voltage

high uniformity and flexibility, can grow directly on the ITO/PET film. An Al film was then pressed onto the perovskite film surface on the ITO/PET to form a flexible generator (Fig. 11.10d and inset). When the Al film reciprocated without bending on the perovskite film, an output pulse voltage of − 0.45 V could be measured in dark conditions, and a continuous DC voltage signal was generated when the movement was steady (Fig. 11.10e and f). The generator made from Al/perovskite on ITO/PET retained 72.5% and 50% of its initial output voltage after 20 and 50 cycles, respectively, indicating its stability. When the perovskite film was bent at 120°, the flexible generator retained 70% of the original output voltage, showing that the flexible generator made from Al/perovskite on ITO/PET possesses great flexibility and durability, revealing its potential for flexible applications.

11.3.2 Friction Energy and Thermal Energy Coupling Energy Device

Due to the thermoelectric effect in semiconductor materials, temperature variations in the operating conditions of a TVNG lead to different electrical performance responses. Zhang's [7] team first proposed the coupling effect between the tribo-electric thermoelectric effect and the tribovoltaic effect. To illustrate these concepts, a sliding system of a metal–semiconductor DC TVNG composed of a copper and n-type silicon wafer was used. In addition to measuring the electrical output of the TVNG, a thermocouple pair and an infrared thermal imager were placed on the copper block and bottom electrode for temperature sampling. To eliminate interference from other factors, the entire experimental system was placed in an opaque acrylic box with copper paint shielding (Fig. 11.11a). When relative sliding occurs at the interface, the frictional heat generated and the uneven temperature distribution in the semiconductor cause the majority carriers to flow from the hot end to the cold end. In the friction state (pink region), the open-circuit voltage (V_{oc}) output signal initially increases with time and then stabilizes at a certain level. V_{oc} also fluctuates with movement, matching the same cycle as the reciprocating motion (Δt). The average V_{oc} in the steady state is about 2.23 mV, with a fluctuation range of 1.54–2.93 mV. Unlike previous studies, when friction stops, V_{oc} does not directly drop to zero but instead gradually declines from around 1–0 mV (green region). Similarly, the short-circuit current (I_{sc}) output follows a similar trend to the voltage. The average I_{sc} in the steady state is approximately 2.16 μA, with a fluctuation range of 1.41–2.84 μA, and the decline point starting from 1.38 μA. The temperature variations over time on the hot side (copper, T_h) and cold side (silicon wafer bottom electrode, T_c) are shown in Fig. 11.11b and c. T_h significantly increases with extended friction time, and T_c also slightly rises due to heat conduction from the high-temperature interface. When motion ceases, the temperatures of both points gradually return to room temperature. Infrared thermal imaging and film documented this process (Fig. 11.11d). Once the metal–semiconductor interface begins sliding, both the tribovoltaic effect and thermoelectric effect occur simultaneously. For the tribovoltaic effect, the frictional energy excites electrons and electron–hole pairs in the space charge region at the metal–semiconductor interface to higher energy levels, where they are directionally separated by the built-in electric field, forming a current toward the metal side (I_t). As for the thermoelectric effect, the carrier concentration at the semiconductor surface, heated by friction, exceeds that inside the material. The majority carriers (electrons in this case) diffuse from the high-temperature region to the low-temperature region, generating a thermoelectric current (I_h). This friction process can thus provide continuous electrical output to an external circuit (Fig. 11.11e). The study suggests that the tribovoltaic effect is an interface effect closely related to the built-in electric field formed at the interface, while the thermoelectric effect is a bulk effect tied to the temperature difference within the material. Unlike the instantaneous tribovoltaic effect, the thermoelectric effect accumulates over time. Introducing the thermoelectric effect offers a better understanding of the tribovoltaic effect and the tribo-thermoelectric coupling effect.

Temperature affects the physical properties of semiconductor materials, and changes in the operating temperature of a Triboelectric Nanogenerator (TVNG) can alter its triboelectric signals. Lin and colleagues conducted a study using a TVNG composed of copper (Cu) and silicon (Si) (Fig. 11.12a). In low-temperature environments (77 K), the TVNG can still operate smoothly. Compared to room temperature, the voltage and current outputs can increase to 1.21 V and 11.38 μA, respectively, while at room temperature, the values are 0.76 V and 4.86 μA, demonstrating a significant enhancement in electrical performance (Fig. 11.12b and c) [8]. This improvement is attributed to the reduced lattice scattering and ionized impurity scattering of the semiconductor material at lower temperatures, which leads to increased carrier mobility. Additionally, the mean free path of the carriers significantly increases, resulting in improved efficiency of the carriers collected by the

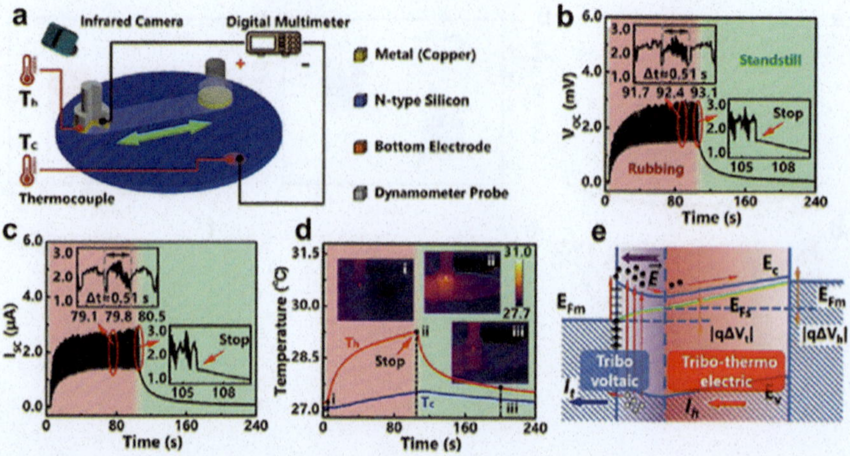

Fig. 11.11 Coupled effects of triboelectric thermoelectric and tribovoltaic effects. **a** 3D schematic of the TVNG experimental setup and structure. **b** Open-circuit voltage output characteristics. **c** Short-circuit current output characteristics. **d** Temperature difference between the hot and cold ends. **e** Energy band diagram showing the effects of triboelectric thermoelectric and tribovoltaic effects on the interface

electrodes, thus enhancing electrical performance. Subsequently, Lin et al. investigated the effects of high-temperature conditions on the electrical performance of the TVNG [9]. Experimental results indicated that an increase in temperature (0–90 °C) leads to an increase in the triboelectric voltage and current at the water/Si and water/metal interfaces during sliding (Fig. 11.12d and e). A new energy band model was proposed to explain the power generation mechanism behind the tribovoltaic effect, which arises from the energy released during the formation of chemical bonds in the contact process between liquids and solids (referred to as "bindington"). As the temperature increases, more "bindingtons" are generated, which excite more electron–hole pairs, thereby increasing carrier concentration (Fig. 11.12f). Consequently, the triboelectric signal is enhanced. Moreover, Zhang et al. [10] reported that by increasing the operating temperature of the TVNG from room temperature to 373 K, the peak power density of the Cu and p-Si significantly improved by several orders of magnitude. The rise in temperature facilitates an increase in the formation rate of Si–O–Cu bonds, resulting in the excitation of more electron–hole pairs at the interface. However, the depletion region of the semiconductor narrows, and an opposing thermoelectric potential due to the thermoelectric effect can lead to a decrease in triboelectric voltage. From the studies mentioned above, it is evident that the TVNG exhibits strong adaptability to both high and low-temperature environments. Furthermore, temperature changes can couple with the tribovoltaic effect to enhance electrical signals, which is advantageous for future applications. Such compact TVNG devices hold great promise for various applications.

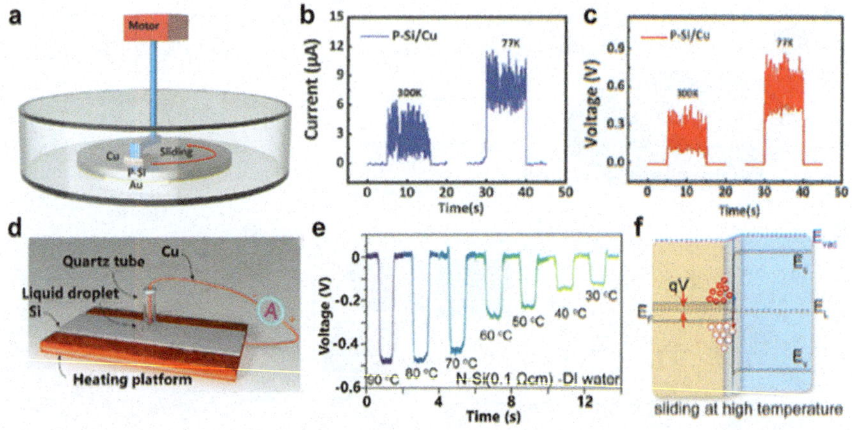

Fig. 11.12 Output characteristics of TVNG at different temperatures. **a–c** Output characteristics of Cu and p-Si at low temperatures. **d–f** Output characteristics of DI water and p-Si at high temperatures

11.3.3 Friction Energy and Droplet Energy Coupling Energy Device

Water is the most abundant natural substance on Earth, making the exploration of energy collection from water crucial for the future development and application of triboelectric nanogenerators (TVNGs). As we know, the electrical performance of traditional triboelectric nanogenerators (TENGs) tends to decrease in liquid environments. However, in the field of TVNG research, TVNGs can not only harvest friction energy from droplets but also operate effortlessly in liquid environments, with enhanced output. This unique power generation capability broadens the application prospects of TVNGs, thanks to the distinctive generating mechanism of the triboelectric effect.

Firstly, water can serve as a triboelectric layer material to generate electricity. Lin et al. were the first to experiment with water as a triboelectric layer in a TVNG. The triboelectric voltage was generated by sliding a water droplet across the surface of silicon. In their setup, a silicon wafer was mounted on a moving platform, while a syringe filled with deionized water was held above the wafer. It is worth noting that the conductive syringe needle did not contact the silicon wafer and was kept at an approximate 0.25 mm distance. Deionized water droplets could be dispensed from the syringe to contact the silicon surface. As the wafer moved with the platform, the needle dragged a DI water droplet along the silicon surface, as shown in Fig. 11.13a and b. This device offers a new approach to liquid–solid TVNGs. Charge transfer occurs due to the triboelectric effect between water molecules and the semiconductor interface, while electron–hole pairs are excited at the silicon surface through bonding interactions at the interface, thus generating an electrical signal under the built-in electric field. Additionally, the electrical performance of TVNGs in high-humidity environments is significantly enhanced, unlike the output characteristics of traditional TENGs. A humidity-enhanced TVNG was proposed by Zhang et al. [11] (Fig. 11.13c). Under high humidity conditions (RH 90%), the charge transfer and peak power density reached 12.6 mC m^{-2} and 1.6 mW m^{-2}, which are nearly 88 and 100 times, respectively, those observed under low humidity (RH 30%) (Fig. 11.13d). XPS characterization suggests that high humidity strengthens the surface states of silicon, which facilitates the directional separation of more charge carriers under the interface electric field, thereby generating DC power. Additionally, water molecules fill gaps in the friction interface, significantly reducing interface contact resistance. This is a coupled effect of enhanced silicon surface states, friction-induced carrier excitation, and reduced contact resistance in high-humidity conditions.

In recent years, researchers have been exploring power generation characteristics by replacing the friction layer material at solid–liquid interfaces with various organic and inorganic solvents, such as acids and bases. Lin and colleagues used an alkaline solution/Si as the tribo-pair material for TVNG, finding that electrical performance increased with stronger alkalinity [12]. The energy produced during the redox reaction between OH^- and the Si interface was attributed to "bindingtons." A higher OH^- concentration or temperature accelerated the generation rate of these "bindingtons," releasing significant energy to excite more charge carriers. Later, Lin et al. [13] introduced different solutions between graphene-silicon friction layers to explore power generation performance (Fig. 11.13e). The solution could

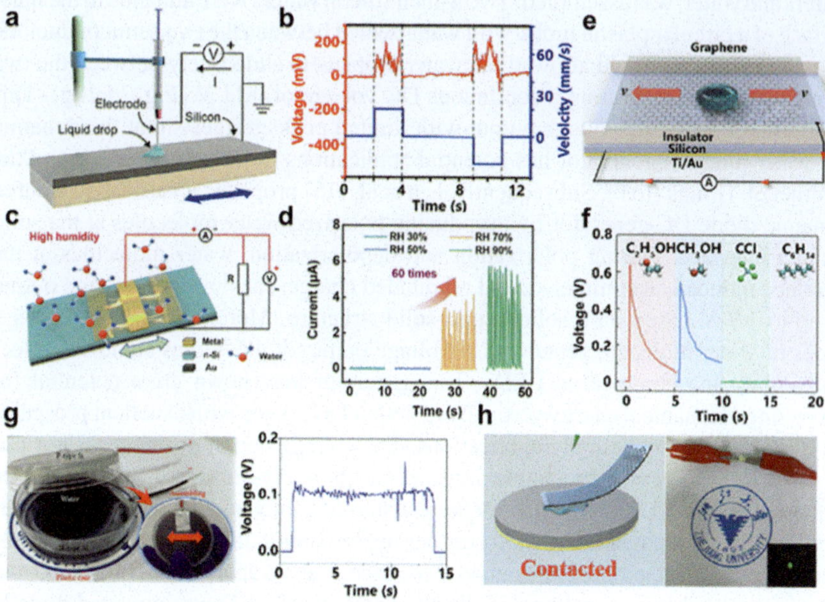

Fig. 11.13 Energy harvesting of TVNG from liquid droplets. a, b Tribovoltaic experiments at the DI water–silicon interface. c, d Humidity-enhanced silicon-based TVNG. e, f Experiments with different liquids at the semiconductor interface. g, h Demonstrations of applications for water/semiconductor TVNG devices

move freely between the two solid interfaces. Adding polar non-electrolyte liquids, such as ethanol (C_2H_5OH) and methanol (CH_3OH), generated DC voltages of 0.63 and 0.54 V, respectively, while non-polar non-electrolyte liquids like carbon tetrachloride (CCl_4) and hexane (C_6H_{14}) produced no voltage (Fig. 11.13f). This was because, in the interfacial electric field, polar molecules undergo a depolarization process, leading to charge carrier movement and generating electrical output.

Lin et al. [14] further experimented by sliding water droplets between P-type and N-type silicon with different Fermi levels, generating a continuous current output. When water molecules contacted silicon, interlayer water molecules polarized instantly, and free charge carriers in the silicon accumulated at the water-semiconductor interface, reaching electrostatic and polarization equilibrium driven by the Fermi level difference between the two semiconductors. As the water droplet moved, this equilibrium was repeatedly broken and re-established, with the induced polarized electrons and holes being released and bouncing back onto the P-type and N-type semiconductors. This created a continuous voltage or current in an external circuit via the dynamic polarized water-semiconductor PN junction. Based on the dynamic water-semiconductor junction, this DC generator, which uses polarized water as a mobile dielectric medium, achieved an open-circuit voltage of up to 0.3 V and a short-circuit current of 0.64 μA. Here, a simple, portable, and sustainable DC power-generating device is further demonstrated. As shown in Fig. 11.13g, this dynamic silicon–water–silicon generator, consisting of P-type and N-type silicon wafers and water, was assembled. Two 2-inch silicon wafers were attached to the inner surface of a circular plastic mold, with water sealed between the two semiconductors. Shaking the plastic mold allowed the water droplets to slide freely between the two semiconductors, generating a continuous DC voltage of 0.11 V (Fig. 11.13g). This continuous and stable voltage output with limited peaks suggests that the dynamic PN water junction generator has potential applications in wearable devices and the Internet of Things (IoT). Subsequently, Lin et al. [15] proposed a vertical-structured dynamic diode DC generator by introducing polarized water molecules at the solid contact interface. During polarization and depolarization, water molecules at the interface induced, accumulated, and rebounded charges, achieving a voltage output of up to 1.7 V based on a solid–liquid-solid structure. Moreover, by integrating a series of water molecule generators, a voltage output of 4.5 V was achieved, effectively powering commercial LEDs. This generator has shown great potential for harvesting mechanical energy vertically to output DC power, with excellent prospects for distributed energy supply applications. The voltage output of the generator can be effectively increased by simple series connection. As illustrated in Fig. 11.13h, voltage outputs of 1.6, 3.1, and 4.5 V were achieved by connecting 1, 2, and 3 generator units in series, respectively. To further explore potential practical applications, three series-connected generator units were used to light up a commercial LED. As shown in the schematic in Fig. 11.13h, the commercial LED was powered directly in series without any rectification circuit or energy storage unit, indicating the feasibility of using it as an integrated power-generating chip or as a direct power source for commercial electronic components.

11.3.4 Friction Energy and Wind Energy Coupling Energy Device

With the rapid development of the Internet of Things (IoT), the use of distributed sensors has become increasingly frequent, leading to a growing demand for small, portable power sources. However, current portable power supplies, such as lithium batteries, have limited capacity and require frequent replacement or recharging. A portable power source capable of generating electricity continuously and in situ would be an ideal solution. In response to this need, researchers have demonstrated a wind-driven semiconductor generator based on a dynamic Schottky junction [16]. This generator is composed of a wind-driven turbine blade, a semiconductor rotor, and a metal stator structure, all supported by a metal framework to form a complete device, as illustrated in Fig. 11.14a. The tight contact between the metal and semiconductor components allows for efficient energy generation. Using wind energy instead of previous manual sliding operations ensures that the pressure between the copper and silicon rod is better suited to produce higher direct current at the milliampere level. Thus, a wind turbine generator based on a dynamic Schottky junction Cu/p-Si structure was realized. Remarkably, this generator produces milliampere-level direct current output that exceeds the lower limit of the simulated current value of 4 mA, making it suitable for long-distance current transmission of analog signals. To explore its practical usability, a Cu/p-Si wind-driven semiconductor generator was fabricated. Continuous output voltage and current lasting over 740 s are shown in Fig. 11.14b. With a rotational speed of 11 rpm s^{-1} and a contact area of 0.45 cm^2, the output voltage stabilized at an average value of 40.1 mV, and the output current at 4.4 mA. This stable direct current voltage and current output clearly demonstrate that our generator can consistently and effectively convert low-frequency wind energy into electrical energy. Additionally, pre- and post-experiment static I–V curves show that there is minimal formation of a natural oxide layer on the Cu and silicon rod structure, and the contact surface remains undamaged, indicating potential for practical application. The generator can also serve as a rotation counter to track the number of revolutions, as shown in Fig. 11.14c. For counting purposes, a polyimide insulating layer was added to the silicon surface to separate the metal plate from the silicon rod surface. Distinct contrasts between current peaks and valleys allow accurate counting, as the number of current peaks corresponds well with the generator's revolutions, demonstrating that our generator achieved reliable rotation counting. Furthermore, the generator successfully powered a graphene/GaN ultraviolet photodetector, as illustrated in Fig. 11.14d. Details on the preparation of the graphene/GaN heterostructure device are provided in the annotations. When exposed to 365 nm UV light, the photodetector—powered directly by the generator—achieved a responsivity of 35.8 A W^{-1} and operated stably, indicating that our generator can indeed serve as a power source for sensors.

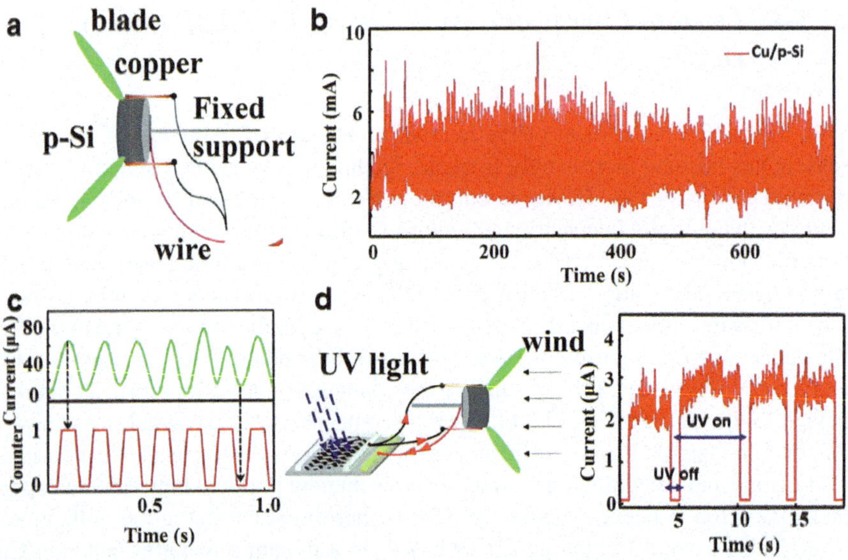

Fig. 11.14 Wind energy harvesting by TVNG. **a** Schematic diagram of the TVNG structure. **b** Durability test. **c** Revolution counter, calculating the rotations of the wind turbine. **d** Driving a graphene/GaN ultraviolet photodetector and photoelectric response under 365 nm UV light

11.4 Applications in Smart Flexible Devices

In the previous two sections, we primarily discussed the applications of triboelectric nanogenerators (TVNG) in inorganic semiconductors. However, the inherent mechanical brittleness of inorganic semiconductor materials limits their development in the field of flexible wearable electronics. In contrast, organic semiconductors, with their superior flexibility, stretchability, and ease of fabrication, enable TVNGs to more effectively harvest mechanical energy from flexible interfaces. This subsection will explore the sensing and power supply applications of triboelectric generators in smart flexible electronic devices, showcasing their immense potential in the wearable electronics domain.

11.4.1 Applications of Flexible TVNG for Sensing

11.4.1.1 Respiratory Rate Monitoring

To demonstrate potential applications of flexible TVNG, Chen et al. proposed a moisture-flexible TVNG with an asymmetric graphene oxide (aGO) and an Al foil interface, which works by a new mechanism and outperforms existing flexible semiconductor-based DC nanogenerators by several orders of magnitude in both current density (81.06 A/m^2) and power density (24.08 W/m^2) and the mechanic-to-electricity conversion efficiency can achieve to 2.29%.[17] To identify the effect of applied pressure, we changed the pressures and set the sliding speed (60 mm/s) and humidity (70%) to be constant. The open-circuit voltage (V_{OC}) is almost stable at 1.35 V at different pressures. While the short-circuit current (I_{SC}) increases linearly from 2.95 μA to 81.06 μA as the pressure increases from 0.1 to 5 N. The real contact area is much smaller than the contact area on the macro-level due to the roughness of the friction surface on the micro-level. Therefore, when increasing the pressure, the real contact area between the aGO and Al wire is enlarged due to the extrusion of the two friction interfaces, causing more electrons/holes to be excited during sliding and increasing I_{SC}. For V_{OC}, the increased real contact area also results in a smaller voltage because of the characteristics of plate capacitors. The linear electric current output can be used as sensing signal.

Integrated flexible TVNG into clothing can fabricate wearable electrical power supplies and self-powered sensors for monitoring the respiratory systems of humans. Five TVNGs (in film form) are connected in series and integrated into clothing to convert human mechanical energy into electrical energy. The mechanical energy from armpit friction can be converted into electricity enough to directly light a commercial LED bulb in the absence of extra rectifying circuits or capacitors. In addition, non-invasive and self-powering respiration monitoring devices are highly practical in future wearable health-monitoring electronics. Considering sweating would increase the body surface humidity, the M-TVNG based sensors would give more comprehensive physical state information of humans by simultaneously detecting the body surface sweat levels and respiratory system. A fibrous M-TVNG is prepared by braiding a naked Al wire and an aGO-coated Al wire together. After being integrated into gauze bound on the chest, the M-TSNG monitors the respiratory system of a subject under "Quietness", "30min running", and "after 15min relaxation" conditions. The chest rises and falls during the respiratory cycles, causing the naked Al wire and aGOC-coated Al wire to slide relative to each other, generating electric signals. Three key parameters need to be focused on: (i) V_0 is the starting voltage, related to the relative humidity of the body surface. (ii) ΔV represents the change in voltage during breathing, related to the respiratory rate (the rate of one breath). (iii) Δt is the interval time between each breath, related to breathing frequency (the breaths per minute). When the subject breathes in the calm state, three parameters were extracted: $V_{01} = 0.030$ V, $\Delta V_1 = 0.091$ V, and $\Delta t_1 = 4.1$s (Fig. 11.5f). The breathing frequency in the case of "Quietness" was calculated to be around 14 breaths min 1. The three parameters in the "30min running" state are $V_{02} = 0.120$ V, $\Delta V_2 = 0.238$ V, and $\Delta t_2 = 2.0$ s. Compared with "Quietness", the starting voltage increased by 4 times from 0.030 V to 0.120 V. Sweating during running and the resulting surface humidity increase of skin is evidently reflected by the starting voltage V_0 increase. The increase of ΔV from 0.091 V to 0.238 V and the calculated breathing frequency to be around 30 breaths min 1, which means the subject breathes fast and deep after running. The small plateau within each breath cycle may result from that the lungs tend to wait a moment after inhaling before exhaling to allow the oxygen to be fully absorbed during hypoxia. The obtained signals are not very regular because of the unsteady state of the body condition after running. After 15min relaxation, $V_{03} = 0.069$ V, $\Delta V_3 = 0.096$ V, and $\Delta t_3 = 4.2$ s were measured. As expected, the starting voltage decreases after relaxation because most sweat evaporates over time. The change in voltage and the interval time between each breath decreased to normal levels after 15 min relaxation. These signals indicate that the subject may need to take a 15-min break after 30-min running to recuperate the respiratory system to the original state. Of course, for different individuals, running and relaxation time can be adjusted according to their personal measured data (Fig. 11.15a).

11.4.1.2 Running Speed Sensor Monitoring

In 2023, Fan et al. proposed a novel flexible textile TVNGwith high direct current output, designed based on a 2D dynamic heterojunction of 2H-MoS_2/Ta_4C_3. During the friction process, a built-in electric field and an additional interfacial electric field are generated through continuous dynamic contact [18]. The open circuit voltage (V_{oc}) of a single DC generator is approximately 0.3 V, while the short circuit current (I_{sc}) is about 8.5 μA. By connecting ten TVNGs in series, a DC output voltage of 3.3 V can be achieved, and a parallel connection yields a DC current of 75 μA. Unlike traditional battery power, self-powered sensors enable the use of TVNGs in wireless wearable electronics. By wrapping the flexible TVNG around the legs, the system effectively responds to the frequency of human movement, allowing for the development of a wireless sensing system to monitor running speed. Testing the relationship between various running speeds and TVNG output voltage signals demonstrated a strong response across different speeds. As industrial production increasingly integrates information technology and automation, safety issues in factory workshops have become pressing challenges, particularly in environments with high personnel density and numerous machines. Therefore, developing a simple, efficient, and self-powered warning system is crucial for maintaining human health and ensuring orderly production. A self-powered wireless warning system is designed based on TVNGs, which includes a power supply system, circuit management system, wireless transmission system, and data processing program. Three groups of ten TVNGs are connected in series and fixed them on the shoulders, abdomen, and legs of the human body, subsequently testing the voltage signals that power the entire warning system at different speeds. Additionally, a warning program is designed based on the relationship between wireless signal strength and distance, tested within a $10 \times 10\ m^2$ space. By measuring signal strength based on proximity and converting these readings into digital signals using algorithms, the mobile application provides danger feedback when an individual enters the designated danger zone. The developed warning system demonstrates excellent applicability and shows significant potential for future applications in big data collection, human–computer interaction, and access control systems (Fig. 11.15b).

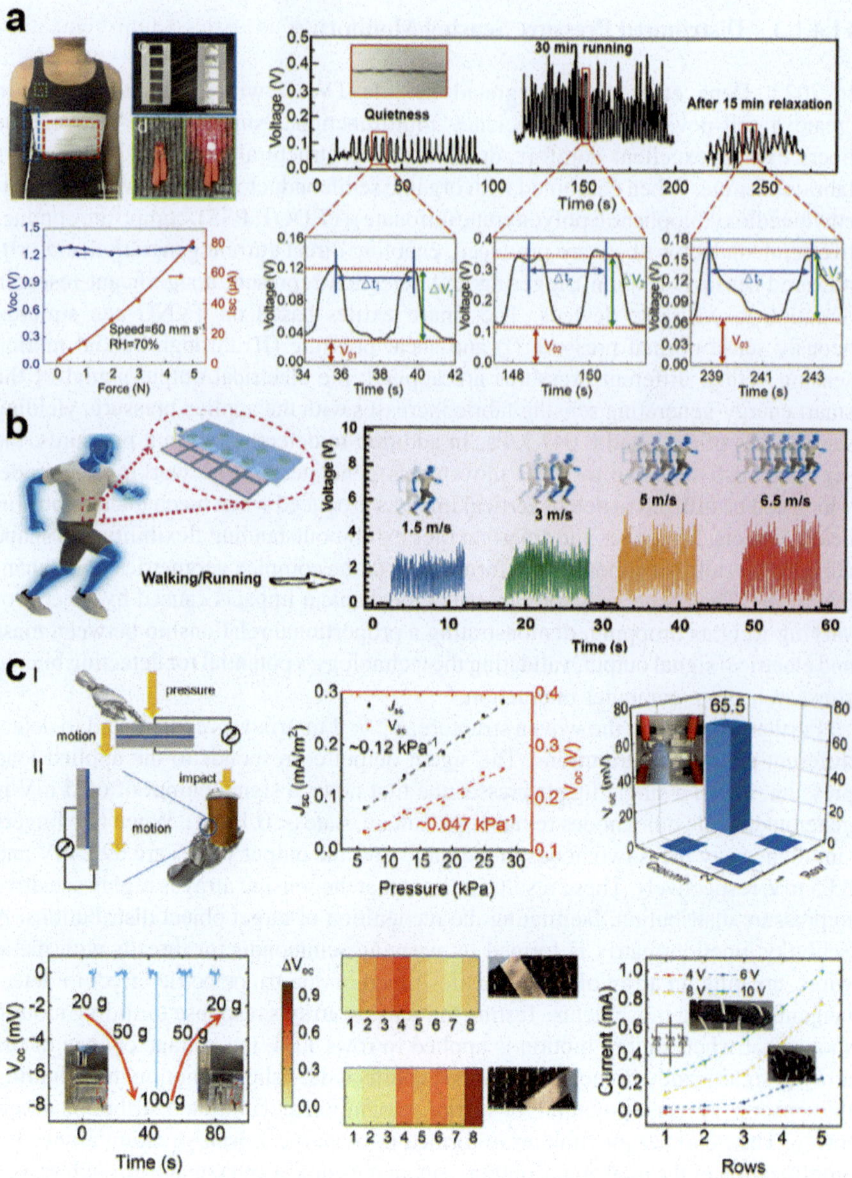

Fig. 11.15 **a** The serial moist TVNG based on Al wire and rGO wires is integrated into clothing. Photograph of circuit design of five M-TVNG in series. The signals for respiratory monitors during "Quietness", "30 min running", and "after 15min relaxation" based on a fibrous M-TVNGs. **b** Image illustration of human walking and running with a self-powered TVNG based on 2H-MoS_2/Ta_4C_3 heterojunction. V_{oc} of at different human motion speeds. **c** The application of tribovoltaic textile based on PEDOT: PSS to sense normal pressure (I) and shear motion (II). Demonstration of the motion sensing array of different distributions and effective in-parallel circuit of woven structure and its output as a function of the number of rows

11.4.1.3 Distributed Pressure Sensing Monitoring

In 2023, Deng et al. [19] integrated flexible TVNG with cellulose textiles to create a self-powered, high-efficiency multifunctional smart textile sensor. The fibers exhibit excellent comfort, durability, and structural integrity, serving as a fabric substrate. When combined with organic semiconductors coated with poly(3,4-ethylenedioxythiophene): poly(styrenesulfonate) (PEDOT: PSS), composite conductive semiconductor fibers are produced, enabling direct current generation and self-powered sensing. Motion triggering and detection represent a significant research direction in wearable devices. The smart textiles based on TVNG can simultaneously sense normal pressure (I) and shear pressure (II) through biaxial motion sensing. When different pressures are applied, the electrical output signal of the smart energy-generating sensing fabric increases with the applied pressure, yielding sensitivities of 0.12 and 0.047 k/Pa. In addition to detecting normal pressures, the sensor is also sensitive to shear movements generated by the working electrode, which can be utilized to detect vertical impacts produced when mechanical limbs lift heavy objects. Moreover, the smart textiles exhibit outstanding flexibility and shape adaptability, allowing them to conform easily to the complex geometries of mechanical limbs. This impact sensor can detect mechanical impacts caused by objects of varying weights dropping, demonstrating a proportional relationship between mass and electrical signal output, validating the technology's potential for detecting biaxial stress in human–computer interaction.

Further inspired by the woven structure, a 2×2 matrix was constructed to detect different pressure distributions. The signal output corresponds to the applied load pressure. When a single finger presses the first node, a signal output of 65.5 mV is generated, while other nodes remain in a "silent" state (< 0.1 mV). When two fingers simultaneously press two nodes in the same row, the output values are 69.2 mV and 53.5 mV, respectively. These results indicate that the sensing array is highly sensitive to pressure distribution, facilitating the recognition of target object distributions. A Schottky junction matrix is formed by weaving semiconductor threads with metal wires, creating an array of smart textiles based on the triboelectric effect to detect tangential sliding movements. Testing the sensing array's response to sliding motion shows that when sliding motion is applied to rows 11.4, the voltage change of the array is significantly higher than in areas unaffected by sliding motion, highlighting the sensor's immense potential for spatial recognition and motion distribution monitoring. The Schottky junction array formed through the woven structure allows for amplification of the total signal output, enhancing current and voltage through woven arrangements. In summary, smart sensing textiles based on the triboelectric effect can achieve self-powered sensing networks without any external circuitry, capable of responding to both normal and tangential biaxial movements, showcasing their extensive application potential in human–computer interaction (Fig. 11.15c).

11.4.2 Applications for Powering Micro-electric Device

High-performance flexible direct current triboelectric nanogenerators (TVNG) provide sustainable energy solutions for various wearable electronic devices due to their excellent mechanical flexibility and high current output characteristics. Multiple studies have shown that flexible TVNGs can significantly enhance voltage and current output under different connection configurations (series or parallel), making them sufficient to power electronic devices.

In 2021, Meng et al. [20] introduced a flexible textile-based triboelectric generator that uses Al and PEDOT: PSS fabrics to construct asymmetric Schottky junctions for direct current generation. Using porous fabric as the substrate enhances the adhesion between the PEDOT layer and the substrate. The coated conductive fabric can be bent and twisted around the fingers. The energy released during the formation of an atom–atom bond at the dynamic Schottky contact between an Al slider and the PEDOT coating can excite non-equilibrium electron–hole pairs, which are subsequently separated by the built-in electric field. This accumulation of potential energy at the interface generates voltage and electric work for the external circuit. This frictional generator can produce approximately 0.45–0.70 V in voltage and around 2.11.170 μA in current when operating as a single unit. Compared to previously reported AC TENGs, the flexible TVNG's high current output and low internal resistance make it easier to integrate with self-powered electronic devices. Testing revealed that even in a bent state, the flexible TVNG maintains good output performance, showcasing remarkable flexibility and durability. This innovative triboelectric textile material demonstrates the practicality of TVNGs for wearable applications. In practice, the power-generating fabric can be integrated into clothing, where friction between an aluminum foil-covered index finger and the PEDOT coating yields an open-circuit voltage (V_{OC}) of about 0.5 V and a short-circuit current (I_{SC}) of approximately 12 μA. A significant feature of this triboelectric textile material is its ability to be connected in series or parallel to enhance voltage and current output. For instance, when seven triboelectric generators are connected in series, the V_{OC} increases to about 3.5 V, while parallel connections yield an I_{SC} of around 70 μA. The electric output of the series-connected TVNG is sufficient to charge batteries and power microelectronic devices. For example, charging a 4.7 μF capacitor to 2.3 V meets the power requirements of an electronic watch. After connecting to the watch, the frictional movement of the textile generator continues to operate the device for approximately one minute. During this time, the voltage across the capacitor stabilizes between 1.2 and 1.5 V, indicating that the energy output from the generator balances the energy consumption of the watch, effectively demonstrating self-powered electronic devices. Compared to traditional TENGs, this system is highly energy-efficient and does not require additional rectification and regulation circuits. This work showcases the first flexible and practically viable TVNG, which has broad prospects in the field of wearable electronics (Fig. 11.16a).

In 2022, You et al. developed an organic flexible triboelectric nanogenerator (TVNG) based on PEDOT and various friction materials, including aluminum alloy,

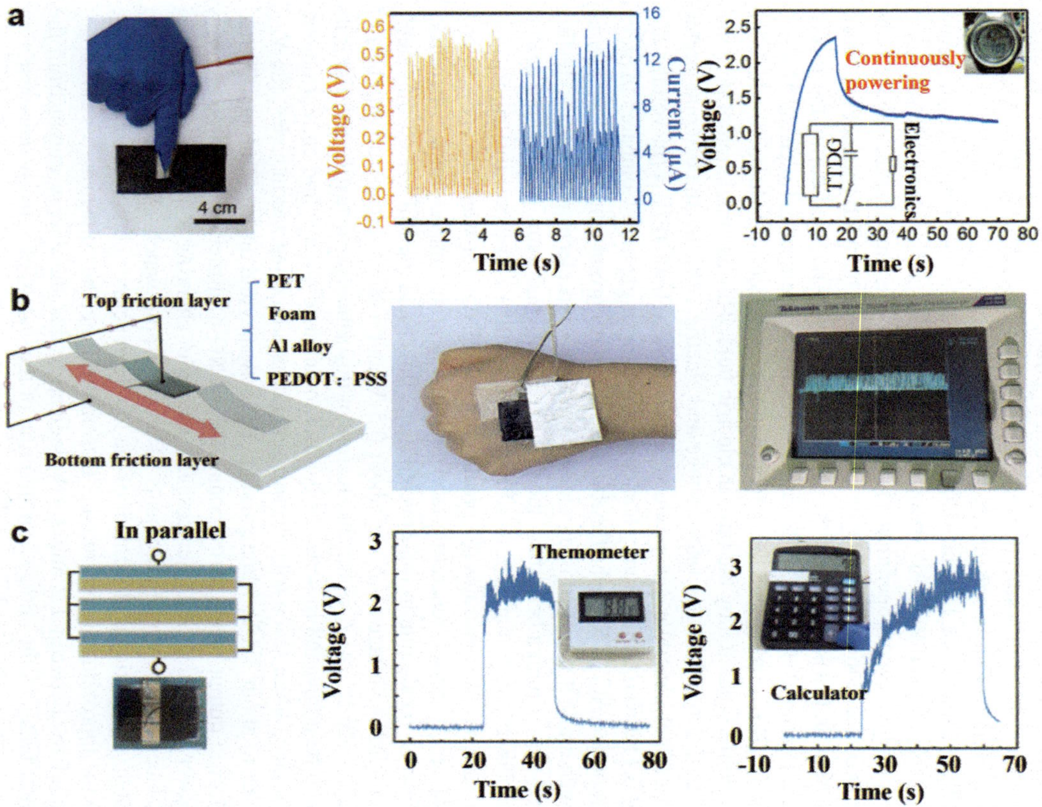

Fig. 11.16 The application of flexible tribovoltaic device for powering. **a** Photo of a textile TVNG integrated on a cloth. V_{OC} and I_{SC} of a single TVNG when sliding an Al-attached finger on the cloth. A digital electronic watch (inset) was powered continuously by the seven TVNGs. **b** The electronic skin based on Al and PEDOT: PSS TVNG and its output voltage output signal. **c** In series stated, the 3 series-connected flexible textile based on organic TVNG drives thermometer and calculator, respectively

silicon, indium tin oxide (ITO), and copper [21]. This innovative design significantly enhanced the durability and flexibility of the TVNG, broadening its applications in wearable electronic devices. The study found that the TVNG based on the PEDOT: PSS and Al friction interface exhibited optimal electrical output performance, with an V_{OC} of about 1 V and a I_{SC} of 309 μA. Additionally, its internal impedance is several orders of magnitude lower than that of inorganic semiconductors, laying the foundation for its high efficiency.

Under sliding speeds of 30 cm/s and loading pressures of 5 N, the output performance of the TVNG based on the PEDOT/aluminum alloy interface was significant. The V_{OC} ranged from 0.11 to 0.63 V as the external resistance varied from 10^3 to 10^9 Ω, while ISC decreased from 24.7 to 2.7 μA. Calculations indicated that maximum power of 5.84 μW (11.67 mW/m^2) was achieved at an external resistance of 2×10^4 Ω, leading to an estimated internal resistance of the TVNG of about 2×10^4 Ω. Further experiments demonstrated that our device could charge a 4.7 μF capacitor to 0.6 V within about 2 s. When the flexible device was bent approximately 100 to 1000 times, noticeable wrinkles appeared, but its performance remained unaffected, further confirming its feasibility as a flexible or wearable electronic sensor or power source. The PEDOT: PSS/aluminum alloy DC TENG can serve as an electronic skin device, sensing human motion signals. When a finger reciprocates over the friction layer, it drives the sliding of the top and bottom friction layers, generating a DC electrical signal (Fig. 11.16b).

In 2023, Liu et al. designed a flexible organic DC textile triboelectric nanogenerator (FT-OTG) made from a blend of 11.11% PEDOT and 88.89% polyvinyl alcohol (PVA), hydrophobic conductive textiles, and aluminum sliders. The operating mechanism of the OTG involves the excitation of abundant electron–hole pairs at the metal–semiconductor interface upon absorbing friction energy [22]. These carriers move directionally, driven by both the electrostatic field and the built-in electric field, resulting in the formation of a current. Due to the flexible substrate, the fabricated DC-OTG exhibits good flexibility and strain resistance. The tensile strength and elongation at break of the mixed film increased by eightfold and 3.2-fold, respectively, with significant improvements in output performance: the open-circuit voltage rose from 0.54 to 2.54 V, and the short-circuit current increased from 3.6 to 15.6 μA.

Based on the DC power generation characteristics of the FT-OTG, several low-power electronic devices can be powered simply by series or parallel connections. For instance, three FT-OTGs can be connected in series, yielding an open-circuit voltage of 6.7 V and a short-circuit current of 6.4 μA. Common portable electronic devices like digital watches, thermometers, and calculators can be directly powered by the series FT-OTG source. When the FT-OTG outputs a DC signal, the portable devices start up and function normally almost immediately, showcasing the outstanding application potential of the FT-OTG as a sustainable bioenergy harvester (Fig. 11.16d).

In summary, flexible triboelectric generators (TVNGs) demonstrate vast potential in the fields of sensing and power supply. These research findings not only provide innovative energy solutions for wearable electronic devices but also advance the development of smart textiles and self-powered sensors. In the future, there will be even greater application opportunities in areas such as health monitoring and smart interaction.

11.5 Applications in Self-powered Sensing and Monitoring

11.5.1 TVNG for Motion Sensing

Sensing and monitoring of mechanical motion are widely applied in research and engineering fields such as manufacturing, automation, and robotics. Most existing motion sensors require additional sensors or external power sources. As the number and density of IoT sensing nodes continue to increase, there is an urgent demand for active sensing technology. The tribovoltaic effect enables the conversion of mechanical energy into electrical energy at frictional interfaces, producing electrical signals without requiring additional power supplies. Based on this principle, TVNG can monitor the mechanical motion of objects in situ. In a mechanical system, motion parameters such as displacement, velocity, acceleration, and angles are crucial for characterizing motion. TVNGs with different structural designs can harvest energy from various types of mechanical motion, such as linear and rotational motion. Additionally, since the output signal of TVNGs correlates directly with the state of the frictional interface and the mechanical motion parameters, TVNGs can be used as motion sensors. Luo et al. demonstrated a motion-sensing technique based on the tribovoltaic effect that can monitor mechanical motion parameters such as speed, displacement, tension, and rotation angle [23].

As shown in Fig. 11.17a, this motion sensor is composed of a TVNG based on a Mxene-silicon heterojunction. Ti_3C_2Tx Mxene powder is pressed onto a copper electrode to form a thin film. After cleaning the oxide layer of crystalline silicon with hydrofluoric acid, approximately 50 nm of gold electrode is deposited on the surface using magnetron sputtering. The top and bottom electrodes are then connected. The Mxene film makes effective contact with the silicon surface under the drive of a linear motor. When sliding friction occurs between the Mxene film and silicon, a DC output is generated in the external circuit based on the tribovoltaic effect. The metallic properties of Mxene provide excellent wear resistance and stable output performance, maintaining output stability even after over 210 min of continuous operation, demonstrating the potential of TVNG in sensing applications. The tribovoltaic effect's in situ generation of electrical signals at the friction interface makes it highly sensitive to motion parameters and allows for intuitive monitoring of an object's motion state. When the Mxene slider slides on the silicon surface at varying speeds, the generated frictional energy and the excited semiconductor carriers differ, leading to distinct electrical signals in the external circuit. TVNG produces an output current proportional to the sliding speed, making it suitable for use as a velocity sensor. Moreover, the tribovoltaic effect enables the generation of accurate, continuous, and real-time DC signals, showing great potential in displacement, tension, angular oscillation, and vibration sensing.

Figure 11.17b shows a multifunctional sensor based on Mxene-Si TVNG. The device converts mechanical energy into potential energy using a spring, generating sustained output. When the Mxene slider is pulled to different distances with an external force FFF, it acquires varying elastic potential energy. After releasing FFF, the slider moves due to the combined elastic force of the spring and its own gravity, with the spring's elastic potential energy converting into the slider's kinetic energy. The frictional energy loss between the Mxene slider and silicon eventually causes the slider to stop after several oscillations. The elastic potential energy of the Mxene slider correlates with the initial velocity upon release, and as the sliding distance increases, the output current increases proportionally. Figure 11.17c illustrates the relationship between the Mxene-Si TVNG output current and the slider's stretching distance. With a minimum displacement interval of 0.25 cm and a sensitivity of 49.16 nA cm^{-1}, the device maintains a good linear relationship between displacement and current, verifying its reliability as a displacement sensor. There is also a linear relationship between the force applied to the slider and the spring's elongation, making the output current (ΔI) of the Mxene-Si TVNG proportional to the spring's tension. Figure 11.17d depicts the observed linear relationship between calculated tension and measured current, confirming the feasibility of constructing a force sensor with a sensitivity of 151.68 nA N^{-1}. Furthermore, as shown in Fig. 11.17e, replacing the spring with a fixed rope transforms the Mxene-Si TVNG into a pendulum device. With varying oscillation angles θ, the maximum velocity of the slider changes accordingly, enabling the device to function as an oscillation sensor. Figure 11.17f demonstrates the changes in output current at different oscillation angles, revealing distinct linear relationships under high (> 25°) and low (< 25°) oscillation conditions. The sensitivity reaches 2.15 nA deg^{-1} for high angles and 0.65 nA deg^{-1} for low angles. Notably, negative signals in oscillation measurements are due to noise and have minimal impact on sensing performance. Hence, the sensor can respond to environmental vibrations as a vibration sensor when mounted on a horizontal support.

The tribovoltaic effect's ability to generate accurate, real-time DC signals also applies to high-precision displacement sensors. Wang et al. developed a micro/nano grid displacement sensor based on a metal-gallium arsenide (GaAs) interface [24]. This sensor comprises grid-patterned copper electrodes on high-resistance glass and GaAs strips coated with copper electrodes on their backs. It achieves micron-level displacement precision.

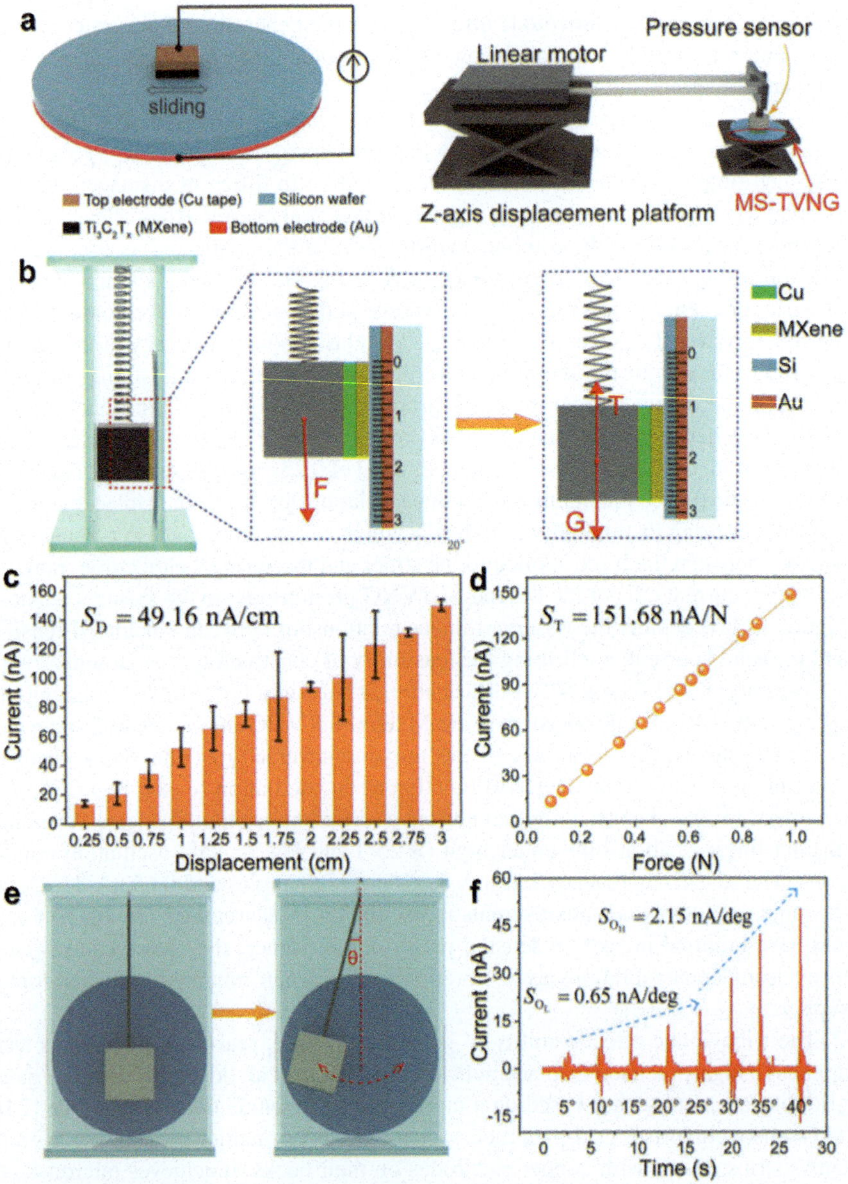

Fig. 11.17 **a** Structural diagram and test setup of Mxene-Si TVNG. **b** Schematic of displacement and force sensing. **c** Output current at different slider stretching distances. **d** Linear fit of output current and spring tension. **e** Schematic of Mxene-Si TVNG oscillation sensor. **f** Output current changes at various oscillation angles

11.5.2 TVNG for Lubrication State Monitoring

Lubrication sensing technology plays a crucial role in industrial and academic research, significantly enhancing mechanical efficiency, reducing energy consumption, and extending equipment life. In recent years, rapid advancements in nanotechnology and materials science have led to breakthroughs in lubrication sensing technology. Superlubricity, characterized by a coefficient of friction (COF) below 0.01, greatly reduces energy loss and improves mechanical efficiency. However, traditional methods of detecting frictional states, such as pressure, temperature, acoustic, and optical sensing, often face limitations in real-time and on-site detection. Therefore, developing a sensing technology capable of real-time and on-site detection of superlubricity is essential. The in-situ generation of electrical signals by tribovoltaic effects at the interface enables intuitive monitoring of the lubrication state.

In Zhang et al.'s work [25], they developed a lubrication sensing technology based on macroscopic superlubricity using a superlubricity tribovoltaic nanogenerator (SL-TENG). This technology utilized hydrogenated diamond-like carbon (DLC) films to achieve ultra-low friction coefficients and wear rates under high contact stress. As shown in Fig. 11.18a, the SL-TENG was prepared by depositing a 2.5 μm-thick hydrogenated DLC film onto a stainless-steel plate using an unbalanced magnetron sputtering method with methane as the hydrogen source. Figure 11.18b illustrates the SL-TENG structure. The device utilized a GCr15 steel ball with a diameter of 6 mm as the upper triboelectric pair and electrode. The steel ball was fixed onto a metal holder insulated from the remaining tribometer (TRB) parts. A copper wire was attached to the bottom of the holder, and the DLC-coated stainless steel plate served as the other triboelectric layer. The steel plate was grounded to avoid electromagnetic interference. The working principle of SL-TENG is based on the tribovoltaic effect, where sliding at a metal–semiconductor interface generates direct current due to non-equilibrium carriers created by mechanical friction. During contact and sliding, electrons flow from the steel ball to the DLC film due to the work function difference, achieving equilibrium. In the sliding process, internal electric fields within the space charge region (SCR) of the DLC layer drive electron–hole pairs to the metal side, producing current in the external circuit. Experiments showed that DLC films achieved a friction coefficient much lower than other materials and nearly zero wear under superlubricity conditions, highlighting their potential for long-term, low-wear applications. Figure 11.18c–e depict the short-circuit current, open-circuit voltage, and power density of SL-TENG in the lubricated state, demonstrating stable COF (< 0.01) over 10,000 cycles. Moreover, environmental conditions like gas atmosphere influenced COF, altering superlubricity states and consequently affecting output signals. As shown in Fig. 11.18f, the SL-TENG exhibited different friction coefficients in nitrogen (N_2) and air environments, leading to variations in output current. This enabled the development of a gas atmosphere monitoring system based on lubrication states. Compared to N_2 environments, the higher COF in air produced greater electrical outputs, allowing macroscale phenomena such as lighting an LED in air but not in N_2.

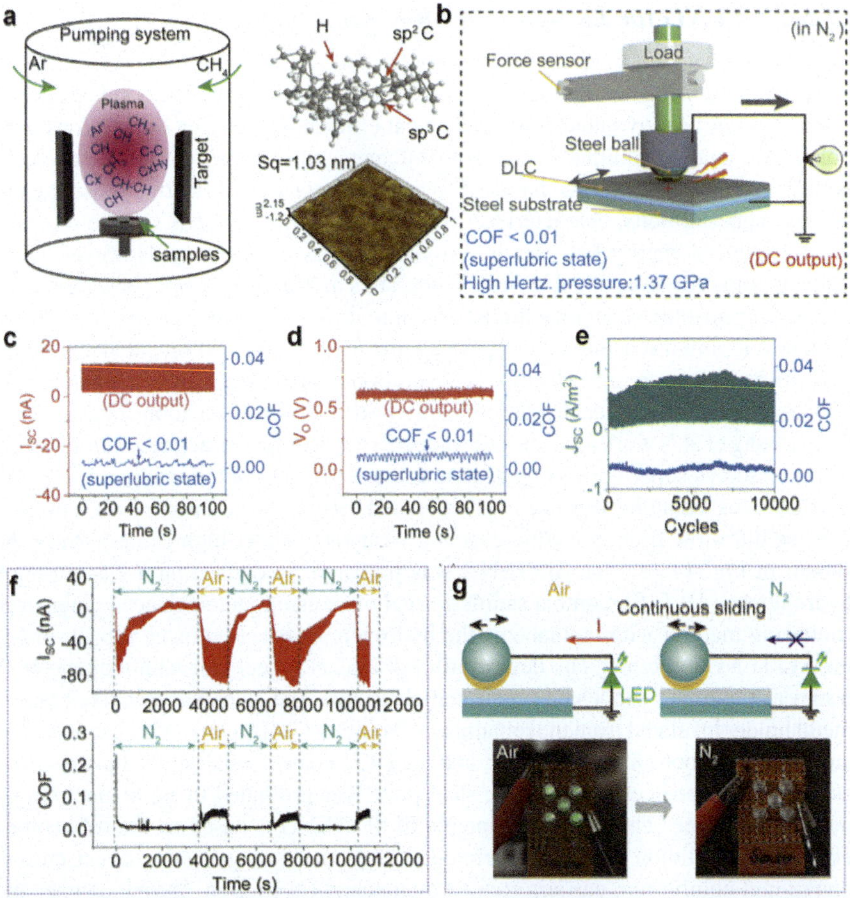

Fig. 11.18 **a** Preparation of hydrogenated DLC films. **b** Structural schematic of SL-TENG. **c** Short-circuit current in superlubricity states. **d** Open-circuit voltage in superlubricity states. **e** Power density and durability in superlubricity states. **f** Short-circuit current and COF in different atmospheric environments. **g** Lubrication state monitoring in various gas environments

The tribovoltaic-effect-based lubrication sensing technology offers an effective means for real-time monitoring of superlubricity. This research not only demonstrates the theoretical potential of the tribovoltaic effect but also its significant practical advantages. With further advancements in materials science and nanotechnology, this technology is expected to find broad applications in industrial IoT. Overall, tribovoltaic-effect-based lubrication sensing provides an efficient and accurate solution for monitoring superlubricity, reducing energy consumption, minimizing equipment wear, and enabling intelligent monitoring and maintenance in industrial IoT systems.

11.5.3 TVNG for Self-powered Sensing

TVNGs utilize tribovoltaic effects to generate direct current (DC), offering advantages like low impedance matching, high current density, and continuous output. These features make TVNGs promising solutions for distributed electronics and power supply challenges. As a result, their application in self-powered sensing has become a hotspot in energy harvesting and sensor technology research.

Qiao et al. [26] proposed an interface lubrication strategy to enhance the current density and lifespan of TVNGs. As shown in Fig. 11.19a, introducing a water-based graphene oxide solution between a copper-based graphene (Cu-G) thin film and a silicon wafer significantly improved TVNG performance. This strategy increased sliding surface carriers to enhance current density and reduced wear between copper and silicon due to its excellent lubrication properties. The TVNG achieved peak current densities up to 775 mA/m^2 and power densities of 54.3 mW/m^2, maintaining high output stability after 30,000 cycles, as illustrated in Fig. 11.19b.

Further experiments explored the effects of parameters like sliding speed, pressure, displacement, graphene oxide solution volume, and external loads on TVNG performance (Fig. 11.19c). Results indicated that higher sliding speed, pressure, or displacement significantly increased current output while voltage remained relatively stable. The current signal's sensitivity to motion parameters demonstrated the potential of TVNGs as motion sensors. Additionally, TVNGs were integrated with alternating current (AC)-based triboelectric nanogenerators (AC-TENGs) to form dual-signal (DS) sensors for applications like bridge vibration and cargo weight monitoring (Fig. 11.19d). These sensors provided AC signals within safe operation zones and triggered alarms with DC signals when exceeding safety limits.

Other advancements include Zhang et al.'s [27] work on a heterojunction TVNG using gallium nitride (GaN) and bismuth telluride (Bi_2Te_3), achieving ultra-high power density. This TVNG generated peak open-circuit voltages of 40 V and power densities of 11.85 W/m^2, representing a 200-fold increase over conventional centimeter-scale SDC-TENGs. Liu et al. [22] introduced an organic tribovoltaic nanogenerator (FT-OTG) using PEDOT:PSS and polyvinyl alcohol (PVA), enhancing both electrical output and mechanical properties. This device achieved significant improvements in durability, tensile strength, and elongation at break, enabling continuous power supply to portable electronics.

Collectively, these studies highlight the diverse applications of tribovoltaic effects in self-powered sensing, offering high performance, low cost, and eco-friendly energy solutions. Future research will focus on improving energy conversion efficiency, reducing costs, and enhancing mechanical stability while expanding application domains in smart electronics and IoT systems.

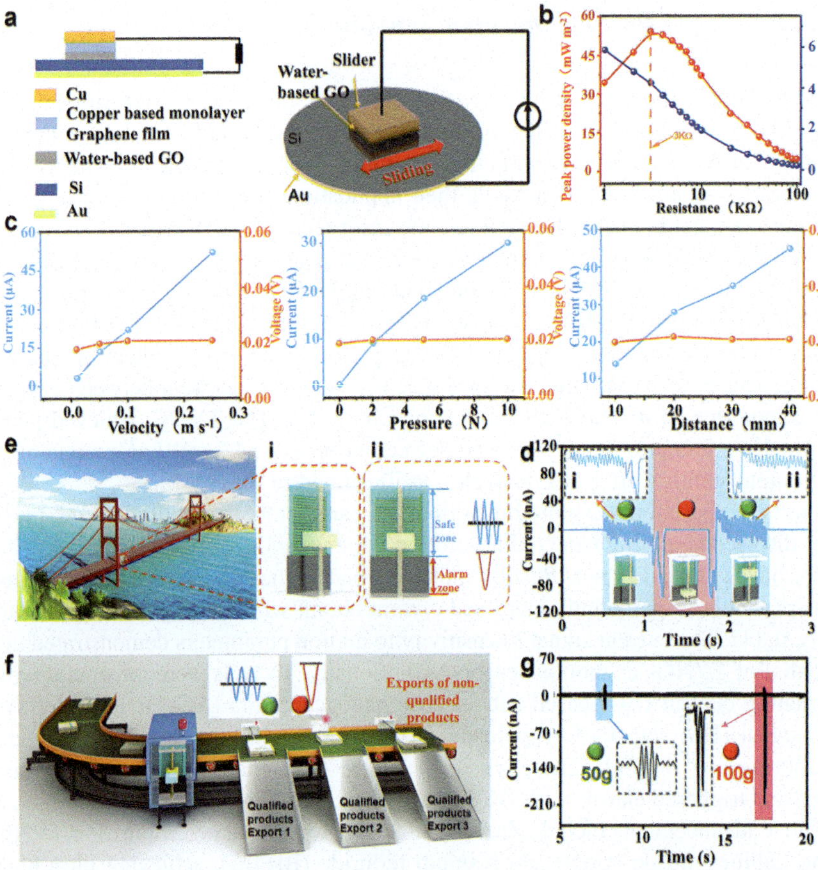

Fig. 11.19 **a** GWS-TVNG structure and circuit. **b** Peak power and current densities of GWS-TVNG. **c** Durability testing on Cu-G thin films over silicon. **d** DS sensor for bridge vibration monitoring. **e** Motion parameter responses. **f** DS sensor for weight monitoring

11.5.4 TVNG for Environmental Monitoring

With the rapid advancement of environmental monitoring and IoT technologies, there is a growing demand for real-time monitoring of environmental parameters such as wind speed, water flow, and atmospheric conditions, especially in the field of energy harvesting and self-powered systems. This section introduces TVNG-based environmental sensing technologies, detailing their fabrication processes and principles, with a focus on their applications in environmental monitoring. The development of tribovoltaic effects offers a novel solution for environmental monitoring. By converting mechanical energy from the environment into electrical energy, these devices can act as power sources for sensors or monitoring systems, enabling self-powered environmental monitoring.

11.5.4.1 TVNG for Wind Speed Sensing and Meteorological Monitoring

Yu et al. [16] developed a wind-driven tribovoltaic generator based on dynamic Schottky junctions, capable of delivering pulse outputs up to 8.4 mA. The structure of the TVNG is shown in Fig. 11.20a. It uses n-type and p-type silicon rods along with copper sheets as base materials, chosen for their semiconductor properties and ability to form Schottky junctions when in contact with metal. The silicon rods are cleaned with acetone, isopropanol, and deionized water to remove native oxide layers, ensuring good contact between the metal and semiconductor. Silver electrodes are formed on the silicon rods and annealed at high temperatures to establish low resistance ohmic contacts. These processed silicon rods are fixed onto windmill blades and paired with copper sheets to form a rotary structure. This design enables energy generation driven by wind-induced rotation.

As shown in Fig. 11.20b, the stable current output of the generator can record the number of windmill rotations, offering potential applications in wind energy harvesting and meteorological monitoring. Additionally, as illustrated in Fig. 11.20c, the generator can directly power a graphene ultraviolet photodetector, showcasing its potential as a sensor power source. This is particularly significant in environmental monitoring, where changes in ultraviolet intensity provide critical information about atmospheric conditions and ozone layer status.

11.5.4.2 TVNG for Water Flow Sensing and Aquarium Bubble Monitoring

Huang et al. [28] reported an immersed TVNG that utilizes the interaction between fluorinated graphite paper (FGP) and water flow to generate electricity. The working principle of the FGP-TVNG involves two mechanisms: triboelectric effects and tribovoltaic effects. The triboelectric effect generates an alternating current (AC) signal through the interaction of FGP with ions in the water, while the tribovoltaic effect produces a direct current (DC) signal via friction between FGP and water. This dual mechanism allows the FGP-TVNG to convert the mechanical energy of water flow into electrical energy. Figure 11.20d illustrates a self-powered water speed sensing system using FGP-TVNG, where plastic markers are employed to estimate water flow velocity. As shown in Fig. 11.20e, the open-circuit voltage and short-circuit current of the FGP sensor exhibit a linear relationship with water flow speed, with a fitting degree R2R^2R2 exceeding 0.99. This demonstrates the exceptional performance of FGP as a self-powered water flow speed sensor, which is crucial for fields such as hydrology, water treatment, and environmental monitoring. Moreover, as depicted in Fig. 11.20f, researchers placed the TVNG in an aquarium to harness energy from interactions between FGP and flowing water or bubbles. A bubble generator simulates bubble production, and as bubbles rise, they induce upward water flow, increasing the output current, as shown in Fig. 11.20g. In addition to water flow energy, FGP can also detect other mechanical energies in the aquarium, such as ultrasonic waves.

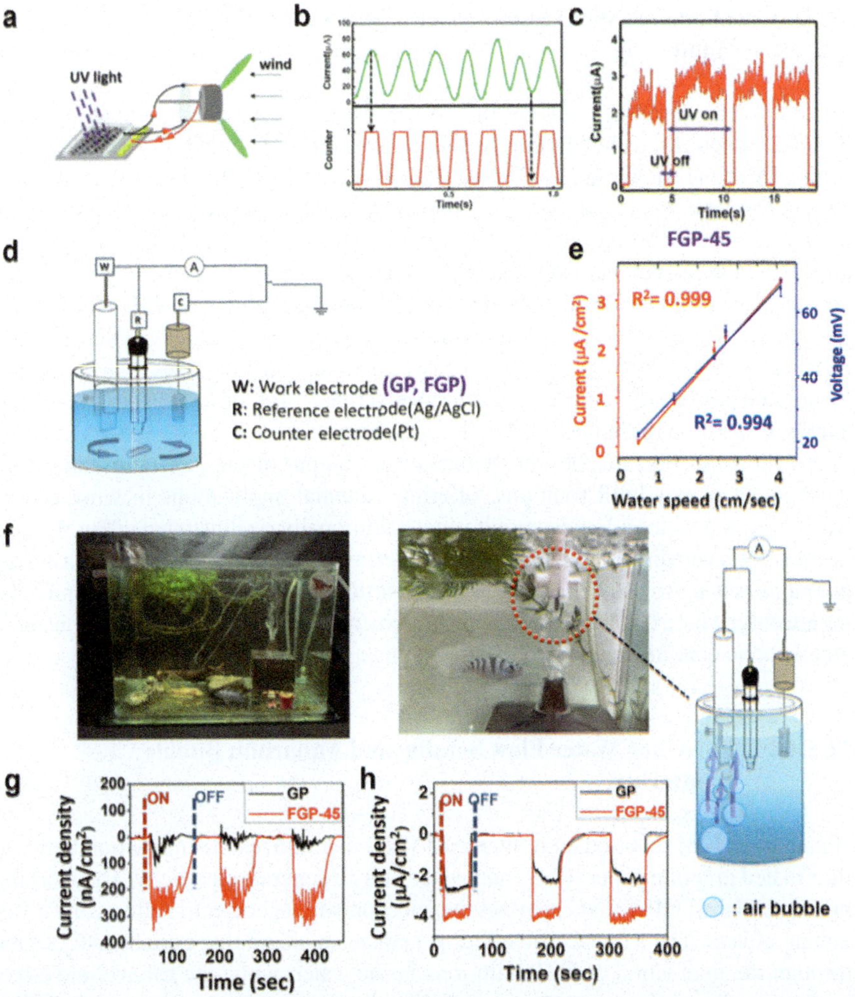

Fig. 11.20 **a** Structure of wind-driven TVNG. **b** TVNG as a counter for wind speed sensing. **c** TVNG powering a graphene/GaN ultraviolet photodetector for UV monitoring. **d** Self-powered water flow velocity sensing system. **e** Open-circuit voltage and short-circuit current response to water flow speed. **f** FGP-based aquarium device for monitoring bubbles and ultrasonic waves. **g** Output current response to bubble generation in water. **h** Output current response to ultrasonic waves in water

Figure 11.20h shows that when an ultrasonic cleaner operating at 100 W emits ultrasonic waves into the water, an increase in output current is observed. This monitoring capability is significant for aquaculture and water quality management.

TVNG-based environmental sensing technology, with its unique energy harvesting and conversion mechanism, offers a novel self-powered solution for environmental monitoring. Wind-driven semiconductor generators can serve as counters to record windmill rotations and directly power graphene ultraviolet photodetectors, demonstrating their potential as sensor power sources. The output current of FGP-TVNG shows a linear relationship with water flow velocity, making it suitable for monitoring water flow. In aquariums, FGP-based devices can detect bubble generation and help remotely monitor this process in real time. With advancements in materials science and nanotechnology, TVNGs are expected to play an increasingly significant role in future environmental monitoring and self-powered sensing applications.

11.6 Conclusion

This chapter summarizes the extensive applications and potential of the TVNG as an innovative energy conversion device, highlighting its remarkable performance across various fields. By efficiently converting mechanical vibrations and frictional energy into electrical energy, TVNG has achieved significant advancements in optimizing energy utilization and enabling self-driven functionality in traditional mechanical systems. In multi-source energy harvesting, TVNG integrates multiple energy forms and couples with other energy conversion mechanisms, significantly improving energy conversion efficiency and supporting the development of sustainable energy technologies. Moreover, the flexible design and high current density of TVNG empower intelligent flexible devices to achieve battery-free operation through self-powering, enhancing their portability and flexibility. In self-powered sensing systems, TVNG offers reliable real-time monitoring capabilities, providing efficient solutions for smart environmental monitoring, health tracking, and industrial automation. In conclusion, TVNG demonstrates exceptional adaptability and efficient energy conversion capabilities in diverse scenarios, offering critical guidance for the design of intelligent systems and the future development of renewable energy technologies. This underscores its broad application prospects in advancing sustainable development and technological innovation.

References

1. Y. Shi, H. Li, X. Fu, R. Luan, Y. Wang, N. Wang, Z. Sun, Y. Niu, C. Wang, C. Zhang, Z.L. Wang, Self-powered difunctional sensors based on sliding contact-electrification and tribovoltaic effects for pneumatic monitoring and controlling. Nano Energy **110**, 108339 (2023)
2. L. Gong, Z. Zhang, W. Yu, J. Zeng, J. Cao, B. Fan, J. Zhao, C. Zhang, Ultra-durable polysilicon based tribovoltaic nanogenerators for bearing in situ rotational speed sensing. Small **20**(50), 2405992 (2024)
3. S. Dong, T. Bu, Z. Wang, Y. Feng, G. Liu, J. Zeng, Z. Wang, J. Cao, Z. Zhang, F. Liu, C. Zhang, Freestanding-mode tribovoltaic nanogenerator for harvesting sliding and rotational mechanical energy. Adv. Energy Mater. **13**(16), 2300079 (2023)

4. Z. Zhang, N. Wu, L. Gong, R. Luan, J. Cao, C. Zhang, An ultrahigh power density and ultralow wear GaN-based tribovoltaic nanogenerator for sliding ball bearing as self-powered wireless sensor node. Adv. Mater. **36**(6), 2310098 (2024)
5. J. Liu, Y. Zhang, J. Chen, R. Bao, K. Jiang, F. Khan, A. Goswami, Z. Li, F. Liu, K. Feng, J. Luo, T. Thundat, Separation and quantum tunneling of photo-generated carriers using a tribo-induced field. Matter. **1**(3), 650–660 (2019)
6. Z. Hao, T. Jiang, Y. Lu, S. Feng, R. Shen, T. Yao, Y. Yan, Y. Yang, Y. Lu, S. Lin, Co-harvesting light and mechanical energy based on dynamic metal/perovskite schottky junction. Matter **1**(3), 639–649 (2019)
7. Z. Zhang, T. He, J. Zhao, G. Liu, Z.L. Wang, C. Zhang, Tribo-thermoelectric and tribovoltaic coupling effect at metal-semiconductor interface. Mater. Today Phys. **16**, 100295 (2021)
8. H. Zheng, R. Shen, H. Zhong, Y. Lu, X. Yu, S. Lin, Dynamic Schottky diode direct-current generator under extremely low temperature. Adv. Funct. Mater. **31**(40), 2105325 (2021)
9. M. Zheng, S. Lin, L. Zhu, Z. Tang, Z.L. Wang, Effects of temperature on the tribovoltaic effect at liquid–solid interfaces. Adv. Mater. Interfaces **9**(3), 2101757 (2021)
10. S. Deng, W. Seh, Q. Zhang, Thermal characteristics of tribovoltaic dynamic Schottky junctions. Nano Energy **114**, 108665 (2023)
11. Z. Wang, L. Gong, S. Dong, B. Fan, Y. Feng, Z. Zhang, C. Zhang, A humidity-enhanced silicon-based semiconductor tribovoltaic direct-current nanogenerator. J. Mater. Chem. A. **10**(47), 25230–25237 (2022)
12. S. Lin, X. Chen, Z.L. Wang, The tribovoltaic effect and electron transfer at a liquid-semiconductor interface. Nano Energy **76**, 105070 (2020)
13. Y. Yan, X. Zhou, S. Feng, Y. Lu, J. Qian, P. Zhang, X. Yu, Y. Zheng, F. Wang, K. Liu, S. Lin, Direct current electricity generation from dynamic polarized water–semiconductor interface. J. Phys. Chem. C **125**(26), 14180–14187 (2021)
14. Y. Lu, Y. Yan, X. Yu, X. Zhou, S. Feng, C. Xu, H. Zheng, Z. Yang, L. Li, K. Liu, S. Lin, Polarized water driven dynamic PN junction-based direct-current generator. Research **2021**, 1–10 (2021)
15. X. Yu, H. Zhong, Y. Lu, C. Liu, Y. Chen, R. Yang, C. Wang, S. Lin, Flexible liquid-based continuous direct-current tribovoltaic generators enable self-powered multi-modal sensing. Adv. Mater. Technol. **8**(21), 2300873 (2023)
16. X. Yu, H. Zheng, Y. Lu, R. Shen, Y. Yan, Z. Hao, Y. Yang, S. Lin, Wind driven semiconductor electricity generator with high direct current output based on a dynamic Schottky junction. RSC Adv. **11**(31), 19106–19112 (2021)
17. J. Chen, P. He, T. Huang, D. Zhang, G. Wang, S. Yang, X. Xie, G. Ding, Boosting carrier transfer at flexible schottky junctions with moisture: A strategy for high-performance wearable direct-current nanogenerators. Nano Energy **90**, 106593 (2021)
18. B. Fan, Z. Wang, G. Liu, Z. Wang, X. Fu, L. Gong, C. Zhang, Robust flexible textile tribovoltaic nanogenerator via a 2D 2H-MoS_2/Ta_4C_3 dynamic heterojunction. Adv. Func. Mater. **33**(33), 2301821 (2023)
19. P. Deng, Y. Wang, R. Yang, Z. He, Y. Tan, Z. Chen, J. Liu, T. Li, Self-powered smart textile based on dynamic schottky diode for human-machine interactions. Adv. Sci. **10**(11), 2207298 (2023)
20. J. Meng, Z.H. Guo, C. Pan, L. Wang, C. Chang, L. Li, X. Pu, Z.L. Wang, Flexible textile direct-current generator based on the tribovoltaic effect at dynamic metal-semiconducting polymer interfaces. ACS Energy Lett. **6**(7), 2442–2450 (2021)
21. Z. You, S. Wang, Z. Li, Y. Zou, T. Lu, F. Wang, B. Hu, X. Wang, L. Li, W. Fang, Y. Liu, High current output direct-current triboelectric nanogenerator based on organic semiconductor heterojunction. Nano Energy **91**, 106667 (2022)
22. G. Liu, R. Luan, Y. Qi, L. Gong, J. Cao, Z. Wang, F. Liu, J. Zeng, X. Huang, Y. Qin, S. Dong, Y. Feng, L.-B. Huang, C. Zhang, Organic tribovoltaic nanogenerator with electrically and mechanically tuned flexible semiconductor textile. Nano Energy **106**, 108075 (2023)
23. X. Luo, L. Liu, Y.-C. Wang, J. Li, A. Berbille, L. Zhu, Z.L. Wang, Tribovoltaic nanogenerators based on MXene–Silicon heterojunctions for highly stable self-powered speed, displacement, tension, oscillation angle, and vibration sensors. Adv. Func. Mater. **32**(23), 2113149 (2022)

24. M. Wang, J. Yang, S. Liu, Y. Meng, Y. Qin, X. Li, Metal-gallium arsenide based tribovoltaic nanogenerators and its application for high-precision self-powered displacement sensors. Adv. Mater. Technol. **8**(3), 2200677 (2023)
25. L. Zhang, H. Cai, L. Xu, L. Ji, D. Wang, Y. Zheng, Y. Feng, X. Sui, Y. Guo, W. Guo, F. Zhou, W. Liu, Z.L. Wang, Macro-superlubric triboelectric nanogenerator based on tribovoltaic effect. Matter **5**(5), 1532–1546 (2022)
26. W. Qiao, Z. Zhao, L. Zhou, D. Liu, S. Li, P. Yang, X. Li, J. Liu, J. Wang, Z.L. Wang, Simultaneously enhancing direct-current density and lifetime of tribovotaic nanogenerator via interface lubrication. Adv. Func. Mater. **32**(46), 2208544 (2022)
27. Z. Zhang, Z. Wang, Y. Chen, Y. Feng, S. Dong, H. Zhou, Z.L. Wang, C. Zhang, Semiconductor contact-electrification-dominated tribovoltaic effect for ultrahigh power generation. Adv. Mater. **34**(20), 2200146 (2022)
28. H.-H. Huang, R.-C. Wang, Y.-J. Chen, Fluorinated graphite paper used for self-powered water speed sensors by immersion-type tribovoltaic effect-dominated triboelectric nanogenerators. Nano Energy **93**, 106887 (2022)

Chapter 12
Applications of Contact-Electro-Catalysis

Abstract With the rapid progression of society and the continuous advancement of industrialization, catalysis has become an indispensable tool in manufacturing, energy conversion and storage, resource recycling, and environmental management. Over the past few decades, electrocatalysis and photocatalysis have been widely recognized as the most promising approaches to support sustainable development. In recent years, emerging catalytic methods such as contact-electro-catalysis (CEC) have also garnered considerable attention and research efforts. CEC is based on the interdisciplinary mechanism of physics and chemistry, and specifically, realized through interfacial electron transfer during the mechano-driven solid–liquid (or liquid–liquid, etc.) contact electrification process, which differs from conventional catalytic principles in terms of both electron and energy sources. This endows CEC with the characteristics of being environmentally friendly, low-energy-consuming, and offering a wider range of catalyst options. Here, in this chapter, we will review the potential applications of CEC that have been proposed thus far, including water treatment, carbon dioxide capture, nitrogen fixation, hydrogen peroxide generation, recycling of lithium-ion cathode materials, and cancer therapy.

12.1 Introduction

With the rapid development of society and the continuous advancement of industrialization, issues such as environmental degradation and energy shortages have become increasingly apparent. As a result, clean energy conversion, including energy storage and reversible energy utilization, has attracted widespread attention [1]. In current research, catalysis plays an irreplaceable role in chemical manufacturing, energy conversion and storage, resource recycling, and environmental management [2]. Catalysis not only significantly improves the efficiency of chemical reactions, reduces energy consumption and byproducts, but also promotes sustainable development, environmental protection, and resource recycling [3, 4]. At present, electrocatalysis and photocatalysis are widely regarded as the most promising methods to address the energy crisis. Additionally, emerging catalytic methods such as piezo-catalysis

Z. L. Wang, *Contact-Electrification of Matter*, Springer Series in Surface Sciences 75,
https://doi.org/10.1007/978-3-031-88504-4_12

and contact-electro-catalysis (CEC) have also received considerable attention and research. Contact-electro-catalysis was first proposed by Wang et al. [5]. CEC is realized by electron transfer during mechano-driven contact electrification process, which differs from conventional catalytic principles in both electron and energy sources [6, 7]. Due to the significant electron transfer during the contact electrification process, and the fact that the charged surface after contact can locally introduce a strong electric field, these electrons combine with water or dissolved oxygen in water to generate active oxygen species such as hydroxyl radicals and superoxide radicals [8]. This promotes subsequent redox reactions and enables various applications in areas such as green energy synthesis, environmental remediation, resource recovery, and biomedicine. Moreover, the selection range for its catalysts is broad [6]. As long as the material exhibits excellent contact electrification effects, such as commercially available polymers, it can be used for CEC. These characteristics give it unique advantages in practical applications.

From the detailed mechanistic analysis of contact electrolysis in Chap. 8, CEC is a catalytic process in which chemical reactions are driven by electron transfer generated through contact electrification. As an emerging catalytic mechanism, it utilizes the transfer of electrons at the interface between water and dielectric materials to generate active radicals from water or oxygen, which then undergo redox reactions. CEC has stronger electrification advantages and a wider range of catalyst options. Due to its environmentally friendly nature, mild reaction conditions, and other characteristics, it holds promise for exploring its applications in other catalytic reactions and even interdisciplinary fields, such as the degradation of organic pollutants, the synthesis of important chemicals, metal ion recovery, and even biomedicine. In conclusion, the field of CEC provides exciting opportunities for the development of mechanochemical catalysis. This makes CEC a promising strategy for future scientific research and industrial applications.

12.2 Low-Carbon Applications (Pollution Degradation, Carbon Dioxide Capture)

Organic pollutants commonly originate from industrial discharges, pharmaceuticals, pesticides, and other human activities. Their toxicity and potential for bioaccumulation pose significant threats to the environment and human health. Effective degradation of these pollutants helps prevent ecosystem disruption and reduces health risks, making it essential for safeguarding environmental health and protecting water and soil resources [9, 10].

Contact Electrocatalysis (CEC) is an emerging catalytic technology that utilizes the contact electrification effect between materials to generate reactive oxygen species (ROS), enabling the efficient degradation of organic pollutants and other hazardous substances. CEC technology leverages mechanical energy sources, such as ultrasound and vibration, which are widely available and cost-effective. It operates effectively across a broad range of temperatures and pH conditions, making it adaptable to various environmental scenarios. Unlike traditional catalytic systems, CEC expands the selection of catalysts to include chemically inert materials such as Polytetrafluoroethylene (PTFE), polydimethylsiloxane (PDMS), and fluorinated ethylene propylene (FEP). These materials maintain their chemical composition before and after the reaction, allowing for repeated recycling and reuse, thereby reducing long-term operational costs, and minimizing environmental impact. The core mechanism of CEC lies in the charge transfer phenomenon generated by physical contact between materials, leading to the formation of highly redox-active radicals, such as hydroxyl radicals (·OH) and superoxide anion radicals ($\cdot O_2^-$). These radicals rapidly oxidize and degrade organic pollutants. CEC holds significant promise for environmental remediation and sustainable development, demonstrating substantial potential in various practical applications.

Using the degradation of dye species as an example, Wang et al. investigated the degradation process of methyl orange (MO). The experimental setup is illustrated in Fig. 12.1a. In their study, 20 mg of FEP powder was added to 50 mL of an aqueous MO solution (5 ppm) and stirred for 48 h to enhance the contact between FEP and water. Subsequently, the prepared suspension was subjected to ultrasonic treatment for 3 h, and the initial light-yellow solution became transparent. To investigate the discoloration process, the corresponding UV–Vis spectra (Fig. 12.1b) revealed that the characteristic absorption peak of MO gradually decreased with increasing ultrasonic treatment time, eventually disappearing after 120 min. These results demonstrated that the MO solution was completely degraded within 3 h of ultrasonic treatment. To further understand the underlying mechanism, trapping experiments indicated that two reactive radicals, ·OH and $\cdot O_2^-$, were involved in the degradation process. Among these, hydroxyl radicals were identified as the primary limiting factor [5]. Wang et al. further demonstrated that ball milling could directly promote frequent contact-separation cycles, with triboelectric materials catalyzing the generation of ROS to interact with organic pollutants in aqueous solutions. A ball milling setup was constructed (Fig. 12.1c), where 50 mL of a 5 ppm MO solution was milled with 100 g of polymer balls for 120 min. The degradation rates of MO were 3.1%, 32.05%, and 99.18% for polypropylene (PP), PDMS, and PTFE, respectively. To investigate the underlying mechanism, ultraviolet–visible spectroscopy, and liquid chromatography-mass spectrometry (LC–MS) were performed, as shown in Fig. 12.1d and e. The characteristic absorbance intensity of MO in the UV–Vis spectrum decreased with increasing milling time, approaching zero after 60 min. The evolution of MO's relative concentration in the presence of different scavengers revealed the contributions of ·OH and superoxide radicals $\cdot O_2^-$ to the degradation process. Additionally, the study investigated changes in MO's relative concentration under varying rotational speeds using PTFE. When the rotational speed was below 100 RPM, MO degradation was negligible. However, at 150 RPM, the relative concentration of MO dropped 19.5-fold, and further increases in rotational speed resulted in even higher degradation rates [11]. To further enhance the catalytic performance of CEC, Dong et al. developed a polymer/metal Janus composite catalyst that can modulate the reaction rates of water oxidation reactions (WOR), and oxygen reduction reactions (ORR) based on its composition (Fig. 12.1f). Cavitation bubbles induced CE under ultrasonic treatment. This resulted in negative charges on the FEP surface, generating an electric field that drove electrons from within the copper (Cu) component to its outer surface, thus facilitating the production of ROS. The efficacy of ROS production was validated by the degradation of MO in aqueous solutions. The FEP/Cu catalyst exhibits the highest MO degradation rate (96.56%), followed by those of FEP (46.21%) and FEP + Cu (21.38%). FEP/Cu exhibited a significant enhancement of 103.80 and 351.64% (Fig. 12.1g) over those of FEP and FEP + Cu, respectively. FEP/Cu shows the highest kinetic rate of $k = 0.03038\ \text{min}^{-1}$ (Fig. 12.1h), illustrating the advantage of FEP/Cu composites over pristine polymers for the generation of ROS [8].

Antibiotic contamination in water bodies poses a significant challenge to environmental management. In a study by Cao et al., an innovative CEC process utilizing reusable PTFE particles as trigger catalysts was applied as an alternative to conventional advanced oxidation processes (AOPs). This approach successfully degraded three common antibiotics—sulfamethoxazole (SMX), ciprofloxacin (CIP), and tetracycline (TET)—from secondary effluent of a wastewater treatment plant, marking the first demonstration of this method in such applications (Fig. 12.2a). In the experiments, PTFE particles at a concentration of 0.8 g/L were added to antibiotic solutions containing 1 mg/L of each antibiotic. After 90 min of operation, removal rates of 50%, 80%, and 90% were achieved for SMX, CIP, and TET, respectively (Fig. 12.2b). The findings revealed that the CEC process generated ·OH and singlet oxygen ($^1O_2$), which played key roles in the antibiotic degradation (Fig. 12.2c). Based on these results, novel ROS generation mechanisms during the CEC process were proposed. To enhance sustainability, polyvinylidene fluoride (PVDF) nanofiber membranes were fabricated using electrospinning for the efficient recovery of PTFE particles. These membranes achieved nearly 100% particle recovery through dead-end filtration with a high water flux of 7.5 $m^3/(m^2\ h)$ (Fig. 12.2d). This study provides a promising strategy for antibiotic removal from wastewater, advancing both CEC applications and environmental remediation techniques [12].

Chlorophenols (CPs) pose significant risks to human health due to their toxicity and carcinogenic properties. The direct oxidative decomposition of CPs often generates more harmful byproducts, leading to secondary pollution. To address this, Li et al. employed commercially viable FEP as a catalyst to degrade pentachlorophenol (PCP) via an ultrasound-activated CEC process (Fig. 12.3a). A mechanism based on contact-induced electron transfer was proposed, which triggers the generation of reactive species for PCP degradation. The experimental results demonstrated that as little as 1.0 mg of FEP was sufficient to achieve complete PCP degradation. After 1 h of treatment with 10.0 mg/L PCP, the degradation efficiency and mineralization efficiency were 24% and 21%, respectively. Extending the treatment time to 3 h significantly improved these results, with a degradation efficiency of 91% and a mineralization efficiency close to 40% (Fig. 12.3b). Identification of ROS and quenching experiments revealed that ·OH, $^1O_2$, and $\cdot O_2^-$ played critical roles in the PCP degradation process (Fig. 12.3c). This study highlights the potential

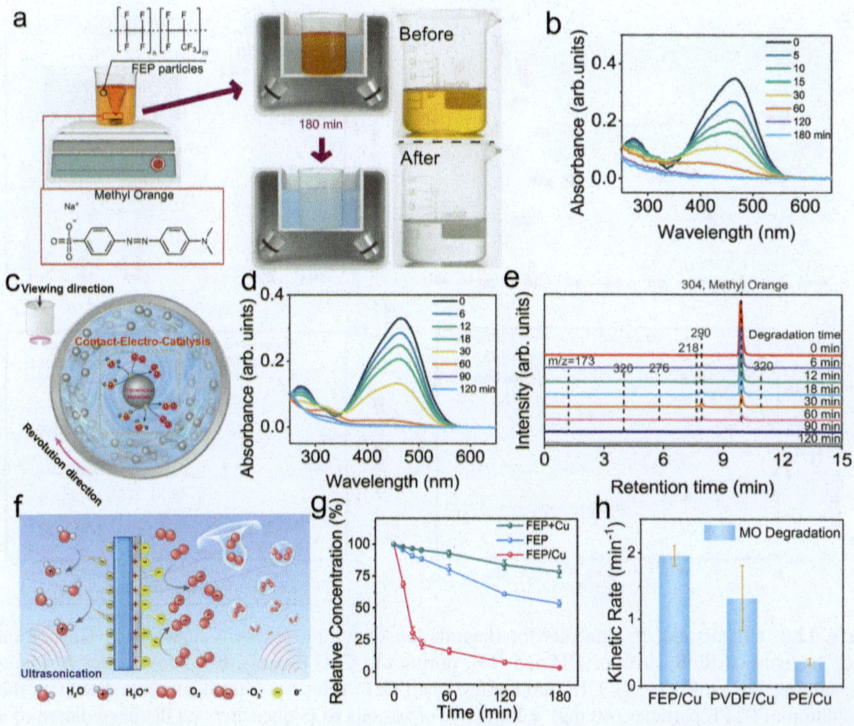

Fig. 12.1 Degradation of methyl orange by contact-electro-catalysis. **a** 3D schematic of the experimental setup and protocol. **b** UV–Vis spectra of MO aqueous solution samples during ultrasonication in presence of FEP powder (20 mg) from 0 to 180 min. **c** Proposed working principle of CEC during ball milling using triboelectric materials. **d** UV–Vis spectra of a 5 ppm MO aqueous solution when PTFE vial and balls were employed for ball milling. **e** Electron transfer principle for Janus composites. **f** Liquid chromatograph of the aqueous MO solution at different degradation time intervals with peaks being identified by mass spectra. Evolution of the **g** relative concentration of methyl orange in the presence of different catalysts (FEP + Cu, FEP, and FEP/Cu). **h** Effect of dielectric-end electrification performance on the kinetic constants of MO degradation. **a, b** Reproduced with permission from Ref. [5]. Copyright 2022, Springer Nature. **c–e** Reproduced with permission from Ref. [11]. Copyright 2024, Springer Nature. **f–h** Reproduced with permission from Ref. [8]. Copyright 2024, American Chemical Society

of ultrasound-activated CEC using FEP as a catalyst for effective and sustainable remediation of chlorophenol-contaminated environments [13].

In summary, CEC presents a novel solution that combines mechanical energy with catalytic principles to enhance ROS production and pollutant degradation. This approach seeks to address the limitations of traditional methods by being adaptable, energy-efficient, and capable of operating under ambient conditions, potentially offering a sustainable alternative for organic degradation in environmental applications.

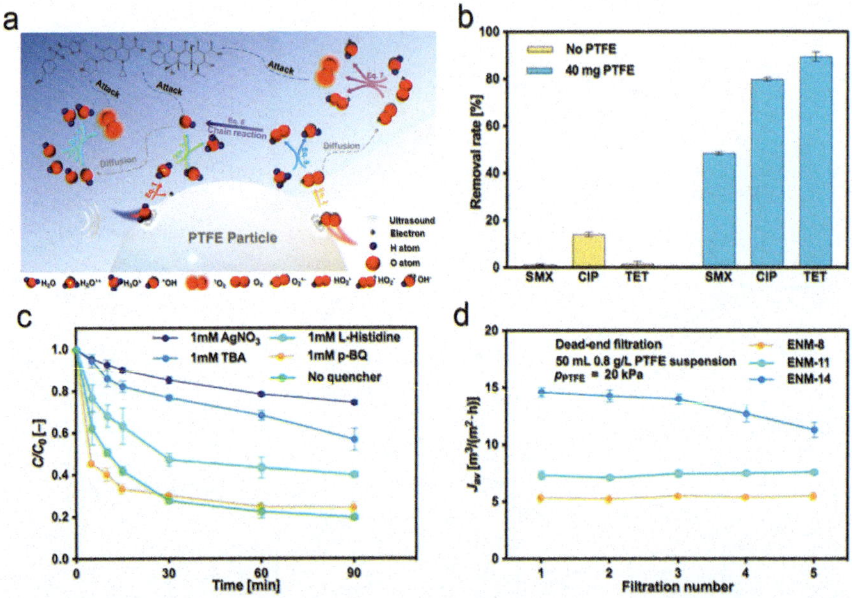

Fig. 12.2 Contact-electro-catalysis for degradation of antibiotics in wastewater. **a** Generation mechanisms of ROS, such as ·OH and $^1O_2$, during the CEC process. **b** Removal rates of 1 mg/L sulfamethoxazole (SMX), CIP, and tetracycline (TET) solutions (50 mL) with and without the addition of PTFE particles (40 mg). **c** Influence of various ROS quenchers on the degradation of a ciprofloxacin (CIP) solution (50 mL, 1 mg/L); here, $AgNO_3$, Tert-Butanol (TBA), L-Histidine, and p-Benzoquinone (p-BQ) were employed to quench e^-, ·OH, $^1O_2$, and O_2^-, respectively. **d** Average water flux (Jav) in the filtrations of PTFE suspension for five cycles. [12] Reproduced with permission from Ref.. Copyright 2024, Chemical Engineering Journal

The rising concentration of carbon dioxide (CO_2) poses significant threats to global climate and human health. As the world strives to tackle challenges related to climate change and energy security, the importance of utilizing CO_2 as a carbon resource has become increasingly evident, offering potential benefits for mitigating climate change and promoting sustainable development [14]. In recent years, various catalytic approaches, including catalytic hydrogenation, photocatalysis, electrocatalysis, photo-electrocatalysis and photothermal catalysis, have been employed to convert CO_2 into high-value products. These processes can yield carbon-containing compounds such as formic acid, methanol, methane, acetic acid, ethylene, and even long-chain hydrocarbons [15, 16]. However, several limitations hinder the practical application of these methods. For instance, photocatalytic processes face challenges in enhancing efficiency, electrocatalytic CO_2 reduction often requires substantial consumption of high-quality electricity. What's more, photocatalysis, electrocatalysis, photothermal catalysis, and photo-electrocatalysis all impose stringent requirements on catalysts, often involving complex preparation processes. In the case of

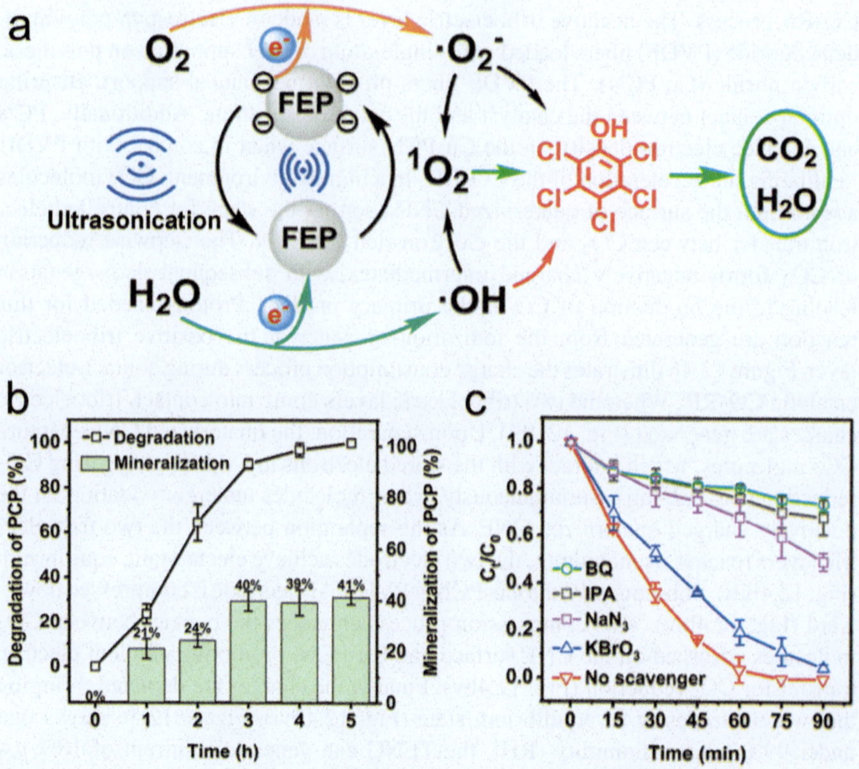

Fig. 12.3 Contact-electro-catalysis for degradation of Pentachlorophenol Using Inert FEP Powders. **a** Possible mechanism of degradation of PCP under US/FEP system. **b** The quenching effects of various scavengers on PCP degradation in the US/FEP system. **c** The quenching effects of various scavengers on PCP degradation in the US/FEP system. Reproduced with permission from Ref. [13]. Copyright 2024, American Chemical Society

catalytic CO_2 hydrogenation, hydrogen gas is an essential feedstock, but its production, storage, and transportation involve complex systemic challenges and a heavy reliance on fossil fuels for energy supply [17].

CEC presents a promising solution to the challenges faced in conventional CO_2 reduction processes. Wang et al. demonstrated the use of CEC for CO_2 reduction, achieving a Faradaic efficiency of 96.24% for CO production with a yield rate of 33 μmol g^{-1} h^{-1}. The underlying mechanism is illustrated in Fig. 12.4. Figure 12.4a shows a schematic of the TENG applied in the CEC-based CO_2 reduction reaction (CO_2RR). The positive triboelectric layer of the TENG utilizes quaternized cellulose nanofibers (CNF). CNF possess a high CO_2 adsorption capacity and have abundant hydroxyl groups which can form hydrogen bonds with water molecules in high-humidity environments to fix water and participates in contact electrification. In such humid conditions, water molecules undergo oxidation on the quaternized CNF surface, producing oxygen and protons. These protons act as intermediates in the

CO_2RR process. The negative triboelectric layer is made of electrospun polyvinylidene fluoride (PVDF) fibers loaded with single-atom copper supported on polymeric carbon nitride (Cu-PCN). The PVDF fibers provide mechanical support, ensuring optimal contact between the catalyst and the catalytic substrate. Additionally, PCN can enhance electron density on the Cu-PCN surface when in contact with PVDF, facilitating the acceleration of the CO_2RR. In a humid environment, CO_2 molecules adsorb onto the surface of quaternized CNFs, setting the stage for controlled electron transfer between CO_2 and the electron-rich Cu-PCN. The stepwise reduction of CO_2 forms negatively charged intermediates, with subsequent deoxygenation leading to the production of CO as the primary product. Protons needed for this reaction are generated from the ionization of water at the positive triboelectric layer. Figure 12.4b illustrates the charge consumption process during contact-electro-catalytic CO_2RR. When the two triboelectric layers come into contact, triboelectric charges are generated (Fig. 12.4bi). Upon separation, the quaternized CNFs capture CO_2 molecules, which interact with the stored electrons in Cu-PCN, triggering CO_2 reduction (Fig. 12.4bii). Simultaneously, water molecules undergo oxidation on the positively charged quaternized CNF. As the separation between the two triboelectric layers reaches its maximum, the two electrodes achieve electrostatic equilibrium (Fig. 12.4biii). Subsequently, the Cu-PCN@PVDF/Al electrode is compressed downward (Fig. 12.4biv). This compression process enhances the contact between CO_2 molecules adsorbed on the CNF surface and Cu-PCN, promoting efficient electron transfer for CO_2 reduction (Fig. 12.4bv). Finally, the charges are depleted, bringing the two electrodes to an equilibrium state (Fig. 12.4bvi). Figure 12.4c shows that under 99% relative humidity (RH), the TENG can generate a current of 18.7 μA and a voltage of 405 V. After 5 h, gas chromatography analysis revealed a distinct CO peak (Fig. 12.4d), indicating that CO_2 was effectively reduced, with CO as the final product. A schematic of the air-based contact-electro-catalytic CO_2 reduction reaction (CO_2RR) is presented in Fig. 12.4e. Atmospheric CO_2 is drawn into the wind-driven TENG device, where mechanical energy is converted into electrical energy through the interaction of the positive and negative triboelectric layers. This energy is then utilized in the CEC process to catalytically reduce CO_2 to CO [18].

This technology utilizes wind energy as the power source to achieve efficient CO_2 reduction, providing a green and sustainable method for lowering atmospheric CO_2 levels. It offers a novel approach to advancing carbon capture efforts, contributing to global sustainability initiatives.

12.3 Synthesis of Vital Chemicals (Hydrogen Peroxide, Ammonia Synthesis)

Hydrogen peroxide (H_2O_2) presents significant potential for environmental remediation and as an alternative to carbon-based energy sources [19]. As a green and mild oxidizing agent, hydrogen peroxide is a crucial chemical raw material with a wide

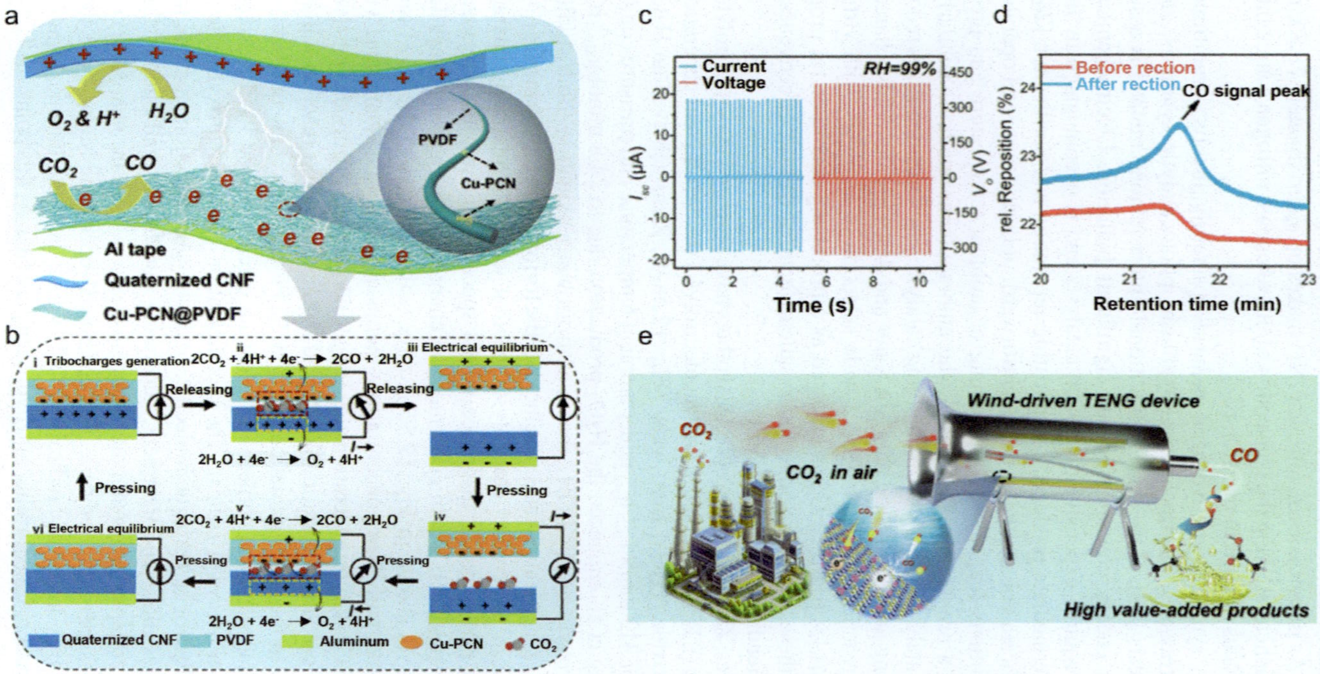

Fig. 12.4 Device design and characterizations of contact-electro-catalytic CO_2RR. **a** Schematic diagram showing the structure of TENG. **b** Schematic diagram of the TENG working cycle, highlighting the integrated TENG charge generation and catalytic charge consumption process during contact-electro-catalytic CO_2RR. **c** Output current and voltage of the TENG at 99% RH. **d** Comparison of CO signal peaks before and after CO_2RR tested by GC. **e** Schematic illustration of contact-electro-catalytic CO_2 reduction in air. Reproduced with permission from Ref. [18]. Copyright 2024, Springer Nature

range of applications in chemical synthesis, fuel energy, and the electronics industry [20, 21]. Additionally, ammonia (NH_3) is one of the most important inorganic chemicals in industry. As is well known, nitrogen (N_2) makes up approximately 78.8% of the Earth's atmosphere, making the conversion of this abundant nitrogen source into ammonia a key focus of current research [22].

Currently, the synthesis of hydrogen peroxide primarily involves traditional anthraquinone processes, electrochemical synthesis, photocatalytic synthesis, and enzymatic synthesis [23]. However, the traditional anthraquinone method is associated with high energy consumption and significant pollution. Electrochemical synthesis often requires expensive palladium catalysts and is accompanied by the generation of intermediate products, complicating separation and purification [24]. In contrast, photocatalytic synthesis is straightforward, pollution-free, and operates under mild conditions (capable of functioning at ambient temperature and pressure), making it a promising alternative to the anthraquinone method. Similarly, contact-electro-catalysis (CEC), as an emerging catalytic mechanism, utilizes redox reactions occurring at the interface between water and dielectric materials, characterized by mild reaction conditions and a wide selection of catalysts, with virtually no limitations [5, 6, 8]. CEC has emerged as a promising candidate for properly addressing these difficulties due to its salient features of cost-effectiveness, environmental friendliness, and scalability.

In recent studies, Zhao [25], Wang [26], and others have proposed the use of polytetrafluoroethylene (PTFE) particles in contact electrolysis for the efficient generation of hydrogen peroxide (H_2O_2) from water (Fig. 12.5a–f). Zhao achieved an H_2O_2 production rate of 313 μmol L^{-1} h^{-1} under ambient temperature and pressure conditions, while Wang et al. enhanced the H_2O_2 production rate to 24.8 mmol g_{at}^{-1} h^{-1} by utilizing hydroxyl radicals (·OH) generated from water oxidation. Both studies achieved high H_2O_2 generation rates. To elucidate the formation mechanism of H_2O_2, the intermediates in the contact electrolysis reaction were systematically characterized. Electron paramagnetic resonance (EPR) spectroscopy was used to identify intermediate products, which confirmed the charge transfer and pathways for H_2O_2 formation. As shown in Fig. 12.5c–e, three active radicals *OH, $*O_2^-$, and $^1O_2$ were generated, with their concentrations increasing as the ultrasonic time was extended. The CEC catalytic generation of hydrogen peroxide can be divided into two pathways: water oxidation reaction (WOR) and oxygen reduction reaction (ORR) [25] (Fig. 12.5f). In the water oxidation reaction, PTFE particles come into contact with water, causing the water molecules to lose an electron to the PTFE surface, which generates a hydroxyl radical. The recombination of two hydroxyl radicals then produces a hydrogen peroxide molecule. In the oxygen reduction reaction, the solution undergoes ultrasonic treatment, generating numerous small bubbles that expand and then collapse. Oxygen molecules gain electrons from the charged PTFE surface, forming superoxide radicals. The generated superoxide radicals are first protonated to form hydroperoxyl radicals (·OOH), which then simultaneously gain a proton and an electron, transforming into hydrogen peroxide. At this point, the PTFE particles return to their original state.

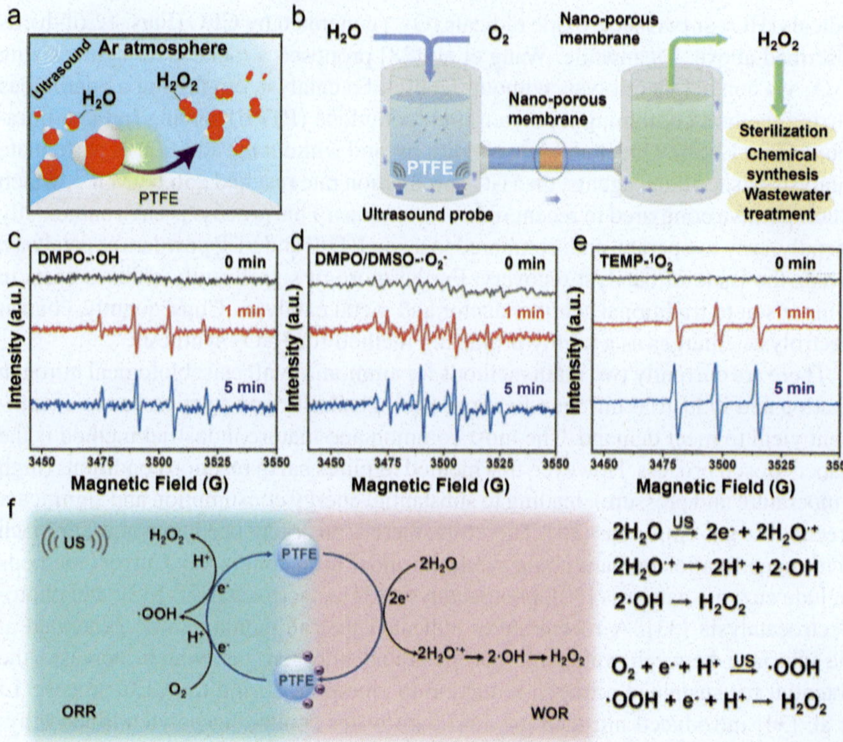

Fig. 12.5 **a** Schematic representation of the overall reaction. Reproduced with permission from Ref. [25]. Copyright 2024, American Chemical Society **b** Expected large application of contact-electro-catalysis technology in the field of efficient production of H_2O_2. **c** EPR spectra for DMPO-*OH, **d** DMPO/DMSO-*O_2, **e** TEMP-$^1O_2$ in the presence of PTFE particles during ultrasonication in air atmosphere. **f** Proposed reaction mechanism of contact-electro-catalysis for H_2O_2 generation. Reproduced with permission from Ref. [25]. Copyright 2023, Wiley

Additionally, an economical H_2O_2 production system based on the principles of contact electrolysis has been designed. The design concept (Fig. 12.5b) aims to enable large-scale production, thereby accelerating the commercialization of contact electrolysis for H_2O_2 production. A simple experimental setup was constructed to verify that this method offers high durability. This also demonstrates that the contact electrolysis method for H_2O_2 production can simplify the preparation process, reduce economic costs, and achieve overall green production.

Additionally, Berbille et al. [27] used fluorinated ethylene propylene (FEP) as a catalyst and employed ultrasound-driven contact electrolysis (CEC) to generate H_2O_2 from air and deionized water (Fig. 12.6a). A detailed mechanistic study of the contact electrolysis synthesis of H_2O_2 was also conducted (Fig. 12.6c). At 20 °C, the kinetics of H_2O_2 production reached 58.87 mmol L^{-1} $g_{at}{}^{-1}$ h^{-1}. Electron paramagnetic resonance (EPR) spectroscopy showed that, during ultrasonic treatment, charged FEP emitted electrons into the solution, suggesting that H_2O_2 was formed from hydroxyl radicals (HO) or two superoxide radicals (O_2^-) generated by CEC (Figs. 12.6f–h), as described above. Meanwhile, Wang et al. [28] proposed a method for synthesizing H_2O_2 via contact electrolysis without the use of a catalyst, employing a continuous stirring method combining a polytetrafluoroethylene (PTFE) stirring rod and ultrasound (Fig. 12.6b). Under ambient conditions, and without the addition of particulate catalysts or sacrificial agents, the H_2O_2 production rate reached 256.6 μM h^{-1}, which is competitive compared to recent redox techniques. This process is environmentally friendly, easy to operate, and, compared to using PTFE and FEP powders as catalysts, avoids the issue of catalyst recovery. Furthermore, it significantly reduces costs in comparison to traditional semiconductor and metal catalysts. Consequently, contact electrolysis emerges as a highly promising method for H_2O_2 synthesis.

There are currently two main methods for ammonia synthesis: biological nitrogen fixation and industrial nitrogen fixation. Biological nitrogen fixation has an insufficient yield to meet demand. The most common and mature industrial method is the Haber–Bosch process. However, this method requires harsh reaction conditions (high temperature and pressure), leading to substantial energy consumption and significant greenhouse gas emissions [29]. Therefore, there is an urgent need for a new approach to nitrogen fixation and ammonia synthesis under mild conditions. Current methods include enzyme catalysis [30], photocatalysis [31], electrocatalysis [32], and photoelectrocatalysis [33]. A recent study indicated that ammonia can be generated at the interface between water microdroplets and nitrogen gas, which increases the potential for sustained nitrogen reduction to ammonia during the CEC process. Li et al. [34] introduced nitrogen gas into pure water containing polytetrafluoroethylene (PTFE) particles at room temperature (Fig. 12.7a) and synthesized ammonia (NH_3) via CEC principles with ultrasound. This process does not require additional electrical energy or radiation and allows continuous ammonia production. Under optimal conditions, the ammonia production rate per gram of PTFE particles was approximately 420 μmol L^{-1} h^{-1}. As shown in Fig. 12.7b, nitrogen gas injection under two hours of ultrasonic treatment was more favorable for ammonia generation compared to air injection. Furthermore, experiments were conducted with varying ratios of N_2–O_2 mixtures. As shown in Fig. 12.7c, reducing the nitrogen content in the N_2–O_2 mixture not only decreased NH_3 production but also led to a reduction in H_2O_2 generation. Figure 12.7d and e, utilizing electron paramagnetic resonance (ESR) spectroscopy, revealed that the main radicals generated in the nitrogen environment were $OH^{\cdot}$ and $H^{\cdot}$. Based on the above experimental results, Li et al. proposed a possible reaction mechanism for ammonia formation, suggesting that the ammonia synthesis reaction triggered by the contact electrolysis process can be divided into three steps.

First, water forms hydrogen ions and hydroxide ions at the interface. At the same time, nitrogen is adsorbed on the surface of PTFE, as shown in reactions (12.1) and (12.2)

$$H_2O \rightarrow H^+ + OH^-, \tag{12.1}$$

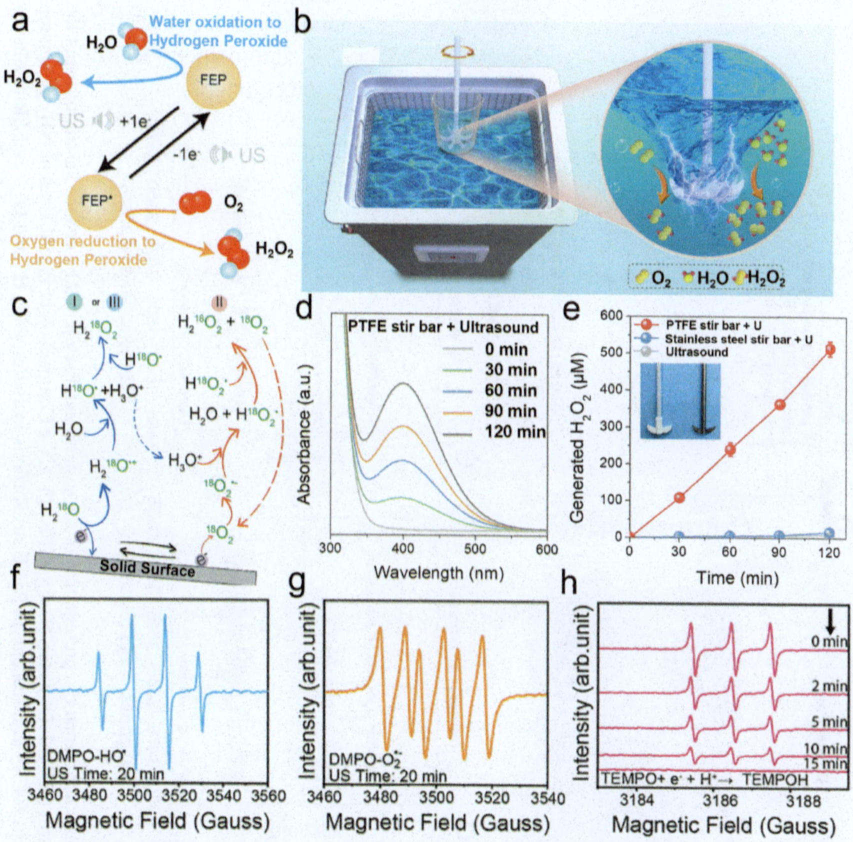

Fig. 12.6 **a** Schematic illustration of generating H_2O_2 from water and oxygen by ultrasonication (US) in presence of FEP **b** Schematic of the experimental setup for H_2O_2 generation. **c** Representation of the reaction of 4-carbonxyphenylboronic acid (4-CPB) with unlabeled (black arrow) and labeled (green arrow) H_2O_2. **d** Absorption spectrum of PTO in the reaction solution of the PTFE stir bar + ultrasound system. **e** Comparison of H_2O_2 generation using PTFE and stainless-steel stir bars. Detection by EPR of **f** DMPO-HO$^\cdot$, **g** DMPO-$O_2^{\cdot -}$, and **h** the loss of signal intensity from TEMPO as it is reduced to TEMPOH by addition of 1 electron and 1 proton during CEC. **a, c, f–h** Reproduced with permission from Ref. [26]. Copyright 2024, Wiley. **b, d, e** Reproduced with permission from Ref. [28]. Copyright 2024, Royal Society of Chemistry

$$\text{PTFE} + N_2 \rightarrow \text{PTFE} - N_2. \quad (12.2)$$

Second, promoted by the strong electric field caused by contact electrification at the interface, the hydroxide ion loses an electron to form a hydroxyl radical. Some electrons are trapped by hydrogen ions to form protons (hydrogen radicals), as shown in reactions (12.3) and (12.4).

$$OH^- \rightarrow OH^\cdot + e^-, \quad (12.3)$$

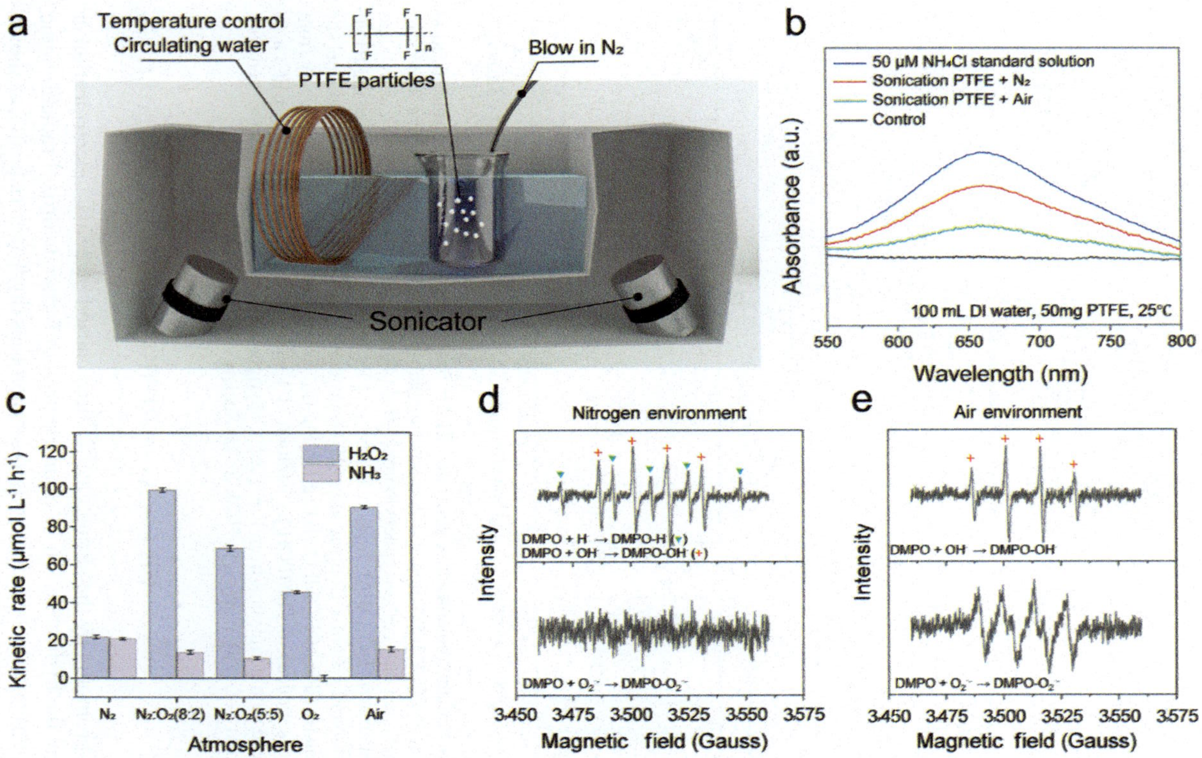

Fig. 12.7 Progress of contact electrocatalysis in ammonia synthesis **a** Schematic representation of the experimental setup. **b** Absorbance variations using the indophenol blue method under sonication with different atmospheres (100 mL DI water, 50 mg PTFE, 25 °C). **c** Comparison of kinetic rates of NH_3 generation in PTFE suspensions with different gas components. **d** ESR results for N_2 as a reactant in PTFE suspensions including $OH^·$, $H^·$, and $O_2^{·-}$. **e** ESR results for N_2–O_2 mixture (8:2) as a reactant in PTFE suspensions including $OH^·$, $H^·$, and $O_2^{·-}$.

$$H^+ + e^- \rightarrow H^{\cdot}. \tag{12.4}$$

Finally, $H^{\cdot}$ primarily reacts with the adsorbed nitrogen (PTFE-N_2) in a series of steps to form NH_3, which escapes from the PTFE to regenerate it, as shown in reaction (12.5). In addition, hydroxyl radicals recombine to hydrogen peroxide, as shown in reaction (12.6).

$$PTFE - N_2 + 6H^{\cdot} \rightarrow 2NH_3 + PTFE, \tag{12.5}$$

$$OH^{\cdot} + OH^{\cdot} \rightarrow H_2O_2. \tag{12.6}$$

In summary, emerging catalytic methods have enabled contact-electro-catalysis to facilitate the synthesis of important chemicals, such as the efficient production of hydrogen peroxide under mild conditions and ammonia synthesis from air and nitrogen. This expands the application scope of contact-electro-catalysis and provides new insights for the industrial production of chemicals.

12.4 Resources Recovery (LIBs, Precious Metals)

In the current landscape of rapid technological innovation and increasingly severe environmental challenges, resource recovery has emerged as a critical area of research and practice. At its core, resource recovery focuses on extracting valuable materials from waste streams, including metals, non-metallic substances, and by-products from industrial activities. This process not only mitigates the environmental impact associated with waste management but also promotes the sustainable use of finite natural resources. By doing so, resource recovery serves a dual purpose: safeguarding the environment and conserving essential resources, thereby contributing to a more sustainable future.

Driven by high-tech industries such as electronics, automotive manufacturing, and renewable energy, the global demand for metal materials continues to rise, making traditional resource extraction methods increasingly unsustainable. Conventional mining and extraction processes have significantly contributed to environmental degradation and resource depletion. Additionally, the rapid growth in demand for electronic devices and Lithium-ion batteries (LIBs) has led to the generation of large volumes of related waste. These wastes contain heavy metals and organic chemicals (such as chlorinated/brominated hydrocarbons), posing substantial risks to human health and the environment. More importantly, the metal content in waste is higher than that found in natural ores [35]. Therefore, the recovery of metals (such as gold, palladium, platinum, lithium, cobalt, nickel, etc.) from waste should be prioritized [36]. Recovering resources from waste offers a dual benefit: it helps to reduce environmental pollution while simultaneously addressing the growing demand for raw materials.

LIBs are essential for portable electronics, EVs, and energy storage, driving high demand for critical metals like lithium, nickel, and cobalt. These metals are often mined in politically unstable regions with limited reserves, causing prices to rise [37, 38]. Additionally, their extraction and processing have significant environmental impacts [39]. The rapid growth of LIBs leads to large volumes of waste batteries which improper disposal releases pollutants that challenge waste management. The content of valuable metals in spent LIBs is higher than natural ores [39]. Therefore, recycling LIBs can not only address the demand for raw materials but also reduce the environmental and health hazards posed by spent batteries. Current recycling methods include pyrometallurgy, hydrometallurgy, and direct recycling. While pyrometallurgy is suitable for large-scale use, it is energy-intensive, costly, and releases toxic gases and greenhouse gases, high lithium loss rate and only yield metal alloys. Direct recycling allows for the reuse of recovered cathode materials in the production of new batteries. However, this method requires strict sorting and pretreatment processes, accurate assessments of the battery's charge state and composition, making it a less flexible technology. Hydrometallurgical recycling is favored for its high metal leaching rates and purity of recovered materials. However, the use of inorganic acids for leaching generates large volumes of corrosive waste liquids and toxic gases. Organic acid leaching, in contrast, can reduce toxic gas emissions and mitigate adverse effects on the environment and operators. Nonetheless, organic acids have lower leaching efficiency and require additional reducing agents (H_2O_2), to accelerate the process, which increases reagent costs. To address this, Li et al. proposed a contact-electro-catalysis method for recycling LIBs, using citric acid as the leaching agent and incorporating inexpensive, recyclable SiO_2 to replace the need for reducing agents, thereby enhancing leaching efficiency [40].

Figure 12.8 illustrates the flow chart of LIBs recovery by CEC leaching. The process begins with pretreating the collected spent batteries to obtain lithium cobalt oxide (LCO) electrode powder. This is followed by CEC leaching, where citric acid serves as the leaching agent, and recyclable SiO_2 is added as a catalyst. The electrode powder is mixed with a solid-to-liquid ratio of 10 g/L, and ultrasonic stimulation is applied. The ultrasound generates cavitation bubbles, which collapse, causing frequent contact between SiO_2 and the aqueous phase. This results in the transfer of electrons from deionized water to the SiO_2 surface. Hydroxyl radicals (·OH) generated by the reaction of water cations with water molecules break the Co–O bond, facilitating the leaching reaction. Electrons on the SiO_2 surface reduce Co^{3+} to Co^{2+}, enabling the leaching of metals. Oxygen captures electrons from the SiO_2 surface, forming superoxide anions, and the leaching of metals from LCO can be described by the reactions in equations. More importantly, hydroxyl radicals can recombine to form hydrogen peroxide, while superoxide radicals can also react with electrons to generate hydrogen peroxide, thereby accelerating the leaching of metals. This process results in a pink solution containing lithium and cobalt ions. ICP-OES analysis indicates that after 6 h of ultrasound at 90 °C, the leaching efficiencies for lithium and cobalt are 100% and 92.19%, respectively. At 70 °C, after 6 h of ultrasound, the leaching efficiencies for lithium, nickel, manganese, and cobalt in

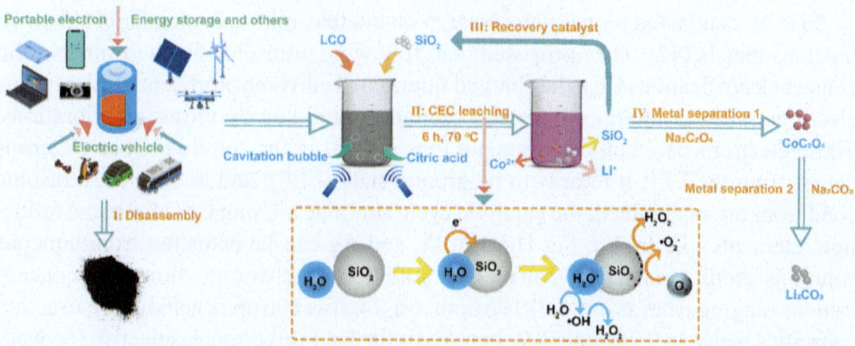

Fig. 12.8 Flow chart of lithium battery recovery by CEC leaching. Reproduced with permission from Ref. [40]. Copyright 2023, Springer Nature

ternary lithium-ion batteries reached 94.56%, 96.62%, 96.54%, and 98.39%, respectively. The catalyst and leachate are then separated by centrifugation. The recovered catalyst can be reused in subsequent leaching cycles. After five cycles, the leaching efficiencies for lithium and cobalt from the recycled SiO_2 are 88.14% and 80.01%, respectively. Finally, the metals are separated from the leachate through precipitation. Cobalt ions are precipitated as cobalt oxalate using sodium oxalate, and lithium ions are precipitated as lithium carbonate using sodium carbonate, achieving metal separation. Economic analysis shows that this method has a profit of $8.986 per kilogram of spent $LiCoO_2$ powder, which is highly favorable compared to other acid leaching methods.

Therefore, CEC proves to be a more environmentally friendly and cost-effective method for recycling lithium-ion batteries. Future efforts to enhance the efficiency of CEC leaching could focus on improving the surface chargeability of SiO_2, increasing the concentration of the leaching solution, and optimizing the overall recovery process.

The rapid development of the electronics industry has led to the constant innovation of devices such as laptops, refrigerators, and smartphones. This rapid market expansion has significantly increased the generation of electronic waste (e-waste), which is now the fastest-growing waste category. E-waste is both a source of hazardous substances and a valuable secondary resource, containing heavy metals and organic chemicals that pose substantial environmental burdens [41], as well as varying levels of precious metals and rare earth elements. Discarded printed circuit boards contain precious metals at concentrations up to ten times higher than those found in high-grade ores. Moreover, global demand for these critical metals continues to rise, leading to steady price increases. Therefore, recycling e-waste can not only address environmental pollution but also alleviate the pressure on raw material supply. However, only 17.4% of these metals are currently recycled, with approximately $57 billion worth of metals wasted annually through landfilling, incineration, and other improper disposal methods [41]. Thus, it is crucial to design green and efficient recycling methods to recover metals.

Su et al. conducted pioneering research on the feasibility of using CEC to extract precious metals [42]. They proposed that, following ultrasonic-assisted water–solid contact electrification (CE), the charged fluorinated ethylene propylene (FEP*) emits electrons into the solution under the excitation provided by ultrasonic vibrations. These electrons can then reduce metal ions present in the solution. Once electrons depart from the FEP, it returns to its ground state (FEP), and as long as ultrasonic conditions are maintained, the catalytic cycle continues. Using CEC for metal reduction, elements like Ir, Au, Pd, Hg, Rh, Pt, and Ag can be extracted from aqueous solutions. Additionally, CEC can recover gold from synthetic solutions with concentrations ranging from as low as 0.196 ppm to as high as 196 ppm, achieving extraction capacities between 0.756 and 722.5 mg/g within 3 h. Furthermore, selective recovery of gold from e-waste was demonstrated as shown in Fig. 12.9a is the flowchart of the recovery process for this method. After pre-treating recovered CPUs to obtain a 10 ppm gold solution, adding 10 mg of FEP under aerobic conditions at 25 °C for 20 h resulted in 94.4% of $AuCl_4^-$ being reduced to elemental gold as depicted in Fig. 12.9b, with minimal reduction of other metal ions (Zn, Fe, Ni, Cu) as illustrated in Fig. 12.9c, indicating high selectivity of the CEC process. A similar approach applied to electroplating waste achieved a 91.5% gold extraction rate. This selectivity is likely due to a kinetic mechanism that prevents the re-dissolution of metals by neutralizing the solution, thus avoiding reversible reactions. Additionally, as shown in Fig. 12.9d, after five cycles, the gold recovery efficiency did not significantly decrease.

Most methods that utilize advanced oxidation processes (AOPs) to produce highly oxidative free radicals for oxidizing and dissolving precious metals require the use of organic solvents, such as acetonitrile (CH_3CN), benzaldehyde (C_6H_5CHO), and dichloromethane (CH_2Cl_2), which pose environmental pollution concerns. Although recent studies have achieved precious metal extraction via AOPs in inorganic solution systems, they require the addition of radical precursors, like persulfate (PDS). To address these issues, Li et al. developed a novel catalyst by loading the dielectric material polytetrafluoroethylene (PTFE) onto ZSM-5, termed PTFE/ZSM-5 (PZ), which enabled uniform dispersion of the catalyst in water [43]. They introduced the Fe^{III}-initiated self-cycling Fenton system (SF-CEC). Unlike traditional CEC, the presence of Fe^{III} activates in-situ generated hydrogen peroxide to produce singlet oxygen active radicals, enhancing the formation of reactive oxygen species (ROS). This system addresses the limitation of CEC in dissolving Pd^{2+} as shown in Fig. 12.10a. The hydrogen peroxide generated by CEC is continuously decomposed by Fe(III) into ROS, which oxidizes Pd loaded on the catalyst. The abundant Cl^- ions in the solution stabilize Pd^{2+} by forming complexes, enabling its continuous dissolution as represented in Fig. 12.10b. Under ambient temperature, atmospheric pressure, and pH 2 conditions, SF-CEC can effectively dissolve Pd, Au, Rh, and Ir as depicted in Fig. 12.10c, demonstrating its broad applicability for dissolving precious metal catalysts.

Therefore, CEC has demonstrated outstanding performance in the recovery of precious metals. By further modifications, using new materials that are less polluting than fluorinated polymers, the efficiency of the CEC process can be enhanced, making it greener and more effective. This opens up new avenues for its industrial application.

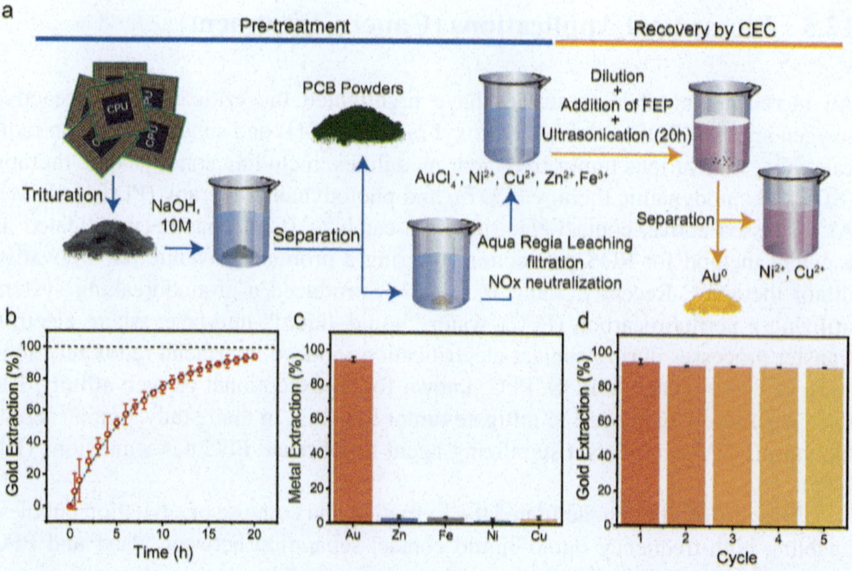

Fig. 12.9 Selective extraction of Gold from the leachate of spent CPUs by CEC. **a** Schematic of the pre-treatment of CPU (blue) and recovery of CEC for extracting gold from spent CPUs. Arrows are guide to eyes. **b** Evolution of gold extraction over time from a leachate of CPU at a concentration of 10 ppm of gold. The doted lines are guides to the eyes. **c** Amount of $AuCl_4^-$, Zn^{2+}, Fe^{2+}, Ni^{2+}, and Cu^{2+} extracted after 20 h. **d** Recycling of FEP particle for 5 cycles of 18 h. Error bars represent standard deviations for 3 reproduced experiments. Reproduced with permission from Ref. [42]. Copyright 2024, Springer Nature

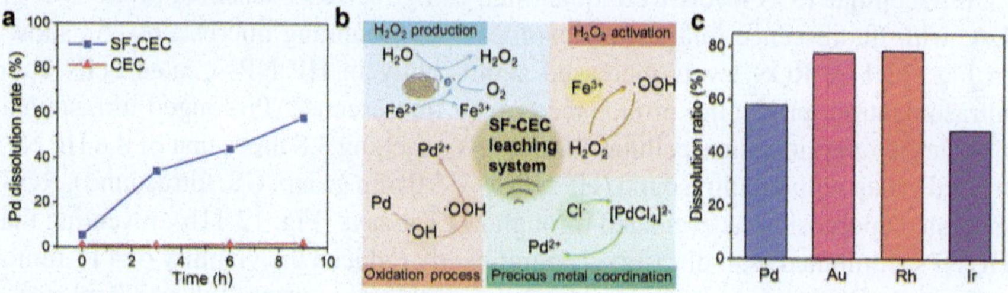

Fig. 12.10 Precious metal dissolution in SF-CEC systems. **a** Testing the dissolution performance of precious metals in Pd-based catalysts using SF-CEC and CEC. The Pd-based catalyst is $Pd-ZnTiO_3$, used for thermal catalytic methanol oxidation. **b** The mechanism of precious metal dissolution from catalysts in the SF-CEC system. **c** Testing the dissolution performance of precious metals in Au, Rh, and Ir-based catalysts using the SF-CEC system. The Au-based catalyst is $Au-TiO_2$, used for photocatalytic methane oxidation. The Rh-based catalyst is Rh-PMA (phosphomolybdic acid), used for thermal catalytic ethanol oxidation. The Ir-based catalyst is IrOx-FTO, used for electrocatalytic water splitting to produce hydrogen. Reproduced with permission from Ref. [43]. Copyright 2024, Wiley

12.5 Biomedical Applications (Cancer Treatment)

An increasing number of studies have highlighted the critical role of reactive oxygen species (ROS), such as hydroxyl radicals (·OH) and superoxide anion radicals ($\cdot O_2^-$), in various tumor treatment modalities, including sonodynamic therapy (SDT), chemodynamic therapy (CDT), and photodynamic therapy (PDT) [44–46]. As discussed above, contact-electrification catalysis (CEC) has been validated as a potent method for ROS production, offering a promising avenue for innovative tumor therapies. Recently, Jiang et al. [47] introduced a groundbreaking system utilizing a perfluorocarbon (PFC)–water "liquid–liquid" interface, where electron transfer processes during contact-electrification-induced interfacial redox reactions were shown to generate ROS. PFC, known for its exceptional oxygen affinity, has been extensively employed to mitigate tumor hypoxia. In their study, human serum albumin (HSA) served as a stabilizing agent to fabricate PFC nanoemulsions (HP NPs).

Ultrasonic oscillation facilitated the formation and collapse of cavitation bubbles, enabling high-frequency liquid–liquid contact separation between water and PFC nanoemulsions. This process promoted electron transfer, wherein electrons moved from water molecules and hydroxide ions (OH^-) to PFC under ultrasonication, producing ·OH. Simultaneously, dissolved oxygen in water gained electrons from negatively charged PFC (PFC*) to form $\cdot O_2^-$. Leveraging this ROS generation mechanism, HSA-modified perfluorotributylamine (PFTBA) nanoemulsions were applied to induce ROS-mediated apoptosis in tumor cells, culminating in the development of a novel therapeutic approach termed "Contact-Electrodynamic Therapy (CEDT)."

Intracellular ROS levels were quantified using the ROS-sensitive probe DCFH-DA, with fluorescence imaging by confocal laser scanning microscopy. As shown in Fig. 12.11a, ROS levels increased significantly in HP NPs-treated cells upon ultrasonication, producing pronounced green fluorescence. Prolonged ultrasonication further amplified intracellular ROS levels, reaching 3.8 times that of the HP NPs treated group without ultrasound (HP NPs + US 0 min group, US: ultrasound). ROS-mediated apoptosis was evaluated through MTT assays (Fig. 12.11b), revealing that HP NPs, combined with ultrasound, significantly reduced the viability of 4T1 tumor cells due to ROS-induced cellular damage. Additionally, extracellular ATP release, a "find me" signal for attracting antigen-presenting cells (APCs), increased markedly in the HP NPs + US group (Fig. 12.11c), indicating immunogenic cell death (ICD).

CEDT-induced ICD was further supported by confocal imaging of high-mobility group box 1 (HMGB1) migration from the nucleus to the extracellular environment and calreticulin exposure on the cell membrane (Fig. 12.11e). The HP NPs + US group exhibited the highest levels of CRT and HMGB1 expression, demonstrating strong green and red fluorescence signals, respectively. These "danger" signals are crucial for enhancing APC-mediated antigen presentation and activating systemic immune responses.

The anti-tumor efficacy of CEDT was evaluated in a subcutaneous 4T1 tumor-bearing mouse model. As illustrated in Fig. 12.12a, mice received either PBS or HP

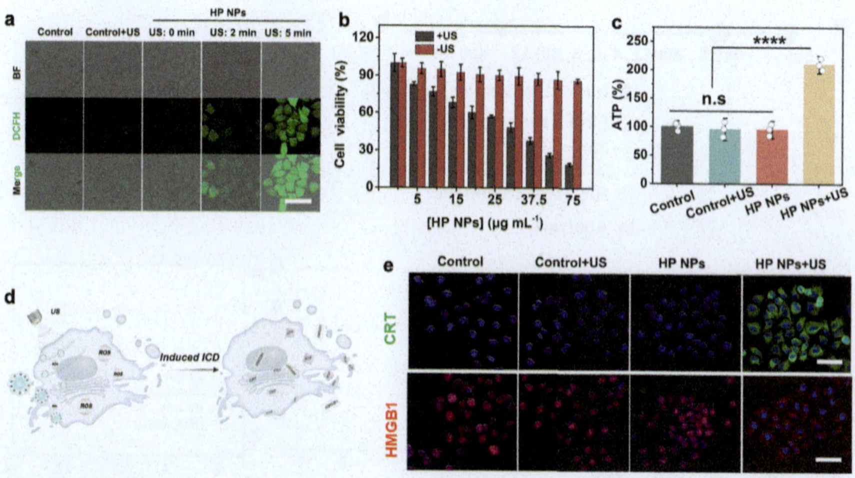

Fig. 12.11 ROS generation performance of HP NPs in cell. **a** Representative confocal fluorescence imaging of ROS generation in 4T1 cells co-incubated with HP NPs under different ultrasonication (US) time. Scale bar: 50 μm. **b** Cell viability analysis of 4T1 incubated with different concentration of HP NPs with and without ultrasonication. **c** Relative percentages of ATP in the extracellular environment of 4T1 cells under different treatment. **d** Schematic representation of HP NPs-induced ICD. **e** Representative fluorescence imaging of HMGB1 and calreticulin expressed by 4T1 cells under different treatment. Scale bar: 50 μm

NPs every two days, with or without ultrasound stimulation (3 kHz, 1 W/cm^2, 50% duty cycle, 10 min). Tumor size and body weight were monitored for 14 days. While tumors in the control groups grew rapidly (Fig. 12.12b–d), HP NPs combined with ultrasound significantly inhibited tumor growth and even eradicated solid tumors. Importantly, the treatment did not cause adverse effects on mouse body weight, indicating good biosafety (Fig. 12.12e).

CEDT presents several distinct advantages, including its independence from endogenous tumor microenvironment substrates like O_2 and H_2O_2 for ·OH production, its noninvasive nature with deep tissue penetration and minimal damage to healthy tissues, and its use of clinically approved PFTBA, which ensures high biosafety and clinical applicability. Moreover, CEDT can activate the immune system to enhance anti-tumor efficacy. This study not only elucidates the mechanism of ROS generation through liquid–liquid interfacial contact-electrification but also introduces a promising tumor therapy with significant clinical potential, free from the constraints of the tumor microenvironment.

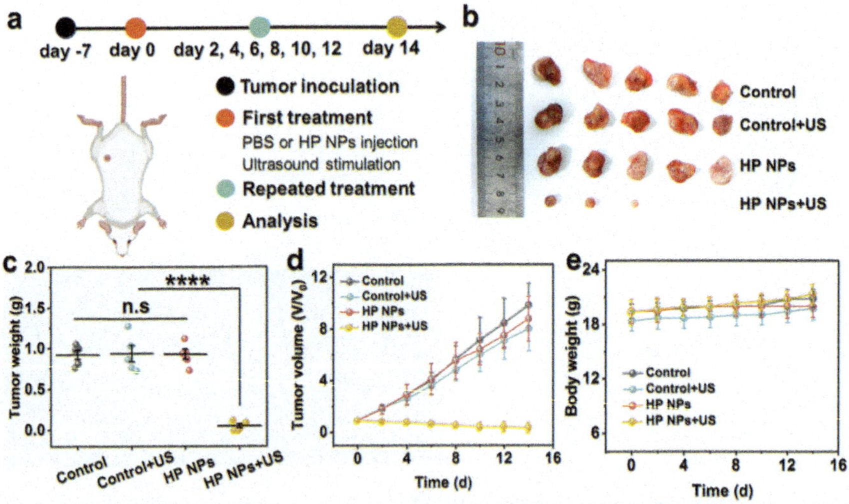

Fig. 12.12 Evaluation of HP NPs-mediated therapy in solid tumors. **a** Schematic illustration of 4T1 tumor-bearing mice model construction and treatment. **b** Photographs of tumors, **c** tumor weight, **d** tumor growth curves, and **e** body weight of mice under different treatment

12.6 Summary and Perspective

Contact-electro-catalysis (CEC) is an emerging catalytic strategy that leverages contact electrification at interfaces to induce electron transfer, subsequently generating catalytically active species to drive redox reactions. This approach has demonstrated excellent performance across a wide range of applications, including pollutant degradation, CO_2 capture, the synthesis of important chemicals (such as hydrogen peroxide and ammonia), resource recovery (e.g., recycling of lithium-ion batteries and precious metals), and even cancer therapy. CEC offers high reaction efficiency under mild conditions, which sets it apart from more demanding catalytic processes. More importantly, a particularly significant advantage of CEC lies in its broad catalyst selectivity. Unlike conventional photocatalysis, electrocatalysis, piezo-catalysis, or tribo-catalysis, which typically rely on semiconductors, metals, or carbon-based materials, CEC can utilize even chemically inert materials like FEP and PTFE to achieve catalytic activity. This flexibility expands the range of potential catalysts significantly, making it easier to tailor catalytic systems to specific applications. Moreover, the scope of CEC extends beyond solid catalysts to liquid–liquid interfaces, such as oil–water boundaries, where liquid–liquid contact electrification can induce electron transfer to drive redox reactions, similar to solid–liquid systems. This opens up new possibilities for catalytic processes at liquid interfaces, where traditional catalytic methods may be less effective. Moreover, CEC is not limited to solid-state catalysts and can be extended to liquid–liquid interfaces, such as oil–water systems. At these interfaces, liquid–liquid contact electrification can induce electron transfer, driving redox reactions similarly to solid–liquid systems. This capability opens up new possibilities for catalytic reactions that take place at liquid interfaces, where traditional catalytic methods may be less effective, thus offering a versatile and innovative pathway for advancing catalytic processes across various fields.

Moreover, due to the relatively low requirements for catalysts and the ability to drive reactions using mechanical energy, CEC presents a promising solution to a wide range of industrial and environmental challenges, particularly in the fields of environmental sustainability and renewable energy. For example, CEC holds great potential in the reduction of CO_2 emissions. By utilizing environmental energy sources such as wind or mechanical vibrations, CEC can effectively convert CO_2 into valuable products such as carbon monoxide (CO), formic acid, and other useful chemicals. This provides an energy-efficient and environmentally friendly alternative to traditional CO_2 capture methods, helping to mitigate climate change and support the development of a circular carbon economy. Additionally, CEC has been proven effective in removing pollutants such as methyl orange and antibiotics, demonstrating its immense potential in wastewater treatment and environmental remediation. In practical applications, CEC can degrade pollutants and microorganisms in water by harnessing mechanical energy from sources like ocean currents and wind. This sustainable wastewater management approach reduces the need for catalysts and non-renewable energy, offering more energy-efficient solutions and providing a more effective and convenient method for producing cleaner water. Furthermore, in response to the dual challenges of resource depletion and increasing e-waste generation, CEC offers a sustainable method for recovering valuable metals. It reduces the environmental impact of mining and lowers the pollution and waste associated with electronic waste. More importantly, CEC can harness mechanical energy from the environment and generate electrical charges through the triboelectric effect, thereby promoting the reduction of water. Compared to traditional methods, which often involve high energy consumption, expensive catalysts, or the release of carbon emissions, CEC-based hydrogen production offers an innovative alternative. By utilizing renewable, low-cost mechanical energy, CEC addresses the limitations of existing methods, providing a more sustainable and cost-effective approach to hydrogen production. As shown in Fig. 12.13, CEC can utilize mechanical energy from the environment, and due to its low catalyst requirements, it has broad application scenarios. Through the commonly observed triboelectric effect, CEC can trigger catalytic reactions and generate active species, making it promising for hydrogen production, plastic degradation, and other fields. In summary, the advancements in CEC research have led to its potential widespread application across various industries, particularly in those focused on sustainability and environmental protection. With continued progress in catalyst design, materials science, and energy integration, CEC has the potential to play a key role in the transition to a more sustainable and circular economy.

Currently, the greatest challenge limiting the development of contact-electro-catalysis (CEC) lies in its stringent requirements for reaction conditions. Effective CEC performance depends on strong contact electrification between the solid and

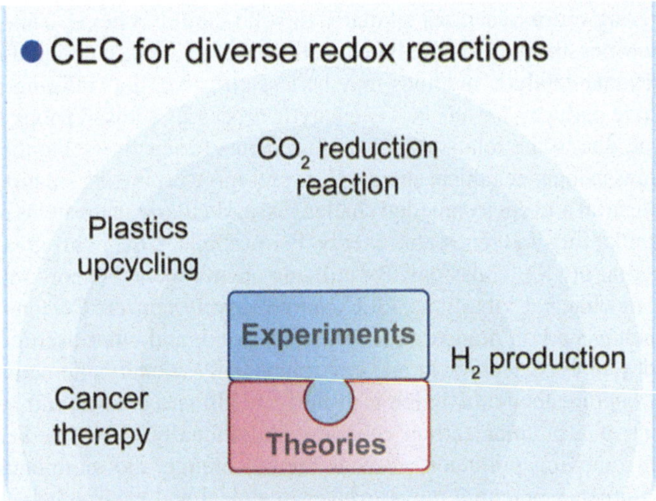

Fig. 12.13 The future application diagram of CEC

liquid phases, which is typically optimized under neutral pH and low ionic concentration. This limitation results in lower efficiency when treating highly concentrated pollutants or operating in solutions with strong acidity or alkalinity. To address this issue, there is a need to explore more suitable catalysts that can meet the demands of specific catalytic reactions. Another potential approach is to incorporate other materials to enhance catalytic performance, thereby expanding the applicability of CEC. Additionally, improving charge density can further enhance catalytic efficiency, thereby reducing the constraints on reaction conditions. These strategies aim to enhance contact electrification capabilities, thereby improving the overall catalytic efficiency of CEC systems. Furthermore, a deeper exploration is needed to clarify the catalytic mechanism.

In summary, the field of contact-electro-catalysis (CEC) offers exciting opportunities to advance mechanochemical catalysis. By modifying catalysts and gaining a deeper understanding of the underlying mechanisms, it is possible to discover novel catalysts and expand the scope of CEC applications. This positions CEC as a promising strategy for future research and industrial applications.

References

1. F. Li, G.F. Han, J.B. Baek, Nanocatalytic materials for energy-related small-molecules conversions: active site design, identification and structure-performance relationship discovery. Acc. Chem. Res. **55**, 110–120 (2022)
2. M.G. Sendeku, T.A. Shifa, F.T. Dajan, K.B. Ibrahim, B. Wu, Y. Yang, E. Moretti, A. Vomiero, F. Wang, Frontiers in photoelectrochemical catalysis: a focus on valuable product synthesis. Adv. Mater. **36**, 2308101 (2024)

3. T. Xu, Z. Wang, W. Zhang, S. An, L. Wei, S. Guo, Y. Huang, S. Jiang, M. Zhu, Y.B. Zhang, W.H. Zhu, Constructing photocatalytic covalent organic frameworks with aliphatic linkers. J. Am. Chem. Soc. **146**, 20107–20115 (2024)
4. A. Shu, C. Qin, M. Li, L. Zhao, Z. Shangguan, Z. Shu, X. Yuan, M. Zhu, Y. Wu, H. Wang, Electric effects reinforce charge carrier behaviour for photocatalysis. Energy Environ. Sci. **17**, 4907–4928 (2024)
5. Z. Wang, A. Berbille, Y. Feng, S. Li, L. Zhu, W. Tang, Z.L. Wang, Contact-electro-catalysis for the degradation of organic pollutants using pristine dielectric powders. Nat. Commun. **13**, 130 (2022)
6. Z. Wang, X. Dong, W. Tang, Z.L. Wang, Contact-electro-catalysis (CEC). Chem. Soc. Rev. **53**, 4349–4373 (2024)
7. Z. Chen, Y. Lu, R. Hong, Z. Liang, L. Wen, X. Liu, Q. Liu, Recent progress of solid–liquid interface-mediated contact-electro-catalysis. Langmuir **40**, 5557–5570 (2024)
8. X. Dong, Z. Wang, Y. Hou, Y. Feng, A. Berbille, H. Li, Z.L. Wang, W. Tang, Regulating contact-electro-catalysis using polymer/metal Janus composite catalysts. J. Am. Chem. Soc. **146**, 28110–28118 (2024)
9. S.B. Grant, J.-D. Saphores, D.L. Feldman, A.J. Hamilton, T.D. Fletcher, P.L.M. Cook, M. Stewardson, B.F. Sanders, L.A. Levin, R.F. Ambrose, A. Deletic, R. Brown, S.C. Jiang, D. Rosso, W.J. Cooper, I. Marusic, Taking the "waste" out of "wastewater" for human water security and ecosystem sustainability. Science **337**, 681–686 (2012)
10. W.-W. Li, H.Q. Yu, B.E. Rittmann, Chemistry: reuse water pollutants. Nature **528**, 29–31 (2015)
11. Z. Wang, X. Dong, X.F. Li, Y. Feng, S. Li, W. Tang, Z.L. Wang, A contact-electro-catalysis process for producing reactive oxygen species by ball milling of triboelectric materials. Nat. Commun. **15**, 757 (2024)
12. D.Q. Cao, R.K. Fang, Y.X. Song, M.G. Ma, H. Li, X.D. Hao, R. Wu, X. Chen, Contact-electro-catalysis for degradation of trace antibiotics in wastewater. Chem. Eng. J. **487**, 150531 (2024)
13. K. Li, Y. Lai, S. Lin, L. Zhou, M. He, H. Lin, Y. Yuan, Contact-electro-catalysis for the degradation of pentachlorophenol using inert fluorinated ethylene propylene powders. ACS EST Eng. **4**, 2485–2494 (2024)
14. C. Zhang, Z. Lin, L. Jiao, H. Jiang, Metal-organic frameworks for electrocatalytic CO_2 reduction: from catalytic site design to microenvironment modulation. Angew. Chem. Int. Ed. **63**(50), e202414506 (2024)
15. C. Hepburn, E. Adlen, J. Beddington, E.A. Carter, S. Fuss, N. Mac Dowell, J.C. Minx, P. Smith, C.K. Williams, The technological and economic prospects for CO_2 utilization and removal. Nature **575**, 87–97 (2019)
16. Z. Liu, K. Wang, Y. Chen, T. Tan, J. Nielsen, Third-generation biorefineries as the means to produce fuels and chemicals from CO_2. Nat. Catal. **3**, 274–288 (2020)
17. Z. Tang, X. Liu, Y. Yang, F. Jin, Recent advances in CO_2 reduction with renewable reductants under hydrothermal conditions: towards efficient and net carbon benefit CO_2 conversion. Chem. Sci. **15**, 9927–9948 (2024)
18. N. Wang, W. Jiang, J. Yang, H. Feng, Y. Zheng, S. Wang, B. Li, J.Z.X. Heng, W.C. Ong, H.R. Tan, Y.W. Zhang, D. Wang, E. Ye, Z. Li, Contact-electro-catalytic CO_2 reduction from ambient air. Nat. Commun. **15**, 5913 (2024)
19. Y. Holade, S. Ghosh, T.W. Napporn, Best practices for hydrogen peroxide (photo)electrosynthesis. Nat. Sustain. **7**, 1085–1087 (2024)
20. Y. Liu, L. Li, Z. Sang, H. Tan, N. Ye, C. Sun, Z. Sun, M. Luo, S. Guo, Enhanced hydrogen peroxide photosynthesis in covalent organic frameworks through induced asymmetric electron distribution. Nat. Synth. **4**, 134–141 (2024)
21. B.H. Lee, H. Shin, A.S. Rasouli, H. Choubisa, P. Ou, R. Dorakhan, I. Grigioni, G. Lee, E. Shirzadi, R.K. Miao, J. Wicks, S. Park, H.S. Lee, J. Zhang, Y. Chen, Z. Chen, D. Sinton, T. Hyeon, Y.-E. Sung, E.H. Sargent, Supramolecular tuning of supported metal phthalocyanine catalysts for hydrogen peroxide electrosynthesis. Nat. Catal. **6**, 234–243 (2023)

22. T.N. Ye, S.W. Park, Y. Lu, J. Li, M. Sasase, M. Kitano, T. Tada, H. Hosono, Vacancy-enabled N_2 activation for ammonia synthesis on an Ni-loaded catalyst. Nature **583**, 391–395 (2020)
23. X. Sun, J. Yang, X. Zeng, L. Guo, C. Bie, Z. Wang, K. Sun, A.K. Sahu, M. Tebyetekerwa, T.E. Rufford, X. Zhang, Pairing oxygen reduction and water oxidation for dual-pathway H_2O_2 Production. Angew. Chem. Int. Ed. **63**(52), e202414417 (2024)
24. F. Liu, P. Zhou, Y. Hou, H. Tan, Y. Liang, J. Liang, Q. Zhang, S. Guo, M. Tong, J. Ni, Covalent organic frameworks for direct photosynthesis of hydrogen peroxide from water, air and sunlight. Nat. Commun. **14**, 4344 (2023)
25. J. Zhao, X. Zhang, J. Xu, W. Tang, Z. Lin Wang, F. Ru Fan, Contact-electro-catalysis for direct synthesis of H_2O_2 under ambient conditions. Angew. Chem. Int. Ed. **62**, e202300604 (2023)
26. Y. Wang, P. Wei, Z. Shen, C. Wang, J. Ding, W. Zhang, X. Jin, C.D. Vecitis, G. Gao, O_2-independent H_2O_2 production via water–polymer contact electrification. Environ. Sci. Technol. **58**, 925–934 (2024)
27. A. Berbille, X. Li, Y. Su, S. Li, X. Zhao, L. Zhu, Z.L. Wang, Mechanism for generating H_2O_2 at water–solid interface by contact-electrification. Adv. Mater. **35**, 2304387 (2023)
28. Y. Wang, Y. Wang, B. Hu, M. Qiu, G. Gao, P. Wei, Catalyst-free contact-electro-catalytic H_2O_2 synthesis via simple combination of a poly(tetrafluoroethylene) stir bar and ultrasound. Chem. Commun. **60**, 7331–7334 (2024)
29. Y. Shi, Z. Zhao, D. Yang, J. Tan, X. Xin, Y. Liu, Z. Jiang, Engineering photocatalytic ammonia synthesis. Chem. Soc. Rev. **52**, 6938–6956 (2023)
30. J.B. Solomon, C.C. Lee, Y.A. Liu, C. Duffin, M.W. Ribbe, Y. Hu, Ammonia synthesis via an engineered nitrogenase assembly pathway in *Escherichia coli*. Nat. Catal. **7**, 1130–1141 (2024)
31. H. Li, R. Chen, L. Sun, Y. Wang, Q. Liu, Q. Zhang, C. Xiao, Y. Xie, Hole polaron-mediated suppression of electron-hole recombination triggers efficient photocatalytic nitrogen fixation. Adv. Mater. **36**, 2408778 (2024)
32. R.G. Fiddian-Green, W. Silen, Mechanisms of disposal of acid and alkali in rabbit duodenum. Am. J. Physiol. **229**, 1641–1648 (1975)
33. Y. Fang, M. Li, Y. Gao, Y. Wen, B. Shan, Static organic p-n junctions in photoelectrodes for solar ammonia production with 86% internal quantum efficiency. Angew. Chem. Int. Ed. **64**(3), e202415729 (2024)
34. J. Li, Y. Xia, X. Song, B. Chen, R.N. Zare, Continuous ammonia synthesis from water and nitrogen via contact electrification. Proc. Natl. Acad. Sci. **121**, e2318408121 (2024)
35. E. Hsu, K. Barmak, A.C. West, A.H.A. Park, Advancements in the treatment and processing of electronic waste with sustainability: a review of metal extraction and recovery technologies. Green Chem. **21**, 919–936 (2019)
36. S.S.V. Vuppaladadiyam, B.S. Thomas, C. Kundu, A.K. Vuppaladadiyam, H. Duan, S. Bhattacharya, Can e-waste recycling provide a solution to the scarcity of rare earth metals? An overview of e-waste recycling methods. Sci. Total. Environ. **924**, 171453 (2024)
37. X. Sun, Z. Liu, F. Zhao, H. Hao, Global competition in the lithium-ion battery supply chain: a novel perspective for criticality analysis. Environ. Sci. Technol. **55**, 12180–12190 (2021)
38. X. Xiao, L. Wang, Y. Wu, Y. Song, Z. Chen, X. He, Cathode regeneration and upcycling of spent LIBs: toward sustainability. Energy Environ. Sci. **16**, 2856–2868 (2023)
39. G. Zhang, X. Yuan, Y. He, H. Wang, T. Zhang, W. Xie, Recent advances in pretreating technology for recycling valuable metals from spent lithium-ion batteries. J. Hazard. Mater. **406**, 124332 (2021)
40. H. Li, A. Berbille, X. Zhao, Z. Wang, W. Tang, Z.L. Wang, A contact-electro-catalytic cathode recycling method for spent lithium-ion batteries. Nat. Energy **8**, 1137–1144 (2023)
41. G. Kwon, K. Yoon, E. Kwon, J. Park, H. Lee, H. Song, Technical advancement in valorization of electronic waste and its contribution to establishing economic value-chain. Chem. Eng. J. **494**, 153154 (2024)
42. Y. Su, A. Berbille, X.F. Li, J. Zhang, M. PourhosseiniAsl, H. Li, Z. Liu, S. Li, J. Liu, L. Zhu, Z.L. Wang, Reduction of precious metal ions in aqueous solutions by contact-electro-catalysis. Nat. Commun. **15**, 4196 (2024)

43. W. Li, J. Tu, J. Sun, Y. Zhang, J. Fang, M. Wang, X. Liu, Z.Q. Tian, F.R. Fan, Boosting reactive oxygen species generation via contact-electro-catalysis with Fe^{III}-initiated self-cycled Fenton system. Angew. Chem. Int. Ed. **64**(1), e202413246 (2024)
44. Y. Zhang, J. Zhao, L. Zhang, Y. Zhao, Y. Zhang, L. Cheng, D. Wang, C. Liu, M. Zhang, K. Fan, M. Zhang, A cascade nanoreactor for enhancing sonodynamic therapy on colorectal cancer via synergistic ROS augment and autophagy blockage. Nano Today **49**, 101798 (2023)
45. J. Kang, H. Jeong, M. Jeong, J. Kim, S. Park, J. Jung, J.M. An, D. Kim, In situ activatable nitrobenzene–cysteine–copper(II) nano-complexes for programmed photodynamic cancer therapy. J. Am. Chem. Soc. **145**, 27587–27600 (2023)
46. H. Li, Y. Liu, B. Huang, C. Zhang, Z. Wang, W. She, Y. Liu, P. Jiang, Highly efficient GSH-responsive "off–on" NIR-II fluorescent fenton nanocatalyst for multimodal imaging-guided photothermal/chemodynamic synergistic cancer therapy. Anal. Chem. **94**, 10470–10478 (2022)
47. H. Li, Z. Wang, X. Chu, Y. Zhao, G. He, Y. Hu, Y. Liu, Z.L. Wang, P. Jiang, Free radicals generated in perfluorocarbon-water (liquid–liquid) interfacial contact electrification and their application in cancer therapy. J. Am. Chem. Soc. **146**, 12087–12099 (2024)

The manufacturer's authorised representative in the EU is Springer Nature Customer Service Centre GmbH, Europaplatz 3, 69115 Heidelberg, Germany. If you have any concerns regarding our products, please contact ProductSafety@springernature.com

Printed and bound by CPI Group (UK) Ltd, Croydon, CR0 4YY
30/09/2026
02225454-0014